Introductory Algebra

An Interactive Approach

Introductory Algebra

An Interactive Approach

SECOND EDITION

Linda Pulsinelli
Patricia Hooper
Department of Mathematics
Western Kentucky University

MACMILLAN PUBLISHING COMPANY
NEW YORK

Collier Macmillan Publishers
London

For our children
Gary and Elizabeth
Jim W., Regina, Bob, and Mel

Printed in the United States of America.

Macmillan Publishing Company
866 Third Avenue, New York, New York 10022

Collier Macmillan Canada, Inc.

Library of Congress Cataloging in Publication Data

Pulsinelli, Linda Ritter.
Introductory algebra.

Includes index.
1. Algebra. I. Hooper, Patricia (Patricia I.)
II. Title.
QA152.2.P85 1987 512 86-2795
ISBN 0-02-396940-7

Printing: 3 4 5 6 7 8 Year: 8 9 0 1 2 3 4 5

ISBN 0-02-396940-7

Preface

As its title indicates, *Introductory Algebra: An Interactive Approach* is intended to teach students the fundamental concepts and skills of algebra by involving them continually in the learning process, allowing them to see their own progress and therefore motivating them to continue.

The vocabulary level of a book of this type is crucial to its success and we have attempted to word our explanations in simple, straightforward language. The writing style is informal without being condescending, and important ideas are summarized in table form rather than paragraph form whenever appropriate.

Our experience with students at this level leads us to believe that they must practice each new skill as soon as it has been presented. For this reason, the text includes several unique features that are designed to provide maximum reinforcement.

Each chapter in the book follows the same basic structure.

Motivational Applied Problem

At the beginning of each chapter we have presented an applied problem that can be solved after the student has mastered the skills in that chapter. Its solution appears within the chapter.

Explanations

We have tried to avoid the "cookbook" approach to algebra by including a straightforward and readable explanation of each new concept. Realizing that students at the introductory level become easily bogged down in reading lengthy explanations, we have attempted to make our explanations as brief as possible without sacrificing rigor.

Highlighting

Definitions, properties, theorems, and formulas are highlighted in boxes throughout the book for easy student reference. In most cases a rephrasing of a generalization in words accompanies the symbolic statement, and it is also highlighted in a box.

Examples

Immediately following the presentation of a new idea, several completely worked-out examples appear together with several partially worked-out examples to be finished by the student. These examples are completed correctly at the end of each section and the student is advised to check his or her work immediately.

Trial Runs

Sprinkled throughout each section are several short Trial Runs, a list of six or eight problems to check on the student's grasp of a new skill. The answers appear at the end of the section.

Exercise Sets
Each section concludes with an extensive Exercise Set in which each odd-numbered problem corresponds closely to the following even-numbered problem.

Stretching the Topics
At the end of each Exercise Set there are several problems designed to challenge the better students by extending to the next level of difficulty the skills learned in the chapter.

Checkups
Following each Exercise Set, a list of about 10 problems checks on the student's mastery of the most important concepts in the section. Each Checkup problem is keyed to comparable examples in the section for restudy if necessary.

Problem Solving
One section of almost every chapter involves switching from words to algebra. By including such a section in each chapter we are attempting to treat problem solving as a natural outgrowth of acquiring algebraic skills.

Chapter Summaries
Each chapter concludes with a summary in which the important ideas are again highlighted, in tables when possible. New concepts are presented in symbolic form and verbal form, accompanied by a typical example.

Speaking the Language of Algebra
Following the summary, we have included a group of sentences to be completed *with words* by the student. Algebra students (especially those in self-paced programs) often lack the opportunity to "speak mathematics." We hope that these short sections will help them develop a better mathematics vocabulary.

Review Exercises
A list of exercises reviewing all the chapter's important concepts serves to give the student an overview of the content. Each problem is keyed to the appropriate section and examples.

Practice Test
A Practice Test is included to help the student prepare for a test over the material in the chapter. Once again, each problem is keyed to the appropriate chapter sections and examples.

Sharpening Your Skills
Finally, we have included a short list of exercises that will provide a cumulative review of concepts and skills from earlier chapters. Retention seems to be a very real problem with students at this level and we hope that these exercises will serve to minimize that problem. Each cumulative review exercise is keyed to the appropriate chapter and section.

Throughout the book we have adhered to a rather standard order of topics, making an attempt to connect new concepts to old ones whenever appropriate. This modified spiraling technique is designed to help students maintain an overview of the content. Success in future courses seems to us to hinge on students' seeing that algebra is a logical progression of ideas rather than a set of unrelated skills to be memorized and forgotten.

Chapter 1 contains a review of the arithmetic of whole numbers, fractions, decimals, and percents. Students who have recently completed an arithmetic course or whose arithmetic skills are not weak may be allowed to skip this chapter. All the properties of real numbers are included in Chapter 2, so the omission of Chapter 1 will not handicap the student who is proficient in computation with fractional and decimal numbers. Problems involving such numbers appear throughout the remaining chapters.

A section on negative exponents is included in Chapter 5 for those instructors who consider this an appropriate point at which to introduce this topic to introductory algebra students. For those who do not, the section can be omitted or inserted at a later time without disturbing the continuity of the remainder of the book.

The answers to the odd-numbered exercises in the Exercise Sets appear in the back of the book together with answers for *all* items in Stretching the Topics, Checkups, Speaking the Language of Algebra, Review Exercises, Practice Tests, and Sharpening Your Skills.

More assistance for students and instructors can be found among the supplementary materials that accompany this book.

Instructor's Manual with Test Bank

The Instructor's Manual contains the answers for all exercises in the Exercise Sets and Stretching the Topics. In addition there are eight Chapter Tests (four open-ended and four multiple choice) for each chapter and three Final Examinations (two open-ended and one multiple choice). Answers to these tests and examinations also appear in the Instructor's Manual. The chapter tests are also available on computer disks.

Student's Solutions Manual

The Student's Solutions Manual, written by Rebecca Stamper, contains step-by-step solutions for the even-numbered exercises in the Exercise Sets and for *all* items in the Review Exercises, Practice Tests, Sharpening Your Skills, and exercises involving word problems. Using the same style as appears in the text, these solutions emphasize the procedure as well as the answer.

Video Tapes

A series of 10 video tapes (each 20 to 30 minutes in length) provides explanations for some of the more difficult topics in the course.

Audio Tapes

A series of 10 audio cassettes (each 20 to 30 minutes in length) also offer explanations for the more difficult topics. Keyed to examples in the text, these cassettes encourage students to work along.

Computerized Test Generator

A set of computer-generated tests is available for producing either a 10-item test for each chapter *or* tests of any length from objective-referenced items. Cumulative tests and final exams may also be constructed using the objective-referenced items.

Interactive Software Program

This computer assisted program is available in Apple and IBM and is a series of lessons including problems at differing levels of difficulty.

Acknowledgments

The writing of this book would not have been possible without the assistance of many people. We express our appreciation to our indefatigable typist Maxine Worthington, to Rebecca Stamper for carefully working all our problems, to our families for tolerating our obsessive work schedules, to our Mathematics Editors Gary Ostedt and Bob Clark for their enthusiastic support, and to our Production Supervisor Elaine Wetterau for her efficiency and expertise.

We also thank our reviewers: Mary Jean Brod, University of Montana; Donald R. Johnson, Scottsdale Community College; Adele Le Gere, Oakton Community College; Gerald J. LePage, Bristol Community College; Lois Miller, El Camino College; Harold M. Nerr, University of Wisconsin; Dan Streeter, Portland State University; and Jack W. Rotman, Lansing Community College; for their careful scrutiny and helpful comments.

L. R. P.
P. I. H.

Contents

1 Working with the Numbers of Arithmetic 1

1.1 Working with Whole Numbers 2
1.2 Working with Fractions 23
1.3 Working with Decimals and Percents 41
1.4 Switching from Word Expressions to Number Expressions 55
Summary 67
Review Exercises 69
Practice Test 71

2 Working with Real Numbers 73

2.1 Understanding Integers 74
2.2 Subtracting Integers 91
2.3 Multiplying and Dividing Integers 99
2.4 Understanding Real Numbers 117
2.5 Switching from Word Expressions to Real Number Expressions 129
Summary 139
Review Exercises 143
Practice Test 147

3 Working with Variable Expressions 149

3.1 Switching from Word Expressions to Algebraic Expressions 150
3.2 Combining Like Terms 163
3.3 Working with Symbols of Grouping 171
Summary 183
Review Exercises 185
Practice Test 187
Sharpening Your Skills after Chapters 1–3 189

4 Solving First-Degree Equations 191

4.1 Solving Equations Using One Property 192
4.2 Solving Equations Using Several Properties 207
4.3 Switching from Word Statements to Equations 223
Summary 233
Review Exercises 235
Practice Test 237
Sharpening Your Skills after Chapters 1–4 239

5 Working with Exponents 241

5.1 Multiplying with Whole-Number Exponents 242
5.2 Dividing with Whole-Number Exponents 255
5.3 Working with Negative Exponents 267
Summary 283
Review Exercises 285
Practice Test 287
Sharpening Your Skills after Chapters 1–5 289

6 Working with Polynomials 291

6.1 Simplifying Polynomials 292
6.2 Multiplying with Monomials 301
6.3 Multiplying Binomials 309
6.4 Dividing Polynomials 319
6.5 Switching from Word Expressions to Polynomials 329
Summary 339
Review Exercises 341
Practice Test 343
Sharpening Your Skills after Chapters 1–6 345

7 Factoring Polynomials and Solving Quadratic Equations 347

7.1 Looking for Common Factors 348
7.2 Factoring the Difference of Two Squares 359
7.3 Factoring Trinomials 367
7.4 Using All Types of Factoring 383
7.5 Using Factoring to Solve Quadratic Equations 389
7.6 Switching from Word Statements to Quadratic Equations 403
Summary 413
Review Exercises 415
Practice Test 417
Sharpening Your Skills after Chapters 1–7 419

8 Working with Rational Expressions 421

8.1 Simplifying Rational Algebraic Expressions 422
8.2 Multiplying and Dividing Rational Algebraic Expressions 437
8.3 Building Rational Algebraic Expressions 447
8.4 Adding and Subtracting Rational Algebraic Expressions 455
8.5 Working with Complex Fractions 469
8.6 Solving Fractional Equations 479
8.7 Switching from Word Statements to Fractional Equations 491
Summary 503
Review Exercises 507
Practice Test 511
Sharpening Your Skills after Chapters 1–8 513

9 Working with Inequalities 515

9.1 Writing and Graphing Inequalities 516
9.2 Solving Inequalities 525
9.3 Combining Variable Inequalities (Optional) 537
9.4 Switching from Word Statements to Inequalities 547
Summary 555
Review Exercises 557
Practice Test 559
Sharpening Your Skills after Chapters 1–9 561

10 Working with Equations in Two Variables 563

10.1 Switching from Word Statements to Equations Containing Two Variables 564
10.2 Graphing First-Degree Equations in Two Variables 575
10.3 Graphing Linear Equations by Other Methods 589
10.4 Graphing Lines Using the Slope 603
Summary 621
Review Exercises 623
Practice Test 625
Sharpening Your Skills after Chapters 1–10 629

11 Solving Systems of Linear Equations 631

11.1 Solving a System of Equations by Graphing 632
11.2 Solving a System of Equations by Substitution 643
11.3 Solving a System of Equations by Elimination (or Addition) 653
11.4 Switching from Word Statements to Systems of Equations 665
Summary 675
Review Exercises 677

	Practice Test	679
	Sharpening Your Skills after Chapters 1–11	681
12	**Working with Square Roots**	**683**
12.1	Working with Radical Expressions	684
12.2	Operating with Radical Expressions	697
12.3	Rationalizing Denominators	709
12.4	Solving More Quadratic Equations	719
12.5	Switching from Word Statements to Quadratic Equations	729
	Summary	737
	Review Exercises	739
	Practice Test	741
	Sharpening Your Skills after Chapters 1–12	743
	Answers to Odd-Numbered Exercises, Stretching the Topics, Checkups, Speaking the Language of Algebra, Review Exercises, Practice Tests, and Sharpening Your Skills	**A1**
	Index	**I1**

Working with the Numbers of Arithmetic

1

Each month, Estelle saves $\frac{1}{10}$ of her paycheck in her savings bank and $\frac{1}{9}$ of her paycheck in her credit union. What fractional part of her paycheck does Estelle save each month?

When you have completed this chapter, you will be able to solve this problem using the numbers of arithmetic. Although we assume that your basic arithmetic facts are in good shape, we must be sure that you recall the techniques required for working with whole numbers, fractions, and decimals.

In this chapter we

1. Perform operations with whole numbers.
2. Observe some properties of zero and 1.
3. Work with fractions.
4. Work with decimals and percents.
5. Switch from word expressions to number expressions.

1.1 Working with Whole Numbers

As a child, you learned to count "one, two, three," and so on. These counting numbers are also called **natural numbers**.

Natural numbers: $\{1, 2, 3, 4, 5, \ldots\}$

The dots after the 5 mean "and so on"; in other words, this set continues indefinitely. If we include the number zero in this set, we obtain a new set of numbers called the **whole numbers**.

Whole numbers: $\{0, 1, 2, 3, 4, 5, \ldots\}$

The Number Line

To picture the whole numbers, we shall use a **number line**. We draw a line and choose a zero point and a length to represent 1 unit. Then all points spaced 1 unit apart to the right of zero are labeled with the whole numbers in order. Many points between the whole numbers are named by fractions and decimal numbers.

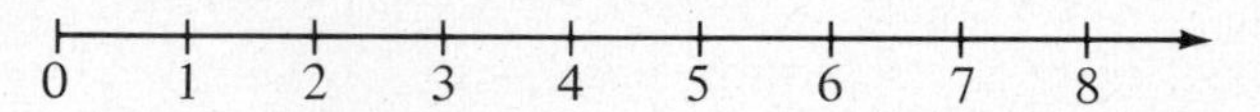

The arrow shows that this line goes on indefinitely, so that numbers such as 24 and 793 also correspond to points.

To show a whole number on the number line, we put a solid dot at the point corresponding to that number. This is called **plotting a point** on the number line. The point is called the **graph** of the number and the number is called the **coordinate** of the point.

Example 1. Graph 4 on the number line.

Solution

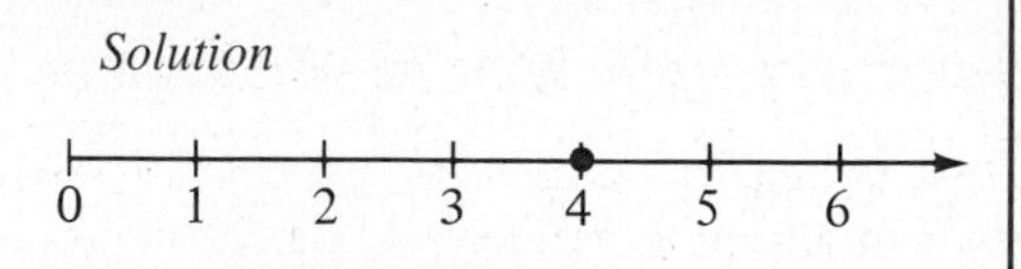

You try Example 2.

Example 2. Graph 0 on the number line.

Solution

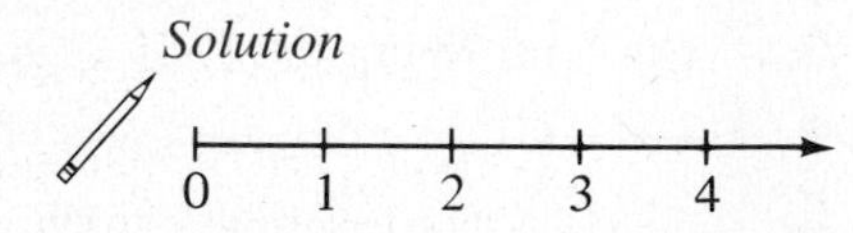

Check your work on page 13. ▶

The number line also gives us a handy way of comparing numbers. For example, if we look at the graph of the number 3, we notice that all the numbers to the right of 3 are greater than 3. All numbers to the left of 3 are less than 3.

Mathematicians use symbols to compare numbers.

Symbol	Meaning	Example
$>$	is greater than	$4 > 1$
$<$	is less than	$6 < 20$

Such statements are called **inequalities** and we summarize them as follows, for any numbers A and B.

Less than:	$A < B$	if A lies to the left of B on the number line
Greater than:	$A > B$	if A lies to the right of B on the number line

Example 3. Place a $<$ or $>$ symbol between the numbers.

12 ____ 3

0 ____ 1

Solution

$12 > 3$ because 12 lies to the right of 3

$0 < 1$ because 0 lies to the left of 1

You try Example 4.

Example 4. Place a $<$ or $>$ symbol between the numbers.

5 ____ 7

110 ____ 99

Solution

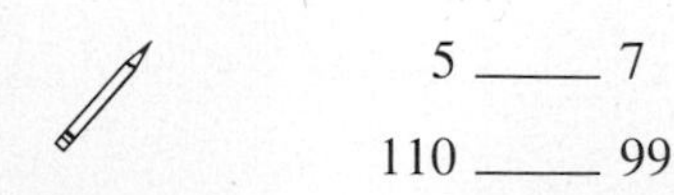

5 ____ 7

110 ____ 99

Check your work on page 13. ▶

Trial Run

1. Graph 6 on the number line.

0

Compare the numbers using the symbols $<$ or $>$.

2. 7 ____ 5 **3.** 0 ____ 2

4. 186 ____ 187 **5.** 2 ____ 0

Answers are on page 15.

Performing Operations with Whole Numbers

The four basic operations of arithmetic are *addition, subtraction, multiplication*, and *division*. It will help you in the study of algebra if you learn to name the parts of each of the four types of arithmetic problems now.

Addition

$$\underset{\text{term}}{\underset{\uparrow}{12}} + \overset{\text{term}}{\overset{\downarrow}{3}} = 15 \leftarrow \text{sum}$$

Subtraction

$$\underset{\text{term}}{\underset{\uparrow}{12}} - \overset{\text{term}}{\overset{\downarrow}{3}} = 9 \leftarrow \text{difference}$$

Multiplication

$$\underset{\substack{\uparrow \\ \text{factor}}}{12} \cdot \overset{\substack{\text{factor} \\ \downarrow}}{3} = 36 \leftarrow \text{product}$$

Division

$$\underset{\substack{\uparrow \\ \text{dividend}}}{12} \div \overset{\substack{\text{divisor} \\ \downarrow}}{3} = 4 \leftarrow \text{quotient}$$

Example 5. Find the sum of 10 and 9.

Solution

$$10 + 9 = 19$$

The sum is 19.

Example 6. Find the product of 10 and 9.

Solution

$$10 \cdot 9 = 90$$

The product is 90.

If you ever used your fingers to find a sum, you should agree that *addition is just a form of counting*. When you learned your basic addition facts in elementary school, you mastered a shortcut for finding sums without counting.

To learn your basic subtraction facts, you discovered that *every subtraction statement corresponds to an addition statement.*

$$15 - 8 = 7 \quad \text{because} \quad 15 = 8 + 7$$

$$9 - 0 = 9 \quad \text{because} \quad 9 = 0 + 9$$

Relating Subtraction to Addition. For whole numbers A, B, and C

$$A - B = C \quad \text{means} \quad A = B + C$$

There are several ways to indicate that two numbers are to be multiplied. For example, to multiply 4 times 3 we may write

$$4 \text{ X } 3 \qquad 4 \cdot 3 \qquad 4(3) \qquad (4)3 \qquad (4)(3)$$

Each of these expressions says the same thing, but we shall not use the first notation 4 X 3, because of the confusion that might occur when we begin to use the letter X to mean something else.

One way to find the product of two numbers is to realize that *multiplication is just repeated addition*. Your knowledge of addition helped you learn your multiplication facts.

$$4 \cdot 3 \quad \text{means} \quad 3 + 3 + 3 + 3 \quad \text{so} \quad 4 \cdot 3 = 12$$

$$3 \cdot 4 \quad \text{means} \quad 4 + 4 + 4 \quad \text{so} \quad 3 \cdot 4 = 12$$

$$2 \cdot 5 \quad \text{means} \quad 5 + 5 \quad \text{so} \quad 2 \cdot 5 = 10$$

Relating Multiplication to Addition. For whole numbers A and B

$$A \cdot B \quad \text{means} \quad \underbrace{B + B \cdot \cdot \cdot \cdot + B}_{A \text{ terms}}$$

Example 7. Write the addition statement that corresponds to the subtraction statement $25 - 12 = 13$.

Solution

$$25 - 12 = 13$$

$$\text{means} \quad 25 = 12 + 13$$

You try Example 8.

Example 8. Write the product $5 \cdot 7$ as a sum.

Solution

$$5 \cdot 7 = \_\_\_\_ + \_\_\_\_ + \_\_\_\_ + \_\_\_\_ + \_\_\_\_$$

Check your answer on page 13. ▶

There are several ways to indicate the operation of division. To show 6 divided by 3, we may write

$$6 \div 3 \qquad 3\overline{)6} \qquad 6/3 \qquad \frac{6}{3}$$

Each of these expressions is equal to 2.

To understand the operation of division, you should remember that *every division statement corresponds to a multiplication statement.*

$$\frac{6}{3} = 2 \quad \text{because} \quad 6 = 3 \cdot 2$$

$$\frac{40}{8} = 5 \quad \text{because} \quad 40 = 8 \cdot 5$$

$$\frac{19}{1} = 19 \quad \text{because} \quad 19 = 1 \cdot 19$$

Relating Division to Multiplication. For whole numbers A, B, and C

$$\frac{A}{B} = C \quad \text{means} \quad A = B \cdot C$$

where B does not equal zero ($B \neq 0$).

Example 9. Write the multiplication statement that corresponds to the division statement $\frac{0}{15} = 0$.

Solution

$$\frac{0}{15} = 0$$

$$\text{means} \quad 0 = 15 \cdot 0$$

You complete Example 10.

Example 10. Write the multiplication statement that corresponds to the division statement $\frac{23}{23} = 1$.

Solution

$$\frac{23}{23} = 1$$

$$\text{means} \quad \_\_\_\_ = \_\_\_\_ \cdot \_\_\_\_$$

Check your answer on page 13. ▶

Trial Run

______ 1. Write the addition statement that corresponds to the subtraction statement $11 - 0 = 11$.

______ 2. Write the product $4 \cdot 2$ as a sum.

_______ **3.** Write the multiplication statement that corresponds to the division statement $\frac{18}{6} = 3$.

_______ **4.** Write the multiplication statement that corresponds to the division statement $\frac{0}{3} = 0$.

Answers are on page 15.

Working with Zero and One

When we operate with the whole numbers zero and 1, we observe some interesting facts. Consider the following sums:

$$13 + 0 = 13$$

$$0 + 7 = 7$$

Notice that

> If zero is added to any whole number, the answer is that whole number.

This law is called the **addition property of zero** (or the **identity property for addition**).

> **Addition Property of Zero.** For any whole number A
>
> $$A + 0 = A$$
>
> $$0 + A = A$$

We call zero the **identity for addition** because when zero is added to any number, the answer is identical to the original number.

Similarly, the differences

$$17 - 0 = 17$$

$$1 - 0 = 1$$

should help you see that

> If zero is subtracted from any whole number, the answer is that whole number.

We call this law the **subtraction property of zero**.

> **Subtraction Property of Zero.** For any whole number A
>
> $$A - 0 = A$$

It is also interesting to note here that if any whole number is subtracted from itself, the answer is zero.

$$7 - 7 = 0 \quad \text{because} \quad 7 = 7 + 0$$

$$96 - 96 = 0 \quad \text{because} \quad 96 = 96 + 0$$

For any whole number

$$A - A = 0$$

What happens when we multiply any number times zero? Recalling that multiplication is repeated addition, look at this product:

$$\begin{aligned} 5 \cdot 0 &= 0 + 0 + 0 + 0 + 0 \\ &= 0 \end{aligned}$$

If any whole number is multiplied times zero, the answer is 0.

This law is called the **multiplication property of zero**.

Multiplication Property of Zero. For any whole number A

$$A \cdot 0 = 0$$

$$0 \cdot A = 0$$

Use the properties of zero to complete Example 11.

Example 11. Complete each statement.

$13 + 0 = \_\_\_\_$ $\qquad 167 - 0 = \_\_\_\_$

$73 + \_\_\_\_ = 73$ $\qquad 29 \cdot 0 = \_\_\_\_$

$19 - \_\_\_\_ = 19$ $\qquad 0 \cdot 15 = \_\_\_\_$

$19 - \_\_\_\_ = 0$ $\qquad 1 \cdot \_\_\_\_ = 0$

Check your work on page 13. ▶

Let's investigate what happens when zero appears as the divisor or dividend in a division problem. First we shall try to divide a whole number *by* zero. Consider the quotient $\frac{6}{0}$. Remember that every division statement corresponds to a multiplication statement. Can we say

$$\frac{6}{0} = 0 \quad \text{because} \quad 6 = 0 \cdot 0?$$

or

$$\frac{6}{0} = 6 \quad \text{because} \quad 6 = 0 \cdot 6?$$

Of course, *neither* of these answers is correct. In fact, there is *no answer* for the problem $\frac{6}{0}$.

On the other hand, for the quotient $\frac{0}{0}$ the answer could be *any* number. For instance,

$$\frac{0}{0} = 3 \quad \text{because} \quad 0 = 0 \cdot 3$$

$$\frac{0}{0} = 0 \quad \text{because} \quad 0 = 0 \cdot 0$$

In this case there are *too many answers*.

Because division *by* zero always gives us no answer or too many answers, we agree that *division by zero is impossible*, or

> Division by zero is undefined.

There is no way to define division by zero to give us *one and only one* answer.

What happens if we try to divide zero by any *other* whole number? You should agree that

$$\frac{0}{6} = 0 \quad \text{because} \quad 0 = 6 \cdot 0$$

$$\frac{0}{1} = 0 \quad \text{because} \quad 0 = 1 \cdot 0$$

$$\frac{0}{795} = 0 \quad \text{because} \quad 0 = 795 \cdot 0$$

> If zero is divided by any *nonzero* number, the answer is zero.

To avoid confusion between division *by* zero and division *of* zero, we summarize the two situations here for any whole number A.

Division by Zero	Division of Zero
$\frac{A}{0}$ is undefined	$\frac{0}{A} = 0$ provided that $A \neq 0$

Use the facts about zero in division to complete Example 12.

Example 12. Complete each statement.

$\frac{0}{36} =$ ____ $\qquad \frac{?}{17} = 0$

$\frac{36}{0} = \_\_\_\_$ $\qquad \frac{0}{?}$ is undefined

$\frac{0}{0} = \_\_\_\_$ $\qquad \frac{17}{?}$ is undefined

Check your work on page 14. ▶

Now let us consider what happens when we multiply any whole number by the whole number 1. Look at this product:

$$5 \cdot 1 = 1 + 1 + 1 + 1 + 1$$

$$5 \cdot 1 = 5$$

If any whole number is multiplied times 1, the answer is that whole number.

This law is called the **multiplication property of 1** (or the **identity property for multiplication**).

Multiplication Property of 1. For any whole number A

$$A \cdot 1 = A$$

$$1 \cdot A = A$$

We call 1 the **identity for multiplication** because when we multiply any number by 1, the answer is identical to the original number.

Similarly, the quotients

$$\frac{26}{1} = 26$$

$$\frac{103}{1} = 103$$

should help you see that

If any whole number is divided by 1, the answer is that whole number.

We call this law the **division property of 1**.

Division Property of 1. For any whole number A

$$\frac{A}{1} = A$$

It is also interesting to note here that if any nonzero whole number is divided by itself, the answer is 1.

$$\frac{13}{13} = 1 \quad \text{because} \quad 13 = 13 \cdot 1$$

$$\frac{149}{149} = 1 \quad \text{because} \quad 149 = 149 \cdot 1$$

> For any whole number
>
> $\frac{A}{A} = 1$ provided that $A \neq 0$

Use the properties of 1 to complete Example 13.

Example 13. Complete each statement.

$78 \cdot 1 =$ ____ $\qquad \frac{?}{1} = 8$

____ $\cdot\ 1 = 16$ $\qquad \frac{33}{33} =$ ____

____ $\cdot\ 10 = 10$ $\qquad \frac{159}{?} = 1$

$\frac{98}{1} =$ ____

Check your answers on page 14. ▶

Trial Run

Complete each statement.

______ **1.** $12 + 0 =$ ____ ______ **2.** $46 -$ ____ $= 46$

______ **3.** $56 -$ ____ $= 0$ ______ **4.** $29 \cdot 1 =$ ____

______ **5.** $\frac{100}{1} =$ ____ ______ **6.** $\frac{?}{17} = 1$

Answers are on page 15.

Working with Exponents

When a whole number is repeated as a factor in a product, there is a handy notation that can be used.

Product	Notation	Read as:
$4 \cdot 4$	4^2	4 squared
$2 \cdot 2 \cdot 2$	2^3	2 cubed
$3 \cdot 3 \cdot 3 \cdot 3 \cdot 3$	3^5	3 to the fifth power

In these expressions the small raised whole number is called the **exponent** and it tells us how many times the other number, called the **base**, is to be used as a factor in a product. The product itself is called a **power**.

Example 14. Find 2^5.

Solution

$$2^5$$
$$= 2 \cdot 2 \cdot 2 \cdot 2 \cdot 2$$
$$= 32$$

You try Example 15.

Example 15. Find 4^3.

Solution

$$4^3$$
$$= \_\_\_\_ \ \_\_\_\_ \ \_\_\_\_$$
$$= \_\_\_\_$$

Check your work on page 14. ▶

Using Symbols of Grouping

When *parentheses* appear in an expression, they indicate the order in which operations are to be performed. Parentheses tell us to "do this first."

Example 16. Find $(3 + 2) \cdot 6$.

Solution

$$\underbrace{(3 + 2)} \cdot 6$$
$$= \quad 5 \quad \cdot 6$$
$$= 30$$

Example 17. Find $3 + (2 \cdot 6)$.

Solution

$$3 + \underbrace{(2 \cdot 6)}$$
$$= 3 + \quad 12$$
$$= 15$$

Notice in Examples 16 and 17 that the location of the parentheses made the problems completely different.

Parentheses ( ), brackets [ ], and braces { } are symbols of grouping that give you directions about the order in which operations are to be done. In problems containing parentheses, braces, and brackets, you must remember to *work inside the innermost symbols of grouping first*.

Example 18. Simplify $[2 + (8 \cdot 9) + 1] + 5$.

Solution

$$[2 + (8 \cdot 9) + 1] + 5$$
$$= [2 + 72 + 1] + 5$$
$$= 75 + 5$$
$$= 80$$

You complete Example 19.

Example 19. Simplify $(3 + 2)(8 + 1)$.

Solution

$$(3 + 2)(8 + 1)$$
$$= 5 \cdot \_\_\_\_$$
$$= \_\_\_\_$$

Check your work on page 14. ▶

The fraction bar acts like parentheses in division problems, telling us that we should simplify the dividend (top) and simplify the divisor (bottom) before performing the division.

Example 20. Simplify $\frac{7 + 8}{2 + 3}$.

Solution

$$\frac{7 + 8}{2 + 3}$$
$$= \frac{15}{5}$$
$$= 3$$

Now you complete Example 21.

Example 21. Simplify $\frac{5 + 19}{2 \cdot 3}$.

Solution

$$\frac{5 + 19}{2 \cdot 3}$$
$$= \frac{?}{6}$$
$$= \_\_\_\_$$

Check your work on page 14. ▶

But suppose that parentheses are not used in a problem which involves several operations. To avoid confusion, mathematicians have agreed that:

Order of Operations. If there are no parentheses to indicate order of operations, first deal with any exponents, then perform multiplications and/or divisions from left to right, and then perform additions and/or subtractions from left to right.

Example 22. Simplify $3^2 + \frac{42}{7}$.

Solution

$$3^2 + \frac{42}{7}$$
$$= 9 + \frac{42}{7}$$
$$= 9 + 6$$
$$= 15$$

You try Example 23.

Example 23. Simplify $5 \cdot 2^3 + 6 \cdot 8$.

Solution

$$5 \cdot 2^3 + 6 \cdot 8$$
$$= 5 \cdot \_\_\_\_ + 6 \cdot 8$$
$$= \_\_\_\_ + \_\_\_\_$$
$$= \_\_\_\_$$

Check your work on page 14. ▶

Example 24. Simplify $6\left(9 + \frac{9}{3}\right) - 2$.

Solution

$$6\left(9 + \frac{9}{3}\right) - 2$$
$$= 6(9 + 3) - 2$$
$$= 6(12) - 2$$
$$= 72 - 2$$
$$= 70$$

You complete Example 25.

Example 25. Simplify $\frac{20 - 3 \cdot 4}{2(5 - 3)}$.

Solution

$$\frac{20 - 3 \cdot 4}{2(5 - 3)}$$
$$= \frac{20 - 12}{2(\_\_\_\_)}$$
$$= \frac{?}{4}$$
$$= \_\_\_\_$$

Check your work on page 14. ▶

Trial Run

Simplify.

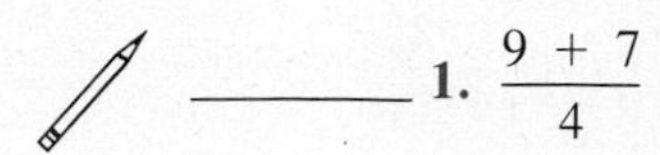

______ 1. $\frac{9 + 7}{4}$

______ 2. $4[5 + (3 \cdot 2)]$

______ 3. $(7 + 2)(3 + 1)$

______ 4. $3^2(7 + 0)$

______ 5. $5[2 + 7 \cdot 0]$

______ 6. $\frac{4 \cdot 9 - 6}{2(8 - 5)}$

______ 7. $5\left(6 + \frac{8}{2}\right) - 3$

______ 8. $7 + \frac{8 + 24}{4} + 2$

Answers are on page 15.

Examples You Completed

Example 2. Graph 0 on the number line.

Solution

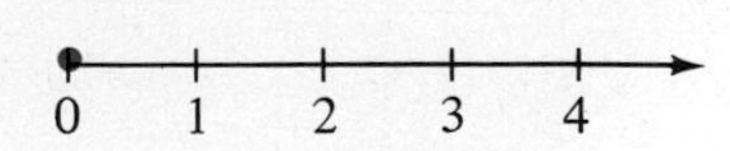

Example 4. Place a < or > symbol between the numbers.

5 ____ 7

110 ____ 99

Solution

$$5 < 7$$

$$110 > 99$$

Example 8. Write the product $5 \cdot 7$ as a sum.

Solution

$$5 \cdot 7 = 7 + 7 + 7 + 7 + 7$$

Example 10. Write the multiplication statement that corresponds to the division statement $\frac{23}{23} = 1$.

Solution

$$\frac{23}{23} = 1$$

means $23 = 23 \cdot 1$

Example 11. Complete each statement.

$13 + 0 = 13$	$167 - 0 = 167$
$73 + 0 = 73$	$29 \cdot 0 = 0$
$19 - 0 = 19$	$0 \cdot 15 = 0$
$19 - 19 = 0$	$1 \cdot 0 = 0$

Example 12. Complete each statement.

$\frac{0}{36} = 0$ $\frac{0}{17} = 0$

$\frac{36}{0} =$ undefined $\frac{0}{0}$ is undefined

$\frac{0}{0} =$ undefined $\frac{17}{0}$ is undefined

Example 13. Complete each statement.

$78 \cdot 1 = 78$ $\frac{8}{1} = 8$

$16 \cdot 1 = 16$ $\frac{33}{33} = 1$

$1 \cdot 10 = 10$ $\frac{159}{159} = 1$

$\frac{98}{1} = 98$

Example 15. Find 4^3.

Solution

$$4^3$$
$$= 4 \cdot 4 \cdot 4$$
$$= 64$$

Example 19. Simplify $(3 + 2)(8 + 1)$.

Solution

$$(3 + 2)(8 + 1)$$
$$= 5 \cdot 9$$
$$= 45$$

Example 21. Simplify $\frac{5 + 19}{2 \cdot 3}$.

Solution

$$\frac{5 + 19}{2 \cdot 3}$$
$$= \frac{24}{6}$$
$$= 4$$

Example 23. Simplify $5 \cdot 2^3 + 6 \cdot 8$.

Solution

$$5 \cdot 2^3 + 6 \cdot 8$$
$$= 5 \cdot 8 + 6 \cdot 8$$
$$= 40 + 48$$
$$= 88$$

Example 25. Simplify $\frac{20 - 3 \cdot 4}{2(5 - 3)}$.

Solution

$$\frac{20 - 3 \cdot 4}{2(5 - 3)}$$
$$= \frac{20 - 12}{2(2)}$$
$$= \frac{8}{4}$$
$$= 2$$

Answers to Trial Runs

page 3 **1.**  **2.** $7 > 5$ **3.** $0 < 2$

4. $186 < 187$ **5.** $2 > 0$

page 5 **1.** $11 = 0 + 11$ **2.** $2 + 2 + 2 + 2$ **3.** $18 = 6 \cdot 3$ **4.** $0 = 3 \cdot 0$

page 10 **1.** $12 + 0 = 12$ **2.** $46 - 0 = 46$ **3.** $56 - 56 = 0$ **4.** $29 \cdot 1 = 29$ **5.** $\frac{100}{1} = 100$

6. $\frac{17}{17} = 1$

page 13 **1.** 4 **2.** 44 **3.** 36 **4.** 63 **5.** 10 **6.** 5 **7.** 47 **8.** 17

Name ______________________________ Date ______________

EXERCISE SET 1.1

1. Graph the points corresponding to these numbers on a number line.
(a) 3 **(b)** 0 **(c)** 11 **(d)** 1

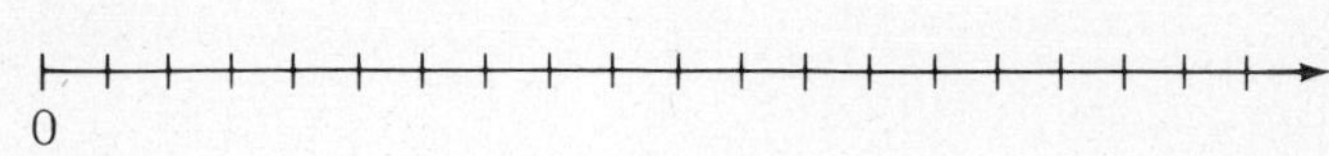

2. Graph the points corresponding to these numbers on a number line.
(a) 5 **(b)** 1 **(c)** 12 **(d)** 0

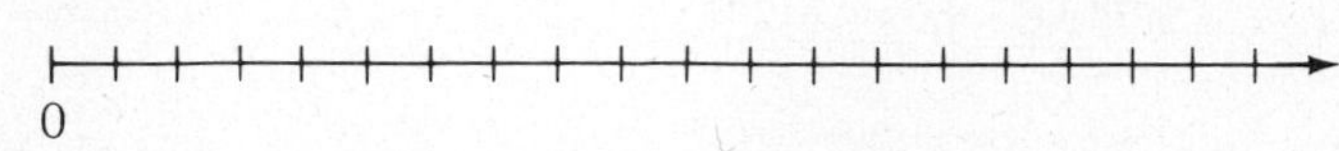

3. Tell what coordinate corresponds to each letter on the number line.

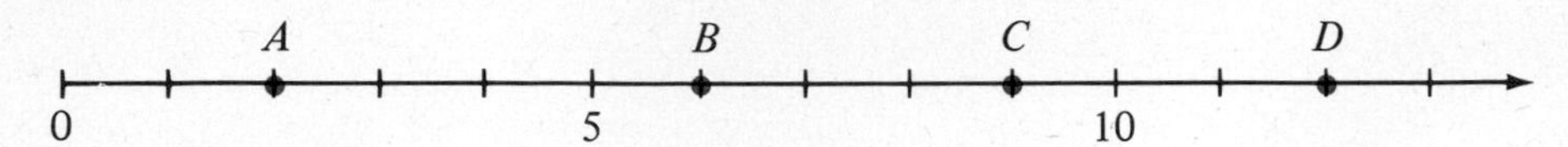

4. Tell what coordinate corresponds to each letter on the number line.

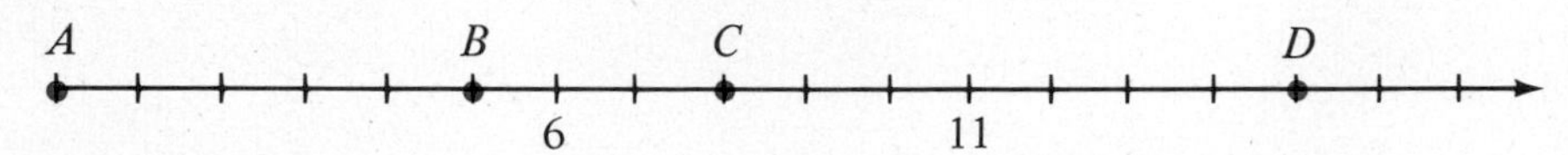

Compare the numbers using the symbols < or >.

5. 12 ____ 7

6. 23 ____ 15

7. 0 ____ 9

8. 1 ____ 10

9. 22 ____ 55

10. 43 ____ 17

______ **11.** Write the addition statement that corresponds to the subtraction statement $32 - 15 = 17$.

______ **12.** Write the addition statement that corresponds to the subtraction statement $28 - 19 = 9$.

______ **13.** Write the multiplication statement that corresponds to the division statement $\frac{35}{7} = 5$.

______ **14.** Write the multiplication statement that corresponds to the division statement $\frac{24}{6} = 4$.

Simplify by performing the operations.

______ **15.** $3 + 2 + 1$

______ **16.** $10 + 6 + 7$

______ **17.** $15 - 3$

______ **18.** $14 - 11$

______ **19.** $2 \cdot 3 \cdot 8$

______ **20.** $5 \cdot 6 \cdot 4$

_______ 21. $8(5 \cdot 4)$

_______ 22. $9(7 \cdot 2)$

_______ 23. $\frac{8 + 6}{7}$

_______ 24. $\frac{9 + 3}{4}$

_______ 25. $6 \cdot 7 + 5$

_______ 26. $4 + 9 \cdot 8$

_______ 27. $17 - (14 - 2)$

_______ 28. $27 - (9 - 3)$

_______ 29. $3 \cdot 5 + 7^2$

_______ 30. $6 \cdot 12 + 5^2$

_______ 31. $(8 + 2^2) - (6 + 1)$

_______ 32. $(13 + 1^3) - (2 + 3)$

_______ 33. $5(9 - 3)$

_______ 34. $7(8 - 1)$

_______ 35. $(9 + 7)(3 + 1)$

_______ 36. $(7 + 2)(3 + 3)$

_______ 37. $6^2 + \frac{5}{5}$

_______ 38. $9^2 + \frac{7}{7}$

_______ 39. $\frac{11 + 23}{13 + 4}$

_______ 40. $\frac{19 + 11}{5 + 10}$

_______ 41. $6 + 2(15 - 4)$

_______ 42. $7 + 3(14 - 1)$

_______ 43. $\frac{37 - 22}{8 - 8}$

_______ 44. $\frac{27 + 13}{11 - 11}$

_______ 45. $2(8 + 3 \cdot 2)$

_______ 46. $3(10 + 5 \cdot 3)$

_______ 47. $4 + [3^2 + 4(1 + 5)]$

_______ 48. $7 + [6^2 + 2(1 + 3)]$

_______ 49. $2^3\left(7 + \frac{10}{2}\right)$

_______ 50. $3^2\left(11 + \frac{16}{4}\right)$

_______ 51. $\frac{7(18 - 2 \cdot 3^2)}{4}$

_______ 52. $\frac{9(12 - 3 \cdot 2^2)}{8}$

_______ 53. $\frac{2^5 + 6 \cdot 5}{7 - \frac{15}{3}}$

_______ 54. $\frac{3^4 - 9 \cdot 3}{\frac{42}{6} + 2}$

_______ 55. $2\left[\frac{14}{7} + 3(8 - 2)\right]$

_______ 56. $3\left[4(9 - 3) + \frac{15}{3}\right]$

_______ 57. $7\left(8 - \frac{2 \cdot 8}{4}\right)$

_______ 58. $6\left(10 - \frac{6 \cdot 9}{27}\right)$

_______ 59. $11\left(2 \cdot 3 - \frac{23 - 5}{3}\right)$

_______ 60. $13\left(\frac{28}{7} - \frac{22 - 6}{4}\right)$

☆ Stretching the Topics

Simplify by performing the operations.

_______ **1.** $2^3\left\{[(2 \cdot 3)^2 + 7(3 + 2)] - \frac{128}{2}\right\}$

_______ **2.** $\dfrac{5\left[2^2(3) - \frac{40}{5}\right]}{3^3 - 7}$

_______ **3.** $15 - \{5[16 - (5 - 2)^2] - 4[(9 - 7)^3 - 3]\}$

Check your answers in the back of your book.

If you can simplify the expressions in **Checkup 1.1**, you are ready to go on to Section 1.2.

Name ______________________ Date ____________

CHECKUP 1.1

Simplify. Perform the operations.

_______ **1.** $(2^5 - 12) \cdot 4$

_______ **2.** $4^3 + (3 \cdot 12)$

_______ **3.** $(8 - 3)(15 - 7)$

_______ **4.** $\dfrac{19 + 3}{17 + 15}$

_______ **5.** $\dfrac{8 + 28}{3 \cdot 2}$

_______ **6.** $[(4 \cdot 6) - 3^2] + \dfrac{48}{6}$

_______ **7.** $4 \cdot 3^2 + 8 \cdot 7$

_______ **8.** $8\left(15 - \dfrac{12}{2}\right) + 13$

_______ **9.** $\dfrac{13(8 - 4 \cdot 2)}{5 \cdot 6 - 4 \cdot 5}$

_______ **10.** $\dfrac{5 + 8 \cdot 4}{28 - 4 \cdot 7}$

Check your answers in the back of your book.

If You Missed Problems:	You Should Review Examples:
1, 2	14–17
3	19
4	20, 13
5	21
6	18, 22
7, 8	23, 24
9, 10	25, 12

1.2 Working with Fractions

So far we have learned the rules for adding, subtracting, multiplying, and dividing whole numbers, but there are many useful numbers that are *not* whole numbers. Numbers such as $\frac{1}{2}$, $\frac{5}{3}$, and $\frac{17}{25}$ are called **fractions**. If the numerator of a fraction is smaller than the denominator, it is called a **proper fraction**. If not, it is called an **improper fraction**. Every improper fraction can be written as a whole number or a **mixed number**. A mixed number contains a whole number part and a fractional part.

Example 1. Complete each statement.

$\frac{1}{2}$ is a(n) ________ fraction.

$\frac{5}{3}$ is a(n) ________ fraction.

Check your answers on page 33. ▶

Example 2. Write $\frac{5}{3}$ as a mixed number.

Solution

$$\frac{5}{3}$$

$$= \frac{3}{3} + \frac{2}{3}$$

$$= 1\frac{2}{3}$$

Reducing Fractions

Sometimes we can find a new fraction that is equivalent to a certain fraction by a process called **reducing the fraction to lowest terms**. To reduce a fraction, we first look for any number(s) that will divide evenly into both the numerator and denominator (with no remainder). Such numbers are called **common factors** for the numerator and denominator. We then divide the numerator and denominator by any common factors.

For instance, to reduce $\frac{21}{33}$ we first look for common factors. One way to do this is to rewrite the numerator and denominator as products of factors.

$$\frac{21}{33} = \frac{3 \cdot 7}{3 \cdot 11}$$ Rewrite numerator and denominator.

$$= \frac{\overset{1}{\cancel{3}} \cdot 7}{\underset{1}{\cancel{3}} \cdot 11}$$ Divide numerator and denominator by the common factor 3. Remember, $\frac{3}{3} = 1$.

$$= \frac{7}{11}$$ Write the equivalent fraction.

Reducing Fractions

1. Write numerator and denominator as products of factors.
2. Divide numerator and denominator by any common factors.

Example 3. Reduce $\frac{24}{84}$.

Solution

$$\frac{24}{84}$$

$$= \frac{2 \cdot 2 \cdot 2 \cdot 3}{2 \cdot 2 \cdot 3 \cdot 7}$$

$$= \frac{\overset{1}{\cancel{2}} \cdot \overset{1}{\cancel{2}} \cdot 2 \cdot \overset{1}{\cancel{3}}}{\underset{1}{\cancel{2}} \cdot \underset{1}{\cancel{2}} \cdot \underset{1}{\cancel{3}} \cdot 7}$$

$$= \frac{2}{7}$$

Now try Example 4.

Example 4. Reduce $\frac{110}{25}$.

Solution

Check your work on page 33. ▶

Trial Run

Describe each fraction as proper or improper.

______ **1.** $\frac{16}{29}$ ______ **2.** $\frac{93}{92}$

Reduce each fraction.

______ **3.** $\frac{28}{44}$ ______ **4.** $\frac{70}{42}$

______ **5.** $\frac{51}{93}$ ______ **6.** $\frac{63}{180}$

Answers are on page 34.

Multiplying and Dividing Fractions

From previous work in arithmetic we remember that to multiply fractions, we multiply the numerators and multiply the denominators.

Multiplying Fractions

$$\frac{A}{B} \cdot \frac{C}{D} = \frac{A \cdot C}{B \cdot D} \qquad B \neq 0, D \neq 0$$

You try multiplying the fractions in Example 5.

Example 5. Find $\frac{1}{3} \cdot \frac{2}{5}$.

Solution

$$\frac{1}{3} \cdot \frac{2}{5}$$

$$= \frac{1 \cdot 2}{3 \cdot 5}$$

$$= \underline{\qquad}$$

Check your work on page 33. ▶

Example 6. Find $\frac{11}{3} \cdot \left(\frac{2}{3}\right)^2$.

Solution

$$\frac{11}{3} \cdot \left(\frac{2}{3}\right)^2$$

$$= \frac{11}{3} \cdot \frac{2}{3} \cdot \frac{2}{3}$$

$$= \frac{11 \cdot 2 \cdot 2}{3 \cdot 3 \cdot 3}$$

Notice that we leave our answers in fractional form. For work in algebra, fractional form is always preferred to mixed number form.

Now let's use three approaches to find the product $\frac{3}{8} \cdot \frac{4}{9}$.

Approach 1

$$\frac{3}{8} \cdot \frac{4}{9} = \frac{3 \cdot 4}{8 \cdot 9}$$

$$= \frac{12}{72}$$

Now reduce.

$$\frac{12}{72} = \frac{\overset{1}{\cancel{12}} \cdot 1}{\underset{1}{\cancel{12}} \cdot 6}$$

$$= \frac{1}{6}$$

Approach 2

$$\frac{3}{8} \cdot \frac{4}{9} = \frac{3 \cdot 4}{8 \cdot 9}$$

Now reduce.

$$\frac{3 \cdot 4}{8 \cdot 9} = \frac{\overset{1}{\cancel{3}} \cdot \overset{1}{\cancel{4}}}{2 \cdot \underset{1}{\cancel{4}} \cdot \underset{1}{\cancel{3}} \cdot 3}$$

$$= \frac{1}{6}$$

Approach 3

First factor numerators and denominators and divide by factors common to any numerator and any denominator.

$$\frac{3}{8} \cdot \frac{4}{9} = \frac{\overset{1}{\cancel{3}}}{\underset{1}{\cancel{4}} \cdot 2} \cdot \frac{\overset{1}{\cancel{4}}}{\underset{1}{\cancel{3}} \cdot 3}$$

$$= \frac{1}{6}$$

Each of these approaches gives the same answer. Since the third approach requires the fewest steps, it will be our preferred method for multiplying fractions. From now on, we shall always *reduce first, then multiply*.

Example 7. Multiply $\frac{15}{36} \cdot \frac{44}{75}$.

Solution

$$\frac{15}{36} \cdot \frac{44}{75}$$

$$= \frac{\overset{1}{\cancel{3}} \cdot \overset{1}{\cancel{5}}}{\underset{1}{\cancel{4}} \cdot 9} \cdot \frac{\overset{1}{\cancel{4}} \cdot 11}{\underset{1}{\cancel{3}} \cdot \underset{1}{\cancel{5}} \cdot 5}$$

$$= \frac{1 \cdot 11}{9 \cdot 5}$$

$$= \frac{11}{45}$$

You complete Example 8.

Example 8. Multiply $\frac{3}{2} \cdot \frac{5}{9} \cdot \frac{4}{15}$.

Solution

$$\frac{3}{2} \cdot \frac{5}{9} \cdot \frac{4}{15}$$

$$= \frac{3}{2} \cdot \frac{5}{3 \cdot 3} \cdot \frac{2 \cdot 2}{3 \cdot 5}$$

Check your work on page 33. ▶

In a division problem, the number being divided is called the *dividend*; the number doing the dividing is called the *divisor*; the answer is called the *quotient*. Recall the steps for dividing fractions.

> **Dividing Fractions**
>
> 1. Keep the dividend as it is.
> 2. Invert the divisor (turn it upside down).
> 3. Multiply, reducing if possible.
>
> $$\frac{A}{B} \div \frac{C}{D} = \frac{A}{B} \cdot \frac{D}{C} \qquad B, C, D \neq 0$$

Example 9. Find $\frac{9}{16} \div \frac{3}{8}$.

Solution

$$\frac{9}{16} \div \frac{3}{8}$$

$$= \frac{9}{16} \cdot \frac{8}{3}$$

$$= \frac{3 \cdot \overset{1}{\cancel{3}}}{2 \cdot \underset{1}{\cancel{8}}} \cdot \frac{\overset{1}{\cancel{8}}}{\underset{1}{\cancel{3}}}$$

$$= \frac{3}{2}$$

Now you complete Example 10.

Example 10. Find $\frac{2}{3} \div 6$.

Solution

$$\frac{2}{3} \div 6$$

$$= \frac{2}{3} \div \frac{6}{1}$$

$$= \frac{2}{3} \cdot \frac{1}{6}$$

Check your work on page 33. ▶

Symbols of grouping continue to tell us what to do first.

Example 11. Find $\frac{1}{7} \div \left(\frac{2}{3} \cdot \frac{9}{14}\right)$.

Solution

$$\frac{1}{7} \div \left(\frac{2}{3} \cdot \frac{9}{14}\right)$$

$$= \frac{1}{7} \div \left(\frac{\overset{1}{\cancel{2}}}{\underset{1}{\cancel{3}}} \cdot \frac{\overset{1}{\cancel{3}} \cdot 3}{\underset{1}{\cancel{2}} \cdot 7}\right)$$

$$= \frac{1}{7} \div \frac{3}{7}$$

$$= \frac{1}{7} \cdot \frac{7}{3}$$

$$= \frac{1}{3}$$

Trial Run

Simplify.

______ **1.** $\frac{20}{21} \cdot \frac{14}{15}$

______ **2.** $\frac{4}{5} \cdot 15$

______ **3.** $\frac{3}{8} \cdot \left(\frac{4}{3}\right)^2$

______ **4.** $\frac{2}{3} \cdot \frac{5}{14} \cdot \frac{33}{25}$

______ **5.** $\frac{8}{17} \div \frac{4}{5}$

______ **6.** $\frac{4}{5} \div 3$

______ **7.** $\frac{1}{15} \div \frac{5}{9}$

______ **8.** $\left(\frac{2}{3} \cdot \frac{15}{22}\right) \div \frac{5}{11}$

Answers are on page 34.

Building Fractions

We have spent some time reducing fractions to simplest form by dividing numerator and denominator by common factors. Now we must learn how to "build up" a fraction so that it will have a specified denominator. We do this by *multiplying* numerator and denominator by the same factor.

Suppose that we wish to write the fraction $\frac{1}{2}$ as a new fraction with a denominator of 10.

$\frac{1}{2} = \frac{?}{10}$	We must find a numerator for the fraction on the right. By what factor must 2 be multiplied to give 10?
$\frac{1}{2} = \frac{?}{2(5)}$	The factor is 5.
$\frac{1}{2} = \frac{1(5)}{2(5)}$	Since we multiplied the old denominator by 5, we must also multiply the old numerator by 5, because $\frac{5}{5} = 1$.
$\frac{1}{2} = \frac{5}{10}$	We have found our new fraction.

Example 12. Find the new fraction.

$$\frac{2}{3} = \frac{?}{21}$$

Solution

$$\frac{2}{3} = \frac{?}{21}$$

$$\frac{2}{3} = \frac{?}{3(7)}$$

$$\frac{2}{3} = \frac{2(7)}{3(7)}$$

$$\frac{2}{3} = \frac{14}{21}$$

Now you complete Example 13.

Example 13. Find the new fraction.

$$\frac{5}{9} = \frac{?}{180}$$

Solution

$$\frac{5}{9} = \frac{?}{180}$$

$$\frac{5}{9} = \frac{?}{9(20)}$$

$$\frac{5}{9} = \frac{5(?)}{9(20)}$$

$$\frac{5}{9} = \frac{?}{180}$$

Check your work on page 33. ▶

Trial Run

Find the new fraction.

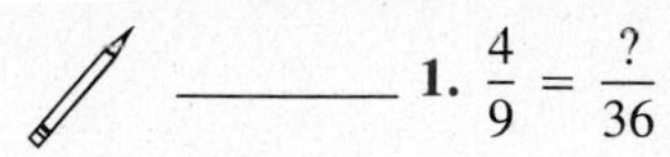

______ 1. $\frac{4}{9} = \frac{?}{36}$

______ 2. $\frac{3}{8} = \frac{?}{24}$

______ 3. $\frac{5}{12} = \frac{?}{84}$

______ 4. $\frac{2}{7} = \frac{?}{91}$

Answers are on page 34.

Adding and Subtracting Fractions with the Same Denominator

To add (or subtract) fractions with the same denominator, we must add (or subtract) the numerators and keep the same denominator.

Adding Fractions	**Subtracting Fractions**
$\frac{A}{C} + \frac{B}{C} = \frac{A + B}{C}$	$\frac{A}{C} - \frac{B}{C} = \frac{A - B}{C}$

Example 14. Find $\frac{1}{9} + \frac{4}{9}$.

Solution

$$\frac{1}{9} + \frac{4}{9}$$

$$= \frac{1 + 4}{9}$$

$$= \frac{5}{9}$$

You try Example 15.

Example 15. Find $\frac{5}{3} - \frac{2}{3}$.

Solution

$$\frac{5}{3} - \frac{2}{3}$$

Did you remember to reduce? Check your work on page 34. ▶

Example 16. Find $\frac{3}{29} + \frac{9}{29} - \frac{12}{29}$.

Solution

$$\frac{3}{29} + \frac{9}{29} - \frac{12}{29}$$

$$= \frac{3 + 9 - 12}{29}$$

$$= \frac{12 - 12}{29}$$

$$= \frac{0}{29}$$

$$= 0$$

You complete Example 17.

Example 17. Find $\frac{8}{45} + \frac{14}{45} - \frac{7}{45}$.

Solution

$$\frac{8}{45} + \frac{14}{45} - \frac{7}{45}$$

$$= \frac{8 + 14 - 7}{45}$$

Did you remember to reduce? Check your work on page 34. ▶

Trial Run

Find the sums or differences.

______ 1. $\frac{3}{8} + \frac{2}{8}$

______ 2. $\frac{7}{15} - \frac{4}{15}$

______ 3. $\frac{16}{21} - \frac{5}{21} - \frac{11}{21}$

______ 4. $\frac{21}{20} + \frac{3}{20} - \frac{7}{20}$

Answers are on page 34.

Adding and Subtracting Fractions with Different Denominators

If we wish to add or subtract fractions with different denominators, we must make all the denominators *match*. First we decide on a good choice for a new denominator and then *build up* each of the original fractions to a new fraction having the new denominator.

Consider the sum

$$\frac{2}{3} + \frac{1}{2} + \frac{5}{6}$$

What would be a good choice for a new denominator? We may choose any number that is a multiple of 3 and 2 and 6. The numbers, 6, 12, 18, 24, 30, and so on, are all good choices, but the *best* choice is the *smallest* number in that list. We shall use 6.

$$\frac{2}{3} + \frac{1}{2} + \frac{5}{6}$$

$$= \frac{?}{6} + \frac{?}{6} + \frac{?}{6}$$ From each old fraction we must build a new fraction with a denominator of 6.

$$= \frac{?}{3 \cdot 2} + \frac{?}{2 \cdot 3} + \frac{?}{6 \cdot 1}$$ Decide by what factor each old denominator was multiplied.

$$= \frac{2 \cdot 2}{3 \cdot 2} + \frac{1 \cdot 3}{2 \cdot 3} + \frac{5 \cdot 1}{6 \cdot 1}$$ Now multiply the old numerator by the same factor.

$$= \frac{4}{6} + \frac{3}{6} + \frac{5}{6}$$ Simplify each numerator and denominator by multiplying the factors.

$$= \frac{4 + 3 + 5}{6}$$ Combine the numerators over the same denominator.

$$= \frac{12}{6}$$ Simplify.

$$= 2$$ Reduce.

In this example, the number 6 is called the **lowest common denominator**.

Lowest Common Denominator (LCD) The lowest common denominator for a set of fractions is the smallest number that is a multiple of each of the denominators of the original fractions.

To find the LCD for a set of fractions, we must consider the factors of the original denominators. *The LCD must contain any factor that appears in any of the original denominators.*

Consider the following sum.

$$\frac{6}{7} + \frac{4}{5} - \frac{3}{10}$$

$$= \frac{6}{7} + \frac{4}{5} - \frac{3}{2 \cdot 5}$$ Factor the original denominators. Select the common denominator. LCD: $7 \cdot 5 \cdot 2$.

$$= \frac{?}{7 \cdot 5 \cdot 2} + \frac{?}{7 \cdot 5 \cdot 2} - \frac{?}{7 \cdot 5 \cdot 2}$$ Notice by what building factor each denominator was multiplied to get the LCD.

$= \frac{6 \cdot 5 \cdot 2}{7 \cdot 5 \cdot 2} + \frac{4 \cdot 7 \cdot 2}{7 \cdot 5 \cdot 2} - \frac{3 \cdot 7}{7 \cdot 5 \cdot 2}$ Multiply each numerator by that building factor.

$= \frac{60}{70} + \frac{56}{70} - \frac{21}{70}$ Simplify each numerator and denominator, by multiplying the factors.

$= \frac{60 + 56 - 21}{70}$ Combine numerators over the common denominator.

$= \frac{95}{70}$ Simplify.

$= \frac{5 \cdot 19}{5 \cdot 14}$ Factor numerator and denominator.

$= \frac{19}{14}$ Reduce the fraction.

Let's summarize the steps we must use to find the LCD.

Finding the LCD

1. Factor the denominators of the original fractions.
2. Decide what different factors appear in the denominators.
3. Notice the most times each factor appears in any single denominator.
4. The LCD is the product of all the different factors, with each factor used the most times that it appears in any single denominator.

Let's look at another sum.

Example 18. Find $\frac{5}{9} + \frac{1}{12}$.

Solution

$\frac{5}{9} + \frac{1}{12}$

$= \frac{5}{3 \cdot 3} + \frac{1}{3 \cdot 2 \cdot 2}$ To form the LCD, we must use any factor the *most* number of times that it appears in any *single* denominator; LCD: $3 \cdot 3 \cdot 2 \cdot 2$.

$= \frac{5 \cdot 2 \cdot 2}{3 \cdot 3 \cdot 2 \cdot 2} + \frac{1 \cdot 3}{3 \cdot 3 \cdot 2 \cdot 2}$ Build each fraction up to a new fraction having the LCD.

$= \frac{20}{36} + \frac{3}{36}$ Simplify numerators and denominators.

$= \frac{20 + 3}{36}$ Combine numerators over one denominator.

$= \frac{23}{36}$ Simplify.

Now you try to complete Example 19.

Example 19. Find $\frac{7}{25} + \frac{3}{2} - \frac{1}{8}$.

Solution

$$\frac{7}{25} + \frac{3}{2} - \frac{1}{8}$$

$$= \frac{7}{5 \cdot 5} + \frac{3}{2} - \frac{1}{2 \cdot 2 \cdot 2}$$

LCD: $5 \cdot 5 \cdot 2 \cdot 2 \cdot 2 = 200$.

$$= \frac{7 \cdot 2 \cdot 2 \cdot 2}{5 \cdot 5 \cdot 2 \cdot 2 \cdot 2} + \frac{3 \cdot 5 \cdot 5 \cdot 2 \cdot 2}{5 \cdot 5 \cdot 2 \cdot 2 \cdot 2} - \frac{1 \cdot 5 \cdot 5}{5 \cdot 5 \cdot 2 \cdot 2 \cdot 2}$$

$$= \frac{?}{200} + \frac{?}{200} - \frac{?}{200}$$

Check your work on page 34. ▶

Trial Run

Find the sum or diffference.

_______ 1. $\frac{3}{10} + \frac{5}{14}$

_______ 2. $\frac{3}{4} + \frac{2}{5} - \frac{5}{6}$

_______ 3. $\frac{5}{2} - \frac{1}{4} - \frac{11}{18}$

_______ 4. $\frac{9}{16} + \frac{7}{36} - \frac{2}{3}$

_______ 5. $\frac{11}{45} - \frac{7}{30}$

_______ 6. $\frac{7}{6} - \frac{1}{3} - \frac{5}{7}$

Answers are on page 34.

▶ Examples You Completed

Example 1. Complete each statement.

$\frac{1}{2}$ is a proper fraction.

$\frac{5}{3}$ is an improper fraction.

Example 4. Reduce $\frac{110}{25}$.

Solution

$$\frac{110}{25}$$
$$= \frac{2 \cdot 5 \cdot 11}{5 \cdot 5}$$
$$= \frac{2 \cdot \overset{1}{\cancel{5}} \cdot 11}{5 \cdot \underset{1}{\cancel{5}}}$$
$$= \frac{22}{5}$$

Example 5. Find $\frac{1}{3} \cdot \frac{2}{5}$.

Solution

$$\frac{1}{3} \cdot \frac{2}{5}$$
$$= \frac{1 \cdot 2}{3 \cdot 5}$$
$$= \frac{2}{15}$$

Example 8. Multiply $\frac{3}{2} \cdot \frac{5}{9} \cdot \frac{4}{15}$.

Solution

$$\frac{3}{2} \cdot \frac{5}{9} \cdot \frac{4}{15}$$
$$= \frac{3}{2} \cdot \frac{5}{3 \cdot 3} \cdot \frac{2 \cdot 2}{3 \cdot 5}$$
$$= \frac{\overset{1}{\cancel{3}}}{\underset{1}{\cancel{2}}} \cdot \frac{\overset{1}{\cancel{5}}}{\underset{1}{\cancel{3}} \cdot 3} \cdot \frac{\overset{1}{\cancel{2}} \cdot 2}{3 \cdot \underset{1}{\cancel{5}}}$$
$$= \frac{2}{9}$$

Example 10. Find $\frac{2}{3} \div 6$.

Solution

$$\frac{2}{3} \div 6$$
$$= \frac{2}{3} \div \frac{6}{1}$$
$$= \frac{2}{3} \cdot \frac{1}{6}$$
$$= \frac{\overset{1}{\cancel{2}}}{3} \cdot \frac{1}{\underset{1}{\cancel{2}} \cdot 3}$$
$$= \frac{1}{9}$$

Example 13. Find the new fraction.

$$\frac{5}{9} = \frac{?}{180}$$

Solution

$$\frac{5}{9} = \frac{?}{180}$$
$$\frac{5}{9} = \frac{?}{9(20)}$$
$$\frac{5}{9} = \frac{5(20)}{9(20)}$$
$$\frac{5}{9} = \frac{100}{180}$$

Example 15. Find $\frac{5}{3} - \frac{2}{3}$.

Solution

$$\begin{aligned} &\frac{5}{3} - \frac{2}{3} \\ &= \frac{5 - 2}{3} \\ &= \frac{3}{3} \\ &= 1 \end{aligned}$$

Example 17. Find $\frac{8}{45} + \frac{14}{45} - \frac{7}{45}$.

Solution

$$\begin{aligned} &\frac{8}{45} + \frac{14}{45} - \frac{7}{45} \\ &= \frac{8 + 14 - 7}{45} \\ &= \frac{22 - 7}{45} \\ &= \frac{15}{45} \\ &= \frac{1}{3} \end{aligned}$$

Example 19. Find $\frac{7}{25} + \frac{3}{2} - \frac{1}{8}$.

Solution

$$\begin{aligned} &\frac{7}{25} + \frac{3}{2} - \frac{1}{8} \\ &= \frac{7}{5 \cdot 5} + \frac{3}{2} - \frac{1}{2 \cdot 2 \cdot 2} \qquad \text{LCD: } 5 \cdot 5 \cdot 2 \cdot 2 \cdot 2 = 200. \\ &= \frac{7 \cdot 2 \cdot 2 \cdot 2}{5 \cdot 5 \cdot 2 \cdot 2 \cdot 2} + \frac{3 \cdot 5 \cdot 5 \cdot 2 \cdot 2}{5 \cdot 5 \cdot 2 \cdot 2 \cdot 2} - \frac{1 \cdot 5 \cdot 5}{5 \cdot 5 \cdot 2 \cdot 2 \cdot 2} \\ &= \frac{56}{200} + \frac{300}{200} - \frac{25}{200} \\ &= \frac{56 + 300 - 25}{200} \\ &= \frac{331}{200} \end{aligned}$$

Answers to Trial Runs

page 24 **1.** Proper **2.** Improper **3.** $\frac{7}{11}$ **4.** $\frac{5}{3}$ **5.** $\frac{17}{31}$ **6.** $\frac{7}{20}$

page 27 **1.** $\frac{8}{9}$ **2.** 12 **3.** $\frac{2}{3}$ **4.** $\frac{11}{35}$ **5.** $\frac{10}{17}$ **6.** $\frac{4}{15}$ **7.** $\frac{3}{25}$ **8.** 1

page 28 **1.** $\frac{16}{36}$ **2.** $\frac{9}{24}$ **3.** $\frac{35}{84}$ **4.** $\frac{26}{91}$

page 29 **1.** $\frac{5}{8}$ **2.** $\frac{1}{5}$ **3.** 0 **4.** $\frac{17}{20}$

page 32 **1.** $\frac{23}{35}$ **2.** $\frac{19}{60}$ **3.** $\frac{59}{36}$ **4.** $\frac{13}{144}$ **5.** $\frac{1}{90}$ **6.** $\frac{5}{42}$

Name ______________________ Date __________

EXERCISE SET 1.2

Reduce the fractions.

______ 1. $\frac{3}{15}$

______ 2. $\frac{6}{18}$

______ 3. $\frac{21}{15}$

______ 4. $\frac{30}{25}$

______ 5. $\frac{9}{16}$

______ 6. $\frac{25}{36}$

______ 7. $\frac{25}{35}$

______ 8. $\frac{49}{63}$

______ 9. $\frac{18}{12}$

______ 10. $\frac{24}{21}$

______ 11. $\frac{7}{49}$

______ 12. $\frac{9}{81}$

Find the product or quotient.

______ 13. $\frac{1}{3} \cdot \frac{2}{7}$

______ 14. $\frac{2}{5} \cdot \frac{1}{3}$

______ 15. $\frac{1}{2} \cdot 6$

______ 16. $\frac{1}{5} \cdot 10$

______ 17. $\left(\frac{2}{3}\right)^2 \cdot \frac{9}{10}$

______ 18. $\left(\frac{3}{5}\right)^2 \cdot \frac{25}{33}$

______ 19. $\frac{2}{3} \cdot \frac{3}{5} \cdot \frac{1}{6}$

______ 20. $\frac{4}{7} \cdot \frac{7}{8} \cdot \frac{1}{3}$

______ 21. $\frac{1}{4} \div \left(\frac{1}{6}\right)^2$

______ 22. $\frac{1}{9} \div \left(\frac{1}{12}\right)^2$

______ 23. $\frac{2}{3} \div \frac{5}{7}$

______ 24. $\frac{2}{5} \div \frac{7}{9}$

______ 25. $\frac{4}{5} \div \frac{12}{35}$

______ 26. $\frac{3}{8} \div \frac{15}{16}$

______ 27. $\frac{3}{5} \div 9$

______ 28. $\frac{7}{8} \div 21$

______ 29. $\frac{20}{7} \div \frac{16}{77}$

______ 30. $\frac{18}{5} \div \frac{27}{55}$

_____ 31. $\frac{8}{25} \div \left(\frac{24}{35} \cdot \frac{14}{27}\right)$

_____ 32. $\frac{10}{66} \cdot \left(\frac{8}{9} \div \frac{25}{33}\right)$

Find the new fractions.

_____ 33. $\frac{6}{7} = \frac{?}{49}$

_____ 34. $\frac{3}{5} = \frac{?}{25}$

_____ 35. $\frac{8}{3} = \frac{?}{36}$

_____ 36. $\frac{9}{4} = \frac{?}{48}$

_____ 37. $\frac{4}{11} = \frac{?}{99}$

_____ 38. $\frac{2}{13} = \frac{?}{65}$

_____ 39. $\frac{17}{15} = \frac{?}{45}$

_____ 40. $\frac{13}{12} = \frac{?}{60}$

_____ 41. $\frac{7}{8} = \frac{?}{72}$

_____ 42. $\frac{7}{10} = \frac{?}{120}$

_____ 43. $9 = \frac{?}{5}$

_____ 44. $7 = \frac{?}{4}$

_____ 45. $\frac{31}{84} = \frac{?}{252}$

_____ 46. $\frac{29}{132} = \frac{?}{396}$

Find the sum or difference and write the answers in reduced form.

_____ 47. $\frac{3}{5} + \frac{1}{5}$

_____ 48. $\frac{4}{5} - \frac{1}{5}$

_____ 49. $\frac{2}{27} + \frac{7}{27}$

_____ 50. $\frac{3}{8} + \frac{1}{8}$

_____ 51. $\frac{7}{13} - \frac{2}{13}$

_____ 52. $\frac{11}{15} - \frac{4}{15}$

_____ 53. $\frac{3}{11} + \frac{2}{11} - \frac{4}{11}$

_____ 54. $\frac{9}{13} + \frac{2}{13} - \frac{10}{13}$

_____ 55. $\frac{9}{11} - \frac{2}{11} + \frac{4}{11}$

_____ 56. $\frac{10}{17} - \frac{5}{17} + \frac{12}{17}$

_____ 57. $\frac{4}{5} - \frac{3}{5} + \frac{2}{5} - \frac{3}{5}$

_____ 58. $\frac{4}{7} - \frac{2}{7} + \frac{3}{7} - \frac{5}{7}$

_____ 59. $\frac{1}{3} + \frac{1}{5}$

_____ 60. $\frac{1}{7} + \frac{1}{9}$

_____ 61. $\frac{2}{3} - \frac{4}{9}$

_____ 62. $\frac{3}{5} - \frac{11}{25}$

______ 63. $3 - \frac{4}{5}$

______ 64. $2 - \frac{3}{7}$

______ 65. $\frac{7}{12} - \frac{11}{28} + \frac{17}{21}$

______ 66. $\frac{3}{10} + \frac{12}{35} + \frac{5}{14}$

______ 67. $\frac{29}{45} - \frac{17}{36} + \frac{7}{20}$

______ 68. $\frac{11}{18} - \frac{37}{63} + \frac{3}{14}$

______ 69. $\frac{8}{15} + \frac{24}{35} + \frac{1}{21}$

______ 70. $\frac{20}{21} - \frac{13}{28} + \frac{7}{12}$

☆ Stretching the Topics

Perform the operations indicated.

______ 1. $\left(\frac{3}{5} \cdot \frac{7}{9}\right) + \left(\frac{9}{10} \cdot \frac{5}{18}\right)$

______ 2. $\frac{4}{5}\left[\left(\frac{3}{4} + \frac{2}{5}\right) - \left(\frac{3}{8} + \frac{5}{12}\right)\right]$

______ 3. $\frac{1}{2}\left\{\left[\frac{5}{6} - \left(\frac{2}{3}\right)^2\right] \div \left(\frac{5}{6} \cdot \frac{7}{3}\right)\right\}$

Check your answers in the back of your book.

If you can complete **Checkup 1.2**, you are ready to go on to Section 1.3.

Name ______________________ Date ____________

CHECKUP 1.2

______ **1.** Reduce $\frac{49}{63}$.

______ **2.** Find the new fraction. $\frac{5}{7} = \frac{?}{56}$

Perform the operations and reduce if possible.

______ **3.** $\frac{5}{9} \cdot \frac{6}{35}$

______ **4.** $\frac{2}{3} \div \frac{10}{9}$

______ **5.** $12 \div \frac{4}{5}$

______ **6.** $\frac{9}{25} \div \left(\frac{12}{35} \cdot \frac{28}{39}\right)$

______ **7.** $\frac{2}{3} - \frac{4}{21}$

______ **8.** $5 - \frac{2}{3}$

______ **9.** $\frac{1}{7} + \frac{3}{8}$

______ **10.** $\frac{8}{15} + \frac{24}{35} - \frac{1}{21}$

Check your answers in the back of your book.

If You Missed Problem:	You Should Review Examples:
1	3
2	12
3	5
4, 5	9, 10
6	11
7–9	14, 15
10	17

1.3 Working with Decimals and Percents

When you receive a check for \$73.28 or when you read that unemployment has reached 10.2%, you are being asked to deal with **decimal numbers**. Every decimal number has two parts, separated by the decimal point. The part to the left of the decimal point is the **whole-number part**, and the part to the right of the decimal point is called the **fractional part**.

Relating Decimal Numbers and Fractions

Each digit in a decimal number has a certain **place value**. The value of each place is *ten times* the value of the place immediately to its *right*. Or, stated in another way, each place has *one-tenth* the value of the place immediately to its *left*.

4	6	9	8	.	2	3	1	5
thousands	hundreds	tens	ones		tenths	hundredths	thousandths	ten thousandths

To read the fractional part of a decimal number, we note the position where the last digit appears. The place value of that position tells us whether we are dealing with tenths or hundredths and so on. Every decimal number can then be rewritten as a fraction. If a decimal number has no fractional part, it is a whole number.

Decimal Number	Meaning	Fraction
0.3	3 tenths	$\frac{3}{10}$
0.19	19 hundredths	$\frac{19}{100}$
0.257	257 thousandths	$\frac{257}{1000}$
73.29	73 and 29 hundredths	$73\frac{29}{100}$ or $\frac{7329}{100}$
16	16	$\frac{16}{1}$

Example 1. Write 0.013 as a fraction.

Solution

$$0.013 = \frac{13}{1000}$$

You try Example 2.

Example 2. Write $\frac{37}{10,000}$ as a decimal number.

Solution: $\frac{37}{10,000}$ means 37 ten thousandths.

$$\frac{37}{10,000} = 0.\ \_\_\_\_\_\_$$

Check your work on page 48. ▶

In the fractional part of a decimal number, zeros may be added *after* the last nonzero digit without changing the value of the number. We may write

$$0.5 = 0.50 = 0.500$$

because

$$\frac{5}{10} = \frac{50}{100} = \frac{500}{1000}$$

Example 3. Rewrite 1.2 as a decimal number that ends in the thousandths place.

Solution: We add the necessary zeros.

$$1.2 = 1.200$$

You try Example 4.

Example 4. Rewrite $4 as a decimal number that ends in the hundredths place.

Solution: We add the necessary zeros.

$4 = $4. ____

Check your work on page 48. ▶

To change any fraction to a decimal number, we must divide the numerator by the denominator, placing a decimal point and adding zeros where necessary in the dividend. For example, let's change $\frac{1}{4}$ to a decimal number.

$4\overline{)1.00}$ Place the decimal point after the 1 and add zeros.

```
   .25
4)1.00    Divide in the usual way.
   8
   20
   20
    0
```

$$\frac{1}{4} = 0.25$$

Example 5. Change $\frac{1}{3}$ to a decimal number.

Solution

```
   .333
3)1.000
   9
   10
    9
   10
    9
    1
```

We shall continue to obtain the digit 3 in the quotient. This **repeating decimal** can be written $0.3\overline{3}$, where the bar shows the repeating digit.

$$\frac{1}{3} = 0.3\overline{3}$$

Now try Example 6.

Example 6. Change $\frac{5}{8}$ to a decimal number.

Solution

$8\overline{)5.000}$

So $\frac{5}{8} =$ ________ .

Check your work on page 48. ▶

Sometimes it is necessary to **round off** decimal numbers to some particular place, and we agree to use the following procedure.

Rounding-Off Decimal Numbers

1. Look at the digit in the place immediately to the right of the place to which you wish to round the number.
2. If the digit in the place to the right is less than 5, leave the digit in the rounding place as it is.
3. If the digit in the place to the right is 5 or more, add 1 to the digit in the rounding place.
4. State that your original decimal number is approximately equal to ($\doteq$) your rounded decimal number.

Example 7. Round 17.261 to the tenths place.

Solution

$$17.261 \doteq 17.3$$

Now you try Example 8.

Example 8. Round 17.261 to the hundredths place.

Solution

$17.261 \doteq$ ______

Check your work on page 48. ▶

Example 9. Round $0.6\overline{6}$ to the thousandths place.

Solution: Remember that this repeating decimal number is

$$0.66666\ldots$$

so in rounded-off form we have

$$0.6\overline{6} \doteq 0.667$$

Now try Example 10.

Example 10. Round $2.7\overline{373}$ to the hundredths place.

Solution

$2.7\overline{373}$

$\doteq$ ______

Check your work on page 48 ▶

Trial Run

______ **1.** Write 0.377 as a fraction.

______ **2.** Write $\frac{49}{100}$ as a decimal number.

______ **3.** Write $\frac{4}{5}$ as a decimal number.

______ **4.** Round 1.3219 to the hundredths place.

______ **5.** Change $\frac{1}{6}$ to a decimal number and round to the thousandths place.

Answers are on page 49.

Operating with Decimal Numbers

To **add** or **subtract** decimal numbers, we follow this procedure.

Adding or Subtracting Decimal Numbers

1. Write the problem vertically, with decimal points lined up beneath each other.
2. Add necessary zeros to make all fractional parts contain the same number of digits.
3. Add or subtract the columns of numbers in the usual way, keeping the decimal point in its original position.

Example 11. Find 3.62 + 2.8 + 0.317.

Solution

$$\begin{array}{r} 3.620 \\ 2.800 \\ +\underline{0.317} \\ 6.737 \end{array}$$

Now try Example 12.

Example 12. Find \$76 − \$15.34.

Solution

$$\begin{array}{r} \$76.00 \\ -\underline{\ 15.34} \\ \$\underline{\qquad} \end{array}$$

Check your work on page 48. ▶

In **multiplying** decimal numbers, the location of the decimal point in the answer depends on the location of the decimal points in the factors being multiplied.

Multiplying Decimal Numbers

1. Multiply the digits (ignoring the decimal points).
2. Add the number of places to the right of the decimal point in all the numbers being multiplied. This sum tells how many places there will be to the right of the decimal point in the product.
3. Locate the decimal point in the answer by "counting off" the places from the right.

Example 13. Find (0.12)(1.8).

Solution

$$\begin{array}{rl} 0.12 & \text{(2 places)} \\ \times\ \underline{1.8} & \text{(1 place)} \\ 96 & \\ \underline{12\ } & \\ .216 & \text{(3 places)} \end{array}$$

$$(0.12)(1.8) = 0.216$$

Example 14. Find (0.182)(0.05).

Solution

$$\begin{array}{rl} 0.182 & \text{(3 places)} \\ \times\ \underline{0.05} & \text{(2 places)} \\ .00910 & \text{(5 places)} \end{array}$$

$$(0.182)(0.05) = 0.00910$$

In Example 14, notice that we added zeros in front of our product in order to reach the correct location of the decimal point.

To **divide** decimal numbers, we always want the divisor to be a *whole number*.

Dividing Decimal Numbers

1. Change the divisor to a whole number by moving the decimal point to the right as many places as necessary.
2. Move the decimal point in the dividend the *same* number of places to the right.
3. Locate the decimal point in the quotient *directly above* the new decimal point location in the dividend.
4. Ignore the decimal points and divide as usual.

Example 15. Find $1.3939 \div 0.53$.

Solution

```
          2.63
 0.53^)1.39^39
       1 06
         33 3
         31 8
          1 59
          1 59
             0
```

$1.3939 \div 0.53 = 2.63$

Example 16. Find $0.464 \div 0.000116$.

Solution

```
                   4000.
 0.000116^)0.464000^
                464
                 00
                 00
                  00
                  00
                   00
                   00
                    0
```

$0.464 \div 0.000116 = 4000$

In Example 16, notice that we added zeros in the dividend so that we could move the decimal point the required number of places to the right.

Trial Run

Perform the operation indicated.

______ **1.** $7.23 + 0.591$

______ **2.** $6.82 + $3 + $9.29

______ **3.** $0.317 - 0.25$

______ **4.** $19 − $18.83

______ **5.** $(10.21)(3.051)$

______ **6.** $(0.0035)(0.28)$

______ **7.** $1.1607 \div 7.3$

______ **8.** $1.065 \div 0.00213$

Answers are on page 49.

Working with Percents

Decimals are used extensively in work with percents. The word **percent** means "per hundred." Understanding this meaning makes it easy to change a percent to a fraction or a decimal number.

Percent	Meaning	Fraction	Decimal Number
31%	31 per hundred	$\frac{31}{100}$	0.31
9%	9 per hundred	$\frac{9}{100}$	0.09
113%	113 per hundred	$\frac{113}{100}$	1.13

Looking closely at the first and last columns in this chart, you should agree to the following method for changing a percent to a decimal number.

Changing a Percent to a Decimal Number. To change a percent to a decimal number, drop the % symbol and move the decimal point *two* places to the *left*.

You try Example 17.

Example 17. Change 69% to a decimal number.

Solution

69% = 0. ____

Check your work on page 48. ▶

Example 18. Change 10.2% to a decimal number.

Solution

10.2% = 0.102

Of course, we can reverse this procedure if we wish to change a decimal number to a percent.

Changing a Decimal Number to a Percent. To change a decimal number to a percent, move the decimal point *two* places to the *right* and attach a % symbol.

Example 19. Change 0.73 to a percent.

Solution

0.73 = 73%

Now try Example 20.

Example 20. Change 1.2 to a percent.

Solution

1.2 = ____ %

Check your work on page 48. ▶

Suppose that your boss promises you a raise that is 8% of your current salary of $300 per week. How can you calculate your dollar raise? You must find 8% of $300. Recall that the word *of* usually tells us to *multiply*. But in order to multiply by a percent, we must change it to a decimal number (or to a fraction).

$$\begin{aligned} &8\% \text{ of } 300 \\ &= 8\% \cdot 300 \\ &= (0.08)(300) \\ &= 24 \end{aligned}$$

Your raise should be $24 per week.

Finding a Percent of a Number. To find a percent of some number, change the percent to a decimal number and multiply.

You try Example 21.

Example 21. Find 110% of 56.

Solution

$$\begin{aligned} &110\% \text{ of } 56 \\ &= 110\% \cdot 56 \\ &= (1.10)(56) \\ &= \_\_\_\_ \end{aligned}$$

110% of 56 is ____

Check your work on page 48. ▶

Example 22. Find 6.5% of $200.

Solution

$$\begin{aligned} &6.5\% \text{ of } 200 \\ &= 6.5\% \cdot 200 \\ &= (0.065)(200) \\ &= 13.0 \end{aligned}$$

6.5% of $200 is $13.

Trial Run

Change each percent to a decimal number.

____ **1.** 53%

____ **2.** 150%

Change each decimal number to a percent.

____ **3.** 0.18

____ **4.** 0.071

Find the percentage.

____ **5.** 12% of 80

____ **6.** 7.8% of $120

Answers are on page 49.

▶ Examples You Completed

Example 2. Write $\frac{37}{10{,}000}$ as a decimal number.

Solution: $\frac{37}{10{,}000}$ means 37 ten thousandths.

$$\frac{37}{10{,}000} = 0.0037$$

Example 4. Rewrite \$4 as a decimal number that ends in the hundredths place.

Solution: We add the necessary zeros.

$$\$4 = \$4.00$$

Example 6. Change $\frac{5}{8}$ to a decimal number.

Solution

$$\begin{array}{r} .625 \\ 8\overline{)5.000} \\ \underline{4\,8} \\ 20 \\ \underline{16} \\ 40 \\ \underline{40} \\ 0 \end{array}$$

So $\frac{5}{8} = 0.625$.

Example 8. Round 17.261 to the hundredths place.

Solution

$$17.261 \doteq 17.26$$

Example 10. Round $2.7\overline{373}$ to the hundredths place.

Solution

$$2.7\overline{373} \doteq 2.74$$

Example 12. Find \$76 − \$15.34.

Solution

$$\begin{array}{r} \$76.00 \\ -\ \ 15.34 \\ \hline \$60.66 \end{array}$$

Example 17. Change 69% to a decimal number.

Solution

$$69\% = 0.69$$

Example 20. Change 1.2 to a percent.

Solution

$$1.2 = 120\%$$

Example 21. Find 110% of 56.

Solution

$$\begin{aligned} &110\% \text{ of } 56 \\ &= 110\% \cdot 56 \\ &= (1.10)(56) \\ &= 61.6 \end{aligned}$$

110% of 56 is 61.6.

Answers to Trial Runs

page 43 **1.** $\frac{377}{1000}$ **2.** 0.49 **3.** 0.8 **4.** 1.32 **5.** 0.167

page 45 **1.** 7.821 **2.** $19.11 **3.** 0.067 **4.** $0.17 **5.** 31.15071 **6.** 0.00098 **7.** 0.159 **8.** 500

page 47 **1.** 0.53 **2.** 1.5 **3.** 18% **4.** 7.1% **5.** 9.6 **6.** $9.36

Name ______________________ Date ____________

EXERCISE SET 1.3

Write the decimal numbers as fractions. Reduce to lowest terms.

_______ **1.** 0.8

_______ **2.** 0.6

_______ **3.** 0.55

_______ **4.** 0.75

_______ **5.** 0.175

_______ **6.** 0.112

_______ **7.** 0.012

_______ **8.** 0.028

_______ **9.** Rewrite 2.3 as a decimal number that ends in the thousandths place.

_______ **10.** Rewrite 4.7 as a decimal number that ends in the thousandths place.

_______ **11.** Rewrite $15 as a decimal number that ends in the hundredths place.

_______ **12.** Rewrite $12 as a decimal number that ends in the hundredths place.

_______ **13.** Rewrite 13.2 as a decimal number that ends in the thousandths place.

_______ **14.** Rewrite the 10.6 as a decimal number that ends in the hundredths place.

Change the fractions to decimal numbers. Round to the nearest hundredths.

_______ **15.** $\frac{3}{4}$

_______ **16.** $\frac{1}{4}$

_______ **17.** $\frac{5}{8}$

_______ **18.** $\frac{7}{16}$

_______ **19.** $\frac{5}{6}$

_______ **20.** $\frac{4}{9}$

_______ **21.** $\frac{8}{15}$

_______ **22.** $\frac{12}{19}$

_______ **23.** $\frac{9}{25}$

_______ **24.** $\frac{18}{25}$

Perform the operations indicated.

_______ **25.** $0.83 + 0.7$

_______ **26.** $49.6 + 3.002$

_______ **27.** $9.75 + $13 + $3.05

_______ **28.** $20 − $15.78

_______ **29.** $9.005 - 3.72$

_______ **30.** $13.003 - 4.139$

_______ **31.** $3.089 + 0.845 - 1.32$

_______ **32.** $50 − $7.85 − $19.46

_______ **33.** $(0.027)(0.78)$

_______ **34.** $(56.9)(3.24)$

_______ **35.** (3.27)(0.008)

_______ **36.** (7.85)(0.007)

_______ **37.** 0.273 ÷ 0.07

_______ **38.** 2.58 ÷ 0.0043

_______ **39.** 1.6849 ÷ 8.3

_______ **40.** 305.2 ÷ 8.72

Change the percents to decimal numbers.

_______ **41.** 19%

_______ **42.** 64%

_______ **43.** 37.1%

_______ **44.** 28.2%

_______ **45.** 150%

_______ **46.** 220%

_______ **47.** 0.5%

_______ **48.** 0.2%

Change the decimal numbers to percents.

_______ **49.** 0.09

_______ **50.** 0.12

_______ **51.** 7.1

_______ **52.** 5.6

_______ **53.** 0.032

_______ **54.** 0.054

_______ **55.** 1.75

_______ **56.** 2.50

Find the following percentages.

_______ **57.** 13% of 470

_______ **58.** 12% of 560

_______ **59.** 9% of 72

_______ **60.** 6% of 84

_______ **61.** 12% of 3240

_______ **62.** 175% of 4760

_______ **63.** 8.3% of 6000

_______ **64.** 9.4% of 5000

_______ **65.** 300% of 74

_______ **66.** 500% of 68

_______ **67.** 10% of 9378

_______ **68.** 20% of 8476

_______ **69.** 25% of 1250

_______ **70.** 50% of 2740

☆ Stretching the Topics

_______ **1.** Find $\frac{0.5[11.36 + (7.2 - 3.065)]}{0.002}$.

_______ **2.** Find $12\frac{1}{2}\%$ of \$5000.

_______ **3.** Find 0.25% of \$600.

Check your answers in the back of your book.

If you can do the problems in **Checkup 1.3**, you are ready to go on to Section 1.4.

Name ______________________ Date ____________

CHECKUP 1.3

_____ **1.** Write 0.024 as a fraction and reduce to lowest terms.

_____ **2.** Rewrite 43.5 as a decimal number that ends in the thousandths place.

_____ **3.** Change $\frac{7}{8}$ to a decimal number.

_____ **4.** Change 3.25% to a decimal number.

_____ **5.** Change 0.125 to a percent.

Perform the operations indicated.

_____ **6.** $2.032 + 7.85 - 4.02$

_____ **7.** $\$25 - \$17.98$

_____ **8.** $(0.32)(0.7)$

_____ **9.** $0.02616 \div 3.27$

_____ **10.** Find 8.5% of 325.

Check your answers in the back of your book.

If You Missed Problems:	You Should Review Examples:
1	1
2	3
3	5, 6
4	17, 18
5	19, 20
6, 7	11, 12
8, 9	13–16
10	21, 22

1.4 Switching from Word Expressions to Number Expressions

It is important for an algebra student to be able to change word statements into number statements. Such an ability will allow us to solve many ''real-life'' problems that could otherwise be solved only by trial and error.

Let's see if we can practice switching some word expressions into number expressions. There are several key phrases that will help you switch from words to numbers.

Words	Numbers
the sum of 7 and 9	$7 + 9$
6 more than 25	$25 + 6$
1.6 increased by 25.4	$1.6 + 25.4$
24 decreased by 8	$24 - 8$
5 fewer than 12	$12 - 5$
0.3 less than 1.7	$1.7 - 0.3$
the product of 6 and 11	$6 \cdot 11$
twice as much as 7.3	$2(7.3)$
14 doubled	$2 \cdot 14$
5 times as large as 10	$5 \cdot 10$
24 divided by 0.8	$\frac{24}{0.8}$
half of $\frac{6}{7}$	$\frac{1}{2} \cdot \frac{6}{7}$

See if you can complete the following table.

Words	Numbers
13 increased by 7	______
____________	$\frac{32}{2.8}$
the sum of 12 and 3	______
the product of 12 and 3	______
5 fewer than 18	______
twice as much as 10.6	______
____________	$42 \div 15$
19 decreased by 15	______
two-thirds of 24	______
3 times as large as 25	______

Check your work on page 59.

Example 1. Nancy's hourly wage is \$3. Next week she will receive a raise of \$1.10 per hour. Write an expression for her new hourly wage.

Solution

\$3 *increased by* \$1.10

\$3 + \$1.10

Complete Example 2.

Example 2. At noon the temperature was 18°. By midnight, the temperature had decreased by 12°. Write an expression for the temperature at midnight.

Solution

18° *decreased by* 12°

18° ____ 12°

Check your work on page 59. ▶

Example 3. Tom weighs 20 pounds more than twice his son's weight. If Tom's son weighs 73 pounds, write an expression for Tom's weight.

Solution

73	son's weight
2(73)	*twice* son's weight
2(73) + 20	20 *more than* twice son's weight
2(73) + 20	Tom's weight

Now let's solve the problem stated at the beginning of this chapter.

Example 4. Each month Estelle saves $\frac{1}{10}$ of her paycheck in her savings bank and $\frac{1}{9}$ of her paycheck in her credit union. What fractional part of her paycheck does Estelle save each month?

Solution: The total of Estelle's savings is the *sum* of the two fractions.

$$\frac{1}{10} + \frac{1}{9}$$

$$= \frac{1}{2 \cdot 5} + \frac{1}{3 \cdot 3} \qquad \text{LCD: } 2 \cdot 5 \cdot 3 \cdot 3 = 90.$$

$$= \frac{1 \cdot 3 \cdot 3}{2 \cdot 5 \cdot 3 \cdot 3} + \frac{1 \cdot 2 \cdot 5}{2 \cdot 5 \cdot 3 \cdot 3}$$

$$= \frac{9}{90} + \frac{10}{90}$$

$$= \frac{19}{90}$$

Estelle saves $\frac{19}{90}$ of her paycheck each month.

Example 5. The sales tax in a certain state is 6.5%. How much tax would a customer pay on a record selling for \$8? How much would the customer pay altogether?

Solution: The tax is 6.5% of \$8.

$$6.5\% \text{ of } 8$$

$$= (0.065)(8)$$

$$= 0.52$$

The tax is \$0.52.

The total paid altogether is the *sum* of the price of the record and the tax.

$$\$8 + 0.52$$
$$= \$8.52$$

Now you complete Example 6.

Example 6. A real estate developer owns a large plot of land with four sides measuring 2.3 miles, 1.4 miles, 4.25 miles, and 3.5 miles. If the developer drives around the boundary of her land, how many miles will she travel?

Solution: The distance traveled is found by ________ the four distances.

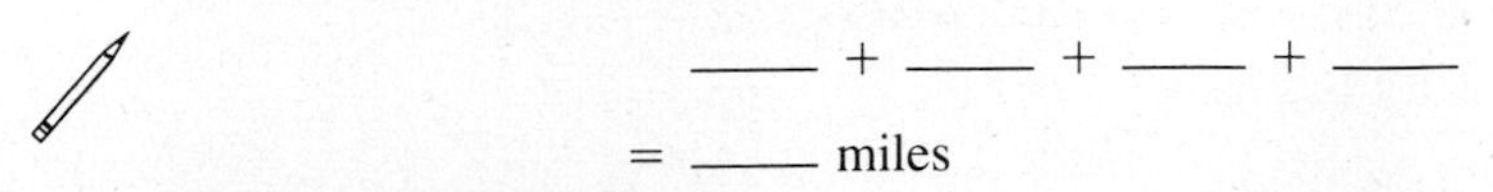

____ + ____ + ____ + ____

= ____ miles

Check your work on page 59. ▶

One type of problem that you will be asked often to solve is that involving some sort of **rate**, such as

miles *per* hour

dollars *per* hour

miles *per* gallon

revolutions *per* minute

Each of these rates is described as "so many of one thing *per* something else" and tells us how many units of the first thing we can expect for *one* unit of the other thing. For instance:

Rate	Meaning
23 miles per gallon	We can travel 23 miles on *one* gallon of gas.
\$3 per hour	We are paid \$3 for every *one* hour worked.
55 miles per hour	We will travel 55 miles in *one* hour of driving.

If we know a rate and if we know how many of the second thing we have, we can find how many of the first thing we have by *multiplying*. For example:

At 55 miles per hour for 3 hours, we will travel $55 \cdot 3 = 165$ miles

At \$3 per hour for 40 hours, we will earn $3 \cdot 40 = 120$ dollars

Example 7. How many revolutions will a record make during a 2-minute song if it is spinning at a rate of 45 revolutions per minute (rpm)?

Solution: We must multiply the rate, 45 rpm, times the number of minutes.

$$45 \cdot 2 = 90 \text{ revolutions}$$

You complete Example 8.

Example 8. Juan drove at a rate of 55 miles per hour (mph) for 3 hours and at a rate of 50 mph for 2 hours. How far did Juan drive?

Solution: We must multiply the first rate, ________ per ________ , times the number of ________ .

$$55 \cdot \_\_\_\_ = \_\_\_\_ \text{ miles}$$

Then we must multiply the second rate, ________ per ________ , times the number of ________ .

$$50 \cdot \_\_\_\_ = \_\_\_\_ \text{ miles}$$

The total distance is the sum of these distances.

$$\_\_\_\_ + \_\_\_\_ = \_\_\_\_ \text{ miles}$$

Check your work on page 59. ▶

Now try Example 9.

Example 9. On a business trip Harvey's car mileage gauge changed from 21623 to 22007. If Harvey's company pays him 25 cents per mile, what will the company pay Harvey for car expenses on this trip?

Solution: How many *miles* did Harvey drive? We must look at the mileage gauge readings.

$$\_\_\_\_ - \_\_\_\_ \text{ is the number of miles}$$

Now we must multiply the rate, ________ per ________ , times the number of ________ .

$$\begin{aligned} &25(\_\_\_\_ - \_\_\_\_) \\ &= 25(\_\_\_\_) \\ &= \_\_\_\_\_\_\_\_ \text{ cents} \\ &= \$\_\_\_\_\_\_\_\_ \end{aligned}$$

Check your work on page 60. ▶

Even when you are no longer a student, you will see the word **average** in the newspaper or on television. You may remember that an average is found by dividing a *total* by the *number of equal parts* that the total is to be broken into.

Example 10. Over a period of 6 months, Angela lost 42 pounds. What was her *average* weight loss each month?

Solution: We wish to break the total of 42 pounds into six equal parts, so we *divide* 42 by 6.

$$\frac{42}{6} = 7 \text{ pounds each month}$$

Angela's average weight loss was 7 pounds each month.

Now you complete Example 11.

Example 11. The Colemans spent \$186 on groceries and \$240 on rent during 4 weeks. What was the Coleman's *average* weekly expense for groceries and rent?

Solution: The Coleman's *total* expense for groceries and rent is ____ + ____ . We wish to break this total into ____ equal parts, so we divide ____ + ____ by ____ .

$$\frac{186 + ?}{4}$$

$$= \frac{?}{4}$$

$$= \$\_\_\_\_$$

The Colemans' average expense for groceries and rent was \$ ____. each week.

Check your work on page 60. ▶

▶ Examples You Completed

Words	Numbers
13 increased by 7	$13 + 7$
32 divided by 2.8	$\frac{32}{2.8}$
the sum of 12 and 3	$12 + 3$
the product of 12 and 3	$12(3)$
5 fewer than 18	$18 - 5$
twice as much as 10.6	$2(10.6)$
42 divided by 15	$42 \div 15$
19 decreased by 15	$19 - 15$
two-thirds of 24	$\frac{2}{3}(24)$
3 times as large as 25	$3(25)$

Example 2. At noon the temperature was 18°. By midnight the temperature had decreased by 12°. Write an expression for the temperature at midnight.

Solution

18° decreased by 12°

$18° - 12°$

Example 6. A real estate developer owns a large plot of land with four sides measuring 2.3 miles, 1.4 miles, 4.25 miles, and 3.5 miles. If the developer drives around the boundary of her land, how many miles will she travel?

Solution: The distance traveled is found by adding the four distances.

$$2.3 + 1.4 + 4.25 + 3.5$$

$$= 11.45 \text{ miles}$$

Example 8. Juan drove at a rate of 55 miles per hour (mph) for 3 hours and at a rate of 50 mph for 2 hours. How far did Juan drive?

Solution: We must multiply the first rate, 55 miles per hour, times the number of hours.

$$55 \cdot 3 = 165 \text{ miles}$$

Then we must multiply the second rate, 50 miles per hour, times the number of hours.

$$50 \cdot 2 = 100 \text{ miles}$$

The total distance is the sum of these distances.

$$165 + 100 = 265 \text{ miles}$$

Example 9. On a business trip Harvey's car mileage gauge changed from 21623 to 22007. If Harvey's company pays him 25 cents per mile, what will the company pay Harvey for car expenses on this trip?

Solution: How many *miles* did Harvey drive? We must look at the mileage gauge readings.

$$22007 - 21623 \text{ is the number of miles}$$

Now we must multiply the rate, 25 cents per mile, times the number of miles.

$$\begin{aligned} &25(22007 - 21623) \\ &= 25(384) \\ &= 9600 \text{ cents} \\ &= \$96 \end{aligned}$$

Example 11. The Colemans spent \$186 on groceries and \$240 on rent during 4 weeks. What was the Coleman's *average* weekly expense for groceries and rent?

Solution: The Coleman's *total* expense for groceries and rent is 186 + 240. We wish to break this total into four equal parts, so we divide 186 + 240 by 4.

$$\begin{aligned} &\frac{186 + 240}{4} \\ &= \frac{426}{4} \\ &= \$106.50 \end{aligned}$$

The Colemans' average expense for groceries and rent was \$106.50 each week.

Name ____________________ Date __________

EXERCISE SET 1.4

Change each word expression to a number expression and simplify.

_______ **1.** Leroy's electric bill this month was $15.93 more than his bill last month. If his bill last month was $33.24, what is his bill this month?

_______ **2.** If one calculator costs $18.95, how much will eight calculators cost?

_______ **3.** If Carl's car averages 19 miles per gallon, how far can he drive when he has 15 gallons of gasoline?

_______ **4.** Sherry, Sandra, and Jo have a job washing the windows of an office building. If the building has 48 windows and they divide the work evenly, how many windows will each wash?

_______ **5.** Each month, Sam's wife earns $125 more than twice Sam's salary. If Sam's monthly salary is $685, what is his wife's monthly salary?

_______ **6.** Debbie scored 10 points higher on her second history test than on her first test. If she scored 76 on the first test, what is her average on the two tests?

_______ **7.** During 6 months, the Sturgis Cap Factory had a profit of $42,000. What was the average profit for 1 month?

_______ **8.** If Mark weighed $235\frac{1}{2}$ pounds before his diet and now weighs $190\frac{3}{4}$ pounds, how much weight did he lose?

_______ **9.** If $\frac{2}{3}$ of 645 homes surveyed had a microwave oven, how many homes in that area had a microwave?

_______ **10.** Chris is putting a string of Christmas lights on his house. If the cord is 20 feet long and has a bulb every $\frac{1}{4}$ of a foot, how many bulbs will he need?

_______ **11.** Rachel knows that it takes $1\frac{3}{4}$ yards of material for each costume she is sewing for the dance recital. If there are 12 dancers in the recital, how many yards of material will she need for the costumes?

_______ **12.** Kelvin has a rectangular garden that measures $24\frac{3}{4}$ feet by $20\frac{1}{3}$ feet. Find the area of the garden.

_______ **13.** Martin spends $\frac{1}{10}$ of his salary each month for gasoline. His car payment is $\frac{1}{5}$ of his salary. What fractional part of his monthly salary is spent on this car?

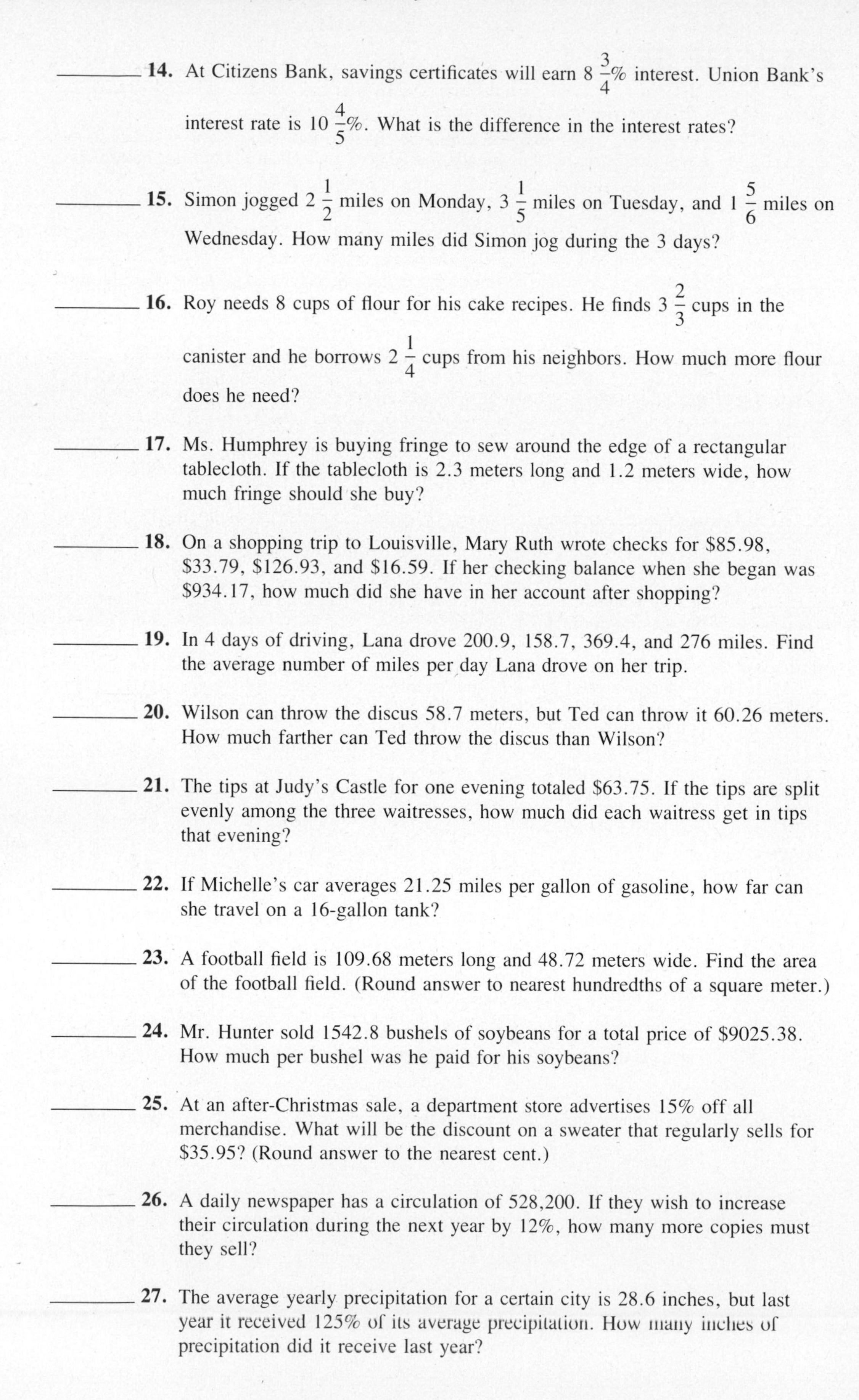

______ **14.** At Citizens Bank, savings certificates will earn $8\frac{3}{4}\%$ interest. Union Bank's interest rate is $10\frac{4}{5}\%$. What is the difference in the interest rates?

______ **15.** Simon jogged $2\frac{1}{2}$ miles on Monday, $3\frac{1}{5}$ miles on Tuesday, and $1\frac{5}{6}$ miles on Wednesday. How many miles did Simon jog during the 3 days?

______ **16.** Roy needs 8 cups of flour for his cake recipes. He finds $3\frac{2}{3}$ cups in the canister and he borrows $2\frac{1}{4}$ cups from his neighbors. How much more flour does he need?

______ **17.** Ms. Humphrey is buying fringe to sew around the edge of a rectangular tablecloth. If the tablecloth is 2.3 meters long and 1.2 meters wide, how much fringe should she buy?

______ **18.** On a shopping trip to Louisville, Mary Ruth wrote checks for $85.98, $33.79, $126.93, and $16.59. If her checking balance when she began was $934.17, how much did she have in her account after shopping?

______ **19.** In 4 days of driving, Lana drove 200.9, 158.7, 369.4, and 276 miles. Find the average number of miles per day Lana drove on her trip.

______ **20.** Wilson can throw the discus 58.7 meters, but Ted can throw it 60.26 meters. How much farther can Ted throw the discus than Wilson?

______ **21.** The tips at Judy's Castle for one evening totaled $63.75. If the tips are split evenly among the three waitresses, how much did each waitress get in tips that evening?

______ **22.** If Michelle's car averages 21.25 miles per gallon of gasoline, how far can she travel on a 16-gallon tank?

______ **23.** A football field is 109.68 meters long and 48.72 meters wide. Find the area of the football field. (Round answer to nearest hundredths of a square meter.)

______ **24.** Mr. Hunter sold 1542.8 bushels of soybeans for a total price of $9025.38. How much per bushel was he paid for his soybeans?

______ **25.** At an after-Christmas sale, a department store advertises 15% off all merchandise. What will be the discount on a sweater that regularly sells for $35.95? (Round answer to the nearest cent.)

______ **26.** A daily newspaper has a circulation of 528,200. If they wish to increase their circulation during the next year by 12%, how many more copies must they sell?

______ **27.** The average yearly precipitation for a certain city is 28.6 inches, but last year it received 125% of its average precipitation. How many inches of precipitation did it receive last year?

_________ **28.** When 3000 athletes were tested for drug usage, 9.4% were found to have used steroids. How many athletes had used steroids?

_________ **29.** A theater seats 2400 persons. If it was filled to 75% capacity, how many seats were filled?

_________ **30.** Seventy-two percent of the employees at a coal mine voted to accept the union contract. If there are 450 miners at that mine, how many voted to accept the contract?

☆ Stretching the Topics

_________ **1.** The Lollipop Tree advertises 15% off on all blouses, 20% off on all jeans, and 25% off on all dresses. If Kathy bought two blouses that regularly sell for $25.00 each, a pair of jeans that regularly sells for $33.95, and a dress that regularly sells for $48.00, find the total amount she owes for the clothing.

_________ **2.** Bruce's car averages 21.5 miles per gallon on the interstate highway and 16.5 miles per gallon on two-lane roads. The trip from his home to college requires that he drive 129 miles on the interstate highway and 66 miles on two-lane roads. If gasoline costs $1.339 per gallon, how much will the trip cost? (Round answer to the nearest cent.)

Check your answers in the back of your book.

If you can complete **Checkup 1.4** correctly, you are ready to do the **Review Exercises for Chapter 1**.

Name ______________________________ Date ____________

✓ CHECKUP 1.4

Change each word expression to a number expression and simplify.

________ **1.** Each spring, Natalie plants $\frac{1}{4}$ of her garden in green beans and $\frac{2}{5}$ in lima beans. What fractional part of her garden is used to raise beans?

________ **2.** In a state that has a 5.5% sales tax, how much sales tax would be paid on an automobile that costs $12,248.82? (Round answer to the nearest cent.)

________ **3.** The three sides of a triangular plot of land measure 13.6 feet, 8.92 feet, and 7.3 feet. Find (to the nearest tenth of a foot) how many feet of fence would be needed to enclose the plot.

________ **4.** If Tom's monthly bills were $325 for rent, $73.86 for utilities, and $195 for groceries, find his average weekly living expenses. (Assume a 4-week month.)

________ **5.** On a business trip Lydia averaged 62 miles per hour for 4 hours on the interstate highway and 53 miles per hour for 3 hours on two-lane roads. If Lydia is paid 25 cents per mile, what will the company pay for her expenses?

Check your answers in the back of your book.

If You Missed Problems:	You Should Review Examples:
1	4
2	5
3	6
4	11
5	8, 9

Summary

In this chapter we reviewed the arithmetic of whole numbers, fractions, and decimal numbers. We observed how the four basic operations of addition, subtraction, multiplication, and division are related to each other, and noted some special properties of zero and 1.

Property	Symbols	Examples
Addition property of zero	$A + 0 = A$	$3 + 0 = 3$
		$0 + 29 = 29$
Subtraction property of zero	$A - 0 = A$	$5 - 0 = 5$
Multiplication property of zero	$A \cdot 0 = 0$	$13 \cdot 0 = 0$
	$0 \cdot A = 0$	$0 \cdot 9 = 0$
Division of zero	$\frac{0}{A} = 0 \quad (A \neq 0)$	$\frac{0}{25} = 0$
Division by zero	$\frac{A}{0}$ is undefined	$\frac{19}{0}$ is undefined
		$\frac{0}{0}$ is undefined
Multiplication property of 1	$A \cdot 1 = A$	$6 \cdot 1 = 6$
	$1 \cdot A = A$	$1 \cdot 77 = 77$
Division property of 1	$\frac{A}{1} = A$	$\frac{769}{1} = 769$

We also noted that $A - A = 0$ and $\frac{A}{A} = 1$ (provided that $A \neq 0$).

The use of **exponents** was discussed as a way to indicate that a base is to be repeated as a factor in a product. For example,

$$5^3 = 5 \cdot 5 \cdot 5 = 125$$

We discovered that **symbols of grouping** (parentheses, braces, brackets, and fraction bars) help us simplify expressions by telling us which operations to perform first. In the absence of parentheses, we agreed to perform operations in the following order.

1. Deal with exponents.
2. Perform multiplications and/or divisions from left to right.
3. Perform additions and/or subtractions from left to right.

In our work with fractions we mastered several important techniques.

In Order to:	We Must:	Examples
Reduce a fraction	Divide numerator and denominator by any common factors.	$\frac{28}{20} = \frac{\overset{1}{\cancel{2}} \cdot \overset{1}{\cancel{2}} \cdot 7}{\underset{1}{\cancel{2}} \cdot \underset{1}{\cancel{2}} \cdot 5} = \frac{7}{5}$
Multiply fractions	Multiply numerators and multiply denominators.	$\frac{5}{9} \cdot \frac{2}{3} = \frac{10}{27}$
Divide fractions	Invert the divisor and multiply.	$\frac{3}{11} \div \frac{1}{3} = \frac{3}{11} \cdot \frac{3}{1} = \frac{9}{11}$
Build fractions	Multiply numerator and denominator by the same factor.	$\frac{3}{4} = \frac{3 \cdot 7}{4 \cdot 7} = \frac{21}{28}$
Add (or subtract) fractions	Be sure that fractions have same denominator. Add (or subtract) numerators and keep same denominator.	$\frac{5}{8} - \frac{2}{8} = \frac{5-2}{8} = \frac{3}{8}$ $\frac{1}{6} + \frac{2}{3} - \frac{3}{4} = \frac{2}{12} + \frac{8}{12} - \frac{9}{12}$ $= \frac{2+8-9}{12} = \frac{1}{12}$

We reviewed the operations of addition, subtraction, multiplication, and division of **decimal numbers** and discovered how **percents** are related to fractions and decimal numbers by the fact that percent means "per hundred." For example,

$$23\% = \frac{23}{100} = 0.23$$

Finally, we learned to use whole numbers, fractions, and decimal numbers to switch from word expressions to numerical expressions.

❑ Speaking the Language of Algebra

1. The set of numbers {0, 1, 2, 3, . . .} is called the set of ________ ________ .
2. When we add two numbers, the result is called the ________ . When we subtract two numbers, the result is called the ________ .
3. In multiplication the numbers being multiplied are called the ________ and the result is called the ________ .
4. In division the number being divided is called the ________ , the number doing the dividing is called the ________ , and the result is called the ________ .
5. If the numerator of a fraction is smaller than the denominator, it is called a ________ fraction. If not, it is called an ________ fraction.
6. To reduce a fraction to lowest terms, we divide the numerator and denominator by any ________ ________ .
7. To divide fractions, we leave the dividend as it is, ________ the divisor, and ________ .
8. Fractions can be added only if they have the ________ ________ .
9. In a decimal number, the part to the left of the decimal point is called the ________ ________ part and the part to the right of the decimal point is called the ________ part.
10. The word "percent" means " ________ ________ ." To change a percent to a decimal number, we drop the % symbol and move the decimal point ________ places to the ________ .

Check your answers in the back of your book.

Name ______________________________ Date ____________

REVIEW EXERCISES for Chapter 1

________ **1.** Use the symbol $<$ or $>$ to complete the statement: 3 _____ 19.

________ **2.** Write the addition statement that corresponds to the subtraction statement $43 - 32 = 11$.

________ **3.** Write the multiplication statement that corresponds to the division statement $\frac{91}{13} = 7$.

Simplify by performing the operations.

________ **4.** $3(5 + 2 \cdot 7)$

________ **5.** $(7 + 2)(8 + 0)$

________ **6.** $5 \cdot 2^3 + 4 \cdot 9$

________ **7.** $[9 + (8 \cdot 7) + 1] + 6$

________ **8.** $7\left(\frac{15}{3} - \frac{5 \cdot 0}{2}\right)$

________ **9.** $\frac{80 - 8 \cdot 6}{9 - 5}$

________ **10.** $\frac{9 \cdot 6 - 3^2}{56 - 7 \cdot 8}$

________ **11.** $\frac{3 \cdot 8 - 6 \cdot 4}{19 - 12}$

________ **12.** $\frac{15}{4} \cdot \left(\frac{2}{5}\right)^2$

________ **13.** $\frac{5}{6} \cdot \frac{7}{25} \cdot \frac{15}{28}$

________ **14.** $\frac{7}{3} \div 63$

________ **15.** $\frac{1}{9} \div \left(\frac{4}{5} \cdot \frac{10}{27}\right)$

________ **16.** $\frac{1}{15} + \frac{9}{25}$

________ **17.** $\frac{10}{21} + \frac{11}{18} - \frac{9}{14}$

________ **18.** Change $\frac{5}{16}$ to a decimal number.

________ **19.** Round 13.072 to the tenths place.

Perform the operations indicated.

________ **20.** $13 + 4.0765 + 0.09$

________ **21.** $\$276.35 - \$98.86$

________ **22.** $(2.75)(0.032)$

________ **23.** $(0.2236) \div (0.043)$

________ **24.** Change 0.09 to a percent.

________ **25.** Change 12.7% to a decimal number.

________ **26.** Find 13.2% of $5000.

Change each word expression to a number expression and simplify.

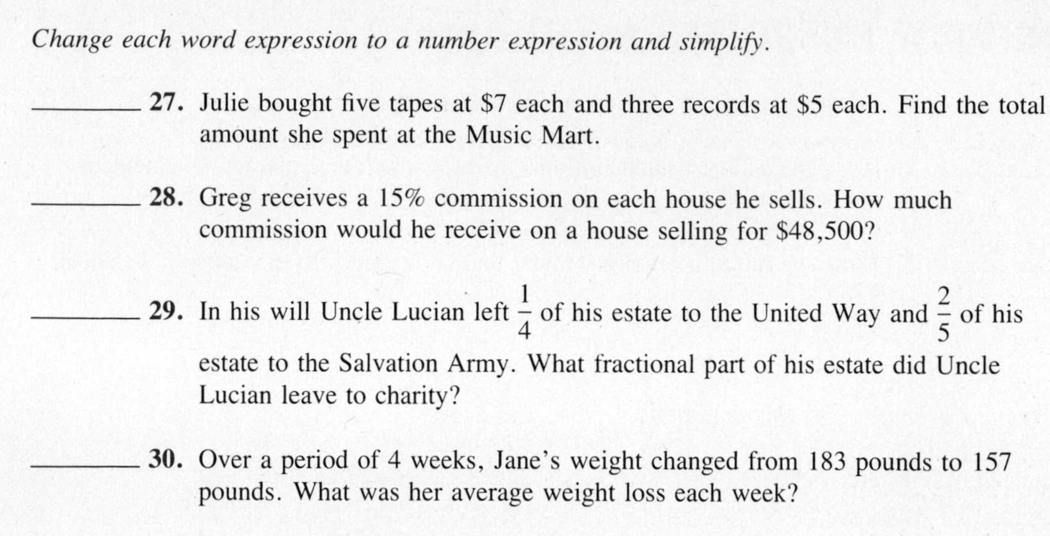

________ **27.** Julie bought five tapes at $7 each and three records at $5 each. Find the total amount she spent at the Music Mart.

________ **28.** Greg receives a 15% commission on each house he sells. How much commission would he receive on a house selling for $48,500?

________ **29.** In his will Uncle Lucian left $\frac{1}{4}$ of his estate to the United Way and $\frac{2}{5}$ of his estate to the Salvation Army. What fractional part of his estate did Uncle Lucian leave to charity?

________ **30.** Over a period of 4 weeks, Jane's weight changed from 183 pounds to 157 pounds. What was her average weight loss each week?

Check your answers in the back of your book.

If You Missed Exercises:	You Should Review Examples:	
1	SECTION 1.1	3, 4
2		7
3		9, 10
4–7		16–19
8–11		20–25
12–15	SECTION 1.2	5–11
16, 17		14–17
18	SECTION 1.3	5, 6
19		7, 8
20, 21		11, 12
22, 23		13–16
24		19, 20
25		17, 18
26		21, 22
27	SECTION 1.4	8
28		5
29		4
30		10

If you have completed the **Review Exercises** and corrected your errors, you are ready to take the **Practice Test for Chapter 1**.

Name ______________________ Date __________

PRACTICE TEST for Chapter 1

		Section	Examples
________	**1.** Write the addition statement that corresponds to the subtraction statement $24 - 13 = 11$.	1.1	7
________	**2.** Write the multiplication statement that corresponds to the division statement $\frac{55}{11} = 5$.	1.1	9, 10

Simplify.

________	**3.** $8(6 + 2 \cdot 9)$	1.1	16, 17
________	**4.** $8 \cdot 3^2 + 5 \cdot 6$	1.1	23
________	**5.** $[8 + (6 \cdot 3) + 2] + 9$	1.1	18
________	**6.** $\frac{8 \cdot 7 - 2^3}{72 - 9 \cdot 8}$	1.1	12, 23, 25
________	**7.** $\frac{9 \cdot 4 - 6 \cdot 6}{27 - 19}$	1.1	12, 25
________	**8.** $\left(\frac{45}{28}\right)\left(\frac{2}{3}\right)^3$	1.2	6
________	**9.** $\frac{8}{9} \div 16$	1.2	10
________	**10.** $\frac{4}{3} \div \left(\frac{5}{18} \cdot \frac{8}{25}\right)$	1.2	11
________	**11.** $\frac{1}{12} + \frac{7}{9}$	1.2	18
________	**12.** $\frac{7}{10} + \frac{9}{12} - \frac{8}{15}$	1.2	19
________	**13.** Change $\frac{3}{8}$ to a decimal number.	1.3	6
________	**14.** Round 7.1354 to the hundredths place.	1.3	8

Perform the operations indicated.

________	**15.** $8 + 11.35 + 0.135$	1.3	11
________	**16.** $(3.29)(0.25)$	1.3	13, 14

________	**17.** (6.3252) ÷ (1.26)	1.3	15, 16
________	**18.** Find 8.5% of $270.	1.3	21, 22

Change each word expression to a number expression and simplify.

________	**19.** Pansy bought 2 pounds of ground beef at $1.79 per pound and 1 pound of ground pork at $1.89 per pound. Find the total amount she spent for the meat.	1.4	8
________	**20.** During the first 4 days of March, Doug jogged 3.7, 4.2, 3.8, and 4.5 miles. Find the average number of miles Doug jogged each day.	1.4	6, 11

Check your answers in the back of your book.

Working with Real Numbers 2

In one series of plays, the State University football team completed two 7-yard passes and ran for 5 yards. Then the quarterback was sacked twice, for a 3-yard loss each time, and the team received a 15-yard penalty. What was the team's total gain or loss?

When you have completed this chapter, you will be able to solve this problem using arithmetic of **real numbers**.

In this chapter we

1. Introduce integers.
2. Add, subtract, multiply, and divide integers.
3. Introduce real numbers.
4. Observe some properties of real numbers.
5. Switch from word expressions to real number expressions.

2.1 Understanding Integers

On a winter day, have you ever watched the thermometer drop to a temperature of 3° below zero? Have you ever seen your favorite football team lose 6 yards on a play? Have you ever borrowed $10?

In each of these situations the set of whole numbers does not provide us with a way to make the necessary measurement. We know that temperatures of 3° above zero and 3° below zero are very different. A gain of 6 yards is very different from a loss of 6 yards, and a credit of $10 is very different from a debt of $10. We can use a **negative sign** to help us, and

3° *below* zero	is written	−3°
a *loss* of 6 yards	is written	−6 yards
a *debt* of 10 dollars	is written	−10 dollars

Locating Integers on the Number Line

On a number line we illustrate these negative numbers by extending our line to the left, marking off each unit in a leftward direction from zero.

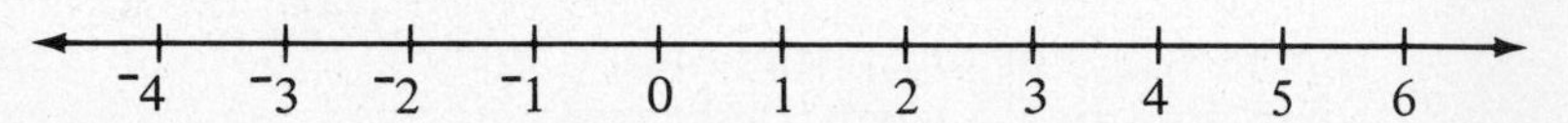

This new set of numbers, which includes all the whole numbers *and* their negative counterparts, is called the set of **integers**.

> Integers: {. . ., −6, −5, −4, −3, −2, −1, 0, 1, 2, 3, 4, . . .}

Notice that this list continues indefinitely in either direction, as shown by the arrows at both ends of the number line. Units to the *right* of zero correspond to the natural numbers (also called **positive integers**) and units to the left of zero correspond to the **negative integers**.

> Positive integers: {1, 2, 3, 4, 5, . . .}
> Negative integers: {. . ., −4, −3, −2, −1}

The set of integers contains all the positive integers, all the negative integers, and the integer zero. Each integer corresponds to one and only one point on the number line.

The number line again gives us a handy way to compare integers. If one integer lies to the *left* of another integer, we say that the first integer is *less than* (<) the second integer. If one integer lies to the *right* of another integer, we say that the first integer is *greater than* (>) the second integer.

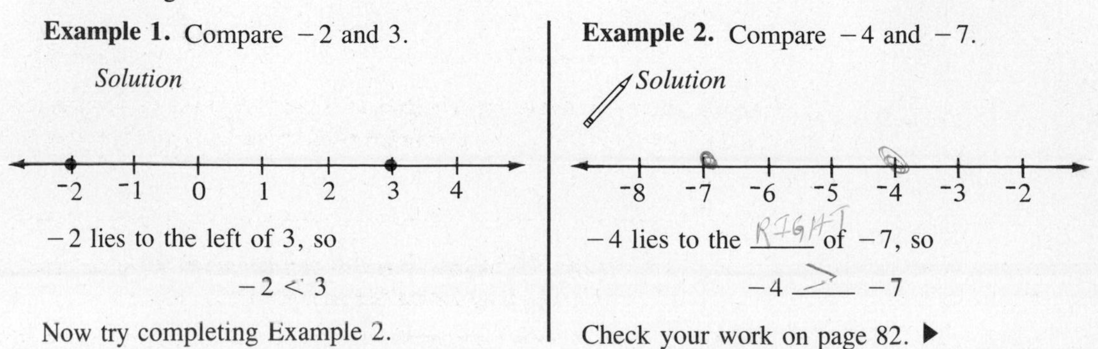

Example 1. Compare −2 and 3.

Solution

−2 lies to the left of 3, so

$$-2 < 3$$

Now try completing Example 2.

Example 2. Compare −4 and −7.

Solution

−4 lies to the ______ of −7, so

−4 ______ −7

Check your work on page 82. ▶

Example 3. Locate -3 and locate 3 on the number line.

Solution

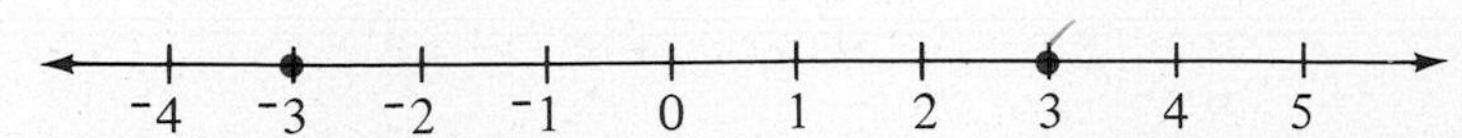

If we were to take a ruler and measure the distance between -3 and 0 on our number line and then measure the distance between 0 and 3 on our number line, what would we observe? Clearly, those distances are the same. Each of the points, -3 and 3, measures 3 units from zero. Their locations are on *opposite* sides of zero, but their distances represent the same number of units. For this reason, we say

-3 is the *opposite* of 3

3 is the *opposite* of -3

When two numbers measure the same distance from zero but the two numbers are on opposite sides of zero, we say that one number is the **opposite** of the other.

Example 4. What is the opposite of 16?

Solution: -16 is the opposite of 16.

You try Example 5.

Example 5. What is the opposite of -5?

Solution: ______ is the opposite of -5.

Check your answer on page 82. ▶

Sometimes we refer to a positive integer, such as 5, as $+5$, but the positive sign is not required. You should understand that if there is no sign in front of a number, it is considered positive. We also agree here that the opposite of 0 is 0.

We can use a *negative sign* in front of a number to denote the opposite of that number.

"The opposite of 3 is -3" can be written $-(3) = -3$

"The opposite of -3 is 3" can be written $-(-3) = 3$

Writing Opposites

Opposite of a:	$-(a)$	$= -a$
Opposite of $-a$:	$-(-a)$	$= a$
Opposite of 0:	$-(0)$	$= 0$

Example 6. Find $-(23)$.

Solution

$$-(23) = -23$$

Example 7. Find $-(-11)$.

Solution

$$-(-11) = 11$$

When we measure the distance between 0 and the point on the number line corresponding to an integer, we are finding the **absolute value** of that integer. The absolute value of a number, a, is symbolized by $|a|$.

Absolute Value. If a is a number, then $|a|$ is the distance between 0 and a on the number line.

Because absolute value represents distance, we note that

> The absolute value of a number is never negative.

Example 8. Find $|-6|$.

Solution

$$|-6| = 6$$

You try Example 9.

Example 9. Find $|23|$.

Solution

$$|23| = \underline{23}$$

Check your work on page 82. ▶

Then try Example 10.

Example 10. Find $|0|$.

Solution

$$|0| = \underline{0}$$

Check your work on page 82. ▶

Example 11. Find $-|-10|$.

Solution

$$-|-10|$$
$$10 = -(10)$$
$$10 = -10$$

Adding Integers

How do we go about adding integers? First, let us recall how to add integers that are positive or zero. Remember, these are the whole numbers 0, 1, 2, 3, 4, . . ., which we have already discussed. Let's use the number line.

Example 12. Use the number line to find $5 + 2$.

Solution: We start at 5 and move 2 units to the right.

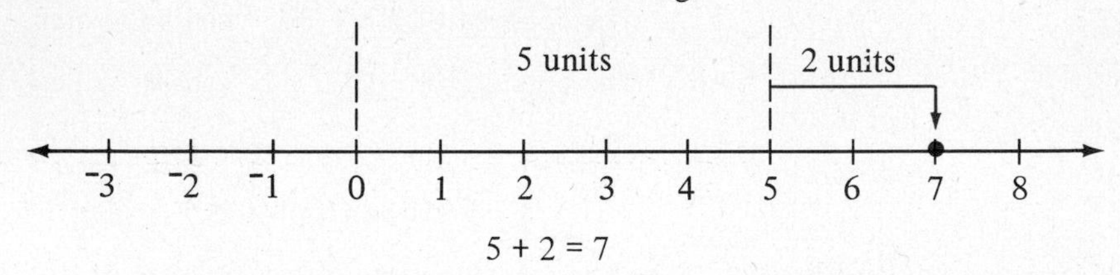

Notice that whenever we add two *positive* integers, the answer will be a *positive* integer.

> To add two *positive* integers, add the units (absolute values) and give the answer a *positive* sign.

What happens when we add two *negative* integers? The number line will be a help here. In locating a negative number, we must move *left* rather than right.

Example 13. Use the number line to add -3 and -2.

Solution: We start at -3 and move 2 units to the *left*.

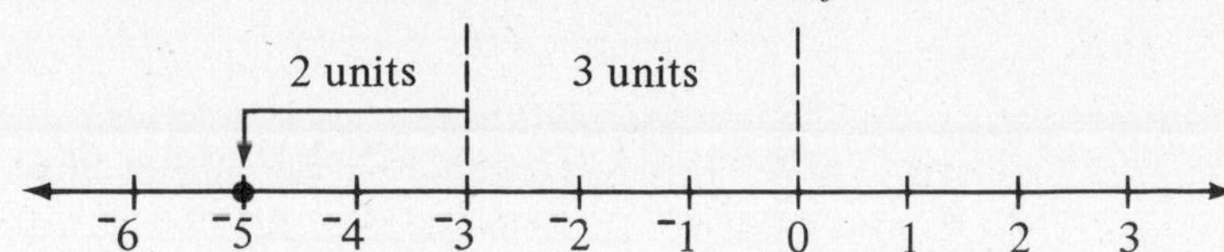

Since we ended up at -5, the sum of -3 and -2 must be -5. We write this as

$$-3 + (-2) = -5$$

Notice that in this example of adding two *negative* numbers, the number of units was the *sum* of the units in the original numbers, but the *sign* of the answer was *negative*. We conclude:

> To add two *negative* integers, add the units (absolute values) and give the answer a *negative* sign.

Example 14. Find $-13 + (-76)$.

Solution: The answer will have a *negative* sign.

$$-13 + (-76) = -89$$

You complete Example 15.

Example 15. Find $[-31 + (-8)] + (-12)$.

Solution: The answer will have a ________ sign.

$[-31 + (-8)] + (-12)$

$=$ _____ $+ (-12)$

$=$ _____

Check your work on page 82. ▶

What happens if we try to add two numbers when one of the numbers is *positive* and one of the numbers is *negative*? The number line will give us a clue if we remember to move *right* for *positive* numbers and move *left* for *negative* numbers.

Example 16. Use the number line to find $-6 + 4$.

Solution: We start at -6 and move 4 units to the *right*.

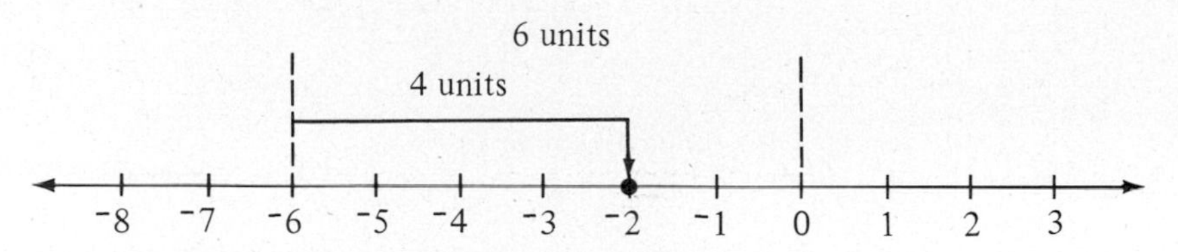

We end up at -2, so

$$-6 + 4 = -2$$

Example 17. Use the number line to find $-3 + 8$.

Solution: We start at -3 and move 8 units to the *right*.

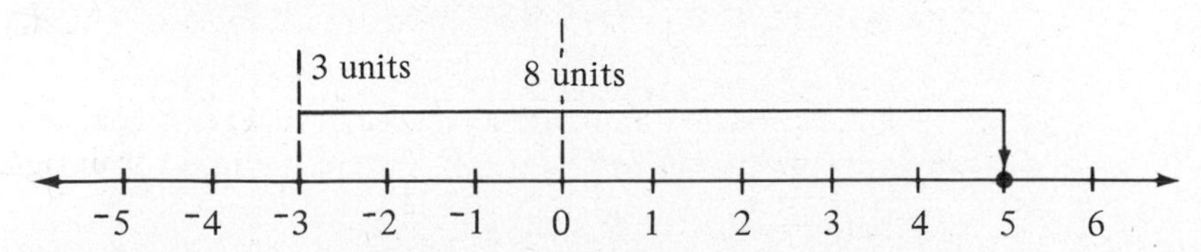

We end up at 5, so

$$-3 + 8 = 5$$

Do you see how we could add two numbers with opposite signs without using the number line? In each example, the *units* in our answer represented the *difference* between our original units. The *sign* of the answer was the sign of the original number having *more* units. Remember, the number of units for an integer is that integer's distance from zero on the number line (absolute value).

> To add a *positive* integer and a *negative* integer, find the *difference* between their units (absolute values) and give the answer the sign of the original integer having more units (larger absolute value).

Example 18. Simplify $-11 + 26$.

Solution: Our answer will have a positive sign. The difference between 11 units and 26 units is 15 units.

$$-11 + 26 = 15$$

You try Example 19.

Example 19. Simplify $-30 + 20$.

Solution: Our answer will have a ________ sign. The difference between 30 units and 20 units is ________ units.

$$-30 + 20 = \underline{\qquad}$$

Check your work on page 82. ▶

As was true for whole numbers, fractions, and decimal numbers, *parentheses* give us directions for the order in which to operate with integers.

Example 20. Simplify $-10 + (-17 + 2)$.

Solution: Parentheses still say "do this first."

$$
\begin{aligned}
& -10 + (-17 + 2) \\
&= -10 + (-15) \\
&= -25
\end{aligned}
$$

You try completing Example 21.

Example 21. Simplify $[5 + (-16)] + 3$.

Solution

$$
\begin{aligned}
& [5 + (-16)] + 3 \\
&= \underline{\qquad} + 3 \\
&= \underline{\qquad}
\end{aligned}
$$

Check your work on page 82. ▶

Example 22. Simplify

$$9\left[\frac{18}{2} + (-7 + 2)\right].$$

Solution

$$
\begin{aligned}
& 9\left[\frac{18}{2} + (-7 + 2)\right] \\
&= 9[9 + (-5)] \\
&= 9 \cdot 4 \\
&= 36
\end{aligned}
$$

You try Example 23.

Example 23. Simplify $\dfrac{7(-6 + 6)}{6 + (-3)}$.

Solution

$$\frac{7(-6 + 6)}{6 + (-3)}$$

Check your work on page 82. ▶

Trial Run

Simplify.

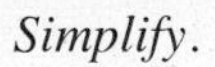

_____ 1. $8 + (-6)$

_____ 2. $-11 + 5$

_____ 3. $-3 + (-3)$

_____ 4. $[-5 + (-6)] + 10$

_____ 5. $(-12 + 2) + (-9)$

_____ 6. $\dfrac{3[9 + (-4)]}{-2 + 7}$

Answers are on page 83.

Using Some Properties of Addition

There are several laws or properties of addition that we may observe from our work with integers. Note these sums.

$$-15 + 8 = -7$$
$$8 + (-15) = -7$$

> It does not matter in which order we add two integers. The sum will be the same.

This law is called the **commutative property for addition**, and we state it in symbols as

> **Commutative Property for Addition.** For any integers A and B
>
> $$A + B = B + A$$

The commutative property gives us two ways in which to write the sum of two integers. Now consider these sums involving *three* integers.

$$(-6 + 8) + (-3) = 2 + (-3) = -1$$
$$-6 + [8 + (-3)] = -6 + 5 = -1$$

In these two sums, the order in which the integers appeared was the same, but the position of the symbols of grouping was different. In both cases the sum was the same.

When adding three integers, it does not matter if we add the first two and then add the third, *or* add the second two and then add the first. The sum will be the same.

This law is called the **associative property for addition**.

Associative Property for Addition. For any integers A, B, and C

$$(A + B) + C = A + (B + C)$$

Complete Examples 24 and 25.

Example 24. Use the commutative property to rewrite the sum $-5 + (-9)$.

Solution

$$-5 + (-9)$$
$$= \underline{-9} + (\underline{-5})$$

Check your answer on page 83. ▶

Example 25. Use the associative property to rewrite the sum $[4 + (-2)] + (-17)$.

Solution

$$[4 + (-2)] + (-17)$$
$$= \underline{4} + [\underline{-2} + (\underline{-17})]$$

Check your answer on page 83. ▶

To add several integers when there are no brackets, the commutative and associative properties of addition allow us to arrange them in any order we choose. It is often easier to *add all the numbers with matching signs first*.

Example 26. Find the sum $-22 + (-11) + 6 + (-30) + 50$.

Solution

$-22 + (-11) + 6 + (-30) + 50$

$= [-22 + (-11) + (-30)] + [6 + 50]$ — Rearrange terms (commutative and associative properties).

$= -63 + 56$ — Add negative integers; add positive integers.

$= -7$ — Find the sum.

You try Example 27.

Example 27. Find the sum $16 + (-23) + 4 + (-8) + (-10)$.

Solution

$16 + (-23) + 4 + (-8) + (-10)$

$= [16 + \underline{4}] + [-23 + (\underline{-8}) + (\underline{-10})]$ — Rearrange terms.

$= \underline{20} + (\underline{-41})$ — Add positive integers; add negative integers.

$= \underline{-21}$ — Find the sum.

Check your work on page 83. ▶

What happens when we add *zero* to an integer? Look at these sums.

$$13 + 0 = 13$$
$$-26 + 0 = -26$$
$$0 + (-93) = -93$$

As was true with whole numbers, when zero is added to any integer, the sum is that integer.

Addition Property of Zero. For any integer A

$$A + 0 = A$$

Recall that zero is called the **identity for addition**.

Consider the result when an integer is added to its *opposite*.

$$9 + (-9) = 0$$
$$-62 + 62 = 0$$

These examples should help you see that

The sum of an integer and its opposite is 0.

The *opposite* of a number is also called its **additive inverse** and we can state this new law as follows.

Property of Opposites *(or Additive Inverses)*. For any integer A

$$A + (-A) = 0$$

Example 28. Find the sum $-15 + 12 + (-11) + 14$.

Solution

$-15 + 12 + (-11) + 14$	
$= [-15 + (-11)] + [12 + 14]$	Rearrange terms.
$= -26 + 26$	Add negative integers; add positive integers.
$= 0$	Property of opposites.

Trial Run

Complete each statement.

1. By the commutative property for addition, $-8 + 7 =$ 7+(-8).

2. By the associative property for addition, $-5 + [3 + (-9)] =$ ______.

Find each sum.

______ 3. $8 + (-6) + (-2)$

______ 4. $-22 + (-7) + 3 + (-20) + 40$

______ 5. $(17 + 0) + [0 + (-3)]$

______ 6. $16 + (-23) + 4 + (-8) + (-10)$

Answers are on page 83.

▶ Examples You Completed

Example 2. Compare -4 and -7.

Solution

Number line: $^-8 \quad ^-7 \quad ^-6 \quad ^-5 \quad ^-4 \quad ^-3 \quad ^-2$ (points marked at $^-7$ and $^-4$)

-4 lies to the right of -7, so

$$-4 > -7$$

Example 5. What is the opposite of -5?

Solution: 5 is the opposite of -5.

Example 9. Find $|23|$.

Solution

$$|23| = 23$$

Example 10. Find $|0|$.

Solution

$$|0| = 0$$

Example 15. Find $[-31 + (-8)] + (-12)$.

Solution: The answer will have a negative sign.

$$[-31 + (-8)] + (-12)$$
$$= -39 + -12$$
$$= -51$$

Example 19. Simplify $-30 + 20$.

Solution: Our answer will have a negative sign. The difference between 30 units and 20 units is 10 units.

$$-30 + 20 = -10$$

Example 21. Simplify $[5 + (-16)] + 3$.

Solution

$$[5 + (-16)] + 3$$
$$= -11 + 3$$
$$= -8$$

Example 23. Simplify $\dfrac{7(-6 + 6)}{6 + (-3)}$.

Solution

$$\frac{7(-6 + 6)}{6 + (-3)}$$
$$= \frac{7(0)}{3}$$
$$= \frac{0}{3}$$
$$= 0$$

Example 24. Use the commutative property to rewrite the sum $-5 + (-9)$.

Solution

$$
\begin{aligned}
& -5 + (-9) \\
&= -9 + (-5)
\end{aligned}
$$

Example 25. Use the associative property to rewrite the sum $[4 + (-2)] + (-17)$.

Solution

$$
\begin{aligned}
& [4 + (-2)] + (-17) \\
&= 4 + [-2 + (-17)]
\end{aligned}
$$

Example 27. Find the sum $16 + (-23) + 4 + (-8) + (-10)$.

Solution

$$
\begin{aligned}
& 16 + (-23) + 4 + (-8) + (-10) \\
&= [16 + 4] + [-23 + (-8) + (-10)] \\
&= 20 + (-41) \\
&= -21
\end{aligned}
$$

Answers to Trial Runs

page 79 **1.** 2 **2.** -6 **3.** -6 **4.** -1 **5.** -19 **6.** 3

page 81 **1.** $7 + (-8)$ **2.** $(-5 + 3) + (-9)$ **3.** 0 **4.** -6 **5.** 14 **6.** -21

Name ______________________ Date ____________

EXERCISE SET 2.1

1. Graph the points corresponding to these integers on a number line.

(a) -7 (b) 4 (c) 0 (d) -3

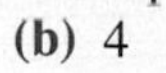

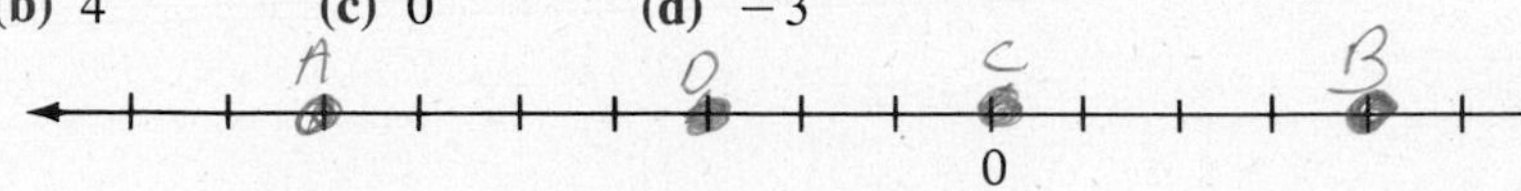

2. Graph the points corresponding to these integers on a number line.

(a) -5 (b) 3 (c) 1 (d) -8

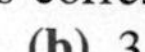

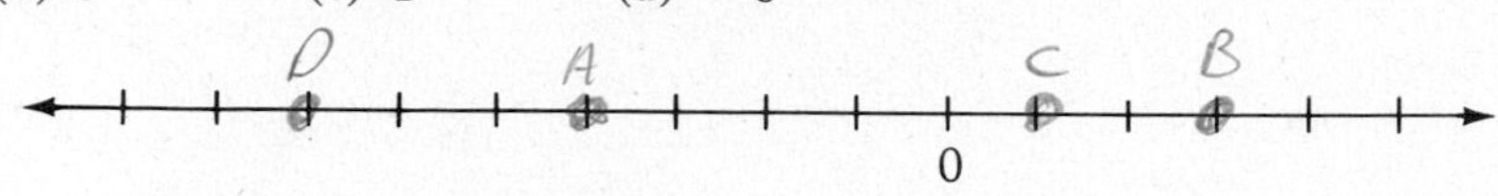

3. Identify by number each letter on the number line.

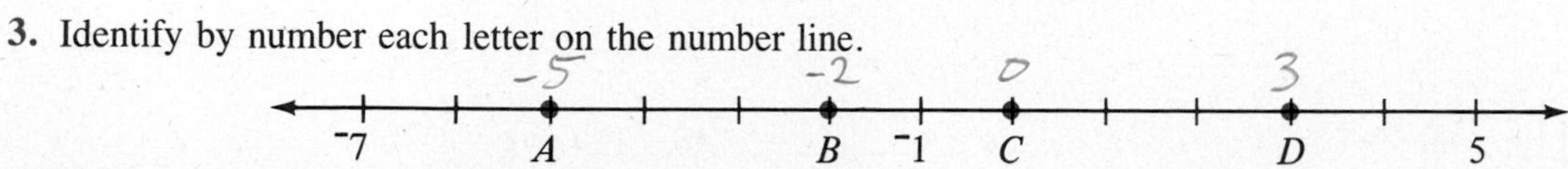

4. Identify by number each letter on the number line.

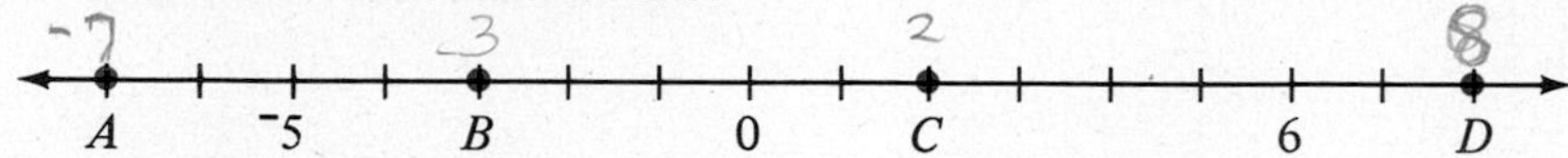

5. Compare each pair of numbers using < or >.

______ (a) 5, 12 ______ (b) -4, -10

______ (c) 8, -11 ______ (d) -3, 7

6. Compare each pair of numbers using < or >.

______ (a) 3, 9 ______ (b) -12, -2

______ (c) 10, -15 ______ (d) -2, 9

7. Use the commutative property for addition to rewrite each sum.

______ (a) $-7 + 2$ ______ (b) $6 + (-9)$

______ (c) $-20 + (-6)$ ______ (d) $8 + 10$

8. Use the commutative property for addition to rewrite each sum.

______ (a) $19 + 0$ ______ (b) $-3 + (-11)$

______ (c) $-12 + 5$ ______ (d) $9 + (-3)$

9. Use the associative property for addition to rewrite each sum.

______ (a) $(6 + 3) + (-2)$ ______ (b) $-5 + [4 + (-3)]$

______ (c) $[(-4) + 9] + (-2)$ ______ (d) $7 + [(-2) + 11]$

10. Use the associative property for addition to rewrite each sum.

______ **(a)** $7 + (2 + 5)$ ______ **(b)** $[(-5) + 8] + 9$

______ **(c)** $13 + [(-2) + (-8)]$ ______ **(d)** $[12 + (-5)] + (-3)$

Simplify by removing absolute value bars.

______ **11.** $|-5|$ ______ **12.** $|-7|$

______ **13.** $-|-8|$ ______ **14.** $-|-17|$

______ **15.** $|9 + (-12)|$ ______ **16.** $|(-10) + 8|$

______ **17.** $|9| + |-3|$ ______ **18.** $|12| + |-5|$

______ **19.** $|19| - |-8|$ ______ **20.** $|15| - |-6|$

Simplify.

______ **21.** $4 + 18$ ______ **22.** $12 + 5$

______ **23.** $-5 + (-7)$ ______ **24.** $-9 + (-12)$

______ **25.** $15 + (-3)$ ______ **26.** $19 + (-4)$

______ **27.** $3 + (-9)$ ______ **28.** $2 + (-15)$

______ **29.** $-11 + 4$ ______ **30.** $-12 + 7$

______ **31.** $-2 + 15$ ______ **32.** $-3 + 22$

______ **33.** $14 + [8 + (-2)]$ ______ **34.** $20 + [10 + (-3)]$

______ **35.** $[9 + (-7)] + 6$ ______ **36.** $[15 + (-8)] + 7$

______ **37.** $[-15 + (-3)] + 1$ ______ **38.** $[-41 + (-9)] + 1$

______ **39.** $-12 + (-7 + 5)$ ______ **40.** $-13 + (-5 + 12)$

______ **41.** $21 + (-12) + 9$ ______ **42.** $32 + (-24) + 7$

______ **43.** $-15 + 7 + (-13)$ ______ **44.** $-26 + 10 + (-4)$

______ **45.** $13 + (-8) + (-9) + 4$ ______ **46.** $18 + (-5) + (-17) + 4$

______ **47.** $-19 + (-9) + 12 + 7 + (-2)$ ______ **48.** $-25 + (-11) + 9 + 8 + (-15)$

______ **49.** $2[8 + (-3)]$ ______ **50.** $3[13 + (-10)]$

______ **51.** $(4.3) + (-17)$ ______ **52.** $(5.6) + (-20)$

______ **53.** $(-12) + (5 \cdot 0)$ ______ **54.** $(-13) + (8 \cdot 0)$

______ **55.** $2[(-8) + 5 \cdot 4]$

______ **56.** $3[(-9) + 3 \cdot 6]$

______ **57.** $\dfrac{24 + (-3)}{-8 + 15}$

______ **58.** $\dfrac{33 + (-7)}{-2 + 15}$

______ **59.** $\dfrac{-3 + 30}{9 + 0}$

______ **60.** $\dfrac{-4 + 20}{8 + 0}$

______ **61.** $\dfrac{-7 + 16}{-4 + 4}$

______ **62.** $\dfrac{-9 + 13}{-5 + 5}$

______ **63.** $-8 + (-9) + \dfrac{20}{2}$

______ **64.** $-10 + (-3) + \dfrac{25}{5}$

______ **65.** $\dfrac{5(-7 + 7)}{9 + (-4)}$

______ **66.** $\dfrac{6(-8 + 8)}{18 + (-9)}$

______ **67.** $-18 + \left(6 + \dfrac{16}{4}\right)$

______ **68.** $-23 + \left(9 + \dfrac{15}{3}\right)$

______ **69.** $3\left[\dfrac{21}{3} + (-9 + 2)\right]$

______ **70.** $4\left[\dfrac{15}{3} + (-6 + 1)\right]$

☆ Stretching the Topics

Simplify.

______ **1.** $5\{(-11) + 6[7 + (-3)]\}$

______ **2.** $-4 + 2\left\{\dfrac{18}{9} + 3[7 + (-1)]\right\}$

______ **3.** $10\left\{2 \cdot 4 + \dfrac{2[13 + (-5)]}{4}\right\}$

Check your answers in the back of your book.

If you can complete **Checkup 2.1**, you are ready to go on to Section 2.2.

Name ____________________ Date ____________

✓ CHECKUP 2.1

______ 1. Compare 3 and -15 using $<$ or $>$.

______ 2. Use the commutative property to rewrite the sum $17 + (-5)$.

______ 3. Use the associative property to rewrite the sum $-9 + (-3 + 7)$.

______ 4. Simplify $-|-12|$ by removing absolute value bars.

Simplify.

______ 5. $-9 + 4$

______ 6. $10 + (-2)$

______ 7. $[-3 + (-5)] + 7$

______ 8. $-8 + (-9) + 5 + (-3)$

______ 9. $6\left[\frac{18}{2} + (-12 + 3)\right]$

______ 10. $\frac{2(-3 + 5)}{6 + (-2)}$

Check your answers in the back of your book.

If You Missed Problems:	You Should Review Examples:
1	1, 2
2, 3	24, 25
4	11
5, 6	18, 19
7	15, 21
8	27
9	22
10	23

2.2 Subtracting Integers

When we discussed the subtraction of whole numbers, we agreed that

$$9 - 3 = 6$$

In the section on the addition of integers, we agreed that

$$9 + (-3) = 6$$

so it seems that $9 - 3$ and $9 + (-3)$ are two different ways of saying the same thing.

Subtraction		*Addition*
$9 - 3$	means the same as	$9 + (-3)$
9 *minus* 3	means the same as	9 *plus* the *opposite* of 3

In fact, this is the way in which *subtraction* is defined using *addition*. Since we know how to *add* integers, subtraction should proceed smoothly.

Example 1. Write $10 - 1$ using addition, and simplify.

Solution

$$10 - 1$$
$$= 10 + (-1)$$
$$= 9$$

Now you complete Example 2.

Example 2. Write $6 - 6$ using addition, and simplify.

Solution

$$6 - 6$$
$$= 6 + (-6)$$
$$= \underline{0}$$

Check your work on page 93. ▶

Every subtraction problem can be written as an addition problem.

> To subtract two integers, take the *opposite* of the integer being subtracted, and *add*.

If A and B represent any integers, we write:

Definition of Subtraction

$$A - B = A + (-B)$$

Example 3. Find $-7 - (-5)$.

Solution

$-7 - (-5)$	Subtraction problem.
$= -7 + 5$	Take the opposite of -5, and *add*.
$= -2$	

You try Example 4.

Example 4. Find $-11 - 19$.

Solution

$-11 - 19$	Subtraction problem.
$= -11 + (\underline{-19})$	Take the opposite of 19, and *add*.
$= \underline{-30}$	

Check your work on page 93. ▶

When expressions contain symbols of grouping, we shall continue to perform the operations within parentheses first and to work from the innermost symbols of grouping outward. If you are impatient and try to hurry through these problems, you will find yourself making mistakes. It is important to get into the habit of dealing carefully with symbols of grouping.

Example 5. Find $(7 - 10) - (8 - 15)$.

Solution

$(7 - 10) - (8 - 15)$	
$= [7 + (-10)] - [8 + (-15)]$	Rewrite inner subtractions as additions.
$= -3 - (-7)$	Simplify within brackets.
$= -3 + 7$	Rewrite subtraction as addition.
$= 4$	Find the sum.

You complete Example 6.

Example 6. Find $8 - (10 - 23)$.

Solution

$8 - (10 - 23)$	
$= 8 - [10 + (\_\_\_\_)]$	Rewrite inner subtraction as addition.
$= 8 - [\_\_\_\_]$	Simplify within brackets.
$= 8 + \_\_\_\_$	Rewrite subtraction as addition.
$= \_\_\_\_$	

Check your work on page 94. ▶

Example 7. Find $(8 - 13) - (10 - 16) - 4$.

Solution

$(8 - 13) - (10 - 16) - 4$	
$= [8 + (-13)] - [10 + (-16)] - 4$	Rewrite inner subtractions as additions.
$= -5 - [-6] - 4$	Simplify within brackets.
$= -5 + 6 + (-4)$	Rewrite subtractions as additions.
$= -5 + (-4) + 6$	Rearrange terms.
$= -9 + 6$	Add negative numbers.
$= -3$	Find the sum.

Trial Run

Simplify.

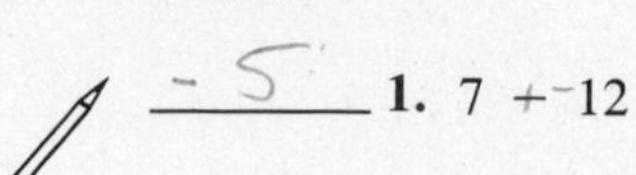

______ **1.** $7 - 12$

______ **2.** $-15 - 8$

______ 3. $-9 - (-17)$ ______ 4. $9 - (3 + 12)$

______ 5. $(3 - 7) - (9 - 12)$ ______ 6. $10 - (9 - 21)$

Answers are on page 94.

A few examples might help us see whether the operation of subtraction is *commutative* or *associative*.

Commutative?

$$-3 - (-7) \stackrel{?}{=} -7 - (-3)$$
$$-3 + 7 \stackrel{?}{=} -7 + 3$$
$$4 \neq -4$$

There is *no* commutative property for subtraction.

Associative?

$$10 - (8 - 3) \stackrel{?}{=} (10 - 8) - 3$$
$$10 - 5 \stackrel{?}{=} 2 - 3$$
$$5 \neq -1$$

There is *no* associative property for subtraction.

▶ Examples You Completed

Example 2. Write $6 - 6$ using addition, and simplify.

Solution

$$6 - 6$$
$$= 6 + (-6)$$
$$= 0$$

Example 4. Find $-11 - 19$.

Solution

$$-11 - 19$$
$$= -11 + (-19)$$
$$= -30$$

Example 6. Find $8 - (10 - 23)$.

Solution

$$
\begin{aligned}
& 8 - (10 - 23) \\
&= 8 - [10 + (-23)] \\
&= 8 - [-13] \\
&= 8 + 13 \\
&= 21
\end{aligned}
$$

Answers to Trial Run

page 92 **1.** -5 **2.** -23 **3.** 8 **4.** -6 **5.** -1 **6.** 22

Name ______________________________ Date ______________

EXERCISE SET 2.2

Write each expression using addition and simplify.

_____ 1. $12 - 9$

_____ 2. $23 - 8$

_____ 3. $11 - 27$

_____ 4. $9 - 18$

_____ 5. $-15 - 7$

_____ 6. $-19 - 8$

_____ 7. $12 - (-3)$

_____ 8. $17 - (-5)$

_____ 9. $-9 - (-7)$

_____ 10. $-11 - (-8)$

Simplify each expression.

_____ 11. $15 + (8 - 13)$

_____ 12. $17 + (9 - 15)$

_____ 13. $21 - (3 - 7)$

_____ 14. $11 - (5 - 12)$

_____ 15. $(8 - 7) - 12$

_____ 16. $(13 - 10) - 20$

_____ 17. $(7 - 14) - 2$

_____ 18. $(9 - 16) - 5$

_____ 19. $(5 - 4) - (-2)$

_____ 20. $(10 - 3) - (-4)$

_____ 21. $-7 - (10 - 5)$

_____ 22. $-13 - (12 - 6)$

_____ 23. $[(-5) - 2] - 3$

_____ 24. $[(-8) - 4] - 9$

_____ 25. $[-5 - (-2)] - 3$

_____ 26. $[-9 - (-4)] - 5$

_____ 27. $[-10 - 5] - (-15)$

_____ 28. $[-12 - 6] - (-18)$

_____ 29. $-13 - [(-3) - 11]$

_____ 30. $-12 - [(-5) - 8]$

_____ 31. $3[(-2) - (-7)]$

_____ 32. $4[(-3) - (-8)]$

_____ 33. $3 \cdot 5 - 17$

_____ 34. $2 \cdot 6 - 21$

_____ 35. $-6 + (8 \cdot 0)$

_____ 36. $-9 - (6 \cdot 0)$

_____ 37. $2(7 \cdot 4 - 8)$

_____ 38. $5(6 \cdot 5 - 10)$

_____ 39. $(12 - 15) - (7 - 13)$

_____ 40. $(8 - 13) - (5 - 9)$

_____ 41. $(-3 - 9) - (7 - 3)$

_____ 42. $(-4 - 11) - (10 - 7)$

_____ 43. $(9 - 15) - (17 - 21) - 5$

_____ 44. $(4 - 12) - (21 - 15) - 7$

_____ 45. $[7 - (-3)] - [-12 - (-9)]$

_____ 46. $[8 - (-5)] - [-16 - (-7)]$

_____ 47. $\dfrac{25 - 3}{4 - (-7)}$

_____ 48. $\dfrac{33 - 9}{6 - (-6)}$

_____ 49. $\dfrac{3 - (-15)}{9 - 0}$

_____ 50. $\dfrac{4 - (-17)}{7 - 0}$

_____ 51. $\dfrac{-3 - (-12)}{5 - 5}$

_____ 52. $\dfrac{17 - (-2)}{7 - 7}$

_____ 53. $-8 - 9 - \dfrac{16}{4}$

_____ 54. $-10 - 7 - \dfrac{20}{10}$

_____ 55. $\dfrac{4(8 - 8)}{-4 - (-9)}$

_____ 56. $\dfrac{3(11 - 11)}{-5 - (-10)}$

_____ 57. $-13 - 8 - \dfrac{10}{2}$

_____ 58. $-17 - 9 - \dfrac{15}{3}$

☆ Stretching the Topics

Simplify.

_____ 1. $-9 - \{10 + 3[1 - (-2)]\}$

_____ 2. $2\{3[10 - (-7)] - 4(-2 + 5)\}$

_____ 3. $9\left[4.1 - \dfrac{14 - (-12)}{6 - (-7)}\right]$

Check your answers in the back of your book.

If you can simplify the expressions in **Checkup 2.2**, you are ready to go on to Section 2.3.

Name ____________________ Date ____________

CHECKUP 2.2

Simplify.

____ **1.** $12 - 5$

____ **2.** $8 - (-13)$

____ **3.** $-17 - 5$

____ **4.** $-19 - (-9)$

____ **5.** $10 - 28$

____ **6.** $-7 - (10 - 23)$

____ **7.** $(4 - 11) - (11 - 19)$

____ **8.** $(-3 - 7) - (9 - 11)$

____ **9.** $(9 - 12) - (21 - 30) - 7$

____ **10.** $[9 - (-5)] - [-13 - (-4)]$

Check your answers in the back of your book.

If You Missed Problems:	You Should Review Examples:
1–5	1–4
6	6
7, 8	5
9, 10	7

2.3 Multiplying and Dividing Integers

Multiplying Integers

When multiplying whole numbers, we observed that multiplication was a way of performing repeated addition. Whenever we multiply two positive integers, we are repeatedly adding positive numbers, so the answer must be *positive*.

$$4 \cdot 3 = 3 + 3 + 3 + 3 = 12$$

$$5 \cdot 1 = 1 + 1 + 1 + 1 + 1 = 5$$

> To find the product of *two positive* integers, multiply the units (absolute values) and give the answer a *positive* sign.

Look at the product of a positive integer and a negative integer, again using repeated addition.

$$5(-3) = -3 + (-3) + (-3) + (-3) + (-3) = -15$$

$$4(-1) = -1 + (-1) + (-1) + (-1) = -4$$

In each case, we were repeatedly adding negative numbers, so the answer was *negative*.

> To find the product of a *positive* integer and a *negative* integer, multiply the units (absolute values) and give the answer a *negative* sign.

Example 1. Find $3(-6)$.

Solution

$$3(-6)$$

$$= -18$$

You do Example 2.

Example 2. Find $-4 \cdot 11$.

Solution

$$-4 \cdot 11$$

$$= \underline{-44}$$

Check your work on page 108. ▶

What happens when we multiply two negative integers? Look at the pattern that occurs when we consider some multiples of -3.

$$4(-3) = -12$$

$$3(-3) = -9$$

$$2(-3) = -6$$

$$1(-3) = -3$$

$$0(-3) = 0$$

As the multipliers of -3 on the left are *decreasing by 1*, notice that the products on the right are *increasing by 3*. To continue this pattern, we must let the multipliers continue to *decrease by 1*. At the same time, the products must continue to *increase by 3*.

$$-1(-3) = 3$$

$$-2(-3) = 6$$

$$-3(-3) = 9$$

$$-4(-3) = 12$$

Observe that the product of two negative numbers turns out to be a positive number.

> To find the product of *two negative* integers, multiply the units (absolute values) and give the answer a *positive* sign.

Example 3. Find $(-1)(-1)$.

Solution

$$(-1)(-1)$$
$$= 1$$

You try Example 4.

Example 4. Find $(-8)(-6)$.

Solution

$$(-8)(-6)$$
$$= \underline{48}$$

Check your work on page 108. ▶

Symbols of grouping still give us direction regarding the order of operations.

Example 5. Find $[-3(-2)](-7)$.

Solution

$$[-3(-2)](-7)$$
$$= [6](-7)$$
$$= -42$$

You complete Example 6.

Example 6. Find $-5(3) + (-2)(-1)$.

Solution

$$-5(3) + (-2)(-1)$$
$$= -15 + \underline{2}$$
$$= \underline{-13}$$

Check your work on page 108. ▶

Example 7. Find $-3[-5 - (10 - 27)]$.

Solution

$-3[-5 - (10 - 27)]$	
$= -3\{-5 - [10 + (-27)]\}$	Rewrite inner subtraction as addition.
$= -3\{-5 - [-17]\}$	Add within brackets.
$= -3\{-5 + 17\}$	Rewrite subtraction as addition.
$= -3 \cdot 12$	Add within braces.
$= -36$	Multiply.

Trial Run

Find the following.

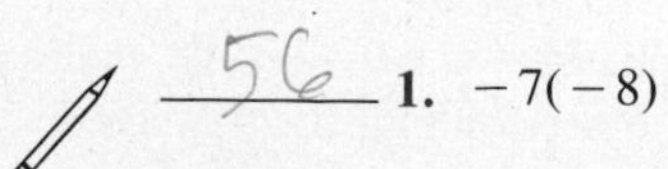

56 **1.** $-7(-8)$

-72 **2.** $-6(12)$

______ 3. $4(-3) + 5$

______ 4. $-5(8 - 17)$

______ 5. $-2[-7 - (13 - 16)]$

______ 6. $[-7(-3)][-1(2)]$

Answers are on page 109.

Using Some Properties of Multiplication

There are several properties of multiplication that are similar to the properties of addition. From the products

$$-6(5) = -30$$

$$5(-6) = -30$$

you might agree that

> It does not matter in which order we multiply two integers. The product is the same.

This law is the **commutative property for multiplication.**

> **Commutative Property for Multiplication.** For integers A and B
>
> $$A \cdot B = B \cdot A$$

Now consider a few products involving three factors.

$$[-2 \cdot 3](-4) = -6(-4) = 24$$

$$-2[3(-4)] = -2(-12) = 24$$

In these products, the order in which the factors appeared was the same, but the position of the symbols of grouping was different. In both cases, the product was the same.

> When multiplying three integers, it does not matter if we multiply the product of the first two times the third, or multiply the first times the product of the second two. The product will be the same.

As you may have guessed, this law is called the **associative property for multiplication.**

> **Associative Property for Multiplication.** For any integers A, B, and C
>
> $$(A \cdot B) \cdot C = A \cdot (B \cdot C)$$

Complete Examples 8 and 9.

Example 8. Use the commutative property to rewrite the product $-2(-11)$.

Solution

$$-2(-11)$$
$$= \underline{\quad\quad}(\underline{\quad\quad})$$

Check your answer on page 108. ▶

Example 9. Use the associative property to rewrite the product $5[-1(-8)]$.

Solution

$$5[-1(-8)]$$
$$= [5(\underline{\quad\quad})] \cdot (\underline{\quad\quad})$$

Check your answer on page 108. ▶

To multiply several integers when there are no brackets, the commutative and associative properties of multiplication allow us to arrange them in any order we choose.

Example 10. Find $3(-2)(-1)(-5)$.

Solution

$3(-2)(-1)(-5)$

$= -6(5)$ Multiply $3(-2)$ and multiply $(-1)(-5)$.

$= -30$ Multiply.

There are two ways to work a problem such as $8(-2 + 5)$.

First method	*Second method*
$8(-2 + 5)$	$8(-2 + 5)$
$= 8 \cdot 3$	$= 8(-2) + 8 \cdot 5$
$= 24$	$= -16 + 40$
	$= 24$

Notice that the results are the same. It seems that the first method is easier, but later we shall have to use the second method, so we note here that

> To multiply a number times the sum of numbers in parentheses, we may multiply the number outside the parentheses times each of the numbers inside the parentheses and add these products together.

This new law is called the **distributive property for multiplication over addition**.

Distributive Property for Multiplication over Addition and Subtraction. For integers A, B, and C

$$A(B + C) = A \cdot B + A \cdot C$$

$$A(B - C) = A \cdot B - A \cdot C$$

Example 11. Use the distributive property to find the product $-5(7 - 1)$.

Solution

$$\begin{aligned} & -5(7 - 1) \\ &= -5(7) - (-5)(1) \\ &= -35 - (-5) \\ &= -35 + 5 \\ &= -30 \end{aligned}$$

Now try Example 12.

Example 12. Use the distributive property to find the product $-10(-17 + 19)$.

Solution: $-10(-17 + 19)$

Check your work on page 108. ▶

What happens when we multiply any integer by 1? Consider these products.

$$-3 \cdot 1 = -3$$

$$1 \cdot 18 = 18$$

As with whole numbers, when 1 is multiplied times any integer, the product is that integer.

Multiplication Property of 1. For any integer A

$$A \cdot 1 = A$$

Recall that 1 is called the **identity for multiplication**.

Trial Run

Complete each statement.

1. According to the commutative property for multiplication, $10(-7) =$ ________.

2. According to the associative property for multiplication, $[6(-5)] \cdot 9 =$ ________.

3. According to the distributive property for multiplication over addition, $-3(2 + 8) =$ ________.

Find each product.

________ **4.** $-10(-3)(-7)$

________ **5.** $6(5)(-1)(-1)$

Answers are on page 109.

Finding Powers of Integers

In Chapter 1 we discovered that **exponents** could be used to write products in which a factor is repeated.

Definition of a^n

$$a^n = \underbrace{a \cdot a \cdot a \cdot \ldots \cdot a}_{n \text{ factors}}$$

This definition also "works" when the base (a) is an *integer* and the exponent (n) is a natural number.

Example 13. Find $(-2)^3$.

Solution

$$\begin{aligned} &(-2)^3 \\ &= (-2)(-2)(-2) \\ &= 4(-2) \\ &= -8 \end{aligned}$$

You try Example 14.

Example 14. Find $(-3)^4$.

Solution

$$\begin{aligned} &(-3)^4 \\ &= (-3)(-3)(-3)(-3) \\ &= 9(\_\_\_\_) \\ &= \_\_\_\_ \end{aligned}$$

Check your work on page 108. ▶

We note here that the sign of a power, a^n, depends on the sign of the base, a, and whether the exponent, n, is *odd* or *even*.

1. If a is *positive*, then a^n is always *positive*.
2. If a is *negative* and n is *even*, then a^n is *positive*.
3. If a is *negative* and n is *odd*, then a^n is *negative*.

For example,

$$(-1)^6 = 1 \quad \text{because} \quad 6 \text{ is even}$$

but

$$(-1)^{13} = -1 \quad \text{because} \quad 13 \text{ is odd}$$

Remember that in our rules for the order of operations, we deal with *exponents* first.

Example 15. Find $-2(-1)^2$.

Solution

$-2(-1)^2$

$= -2(1)$ Find the power $(-1)^2$.

$= -2$ Multiply.

You try Example 16.

Example 16. Find $(-2)^3(-5)^2$.

Solution

$$\begin{aligned} &(-2)^3(-5)^2 \\ &= (-2)(-2)(-2)(-5)(-5) \\ &= (-8)(\_\_\_\_) \\ &= \_\_\_\_ \end{aligned}$$

Check your work on page 108. ▶

Trial Run

Simplify.

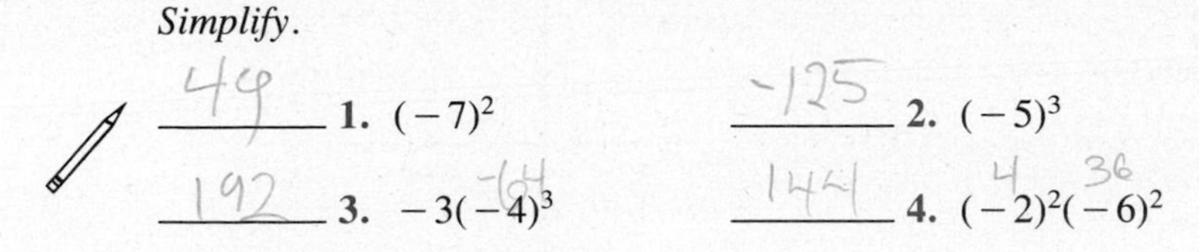

_____ 1. $(-7)^2$ _____ 2. $(-5)^3$

_____ 3. $-3(-4)^3$ _____ 4. $(-2)^2(-6)^2$

Answers are on page 109.

Dividing Integers

Remember that every division statement corresponds to a multiplication statement. To work a division problem, we may look at the corresponding multiplication problem.

Example 17. Find $\frac{123}{3}$ and write the corresponding multiplication statement.

Solution

$$\frac{123}{3} = 41$$

because $3 \cdot 41 = 123$

You try Example 18.

Example 18. Find $\frac{99}{9}$ and write the corresponding multiplication statement.

Solution

$$\frac{99}{9} = \_\_\_\_$$

because $9 \cdot \_\_\_\_ = 99$

Check your work on page 108. ▶

In each of these examples we found that the quotient of *two positive* integers was a *positive integer*.

> To divide a *positive* integer by a *positive* integer, divide the units (absolute values) and give the answer a *positive* sign.

How do we find the quotient of two *negative* integers? We can use multiplication of integers to help us see that

$$\frac{-36}{-9} = 4 \quad \text{because} \quad -9(4) = -36$$

and that

$$\frac{-18}{-6} = 3 \quad \text{because} \quad -6(3) = -18$$

In each case, we found that the quotient of *two negative* integers was a *positive* integer.

> To divide a *negative* integer by a *negative* integer, divide the units (absolute values) and give the answer a *positive* sign.

Example 19. Simplify $\frac{-75}{-3}$.

Solution

$$\frac{-75}{-3} = 25$$

because $-3(25) = -75$

Complete Example 20.

Example 20. Simplify $\frac{-6}{-6}$.

Solution

$$\frac{-6}{-6} = \_\_\_\_$$

because $-6(\_\_\_\_) = -6$

Check your work on page 108. ▶

What happens if the dividend and divisor have *different* signs? Again we must remember our multiplication facts for integers. You should agree that

$$\frac{-35}{7} = -5 \quad \text{because} \quad 7(-5) = -35$$

and

$$\frac{39}{-3} = -13 \quad \text{because} \quad -3(-13) = 39$$

In both cases, we found that the quotient of a *positive* integer and a *negative* integer was a *negative* integer.

> To divide a *positive* integer by a *negative* integer (or a *negative* integer by a *positive* integer), divide the units (absolute values) and give the answer a *negative* sign.

Example 21. Simplify $\frac{-16}{2}$.

Solution

$$\frac{-16}{2}$$
$$= -8$$

Complete Example 22.

Example 22. Simplify $\frac{17}{-17}$.

Solution

$$\frac{17}{-17}$$
$$= \_\_\_\_$$

Check your work on page 109. ▶

Some people prefer to learn the rules for multiplying and dividing integers in a slightly different form.

> **Multiplying or Dividing Two Integers**
>
> 1. If the signs are the *same*, the answer will be *positive*.
> 2. If the signs are *different*, the answer will be *negative*.

The rules for working with zero in a division problem continue to apply here.

Example 23. Simplify $\frac{0}{-13}$.

Solution

$$\frac{0}{-13} = 0$$

because $-13(0) = 0$

Example 24. Simplify $\frac{-37}{0}$.

Solution: $\frac{-37}{0}$ is undefined.

By using symbols of grouping we may combine *all* the integer operations we have learned.

Example 25. Simplify $\dfrac{3(-1) - \frac{24}{-8}}{17 + (-2)(5)}$.

Solution

$$\dfrac{3(-1) - \frac{24}{-8}}{17 + (-2)(5)}$$

$$= \frac{-3 - (-3)}{17 + (-10)}$$

$$= \frac{-3 + 3}{7}$$

$$= \frac{0}{7}$$

$$= 0$$

Now you complete Example 26.

Example 26. Simplify $\frac{5[18 + 2(-3)]}{-6}$.

Solution

$$\frac{5[18 + 2(-3)]}{-6}$$

$$= \frac{5[18 + (-6)]}{-6}$$

$$= \frac{5[\quad]}{-6}$$

$$= \frac{\quad}{-6}$$

$= \_\_\_\_\_$

Check your work on page 109. ▶

Trial Run

Simplify.

______ 1. $\frac{-16}{2}$

______ 2. $\frac{-24}{-8}$

______ 3. $\frac{-9}{0}$

______ 4. $\frac{-4 + 10}{-3}$

______ 5. $\frac{3(-6) + 8}{5}$

______ 6. $\frac{-3[2(-4) + (-2)]}{-11 + 2}$

Answers are on page 109.

We mention here that

There is *no* commutative property for division.
There is *no* associative property for division.

The following quotients should convince you.

$$\frac{6}{3} \neq \frac{3}{6} \qquad -16 \div (4 \div 2) \neq (-16 \div 4) \div 2$$

$$2 \neq \frac{1}{2} \qquad -16 \div 2 \neq -4 \div 2$$

$$-8 \neq -2$$

▶ Examples You Completed

Example 2. Find $-4 \cdot 11$.

Solution

$$-4 \cdot 11$$
$$= -44$$

Example 4. Find $(-8)(-6)$.

Solution

$$(-8)(-6)$$
$$= 48$$

Example 6. Find $-5(3) + (-2)(-1)$.

Solution

$$-5(3) + (-2)(-1)$$
$$= -15 + 2$$
$$= -13$$

Example 8. Use the commutative property to rewrite the product $-2(-11)$.

Solution

$$-2(-11)$$
$$= -11(-2)$$

Example 9. Use the associative property to rewrite the product $5[-1(-8)]$.

Solution

$$5[-1(-8)]$$
$$= [5(-1)] \cdot (-8)$$

Example 12. Use the distributive property to find the product $-10(-17 + 19)$.

Solution

$$-10(-17 + 19)$$
$$= -10(-17) + (-10)(19)$$
$$= 170 + (-190)$$
$$= -20$$

Example 14. Find $(-3)^4$.

Solution

$$(-3)^4$$
$$= (-3)(-3)(-3)(-3)$$
$$= 9(9)$$
$$= 81$$

Example 16. Find $(-2)^3(-5)^2$.

Solution

$$(-2)^3(-5)^2$$
$$= (-2)(-2)(-2)(-5)(-5)$$
$$= (-8)(25)$$
$$= -200$$

Example 18. Find $\frac{99}{9}$ and write the corresponding multiplication statement.

Solution

$$\frac{99}{9} = 11$$

because $9 \cdot 11 = 99$

Example 20. Simplify $\frac{-6}{-6}$.

Solution

$$\frac{-6}{-6} = 1$$

because $-6(1) = -6$

Example 22. Simplify $\frac{17}{-17}$.

Solution

$$\frac{17}{-17}$$
$$= -1$$

Example 26. Simplify $\frac{5[18 + 2(-3)]}{-6}$.

Solution

$$\frac{5[18 + 2(-3)]}{-6}$$
$$= \frac{5[18 + (-6)]}{-6}$$
$$= \frac{5(12)}{-6}$$
$$= \frac{60}{-6}$$
$$= -10$$

Answers to Trial Runs

page 100 **1.** 56 **2.** -72 **3.** -7 **4.** 45 **5.** 8 **6.** -42

page 103 **1.** $-7(10)$ **2.** $6[-5(9)]$ **3.** $-3(2) + (-3)(8)$ **4.** -210 **5.** 30

page 105 **1.** 49 **2.** -125 **3.** 192 **4.** 144

page 107 **1.** -8 **2.** 3 **3.** Undefined **4.** -2 **5.** -2 **6.** -2

Name ______________________ Date ____________

EXERCISE SET 2.3

1. Use the commutative property for multiplication to rewrite the products.

____ (a) $-9(-8)$ ____ (b) $5(-12)$

____ (c) $0 \cdot 6$ ____ (d) $-3(7)$

2. Use the commutative property for multiplication to rewrite the products.

____ (a) $-5(0)$ ____ (b) $-7(10)$

____ (c) $12(-8)$ ____ (d) $-3(-4)$

3. Use the associative property for multiplication to rewrite the products.

____ (a) $(3 \cdot 4)2$ ____ (b) $6[5(-7)]$

____ (c) $-7(-3 \cdot 10)$ ____ (d) $[9(-2)](-8)$

4. Use the associative property for multiplication to rewrite the products.

____ (a) $(5 \cdot 7)9$ ____ (b) $12[7(-9)]$

____ (c) $-8(-2 \cdot 5)$ ____ (d) $[3(-11)](-5)$

5. Use the distributive property for multiplication over addition and subtraction to rewrite the products.

____ (a) $-3(7 + 2)$ ____ (b) $5(-8 - 3)$

____ (c) $4[7 - (-2)]$ ____ (d) $-10(-9 + 8)$

6. Use the distributive property for multiplication over addition and subtraction to rewrite the products.

____ (a) $-9(8 + 10)$ ____ (b) $2(-11 - 4)$

____ (c) $5[6 - (-3)]$ ____ (d) $-4(-1 + 7)$

Simplify.

____ 7. $5(-6)$ ____ 8. $3(-4)$

____ 9. $-6(-7)$ ____ 10. $-4(-8)$

____ 11. $-3(8)$ ____ 12. $-7(9)$

____ 13. $-10(3)(-5)$ ____ 14. $-5(7)(-3)$

____ 15. $4(-3)(0)(9)$ ____ 16. $6(-5)(0)(11)$

____ 17. $-9(3 - 8)$ ____ 18. $-10(5 - 12)$

______ 19. $6(-10) - 20$

______ 20. $8(-9) - 15$

______ 21. $23 + 7(-2)$

______ 22. $29 + 3(-5)$

______ 23. $(-8)^2$

______ 24. $(-2)^4$

______ 25. $(-3)^3$

______ 26. $(-4)^3$

______ 27. $(-2)^3(-6)^2$

______ 28. $(-5)^3(-1)^2$

______ 29. $\frac{24}{-6}$

______ 30. $\frac{20}{-4}$

______ 31. $\frac{-16}{8}$

______ 32. $\frac{-15}{5}$

______ 33. $\frac{-35}{-7}$

______ 34. $\frac{-48}{-6}$

______ 35. $\frac{11(-4)}{22}$

______ 36. $\frac{12(-3)}{4}$

______ 37. $\frac{15}{-3} + \frac{-12}{6}$

______ 38. $\frac{21}{-7} + \frac{-28}{4}$

______ 39. $\frac{-20}{5} - \frac{-30}{-6}$

______ 40. $\frac{-45}{9} + \frac{-55}{-11}$

______ 41. $2\left(\frac{-18}{3}\right)$

______ 42. $5\left(\frac{-54}{9}\right)$

______ 43. $\frac{-20 + 15}{7 - 7}$

______ 44. $\frac{-35 + 20}{5 - 5}$

______ 45. $\frac{-40 + 15}{16 + (+9)}$

______ 46. $\frac{-17 + 10}{4 + (+3)}$

______ 47. $\frac{3(-8) + (-2)(-4)}{4(-4)}$

______ 48. $\frac{5(-9) + (-3)(-2)}{3(-13)}$

______ 49. $\frac{9(-1) + \frac{14}{-2}}{4}$

______ 50. $\frac{3(-9) + \frac{5}{-1}}{8}$

______ 51. $\frac{7[18 + 5(-2)]}{-8}$

______ 52. $\frac{9[21 + 7(-2)]}{-21}$

______ 53. $\frac{-4[3(-2) + (+7)]}{13 + 11}$

______ 54. $\frac{-5[6(-3) + (+20)]}{17 + 12}$

______ 55. $\frac{20(-4)}{10} + \frac{(21 - 6)}{-3}$

______ 56. $\frac{36(-2)}{12} + \frac{27 - 9}{-6}$

______ **57.** $\dfrac{-9(7-1)+8(-3)}{-5(-8+6)}$

______ **58.** $\dfrac{-8(10-3)+9(-4)}{2(-9+4)}$

______ **59.** $-3\left[\dfrac{-14}{7}+3(7-12)\right]$

______ **60.** $-5\left[\dfrac{-15}{3}+2(9-12)\right]$

______ **61.** $10\left[3(6)+\dfrac{13+(-5)}{4}\right]$

______ **62.** $5\left[-5(7)+\dfrac{14+(-5)}{3}\right]$

______ **63.** $\dfrac{\dfrac{-15}{5}+\dfrac{0}{7}}{\dfrac{25}{-5}+2}$

______ **64.** $\dfrac{\dfrac{-18}{6}+\dfrac{0}{9}}{\dfrac{-30}{5}+3}$

______ **65.** $4\left[-9+\dfrac{7+(-3)}{6-8}\right]-3^3$

______ **66.** $5\left[-10+\dfrac{8+(-3)}{15-20}\right]-6^2$

☆ Stretching the Topics

Simplify.

______ **1.** $-7\left[-2(5)+\dfrac{3^2-(-2)^2}{6-11}\right]$

______ **2.** $(-2)^2\{4(-1)^3+3[8+(+12)]\}$

______ **3.** $3^3[7+(-2)^3]-[(-1)^5(-3)-12]-(8-12)^2$

Check your answers in the back of your book.

If you can simplify the expressions in **Checkup 2.3**, you are ready to go on to Section 2.4.

Name ______________________________ Date ______________

CHECKUP 2.3

______ 1. Use the commutative property for multiplication to rewrite the product $(-3)(5)$.

______ 2. Use the associative property for multiplication to rewrite the product $-3[2(-7)]$.

______ 3. Use the distributive property for multiplication over addition to rewrite the product $-5(-2+8)$.

Simplify.

______ 4. $6(-5)$

______ 5. $\frac{-35}{7}$

______ 6. $\frac{-50}{(-5)^2}$

______ 7. $[-4(-3)](-5)$

______ 8. $-2[-3+(7-12)]$

______ 9. $\frac{5[13+3(-1)]}{-25}$

______ 10. $\frac{-3\left[2(-1)+\frac{24}{-6}\right]}{-10+19}$

Check your answers in the back of your book.

If You Missed Problems:	You Should Review Examples:
1	8
2	9
3	11, 12
4	1–4
5, 6	13, 19–22
7, 8	5–7
9, 10	25, 26

2.4 Understanding Real Numbers

So far we have discussed several important sets of numbers.

Natural numbers: $\{1, 2, 3, \ldots\}$

Whole numbers: $\{0, 1, 2, 3, \ldots\}$

Integers: $\{\ldots, -3, -2, -1, 0, 1, 2, 3, \ldots\}$

We have learned how to add, subtract, multiply, and divide such numbers. But in Chapter 1 we also discussed operations with fractional and decimal numbers that do *not* belong to these sets. In fact, we must define two more sets of numbers in order to include fractions and decimal numbers in our study of algebra.

Rational Numbers. A rational number is any number that can be written as a *fraction* with an *integer* as the numerator and a *nonzero integer* as the denominator.

Rational Number	Fraction Form
$\frac{17}{37}$	$\frac{17}{37}$
$\frac{-5}{3}$	$\frac{-5}{3}$
$3\frac{1}{2}$	$\frac{7}{2}$
0.9	$\frac{9}{10}$
$\sqrt{25}$	$\frac{5}{1}$
-0.73	$\frac{-73}{100}$
4.32	$\frac{432}{100}$
2	$\frac{2}{1}$
-3	$\frac{-3}{1}$
0	$\frac{0}{1}$

You should agree that every rational number corresponds to a point on the number line.

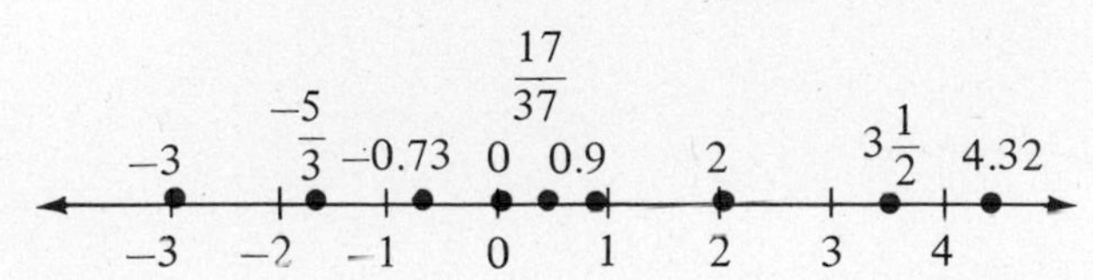

Notice also that *every integer is a rational number*.

It may seem to you that all numbers must be rational. Indeed, most numbers that you work with every day *are* rational. However, the numbers π, $\sqrt{2}$, and $\sqrt{3}$, which you may have seen before, are examples of numbers that are *not* rational because they cannot be written as fractions of integers. Such numbers are called **irrational numbers**; they will be discussed more completely in Chapter 12.

Irrational Numbers. An irrational number is a number that corresponds to a point on the number line but is *not* a rational number.

The set that includes *all* the numbers in the sets we have discussed is the set of **real numbers**.

Real Numbers. A real number is any number that corresponds to a point on the number line.

We can illustrate the relationships among our sets of numbers in the following chart.

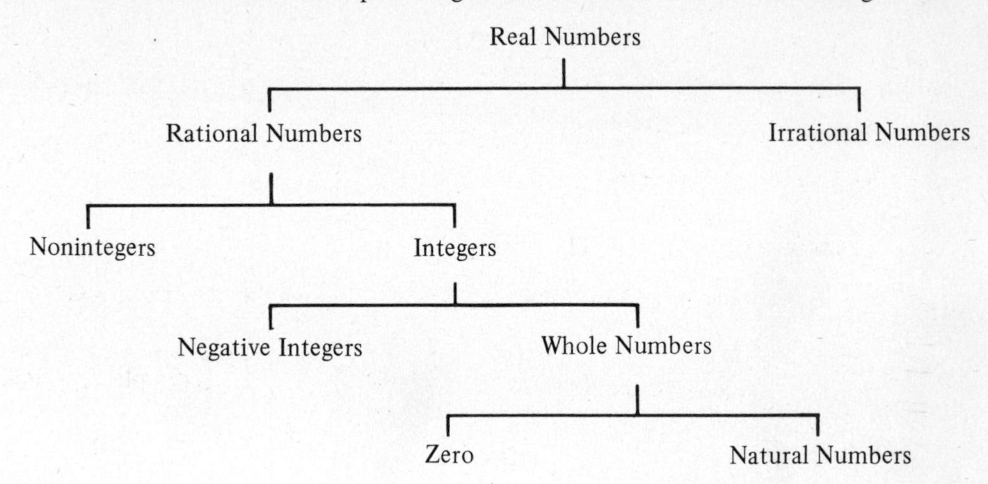

You will be relieved to know that *all the properties and definitions that you learned for operating with integers will continue to work for the real numbers*.

Example 1. What is the opposite of $\frac{5}{6}$?

Solution: $\frac{-5}{6}$ is the opposite of $\frac{5}{6}$.

You try Example 2.

Example 2. What is the opposite of -2.3?

Solution: 2.3 is the opposite of -2.3.

Check your answer on page 123. ▶

Example 3. Find $|-7.35|$.

Solution

$$|-7.35| = 7.35$$

Now you try Example 4.

Example 4. Find $-\left|\frac{-7}{3}\right|$.

Solution

$$-\left|\frac{-7}{3}\right|$$

$$= -\left(\frac{7}{3}\right)$$

$$= \frac{-7}{3}$$

Check your work on page 123. ▶

Example 5. Simplify $\frac{-6}{11} + \frac{2}{11} + \frac{-5}{11}$.

Solution

$$\frac{-6}{11} + \frac{2}{11} + \frac{-5}{11}$$

$$= \frac{-6}{11} + \frac{-5}{11} + \frac{2}{11}$$

$$= \frac{-6 + (-5) + 2}{11}$$

$$= \frac{-11 + 2}{11}$$

$$= \frac{-9}{11}$$

Now you try Example 6.

Example 6. Simplify $-2.3 + (-5.1) + 12.8$.

Solution

$$-2.3 + (-5.1) + 12.8$$

$$= -7.4 + 12.8$$

$$= 5.4$$

Check your work on page 124. ▶

Example 7. Simplify $\frac{5}{3} - \frac{11}{2}$.

Solution

$$\frac{5}{3} - \frac{11}{2}$$

$$= \frac{5}{3} + \frac{-11}{2} \qquad \text{LCD: } 3 \cdot 2.$$

$$= \frac{5 \cdot 2}{3 \cdot 2} + \frac{-11 \cdot 3}{3 \cdot 2}$$

$$= \frac{10}{6} + \frac{-33}{6}$$

$$= \frac{10 + (-33)}{6}$$

$$= \frac{-23}{6}$$

You complete Example 8.

Example 8. Simplify $\frac{-1}{2} - \frac{1}{5}$.

Solution

$$\frac{-1}{2} - \frac{1}{5}$$

$$= \frac{-1}{2} + \frac{-1}{5} \qquad \text{LCD: } 2 \cdot 5.$$

$$= \frac{-1 \cdot 5}{2 \cdot 5} + \frac{-1 \cdot 2}{2 \cdot 5}$$

$$\frac{-5}{10} + \frac{-2}{10}$$

$$\frac{-5 + -2}{10}$$

$$\frac{-7}{10}$$

Check your work on page 124. ▶

Example 9. Simplify $2.3 - (3.51 - 7.6)$.

Solution

$$\begin{aligned} &2.3 - (3.51 - 7.6) \\ &= 2.3 - [3.51 + (-7.6)] \\ &= 2.3 - (-4.09) \\ &= 2.3 + 4.09 \\ &= 6.39 \end{aligned}$$

You complete Example 10.

Example 10. Simplify $(0.37 - 3.2) - (6.1 - 0.09)$.

Solution

$$\begin{aligned} &(0.37 - 3.2) - (6.1 - 0.09) \\ &= [0.37 + (-3.2)] - [(6.1 + (-0.09)] \end{aligned}$$

Check your work on page 124. ▶

Example 11. Find $\frac{-2}{3} \cdot \frac{5}{6}$.

Solution

$$\begin{aligned} &\frac{-2}{3} \cdot \frac{5}{6} \\ &= \frac{-1 \cdot \overset{1}{\cancel{2}}}{3} \cdot \frac{5}{\underset{1}{\cancel{2}} \cdot 3} \\ &= \frac{-1 \cdot 5}{3 \cdot 3} \\ &= \frac{-5}{9} \end{aligned}$$

Now you try Example 12.

Example 12. Find $\frac{-4}{9} \cdot \frac{3}{7} \cdot \frac{-21}{2}$.

Solution

Check your work on page 124. ▶

Example 13. Find $-3(2.7 - 8.2)$.

Solution

$$\begin{aligned} &-3(2.7 - 8.2) \\ &= -3[2.7 + (-8.2)] \\ &= -3(-5.5) \\ &= 16.5 \end{aligned}$$

You complete Example 14.

Example 14. Find $0.5[-3.9 - (6 + 9.1)]$.

Solution

$$\begin{aligned} &0.5[-3.9 - (6 + 9.1)] \\ &= 0.5[-3.9 - (15.1)] \\ &= 0.5[-3.9 + (-15.1)] \\ &= 0.5[\_\_\_\_] \\ &= \_\_\_\_ \end{aligned}$$

Check your work on page 124. ▶

Example 15. Find $\left(\frac{-2}{3}\right)^3$.

Solution

$$\left(\frac{-2}{3}\right)^3$$
$$= \left(\frac{-2}{3}\right)\left(\frac{-2}{3}\right)\left(\frac{-2}{3}\right)$$
$$= \frac{(-2)(-2)(-2)}{3 \cdot 3 \cdot 3}$$
$$= \frac{-8}{27}$$

You try Example 16.

Example 16. Find $(-1.1)^2$.

Solution

$$(-1.1)^2$$
$$= (-1.1)(-1.1)$$
$$= \underline{\qquad}$$

Check your work on page 124. ▶

Example 17. Find $-4.44 \div 3.7$.

Solution: The answer will be *negative*.

$$\begin{array}{r} 1.2 \\ 3.7_{\wedge}\overline{)4.4_{\wedge}4} \\ \underline{3\,7} \\ 7\,4 \\ \underline{7\,4} \\ 0 \end{array}$$

$$-4.44 \div 3.7 = -1.2$$

Example 18. Find $\frac{-5}{7} \cdot \frac{3}{10} \div \frac{-9}{20}$.

Solution

$$\frac{-5}{7} \cdot \frac{3}{10} \div \frac{-9}{20}$$
$$= \frac{-5}{7} \cdot \frac{3}{10} \cdot \frac{20}{-9}$$
$$= \frac{-\cancel{1} \cdot 5}{7} \cdot \frac{\cancel{3}}{\cancel{2} \cdot \cancel{5}} \cdot \frac{2 \cdot \cancel{2} \cdot \cancel{5}}{-\cancel{1} \cdot \cancel{3} \cdot 3}$$
$$= \frac{2 \cdot 5}{7 \cdot 3}$$
$$= \frac{10}{21}$$

Trial Run

Simplify.

_______ 1. $\frac{5}{8} - \frac{1}{2} - \frac{3}{4}$

_______ 2. $8.26 - (-3.5)$

_______ 3. $\frac{6}{7}\left(\frac{1}{7} + \frac{-3}{7}\right)$

_______ 4. $-1.7(-0.2)$

_____ **5.** $-0.1[6.8 - (2 + 1.9)]$

_____ **6.** $\frac{-8}{9} \cdot \frac{3}{4}$

_____ **7.** $\frac{-2.5}{-0.5}$

_____ **8.** $\frac{3}{4} \cdot \frac{-5}{3} \div 2$

Answers are on page 124.

In working many examples throughout this chapter in which you added and multiplied real numbers, you have relied on two properties for real numbers of which you were probably unaware.

When two real numbers are added, their sum is a unique real number.

When two real numbers are multiplied, their product is a unique real number.

The fact that sums and products of real numbers are *unique* real numbers means that there is *one and only one* answer when two certain numbers are added or multiplied. Think how confusing mathematics would be if this were not so!

These laws are called the **closure properties** for real numbers.

Closure Properties. For real numbers A and B

$A + B$ is real and unique

$A \cdot B$ is real and unique

We noted earlier that every real number has an *opposite* and that the sum of every number and its opposite is zero, the *additive identity*. A similar property exists for multiplication. Consider these products.

$$2 \cdot \frac{1}{2} = \frac{\overset{1}{\cancel{2}}}{1} \cdot \frac{1}{\underset{1}{\cancel{2}}} = 1$$

$$\frac{5}{3} \cdot \frac{3}{5} = \frac{\overset{1}{\cancel{5}}}{\underset{1}{\cancel{3}}} \cdot \frac{\overset{1}{\cancel{3}}}{\underset{1}{\cancel{5}}} = 1$$

$$\frac{-1}{4} \cdot \frac{-4}{1} = \frac{-1}{\underset{1}{\cancel{4}}} \cdot \frac{\overset{-1}{-\cancel{4}}}{1} = 1$$

In each product the second number is called the **reciprocal** of the first number, and their product is 1, the *identity for multiplication*. Every real number (except zero) has its own reciprocal (also called its **multiplicative inverse**).

Property of Reciprocals. For any real number A $(A \neq 0)$

$$A \cdot \frac{1}{A} = 1$$

Example 19. Find the reciprocal of $\frac{-8}{5}$.

Solution: The reciprocal of $\frac{-8}{5}$ is $\frac{5}{-8}$ because

$$\frac{-8}{5} \cdot \frac{5}{-8} = 1$$

You complete Example 20.

Example 20. Find the reciprocal of $\frac{1}{\pi}$.

Solution: The reciprocal of $\frac{1}{\pi}$ is ____ because

$$\frac{1}{\pi} \cdot \_\_\_\_ = 1$$

Check your work on page 124. ▶

The fact that every real number (except 0) has a reciprocal allows us to define the operation of division in terms of the operation of multiplication.

Definition of Division. For real numbers A and B $(B \neq 0)$

$$\frac{A}{B} = A \cdot \frac{1}{B}$$

Example 21. Rewrite the quotient $\frac{-63}{7}$ as a product.

Solution

$$\frac{-63}{7} = -63\left(\frac{1}{7}\right)$$

Example 22. Rewrite the product $\frac{1}{4}\pi$ as a quotient.

Solution

$$\frac{1}{4}\pi = \frac{\pi}{4}$$

▶ Examples You Completed

Example 2. What is the opposite of -2.3?

Solution: 2.3 is the opposite of -2.3.

Example 4. Find $-\left|\frac{-7}{3}\right|$.

Solution

$$-\left|\frac{-7}{3}\right|$$

$$= -\left(\frac{7}{3}\right)$$

$$= \frac{-7}{3}$$

Example 6. Simplify $-2.3 + (-5.1) + 12.8$.

Solution

$$
\begin{aligned}
&-2.3 + (-5.1) + 12.8 \\
&= -7.4 + 12.8 \\
&= 5.4
\end{aligned}
$$

Example 10. Simplify $(0.37 - 3.2) - (6.1 - 0.09)$.

Solution

$$
\begin{aligned}
&(0.37 - 3.2) - (6.1 - 0.09) \\
&= [0.37 + (-3.2)] - [6.1 + (-0.09)] \\
&= -2.83 - 6.01 \\
&= -2.83 + (-6.01) \\
&= -8.84
\end{aligned}
$$

Example 8. Simplify $\frac{-1}{2} - \frac{1}{5}$.

Solution

$$
\begin{aligned}
&\frac{-1}{2} - \frac{1}{5} \\
&= \frac{-1}{2} + \frac{-1}{5} \qquad \text{LCD: } 2 \cdot 5. \\
&= \frac{-1 \cdot 5}{2 \cdot 5} + \frac{1 \cdot 2}{2 \cdot 5} \\
&= \frac{-5}{10} + \frac{-2}{10} \\
&= \frac{-5 + (-2)}{10} \\
&= \frac{-7}{10}
\end{aligned}
$$

Example 12. Find $\frac{-4}{9} \cdot \frac{3}{7} \cdot \frac{-21}{2}$.

Solution

$$
\begin{aligned}
&\frac{-4}{9} \cdot \frac{3}{7} \cdot \frac{-21}{2} \\
&= \frac{-1 \cdot \overset{1}{\cancel{2}} \cdot 2}{\underset{1}{\cancel{3}} \cdot \underset{1}{\cancel{3}}} \cdot \frac{\overset{1}{\cancel{3}}}{\underset{1}{\cancel{7}}} \cdot \frac{-1 \cdot \overset{1}{\cancel{3}} \cdot \overset{1}{\cancel{7}}}{\underset{1}{\cancel{2}}} \\
&= 2
\end{aligned}
$$

Example 14. Find $0.5[-3.9 - (6 + 9.1)]$.

Solution

$$
\begin{aligned}
&0.5[-3.9 - (6 + 9.1)] \\
&= 0.5[-3.9 - (15.1)] \\
&= 0.5[-3.9 + (-15.1)] \\
&= 0.5[-19] \\
&= -9.5
\end{aligned}
$$

Example 16. Find $(-1.1)^2$.

Solution

$$
\begin{aligned}
&(-1.1)^2 \\
&= (-1.1)(-1.1) \\
&= 1.21
\end{aligned}
$$

Example 20. Find the reciprocal of $\frac{1}{\pi}$.

Solution: The reciprocal of $\frac{1}{\pi}$ is π because

$$\frac{1}{\pi} \cdot \pi = 1$$

Answers to Trial Run

page 121 **1.** $\frac{-5}{8}$ **2.** 11.76 **3.** $\frac{8}{7}$ **4.** 0.34 **5.** -0.29 **6.** $\frac{-2}{3}$ **7.** 5 **8.** $\frac{-5}{8}$

Name ____________________ Date __________

EXERCISE SET 2.4

Simplify.

______ 1. $|-8.3|$

______ 2. $|-12.7|$

______ 3. $|3.2 - 8.9|$

______ 4. $|4.5 - 7.3|$

______ 5. $-\left|\frac{-2}{3}\right|$

______ 6. $-\left|\frac{-7}{8}\right|$

______ 7. $\left|\frac{1}{8}\right| - \left|\frac{-7}{8}\right|$

______ 8. $\left|\frac{3}{11}\right| - \left|\frac{-7}{11}\right|$

______ 9. $\frac{19}{21} - \left(\frac{18}{21} - \frac{7}{21}\right)$

______ 10. $\frac{16}{27} + \left(\frac{13}{27} - \frac{17}{27}\right)$

______ 11. $\frac{-5}{6} + \frac{2}{5} - \frac{1}{2}$

______ 12. $\frac{-10}{21} + \frac{11}{18} - \frac{9}{14}$

______ 13. $-13.5 + 7.5$

______ 14. $-18.6 + 21.6$

______ 15. $3.72 - (0.49 - 1.08)$

______ 16. $12.35 - (4.06 - 0.04)$

______ 17. $\frac{-4}{5} \cdot \frac{15}{28}$

______ 18. $\frac{-7}{8} \cdot \frac{12}{35}$

______ 19. $(-8.62)(1.03)$

______ 20. $(-2.74)(2.03)$

______ 21. $(-4.3)(-7.001)(1000)$

______ 22. $(-3.2)(-4.005)(1000)$

______ 23. $\left(\frac{-3}{4}\right)^3$

______ 24. $\left(\frac{-2}{5}\right)^3$

______ 25. $(-1.2)^2$

______ 26. $(-2.3)^2$

______ 27. $\left(\frac{1}{5}\right)^4$

______ 28. $\left(\frac{1}{3}\right)^5$

______ 29. $-0.035 \div 0.05$

______ 30. $-0.72 \div 0.06$

______ 31. $-1.606 \div (-7.3)$

______ 32. $-2.697 \div (-4.65)$

______ 33. $\frac{-12}{75} \cdot \frac{-75}{84}$

______ 34. $\frac{-27}{50} \cdot \frac{-70}{81}$

______ 35. $\frac{3}{5} \div \frac{6}{25}$

______ 36. $\frac{5}{7} \div \frac{20}{63}$

______ 37. $-0.3[-6.2 - (8 - 9.3)]$

______ 38. $-0.4[-5.3 - (7 - 8.6)]$

______ 39. $\frac{-3}{8} \cdot \frac{4}{9} \div \frac{-5}{12}$

______ 40. $\frac{-3}{7} \cdot \frac{14}{15} \div \frac{-4}{25}$

☆ Stretching the Topics

Simplify.

_______ **1.** $\left[\left(\frac{1}{2}-\frac{2}{3}\right)\div\left(\frac{5}{6}-\frac{5}{8}\right)\right]^2$

_______ **2.** $(-0.1)^3[3.76(0.2)^2+(-0.3)^3(-5)]$

_______ **3.** $\left(\frac{3}{4}\right)^2-\left(\frac{7}{12}\div\frac{2}{3}\right)+\left(\frac{2}{9}\cdot\frac{3}{4}\right)$

Check your answers in the back of your book.

If you can do the problems in **Checkup 2.4**, you are ready to go to Section 2.5.

Name ______________________ Date ____________

CHECKUP 2.4

Simplify.

______ 1. $\left|\frac{-4}{5}\right|$

______ 2. $\frac{-2}{3} + \frac{1}{2} + \frac{-1}{6}$

______ 3. $-0.3 - 0.25 + 2.26$

______ 4. $\frac{3}{5} - \frac{7}{6}$

______ 5. $\frac{-3}{5} \cdot \frac{-7}{36} \div \frac{14}{3}$

______ 6. $-5(0.23 - 1.09)$

______ 7. $\left(\frac{-3}{5}\right)^3$

______ 8. $2.5(-0.2)^2$

______ 9. $-0.1764 \div 0.063$

______ 10. $0.2[-1.6 - (8.2 - 1.9)]$

Check your answers in the back of your book.

If You Missed Problems:	You Should Review Examples:
1	3, 4
2, 3	5, 6
4	7, 8
5	11, 12
6	13
7, 8	15, 16
9	17
10	14

2.5 Switching from Word Expressions to Real Number Expressions

Now that we understand how to operate with all the real numbers, we must practice switching from words to expressions involving positive and negative numbers.

Example 1. Yesterday, Amy's checking account showed a balance of \$67.39. Today she wrote checks for \$31.93 and \$17.29, deposited \$100, and wrote another check for \$125. What is the condition of her account now?

Solution: Remember that deposits will be positive numbers, but checks will be negative numbers.

67.39	Amy's balance
$-31.93 + (-17.29) + (-125)$	Amy's checks
100	Amy's deposit

We must find the *sum* of these quantities.

$$67.39 + (-31.93) + (-17.29) + (-125) + 100$$

$$= 67.39 + 100 + (-31.93) + (-17.29) + (-125)$$ Rearrange terms.

$$= 167.39 + (-174.22)$$ Add positive numbers; add negative numbers.

$$= -6.83$$ Find the sum.

Amy has *overdrawn* her account by \$6.83.

Example 2. In one series of plays, the State University football team completed two 7-yard passes and ran for 5 yards. Then the quarterback was sacked twice, for a 3-yard loss each time, and the team received a 15-yard penalty. What was the team's total gain or loss?

Solution: Gains should be represented by ________ numbers, but losses should be represented by ________ numbers.

2(____)	gains from passes
____	gain from run
2(____)	losses from sacks
____	loss from penalty

We can find the team's total gain or loss by ________ these numbers.

2(7) + ____ + 2(−3) + (____)

The team had a ____-yard ________ .

Check your work on page 131. ▶

Example 3. On Monday morning, a share of Intral stock was selling for \$27. The following changes occurred in the price of the stock during the week. On Monday the price rose by $\frac{1}{8}$;

on Tuesday it rose by $\frac{1}{4}$; on Wednesday it fell by $\frac{5}{8}$; on Thursday it rose by $\frac{1}{8}$; and on Friday it fell by $\frac{1}{8}$. What was the price of a share of Intral stock on Friday afternoon?

Solution: A rise in price should be represented by a *positive* number, but a drop in price should be represented by a *negative* number.

$$27 + \frac{1}{8} + \frac{1}{4} + \frac{-5}{8} + \frac{1}{8} + \frac{-1}{8}$$

$$= 27 + \frac{1}{8} + \frac{1}{4} + \frac{1}{8} + \frac{-5}{8} + \frac{-1}{8} \qquad \text{LCD: 8.}$$

$$= \frac{216}{8} + \frac{1}{8} + \frac{2}{8} + \frac{1}{8} + \frac{-5}{8} + \frac{-1}{8}$$

$$= \frac{216 + 1 + 2 + 1 + (-5) + (-1)}{8}$$

$$= \frac{220 + (-6)}{8}$$

$$= \frac{214}{8}$$

$$= 26\frac{6}{8}$$

$$= 26\frac{3}{4}$$

The price of a share of Intral stock was \$26.75 on Friday.

Now you complete Example 4.

Example 4. Last week the daily temperatures at noon in Anytown were $-4°$, $-3°$, $0°$, $8°$, $11°$, $-2°$, and $-3°$. What was the *average* noon temperature last week?

Solution: The *total* of last week's noon temperatures is found by adding all the temperatures.

$-4° + (\_\_\_\_) + \_\_\_\_ + \_\_\_\_ + \_\_\_\_ + (\_\_\_\_) + (-3°)$

We must break this total into ____ parts, so we must *divide* the total by ____ .

$$\frac{-4° + (-3°) + 0° + 8° + 11° + (-2°) + (-3°)}{7}$$

The average noon temperature was ____ .

Check your work on page 131. ▶

▶ Examples You Completed

Example 2. In one series of plays, the State University football team completed two 7-yard passes and ran for 5 yards. Then the quarterback was sacked twice, for a 3 yard loss each time, and the team received a 15-yard penalty. What was the team's gain or loss?

Solution: Gains should be represented by positive numbers, but losses should be represented by negative numbers.

$2(7)$	gains from passes
5	gain from run
$2(-3)$	losses from sacks
-15	loss from penalty

We can find the team's total gain or loss by adding these numbers.

$$\begin{aligned} & 2(7) + 5 + 2(-3) + (-15) \\ &= 14 + 5 + (-6) + (-15) \\ &= 19 + (-21) \\ &= -2 \end{aligned}$$

The team had a 2-yard loss.

Example 4. Last week the daily temperatures at noon in Anytown were $-4°$, $-3°$, $0°$, $8°$, $11°$, $-2°$, and $-3°$. What was the *average* noon temperature last week?

Solution: The *total* of last week's noon temperatures is found by adding all the temperatures.

$$-4° + (-3°) + 0° + 8° + 11° + (-2°) + (-3°)$$

We must break this total into seven parts, so we must divide the total by 7.

$$\begin{aligned} & \frac{-4° + (-3°) + 0° + 8° + 11° + (-2°) + (-3°)}{7} \\ &= \frac{-4° + (-3°) + (-2°) + (-3°) + 0° + 8° + 11°}{7} \\ &= \frac{-12° + 19°}{7} \\ &= \frac{7°}{7} \\ &= 1° \end{aligned}$$

The average noon temperature was 1°.

Name ______________________________ Date ____________

EXERCISE SET 2.5

Change each word expression to a number expression and simplify.

_______ **1.** Darlene scored 87 on her first math test. If she scored 14 points lower on her next test, what score did she receive on her second test?

_______ **2.** At a summer sale, Bob bought a \$39 stepladder, which had been reduced by \$14. What was the sale price of the stepladder?

_______ **3.** The Momans used 312 fewer kilowatt-hours in March then they did in February. If they used 1605 kilowatt-hours in February, how many kilowatt-hours did they use in March?

_______ **4.** At the Key Market, Royce's total grocery bill was \$63.78, but after the total value of his coupons was deducted, he paid only \$55.28. Find how much he saved by using the coupons.

_______ **5.** A dinner for two at the Chicken King costs \$5.86. If Jim and Maria split the expenses evenly, how much will each pay?

_______ **6.** If Carol's car averages 19.75 miles per gallon, how many gallons will she need to drive 316 miles?

_______ **7.** If Sarah drives an average speed of 53 miles per hour, how far can she drive in 4.5 hours?

_______ **8.** Carol bought 15.6 gallons of gasoline at \$1.69 per gallon. Find the cost of the gasoline.

_______ **9.** Ms. Fletcher uses $\frac{2}{3}$ cup of fertilizer for each shrub in her yard. If she has 25 shrubs, how many cups of fertilizer does she need?

_______ **10.** Pansy uses $2\frac{1}{2}$ cups of flour for each jam cake she bakes. If she is baking seven jam cakes for the Junior League Bake Sale, how many cups of flour does she need?

_______ **11.** Lester played the 25-cent slot machine seven times, winning the 75-cent jackpot three times. How much did Lester win or lose?

_______ **12.** If Harvey works 8 hours at \$9.85 an hour and has \$12.73 deducted for taxes, what will be his take-home pay?

_______ **13.** If taxi fare is \$1.75 for the first mile and \$1.10 for each additional mile, what will be the fare for an 8-mile trip?

_______ **14.** A club sold 150 dance tickets for \$4.25 each and had expenses of \$275.85. What was the club's profit?

_______ **15.** Miranda borrowed \$5 from her brother twice last month. This month she paid him \$7 and borrowed \$3. Where does Miranda's account with her brother stand now?

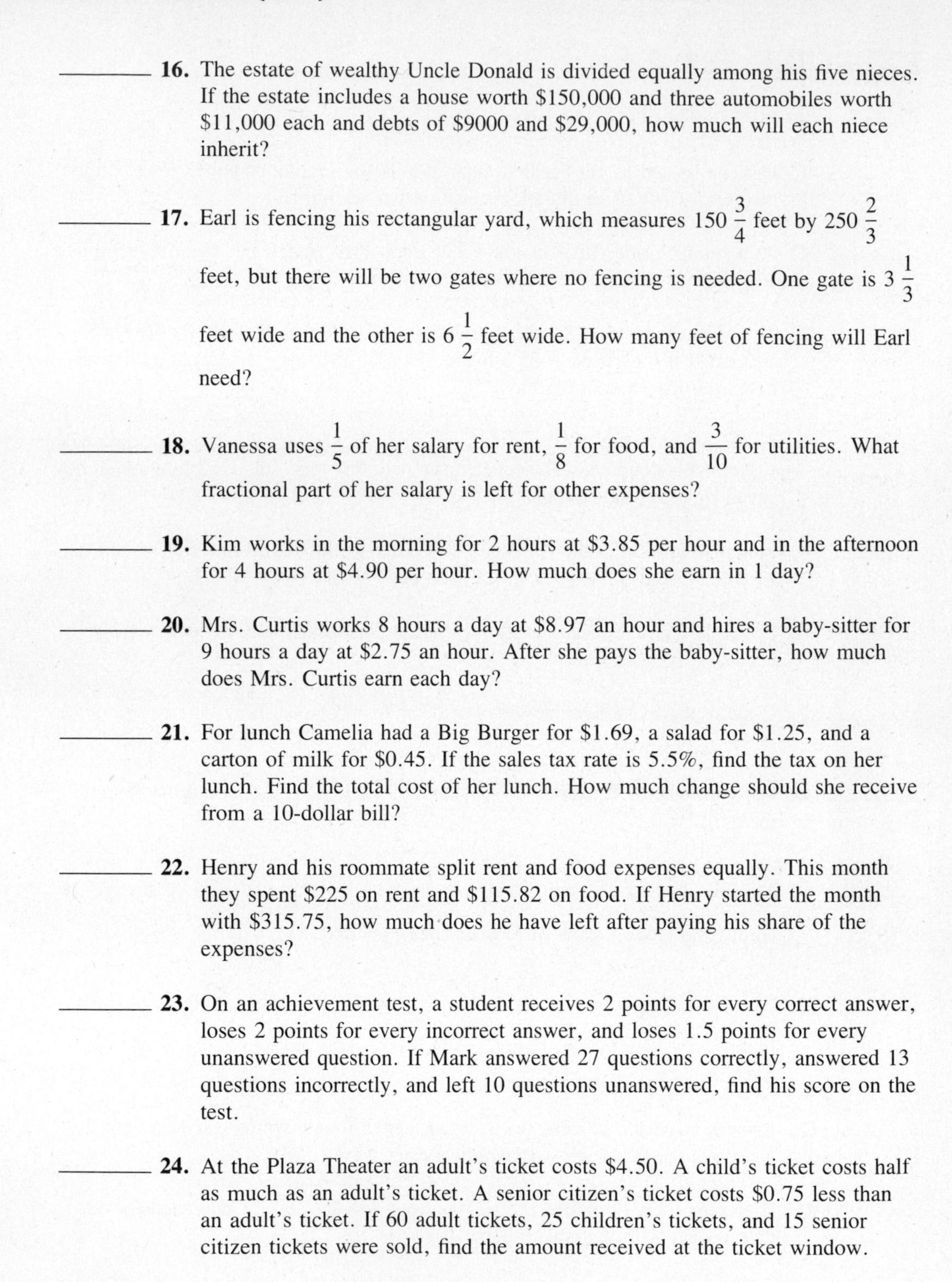

_______ **16.** The estate of wealthy Uncle Donald is divided equally among his five nieces. If the estate includes a house worth \$150,000 and three automobiles worth \$11,000 each and debts of \$9000 and \$29,000, how much will each niece inherit?

_______ **17.** Earl is fencing his rectangular yard, which measures $150\frac{3}{4}$ feet by $250\frac{2}{3}$ feet, but there will be two gates where no fencing is needed. One gate is $3\frac{1}{3}$ feet wide and the other is $6\frac{1}{2}$ feet wide. How many feet of fencing will Earl need?

_______ **18.** Vanessa uses $\frac{1}{5}$ of her salary for rent, $\frac{1}{8}$ for food, and $\frac{3}{10}$ for utilities. What fractional part of her salary is left for other expenses?

_______ **19.** Kim works in the morning for 2 hours at \$3.85 per hour and in the afternoon for 4 hours at \$4.90 per hour. How much does she earn in 1 day?

_______ **20.** Mrs. Curtis works 8 hours a day at \$8.97 an hour and hires a baby-sitter for 9 hours a day at \$2.75 an hour. After she pays the baby-sitter, how much does Mrs. Curtis earn each day?

_______ **21.** For lunch Camelia had a Big Burger for \$1.69, a salad for \$1.25, and a carton of milk for \$0.45. If the sales tax rate is 5.5%, find the tax on her lunch. Find the total cost of her lunch. How much change should she receive from a 10-dollar bill?

_______ **22.** Henry and his roommate split rent and food expenses equally. This month they spent \$225 on rent and \$115.82 on food. If Henry started the month with \$315.75, how much does he have left after paying his share of the expenses?

_______ **23.** On an achievement test, a student receives 2 points for every correct answer, loses 2 points for every incorrect answer, and loses 1.5 points for every unanswered question. If Mark answered 27 questions correctly, answered 13 questions incorrectly, and left 10 questions unanswered, find his score on the test.

_______ **24.** At the Plaza Theater an adult's ticket costs \$4.50. A child's ticket costs half as much as an adult's ticket. A senior citizen's ticket costs \$0.75 less than an adult's ticket. If 60 adult tickets, 25 children's tickets, and 15 senior citizen tickets were sold, find the amount received at the ticket window.

☆ Stretching the Topics

_______ **1.** Profits or losses are split equally among the five owners of the Pizza Shack. During the past weeks, the shack took in $1873.95. Supplies cost $928.17, eight workers were paid $150 each, and advertising cost $53.75. What was each owner's share of the profits or losses?

_______ **2.** June Ricketts bought a lot and built a house on it. The total cost was $95,000. She sold the property a year later for $116,000. During the year she spent $787 for taxes and $585 for insurance. She paid 6% of the selling price to a real estate agent for selling the property. How much profit did she make on the transaction?

Check your answers in the back of your book.

If you can do the problems in **Checkup 2.5**, you are ready to do the **Review Exercises for Chapter 2.**

Name ______________________________ Date ____________

✓ CHECKUP 2.5

Change each word expression to a number expression and simplify.

_______ **1.** Atha has a balance of $785.28 in her checking account. If she writes checks at the grocery for $119.61 and at the cleaners for $28.75, what is her balance?

_______ **2.** If a ballon rises from the ground at a rate of 3 feet per second for 20 seconds and then falls at a rate of 2 feet per second for 10 seconds, what is its final height?

_______ **3.** During 1 year a farm with three equal-sharing owners sold products totaling $625,823 and had expenses of $125,729. What is each owner's share of the profit? (Round to the nearest cent.)

_______ **4.** In a certain town the Ohio River reached flood stage at 48 feet. During the next 5 hours, readings indicated that the river rose $\frac{2}{9}$ foot, rose $\frac{1}{8}$ foot, remained steady, fell $\frac{1}{6}$ foot, and fell $\frac{2}{3}$ foot. What was the river stage after the last reading?

_______ **5.** At the local Bingo game, Betty lost $5 per week for 3 weeks, won $7 per week for 2 weeks, and lost $2.50 per week for 4 weeks. Find her average gain or loss per week.

Check your answers in the back of your book.

If You Missed Problem:	You Should Review Examples:
1	1
2	2
3	4
4	3
5	2, 4

Summary

In this chapter we learned that all the numbers that we have discussed belong to the set of **real numbers**. Every number that corresponds to a point on the number line is a real number.

Number Set	Description	Examples
Natural numbers	$\{1, 2, 3, \ldots\}$	7 is a natural number
Whole numbers	$\{0, 1, 2, 3, \ldots\}$	0 is a whole number 195 is a whole number
Integers	$\{\ldots, -3, -2, -1, 0, 1, 2, 3, \ldots\}$	11 is an integer -29 is an integer
Rational numbers	Numbers that can be written as fractions of integers	$\frac{5}{9}$ is a rational number 2.7 is a rational number -16 is a rational number $\sqrt{9}$ is a rational number
Irrational numbers	Numbers on the number line that are *not* rational	$\sqrt{5}$ is an irrational number π is an irrational number
Real numbers	Numbers on the number line	6.3 is a real number -82 is a real number $\sqrt{2}$ is a real number

We developed several definitions and properties that apply to all real numbers. Suppose that we let A, B, and C represent any real numbers.

Symbol	Words	Meaning	Examples
$A < B$	A is less than B	A lies to the left of B on the number line	$0 < 3$ $-7 < -4$
$A > B$	A is greater than B	A lies to the right of B on the number line	$9 > 1$ $-3.7 > -4$
A^n	A to the nth power	$A^n = \underbrace{A \cdot A \cdot \ldots \cdot A}_{n \text{ factors}}$	$3^4 = 3 \cdot 3 \cdot 3 \cdot 3$ $\left(\frac{1}{2}\right)^3 = \frac{1}{2} \cdot \frac{1}{2} \cdot \frac{1}{2}$ $(-1.5)^2 = (-1.5)(-1.5)$
$\|A\|$	Absolute value of A	The distance between 0 and A on the number line	$\|-3\| = 3$ $\left\|\frac{17}{5}\right\| = \frac{17}{5}$ $\|-6.84\| = 6.84$

Property Name	Statement	Examples
Closure property for addition	$A + B$ is real and unique	$3 + 2.8$ is real $\frac{-1}{6} + \frac{2}{5}$ is real
Closure property for multiplication	$A \cdot B$ is real and unique	$-4.1(10)$ is real $\frac{1}{5} \cdot \frac{2}{9}$ is real
Commutative property for addition	$A + B = B + A$	$3 + (-6) = -6 + 3$ $\frac{1}{8} + \frac{5}{4} = \frac{5}{4} + \frac{1}{8}$
Commutative property for multiplication	$A \cdot B = B \cdot A$	$0.6(2.9) = 2.9(0.6)$ $\frac{-5}{3} \cdot 6 = 6 \cdot \frac{-5}{3}$
Associative property for addition	$(A + B) + C = A + (B + C)$	$(6 + 3.1) + 0.9 = 6 + (3.1 + 0.9)$ $\left(\frac{5}{7} + \frac{1}{2}\right) + \frac{-1}{2} = \frac{5}{7} + \left(\frac{1}{2} + \frac{-1}{2}\right)$
Associative property for multiplication	$(A \cdot B)C = A(B \cdot C)$	$(3 \cdot 2)\frac{1}{2} = 3\left(2 \cdot \frac{1}{2}\right)$ $(-0.9 \cdot 5)2 = -0.9(5 \cdot 2)$
Distributive properties	$A(B + C) = A \cdot B + A \cdot C$ $A(B - C) = A \cdot B - A \cdot C$	$3(2 + 5) = 3 \cdot 2 + 3 \cdot 5$ $\frac{1}{2}(8 - 0.4) = \frac{1}{2} \cdot 8 - \frac{1}{2}(0.4)$
Opposites property	$A + (-A) = 0$	$-3.7 + 3.7 = 0$ $\frac{5}{6} + \left(\frac{-5}{6}\right) = 0$
Reciprocals property	$A \cdot \frac{1}{A} = 1 \quad (A \neq 0)$	$\frac{2}{3} \cdot \frac{3}{2} = 1$ $-5\left(-\frac{1}{5}\right) = 1$

We also continued to use the important properties of zero and 1.

Property Name	Statement	Example
Addition property of zero	$A + 0 = A$	$2.3 + 0 = 2.3$
Subtraction property of zero	$A - 0 = A$	$\frac{17}{39} - 0 = \frac{17}{39}$
Multiplication property of 1	$A \cdot 1 = A$	$-629 \cdot 1 = -629$
Multiplication property of zero	$A \cdot 0 = 0$	$\frac{-2}{3} \cdot 0 = 0$
Division property of 1	$\frac{A}{1} = A$	$\frac{56}{1} = 56$
Division of zero	$\frac{0}{A} = 0 \quad (A \neq 0)$	$\frac{0}{6.91} = 0$
Division by zero	$\frac{A}{0}$ is undefined	$\frac{-0.23}{0}$ is undefined

❑ Speaking the Language of Algebra

1. The set containing all the whole numbers and their opposites is called the set of __________ .

2. Every rational number can be expressed as a fraction with a numerator that is an __________ and a denominator that is a __________ __________ .

3. A real number that is *not* rational is called an __________ number.

4. The symbol $|a|$ represents the __________ __________ __________ __________ . On the number line, $|a|$ is found as the __________ between __________ and a.

5. The absolute value of a real number is never __________ .

6. On the number line, points to the right of zero correspond to __________ real numbers, and points to the left of zero correspond to __________ real numbers.

7. The sum of a real number and its opposite is always __________ .

8. The operations of __________ and __________ are commutative and associative, but the operations of __________ and __________ are not.

9. To find the difference $A - B$, we must find the sum of A and the __________ of B.

10. The sum of two positive numbers is a __________ number. The sum of two negative numbers is a __________ number.

11. The product of two positive numbers is a __________ number. The product of two negative numbers is a __________ number. The product of a positive number and a negative number is a __________ number.

12. The expression A^n tells us to use the base, __________ , as a factor in a product __________ times.

Check your answers in the back of your book.

Name ______________________ Date __________

REVIEW EXERCISES for Chapter 2

1. Graph the point corresponding to each number on the number line.

(a) -3 **(b)** 0 **(c)** $\frac{5}{2}$ **(d)** -0.6

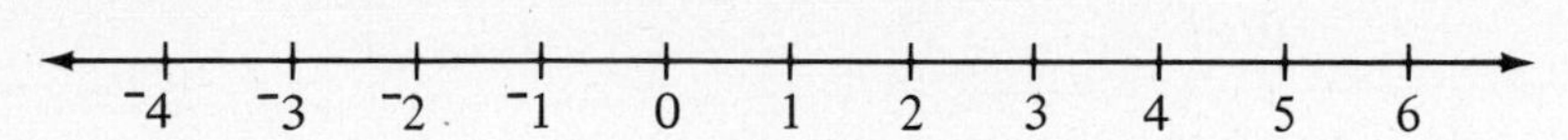

_______ **2.** Use the commutative property for addition to rewrite the sum $-5 + (-6)$.

_______ **3.** Use the associative property for multiplication to rewrite the sum $[-3 + (-2)] + 5$.

_______ **4.** Use the commutative property for multiplication to rewrite the product $-4 \cdot 5$.

_______ **5.** Use the associative property for multiplication to rewrite the product $-7(-8 \cdot 9)$.

_______ **6.** Use the distributive property for multiplication over addition to rewrite $-2(-6 + 9)$.

7. Compare each of the following pairs using < or >.

_______ **(a)** $4, 14$ _______ **(b)** $-5, -11$

_______ **(c)** $9, -12$ _______ **(d)** $-9, 3$

Simplify.

_______ **8.** $|-13|$ _______ **9.** $-6 + (-12)$

_______ **10.** $8 + (-5)$ _______ **11.** $-6 + 3 + 8 + (-10)$

_______ **12.** $-3 + \left(-5 + \frac{24}{3}\right)$ _______ **13.** $32 - (-6)$

_______ **14.** $-9 - 7$ _______ **15.** $[7 + (-2)] - (-5 + 1)$

_______ **16.** $(-8)(-3)(5)$ _______ **17.** $-4(5) + (-3)(-7)$

_______ **18.** $7[4(8 - 5) - (-3)]$ _______ **19.** $(-3)^2(-1)^3$

_______ **20.** $\frac{-56}{8}$ _______ **21.** $\frac{18}{2} - \frac{-15}{3}$

_______ **22.** $\frac{-9 - (-3)}{8 - 10}$ _______ **23.** $\frac{3(-4) + (-6)(-2)}{-7 - 2}$

_______ **24.** $\frac{-3(8 - 2) - 2(-1)}{7 + (1 - 10)}$ _______ **25.** $\frac{\frac{-24}{3} + \frac{0}{2}}{\frac{10}{5} - \frac{8}{4}}$

_______ **26.** $-3.6 + (-4.02) + 9$

_______ **27.** $\frac{3}{4} - \frac{7}{6} - \frac{2}{3}$

_______ **28.** $0.7[-4.6 - (4 - 8.3)]$

_______ **29.** $14.77 \div (-0.35)$

_______ **30.** $\left(\frac{-2}{3}\right)^3 \div \frac{16}{45}$

_______ **31.** At the track Eddie bet \$2 on each of 10 races. Twice he won \$4 and once he won \$15. How much did Eddie win or lose at the track?

_______ **32.** Erica earns \$253.93 per week, but deductions of \$25.19, \$15.32, and \$6 are made before her check is written. She also earned a \$25 bonus for an efficiency suggestion she made last week. What is the amount of her check?

_______ **33.** One day recently the stock market reported that at the beginning of the day Xerox stock shares sold for \$102 $\frac{1}{8}$. During the day the price rose by $\frac{1}{2}$, rose by $\frac{1}{4}$, then fell by $\frac{5}{8}$. What was the price of a share of stock at the close of the day?

_______ **34.** A window-washing crew washed $\frac{1}{9}$ of the windows in an office building during the first week, $\frac{3}{8}$ of the windows the second week, and $\frac{1}{6}$ of the windows the third week. What fractional part of the windows remain to be washed?

_______ **35.** During the first 6 months of this year the rainfall was 1.2 inches above normal for 2 months, 2 inches above normal 1 month, and 0.8 inch below normal for 3 months. Find the average number of inches the precipitation was above or below normal for the first 6 months.

Check your answers in the back of your book.

If You Missed Exercises:	You Should Review Examples:	
1	SECTION 2.1	3
2, 3		24, 25
4, 5	SECTION 2.3	8, 9
6		11, 12
7	SECTION 2.1	1, 2
8		8
9, 10		14, 19
11		26
12		20, 22
13, 14	SECTION 2.2	3, 4
15		5
16, 17	SECTION 2.3	5, 6
18		7
19		16
20, 21		21
22–25		25, 26
26, 27	SECTION 2.4	6–8
28–30		14–18
31	SECTION 2.5	2
32		1
33, 34		3
35		4

If you completed the **Review Exercises** and corrected your errors, you are ready to take the **Practice Test for Chapter 2.**

Name ____________________ Date __________

PRACTICE TEST for Chapter 2

	Section	Examples
1. Graph the point corresponding to each number on the number line. **(a)** $\frac{-2}{3}$ **(b)** 0.7 **(c)** -1.6 **(d)** $\frac{9}{4}$	2.1	3

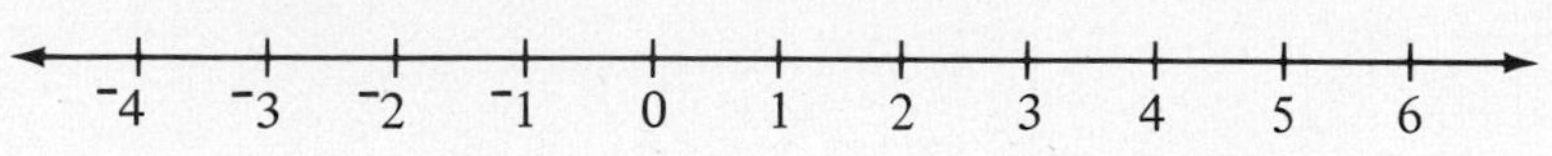

	Section	Examples
______ **2.** Use the commutative property for multiplication to rewrite $-3(5)$.	2.3	4
______ **3.** Use the associative property for addition to rewrite $-4 + [9 + (-2)]$.	2.1	25
______ **4.** Use the distributive property for multiplication over addition to rewrite $-4[11 + (-2)]$.	2.3	6
5. Compare each pair of numbers using < or >.	2.1	1, 2
______ **(a)** $-15, -1$ ______ **(b)** $0, -3$		
______ **(c)** 6, 5.9 ______ **(d)** $\frac{1}{3}, \frac{1}{2}$		

Simplify.

	Section	Examples
______ **6.** $-\lvert -8 \rvert$	2.1	11
______ **7.** $3 + (-11) + (-12) + 7$	2.1	26
______ **8.** $-15 - (-22)$	2.2	3
______ **9.** $[-10 + 6] - [5 + (-9)]$	2.2	5
______ **10.** $(-7)(2)(-3)$	2.3	5
______ **11.** $5[-2(7 - 3) - (-2)]$	2.3	7
______ **12.** $(-2)^4(-1)^5$	2.3	16
______ **13.** $\frac{-24}{6} - \frac{10}{-5}$	2.3	21, 22
______ **14.** $\frac{-11 - 12}{-5 + 8}$	2.3	23, 24
______ **15.** $\frac{-6(-3) + (18)(-1)}{-9 + 4}$	2.3	23, 24
______ **16.** $\frac{-2(11 - 9) - 5(-2)}{6 + (5 - 11)}$	2.3	23, 24
______ **17.** $5.7 + (-8.2) - 2.95$	2.4	6

	Problem	Section	Example
________	**18.** $\frac{3}{8} + \frac{-2}{3} - \frac{7}{6}$	2.4	7, 8
________	**19.** $0.3[-2.3 + (7 - 6.1)]$	2.4	14
________	**20.** $-32.2 \div 0.23$	2.4	17
________	**21.** $\left(\frac{-3}{4}\right)^2 \div \frac{40}{21}$	2.4	15, 18
________	**22.** After attending eight weekly meetings of Weight Worriers, Clyde's record showed that he had lost 2.5 pounds twice, gained 1.5 pounds twice, lost 4 pounds three times, and stayed the same once. What was Clyde's average weekly loss?	2.5	4
________	**23.** Martha opened a quart of milk in the morning. She used $\frac{1}{2}$ cup on her cereal, toook $\frac{3}{4}$ cup to work in her thermos, made pudding with $1\frac{3}{4}$ cups, and drank $\frac{2}{3}$ cup with supper. How much milk was left after supper? (*Remember:* 1 quart = 4 cups.)	2.5	3

Check your answers in the back of your book.

Working with Variable Expressions 3

The Internal Revenue Service allows a taxpayer to deduct $1000 for every dependent claimed. For an income of $12,000, write an expression describing the remaining taxable income after a taxpayer has made the deduction for *d* dependents.

Algebra can be thought of as the bridge between arithmetic and higher mathematics. A person with even a limited knowledge of algebra can solve many everyday problems that otherwise could be solved only by trial and error. In algebra we must work with **constants**, such as the numbers we have already studied, but we must also work with symbols that stand for numbers. Such symbols are called **variables** and are usually represented by letters.

In this chapter we learn how to

1. Switch from word expressions to algebraic expressions.
2. Evaluate algebraic expressions.
3. Add and subtract algebraic expressions.
4. Multiply constants times algebraic expressions.

3.1 Switching from Word Expressions to Algebraic Expressions

Combinations of constants and/or variables involving the operations of addition, subtraction, multiplication, and division are called **algebraic expressions**.

Words	Symbols
add 4 and the variable x	$4 + x$
subtract 9 from the variable y	$y - 9$
divide the variable a by 2	$\frac{a}{2}$
multiply 3 times the variable x	$3 \cdot x$ or $3x$
the sum of x and 5	$x + 5$
the product of 5 and q	$5 \cdot q$ or $5q$
a decreased by 1.5	$a - 1.5$
twice the variable x	$2 \cdot x$ or $2x$
7 less than n	$n - 7$
add 5 and y; then divide by 6	$\frac{5 + y}{6}$
7 times the sum of b and 3	$7(b + 3)$

Let's try switching from word expressions to algebraic expressions.

Example 1. The product of 7 and y.

Solution

$$7 \cdot y \quad \text{or} \quad 7y$$

Notice that the symbol for multiplication may be left out. We agree that a number written next to a variable will always mean *multiply*.

Example 2. 32 decreased by x.

Solution

$$32 - x \quad \text{or} \quad 32 + (-x)$$

Notice that when dealing with variables, we can use the same definition for subtraction that we use with integers.

You try Examples 3 and 4.

Example 3. From 17, subtract the product of 9 and y.

Solution

17 − 9y

Check your work on page 155. ▶

Example 4. Find the sum of 13 and twice x.

Solution

13 + 2x

Check your work on page 155. ▶

Example 5. Subtract x from 11, then divide by -4.

Solution

$$\frac{11 - x}{-4}$$

You try Example 6.

Example 6. Three times the sum of y and 5.

Solution

3(y + 5)

Check your work on page 155. ▶

Trial Run

Write an algebraic expression that means:

______ **1.** Add 7 and x.

______ **2.** Multiply -5 times y.

______ **3.** Subtract 8 from a.

______ **4.** Twice the sum of x and 13.

______ **5.** Subtract x from 15; then multiply by -3.

______ **6.** Add -12 to y; then divide by 10.

Answers are on page 155.

Evaluating Algebraic Expressions

In every algebraic expression containing variables, the number value of the expression depends on what number the variable stands for. For instance, to find the value of the expression $9 + x$, we must know what number x represents.

$$\text{if } x = 2 \text{ then } 9 + x = 9 + 2 = 11$$

$$\text{if } x = -4 \text{ then } 9 + x = 9 + (-4) = 5$$

$$\text{if } x = -15 \text{ then } 9 + x = 9 + (-15) = -6$$

This process is called **evaluating an expression**. To evaluate an algebraic expression for some value of the variable, we *substitute* (or "plug in") the value we are given and then perform the necessary arithmetic.

Example 7. Evaluate $x - 32y$ when $x = 4$ and $y = -1$.

Solution: If $x = 4$ and $y = -1$, then

$$\begin{aligned} x - 32y &= 4 - 32(-1) \\ &= 4 - (-32) \\ &= 4 + 32 \\ &= 36 \end{aligned}$$

Now you try Example 8.

Example 8. Evaluate $7y$ when $y = 3$; when $y = -1.1$; when $y = 0$.

Solution

If $y = 3$, then

$7y = 7(\_\_\_\_) = \_\_\_\_$

If $y = -1.1$, then

$7y = 7(\_\_\_\_) = \_\_\_\_$

If $y = 0$, then

$7y = 7(\_\_\_\_) = \_\_\_\_$

Check your work on page 155. ▶

Example 9. Evaluate $\frac{2b + a}{5}$ when $b = -3$ and $a = 6$.

Solution: If $b = -3$ and $a = 6$, then

$$\frac{2b + a}{5} = \frac{2(-3) + 6}{5}$$
$$= \frac{-6 + 6}{5}$$
$$= \frac{0}{5}$$
$$= 0$$

You try Example 10.

Example 10. Evaluate $4(11 - xy)$ when $x = 5$ and $y = -2$.

Solution: If $x = 5$ and $y = -2$, then

$$4(11 - xy) = 4[11 - 5(-2)]$$
$$= 4[11 + \_\_\_\_]$$
$$= 4[\_\_\_\_]$$
$$= \_\_\_\_$$

Check your work on page 155. ▶

Trial Run

Evaluate each expression.

______ **1.** $6x$ when $x = -2$; when $x = 0$; when $x = 7$.

______ **2.** $x - 23$ when $x = 50$; when $x = 10$; when $x = -5$.

______ **3.** $\frac{60}{y}$ when $y = 6$; when $y = -12$; when $y = 0$.

______ **4.** $\frac{a + b}{-6}$ when $a = 10$ and $b = 8$; when $a = -11$ and $b = 5$; when $a = -2$ and $b = 2$.

______ **5.** $3(13 - m)$ when $m = 3$; when $m = -2$; when $m = 13$.

Answers are on page 155.

Using Algebraic Expressions

Variables and algebraic expressions give us a handy way of dealing with word expressions in which the value of one or more of the quantities is not known.

For example, suppose that movie tickets cost \$3.50 each. We could use a variable expression to represent the cost of *any number* of tickets by letting x stand for the number of tickets sold. Then $3.50x$ would represent the cost of x tickets. For example,

$$\begin{array}{llll} \text{if } x = 2 & \text{then } 3.50x = 3.50(2) & = \$7 \\ \text{if } x = 10 & \text{then } 3.50x = 3.50(10) & = \$35 \\ \text{if } x = 173 & \text{then } 3.50x = 3.50(173) & = \$605.50 \end{array}$$

You try Example 11.

Example 11. If part-time students at a university are charged \$25 per credit hour, write an algebraic expression describing the cost of h credit hours. Then find the cost of 3 hours; of 8 hours.

Solution: Since 1 hour costs \$25, we know that h hours will cost ____ dollars and

if $h = 3$ hours then ____ = ____(3) = ____ dollars

if $h = 8$ hours then ____ = ____(8) = ____ dollars

Check your work on page 155. ▶

Example 12. If the population of a town is 25,230, write an algebraic expression for the number of people in the town after a change in population of x people has occurred. Then find the new population if 2900 people move in; if 1600 people move out.

Solution: Since the current population is 25,230, the new population will be $25{,}230 + x$.

$$\begin{array}{ll} \text{if } x = 2900 & \text{then } 25{,}230 + x = 25{,}230 + 2900 = 28{,}130 \\ \text{if } x = -1600 & \text{then } 25{,}230 + x = 25{,}230 + (-1600) = 23{,}630 \end{array}$$

Notice that we represented the *increase* by $+2900$ and the *decrease* by -1600.

Now you try Example 13.

Example 13. A gasoline distributor supplies 10 local gas stations with equal allotments of gasoline each week. Write an algebraic expression to describe each station's share of the distributor's total number of gallons, G. Then find each station's share if the distributor has 60,000 gallons; 85,000 gallons; 0 gallons.

Solution: Since each station receives an equal share of the distributor's gasoline, we must *divide* the total number of gallons, G, by 10. Each station's share will be $\frac{\quad}{10}$.

if $G = 60{,}000$ then $\frac{G}{10} = \frac{\quad}{10} =$ ____ gallons

if $G = 85{,}000$ then $\frac{G}{10} = \frac{\quad}{10} =$ ____ gallons

if $G = 0$ then $\frac{G}{10} = \frac{\quad}{10} =$ ____ gallons

Check your work on page 155. ▶

Example 14. The Internal Revenue Service allows a taxpayer to deduct \$1000 for each dependent claimed. For an income of \$12,000, write an expression describing the remaining taxable income after a taxpayer has made the deduction for d dependents. Then find the remaining taxable income if a taxpayer claims one dependent; five dependents.

Solution: If \$1000 may be deducted for each dependent, then for d dependents, the total deduction is $\$1000 \cdot d$ or $1000 \cdot d$ dollars. We must *subtract* that deduction from the \$12,000 income, so the remaining taxable income is

$$12{,}000 - 1000 \cdot d$$

$$\text{if } d = 1 \quad \text{then} \quad 12{,}000 - 1000 \cdot d = 12{,}000 - 1000(1)$$
$$= 12{,}000 - 1000$$
$$= 12{,}000 + (-1000) = 11{,}000 \text{ dollars}$$

$$\text{if } d = 5 \quad \text{then} \quad 12{,}000 - 1000 \cdot d = 12{,}000 - 1000(5)$$
$$= 12{,}000 - 5000$$
$$= 12{,}000 + (-5000) = 7000 \text{ dollars}$$

Trial Run

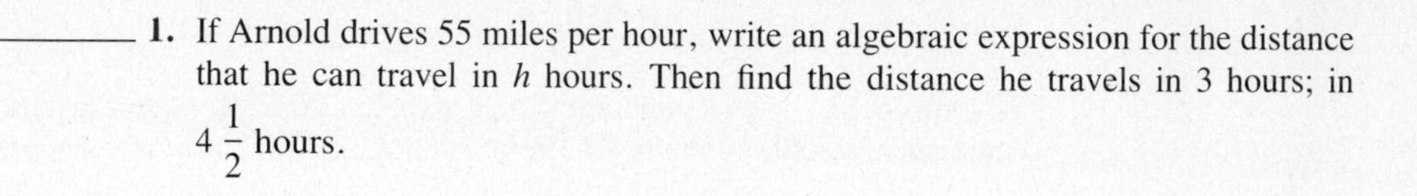

_______ **1.** If Arnold drives 55 miles per hour, write an algebraic expression for the distance that he can travel in h hours. Then find the distance he travels in 3 hours; in $4\frac{1}{2}$ hours.

_______ **2.** If there are 1252 students in a high school, write algebraic expression for the number of students after a change in enrollment of x students has occurred. Then find the enrollment if 142 new students enroll; if 93 students drop out.

_______ **3.** The meat from a beef calf is to be evenly divided among five families. Write an algebraic expression for each family's share if the total weight of the calf is T pounds. Then find each family's share if the calf weighs 725 pounds; if the calf weighs 931 pounds.

_______ **4.** The cost of renting a garden tiller is \$25, plus \$8.50 for each day it is used. Write an algebraic expression for the cost of renting a tiller for d days. Then find the cost for 2 days; for 5 days.

Answers are on page 155.

▶ Examples You Completed

Example 3. From 17, subtract the product of 9 and y.

Solution

$$17 - 9y$$

Example 4. Find the sum of 13 and twice x.

Solution

$$13 + 2x$$

Example 6. Three times the sum of y and 5.

Solution

$$3(y + 5)$$

Example 8. Evaluate $7y$ when $y = 3$; when $y = -1.1$; when $y = 0$.

Solution

If $y = 3$, then $7y = 7(3) = 21$

If $y = -1.1$, then $7y = 7(-1.1) = -7.7$

If $y = 0$, then $7y = 7(0) = 0$.

Example 10. Evaluate $4(11 - xy)$ when $x = 5$ and $y = -2$.

Solution: If $x = 5$ and $y = -2$, then

$$\begin{aligned} 4(11 - xy) &= 4[11 - 5(-2)] \\ &= 4(11 + 10) \\ &= 4(21) \\ &= 84 \end{aligned}$$

Example 11. If part-time students at a university are charged \$25 per credit hour, write an algebraic expression describing the cost of h credit hours. Then find the cost of 3 hours; of 8 hours.

Solution: Since 1 hour costs \$25, we know h hours will cost $25 \cdot h$ dollars and

if $h = 3$ hours then $25h = 25(3) = 75$ dollars

if $h = 8$ hours then $25h = 25(8) = 200$ dollars

Example 13. A gasoline distributor supplies 10 local gas stations with equal allotments of gasoline each week. Write an algebraic expression to describe each station's share of the distributor's total number of gallons, G. Then find each station's share if the distributor has 60,000 gallons; 85,000 gallons; 0 gallons.

Solution: Since each station receives an equal share of the distributor's gasoline, we must divide the total number of gallons, G, by 10. Each station's share will be $\frac{G}{10}$.

if $G = 60{,}000$ then $\frac{G}{10} = \frac{60{,}000}{10} = 6000$ gallons

if $G = 85{,}000$ then $\frac{G}{10} = \frac{85{,}000}{10} = 8500$ gallons

if $G = 0$ then $\frac{G}{10} = \frac{0}{10} = 0$ gallons

Answers to Trial Runs

page 151 **1.** $7 + x$ **2.** $-5y$ **3.** $a - 8$ **4.** $2(x + 13)$ **5.** $-3(15 - x)$ **6.** $\frac{-12 + y}{10}$

page 152 **1.** -12; 0; 42 **2.** 27; -13; -28 **3.** 10; -5; undefined **4.** -3, 1; 0 **5.** 30; 45; 0

page 154 **1.** $55h$; 165; 247.5 **2.** $1252 + x$; 1394; 1159 **3.** $\frac{T}{5}$; 145; 186.2 **4.** $25 + 8.50d$; \$42; \$67.50

Name ______________________ Date ____________

EXERCISE SET 3.1

Write an algebraic expression for each word statement or expression.

______ **1.** Add 10 to y.

______ **2.** Add -5 to x.

______ **3.** Multiply -3 times m.

______ **4.** Multiply 8 times n.

______ **5.** Subtract 6.5 from a.

______ **6.** Subtract 1.3 from b.

______ **7.** Divide t by -15.

______ **8.** Divide s by 12.

______ **9.** The sum of x and -9.

______ **10.** The sum of y and 7.

______ **11.** The product of q and 3.2.

______ **12.** The product of m and -1.3.

______ **13.** n decreased by 6.3.

______ **14.** d decreased by -9.5.

______ **15.** Twice x, divided by 7.

______ **16.** Five times y, divided by -11.

______ **17.** 9 more than the product of $\frac{1}{6}$ and k.

______ **18.** 10 more than the product of $\frac{2}{3}$ and m.

______ **19.** 5 less than the quotient of p divided by 7.

______ **20.** 13 less than the quotient of p divided by -3.

______ **21.** 4 times x is subtracted from 9.

______ **22.** 8 times y is subtracted from 23.

______ **23.** -4 times the sum of x and 1.5.

______ **24.** 9 times the sum of y and 2.5.

Evaluate each expression.

_______ **25.** $y - 3$ when $y = 10$; when $y = -3$

_______ **26.** $y - 8$ when $y = 15$; when $y = -8$

_______ **27.** $\frac{m}{5}$ when $m = -15$; when $m = 0$

_______ **28.** $\frac{m}{9}$ when $m = 36$; when $m = -18$

_______ **29.** $-7a$ when $a = -6$; when $a = 7$

_______ **30.** $-9a$ when $a = 10$; when $a = -3$

_______ **31.** $4(x - y)$ when $x = -3$ and $y = 5$; when $x = 5$ and $y = -5$

_______ **32.** $5(x - y)$ when $x = -1$ and $y = 12$; when $x = 3$ and $y = -3$

_______ **33.** $2x + 4$ when $x = -4$; when $x = -2$

_______ **34.** $5x + 10$ when $x = -3$; when $x = 0$

_______ **35.** $\frac{a - b}{7}$ when $a = 13$ and $b = 6$; when $a = -14$ and $b = 7$

_______ **36.** $\frac{a - b}{5}$ when $a = 14$ and $b = 9$; when $a = -10$ and $b = 15$

Write an algebraic expression for each word expression and evaluate for the given values of the variable.

_______ **37.** If the Betas are selling lottery tickets on a car at \$2.50 per ticket, write an algebraic expression for the cost of y tickets. Find the cost of buying 10 tickets; 25 tickets.

_______ **38.** The cost of a chartered bus to the ballgame is \$570, to be divided evenly among the number of students who ride the bus. Write an algebraic expression for cost to each passenger if x students ride the bus. Find the cost for each student if there are 60 passengers; 30 passengers.

_______ **39.** Write an algebraic expression for the number of liters remaining in a 75-liter tank if x liters have already been used. Find how many liters remain when 8.7 have been used; 13.2 have been used.

_______ **40.** Irene has a part-time job paying \$3.50 per hour. Write an algebraic expression for her earnings after she has worked x hours. Find her earnings when she works $14\frac{1}{4}$ hours; 22 hours.

_______ **41.** An oil tank that contains 200 barrels of oil begins to leak. Three barrels a day are leaking from the tank. Write an algebraic expression for the amount

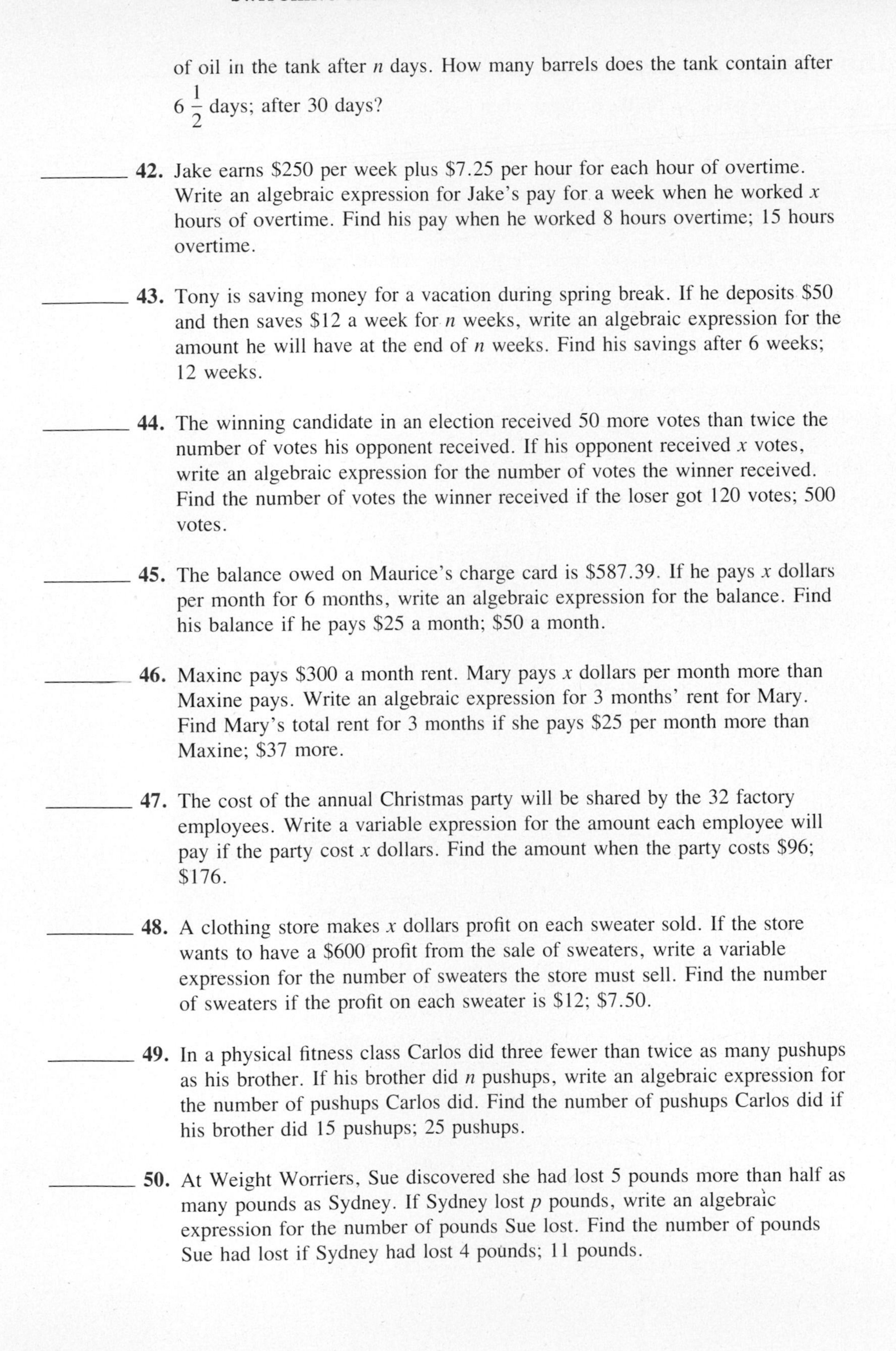

of oil in the tank after n days. How many barrels does the tank contain after $6\frac{1}{2}$ days; after 30 days?

________ **42.** Jake earns $250 per week plus $7.25 per hour for each hour of overtime. Write an algebraic expression for Jake's pay for a week when he worked x hours of overtime. Find his pay when he worked 8 hours overtime; 15 hours overtime.

________ **43.** Tony is saving money for a vacation during spring break. If he deposits $50 and then saves $12 a week for n weeks, write an algebraic expression for the amount he will have at the end of n weeks. Find his savings after 6 weeks; 12 weeks.

________ **44.** The winning candidate in an election received 50 more votes than twice the number of votes his opponent received. If his opponent received x votes, write an algebraic expression for the number of votes the winner received. Find the number of votes the winner received if the loser got 120 votes; 500 votes.

________ **45.** The balance owed on Maurice's charge card is $587.39. If he pays x dollars per month for 6 months, write an algebraic expression for the balance. Find his balance if he pays $25 a month; $50 a month.

________ **46.** Maxine pays $300 a month rent. Mary pays x dollars per month more than Maxine pays. Write an algebraic expression for 3 months' rent for Mary. Find Mary's total rent for 3 months if she pays $25 per month more than Maxine; $37 more.

________ **47.** The cost of the annual Christmas party will be shared by the 32 factory employees. Write a variable expression for the amount each employee will pay if the party cost x dollars. Find the amount when the party costs $96; $176.

________ **48.** A clothing store makes x dollars profit on each sweater sold. If the store wants to have a $600 profit from the sale of sweaters, write a variable expression for the number of sweaters the store must sell. Find the number of sweaters if the profit on each sweater is $12; $7.50.

________ **49.** In a physical fitness class Carlos did three fewer than twice as many pushups as his brother. If his brother did n pushups, write an algebraic expression for the number of pushups Carlos did. Find the number of pushups Carlos did if his brother did 15 pushups; 25 pushups.

________ **50.** At Weight Worriers, Sue discovered she had lost 5 pounds more than half as many pounds as Sydney. If Sydney lost p pounds, write an algebraic expression for the number of pounds Sue lost. Find the number of pounds Sue had lost if Sydney had lost 4 pounds; 11 pounds.

☆ Stretching the Topics

______ **1.** Write an algebraic expression for the quotient when 3 times x subtracted from 4 times y is divided by 15.

______ **2.** Evaluate $\frac{7(2x - 5y)}{4z + y}$ when $x = -2$, $y = 1$, and $z = 2$.

______ **3.** The total area of Kym's house is 2000 square feet. During remodeling, a sewing room with a total area of x square feet was removed. A family room was added which has a total area of 108 square feet more than $\frac{2}{3}$ the area of the sewing room. Write a variable expression for the area of Kym's house after remodeling. What will be the total area if the area of the sewing room was 72 square feet?

Check your answers in the back of your book.

If you can do **Checkup 3.1**, you are ready to go to Section 3.2.

Name ______________________________ Date ______________

✓ CHECKUP 3.1

Write an algebraic expression for each word expression.

________ **1.** The sum of -3 and x.

________ **2.** Eleven decreased by a.

________ **3.** The product of m and -6.

________ **4.** Five more than twice b.

Evaluate each expression.

________ **5.** $x + 9$ when $x = -12$; when $x = -9$

________ **6.** $\frac{24}{n}$ when $n = -6$; when $n = 0$

________ **7.** $3(y - 5)$ when $y = -2$; when $y = 5$

________ **8.** $\frac{m - n}{3}$ when $m = 13$ and $n = 10$; when $m = 3$ and $n = -6$

________ **9.** The total budget for the athletic department will be divided equally among the three major sports. Write an algebraic expression for the share of each sport if the budget is x dollars. Find each sport's share if the budget is \$150,000; \$360,000.

________ **10.** A building contractor must pay a penalty of 6% of \$2500 plus an additional \$500 a day for each day over the deadline for completion of a building. Write an algebraic expression for the amount of the penalty after x days. Find the amount if the building is completed 3 days late; 15 days late.

Check your answers in the back of your book.

If You Missed Problems:	You Should Review Examples:
1	4
2	2
3	1
4	4
5, 6	7, 8
7	10
8	9
9	13
10	14

3.2 Combining Like Terms

We turn our attention now to learning ways of making algebraic expressions as simple as possible. This process is called **simplifying** algebraic expressions.

Recognizing Like Terms

Let's look at some algebraic expressions and learn to call their parts by their correct mathematical names. In the expression $\mathbf{6x + 5x + 2x}$ we say that $6x$ is one **term**, $5x$ is another *term*, and $2x$ is another *term*.

> A **term** is a constant alone, or a product of a constant and one (or more) variables. Terms are separated from each other by addition signs.

Example 1. Identify the terms in the expression

$$4x + 7 - 3x$$

Solution: We may write this expression as the sum

$$4x + 7 + (-3x)$$

Then we see that

$4x$ is a term

7 is a term

$-3x$ is a term

You try Example 2.

Example 2. Identify the terms in the expression

$$5x - 3y - 10$$

Solution: We may write this expression as the sum

$$5x + (\_\_\_\_) + (\_\_\_\_)$$

Then we see that

____ is a term

____ is a term

____ is a term

Check your work on page 166. ▶

Terms that contain a number alone are called **constant terms**. Terms that contain a variable are called **variable terms**. The number part of a variable term is called the **numerical coefficient** for the variable in that term.

Expression	Constant Term	Variable Terms	Numerical Coefficients
$x + 3$	3	x (or $1x$)	1
$5x - 3y - 10$	-10	$5x$ $-3y$	5 -3
$8x - 5x - x$	None (or 0)	$8x$ $-5x$ $-x$ (or $-1x$)	8 -5 -1

Example 3. Identify the numerical coefficient for each variable term of the expression $11x - 9x + 2$.

Solution: Note that 2 is the constant term.

For $11x$, 11 is the numerical coefficient of x.
For $-9x$, -9 is the numerical coefficient of x.

Now you try Example 4.

Example 4. Identify the numerical coefficient for each variable term of the expression $-8a - b + c$.

Solution

For $-8a$, ______ is the numerical coefficient of ______ .
For $-b$, -1 is the numerical coefficient of ______ .
For c, ______ is the numerical coefficient of c.

Check your work on page 166. ▶

Variable terms are called **like terms** if their variable parts are *exactly* the same. All constant terms are like terms. In the expression $6x + 5x + 2x$ the terms are all *like terms*, because the variable part of each term is x. But in the expression $-8a - b + c$ the terms are all *unlike terms*, because their variable parts are different.

Example 5. Identify the like terms in the expression

$$6x + 7y - 2 - 5x - 2y + 8$$

Solution

$6x$ and $-5x$ are like terms.

$7y$ and $-2y$ are like terms.

-2 and 8 are like terms.

You try Example 6.

Example 6. Identify the like terms in the expression

$$6A + 3a - 5A - a$$

Solution

$6A$ and ______ are like terms.

______ and ______ are like terms.

Check your work on page 166. ▶

Trial Run

______ 1. What are the terms in the expression $3x - 2y - 7$?

______ 2. In the expression $7m - 9n + 11$, identify the constant term and the variable terms.

______ 3. Identify the numerical coefficient of each variable term in the expression $-3a + b - 9$.

______ 4. Identify like terms in the expression $-10x + 5 + 6x - 9$.

Answers are on page 166.

Adding Terms

If you were asked, without any hints, to simplify the sum $6x + 5x + 2x$, what would your answer be? If you think the answer is $13x$, you are absolutely correct.

The key is to look for like terms. A sum of terms can be simplified *only if* it contains *like terms*. The process of adding like terms together is called **combining like terms**. The *distributive property* gives us the method for combining like terms.

$6x + 5x + 2x$	
$= (6 + 5 + 2)x$	Use distributive property.
$= 13x$	Simplify the sum.

It is not necessary to show the step in which you use the distributive property if you realize that the process can be described as follows.

> **Combining Like Terms.** To combine like terms, we add the numerical coefficients and keep the same variable part.

Combine like terms in this expression.

$5x + 12y + 3 + 8x - 6y - 1$	Notice which terms are like terms.
$5x + 8x + 12y - 6y + 3 - 1$	Rearrange so that like terms are next to each other.
$13x + 6y + 2$	Combine the coefficients of the like terms.

Try combining like terms in Example 7.

Example 7. Combine like terms in

$$17x + x - 20x$$

Solution

$$17x + x - 20x$$
$$= 17x + 1x + (\underline{\quad}x)$$
$$= \underline{\quad}x$$

Check your work on page 166. ▶

Example 8. Combine like terms in

$$x - (-2y) + 5 - 2x - 9y$$

Solution

$$x - (-2y) + 5 - 2x - 9y$$
$$= x + 2y + 5 + (-2x) + (-9y)$$
$$= 1x + (-2x) + 2y + (-9y) + 5$$
$$= -1x + (-7y) + 5$$
$$= -x - 7y + 5$$

Notice that it is customary to write simplified expressions with just *one sign* between the terms. Also notice that we usually write

$$1x \quad \text{as} \quad x$$
$$-1x \quad \text{as} \quad -x$$

We agree that 1 or -1 is ''understood'' to be the numerical coefficient in such cases.

Trial Run

Combine like terms.

_______ 1. $3x + x + 12x$

_______ 2. $-4x + 5 + 7x - 13$

________ **3.** $-12x + 7y + 3x - 15y$ ________ **4.** $6A + 3a - 5A - a$

________ **5.** $3a + 2b - 5a - 7b - 1$ ________ **6.** $3x - 4y + 9z - 2x + 4y + 6$

Answers are on page 166.

▶ Examples You Completed

Example 2. Identify the terms in the expression

$$5x - 3y - 10$$

Solution: We may write this expression as the sum

$$5x + (-3y) + (-10)$$

Then we see that

$5x$ is a term

$-3y$ is a term

-10 is a term

Example 4. Identify the numerical coefficient for each variable term of the expression $-8a - b + c$.

Solution

For $-8a$, -8 is the numerical coefficient of a.
For $-b$, -1 is the numerical coefficient of b.
For c, 1 is the numerical coefficient of c.

Example 6. Identify the like terms in the expression

$$6A + 3a - 5A - a.$$

Solution

$6A$ and $-5A$ are like terms.

$3a$ and $-a$ are like terms.

Example 7. Combine like terms in

$$17x + x - 20x$$

Solution

$$17x + x - 20x$$
$$= 17x + 1x + (-20x)$$
$$= -2x$$

Answers to Trial Runs

page 164 **1.** $3x, -2y, -7$ **2.** $7m$ and $-9n$ are variable terms; 11 is the constant term
3. -3 is the numerical coefficient of a; 1 is the numerical coefficient of b
4. $-10x$ and $6x$; 5 and -9

page 165 **1.** $14x$ **2.** $3x - 8$ **3.** $-9x - 8y$ **4.** $A + 2a$ **5.** $-2a - 5b - 1$ **6.** $x + 9z + 6$

Name ________________________________ Date ______________

EXERCISE SET 3.2

Identify the terms in the following expressions.

______ **1.** $3x - 5$

______ **2.** $4x - 7$

______ **3.** $\frac{1}{2}a - 7b + 9$

______ **4.** $\frac{1}{3}a - 2b - 2$

______ **5.** $1 - 5m + 6m$

______ **6.** $-1 - 9m + 4n$

______ **7.** $13x - 5y + 2z - 7$

______ **8.** $14x - 12y + 5z - 3$

Identify the constant terms and variable terms.

______ **9.** $4x - 19$

______ **10.** $7x + 23$

______ **11.** $2x + 3y$

______ **12.** $6x - 5y$

______ **13.** $-4a - 7b + 5.2$

______ **14.** $-5a + 11b - 8.4$

______ **15.** $1 + 7m - 5n + 14p - 4$

______ **16.** $2 + 5m - 9n + 3p - 5$

Identify the numerical coefficient of each variable term.

______ **17.** $3x + 7$

______ **18.** $5x - 13$

______ **19.** $\frac{1}{4}x - 12y$

______ **20.** $-4x + \frac{1}{5}y$

______ **21.** $3a + 4b - 2$

______ **22.** $8a - 5b + 9$

______ **23.** $0.3m - n - 2p$

______ **24.** $10m + n - 0.4p$

Identify like terms.

______ **25.** $8a - 3a$

______ **26.** $7a - 5a$

______ **27.** $3h - 2k + 5h + 5k$

______ **28.** $2h - 5k + 6h - 8k$

______ **29.** $\frac{2}{5}a - 6 + 3a - 9$

______ **30.** $\frac{2}{3}a - 8 + 2a - 15$

______ **31.** $3x - 0.24y + 1 + 4y - 1.2x - 7$

______ **32.** $x - 2.5y + 2 - 5y + 0.4x + 3$

Simplify.

______ **33.** $3a - 5a$

______ **34.** $5a - 7a$

______ **35.** $2x - 8x + 9x$

______ **36.** $3x - 11x + 10x$

_______ 37. $9 - 0.2y - 0.3y$

_______ 38. $7 - 0.4y - 0.5y$

_______ 39. $4m - 6 + 11m + 9$

_______ 40. $5m - 9 + 8m + 1$

_______ 41. $3h - k + 5 + 4h + 2k - 9$

_______ 42. $2h - 5k - 1 + 7h + 3k - 11$

_______ 43. $4x - 3y + 13z + 7x + 8y - 6z$

_______ 44. $2x + 5y - 6z + 3x - y - 3z$

_______ 45. $a + 10b - 9a + 7$

_______ 46. $2a - 11b - 6a + 3$

_______ 47. $\frac{2}{3}m - 5n + 4p + \frac{1}{3}m - 6n$

_______ 48. $\frac{3}{4}m - 3n + p + \frac{1}{4}m - 8n$

_______ 49. $5u - 3v + w + 2v - 5u - 7w$

_______ 50. $6u - 2v + 4w + 2v - 4u + 5w$

☆ Stretching the Topics

Simplify.

_______ 1. $0.7a - 3.21b - 0.23c - 1.5a + 5b + 0.45c$

_______ 2. $\frac{1}{2}x - 5y + \frac{4}{5}x - \frac{2}{3}y + 2z - 2x - \frac{1}{4}z$

_______ 3. $-25m + 16 - 73n + 24p - 58 + 32m + 93n - 29p + 42 + 13m$

Check your answers in the back of your book.

If you can complete **Checkup 3.2**, you are ready to go to Section 3.3.

Name ______________________ Date __________

CHECKUP 3.2

______ **1.** Identify the terms in the expression $3 - 7x + 2y$.

______ **2.** Identify the constant terms and the variable terms in the expression $-4a + 11 + 7a - 13$.

______ **3.** Identify the numerical coefficient of each variable term in the expression $3x - y - 4z + 7$.

______ **4.** Identify the like terms in the expression $5x - 2y + 4 - 7x + 2y - 1$.

Simplify.

______ **5.** $3a - 7a$

______ **6.** $x - 2x + 5x$

______ **7.** $5 - 3y + 11y$

______ **8.** $4m - 5 - 3m + 7$

______ **9.** $2x + 7y - 3x - 6$

______ **10.** $4a - 2b + c - 3b - 4a + c$

Check your answers in the back of your book.

If You Missed Problems:	You Should Review Examples:
1	1, 2
2, 3	3, 4
4	5, 6
5, 6	7
7–10	8

3.3 Working with Symbols of Grouping

Parentheses and brackets provide directions for operating with expressions, and they must be treated carefully.

Multiplying or Dividing a Term by a Constant

An expression such as $3(5x)$ means to multiply 3 times $5x$. The associative property for multiplication permits us to write

$$3(5x) = (3 \cdot 5)x \quad \text{Use the associative property.}$$
$$= 15x \quad \text{Simplify the product.}$$

> To multiply a constant times a variable term, multiply the constant times the numerical coefficient and keep the variable part.

Example 1. Multiply $3(-7x)$.

Solution

$$3(-7x) = [3\,(-7)]x$$
$$= -21x$$

You try Example 2.

Example 2. Multiply $-2(-8a)$.

Solution

16A

Check your work on page 176. ▶

Example 3. Multiply $(-2)(3)(5x)$.

Solution

$$(-2)(3)(5x)$$
$$= (-2 \cdot 3)(5x)$$
$$= -6(5x)$$
$$= (-6 \cdot 5)x$$
$$= -30x$$

You try Example 4.

Example 4. Multiply $7(-4)(-x)$.

Solution

$$7(-4)(-x)$$
$$= 7(-4)(-1x)$$

-28
28X

Check your work on page 176. ▶

Example 5. Simplify $4x - 2(3x)$.

Solution

$$4x - 2(3x)$$
$$= 4x - (2 \cdot 3)x$$
$$= 4x - 6x$$
$$= 4x + (-6x)$$
$$= -2x$$

Now you try Example 6.

Example 6. Simplify $2(5a) + 8a$.

Solution

10A+8A
18A

Check your work on page 176. ▶

To divide a variable term by a constant, we can rewrite the quotient.

$$\frac{45x}{5} = \frac{45}{5} \cdot x = 9x$$

$$\frac{-39x}{3} = \frac{-39}{3} \cdot x = -13x$$

> To divide a variable term by a constant, divide the numerical coefficient by the constant and keep the variable part.

Example 7. Divide $\frac{16x}{-2}$.

Solution

$$\frac{16x}{-2}$$

$$= \frac{16}{-2} \cdot x$$

$$= -8x$$

You try Example 8.

Example 8. Divide $\frac{-28x}{-4}$.

Solution

Check your work on page 176. ▶

Example 9. Divide $\frac{-x}{-1}$.

Solution

$$\frac{-x}{-1}$$

$$= \frac{-1x}{-1}$$

$$= \frac{-1}{-1} \cdot x$$

$$= 1x$$

$$= x$$

You try Example 10.

Example 10. Divide $\frac{7a}{7}$.

Solution

Check your work on page 176. ▶

Trial Run

Simplify.

______ 1. $4(5x)$

______ 2. $-3(-8x)$

______ 3. $-11(3x)$

______ 4. $(-2)(5)(7x)$

________ **5.** $3(-2x) + 10x$

________ **6.** $\dfrac{-32x}{-4}$

________ **7.** $\dfrac{x}{-1}$

________ **8.** $\dfrac{-6a}{6}$

Answers are on page 177.

Multiplying a Constant Times a Sum

In working with variable expressions, the **distributive property for multiplication over addition** will be one of our most valuable tools.

> **Distributive Property**
>
> $A(B + C) = A \cdot B + A \cdot C$

Example 11. Simplify $4(x + 2)$.

Solution

$$4(x + 2)$$
$$= 4 \cdot x + 4 \cdot 2$$
$$= 4x + 8$$

Example 12. Simplify $\frac{1}{2}(2x - 6y)$.

Solution

$$\frac{1}{2}(2x - 6y)$$
$$= \frac{1}{2} \cdot 2x - \frac{1}{2} \cdot 6y$$
$$= \left(\frac{1}{2} \cdot 2\right)x - \left(\frac{1}{2} \cdot 6\right)y$$
$$= 1x - 3y$$
$$= x - 3y$$

Perhaps it is time to observe some patterns and take some shortcuts. Let's work some more examples and use *arrows* to indicate our thoughts.

Example 13. Simplify $-4(3x - 7y)$.

Solution

$-4(-7y)$

$-4 \cdot 3x$

$$-4(3x - 7y) = -4(3x + -7y)$$
$$= -12x + 28y$$

You try arrows in Example 14.

Example 14. Simplify $5(6x - 3y)$.

Solution

$$5(6x - 3y) = 5(6x - 3y)$$
$$= \underline{\ 30\ }x - \underline{\ 15\ }y$$

Check your work on page 176. ▶

Trial Run

Simplify.

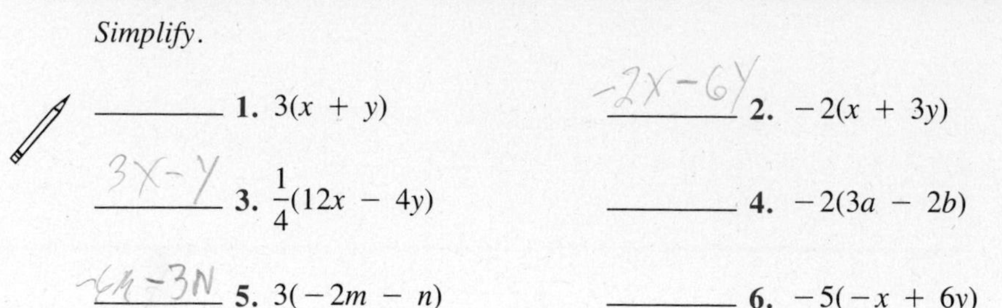

______ 1. $3(x + y)$ ______ 2. $-2(x + 3y)$

______ 3. $\frac{1}{4}(12x - 4y)$ ______ 4. $-2(3a - 2b)$

______ 5. $3(-2m - n)$ ______ 6. $-5(-x + 6y)$

Answers are on page 177.

Using Symbols of Grouping

The algebraic expression $(4x + 5y)$ can be thought of as an example of the distributive property if we agree that 1 is understood to be in front of the parentheses. In other words,

$$(4x + 5y) = 1(4x + 5y)$$

$1 \cdot 5y$

$1 \cdot 4x$

$$= 1(4x + 5y)$$
$$= 4x + 5y$$

We see, then, that to simplify $(4x + 5y)$, we merely "drop the parentheses."

Similarly, we should agree that we can use the indicated steps to simplify the following expression.

$(4x + 5y) - (3x + 2y)$

$= 1(4x + 5y) - 1(3x + 2y)$	Insert 1 before each quantity.
$= 4x + 5y - 3x - 2y$	Drop the first pair of parentheses. Multiply the second quantity by -1.
$= 4x - 3x + 5y - 2y$	Rearrange terms.
$= x + 3y$	Combine like terms.

Now try Example 15.

Example 15. Simplify

$$(8x + 7) - (2x + 3)$$

Solution

$$(8x + 7) - (2x + 3)$$
$$= 1(8x + 7) - 1(2x + 3)$$

Check your work on page 177. ▶

Example 16. Simplify

$$(9a - 6c) - (10a - 5c)$$

Solution

$$(9a - 6c) - (10a - 5c)$$
$$= 1(9a - 6c) - 1(10a - 5c)$$
$$= 9a - 6c - 10a + 5c$$
$$= 9a - 10a - 6c + 5c$$
$$= -a - c$$

Example 17. Simplify

$$-2(5a - 3b) - 7(a - 3b)$$

Solution

$$\begin{aligned} & -2(5a - 3b) - 7(a - 3b) \\ &= -10a + 6b - 7a + 21b \\ &= -10a - 7a + 6b + 21b \\ &= -17a + 27b \end{aligned}$$

Now try Example 18.

Example 18. Simplify

$$3(x - 5y) + 4(-x + 3y)$$

Solution

Check your work on page 177. ▶

In simplifying expressions containing brackets and parentheses, we shall continue the practice of working with the innermost symbols of grouping first. We shall also use our shortcuts.

Example 19. Simplify $2[5x - 3(x - 1)]$.

Solution

$2[5x - 3(x - 1)]$	
$= 2[5x - 3x + 3]$	Remove innermost parentheses (distributive property).
$= 2[2x + 3]$	Find the sum within brackets.
$= 4x + 6$	Remove brackets (distributive property).

Example 20. Simplify $-4[2(a - 3b) - (-3a - 2b)]$.

Solution

$-4[2(a - 3b) - (-3a - 2b)]$	
$= -4[2(a - 3b) - 1(-3a - 2b)]$	Insert 1 before second pair of parentheses.
$= -4[2a - 6b + 3a + 2b]$	Remove parentheses (distributive property).
$= -4[2a + 3a - 6b + 2b]$	Rearrange terms in brackets.
$= -4[5a - 4b]$	Combine like terms within brackets.
$= -20a + 16b$	Remove brackets (distributive property).

Trial Run

Simplify.

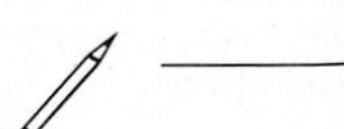

_______ **1.** $-(3x - 5y)$

_______ **2.** $-(-2a - 11b + 2)$

_______ **3.** $(3x + 6) - (2x + 3)$

_______ **4.** $(8a - 5b) - (-a - 3b)$

_______ **5.** $4(2m - 3) - 2(3m + 1)$

_______ **6.** $3[2x - (x - 2)]$

_______ **7.** $3[-2(3x - 4) - (x + 1)]$

_______ **8.** $5[3x - (x + 2)] - 2[2(x - 1) - 4]$

Answers are on page 177.

▶ Examples You Completed

Example 2. Multiply $-2(-8a)$.

Solution

$$-2(-8a) = [-2(-8)]a$$
$$= 16a$$

Example 4. Multiply $7(-4)(-x)$.

Solution

$$7(-4)(-x)$$
$$= 7(-4)(-1x)$$
$$= [7(-4)](-1x)$$
$$= (-28)(-1x)$$
$$= [-28(-1)]x$$
$$= 28x$$

Example 6. Simplify $2(5a) + 8a$.

Solution

$$2(5a) + 8a$$
$$= (2 \cdot 5)a + 8a$$
$$= 10a + 8a$$
$$= 18a$$

Example 8. Divide $\frac{-28x}{-4}$.

Solution

$$\frac{-28x}{-4}$$
$$= \frac{-28}{-4} \cdot x$$
$$= 7x$$

Example 10. Divide $\frac{7a}{7}$.

Solution

$$\frac{7a}{7}$$
$$= \frac{7}{7} \cdot a$$
$$= 1a$$
$$= a$$

Example 14. Simplify $5(6x - 3y)$.

Solution

$$5(6x - 3y) = 5(6x - 3y)$$
$$= 30x - 15y$$

Example 15. Simplify $(8x + 7) - (2x + 3)$.

Solution

$$
\begin{aligned}
&(8x + 7) - (2x + 3) \\
&= 1(8x + 7) - 1(2x + 3) \\
&= 8x + 7 - 2x - 3 \\
&= 8x - 2x + 7 - 3 \\
&= 6x + 4
\end{aligned}
$$

Example 18. Simplify

$$3(x - 5y) + 4(-x + 3y).$$

Solution

$$
\begin{aligned}
&3(x - 5y) + 4(-x + 3y) \\
&= 3x - 15y - 4x + 12y \\
&= 3x - 4x - 15y + 12y \\
&= -x - 3y
\end{aligned}
$$

Answers to Trial Runs

page 172 **1.** $20x$ **2.** $24x$ **3.** $-33x$ **4.** $-70x$ **5.** $4x$ **6.** $8x$ **7.** $-x$ **8.** $-a$

page 174 **1.** $3x + 3y$ **2.** $-2x - 6y$ **3.** $3x - y$ **4.** $-6a + 4b$ **5.** $-6m - 3n$ **6.** $5x - 30y$

page 175 **1.** $-3x + 5y$ **2.** $2a + 11b - 2$ **3.** $x + 3$ **4.** $9a - 2b$ **5.** $2m - 14$ **6.** $3x + 6$ **7.** $-21x + 21$ **8.** $6x + 2$

Name ______________________ Date ____________

EXERCISE SET 3.3

Simplify.

______ 1. $2(4x)$

______ 2. $3(7x)$

______ 3. $3(-0.5)$

______ 4. $4(-0.2)$

______ 5. $-12\left(\frac{3}{4}\right)$

______ 6. $-9\left(\frac{2}{3}m\right)$

______ 7. $-3(-2x)$

______ 8. $-2(-4x)$

______ 9. $(-2)\left(\frac{1}{2}\right)(3y)$

______ 10. $(-3)\left(\frac{1}{3}\right)(2y)$

______ 11. $(-3)(2)(-5k)$

______ 12. $(-4)(3)(-3k)$

______ 13. $\frac{20x}{-4}$

______ 14. $\frac{27x}{-3}$

______ 15. $\frac{-45x}{-9}$

______ 16. $\frac{-34x}{-2}$

______ 17. $\frac{5a}{5}$

______ 18. $\frac{7a}{7}$

______ 19. $\frac{-m}{-1}$

______ 20. $\frac{m}{-1}$

______ 21. $\frac{-10y}{2}$

______ 22. $\frac{-15y}{3}$

______ 23. $\frac{-7n}{7}$

______ 24. $\frac{-15n}{15}$

Simplify the expressions by removing grouping symbols and combining like terms.

______ 25. $5x + 3(2x)$

______ 26. $4x + 3(3x)$

______ 27. $2(5x) - 3x$

______ 28. $3(4x) - 7x$

______ 29. $\frac{1}{4}(a - b)$

______ 30. $\frac{1}{5}(a - b)$

______ 31. $-3(x + 2y)$

______ 32. $-4(x + 5y)$

______ 33. $1.5(6x - 2y)$

______ 34. $0.5(2x - 4y)$

______ 35. $-4(5a - 7b)$

______ 36. $-6(a - 2b)$

______ 37. $9(-2m - 3n)$

______ 38. $8(-3m - 5n)$

_______ **39.** $-3(-x + 4y)$

_______ **40.** $-7(-x + 2y)$

_______ **41.** $-\left(\frac{1}{2}x - 5y\right)$

_______ **42.** $-\left(\frac{1}{3}x - 2y\right)$

_______ **43.** $-(-3a - 7b + 5.4)$

_______ **44.** $-(-4a - 6b + 3.2)$

_______ **45.** $(3x + 5) - (2x + 4)$

_______ **46.** $(5x + 3) - (3x + 1)$

_______ **47.** $(12a + 3b) - (5a - 2b)$

_______ **48.** $(9a + 4b) - (7a - b)$

_______ **49.** $(-4m - 3n) - (2m - 3n)$

_______ **50.** $(-5m - 2n) - (m - 2n)$

_______ **51.** $(7x - 3y) - (-2x + y)$

_______ **52.** $(8x - 5y) - (-3x + 2y)$

_______ **53.** $3(5m - 4) - 4(m + 2)$

_______ **54.** $5(2m - 1) - 2(m + 5)$

_______ **55.** $-3(x - 2y) + 4(-x + y)$

_______ **56.** $-2(2x - y) + 3(-x + 2y)$

_______ **57.** $4[2a - (a - 3.5)]$

_______ **58.** $5[a - (2a - 1.4)]$

_______ **59.** $5[(3x - 2y) - (x + y)]$

_______ **60.** $7[(2x - 3y) - (x + 2y)]$

_______ **61.** $-2[5(a - 2b) + 7(2a + b)]$

_______ **62.** $-3[4(2a - b) + 5(a + 3b)]$

_______ **63.** $4[-3(x - 4) - 2(x + 6)]$

_______ **64.** $5[-6(x - 1) - 3(x + 2)]$

☆ Stretching the Topics

Simplify.

_______ **1.** $0.4(2x - 3y) - 0.5(-y - 3) + 1.2(2x - 1)$

_______ **2.** $-3[(2a - b) - 5(4a + b)] + 7[-2(3a + b) + (b - 3a)]$

_______ **3.** $9 - 3\{4x - 7[5x - (x - 2) + 6] - [2x + (4x - 5)] + 9\}$

Check your answers in the back of your book.

If you can simplify the expressions in **Checkup 3.3**, you are ready to do the **Review Exercises for Chapter 3**.

Name ______________________ Date __________

CHECKUP 3.3

Simplify.

______ 1. $4(-8x)$

______ 2. $(-2)(4)(-5y)$

______ 3. $\frac{48x}{-12}$

______ 4. $\frac{10a}{-10}$

______ 5. $7x - 3(4x)$

______ 6. $5(-2x + 3y)$

______ 7. $(4x - 5) - (2x + 3)$

______ 8. $2(3a - 6b) - 3(-2a + b)$

______ 9. $-6[4a - (3a + 1)]$

______ 10. $7[3(2m - n) - 2(3m - 5n)]$

Check your answers in the back of your book.

If You Missed Problem:	You Should Review Examples:
1	1, 2
2	3, 4
3	7, 8
4	7, 10
5	5, 6
6	11–14
7	15, 16
8	17, 18
9	19
10	20

Summary

In this chapter we learned to recognize, evaluate, and simplify algebraic expressions.

In Order to:	We Must:	Example
Evaluate an algebraic expression for some value of the variable	Substitute the given value for the variable in the expression.	If $x = 2$, then $1 - 3x$ $= 1 - 3(2)$ $= 1 - 6$ $= -5$
Combine like terms in an expression	Add the numerical coefficients and keep the same variable part.	$3x + 2y - 7x$ $= 3x - 7x + 2y$ $= -4x + 2y$
Multiply a constant times a variable term	Multiply the constant times the numerical coefficient and keep the same variable part.	$-5(2x)$ $= (-5 \cdot 2)x$ $= -10x$
Divide a variable term by a constant	Divide the numerical coefficient by the constant and keep the same variable part.	$\frac{-21a}{-7}$ $= \frac{-21}{-7} \cdot a$ $= 3a$
Multiply a constant times a sum of terms	Use the distributive property and simplify.	$3(-4x + y)$ $= 3(-4x) + 3y$ $= -12x + 3y$

❑ Speaking the Language of Algebra

1. In the expression $2x - 3y + 5$, 2 is the ________ ________ of x, -3 is the ________ ________ of y, and 5 is the ________ term.
2. To combine like terms, we add the ________ ________ and keep the same variable part.
3. In the expression $x + y + 5x - 6y$, we say that x and $5x$ are ________ ________ because their variable parts are exactly the same.
4. When we rewrite $2(x + 5)$ as $2x + 10$, we are using the ________ property.
5. When we begin to remove symbols of grouping, we always work with the ________ symbols of grouping first.

Check your answers in the back of your book.

Name ____________________ Date ____________

REVIEW EXERCISES for Chapter 3

Write an algebraic expression for each.

____ **1.** Multiply 6 times x.

____ **2.** The sum of y and 9.

____ **3.** Ten more than 3 times m.

____ **4.** Four times k is subtracted from 23.7.

____ **5.** Nine times the sum of x and 7.

Evaluate each expression.

____ **6.** $x - 9$ when $x = 2$; when $x = -12$

____ **7.** $\frac{56}{n}$ when $n = -7$; when $n = 0$

____ **8.** $3(9 - x)$ when $x = 11$; when $x = 9$

____ **9.** $2x - 7y$ when $x = 5$, $y = -1$

____ **10.** $\frac{4a - 12}{3}$ when $a = 3$; when $a = 0$

____ **11.** A hiker left Horseshoe Cave walking along the trail at 4 miles per hour. Write an algebraic expression for the distance traveled after h hours. Find the distance traveled after 3.5 hours.

____ **12.** Aunt Vina left her entire estate to her three nephews. If the estate is valued at x dollars, write an algebraic expression for each nephew's share. Find the value of each nephew's share if the estate is worth $210,000.

____ **13.** The cost of renting a cabin at the lake is $200 for the first week and $19.50 for each additional day. Write an algebraic expression for the total cost of the cabin after 1 week and d additional days. Find the total cost for staying 1 week and 3 days.

Identify the constant terms and the variable terms.

____ **14.** $3x - 12$

____ **15.** $-2x + 7y$

____ **16.** $4a - 3b + 2$

____ **17.** $3m - 2 + 4n - 9 - p$

Identify the numerical coefficient of each variable term.

____ **18.** $3x - 9$

____ **19.** $7x - 2y$

____ **20.** $-3a + b - 1$

____ **21.** $m - n + 2p$

Simplify each expression and combine like terms.

______ 22. $2x - 9x + 5x$

______ 23. $3m - 4 + 2m + 6$

______ 24. $2a - 13b - 9a + 3$

______ 25. $-4x - 3y + 7z + 8x + 3y - 2z$

______ 26. $4(-5x)$

______ 27. $(-2)(-3)(4y)$

______ 28. $\frac{54x}{9}$

______ 29. $\frac{-a}{-1}$

______ 30. $-3(5x) + 12x$

______ 31. $7(4a - 5b)$

______ 32. $-(7x - 2y)$

______ 33. $(-7x + 2) + (3x + 5)$

______ 34. $(-9m - 3n) - (7m - 3n)$

______ 35. $3[4a - (7a - 2)]$

______ 36. $5[-3(x + y) + 7(x - y)]$

Check your answers in the back of your book.

If You Missed Exercises:	You Should Review Examples:	
1–5	SECTION 3.1	1–4
6–10		7–10
11		11
12		13
13		14
14–21	SECTION 3.2	1–4
22–25		7, 8
26, 27	SECTION 3.3	1–4
28, 29		7–10
30		5, 6
31, 32		11–14
33, 34		17, 18
35, 36		19, 20

If you have completed the **Review Exercises** and have corrected your errors, you are ready to take the **Practice Test for Chapter 3**.

Name ____________________ Date __________

PRACTICE TEST for Chapter 3

	Section	Examples
Write an algebraic expression for each.		
______ **1.** Decrease m by 5.	3.1	2
______ **2.** Nine more than the product of x and 3.	3.1	3
______ **3.** Ten times the sum of b and 4.6.	3.1	6
Evaluate the expressions.		
______ **4.** $6x - 4$ when $x = 1$; when $x = -2$	3.1	7, 8
______ **5.** $-2(7 - x)$ when $x = 0$; when $x = -1$	3.1	10
______ **6.** $\frac{2r - 4s}{6}$ when $r = 5$, $s = -7$	3.1	9
______ **7.** If a hotel room costs $33.79 per night (including tax), write an algebraic expression for the cost of staying n nights. Find the cost of staying 5 nights.	3.1	11
______ **8.** A cable TV subscription salesperson is paid $200 per week plus $4.10 for every new subscriber. Write an algebraic expression for the weekly pay of a salesperson enrolling x new subscribers. Find the weekly pay if 30 new subscribers are enrolled.	3.1	14
______ **9.** The $180 rental of a banquet room is to be shared equally by the English Club members attending the banquet. Write an algebraic expression for each member's share if m members attend the banquet. Find each member's share if 45 members attend.	3.1	13
Simplify each expression and combine like terms.		
______ **10.** $5n - 6n - 3n$	3.2	7
______ **11.** $4x - 10 - 7x + 3$	3.2	8
______ **12.** $-5x - 2y + 3z + 5x - 2y - z$	3.2	8
______ **13.** $(-8)(-2)(-y)$	3.3	3, 4
______ **14.** $\frac{-18x}{-3}$	3.3	8
______ **15.** $-5(4n) + 8n$	3.3	6
______ **16.** $\frac{2}{3}(9x + 3)$	3.3	12

______ 17. $-6(3x - 2y)$	3.3	13
______ 18. $y - (7x - 3y)$	3.3	16
______ 19. $4[2a - (5a - 3)]$	3.3	19
______ 20. $7[-2(x - y) - 4(x + y)]$	3.3	20

Check your answers in the back of your book.

Name ________________________________ Date ______________

SHARPENING YOUR SKILLS after Chapter 1–3

		SECTION
______	1. Change 0.875 to a fraction and reduce.	1.4
______	2. Change $\frac{5}{6}$ to a decimal and round to hundredths.	1.4
______	3. Find the value of $\left(\frac{2}{3}\right)^3$.	1.3
______	4. Find 9% of 35.	1.4
______	5. Complete the statement $2 \cdot 3 - 7$ ____ -8 using $<$, $>$, or $=$.	2.1

Simplify.

______	6. $\dfrac{13(9 - 2)}{5 - \dfrac{15}{3}}$	1.2
______	7. $\frac{5}{8} + \frac{9}{10}$	1.3
______	8. $(2.36)(0.24)$	2.4
______	9. $(23 - 9) - (11 - 6)$	1.2
______	10. $3\left(2 \cdot 3 + \frac{15 - 3}{4}\right)$	1.2
______	11. $-4(7 - 10)$	2.3
______	12. $\dfrac{-2[3(-1) - (-9)]}{-13 + 10}$	2.3
______	13. $(-2)^4$	2.3
______	14. $\lvert -9 \rvert - \lvert 3 \rvert$	2.1
______	15. Write a number expression for the word expression and simplify. The sum of -13 and 12 is divided by 7.	2.5

Check your answers in the back of your book.

Solving First-Degree Equations 4

Charolotte now earns $6 per hour as a grounds keeper working 35 hours per week. She has been offered another job paying $280 per week for the same work. How much of an increase in her hourly wage must her present employer give Charolotte to meet the better offer?

Before you finish this chapter, you will be able to solve this problem using an **equation**. Equations are mathematical sentences which say that one quantity represents the same number as another quantity.

In this chapter we learn how to

1. Solve equations using one property.
2. Solve equations using several properties.
3. Switch from word statements to equations.

4.1 Solving Equations Using One Property

Equations may or may not contain variables. Using the symbol = to mean "is equal to" or "equals," we may write

$$2 + 3 = 6 - 1 \qquad x + 3 = 3 + x$$

$$2(5 - 7) = -4 \qquad 2(x + 5) = 2x + 10$$

and we know that each of these equations is *always* true. Such statements are called **identities**, but they are not our primary concern. Instead, we shall concentrate on statements which are sometimes, but not always, true.

Satisfying Equations

If an equation containing variables is true only for certain values of the variable, we call it a **conditional equation**.

> The **solution** for any equation is the value of the variable that makes the equation a true statement.

For example,

Equation	Solution	Reason
$x + 3 = 5$	$x = 2$	because $2 + 3 = 5$
$17 - y = 11$	$y = 6$	because $17 - 6 = 11$
$5x = 20$	$x = 4$	because $5 \cdot 4 = 20$
$\frac{x}{3} = 4$	$x = 12$	because $\frac{12}{3} = 4$

If we substitute a number for the variable in an equation and obtain a true statement, we say that the number *satisfies* the equation and that the number is a solution for the equation. In this chapter we are concerned with equations containing one variable with no exponent except 1. Such equations are called **first-degree equations**.

Example 1. Is -1 a solution for $1 - 3x = 4$?

Solution: Substituting -1 for x we have

$$1 - 3(-1) \stackrel{?}{=} 4$$

$$1 + 3 \stackrel{?}{=} 4$$

$$4 = 4$$

so -1 is a solution.

You complete Example 2.

Example 2. Is 3 a solution for $2x + 1 = 10 - x$?

Solution: Substituting 3 for x, we have

$$2(\_\_) + 1 \stackrel{?}{=} 10 - \_\_$$

$$\_\_\_\_\_\_ \stackrel{?}{=} 7$$

$$\_\_\_\_\_\_ = 7$$

so ____________________ .

Check your work on page 201. ▶

Example 3. Is -3 a solution for $4(5 - x) = 8$?

Solution: Substituting -3 for x, we have

$$4[5 - (-3)] \stackrel{?}{=} 8$$
$$4[5 + 3] \stackrel{?}{=} 8$$
$$4(8) \stackrel{?}{=} 8$$
$$32 \neq 8$$

so -3 is *not* a solution.

You try Example 4.

Example 4. Is -5 a solution for $\frac{2(y + 2)}{3} = -2$?

Solution: Substituting -5 for y, we have

$$\frac{2[(\quad) + 2]}{3} \stackrel{?}{=} -2$$

so ______________ .

Check your work on page 201. ▶

Trial Run

______ **1.** Is 3 a solution for $3x - 1 = 8$?

______ **2.** Is -5 a solution for $6 - 2y = 16$?

______ **3.** Is 9 a solution for $\frac{m}{3} + 2 = -7$?

______ **4.** Is 0 a solution for $2(a - 5) = -10$?

______ **5.** Is -1 a solution for $2x + 8 = x + 7$?

Answers are on page 202.

Finding Solutions by Addition and Subtraction

As we have already discovered, it is sometimes possible to find a solution just by looking at an equation. However, there are many equations that are too complicated to be solved by inspection. We need some steps that will help us find solutions in an orderly and accurate way.

We would like to have the variable (with a numerical coefficient of 1) all by itself on one side of the equation, and some number (the solution) on the other side. We wish to **isolate the variable**.

To isolate x in the equation

$$x + 6 = 10$$

we must get rid of the 6 being added to x. By the property of opposites we know that $6 + (-6) = 0$, so we can get rid of the 6 by *subtracting* 6 (or adding -6) on the left-hand side. But we cannot do something to one side of an equation without doing the same thing to the other side. If we subtract 6 (or add -6) on the left, we *must* also subtract 6 (or add -6) on the right.

$x + 6 = 10$	
$x + 6 - 6 = 10 - 6$	Subtract 6 from both sides.
$x + 0 = 4$	Simplify the sums.
$x = 4$	Simplify (addition property of 0).

To solve the equation

$$x - 9 = 3$$

we can get rid of the 9 being subtracted from x by *adding* 9 on the left. But we must also add 9 on the right at the same time. Here are the steps.

$x - 9 = 3$	
$x - 9 + 9 = 3 + 9$	Add 9 to both sides.
$x + 0 = 12$	Simplify the sums.
$x = 12$	Simplify (addition property of 0).

> We may add the same quantity to both sides of an equation. We may subtract the same quantity from both sides of an equation. In each case the resulting equation has the same solution as the original equation.

This property of equations can be stated in symbols.

Addition Property of Equality

If $\quad A = B$

then $\quad A + C = B + C$

and $\quad A - C = B - C$

Remember that we can always check our solution by substituting it into the original equation to see if it makes the original statement true. This is called **checking the solution**.

Example 5. Solve $x + 9 = 32$ and check.

Solution

$x + 9 = 32$	
$x + 9 - 9 = 32 - 9$	Subtract 9 from both sides.
$x + 0 = 23$	Simplify the sums.
$x = 23$	Simplify (addition property of 0).

CHECK:

$x + 9 = 32$	Original equation.
$23 + 9 \stackrel{?}{=} 32$	Substitute 23 for x.
$32 = 32$	Simplify.

Now you provide the reason for each step in Example 6.

Example 6. Solve $x - 15.2 = 100$ and check.

Solution

$$x - 15.2 = 100$$

$$x - 15.2 + 15.2 = 100 + 15.2 \qquad \text{\_\_\_\_\_\_\_\_\_\_\_\_\_\_\_\_}$$

$$x + 0 = 115.2 \qquad \text{\_\_\_\_\_\_\_\_\_\_\_\_\_\_\_\_}$$

$$x = 115.2 \qquad \text{\_\_\_\_\_\_\_\_\_\_\_\_\_\_\_\_}$$

CHECK:

$$x - 15.2 = 100 \qquad \text{\_\_\_\_\_\_\_\_\_\_\_\_\_\_\_\_}$$

$$115.2 - 15.2 \stackrel{?}{=} 100 \qquad \text{\_\_\_\_\_\_\_\_\_\_\_\_\_\_\_\_}$$

$$100 = 100 \qquad \text{\_\_\_\_\_\_\_\_\_\_\_\_\_\_\_\_}$$

Check your reasons on page 201. ▶

Example 7. Solve $-5 + y = -3$ and check.

Solution

$$-5 + y = -3$$

$$5 - 5 + y = 5 - 3$$

$$0 + y = 2$$

$$y = 2$$

CHECK:

$$-5 + y = -3$$

$$-5 + 2 \stackrel{?}{=} -3$$

$$-3 = -3$$

You try Example 8.

Example 8. Solve $y + 7 = -2$ and check.

Solution

$$y + 7 = -2$$

CHECK:

$$y + 7 = -2$$

Check your work on page 201. ▶

Trial Run

Solve and check.

______ 1. $x + 3 = 8$

______ 2. $6 = y - 4$

______ 3. $3 + m = 3$

______ 4. $-5 = a + 4$

______ 5. $-8 + x = -4$

______ 6. $9 = -2 + x$

Answers are on page 202.

Finding Solutions by Multiplication and Division

Consider the equation

$$3x = 15$$

We must decide how to *isolate the variable* by getting rid of the 3 that is being multiplied times the variable. We would like to operate on both sides of the equation so that we are left with $1x$ or x by itself.

Look at what happens if we *divide* both sides of this equation by 3.

$$3x = 15$$

$$\frac{3x}{3} = \frac{15}{3}$$

$$\frac{3}{3} \cdot x = \frac{15}{3}$$

$$1x = 5$$

$$x = 5$$

Let's check this solution in the original equation.

$$3x = 15$$

$$3(5) \stackrel{?}{=} 15$$

$$15 = 15$$

Look at what happens if we *multiply* both sides of this equation by $\frac{1}{3}$.

$$3x = 15$$

$$\frac{1}{3}(3x) = \frac{1}{3}(15)$$

$$\left(\frac{1}{3} \cdot 3\right) x = \frac{1}{3}(15)$$

$$1x = 5$$

$$x = 5$$

Notice that both methods give the same solution for the equation.

To solve an equation in which the variable is being multiplied by a constant, we may:

1. Divide both sides by the coefficient of the variable; or
2. Multiply both sides by the *reciprocal* of the coefficient of the variable.

This property of equations may be stated in symbols.

Multiplication Property of Equality

If $A = B$

then $AC = BC$

and $\frac{A}{C} = \frac{B}{C} \quad (C \neq 0)$

Example 9. Solve $-6x = 24$ and check.

Solution

$$-6x = 24$$

$$\frac{-6x}{-6} = \frac{24}{-6} \qquad \text{Divide both sides by } -6.$$

$$\frac{-6}{-6} \cdot x = \frac{24}{-6} \qquad \text{Rewrite.}$$

$$1x = -4 \qquad \text{Simplify (remember: } \frac{A}{A} = 1\text{).}$$

$$x = -4 \qquad \text{Simplify (multiplication property of 1).}$$

CHECK:

$-6x = 24$	Original equation.
$-6(-4) \stackrel{?}{=} 24$	Substitute -4 for x.
$24 = 24$	Find the product.

Now you provide the reason for each step in Example 10.

Example 10. Solve $-24 = \dfrac{-12}{5}x$ and check.

Solution

$$-24 = \frac{-12}{5}x$$

$$\frac{-5}{12} \cdot \frac{-24}{1} = \frac{-5}{12}\left(\frac{-12}{5}x\right) \qquad \underline{\hspace{6cm}}$$

$$\frac{-5}{\cancel{12}_{1}} \cdot \frac{\cancel{-24}^{-2}}{1} = \left(\frac{\cancel{-5}^{-1}}{\cancel{12}_{1}} \cdot \frac{\cancel{-12}^{-1}}{\cancel{5}_{1}}\right)x \qquad \text{Rewrite right side (associative property).}$$

$$10 = 1x \qquad \underline{\hspace{6cm}}$$

$$10 = x \qquad \underline{\hspace{6cm}}$$

CHECK:

$$-24 = \frac{-12}{5}x \qquad \underline{\hspace{6cm}}$$

$$-24 \stackrel{?}{=} \frac{-12}{5}(10) \qquad \underline{\hspace{6cm}}$$

$$-24 = -24 \qquad \underline{\hspace{6cm}}$$

Check your reasons on page 201. ▶

Example 11. Solve $-x = 13$ and check.

Solution

$$-x = 13$$

$$-1x = 13$$

$$\frac{-1x}{-1} = \frac{13}{-1}$$

$$\frac{-1}{-1} \cdot x = \frac{13}{-1}$$

$$x = -13$$

CHECK:

$$-x = 13$$

$$-(-13) \stackrel{?}{=} 13$$

$$13 = 13$$

Trial Run

Solve and check.

________ 1. $4x = 24$

________ 2. $25 = -5x$

________ 3. $3x = 0$

________ 4. $-x = 7$

________ 5. $\frac{1}{3}a = 6$

________ 6. $\frac{-7}{2}x = 21$

Answers are on page 202.

We now know how to solve an equation in which the variable is being *multiplied* by a constant. But how can we get rid of a number that is *dividing* the variable? Consider the equation

$$\frac{x}{2} = 9$$

$\frac{1}{2}x = 9$	Rewrite the quotient as a product (definition of division).
$\frac{2}{1}\left(\frac{1}{2}x\right) = \frac{2}{1} \cdot 9$	Multiply both sides by the reciprocal of the coefficient of x.
$\left(\frac{2}{1} \cdot \frac{1}{2}\right)x = 2 \cdot 9$	Associative property.
$1 \cdot x = 18$	Find the products.
$x = 18$	Simplify.

This method can be shortened if we realize that we can use multiplication to ''un-do'' division.

$$\frac{x}{2} = 9$$

$\frac{2}{1}\left(\frac{x}{2}\right) = 2 \cdot 9$	Multiply both sides by the denominator, 2.
$x = 18$	Find the products.

In each method we solved the equation by the *multiplication property of equality*. From the examples below you will see that we can choose whichever method seems more appropriate.

Example 12. Solve $\frac{-x}{3} = 21$ and check.

Solution: Since the constant -1 is multiplying x *and* the constant 3 is dividing x, we choose to rewrite the left side before solving the equation.

$$\frac{-x}{3} = 21$$

$$\frac{-1}{3}x = 21 \qquad \text{Rewrite the quotient as a product.}$$

$$\frac{-3}{1}\left(\frac{-1}{3}\right)x = -3(21) \qquad \text{Multiply both sides by the reciprocal of the coefficient of } x.$$

$$\left(\frac{-3}{1} \cdot \frac{-1}{3}\right)x = -3(21) \qquad \text{Use the associative property.}$$

$$x = -63 \qquad \text{Find the products.}$$

CHECK:

$$\frac{-x}{3} = 21 \qquad \text{Original equation.}$$

$$\frac{-(-63)}{3} \stackrel{?}{=} 21 \qquad \text{Substitute } -63 \text{ for } x.$$

$$\frac{63}{3} \stackrel{?}{=} 21 \qquad \text{Remove parentheses.}$$

$$21 = 21 \qquad \text{Find the quotient.}$$

Example 13. Solve $\frac{a}{2.5} = -8$ and check.

Solution: Since the variable a is only being divided here by 2.5, we choose to use the shorter method.

$$\frac{a}{2.5} = -8$$

$$\frac{2.5}{1}\left(\frac{a}{2.5}\right) = 2.5(-8) \qquad \text{Multiply both sides by the denominator 2.5.}$$

$$a = -20 \qquad \text{Find the products.}$$

CHECK:

$$\frac{a}{2.5} = -8 \qquad \text{Original equation.}$$

$$\frac{-20}{2.5} \stackrel{?}{=} -8 \qquad \text{Substitute } -20 \text{ for } a.$$

$$-8 = -8 \qquad \text{Find the quotient.}$$

Trial Run

Solve and check.

_______ 1. $\frac{x}{3} = -2$

_______ 2. $4 = \frac{2x}{7}$

_______ 3. $\frac{-x}{2.4} = 5$

_______ 4. $\frac{x}{6} = 0$

Answers are on page 202.

Summarizing, we have discovered that in finding the solution of an equation we are permitted to

1. Add the same quantity to both sides.
2. Subtract the same quantity from both sides.
3. Multiply both sides by the same quantity (not zero).
4. Divide both sides by the same quantity (not zero).

We can use these properties to "un-do" an operation that is being performed on the variable in order to isolate that variable.

To "Un-do":	We Use:	Example	Method
Addition	Subtraction	$x + 3 = 6$	Subtract 3
Subtraction	Addition	$x - 3 = 6$	Add 3
Multiplication	Division	$3x = 6$	Divide by 3
Division	Multiplication	$\frac{x}{3} = 6$	Multiply by 3

▶ Examples You Completed

Example 2. Is 3 a solution for $2x + 1 = 10 - x$?

Solution: Substituting 3 for x, we have

$$2(3) + 1 \overset{?}{=} 10 - 3$$
$$6 + 1 \overset{?}{=} 7$$
$$7 = 7$$

so 3 is a solution.

Example 4. Is -5 a solution for $\frac{2(y + 2)}{3} = -2$?

Solution: Substituting -5 for y, we have

$$\frac{2[(-5) + 2]}{3} \overset{?}{=} -2$$
$$\frac{2(-3)}{3} \overset{?}{=} -2$$
$$2(-1) \overset{?}{=} -2$$
$$-2 = -2$$

so -5 is a solution.

Example 6. Solve $x - 15.2 = 100$ and check.

Solution

$x - 15.2 = 100$	
$x - 15.2 + 15.2 = 100 + 15.2$	Add 15.2 to both sides.
$x + 0 = 115.2$	Simplify.
$x = 115.2$	Simplify.

CHECK:

$x - 15.2 = 100$	Original equation.
$115.2 - 15.2 \overset{?}{=} 100$	Substitute 115 for x.
$100 = 100$	Simplify.

Example 8. Solve $y + 7 = -2$ and check.

Solution

$$y + 7 = -2$$
$$y + 7 - 7 = -2 - 7$$
$$y + 0 = -9$$
$$y = -9$$

CHECK:

$$y + 7 = -2$$
$$-9 + 7 \overset{?}{=} -2$$
$$-2 = -2$$

Example 10. Solve $-24 = \frac{-12}{5}x$ and check.

Solution

$-24 = \frac{-12}{5}x$	
$\frac{-5}{12} \cdot \frac{-24}{1} = \frac{-5}{12} \cdot \frac{-12}{5}x$	Multiply both sides by $\frac{-5}{12}$.
$\frac{-5}{12} \cdot \frac{-24}{1} = \left(\frac{-5}{12} \cdot \frac{-12}{5}\right)x$	Rewrite right side (associative property)
$10 = 1x$	Simplify.
$10 = x$	Simplify.

CHECK:

$-24 = \frac{-12}{5}x$	Original equation.
$-24 \stackrel{?}{=} \frac{-12}{5}(10)$	Substitute 10 for x.
$-24 = -24$	Find the product.

Answers to Trial Runs

page 193 **1.** Yes **2.** Yes **3.** No **4.** Yes **5.** Yes

page 195 **1.** $x = 5$ **2.** $y = 10$ **3.** $m = 0$ **4.** $a = -9$ **5.** $x = 4$ **6.** $x = 11$

page 198 **1.** $x = 6$ **2.** $x = -5$ **3.** $x = 0$ **4.** $x = -7$ **5.** $a = 18$ **6.** $x = -6$

page 200 **1.** $x = -6$ **2.** $x = 14$ **3.** $x = -12$ **4.** $x = 0$

Name ______________________________ Date ______________

EXERCISE SET 4.1

Decide if the given value of the variable is a solution for the equation.

______ **1.** $5x - 1 = 9; x = 2$

______ **2.** $4x - 3 = 21; x = 6$

______ **3.** $9 - 2y = 10; y = \frac{-1}{2}$

______ **4.** $8 - 2y = 9; y = \frac{-1}{2}$

______ **5.** $\frac{m}{2} + 7 = 13; m = 12$

______ **6.** $\frac{m}{4} + 2 = 8; m = 24$

______ **7.** $3(a - 6) = 3; a = 6$

______ **8.** $4(a - 5) = 16; a = 5$

______ **9.** $7x + 1.4 = 0; x = -0.2$

______ **10.** $5x + 1.5 = 0; x = -0.3$

______ **11.** $4x - 5 = 3x + 2; x = -7$

______ **12.** $6x - 7 = 5x + 1; x = -8$

Solve and check.

______ **13.** $x + 5 = 12$

______ **14.** $x + 9 = 16$

______ **15.** $x - 9 = 3$

______ **16.** $x - 8 = 7$

______ **17.** $1.6 + y = 4.2$

______ **18.** $1.9 + y = 5.3$

______ **19.** $y + 2 = -6$

______ **20.** $y + 3 = -8$

______ **21.** $m - 11 = -6$

______ **22.** $m - 9 = -8$

______ **23.** $-4 + a = 6$

______ **24.** $-6 + a = 12$

______ **25.** $12 = k + 6$

______ **26.** $13 = k + 9$

______ **27.** $8 + x = 8$

______ **28.** $-9 + x = -9$

______ **29.** $\frac{-1}{4} = m + \frac{9}{4}$

______ **30.** $\frac{-1}{3} = m + \frac{7}{3}$

______ **31.** $y + 12 = 4$

______ **32.** $y + 9 = 2$

______ **33.** $a - 7 = -14$

______ **34.** $a - 11 = -23$

______ **35.** $-13 = x - 13$

______ **36.** $-15 = x - 15$

______ **37.** $3x = 21$

______ **38.** $4x = 28$

______ **39.** $\frac{1}{2}x = -18$

______ **40.** $\frac{1}{3}x = -27$

______ **41.** $-2x = 35$

______ **42.** $-5x = 46$

______ **43.** $-3.6 = -6x$

______ **44.** $-4.8 = -6x$

_______ **45.** $4x = 0$

_______ **46.** $7x = 0$

_______ **47.** $-x = 6.5$

_______ **48.** $-x = 9.2$

_______ **49.** $\frac{x}{2} = 6$

_______ **50.** $\frac{x}{4} = 12$

_______ **51.** $-3 = \frac{3}{4}x$

_______ **52.** $-9 = \frac{3}{2}x$

_______ **53.** $\frac{-x}{7} = 8$

_______ **54.** $\frac{-x}{8} = 9$

_______ **55.** $\frac{1}{5}x = 0$

_______ **56.** $\frac{1}{9}x = 0$

_______ **57.** $\frac{x}{5} = -8$

_______ **58.** $\frac{x}{11} = -4$

_______ **59.** $-9 = \frac{-x}{7}$

_______ **60.** $-7 = \frac{-x}{6}$

_______ **61.** $\frac{-4x}{3} = 8$

_______ **62.** $\frac{-2x}{5} = 6$

_______ **63.** $\frac{8x}{9} = -12$

_______ **64.** $\frac{6x}{11} = -3$

☆ Stretching the Topics

Solve.

_______ **1.** $\frac{-2}{3} = x - \frac{5}{4}$

_______ **2.** $0.32x = -0.48$

_______ **3.** $y + c = d$, for y.

_______ **4.** $ay = d$, for y.

Check your answers in the back of your book.

If you can now solve the equations in **Checkup 4.1**, you are ready to go on to Section 4.2.

Name ______________________ Date ____________

CHECKUP 4.1

______ 1. Is -2 a solution for $7 + 3x = 13$?

Solve.

______ 2. $x + 9 = 8$

______ 3. $x - 9 = -2$

______ 4. $1.4 = x + 0.2$

______ 5. $x + 11 = 11$

______ 6. $6x = 54$

______ 7. $42 = \frac{-7}{2}x$

______ 8. $4x = 0$

______ 9. $\frac{1}{8}x = 3$

______ 10. $-10 = \frac{x}{6}$

Check your answers in the back of your book.

If You Missed Problems:	You Should Review Examples:
1	1–4
2–5	5–8
6–8	9–11
9, 10	12–14

4.2 Solving Equations Using Several Properties

Many equations involve more than one operation. Equations such as

$$3x + 1 = 7 \qquad \text{and} \qquad -7 - \frac{x}{6} = -2$$

require more than one step to isolate the variable.

Undoing Two Operations

In the equation $3x + 1 = 7$, we must undo the operations of addition (get rid of the 1) and multiplication (get rid of the 3). Let's start by subtracting 1 from both sides of the equation.

$$3x + 1 = 7$$
$$3x + 1 - 1 = 7 - 1$$
$$3x = 6$$

Now we can divide both sides by 3.

$$\frac{3x}{3} = \frac{6}{3}$$
$$x = 2$$

Again we can check this solution.

$$3x + 1 = 7$$
$$3(2) + 1 \stackrel{?}{=} 7$$
$$6 + 1 \stackrel{?}{=} 7$$
$$7 = 7$$

When solving equations with more than one operation, we follow this general procedure:

> If there are no parentheses in the equation, first perform the necessary addition or subtraction, and then perform the necessary multiplication or division.

Example 1. Solve $-5x + 1.3 = 7.8$ and check.

Solution

$-5x + 1.3 = 7.8$	
$-5x + 1.3 - 1.3 = 7.8 - 1.3$	Subtract 1.3 (to undo addition).
$-5x = 6.5$	Find the sums.
$\frac{-5x}{-5} = \frac{6.5}{-5}$	Divide by -5 (to undo multiplication).
$x = -1.3$	Find the quotients.

CHECK:

$-5x + 1.3 = 7.8$	Original equation.
$-5(-1.3) + 1.3 \stackrel{?}{=} 7.8$	Substitute -1.3 for x.
$6.5 + 1.3 \stackrel{?}{=} 7.8$	Remove parentheses.
$7.8 = 7.8$	Find the sum.

Example 2. Solve $-7 = \frac{x}{6} - 2$.

Solution

$-7 = \frac{x}{6} - 2$	
$-7 + 2 = \frac{x}{6} - 2 + 2$	Add 2 (to undo subtraction).
$-5 = \frac{x}{6}$	Find the sums.
$6(-5) = \frac{6}{1}\left(\frac{x}{6}\right)$	Multiply by 6 (to undo division).
$-30 = x$	Find the products.

Example 3. Solve $\frac{2x}{3} - 1 = 5$ and check.

Solution

$\frac{2x}{3} - 1 = 5$	
$\frac{2x}{3} - 1 + 1 = 5 + 1$	Add 1 (to undo subtraction).
$\frac{2x}{3} = 6$	Find the sums.
$\frac{2}{3}x = 6$	Rewrite the variable term.
$\frac{3}{2}\left(\frac{2}{3}x\right) = \frac{3}{2}\left(\frac{6}{1}\right)$	Multiply by $\frac{3}{2}$ (reciprocal of variable coefficient).
$x = 9$	Find the products.

CHECK:

$\frac{2x}{3} - 1 = 5$	Original equation.
$\frac{2(9)}{3} - 1 \stackrel{?}{=} 5$	Substitute 9 for x.
$\frac{18}{3} - 1 \stackrel{?}{=} 5$	Find the product.
$6 - 1 \stackrel{?}{=} 5$	Find one quotient.
$5 = 5$	Find the difference.

Trial Run

Solve and check.

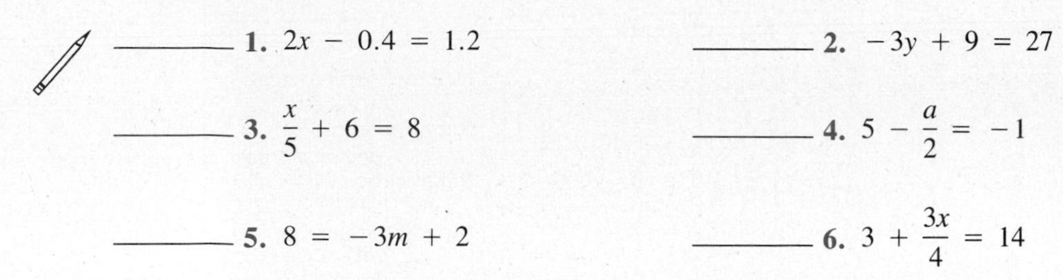

_______ 1. $2x - 0.4 = 1.2$

_______ 2. $-3y + 9 = 27$

_______ 3. $\frac{x}{5} + 6 = 8$

_______ 4. $5 - \frac{a}{2} = -1$

_______ 5. $8 = -3m + 2$

_______ 6. $3 + \frac{3x}{4} = 14$

Answers are on page 216.

Working with Symbols of Grouping in Equations

Parentheses and other symbols of grouping provide us with important directions. As before, we shall deal with symbols of grouping first, then combine like terms, and then solve the equation.

Let's find the solution for

$$2(x + 3) + (x + 5) = 17$$

$2x + 6 + x + 5 = 17$	Remove parentheses (distributive property).
$3x + 11 = 17$	Combine like terms.
$3x + 11 - 11 = 17 - 11$	Subtract 11.
$3x = 6$	Simplify sums.
$\frac{3x}{3} = \frac{6}{3}$	Divide by 3.
$x = 2$	Simplify quotients.

We shall check our solution by substituting 2 for x in the *original equation*.

$$2(x + 3) + (x + 5) = 17$$
$$2(2 + 3) + (2 + 5) \stackrel{?}{=} 17$$
$$2(5) + (7) \stackrel{?}{=} 17$$
$$10 + 7 \stackrel{?}{=} 17$$
$$17 = 17$$

When solving equations involving more than one operation and symbols of grouping, we should follow these steps:

1. Remove symbols of grouping.
2. Combine like terms.
3. Perform necessary addition or subtraction to move constant terms to one side, keeping the variable term on the other side.
4. Perform necessary multiplication or division to isolate the variable.
5. Check the solution in the original equation.

Example 4. Solve $2(x + 7) - 9 = 31$ and check.

Solution

		CHECK:
$2(x + 7) - 9 = 31$		$2(x + 7) - 9 = 31$
$2x + 14 - 9 = 31$	Remove parentheses.	$2(13 + 7) - 9 \stackrel{?}{=} 31$
$2x + 5 = 31$	Simplify sums.	$2(20) - 9 \stackrel{?}{=} 31$
$2x + 5 - 5 = 31 - 5$	Subtract 5.	$40 - 9 \stackrel{?}{=} 31$
$2x = 26$	Simplify sums.	$31 = 31$
$\frac{2x}{2} = \frac{26}{2}$	Divide by 2.	
$x = 13$	Simplify quotients.	

Complete the work in Examples 5 and 6.

Example 5. Solve $1 = 3(2x - 7) - 5(3x + 1)$.

Solution

$$1 = 3(2x + 7) - 5(3x + 1)$$

$$1 = 6x - 21 - 15x - 5$$

Check your work on page 215. ▶

Example 6. Solve $3[x + 2(x - 7)] = -60$.

Solution

$$3[x + 2(x - 7)] = -60$$

$$3[x + 2x - 14] = -60$$

$$3[3x + 14] = -60$$

Check your work on page 215. ▶

Trial Run

Solve.

_______ **1.** $3(x - 2) + 7 = 22$

_______ **2.** $2(x - 6) - (x + 8) = -12$

_______ **3.** $11 = 3(3x - 2) - 5(x - 1)$

_______ **4.** $4[2x + 3(x - 2)] = -4$

Answers are on page 216.

Solving Equations with Variables on Both Sides

In an equation such as

$$3x + 1 = x + 5$$

we must find some way to *get the variable terms together on one side of the equation and the constant terms together on the other side*. We shall do this by addition and subtraction.

$3x + 1 = x + 5$	
$3x - x + 1 = x - x + 5$	Subtract x from both sides.
$2x + 1 = 5$	Combine like terms.
$2x + 1 - 1 = 5 - 1$	Subtract 1.
$2x = 4$	Simplify.
$\frac{2x}{2} = \frac{4}{2}$	Divide by 2.
$x = 2$	Simplify.

We shall check our solution by substituting 2 for x in the original equation.

$$3x + 1 = x + 5$$
$$3(2) + 1 \stackrel{?}{=} 2 + 5$$
$$6 + 1 \stackrel{?}{=} 2 + 5$$
$$7 = 7$$

Example 7. Solve $0.7x - 0.3 = 3.3 - 0.2x$.

Solution

$$0.7x - 0.3 = 3.3 - 0.2x$$
$$0.7x - 0.3 + 0.2x = 3.3 - 0.2x + 0.2x$$
$$0.7x + 0.2x - 0.3 = 3.3$$
$$0.9x - 0.3 = 3.3$$
$$0.9x - 0.3 + 0.3 = 3.3 + 0.3$$
$$0.9x = 3.6$$
$$\frac{0.9x}{0.9} = \frac{3.6}{0.9}$$
$$x = 4$$

Example 8. Solve $5x + 6 = 12x + 41$.

Solution

$$5x + 6 = 12x + 41$$
$$5x + 6 - 5x = 12x + 41 - 5x$$
$$5x - 5x + 6 = 12x - 5x + 41$$
$$6 = 7x + 41$$
$$6 - 41 = 7x + 41 - 41$$
$$-35 = 7x$$
$$\frac{-35}{7} = \frac{7x}{7}$$
$$-5 = x.$$

From Examples 7 and 8, we notice that we may move the variable terms to *either* side of the equation. *You* must decide on which side you wish the variable terms to appear, and then move the constants to the *other* side.

Trial Run

Solve.

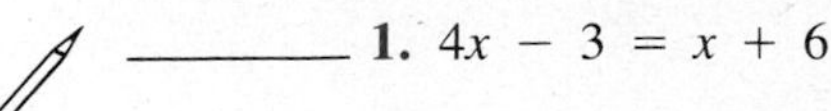

______ **1.** $4x - 3 = x + 6$

______ **2.** $3x + 9 = 8x + 6$

_______ 3. $-3x + 7 = 2x - 18$ _______ 4. $9 + 2x = -x - 3$

Answers are on page 216.

Equations with variables on both sides that also contain symbols of grouping are solved in the same basic way.

Example 9. Solve $2(x + 3) + 5(x + 1) = 3x - 9$ and check.

Solution

$$2(x + 3) + 5(x + 1) = 3x - 9$$

$$2x + 6 + 5x + 5 = 3x - 9 \quad \text{Remove parentheses.}$$

$$7x + 11 = 3x - 9 \quad \text{Combine like terms.}$$

$$7x + 11 - 3x = 3x - 9 - 3x \quad \text{Subtract } 3x \text{ from both sides.}$$

$$4x + 11 = -9 \quad \text{Combine like terms.}$$

$$4x + 11 - 11 = -9 - 11 \quad \text{Subtract 11 from both sides.}$$

$$4x = -20 \quad \text{Simplify.}$$

$$\frac{4x}{4} = \frac{-20}{4} \quad \text{Divide both sides by 4.}$$

$$x = -5$$

CHECK:

$$2(x + 3) + 5(x + 1) = 3x - 9$$

$$2(-5 + 3) + 5(-5 + 1) \stackrel{?}{=} 3(-5) - 9$$

$$2(-2) + 5(-4) \stackrel{?}{=} -15 - 9$$

$$-4 - 20 \stackrel{?}{=} -24$$

$$-24 = -24$$

Now you complete the work for Examples 10 and 11.

Example 10. Solve $5(3 - 2x) + 4 = 3(4 - x)$.

Solution

$$5(3 - 2x) + 4 = 3(4 - x)$$
$$15 - 10x + 4 = 12 - 3x$$
$$19 - 10x = 12 - 3x$$

Check your work on page 216. ▶

Example 11. Solve $2[x + 3(5 - x)] = 7(x + 1) - 21$

Solution

$$2[x + 3(5 - x)] = 7(x + 1) - 21$$
$$2[x + 15 - 3x] = 7x + 7 - 21$$
$$2[15 - 2x] = 7x - 14$$

Check your work on page 216. ▶

Trial Run

Solve.

_______ **1.** $2(x + 3) + 5(x - 1) = 3x - 7$

_______ **2.** $4(2x - 3) + 2 = 2(5 - x)$

_______ **3.** $3 - 2(4 - 5x) = 4(2x - 1) + 7$

_______ **4.** $2[y + 5(y - 3)] = 4(y - 3) - 10$

Answers are on page 216.

Let's summarize the general method for solving first-degree equations.

Steps for Solving Equations

1. Remove symbols of grouping.
2. Combine like terms on each side.
3. Move all variable terms to one side.
4. Move all constant terms to the other side.
5. Isolate the variable to obtain the solution.
6. Check your solution in the original equation.

Solving an Equation for a Certain Variable

Often an equation or formula contains more than one variable. To "solve" such an equation, we must be told which variable is to be isolated. Then we can solve for that variable by the methods we have already learned.

Example 12. The perimeter P of a rectangle with length l and width w is given by the formula $P = 2l + 2w$. Solve this formula for l.

Solution: We wish to isolate l.

$$P = 2l + 2w$$

$$P - 2w = 2l + 2w - 2w \qquad \text{Subtract } 2w \text{ from both sides (to isolate } l\text{-term).}$$

$$P - 2w = 2l \qquad \text{Simplify.}$$

$$\frac{P - 2w}{2} = \frac{2l}{2} \qquad \text{Divide both sides by 2 (to isolate } l\text{).}$$

$$\frac{P - 2w}{2} = l \qquad \text{Simplify.}$$

Example 13. If P dollars are deposited in a savings account paying a simple annual interest rate of r percent, the total amount A after t years is given by the formula $A = P(1 + rt)$. Solve this formula for t.

Solution: We wish to isolate t.

$$A = P(1 + rt)$$

$$A = P + Prt \qquad \text{Remove parentheses.}$$

$$A - P = P + Prt - P \qquad \text{Subtract } P \text{ from both sides (to isolate } t\text{-term).}$$

$$A - P = Prt \qquad \text{Simplify.}$$

$$\frac{A - P}{Pr} = \frac{Prt}{Pr} \qquad \text{Divide both sides by } Pr \text{ (to isolate } t\text{).}$$

$$\frac{A - P}{Pr} = t \qquad \text{Simplify.}$$

Example 14. Solve the equation $3x + \frac{1}{4}y = 2$ for y.

Solution: We wish to isolate y.

$$3x + \frac{1}{4}y = 2$$

$$-3x + 3x + \frac{1}{4}y = -3x + 2 \qquad \text{Subtract } 3x \text{ from both sides (to isolate } y\text{-term).}$$

$$\frac{1}{4}y = -3x + 2 \qquad \text{Simplify.}$$

$$\frac{4}{1} \cdot \frac{1}{4}y = 4(-3x + 2) \qquad \text{Multiply both sides by 4 (to isolate } y\text{).}$$

$$y = -12x + 8 \qquad \text{Find the products.}$$

Trial Run

Solve for the specified variable.

_______ **1.** $A = \frac{b \cdot h}{2}$ for h

_______ **2.** $xy = 3$ for y

_______ **3.** $C = 2\pi r$ for r

_______ **4.** $2A = h(B + b)$ for B

Answers are on page 216.

Examples You Completed

Example 5. Solve $1 = 3(2x - 7) - 5(3x + 1)$.

Solution

$$1 = 3(2x - 7) - 5(3x + 1)$$
$$1 = 6x - 21 - 15x - 5$$
$$1 = -9x - 26$$
$$1 + 26 = -9x - 26 + 26$$
$$27 = -9x$$
$$\frac{27}{-9} = \frac{-9x}{-9}$$
$$-3 = x$$

Example 6. Solve $3[x + 2(x - 7)] = -60$.

Solution

$$3[x + 2(x - 7)] = -60$$
$$3[x + 2x - 14] = -60$$
$$3[3x - 14] = -60$$
$$9x - 42 = -60$$
$$9x - 42 + 42 = -60 + 42$$
$$9x = -18$$
$$\frac{9x}{9} = \frac{-18}{9}$$
$$x = -2$$

Example 10. Solve $5(3 - 2x) + 4 = 3(4 - x)$.

Solution

$$5(3 - 2x) + 4 = 3(4 - x)$$
$$15 - 10x + 4 = 12 - 3x$$
$$19 - 10x = 12 - 3x$$
$$19 - 10x + 10x = 12 - 3x + 10x$$
$$19 = 12 + 7x$$
$$19 - 12 = 12 + 7x - 12$$
$$7 = 7x$$
$$\frac{7}{7} = \frac{7x}{7}$$
$$1 = x$$

Example 11. Solve $2[x + 3(5 - x)] = 7(x + 1) - 21$.

Solution

$$2[x + 3(5 - x)] = 7(x + 1) - 21$$
$$2[x + 15 - 3x] = 7x + 7 - 21$$
$$2[15 - 2x] = 7x - 14$$
$$30 - 4x = 7x - 14$$
$$30 - 4x + 4x = 7x - 14 + 4x$$
$$30 = 11x - 14$$
$$30 + 14 = 11x - 14 + 14$$
$$44 = 11x$$
$$\frac{44}{11} = \frac{11x}{11}$$
$$4 = x$$

Answers to Trial Runs

page 209 **1.** $x = 0.8$ **2.** $y = -6$ **3.** $x = 10$ **4.** $a = 12$ **5.** $m = -2$ **6.** $x = \frac{44}{3}$

page 210 **1.** $x = 7$ **2.** $x = 8$ **3.** $x = 3$ **4.** $x = 1$

page 211 **1.** $x = 3$ **2.** $x = \frac{3}{5}$ **3.** $x = 5$ **4.** $x = -4$

page 213 **1.** $x = -2$ **2.** $x = 2$ **3.** $x = 4$ **4.** $y = 1$

page 215 **1.** $h = \frac{2A}{b}$ **2.** $y = \frac{3}{x}$ **3.** $r = \frac{C}{2\pi}$ **4.** $B = \frac{2A - hb}{h}$

Name ________________________ Date ____________

EXERCISE SET 4.2

Solve the equations.

______ **1.** $3x + 5 = 26$	______ **2.** $4x + 6 = 26$
______ **3.** $2x - 9 = 11$	______ **4.** $7x - 9 = 5$
______ **5.** $5y + 3 = -22$	______ **6.** $2y + 8 = -12$
______ **7.** $0.8x - 1.5 = -0.7$	______ **8.** $0.6x + 2.3 = 1.1$
______ **9.** $-4a + 9 = 45$	______ **10.** $-5a + 10 = 55$
______ **11.** $13 + 2m = 0$	______ **12.** $20 + 3m = 0$
______ **13.** $-x - 7 = 6$	______ **14.** $-x - 8 = 9$
______ **15.** $7\frac{1}{2} - 2x = 13\frac{1}{2}$	______ **16.** $6\frac{3}{4} - 5x = 41\frac{3}{4}$
______ **17.** $24 = 3x - 15$	______ **18.** $17 = 6x - 7$
______ **19.** $4y + \frac{2}{5} = \frac{2}{5}$	______ **20.** $7y + \frac{1}{9} = \frac{1}{9}$
______ **21.** $-23 = 9 - 4a$	______ **22.** $-21 = 15 - 9a$
______ **23.** $-1.7x + 17 = 0$	______ **24.** $-0.4x + 3.2 = 0$
______ **25.** $5 = \frac{3}{2}x - 4$	______ **26.** $2 = \frac{5}{3}x - 8$
______ **27.** $-9 = \frac{1}{5}x - 9$	______ **28.** $-7 = \frac{1}{8}x - 7$
______ **29.** $8 - \frac{2}{5}x = 12$	______ **30.** $9 - \frac{3}{2}x = 15$
______ **31.** $\frac{x}{6} + 3 = 5$	______ **32.** $\frac{x}{7} + 4 = 9$
______ **33.** $0.7 + \frac{a}{5} = -0.1$	______ **34.** $0.6 + \frac{a}{3} = -0.2$
______ **35.** $13 = \frac{y}{4} + 9$	______ **36.** $17 = \frac{y}{5} + 8$
______ **37.** $6 - \frac{x}{4} = 7$	______ **38.** $9 - \frac{x}{2} = 12$
______ **39.** $(x + 3) + (x - 5) = 8$	______ **40.** $(x + 5) + (x - 6) = 9$

_____ 41. $12 = 4\left(\frac{1}{2}x - 7\right) + 8$

_____ 42. $12 = 6\left(\frac{1}{2}x - 6\right) + 9$

_____ 43. $2(x - 8) - (x + 3) = -22$

_____ 44. $3(x - 4) - (2x + 7) = -26$

_____ 45. $3 = 4(x - 5) - 2(x + 6)$

_____ 46. $2 = 5(x - 5) - 3(x - 2)$

_____ 47. $2[3x + 3(x - 1)] = 6$

_____ 48. $3[4x + 3(x - 2)] = 24$

_____ 49. $5[2(x - 4) - 2] = 0$

_____ 50. $6[3(x - 2) + 12] = 0$

_____ 51. $3x + 2 = -x + 10$

_____ 52. $5x + 3 = -3x + 15$

_____ 53. $6x - 0.5 = 3x + 1.6$

_____ 54. $7x - 0.5 = 4x + 0.7$

_____ 55. $11x - 3 = 5x - 15$

_____ 56. $12x - 9 = 7x - 14$

_____ 57. $3x - 9 = 6 - 4x$

_____ 58. $3x - 9 = 12 - 7x$

_____ 59. $5x + 7 = 9x + 23$

_____ 60. $5x + 5 = 11x + 32$

_____ 61. $x + 3 = -5x - 9$

_____ 62. $7x + 6 = -9x - 4$

_____ 63. $3x - 8 = 5x$

_____ 64. $7x - 15 = 12x$

_____ 65. $7.2x + 8 = 8 - 3.8x$

_____ 66. $9.6x + 12 = 12 - 7.4x$

_____ 67. $2(x + 1) + 3(x - 2) = 2x - 37$

_____ 68. $4(x + 1) + (x - 5) = 3x - 19$

_____ 69. $5(2x - 4) - 2 = 2(7 - x)$

_____ 70. $3(2x - 7) - 7 = 5(1 - x)$

_____ 71. $x - (-3x + 16) = 3(x - 12) + 18$

_____ 72. $-4 - 6x = 2(-3 + 4x) + 4(3 - x)$

_____ 73. $3[2x - 5(x - 1)] = 6 - [3 - (x + 2)]$

_____ 74. $4[x - 3(x - 2)] = 9 - [5 - (x - 1)]$

_____ 75. $2[x + 3(x - 2)] = 5(x - 4) - 1$

_____ 76. $3[2x + (x - 5)] = 2(x - 7) + 6$

Solve for the variable specified.

_____ 77. $C = \pi d$ for d

_____ 78. $A = lw$ for w

_____ 79. $x + 2y = 3$ for y

_____ 80. $2x + 3y = 5$ for x

_____ 81. $I = prt$ for t

_____ 82. $V = lwh$ for h

_____ 83. $P = a + b + c$ for a

_____ 84. $M = x + y + z$ for z

_____ 85. $K = c(1 + 2p)$ for p

_____ 86. $Q = m(1 + 3l)$ for l

_____ 87. $2x - 2b = -5(b - x)$ for x.

_____ 88. $6x - 8b = -2(b + x)$ for x.

_____ 89. $l = a + d(n - 1)$ for n.

_____ 90. $E = I(R + r)$ for r.

☆ Stretching the Topics

Solve.

_______ **1.** $2.3 + 0.2x = -0.9 - (9.4x + 1.6)$

_______ **2.** $3(x - 2) = 2(-x - 6) + 3x - (-x - 4)$

_______ **3.** $V = \frac{1}{6} h (B + b + 4M)$, for M.

Check your answers in the back of your book.

If you can solve the equations in **Checkup 4.2**, you are ready to go on to Section 4.3.

3(x − 4) − (2x + 7) = 26
3x + −12 + −2x + −7 =
x − 19 = −26
+19 + 19
−x = −7

3(−7 − 4) − (30 + 7) = 26

3[4x + 3(x + 2)] = 24
4x + 3x + −6 = 24
12x + 9x + −18 = 24
21x + −18 = 24
+18 +18
= 42
21
x = 2

3[4·2 + 3(2 − 2)] = 24
8·3 = 24
24 = 24

M = x + y + z
−x −y
−x =
−y
−x + −y

Name ______________________ Date __________

CHECKUP 4.2

Solve.

______ **1.** $3x - 7 = 8$

______ **2.** $0.7 = 0.8x + 1.5$

______ **3.** $-2y - 5 = 7$

______ **4.** $8 - \frac{x}{3} = 9$

______ **5.** $\frac{3}{2}x + 1 = 5$

______ **6.** $2(x - 3) - 4(x + 2) = 2$

______ **7.** $3[2x - (x + 4)] = 6$

______ **8.** $y + 5 = 7y + 8$

______ **9.** $3(3y - 1) - 9 = 5(4 + y)$

______ **10.** $A = P(1 + rt)$ for r

Check your answers in the back of your book.

If You Missed Problems:	You Should Review Examples:
1–4	1, 2
5	3
6, 7	4–6
8	8
9	10
10	14

4.3 Switching from Word Statements to Equations

In Chapter 3 we learned to use variables and constants to switch from words to algebraic expressions. Now we must learn to switch from word statements to equations.

If 1 credit hour at a university costs \$25, we know the cost of h hours can be written as $25 \cdot h$. Suppose that a student has saved \$150 for part-time courses. How can we figure out how many credit hours that student can take?

We must solve this equation:

$$25h = 150$$

$$\frac{25h}{25} = \frac{150}{25}$$

$$h = 6$$

The student can take 6 credit hours.

Switching problems from words to equations should be done in an orderly way, using these steps.

1. Read the problem carefully to be sure you understand what you are being asked to find.
2. Use a drawing if it helps you visualize the problem.
3. Write down what you want the variable to stand for.
4. Write down any expressions containing the variable.
5. Write an equation from the information in the problem.
6. Solve the equation. Answer the question in a sentence.
7. Check your solution in the original problem, being sure that your solution makes sense.

Example 1. Gary has \$1760 in his savings account now. If he deposits \$40 per week in this account, for *how many weeks* must he continue to save if he wants a total of \$5000 in his account?

Solution: We have italicized what you are being asked to find here.

Let w = number of weeks

$40w$ = dollars saved in w weeks

$1760 + 40w$ = total dollars saved

Our equation is

$$1760 + 40w = 5000$$

and we solve it by the usual methods.

$$1760 + 40w = 5000$$

$$1760 + 40w - 1760 = 5000 - 1760$$

$$40w = 3240$$

$$\frac{40w}{40} = \frac{3240}{40}$$

$$w = 81$$

Gary must save for 81 weeks. Let's check. If Gary saves \$40 each week for 81 weeks, he will save \$40(81) = \$3240. Since he has already saved \$1760, he will have saved a total of \$1760 + \$3240 = \$5000. Our solution is correct.

Example 2. Beverly has 36 yards of fencing to use for enclosing a rectangular garden. If she wishes the garden to be twice as long as it is wide, *what should the garden's dimensions be*?

Solution: A drawing might help us here.

Let x = width (yards)

$2x$ = length (yards)

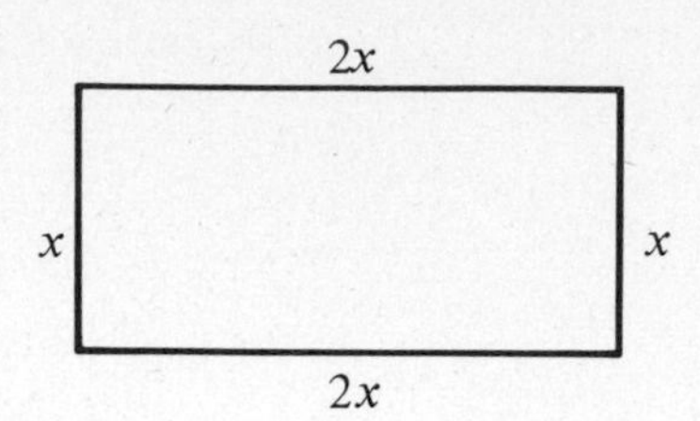

We can express the amount of fencing (which is the perimeter of the rectangle) by

$$P = 2l + 2w$$

Our equation is

$$2(2x) + 2x = 36$$
$$4x + 2x = 36$$
$$6x = 36$$
$$\frac{6x}{6} = \frac{36}{6}$$
$$x = 6$$

In this problem we are asked to find both dimensions of the garden.

Width: $x = 6$

Length: $2x = 12$

The width is 6 yards and the length is 12 yards.

CHECK:

12 yards

6 yards | 6 yards

12 yards

Fencing: 6 + 12 + 6 + 12
= 36 yards

Now you try to complete Example 3.

Example 3. If Mandy earns \$3.65 per hour, *how many hours* did she work in 1 week if she earned \$113.15?

Solution

3.65H = 113.15

Let ____ = number of hours worked in 1 week

____ = dollars earned in 1 week

Our equation is

____ = 113.15

Mandy worked ____ hours.

Check your work on page 227. ▶

Now let's return to the problem stated at the beginning of this chapter and see if we can solve it.

Example 4. Charlotte now earns \$6 per hour as a grounds keeper working 35 hours per week. She has been offered a similar job paying \$280 per week. *How much of an increase* in her hourly wage must her present employer give Charlotte to meet the better offer?

Solution

$$\begin{aligned} \text{Let } x &= \text{increase in hourly wage} \quad \text{(dollars)} \\ 6 + x &= \text{new hourly wage} \quad \text{(dollars)} \\ 35(6 + x) &= \text{new weekly earnings} \quad \text{(dollars)} \end{aligned}$$

Our equation is

$$\begin{aligned} 35(6 + x) &= 280 \\ 210 + 35x &= 280 \\ 210 + 35x - 210 &= 280 - 210 \\ 35x &= 70 \\ \frac{35x}{35} &= \frac{70}{35} \\ x &= 2 \end{aligned}$$

Charlotte must receive an hourly increase of \$2.

CHECK: Charlotte now earns \$6 per hour. If her hourly wage is increased by \$2, she will earn \$8 per hour. If she works 35 hours at \$8 per hour, she will earn

$$35(8) = 280 \text{ dollars}$$

Example 5. Three roommates share monthly rent and food expenses equally. If each person's share of the rent is \$65, *how much can they spend altogether on food* so that each roommate pays \$150 for total monthly expenses?

Solution

$$\begin{aligned} \text{Let } F &= \text{total spent on food} \quad \text{(dollars)} \\ \frac{F}{3} &= \text{each roommate's share of food cost} \quad \text{(dollars)} \\ 65 &= \text{each roommate's share of rent} \quad \text{(dollars)} \\ \frac{F}{3} + 65 &= \text{each roommate's share of monthly expenses} \quad \text{(dollars)} \end{aligned}$$

Our equation is

$$\begin{aligned} \frac{F}{3} + 65 &= 150 \\ \frac{F}{3} + 65 - 65 &= 150 - 65 \\ \frac{F}{3} &= 85 \\ 3 \cdot \frac{F}{3} &= 3(85) \\ F &= 255 \end{aligned}$$

They can spend \$255 altogether on food.

CHECK: If they spend \$255 each month on food, each roommate's share of food expenses is

$$\frac{\$255}{3} = \$85$$

Since each roommate also pays \$65 for rent, each pays a total each month of

$$\$85 + \$65 = \$150$$

Complete Example 6.

Example 6. After a session of Weight Worriers, John announced "I lost 3 pounds more than twice as much as Carla lost this month. I lost 17 pounds." *How many pounds* did Carla lose?

Solution

Let C = number of pounds Carla lost

2C = twice Carla's loss (pounds)

3+2C = 3 pounds more than twice Carla's loss (pounds)

____ = John's loss (pounds)

We must solve the equation:

2C+3 = 17

Carla lost 7 pounds.

CHECK:

Check your solution on page 227. ▶

Example 7. In 1986, the enrollment at State University increased by 3% to 12,360 students. *What was the enrollment* in 1985?

Solution

Let x = enrollment in 1985

3% of $x = 0.03x$ = increase in enrollment

The 1986 enrollment is the *sum* of the 1985 enrollment and the increase.

$x + 0.03x$ = enrollment in 1986

Our equation is

$$x + 0.03x = 12{,}360$$

$$1x + 0.03x = 12{,}360 \qquad \text{Rewrite } x \text{ as } 1x.$$

$$1.03x = 12{,}360 \qquad \text{Combine like terms.}$$

$$\frac{1.03x}{1.03} = \frac{12{,}360}{1.03} \qquad \text{Divide both sides by 1.03.}$$

$$x = 12{,}000$$

The 1985 enrollment was 12,000 students.

▶ Examples You Completed

Example 3. If Mandy earns \$3.65 per hour, *how many hours* did she work in 1 week if she earned \$113.15?

Solution

Let h = number of hours worked in 1 week

$3.65h$ = dollars earned in 1 week

Our equation is

$$3.65h = 113.15$$

$$\frac{3.65h}{3.65} = \frac{113.15}{3.65}$$

$$h = 31$$

Mandy worked 31 hours.

Example 6. John lost 3 pounds more than twice as much as Carla lost. John lost 17 pounds. How many pounds did Carla lose?

Solution

Let x = number of pounds Carla lost

$2x$ = twice Carla's loss

$2x + 3$ = 3 pounds more than twice Carla's loss

$2x + 3$ = John's loss

$$2x + 3 = 17$$

$$2x + 3 - 3 = 17 - 3$$

$$2x = 14$$

$$\frac{2x}{2} = \frac{14}{2}$$

$$x = 7$$

Carla lost 7 pounds.

CHECK: Carla lost 7 pounds. Twice what Carla lost is 14 pounds, and 3 pounds more than twice Carla's loss is 14 + 3 or 17 pounds. John lost 17 pounds.

Name ______________________________ Date ____________

EXERCISE SET 4.3

_______ **1.** The Waller twins' combined weight at birth was 12 pounds. If Lyle weighed 2 more pounds than Lydia, how much did each weigh?

_______ **2.** If Carl finds he can save $12 a week, how many weeks will it take him to save $252?

_______ **3.** If Henry drives at a rate of 55 mph, how long will it take for him to travel 495 miles?

_______ **4.** Cheri bought a shirt and a blazer. The cost of the blazer was $15 less than 3 times the cost of the shirt. If the blazer cost $47.97, what was the cost of the shirt?

_______ **5.** Julia and Sara are selling tickets for a concert to benefit the Special Olympics. Sara has sold 4 times as many tickets as Julia and together they have sold 75 tickets. How many tickets has each sold?

_______ **6.** Carol is building a bookcase that will have to hold 162 books. How many shelves will the bookcase have if each shelf will hold 18 books?

_______ **7.** The area of a rectangle is the product of the length and the width. Find the width of a rectangle with a length of 15 meters and an area of 195 square meters.

_______ **8.** The restaurant manager tells the chairperson of the banquet committee that she will charge $200 for the use of the room plus $8.75 per person for the dinner. If the total cost of the banquet is $900, how many persons will be attending?

_______ **9.** At the Bargain Barn Restaurant, a hamburger costs $0.15 less than twice the cost of french fries. If a hamburger costs $0.95, what is the cost of the fries?

_______ **10.** The width of a room is 5 feet more than one-half the length. If a paperhanger needs 70 feet of ceiling border, what are the dimensions of the room?

_______ **11.** Four neighbors share equally in the expenses for maintaining a nearby empty lot. If each neighbor's share of the mowing is $17 per month, how much can they spend on shrubs and flowers so that each neighbor's share is $26 per month?

_______ **12.** Mickey sells garden tillers. He earns $75 for each tiller he sells. How many must he sell each month so that he will have a monthly salary of at least $1400?

_______ **13.** If Ella loses 16 pounds, 9 pounds, and 11 pounds during the first 3 months of her diet, how much will she have to lose during the fourth month so that she will have an average weight loss of 12 pounds per month?

_______ **14.** Tyrone is buying a bedroom suite on the installment plan. He paid $150 as a down payment and is paying $55.75 a month until it is paid for. If the bedroom suite costs $819 (including finance charges), for how many months will he be making payments?

_______ 15. If Maxine has 134 inches of framing, what will be the dimensions of the largest picture frame she can make if she wants the length to be 5 inches more than the width?

_______ 16. Bernard is saving money for a vacation during spring break. He deposits $50 and then plans to save a certain amount each week for the next 8 weeks. If he estimates he will need $290 for his vacation, how much must he save each week?

_______ 17. Sidney spends 30% of his monthly salary for rent and 25% for food. If the total for these two expenditures is $660, find Sidney's monthly salary.

_______ 18. Mr. McDonald has 120 feet of fencing to build a loading pen for cattle. If the pen must be 3 times as long as it is wide, find the width of the pen.

_______ 19. Lucindy is using her savings to attend an exercise salon twice a week to improve her figure. She must pay $6 each time she attends. How many weeks can she attend if she has only $792 in her savings account?

_______ 20. If Ira has scored 80, 62, 75, and 69 on his first four mathematics tests, what score must he earn on his fifth test to have a test average of 70?

_______ 21. The Miller family decided to save $247.50 monthly. This represented $12\frac{1}{2}\%$ of the family income per month. What was the family's monthly income?

_______ 22. Each consumer receives a discount of 5% if gas bills are paid within 10 days after billing. If Mr. Bryant saved $3.29 by paying promptly, what was the original amount of his gas bill?

_______ 23. A business firm increased its capital by 25% in 1 year. If the firm's capital this year was $525,000, what was its capital last year?

_______ 24. A manufacturing company has decreased its inventory by 8%. If they now have 2070 items in stock, how many items did they have before the decrease?

_______ 25. Lynn is selling a china cabinet he built in a shop class. He wishes to make a profit of 40% of the cost. If he sells the china cabinet for $581, what was the cost of building the cabinet?

☆ Stretching the Topics

_______ 1. A board 50 inches in length is to be sawed into four pieces so that each piece will be 1 inch shorter than the preceding piece. Find the length of each piece.

_______ 2. A business firm increased its capital by 25% and then invested 40% of the total capital in real estate. The amount of capital remaining on hand was $32,250. What was the original capital?

Check your answers in the back of your book.

If you can do the problems in **Checkup 4.3**, you are ready to do the **Review Exercises for Chapter 4**.

Name ____________________ Date __________

CHECKUP 4.3

Solve.

______ 1. Maurice buys a stereo set for \$500. He makes a \$150 down payment and pays the balance by making payments of \$25 a week. For how many weeks will he be making payments.?

______ 2. Jo Ann is buying framing for a rectangular picture that is three times as long as it is wide. The salesperson tells her that she needs 72 inches of framing. What are the dimensions of the frame?

______ 3. Tickets to Opryland are \$8.25 each. How many members of Pi Mu Epsilon can go if the organization has \$198?

______ 4. In physical fitness class, Ramon did 25 more than one-half as many pushups as his father. If Ramon did 65 pushups, how many did his father do?

______ 5. In 1986, the corn production on Mr. Kurtz's farm increased by 7%, to 192.6 bushels per acre. What was the production per acre in 1985?

Check your answers in the back of your book.

If You Missed Problem:	You Should Review Example:
1	1
2	2
3	3
4	6
5	7

Summary

In this chapter we learned that **equations** are mathematical sentences which state that one quantity represents the same number as another quantity. The **solution** for an equation containing a variable is the value of the variable that makes the original equation a true statement.

To solve an equation we learned to **isolate the variable** using the following steps:

1. Remove symbols of grouping.
2. Combine like terms on each side.
3. Perform additions and/or subtractions needed to move variable terms to one side and constant terms to the other side.
4. Perform multiplications and/or divisions needed to isolate the variable.
5. Check the solution by substituting it into the original equation.

We discovered that these same steps can be used to solve for a certain variable in a formula containing several variables. Then we practiced switching from word statements to equations containing variables, using an orderly approach.

❑ Speaking the Language of Algebra

1. An equation is a ________ which states that one quantity represents the same number as another quantity.

2. A value of the variable that makes an equation true is called a ________ for the equation, and we say that the value ________ the equation.

3. In solving an equation, our goal is to ________ the variable.

4. To check a solution, we must ________ it for the variable in the ________ equation.

Check your answers in the back of your book.

Name ______________________ Date __________

REVIEW EXERCISES for Chapter 4

_______ **1.** Is 6 a solution for $19 - 2x = 7$?

_______ **2.** Is 0.5 a solution for $\frac{y}{5} - 3 = -4.5$?

_______ **3.** Is 1 a solution for $7(2a - 3) = 7$?

_______ **4.** Is 10 a solution for $3m - 9 = m + 11$?

Solve and check.

_______ **5.** $x - 7 = 3$

_______ **6.** $5 + y = -9$

_______ **7.** $-3 = m - 7$

_______ **8.** $a + \frac{3}{2} = \frac{7}{2}$

_______ **9.** $0.9x = 7.2$

_______ **10.** $2x = -36$

_______ **11.** $-4x = 28$

_______ **12.** $5x = 0$

_______ **13.** $\frac{x}{3} = -1$

_______ **14.** $2 = \frac{y}{-7}$

_______ **15.** $4x - 7 = 31$

_______ **16.** $8 - 2x = 6$

_______ **17.** $-15 = 12 + 9x$

_______ **18.** $\frac{x}{4} - 2 = -1$

_______ **19.** $\frac{-7}{3}x = 14$

_______ **20.** $9 = \frac{5}{2}x - 1$

_______ **21.** $8 - \frac{3x}{4} = 5$

_______ **22.** $0.2(x - 4) - (0.5x + 0.2) = 2$

_______ **23.** $8 = -2[3(y - 4) - 7]$

_______ **24.** $7x - 5 = 3x + 12$

_______ **25.** $4a - 7 = 3 - a$

_______ **26.** $-4x - 9 = 10 - 3x$

_______ **27.** $18 + 9x = -2x - 15$

_______ **28.** $2(x + 2) + 6 = 5(2 + x)$

_______ **29.** $3(x - 7) - (x + 1) = 0$

_______ **30.** $P = 2a + c$, for a

_______ **31.** Dawn has 5 times as many cassettes as Eddie has. Together they have 54 cassettes. How many cassettes does Dawn have?

_______ **32.** The length of a rectangle is 3 times the width. If the perimeter of the rectangle is 32 feet, find its dimensions.

_______ **33.** Raymond earns \$40 a day plus \$7.35 an hour for each hour he works overtime. If last Tuesday, Raymond earned \$69.40, how many hours of overtime did he work?

Check your answers in the back of your book.

If You Missed Exercises:	You Should Review Examples:	
1–4	SECTION 4.1	1–4
5–8		5–8
9–12		9
13, 14		12, 13
15–21	SECTION 4.2	1–3
22, 23		4–6
24–27		7, 8
28–29		9–11
30		13
31–33	SECTION 4.3	1–6

If you have completed the **Review Exercises** and corrected your errors, you are ready to take the **Practice Test for Chapter 4**.

Name ______________________________ Date ______________

PRACTICE TEST for Chapter 4

	Problem	SECTION	EXAMPLE
______	**1.** Is 2 a solution for $7a - 2 = 2(a + 4)$?	4.1	3
______	**2.** Is 1 a solution for $2 - 7x = 6x - 1$?	4.1	2

Solve.

	Problem	SECTION	EXAMPLE
______	**3.** $y - 8 = 5$	4.1	6
______	**4.** $0.3x = 3.9$	4.1	9
______	**5.** $\frac{x}{-5} = 2$	4.1	13
______	**6.** $6x + 7 = 31$	4.2	1
______	**7.** $\frac{5}{4}x - 3 = 7$	4.2	3
______	**8.** $4(x - 3) - (6x + 1) = 1$	4.2	5
______	**9.** $11 - x = 3x - 7$	4.2	8
______	**10.** $5(x - 1) + 6 = 3(4 - 2x)$	4.2	10
______	**11.** $8\left(\frac{1}{2}x - 2\right) - (5x - 1) = 0$	4.2	9
______	**12.** Solve $2x + \frac{1}{5}y = 3$ for y.	4.2	15
______	**13.** Martha sold 4 times as many tickets to the Chili Supper as Harvey sold. Together they sold 185 tickets. How many tickets did each person sell?	4.3	6
______	**14.** The width of a rectangular photograph is 1 inch more than twice the length. If the perimeter of the photograph is 14 inches, find the dimensions of the photograph.	4.3	2
______	**15.** A limousine rental agency charges customers \$50 per day plus \$2 per mile driven. If Adolph was charged \$216 for 1 day's rental of a limousine, how far did he drive?	4.3	1

Check your answers in the back of your book.

Name ______________________ Date ____________

SHARPENING YOUR SKILLS after Chapters 1–4

Simplify. SECTION

		SECTION
______	**1.** $(-25.8) \div (0.3)$	2.4
______	**2.** $\frac{28}{15} \cdot \frac{6}{21}$	1.3
______	**3.** $-8\left[\frac{-15}{3} + 2(7 - 11)\right]$	2.3
______	**4.** $\frac{5}{8} - \left(\frac{-1}{2} - \frac{3}{4}\right)$	2.4
______	**5.** $-3.92 - 7.45$	2.4
______	**6.** $\lvert 3 - 6\rvert + \lvert -15 + 2\rvert$	2.1
______	**7.** $(-3)^3 - 4(-5)$	2.3
______	**8.** $3x - 7y - 5x + 4y$	3.2
______	**9.** $-5(2x) - (-4x)$	3.3
______	**10.** $4[-2(x - y) + 5(x + y)]$	3.3
______	**11.** Evaluate $7(3 - a)$ when $a = -4$.	3.1
______	**12.** Identify the numerical coefficient of y in the expression $5x - 3y + z$.	3.2
______	**13.** Compare $\frac{2}{3}$ and $\frac{3}{4}$ using $<$ or $>$.	2.1
______	**14.** Evaluate $\frac{3m - 2n}{4}$ when $m = -8$ and $n = 8$.	3.1
______	**15.** A building contractor must pay a penalty of \$500 a day for each day over the deadline for completion of a building. Write a variable expression for the amount of penalty after n days. Find the amount if the building is completed 12 days late.	3.1

Check your answers in the back of your book.

Working with Exponents 5

Mary has a square mirror measuring *s* inches on each side.

After framing, each side is $1\frac{1}{4}$ times as long as it was.

Write an algebraic expression for the *area* of the framed mirror.

In Chapter 1 we learned to use exponents to show that a number was repeated as a factor in a product. Now we turn our attention to variable expressions involving exponents. In this chapter we

1. Define whole-number exponents.
2. Use the laws of exponents.
3. Work with negative exponents.

5.1 Multiplying with Whole-Number Exponents

Using Exponents

Sometimes it is necessary to deal with products in which the same factor is repeated. Mathematicians have invented notation to handle such products.

Product	Notation	Read as:
$3 \cdot 3$	3^2	3 squared
$x \cdot x$	x^2	x squared
$a \cdot a \cdot a$	a^3	a cubed
$b \cdot b \cdot b \cdot b \cdot b$	b^5	b to the fifth
$x \cdot x \cdot y \cdot y \cdot y \cdot y$	x^2y^4	x squared times y to the fourth

Such expressions (sometimes called **powers**) always contain two important pieces of information. The small raised positive integer is called the **exponent** and tells us how many times the other number, called the **base**, is to be used as a factor in a product.

$$\underset{\text{base}}{\underset{\uparrow}{x}}{}^{5} \longleftarrow \text{exponent} \quad = \underbrace{x \cdot x \cdot x \cdot x \cdot x}_{5 \text{ factors}}$$

If n is a positive integer,

$$\underset{\text{base}}{\underset{\uparrow}{a}}{}^{n} \longleftarrow \text{exponent} \quad = \underbrace{a \cdot a \cdot \ldots \cdot a}_{n \text{ factors}}$$

Example 1. Find $\left(\frac{1}{4}\right)^3$.

Solution

$$\left(\frac{1}{4}\right)^3$$

$$= \frac{1}{4} \cdot \frac{1}{4} \cdot \frac{1}{4}$$

$$= \frac{1}{64}$$

You complete Example 2.

Example 2. Find $(-2)^5$.

Solution

$$(-2)^5$$

$$= (-2)(-2)(-2)(\_\_\_\_)(\_\_\_\_)$$

$$= \_\_\_\_$$

Check your work on page 249. ▶

Example 3. Use exponents to rewrite $a \cdot a \cdot a \cdot x \cdot x$.

Solution

$$a \cdot a \cdot a \cdot x \cdot x$$

$$= a^3x^2$$

Now complete Example 4.

Example 4. Write x^4y without exponents.

Solution

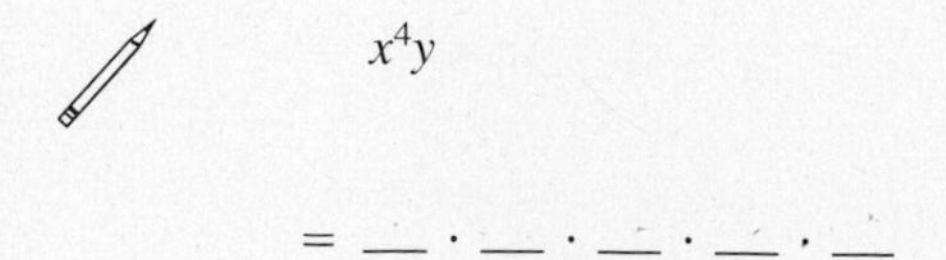

$$x^4y$$

$$= \_\_ \cdot \_\_ \cdot \_\_ \cdot \_\_ \cdot \_\_$$

Check your work on page 249. ▶

Unless parentheses indicate otherwise, an exponent belongs *only* with the number to which it is attached.

Complete Example 5.

Example 5. Write $2x^3$ without exponents.

Solution

$$2x^3$$
$$= 2 \cdot \_\_\_\_ \cdot \_\_\_\_ \cdot \_\_\_\_$$

Check your work on page 249. ▶

Example 6. Write $(2x)^3$ without parentheses.

Solution

$$(2x)^3$$
$$= (2x)(2x)(2x)$$
$$= 2 \cdot x \cdot 2 \cdot x \cdot 2 \cdot x$$
$$= 2 \cdot 2 \cdot 2 \cdot x \cdot x \cdot x$$
$$= 8x^3$$

In Example 6, notice that we regrouped the factors using the commutative and associative properties for multiplication. Notice also that $2x^3$ and $(2x)^3$ do *not* represent the same number.

Example 7. Write $-x^4$ without exponents.

Solution

$$-x^4$$
$$= -1 \cdot x^4$$
$$= -1 \cdot x \cdot x \cdot x \cdot x$$

Example 8. Write $(-x)^4$ without exponents.

Solution

$$(-x)^4$$
$$= (-1x)^4$$
$$= (-1x)(-1x)(-1x)(-1x)$$
$$= (-1)(-1)(-1)(-1) \cdot x \cdot x \cdot x \cdot x$$
$$= 1 \cdot x \cdot x \cdot x \cdot x$$
$$= x \cdot x \cdot x \cdot x$$

Example 9. Evaluate a^3b^5 when $a = -1$ and $b = 2$.

Solution

$$a^3b^5$$
$$= (-1)^3(2)^5$$
$$= (-1)(-1)(-1)(2)(2)(2)(2)(2)$$
$$= (-1)(32)$$
$$= -32$$

Try to complete Example 10.

Example 10. Evaluate $\left(\frac{1}{2}x^2y\right)^3$ when $x = -1$ and $y = 4$.

Solution

$$\left(\frac{1}{2}x^2y\right)^3$$
$$= \left[\frac{1}{2}(-1)^2(4)\right]^3$$
$$= \left[\frac{1}{2} \cdot 1 \cdot 4\right]^3$$
$$= [\_\_\_\_]^3$$
$$= \_\_\_\_$$

Check your work on page 249. ▶

We note here that zero and one can be used as exponents if we follow these rules.

One as an Exponent	**Zero as an Exponent**
$a^1 = a$	$a^0 = 1$ provided that $a \neq 0$

Example 11. Find $(3y)^0$.

Solution

$$(3y)^0$$
$$= 1$$

Example 12. Find $3y^0$.

Solution

$$3y^0$$
$$= 3 \cdot 1$$
$$= 3$$

Example 13. Find $5x^0y^1$.

Solution

$$5x^0y^1$$
$$= 5 \cdot 1 \cdot y$$
$$= 5y$$

Now you try Example 14.

Example 14. Find $4(7xyz)^0$.

Solution

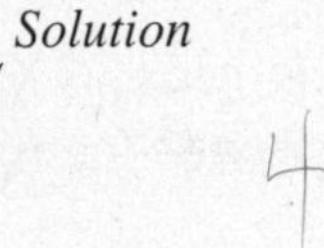

Check your work on page 250. ▶

Trial Run

______ **1.** Use exponents to rewrite $2aaabb$.

______ **2.** Write $-y^2$ without exponents.

______ **3.** Write $(3x)^3$ without parentheses.

______ **4.** Evaluate $(-2)^4$.

______ **5.** Evaluate $(-3x)^2$ when $x = 2$.

______ **6.** Evaluate $-3x^2$ when $x = 2$.

______ **7.** Evaluate $(2x^5y)^3$ when $x = -1$ and $y = 3$.

______ **8.** Find $3x^1y^0$.

Answers are on page 250.

Multiplying Powers

Suppose that we consider some products of powers of the same base.

$$3^2 \cdot 3^4 = 3 \cdot 3 \cdot 3 \cdot 3 \cdot 3 \cdot 3 \qquad = 3^6$$

$$x^3 \cdot x = x \cdot x \cdot x \cdot x \qquad = x^4$$

$$y^4 \cdot y^3 = y \cdot y \cdot y \cdot y \cdot y \cdot y \cdot y = y^7$$

$$a^2 \cdot a^2 = a \cdot a \cdot a \cdot a \qquad = a^4$$

Do you see a rule that we could use to multiply powers? The base in each answer is the original base, and the exponent in each answer is the *sum* of the original exponents.

$$3^2 \cdot 3^4 = 3^{2+4} = 3^6$$

$$x^2 \cdot x = x^{3+1} = x^4$$

$$y^3 \cdot y^4 = y^{3+4} = y^7$$

$$a^2 \cdot a^2 = a^{2+2} = a^4$$

To multiply powers of the *same* base, we keep the same base and *add* the exponents.

This rule is called the **First Law of Exponents**, and we may state it in general, using a to represent any base and m and n to represent any whole numbers.

First Law of Exponents

$$a^m \cdot a^n = a^{m+n}$$

Example 15. Simplify $x^6 \cdot x^9$.

Solution

$$x^6 \cdot x^9$$

$$= x^{6+9}$$

$$= x^{15}$$

You try Example 16.

Example 16. Simplify $y \cdot y^3 \cdot y^5$.

Solution

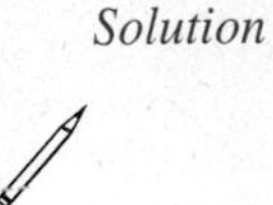

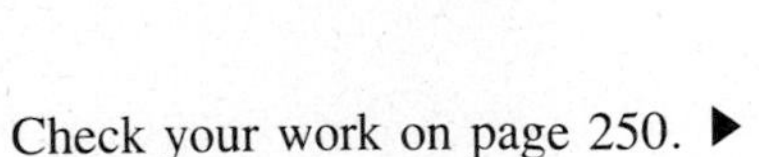

Check your work on page 250. ▶

Example 17. Simplify $x^2 \cdot y$.

Solution

$$x^2 \cdot y$$

$$= x^2y$$

The bases are different, so we cannot add exponents.

Example 18. Simplify $a^{11} \cdot a^0$.

Solution

$$a^{11} \cdot a^0$$

$$= a^{11+0}$$

$$= a^{11}$$

Example 18 should help you see why a^0 must be defined to be 1. The only number that leaves another number unchanged in multiplication is 1, the identity for multiplication. Remember:

$$a^0 = 1 \quad \text{provided that} \quad a \neq 0$$

If a product involves powers of several bases, we may use the associative and commutative properties of multiplication to rearrange the factors and then use the first law of exponents to multiply powers of the same base.

Example 19. Simplify $2x^2y \cdot xy^3$.

Solution

$$2x^2y \cdot xy^3$$
$$= 2 \cdot x^2 \cdot x \cdot y \cdot y^3$$
$$= 2x^{2+1} \cdot y^{1+3}$$
$$= 2x^3y^4$$

You complete Example 20.

Example 20. Simplify $abc \cdot a^2bc^4$.

Solution

$$abc \cdot a^2bc^4$$
$$= a \cdot a^2 \cdot b \cdot b \cdot c \cdot c^4$$

Check your work on page 250. ▶

Trial Run

Find the products.

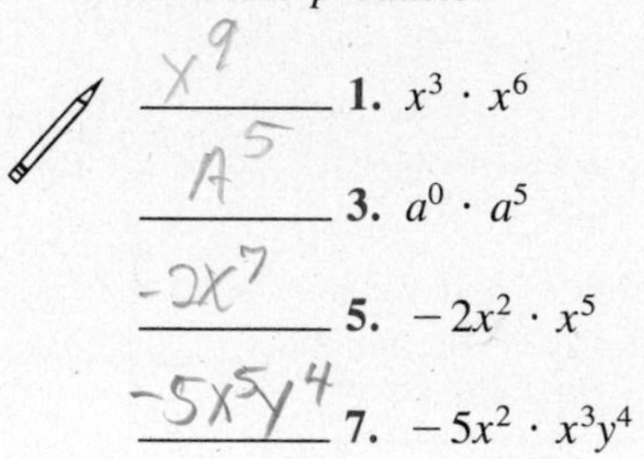

1. $x^3 \cdot x^6$
2. $x^3 \cdot y^2$
3. $a^0 \cdot a^5$
4. $y^2 \cdot y \cdot y^3$
5. $-2x^2 \cdot x^5$
6. $3a \cdot a^2 \cdot a^6$
7. $-5x^2 \cdot x^3y^4$
8. $xy^5z^3 \cdot x^3yz^2$

Answers are on page 250.

Raising a Power to a Power

How would you simplify an expression such as $(x^2)^3$? We know that the exponent of 3 tells us to use the base x^2 as a factor in a product 3 times. Using what we know about the First Law of Exponents, we can simplify this expression and others.

$$(x^2)^3 = x^2 \cdot x^2 \cdot x^2 = x^6$$
$$(y^5)^2 = y^5 \cdot y^5 = y^{10}$$
$$(a^4)^5 = a^4 \cdot a^4 \cdot a^4 \cdot a^4 \cdot a^4 = a^{20}$$
$$(x^3)^3 = x^3 \cdot x^3 \cdot x^3 = x^9$$

This is called "raising a power to a power." You should see that it can be done by multiplying exponents.

$$(x^2)^3 = x^{2 \cdot 3} = x^6$$
$$(y^5)^2 = y^{5 \cdot 2} = y^{10}$$
$$(a^4)^5 = a^{4 \cdot 5} = a^{20}$$
$$(x^3)^3 = x^{3 \cdot 3} = x^9$$

To raise a power to a power, we keep the base and *multiply* the exponents.

This is called the **Second Law of Exponents** and we state it in general, using a to represent any base and m and n to represent any whole numbers.

> **Second Law of Exponents**
>
> $(a^m)^n = a^{m \cdot n}$

Example 21. Simplify $(x^7)^2$.

Solution

$$(x^7)^2$$
$$= x^{7 \cdot 2}$$
$$= x^{14}$$

Now you do Example 22.

Example 22. Simplify $(a^9)^3$.

Solution

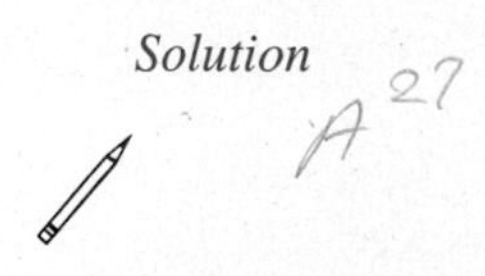

Check your work on page 250. ▶

Example 23. Simplify $(y^2)^3 \cdot (y^4)^2$.

Solution

$$(y^2)^3 \cdot (y^4)^2$$
$$= y^{2 \cdot 3} \cdot y^{4 \cdot 2}$$
$$= y^6 \cdot y^8$$
$$= y^{6+8}$$
$$= y^{14}$$

Now you try Example 24.

Example 24. Simplify $(a^5)^2 \cdot (a^3)^4$.

Solution

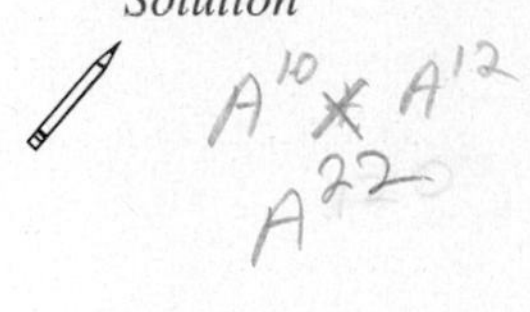

Check your work on page 250. ▶

Raising a Product to a Power

To "raise a product to a power," we again use what we know about exponents.

$$(xy)^3 = (xy)(xy)(xy) = x \cdot x \cdot x \cdot y \cdot y \cdot y = x^3y^3$$

$$(2b)^5 = (2b)(2b)(2b)(2b)(2b) = 2 \cdot 2 \cdot 2 \cdot 2 \cdot 2 \cdot b \cdot b \cdot b \cdot b \cdot b = 2^5b^5$$

Do you see how to raise a product to a power?

> To raise a product to a power, we raise each factor to that power.

This rule is called the **Third Law of Exponents** and can be written in general as

> **Third Law of Exponents**
>
> $(ab)^n = a^nb^n$

Example 25. Simplify $(2x)^7$.

Solution

$$(2x)^7$$
$$= 2^7x^7$$
$$= 128x^7$$

Now you try Example 26.

Example 26. Simplify $\left(\frac{1}{4}y\right)^3$.

Solution

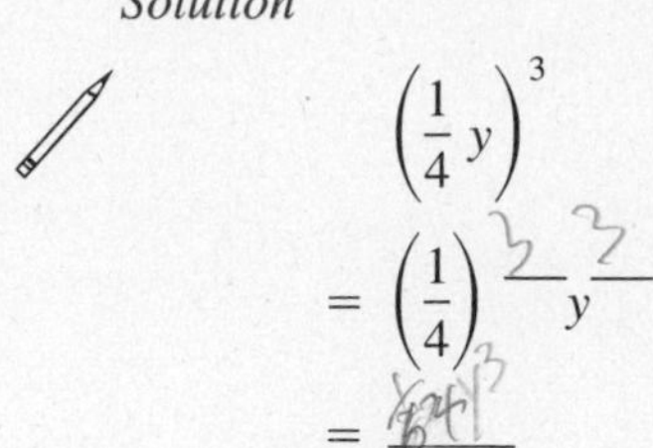

$$\left(\frac{1}{4}y\right)^3$$
$$= \left(\frac{1}{4}\right)^{\_\_} y^{\_\_}$$
$$= \_\_\_\_$$

Check your work on page 250. ▶

Students often make errors in raising products of *negative constants* and variables to a certain power. The safest way to handle such powers is to be sure to write the step in which each factor is raised to the power. Look at this power.

$(-x)^6 = (-1 \cdot x)^6$	Rewrite $-x$ as $-1x$.
$= (-1)^6x^6$	Raise each factor to the power.
$= 1 \cdot x^6$	Find the power of the constant.
$= x^6$	Use multiplication property of 1.

Writing these steps may seem tedious, but they will assure accuracy, especially when problems become more involved later in this chapter.

Example 27. Simplify $(-3y)^5$.

Solution

$$(-3y)^5$$
$$= (-3)^5y^5$$
$$= -243y^5$$

You complete Example 28.

Example 28. Simplify $(-3y)^4$.

Solution

$$(-3y)^4$$
$$= (-3)^{\_\_}y^{\_\_}$$
$$= \_\_\_\_$$

Check your work on page 250. ▶

Example 29. Simplify $(5x^3y^4)^2$.

Solution

$(5x^3y^4)^2$	
$= 5^2(x^3)^2(y^4)^2$	Raise each factor to the power.
$= 5^2x^6y^8$	Raise each power to the power.
$= 25x^6y^8$	Simplify the constant.

Now you complete Example 30.

Example 30. Simplify $x(x^2z)^5$.

Solution

$x(x^2z)^5$	
$= x \cdot (x^2)^5 \cdot z^5$	Raise each factor to the power.
$= x \cdot x^{\_\_} \cdot z^5$	Raise each power to the power
$= \_\_\_\_$	Multiply p[illegible] of the same base.

Check your work on page 250. ▶

Trial Run

Simplify.

______ **1.** $(a^3)^4$

______ **2.** $(x^{10})^8(x^2)^5$

______ **3.** $(xy)^5$

______ **4.** $(4y)^3$

______ **5.** $(-2y)^4$

______ **6.** $(a^2b)^3$

______ **7.** $(3x^2y^4)^2$

______ **8.** $x(x^2y^3)^5$

Answers are on page 250.

Examples You Completed

Example 2. Find $(-2)^5$.

Solution

$$(-2)^5$$
$$= (-2)(-2)(-2)(-2)(-2)$$
$$= -32$$

Example 4. Write x^4y without exponents.

Solution

$$x^4y$$
$$= x \cdot x \cdot x \cdot x \cdot y$$

Example 5. Write $2x^3$ without exponents.

Solution

$$2x^3$$
$$= 2 \cdot x \cdot x \cdot x$$

Example 10. Evaluate $\left(\frac{1}{2}x^2y\right)^3$ when $x = -1$ and $y = 4$.

Solution

$$\left(\frac{1}{2}x^2y\right)^3$$
$$= \left[\frac{1}{2}(-1)^2(4)\right]^3$$
$$= \left[\frac{1}{2} \cdot 1 \cdot 4\right]^3$$
$$= [2]^3$$
$$= 2 \cdot 2 \cdot 2$$
$$= 8$$

Example 14. Find $4(7xyz)^0$.

Solution

$$\begin{aligned} &4(7xyz)^0 \\ &= 4(1) \\ &= 4 \end{aligned}$$

Example 16. Simplify $y \cdot y^3 \cdot y^5$.

Solution

$$\begin{aligned} &y \cdot y^3 \cdot y^5 \\ &= y^{1+3+5} \\ &= y^9 \end{aligned}$$

Example 20. Simplify $abc \cdot a^2bc^4$.

Solution

$$\begin{aligned} &abc \cdot a^2bc^4 \\ &= a \cdot a^2 \cdot b \cdot b \cdot c \cdot c^4 \\ &= a^{1+2}b^{1+1}c^{1+4} \\ &= a^3b^2c^5 \end{aligned}$$

Example 22. Simplify $(a^9)^3$.

Solution

$$\begin{aligned} &(a^9)^3 \\ &= a^{9 \cdot 3} \\ &= a^{27} \end{aligned}$$

Example 24. Simplify $(a^5)^2 \cdot (a^3)^4$

Solution

$$\begin{aligned} &(a^5)^2 \cdot (a^3)^4 \\ &= a^{5 \cdot 2} \cdot a^{3 \cdot 4} \\ &= a^{10} \cdot a^{12} \\ &= a^{10+12} \\ &= a^{22} \end{aligned}$$

Example 26. Simplify $\left(\frac{1}{4}y\right)^3$.

Solution

$$\begin{aligned} &\left(\frac{1}{4}y\right)^3 \\ &= \left(\frac{1}{4}\right)^3 y^3 \\ &= \frac{1}{64}y^3 \end{aligned}$$

Example 28. Simplify $(-3y)^4$.

Solution

$$\begin{aligned} &(-3y)^4 \\ &= (-3)^4y^4 \\ &= 81y^4 \end{aligned}$$

Example 30. Simplify $x(x^2z)^5$.

Solution

$$\begin{aligned} &x(x^2z)^5 \\ &= x \cdot (x^2)^5 \cdot z^5 \\ &= x \cdot x^{10} \cdot z^5 \\ &= x^{11}z^5 \end{aligned}$$

Answers to Trial Runs

page 244 **1.** $2a^3b^2$ **2.** $-1 \cdot y \cdot y$ **3.** $27x^3$ **4.** 16 **5.** 36 **6.** -12 **7.** -216 **8.** $3x$

page 246 **1.** x^9 **2.** x^3y^2 **3.** a^5 **4.** y^6 **5.** $-2x^7$ **6.** $3a^9$ **7.** $-5x^5y^4$ **8.** $x^4y^6z^5$

page 249 **1.** a^{12} **2.** x^{90} **3.** x^5y^5 **4.** $64y^3$ **5.** $16y^4$ **6.** a^6b^3 **7.** $9x^4y^8$ **8.** $x^{11}y^{15}$

IF BASE IS NEG IT WILL BE INCLOSED IN ( )
IF BASE IS NEG IT WILL BE INCLOSED IN ( )
IF BASE IS NEG IT WILL BE INCLOSED IN ( )

Name ______________________________ Date ____________

EXERCISE SET 5.1

Use exponents to rewrite each expression.

______ 1. $3xxyy$

______ 2. $-2xxxxyy$

______ 3. $-a \cdot a \cdot a \cdot a$

______ 4. $-b \cdot b \cdot b \cdot b \cdot b$

______ 5. $(-2x)(-2x)(-2x)$

______ 6. $(3x)(3x)(3x)(3x)$

______ 7. $3x \cdot x \cdot x - 4y \cdot y \cdot y$

______ 8. $x \cdot x \cdot x \cdot x + 2y \cdot y$

Write the expressions without exponents.

______ 9. $8a^3$

______ 10. $5a^4$

______ 11. $(3a)^3$

______ 12. $(2a)^5$

______ 13. $-x^4$

______ 14. $-3x^3$

______ 15. $\frac{2}{3}x^3y^3$

______ 16. $\frac{4}{7}x^4y^2$

Evaluate the expressions.

______ 17. $(-3)^3$

______ 18. $(-2)^5$

______ 19. -2^4

______ 20. -3^4

______ 21. $(2x)^2$ when $x = -1$

______ 22. $(-3x)^2$ when $x = -3$

______ 23. $(xy^2)^3$ when $x = 1$ and $y = -2$

______ 24. $(3x^5y)^2$ when $x = 1$ and $y = 2$

______ 25. $(0.2xy^3)^0$ when $x = 5$ and $y = 3$

______ 26. $(0.3xy^5)^0$ when $x = 2$ and $y = -1$

Simplify.

______ 27. $x^2 \cdot x^3$

______ 28. $a^5 \cdot a^2$

______ 29. $a^0 \cdot a^7$

______ 30. $a^9 \cdot a^0$

______ 31. $y^3 \cdot y \cdot y^5$

______ 32. $y^2 \cdot y \cdot y^7$

______ 33. $-2a^5 \cdot a^7$

______ 34. $-3a^4 \cdot a^6$

______ 35. $x^2y^3 \cdot x^5y^2$

______ 36. $x^3y^4 \cdot x^2y^5$

______ 37. $2m^2 \cdot m \cdot m^5$

______ 38. $3m^4 \cdot m^3 \cdot m$

______ 39. $\frac{1}{2}x^4y^3 \cdot x^2y \cdot x^3$

______ 40. $\frac{-1}{5}x^5y^2 \cdot xy^2 \cdot y$

_______ **41.** $0.5x^2y^3z \cdot x^2yz^4$

_______ **42.** $0.9x^2y^3z^5 \cdot xyz^2$

_______ **43.** $3x^2y \cdot x^7z^2$

_______ **44.** $5xy^3 \cdot x^5z^3$

_______ **45.** $(a^2)^3$

_______ **46.** $(a^3)^2$

_______ **47.** $(xy)^5$

_______ **48.** $(xy)^7$

_______ **49.** $(7x)^2$

_______ **50.** $(8y)^2$

_______ **51.** $\left(\frac{-1}{3}y^2\right)^2$

_______ **52.** $\left(\frac{-1}{5}y^2\right)^2$

_______ **53.** $(-4x)^3$

_______ **54.** $(-3x)^3$

_______ **55.** $(x^2)^3 \cdot (x^3)^4$

_______ **56.** $(x^4)^2 \cdot (x^2)^5$

_______ **57.** $(9x^2y^3)^2$

_______ **58.** $(8x^2y^4)^2$

_______ **59.** $(-2xy^3)^4$

_______ **60.** $(-3x^2y)^5$

_______ **61.** $\frac{1}{4}x^2 \cdot (xy^2)^3$

_______ **62.** $\frac{1}{3}x^2 \cdot (x^2y)^4$

☆ Stretching the Topics

Simplify.

_______ **1.** $\left(\frac{-1}{2}x^3y^2z\right)^3 \cdot (xy^2z^3)^5$

_______ **2.** $7xy(-2xy)(-3x^2y)$

_______ **3.** $9x^2y^3z\left(\frac{-1}{3}x^5yz^4\right)^2$

_______ **4.** $(2ab^2)^3 \cdot (-a^3b^2c)^4 \cdot (3a^2c^0)^2$

Check your answers in the back of your book.

If you can complete **Checkup 5.1**, you are ready to go on to Section 5.2.

Name ______________________ Date ____________

CHECKUP 5.1

_______ **1.** Evaluate $(-2)^4$.

_______ **2.** Use exponents to write $\frac{-1}{3} x \cdot x \cdot x \cdot x$.

_______ **3.** Write $0.5x^3y^2$ without exponents.

_______ **4.** Evaluate $(-3x^2y)^3$ when $x = -1$ and $y = 2$.

Simplify.

_______ **5.** $x^6 \cdot x^2$

_______ **6.** $-y^3 \cdot y^2 \cdot y$

_______ **7.** $-5x^2y^3 \cdot xy^4$

_______ **8.** $(a^3)^2 \cdot (a^4)^3$

_______ **9.** $\left(\frac{2}{3}x\right)^2$

_______ **10.** $(-2xy)^3$

Check your answers in the back of your book.

If You Missed Problems:	You Should Review Examples:
1	1, 2
2	3, 4
3	5–8
4	9, 10
5, 6	15, 16
7	19, 20
8	23, 24
9, 10	25–29

5.2 Dividing with Whole-Number Exponents

To understand how to simplify quotients of powers, we can recall our earlier work with fractions.

Dividing Powers of the Same Base

Let's look at some quotients of powers of the same base. We shall rewrite each power using the definition of exponents.

$$\frac{6^5}{6^2} = \frac{6 \cdot 6 \cdot 6 \cdot 6 \cdot 6}{6 \cdot 6} = \frac{6 \cdot 6}{6 \cdot 6} \cdot 6 \cdot 6 \cdot 6 = 1 \cdot 6 \cdot 6 \cdot 6 = 6^3$$

$$\frac{x^7}{x^3} = \frac{x \cdot x \cdot x \cdot x \cdot x \cdot x \cdot x}{x \cdot x \cdot x} = \frac{x \cdot x \cdot x}{x \cdot x \cdot x} \cdot x \cdot x \cdot x \cdot x = 1 \cdot x \cdot x \cdot x \cdot x = x^4$$

$$\frac{a^3}{a^2} = \frac{a \cdot a \cdot a}{a \cdot a} = \frac{a \cdot a}{a \cdot a} \cdot a = 1 \cdot a = a$$

$$\frac{x^3}{x^6} = \frac{x \cdot x \cdot x}{x \cdot x \cdot x \cdot x \cdot x \cdot x} = \frac{x \cdot x \cdot x}{x \cdot x \cdot x} \cdot \frac{1}{x \cdot x \cdot x} = 1 \cdot \frac{1}{x \cdot x \cdot x} = \frac{1}{x^3}$$

Can you spot a pattern in these quotients? In each case, the exponent on the base in the answer was the *difference* between the original exponents. If the exponent in the numerator was larger than the exponent in the denominator, the power ended up in the numerator. If the exponent in the denominator was larger than the exponent in the numerator, the power ended up in the denominator.

This is the **Fourth Law of Exponents**, and it can be stated in general as follows:

Fourth Law of Exponents

$$\frac{a^m}{a^n} = a^{m-n} \quad \text{if } m \text{ is larger than or equal to } n$$

$$\frac{a^m}{a^n} = \frac{1}{a^{n-m}} \quad \text{if } n \text{ is larger than } m$$

provided that $x \neq 0$.

The statement that $x \neq 0$ arises from the fact that *division by zero is undefined*. Recall from Chapter 2 that a divisor can never be zero.

Example 1. Find $\frac{x^{17}}{x^{10}}$.

Solution

$$\frac{x^{17}}{x^{10}}$$

$$= x^{17-10}$$

$$= x^7$$

You try Example 2.

Example 2. Find $\frac{y^7}{y}$.

Solution

Check your work on page 260. ▶

Example 3. Find $\frac{x^{13}}{x^{13}}$.

Solution

$$\frac{x^{13}}{x^{13}}$$

$$= x^{13-13}$$

$$= x^0$$

$$= 1$$

Remember that $x^0 = 1$.

You complete Example 4.

Example 4. Find $\frac{x^7}{x^{12}}$.

Solution

$$\frac{x^7}{x^{12}}$$

$$= \frac{1}{x^{12-7}}$$

$$= \frac{1}{x^5}$$

Check your answers on page 260. ▶

If a quotient contains constant factors as well as variable factors, we can reduce the constant parts as we reduce a numerical fraction and use the Fourth Law of Exponents with the variable parts.

Example 5. Find $\frac{-42x^{10}}{7x}$.

Solution

$$\frac{-42x^{10}}{7x}$$

$$= \frac{-42}{7} \cdot \frac{x^{10}}{x}$$

$$= -6x^{10-1}$$

$$= -6x^9$$

Example 6. Find $\frac{3x^2}{12x^4}$.

Solution

$$\frac{3x^2}{12x^4}$$

$$= \frac{3}{12} \cdot \frac{x^2}{x^4}$$

$$= \frac{1}{4} \cdot \frac{1}{x^{4-2}}$$

$$= \frac{1}{4} \cdot \frac{1}{x^2}$$

$$= \frac{1}{4x^2}$$

If a quotient contains more than one variable, we reduce the constant parts first, then simplify each of the variable parts by subtracting exponents.

Example 7. Find $\dfrac{-2x^5y^2}{4x^4y^7}$.

Solution

$$\frac{-2x^5y^2}{4x^4y^7}$$

$$= \frac{-2}{4} \cdot \frac{x^5}{x^4} \cdot \frac{y^2}{y^7}$$

$$= \frac{-1}{2} \cdot x^{5-4} \cdot \frac{1}{y^{7-2}}$$

$$= \frac{-1}{2} \cdot x \cdot \frac{1}{y^5}$$

$$= \frac{-x}{2y^5}$$

You complete Example 8.

Example 8. Find $\dfrac{-x^7y^3z}{x^5y^8z}$.

Solution

$$\frac{-x^7y^3z}{x^5y^8z}$$

$$= -1 \cdot \frac{x^7}{x^5} \cdot \frac{y^3}{y^8} \cdot \frac{z}{z}$$

$$= -1 \cdot x^{7-5} \cdot \frac{1}{y^{8-3}} \cdot z^{1-1}$$

Check your work on page 261. ▶

Example 9. Find $\dfrac{(x^2y)^3}{(-xy)^2}$.

Solution

$$\frac{(x^2y)^3}{(-xy)^2}$$

$$= \frac{(x^2)^3\, y^3}{(-1)^2\, x^2y^2}$$ Use Third Law of Exponents in numerator and in denominator.

$$= \frac{x^6y^3}{1 \cdot x^2y^2}$$ Use Second Law of Exponents in numerator.

$$= x^{6-2}y^{3-2}$$ Use Fourth Law of Exponents.

$$= x^4y$$ Simplify.

Trial Run

Find the quotients.

_______ 1. $\dfrac{x^9}{x^2}$

_______ 2. $\dfrac{x^4}{x^5}$

_______ 3. $\dfrac{3y^3}{9y^7}$

_______ 4. $\dfrac{-24a^2b^5}{8a^3b^4}$

______ 5. $\dfrac{64a^7b^2c^3}{56a^2bc}$

______ 6. $\dfrac{(2xy^2)^4}{(-4xy^3)^3}$

Answers are on page 261.

Raising a Quotient to a Power

To "raise a quotient to a power," we can again use what we know about exponents.

$$\left(\frac{x}{y}\right)^2 = \frac{x}{y}\cdot\frac{x}{y} = \frac{x\cdot x}{y\cdot y} = \frac{x^2}{y^2}$$

$$\left(\frac{b}{2}\right)^3 = \frac{b}{2}\cdot\frac{b}{2}\cdot\frac{b}{2} = \frac{b\cdot b\cdot b}{2\cdot 2\cdot 2} = \frac{b^3}{2^3}$$

These examples should help you agree that

> To raise a quotient to a power, we raise the dividend (numerator) to that power and raise the divisor (denominator) to that power.

This rule is called the **Fifth Law of Exponents** and can be written in general as

Fifth Law of Exponents

$$\left(\frac{a}{b}\right)^n = \frac{a^n}{b^n} \quad \text{provided that } b \neq 0.$$

Example 10. Find $\left(\dfrac{x}{3}\right)^4$.

Solution

$$\left(\frac{x}{3}\right)^4$$

$$= \frac{x^4}{3^4}$$

$$= \frac{x^4}{81}$$

You complete Example 11.

Example 11. Find $\left(\dfrac{-2}{y}\right)^5$.

Solution

$$\left(\frac{-2}{y}\right)^5$$

$$= \frac{(-2)^5}{y^5}$$

$$= \frac{\quad}{y^5}$$

Check your answer on page 261. ▶

Example 12. Find $\left(\dfrac{-2a^3}{3c}\right)^4$.

Solution

$$\left(\frac{-2a^3}{3c}\right)^4$$

$$= \frac{(-2a^3)^4}{(3c)^4} \qquad \text{Raise numerator and denominator to power.}$$

$$= \frac{(-2)^4(a^3)^4}{3^4 \cdot c^4} \qquad \text{Raise products to power.}$$

$$= \frac{16a^{12}}{81c^4} \qquad \text{Raise power to power and simplify constants.}$$

To find a power of a quotient containing common variables in the numerator and denominator, we should simplify the quotient *first* before we raise it to the power.

Example 13. Find $\left(\dfrac{14a^2b^3}{7ab^9}\right)^4$.

Solution

$$\left(\frac{14a^2b^3}{7ab^9}\right)^4$$

$$= \left(\frac{14}{7} \cdot \frac{a^2}{a} \cdot \frac{b^3}{b^9}\right)^4 \qquad \text{Rewrite the quotient.}$$

$$= \left(2 \cdot a^{2-1} \cdot \frac{1}{b^{9-3}}\right)^4 \qquad \text{Use the Fourth Law of Exponents inside parentheses.}$$

$$= \left(\frac{2a}{b^6}\right)^4 \qquad \text{Simplify inside parentheses.}$$

$$= \frac{(2a)^4}{(b^6)^4} \qquad \text{Raise numerator and denominator to power.}$$

$$= \frac{2^4a^4}{b^{24}} \qquad \text{Raise products to power and raise power to power.}$$

$$= \frac{16a^4}{b^{24}} \qquad \text{Simplify constant.}$$

Let's try the mirror problem stated at the beginning of this chapter.

Example 14. Mary has a square mirror measuring s inches on each side. After framing, each side is $1\frac{1}{4}$ times as long as it was. Write an algebraic expression for the *area* of the framed mirror.

Solution: Perhaps an illustration would help here.

s s $\frac{5}{4}s$ $\frac{5}{4}s$

Each side is $1\frac{1}{4} \cdot s$ or $\frac{5}{4}s$. The *area* of a square is found by *squaring* the side.

$$\text{Area: } \left(\frac{5}{4}s\right)^2$$

$$= \left(\frac{5s}{4}\right)^2 \qquad \text{Rewrite within parentheses.}$$

$$= \frac{(5s)^2}{4^2} \qquad \text{Raise numerator and denominator to the power.}$$

$$= \frac{25s^2}{16}$$

The area of the framed mirror is $\frac{25}{16}s^2$. Note that since the unframed mirror had an area of s^2, the area of the framed mirror is $\frac{25}{16}\left(\text{or } 1\frac{9}{16}\right)$ times the area of the unframed mirror.

Trial Run

Find the powers.

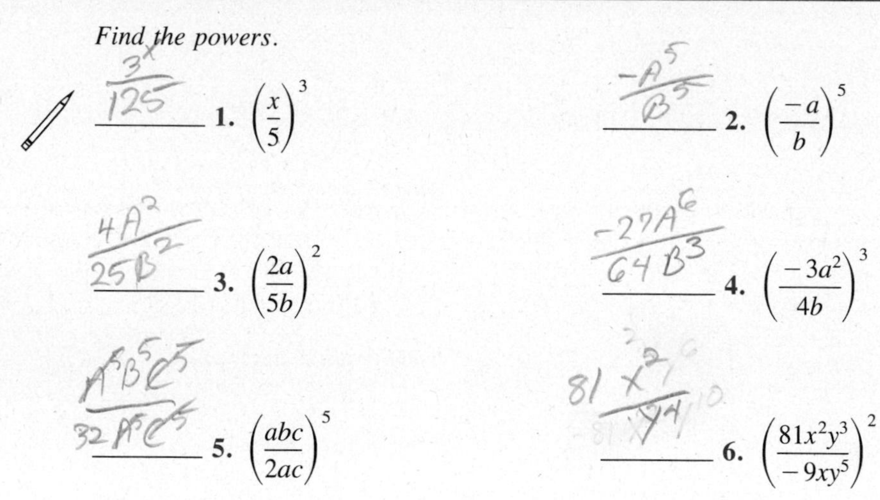

_______ 1. $\left(\frac{x}{5}\right)^3$

_______ 2. $\left(\frac{-a}{b}\right)^5$

_______ 3. $\left(\frac{2a}{5b}\right)^2$

_______ 4. $\left(\frac{-3a^2}{4b}\right)^3$

_______ 5. $\left(\frac{abc}{2ac}\right)^5$

_______ 6. $\left(\frac{81x^2y^3}{-9xy^5}\right)^2$

Answers are on page 261.

▶ Examples You Completed

Example 2. Find $\frac{y^7}{y}$.

Solution

$$\frac{y^7}{y} = \frac{y^7}{y^1}$$

$$= y^{7-1}$$

$$= y^6$$

Example 4. Find $\frac{x^7}{x^{12}}$.

Solution

$$\frac{x^7}{x^{12}}$$

$$= \frac{1}{x^{12-7}}$$

$$= \frac{1}{x^5}$$

Example 8. Find $\frac{-x^7y^3z}{x^5y^8z}$.

Solution

$$\frac{-x^7y^3z}{x^5y^8z}$$

$$= -1 \cdot \frac{x^7}{x^5} \cdot \frac{y^3}{y^8} \cdot \frac{z}{z}$$

$$= -1 \cdot x^{7-5} \cdot \frac{1}{y^{8-3}} \cdot z^{1-1}$$

$$= -1 \cdot x^2 \cdot \frac{1}{y^5} \cdot z^0$$

$$= \frac{-x^2}{y^5}$$

Example 11. Find $\left(\frac{-2}{y}\right)^5$.

Solution

$$\left(\frac{-2}{y}\right)^5$$

$$= \frac{(-2)^5}{y^5}$$

$$= \frac{-32}{y^5}$$

Answers to Trial Runs

page 257 **1.** x^7 **2.** $\frac{1}{x}$ **3.** $\frac{1}{3y^4}$ **4.** $\frac{-3b}{a}$ **5.** $\frac{8a^5bc^2}{7}$ **6.** $\frac{-x}{4y}$

page 260 **1.** $\frac{x^3}{125}$ **2.** $\frac{-a^5}{b^5}$ **3.** $\frac{4a^2}{25b^2}$ **4.** $\frac{-27a^6}{64b^3}$ **5.** $\frac{b^5}{32}$ **6.** $\frac{81x^2}{y^4}$

For handy reference, we shall again state the definitions and laws that we use in working with exponents. Remember that a stands for any base, and m and n represent whole numbers.

	Symbols
Definition of a^n	$a^n = \underbrace{a \cdot a \cdot \ldots \cdot a}_{n \text{ factors}}$
Definition of a^0	$a^0 = 1 \ (a \neq 0)$
First Law of Exponents	$a^m \cdot a^n = a^{m+n}$
Second Law of Exponents	$(a^m)^n = a^{m \cdot n}$
Third Law of Exponents	$(a \cdot b)^n = a^n \cdot b^n$
Fourth Law of Exponents	$\frac{a^m}{a^n} = a^{m-n}$ if $m \geq n$
	$\frac{a^m}{a^n} = \frac{1}{a^{n-m}}$ if $n > m$
Fifth Law of Exponents	$\left(\frac{a}{b}\right)^n = \frac{a^n}{b^n}$

Name ______________________ Date __________

EXERCISE SET 5.2

Find the quotients.

______ 1. $\frac{x^{12}}{x^5}$

______ 2. $\frac{x^{13}}{x^7}$

______ 3. $\frac{a^5}{a}$

______ 4. $\frac{a^9}{a}$

______ 5. $\frac{m^6}{m^6}$

______ 6. $\frac{m^{12}}{m^{12}}$

______ 7. $\frac{y^9}{y^{11}}$

______ 8. $\frac{y^7}{y^{10}}$

______ 9. $\frac{32x^5}{8x^2}$

______ 10. $\frac{35x^7}{5x^5}$

______ 11. $\frac{-9y^7}{3y^2}$

______ 12. $\frac{-12y^8}{3y^3}$

______ 13. $\frac{5x^2}{15x^4}$

______ 14. $\frac{6x^3}{18x^7}$

______ 15. $\frac{-21a^3}{7a^5}$

______ 16. $\frac{-60a^5}{12a^{10}}$

______ 17. $\frac{-5x^3y^2}{25xy^5}$

______ 18. $\frac{-6x^5y^3}{36xy^4}$

______ 19. $\frac{49a^3b^5}{7a^4b^7}$

______ 20. $\frac{64a^4b^7}{8a^5b^9}$

______ 21. $\frac{-x^5y^3z}{x^4y^9z}$

______ 22. $\frac{-x^7yz^3}{x^5yz^7}$

______ 23. $\frac{-12x^2yz^3}{6x^2y^3z}$

______ 24. $\frac{-15x^4y^3z^5}{5x^4yz^7}$

______ 25. $\frac{63a^5b^3c^3}{54a^2bc^5}$

______ 26. $\frac{56a^{10}b^5c^4}{42a^6b^9c^7}$

______ 27. $\frac{(xy^2)^5}{(-xy)^6}$

______ 28. $\frac{(x^3y^5)^4}{(-x^2y^2)^7}$

______ 29. $\frac{(-3xy^3)^3}{(9x^2y)^2}$

______ 30. $\frac{(-4xy^2)^2}{(2x^3y)^3}$

Find the powers.

______ 31. $\left(\frac{x}{2}\right)^3$

______ 32. $\left(\frac{x}{3}\right)^2$

_______ 33. $\left(\frac{-3}{y}\right)^3$

_______ 34. $\left(\frac{-1}{y}\right)^7$

_______ 35. $\left(\frac{x^2}{y}\right)^4$

_______ 36. $\left(\frac{x}{y^3}\right)^2$

_______ 37. $\left(\frac{x^2}{-y^3}\right)^5$

_______ 38. $\left(\frac{-x}{y^2}\right)^3$

_______ 39. $\left(\frac{2x^2}{y}\right)^4$

_______ 40. $\left(\frac{5x^3}{y^2}\right)^2$

_______ 41. $\left(\frac{-5a^3}{2b^2}\right)^2$

_______ 42. $\left(\frac{-4a^5}{3b^3}\right)^2$

_______ 43. $\left(\frac{-2x^2y}{3z}\right)^5$

_______ 44. $\left(\frac{-3xy^3}{2z}\right)^3$

_______ 45. $\left(\frac{15ab^5}{5a^3b^2}\right)^2$

_______ 46. $\left(\frac{12a^3b}{4ab^5}\right)^3$

_______ 47. $\left(\frac{-9xy^2z^5}{27x^3y^2z^2}\right)^3$

_______ 48. $\left(\frac{-5x^5y^3z}{45x^3y^3z^4}\right)^2$

_______ **49.** Mr. Jones has a square garden measuring x feet on each side. He wishes to enlarge the garden by doubling each side. Write an expression for the area of the enlarged garden.

_______ **50.** Hillary has a square tablecloth measuring x inches on each side. After bordering the tablecloth with lace, each side is $1\frac{1}{5}$ times as long as it was. Write an algebraic expression for the area of the bordered tablecloth.

☆ Stretching the Topics

Simplify.

_______ 1. $\left(\frac{-x^3y^2z}{4x^4yz^3}\right)^2 \cdot \left(\frac{-2x^5y^2z^3}{x^4y^2z^6}\right)^3$

_______ 2. $\frac{(a^2b^3)^2}{c^9} \cdot \frac{(ac^5)^3}{(-b^3)^5} \cdot \left(\frac{-b^5}{a^3c^3}\right)^2$

_______ 3. Felicia and Holly each have a picture measuring x inches on each side. Felicia framed her picture after cutting off the border. After framing, each side was $\frac{2}{3}$ as long as it was before framing. Holly framed her picture using a mat. After framing, each side was $1\frac{1}{6}$ as long as it was before framing. Find the difference between the areas of the framed pictures.

Check your answers in the back of your book.

If you can complete **Checkup 5.2**, you are ready to go on to Section 5.3.

Name ______________________ Date ____________

CHECKUP 5.2

Find the quotients.

_______ **1.** $\dfrac{x^{15}}{x^{10}}$

_______ **2.** $\dfrac{-a^7}{a^{10}}$

_______ **3.** $\dfrac{21x^9}{7x^{11}}$

_______ **4.** $\dfrac{-8a^5b^3}{16a^3b^7}$

_______ **5.** $\dfrac{x^5y^3z^2}{x^4y^3z^5}$

_______ **6.** $\dfrac{(xy^2)^5}{(-x^2y^3)^2}$

Find the powers.

_______ **7.** $\left(\dfrac{x}{2}\right)^5$

_______ **8.** $\left(\dfrac{-2}{a}\right)^4$

_______ **9.** $\left(\dfrac{-x^4}{2y}\right)^3$

_______ **10.** $\left(\dfrac{20a^4b^2}{5ab^4}\right)^3$

Check your answers in the back of your book.

If You Missed Problems:	You Should Review Examples:
1, 2	1–4
3–5	5–8
6	9
7, 8	10, 11
9	12
10	13

5.3 Working with Negative Exponents

So far we have used only whole numbers {0, 1, 2, 3, 4, . . .} as exponents, and you may have wondered if negative integers {. . ., −4, −3, −2, −1} could ever appear as exponents. The answer is yes, but let's see how to give meaning to expressions such as

$$2^{-1} \quad \text{or} \quad x^{-2} \quad \text{or} \quad y^{-3}$$

However we decide to define negative exponents, we must agree that we do not wish to lose all the laws of exponents that we have already learned. We would like negative exponents to continue to obey *all* those laws.

Let's see if we can find some meaning for negative exponents by considering the expression x^{-2}.

$$\text{By the First Law of Exponents:} \quad x^2 \cdot x^{-2} = x^{2-2} = x^0 = 1$$

$$\text{By the Property of Reciprocals:} \quad x^2 \cdot \frac{1}{x^2} = \frac{x^2}{1} \cdot \frac{1}{x^2} = 1$$

Since both products yield the same result, it makes sense to conclude that x^{-2} is the *reciprocal* of x^2, or

$$x^{-2} = \frac{1}{x^2}$$

Using similar reasoning with other negative integer exponents, you should also agree that

$$y^{-3} = \frac{1}{y^3} \qquad z^{-8} = \frac{1}{z^8}$$

$$x^{-5} = \frac{1}{x^5} \qquad 2^{-1} = \frac{1}{2^1} = \frac{1}{2}$$

Now we are ready to state the general definition for a base raised to a negative power.

Definition of Negative Exponent

$$a^{-n} = \frac{1}{a^n} \qquad \text{provided that } a \neq 0$$

It is important to realize that a^n and a^{-n} are *reciprocals* for each other. In particular, a^{-1} is the reciprocal of a; in other words, $a^{-1} = \frac{1}{a}$.

See if you can complete Example 1.

Example 1. Write 2^{-5}, y^{-4}, x^{-1}, $(-3)^{-2}$ without negative exponents.

Solution

Using Negative Exponents	Using Positive Exponents
2^{-5}	$\frac{1}{2^5} = \frac{1}{32}$
y^{-4}	____
x^{-1}	____
$(-3)^{-2}$	$\frac{1}{(-3)^2} = \frac{1}{—}$

Check your work on page 276. ▶

Example 2. Write $\left(\frac{2}{3}\right)^{-1}$ without negative exponents.

Solution: $\left(\frac{2}{3}\right)^{-1}$ is the *reciprocal* of $\frac{2}{3}$. Therefore,

$$\left(\frac{2}{3}\right)^{-1} = \frac{3}{2}$$

Example 3. Write $(2x)^{-1}$ without negative exponents.

Solution

$$(2x)^{-1} = \frac{1}{(2x)^1}$$
$$= \frac{1}{2x}$$

Example 4. Write $2x^{-1}$ without negative exponents.

Solution

$$2x^{-1} = 2 \cdot \frac{1}{x^1}$$
$$= \frac{2}{1} \cdot \frac{1}{x}$$
$$= \frac{2}{x}$$

We should notice two important facts from these examples.

1. *A negative exponent has no effect on the sign of the base.* The negative exponent merely says "rewrite the expression as 1 divided by the base raised to the opposite (positive) power." The sign of the answer is determined *after* the expression is rewritten with a positive exponent. So

$$(-1)^{-4} = \frac{1}{(-1)^4} = \frac{1}{1} = 1 \qquad \text{because} \quad (-1)^4 = 1$$

$$(-1)^{-5} = \frac{1}{(-1)^5} = \frac{1}{-1} = -1 \quad \text{because} \quad (-1)^5 = -1$$

2. *A negative exponent applies only to the base to which it is attached, unless parentheses indicate otherwise.* Recall that this fact was true for positive exponents also. So

$$3 \cdot 5^2 = 3 \cdot 25 = 75 \quad \text{but} \quad (3 \cdot 5)^2 = 15^2 = 225$$

$$2x^{-3} = 2 \cdot \frac{1}{x^3} = \frac{2}{x^3} \quad \text{but} \quad (2x)^{-3} = \frac{1}{(2x)^3} = \frac{1}{2^3x^3} = \frac{1}{8x^3}$$

Trial Run

Write each expression without negative exponents.

_____ 1. $(5)^{-2}$ _____ 2. $(-2)^{-3}$

_____ 3. $\left(\frac{3}{4}\right)^{-1}$ _____ 4. $(3a)^{-1}$

_____ 5. $5x^{-2}$ _____ 6. $(5x)^{-2}$

Answers are on page 278.

Negative exponents obey all our laws of exponents. Perhaps you should look back at those laws stated earlier in this chapter before trying to simplify some expressions. Let's agree to use positive exponents in our answers.

The **First Law of Exponents** tells us how to multiply factors that are powers of the same base.

Example 5. Simplify $x^2 \cdot x^{-7}$.

Solution

$$\begin{aligned} x^2 \cdot x^{-7} &= x^{2+(-7)} \\ &= x^{-5} \\ &= \frac{1}{x^5} \end{aligned}$$

Let's rework Example 5 using the definition of a negative exponent, just to verify that the First Law of Exponents does work with negative exponents.

$$\begin{aligned} x^2 \cdot x^{-7} &= \frac{x^2}{1} \cdot \frac{1}{x^7} \\ &= \frac{x^2}{x^7} \\ &= \frac{1}{x^5} \end{aligned}$$

Now complete Example 6.

Example 6. Simplify $a^{-3} \cdot a^{-4}$.

Solution

$$\begin{aligned} a^{-3} \cdot a^{-4} &= a^{-3+(-4)} \\ &= a^{\_\_\_} \\ &= \frac{1}{\_\_\_} \end{aligned}$$

Check your work on page 276. ▶

Example 7. Simplify $(2x^{-3})(x^{-1})$.

Solution

$$\begin{aligned} (2x^{-3})(x^{-1}) &= 2 \cdot x^{-3} \cdot x^{-1} \\ &= 2x^{-3+(-1)} \\ &= 2x^{-4} \\ &= \frac{2}{1} \cdot \frac{1}{x^4} \\ &= \frac{2}{x^4} \end{aligned}$$

Example 8. Simplify $x^{-3} \cdot x^5 \cdot x^{-1}$.

Solution

$$\begin{aligned} x^{-3} \cdot x^5 \cdot x^{-1} &= x^{-3+5+(-1)} \\ &= x^1 \\ &= x \end{aligned}$$

Trial Run

Simplify. Write answers without negative exponents.

_______ **1.** $a^4 \cdot a^{-1}$

_______ **2.** $x^{-3} \cdot x^{-5} \cdot x$

_______ **3.** $(3x^{-6})(x^2)$

_______ **4.** $(-5x^3)(x^{-3})$

Answers are on page 278.

We can use the **Second Law of Exponents** to raise a power to a power.

Simplify $(x^{-3})^2$.

$$\begin{aligned} (x^{-3})^2 &= x^{-3(2)} \\ &= x^{-6} \\ &= \frac{1}{x^6} \end{aligned}$$

Let's rework Example 9 using the definition of a negative exponent, just to verify that the Second Law of Exponents does work with negative exponents.

$$\begin{aligned} (x^{-3})^2 &= \left(\frac{1}{x^3}\right)^2 \\ &= \frac{1^2}{(x^3)^2} \\ &= \frac{1}{x^6} \end{aligned}$$

Now you try Example 10.

Example 10. Simplify $(x^{-5})^{-2}$.

Solution

Check your work on page 277. ▶

Example 11. Simplify $(x^{10})^{-1} \cdot (y^{-4})^{-2}$.

Solution

$$\begin{aligned} (x^{10})^{-1}(y^{-4})^{-2} &= x^{10(-1)}\, y^{-4(-2)} \\ &= x^{-10} \cdot y^8 \\ &= \frac{1}{x^{10}} \cdot \frac{y^8}{1} \\ &= \frac{y^8}{x^{10}} \end{aligned}$$

Example 12. Simplify $(x^{-1})^9 \cdot (x^2)^{-3}$.

Solution

$$\begin{aligned}(x^{-1})^9 \cdot (x^2)^{-3} &= x^{-1(9)} \cdot x^{2(-3)}\\ &= x^{-9} \cdot x^{-6}\\ &= x^{-9+(-6)}\\ &= x^{-15}\\ &= \frac{1}{x^{15}}\end{aligned}$$

Now you try Example 13.

Example 13. Simplify $x^5 \cdot (x^{-3})^2$.

Solution

Check your work on page 277. ▶

The **Third Law of Exponents** gives us a method for raising a product to a power.

Example 14. Simplify $(2a)^{-3}$.

Solution

$$\begin{aligned}&(2a)^{-3}\\ &= 2^{-3}a^{-3}\\ &= \frac{1}{2^3} \cdot \frac{1}{a^3}\\ &= \frac{1}{8} \cdot \frac{1}{a^3}\\ &= \frac{1}{8a^3}\end{aligned}$$

Let's rework Example 14 using the definition of a negative exponent to verify that the Third Law of Exponents works for negative exponents.

$$\begin{aligned}(2a)^{-3} &= \frac{1}{(2a)^3}\\ &= \frac{1}{2^3a^3}\\ &= \frac{1}{8a^3}\end{aligned}$$

Example 15. Simplify $(-5x)^{-2}$.

Solution

$$\begin{aligned}(-5x)^{-2} &= (-5)^{-2} \cdot x^{-2}\\ &= \frac{1}{(-5)^2} \cdot \frac{1}{x^2}\\ &= \frac{1}{25} \cdot \frac{1}{x^2}\\ &= \frac{1}{25x^2}\end{aligned}$$

Now you complete Example 16.

Example 16. Simplify $(3x^{-1})^4$.

Solution

$$\begin{aligned}(3x^{-1})^4 &= 3^4(x^{-1})^4\\ &= 3^4x^{\_\_}\\ &= 81 \cdot \frac{1}{x^{\_}}\\ &= \_\_\_\_\end{aligned}$$

Check your work on page 277. ▶

Example 17. Simplify $(2^{-1}x^{-3}y^{-5})^{-2}$.

Solution

$$\begin{aligned}(2^{-1}x^{-3}y^{-5})^{-2} &= (2^{-1})^{-2}(x^{-3})^{-2}(y^{-5})^{-2} \\ &= 2^2x^6y^{10} \\ &= 4x^6y^{10}\end{aligned}$$

From this example, notice that a step can be skipped if you realize that you must multiply the exponent on each factor within the parentheses times the exponent on the entire quantity. You may be able to do this computation in your head if you are careful.

Example 18. Simplify $(2^{-2}x^4z^{-1})^{-3}$.

Solution

$$\begin{aligned}(2^{-2}x^4z^{-1})^{-3} &= 2^6x^{-12}z^3 \\ &= 64 \cdot \frac{1}{x^{12}} \cdot z^3 \\ &= \frac{64z^3}{x^{12}}\end{aligned}$$

Now complete Example 19.

Example 19. Simplify $(2^{-1}x^2)^{-4}(2x^{-2})^{-3}$.

Solution

$$\begin{aligned}&(2^{-1}x^2)^{-4}(2x^{-2})^{-3} \\ &= 2^4x^{-8} \cdot 2^{-3}x^6 \\ &= 2^4 \cdot 2^{-3} \cdot x^{-8} \cdot x^6\end{aligned}$$

Check your work on page 277. ▶

Trial Run

Simplify. Write answers without negative exponents.

_______ 1. $(x^{-4})^2$

_______ 2. $(a^{-3})^{-4}$

_______ 3. $(y^5)^{-1}(y^{-3})^{-2}$

_______ 4. $3x^{-2}(x^{-1})^3$

_______ 5. $(3x)^{-2}$

_______ 6. $(2a^{-2})^3$

_______ 7. $(3^{-1}y^4)^{-3}$

_______ 8. $(2a^2b^{-3})^2$

Answers are on page 278.

The **Fourth Law of Exponents** involves division. Now that we are allowed to use negative exponents, we need only one version of that law.

> **Fourth Law of Exponents**
>
> $$\frac{a^m}{a^n} = a^{m-n} \qquad \text{provided that } a \neq 0$$

Let's see why our new version for this law works whether the exponent in the denominator is less than or greater than the exponent in the numerator.

Using Old Version	Using New Version
$\frac{x^5}{x^2} = x^{5-2}$	$\frac{x^5}{x^2} = x^{5-2}$
$= x^3$	$= x^3$
$\frac{a^6}{a^{10}} = \frac{1}{a^{10-6}}$	$\frac{a^6}{a^{10}} = a^{6-10}$
$= \frac{1}{a^4}$	$= a^{-4}$
	$= \frac{1}{a^4}$

Since the results are the same, we agree to use the new version of the Fourth Law of Exponents.

Example 20. Simplify $\frac{x^9}{x^{-6}}$.

Solution

$$\frac{x^9}{x^{-6}} = x^{9-(-6)}$$

$$= x^{9+6}$$

$$= x^{15}$$

You try Example 21.

Example 21. Simplify $\frac{10x^{-8}}{2x^{-2}}$.

Solution

$$\frac{10x^{-8}}{2x^{-2}} = \frac{10}{2} \cdot x^{-8-(-2)}$$

Check your work on page 277. ▶

Example 22. Simplify $\frac{2x^{-3}y^{-4}}{6x^8y^{-5}}$.

Solution

$$\frac{2x^{-3}y^{-4}}{6x^8y^{-5}} = \frac{2}{6} \cdot x^{-3-8} \cdot y^{-4-(-5)}$$

$$= \frac{1}{3} \cdot x^{-11} \cdot y^{-4+5}$$

$$= \frac{1}{3} \cdot x^{-11} \cdot y$$

$$= \frac{1}{3} \cdot \frac{1}{x^{11}} \cdot \frac{y}{1}$$

$$= \frac{y}{3x^{11}}$$

You complete Example 23.

Example 23. Simplify $\frac{(-9x^{-3})(x^{-5})}{15x^{-10}}$.

Solution

$$\frac{(-9x^{-3})(x^{-5})}{15x^{-10}}$$

$$= \frac{-9}{15} \cdot \frac{x^{\underline{\quad}}}{x^{-10}}$$

Check your work on page 277. ▶

Trial Run

Simplify. Write answers without negative exponents.

______ 1. $\dfrac{x^{10}}{x^{-6}}$

______ 2. $\dfrac{a^{-4}}{a^4}$

______ 3. $\dfrac{12y^{-7}}{6y^{-3}}$

______ 4. $\dfrac{3a^{-2}b^{-3}}{9a^3b^{-6}}$

______ 5. $\dfrac{(-3x^{-2})(x^{-3})}{9x^{-11}}$

______ 6. $\dfrac{(2x^3)^{-2}}{x^{-10}}$

Answers are on page 278.

The **Fifth Law of Exponents** provides the rule for raising a quotient (or fraction) to a power.

Example 24. Simplify $\left(\dfrac{x}{3}\right)^{-2}$.

Solution

$$\begin{aligned}\left(\frac{x}{3}\right)^{-2} &= \frac{x^{-2}}{3^{-2}}\\ &= x^{-2} \div 3^{-2}\\ &= \frac{1}{x^2} \div \frac{1}{3^2}\\ &= \frac{1}{x^2} \cdot \frac{3^2}{1}\\ &= \frac{3^2}{x^2}\\ &= \frac{9}{x^2}\end{aligned}$$

Example 25. Simplify $\left(\dfrac{2a^2}{b^{-5}}\right)^{-4}$.

Solution

$$\begin{aligned}\left(\frac{2a^2}{b^{-5}}\right)^{-4} &= \frac{2^{-4}a^{2(-4)}}{b^{-5(-4)}}\\ &= \frac{2^{-4}a^{-8}}{b^{20}}\\ &= \frac{1}{2^4} \cdot \frac{1}{a^8} \cdot \frac{1}{b^{20}}\\ &= \frac{1}{16a^8b^{20}}\end{aligned}$$

Notice from Examples 24 and 25 that we must first use the laws of exponents to simplify an expression. After simplifying, we may write our answers using positive exponents if we observe the following:

Observation	Examples
A factor in the numerator with a negative exponent should be moved to the denominator and the exponent will be positive.	$\frac{5x^{-3}}{4} = \frac{5}{4x^3}$
A factor in the denominator with a negative exponent should be moved to the numerator and the exponent will be positive.	$\frac{7}{x^{-1}} = 7x$ $\frac{x^{-2}}{3^{-2}} = \frac{3^2}{x^2} = \frac{9}{x^2}$
A factor with a positive exponent stays where it is.	$\frac{2^{-4}a^{-8}}{b^{20}} = \frac{1}{2^4a^8b^{20}} = \frac{1}{16a^8b^{20}}$

These observations make it much easier to rewrite a simplified fraction using positive exponents. Just be sure you have used the laws of exponents to simplify the expression first.

Example 26. Simplify $\left(\frac{2y^{-5}}{3y^3}\right)^{-2}$.

Solution

$$\left(\frac{2y^{-5}}{3y^3}\right)^{-2} = \left(\frac{2y^{-5-3}}{3}\right)^{-2}$$ Simplify the quotient inside parentheses by subtracting exponents.

$$= \left(\frac{2y^{-8}}{3}\right)^{-2}$$ Simplify the exponent.

$$= \frac{2^{-2}y^{16}}{3^{-2}}$$ Raise the numerator and denominator to the power.

$$= \frac{3^2y^{16}}{2^2}$$ Move factors to have positive exponents.

$$= \frac{9y^{16}}{4}$$ Simplify the constants.

Example 27. Simplify $\left(\frac{-x^3}{y^{-2}}\right)^{-3}$

Solution

$$\left(\frac{-x^3}{y^{-2}}\right)^{-3} = \left(\frac{-1 \cdot x^3}{y^{-2}}\right)^{-3}$$ Rewrite the numerator.

$$= \frac{(-1)^{-3}x^{-9}}{y^6}$$ Raise the numerator and denominator to the power.

$$= \frac{1}{(-1)^3x^9 \cdot y^6}$$ Move factors to have positive exponents.

$$= \frac{1}{-1 \cdot x^9y^6}$$ Simplify $(-1)^3$.

$$= \frac{-1}{x^9y^6}$$ Let the numerator "carry the sign."

Trial Run

Simplify. Write answers without negative exponents.

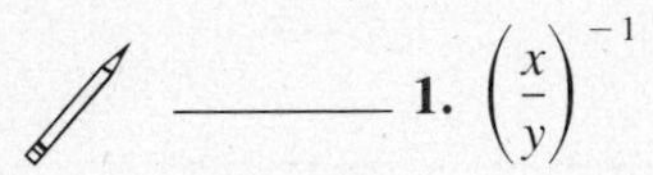

_______ 1. $\left(\frac{x}{y}\right)^{-1}$

_______ 2. $\left(\frac{a^{-2}}{b^3}\right)^{-3}$

_______ 3. $\left(\frac{x^{-4}}{y^{-2}}\right)^{-1}$

_______ 4. $\left(\frac{3x^2}{y^{-3}}\right)^{-2}$

Answers are on page 278.

Examples You Completed

Example 1. Write 2^{-5}, y^{-4}, x^{-1}, and $(-3)^{-2}$ without negative exponents.

Solution

Using Negative Exponents	Using Positive Exponents
2^{-5}	$\frac{1}{2^5} = \frac{1}{32}$
y^{-4}	$\frac{1}{y^4}$
x^{-1}	$\frac{1}{x}$
$(-3)^{-2}$	$\frac{1}{(-3)^2} = \frac{1}{9}$

Example 6. Simplify $a^{-3} \cdot a^{-4}$.

Solution

$$a^{-3} \cdot a^{-4} = a^{-3+(-4)}$$
$$= a^{-7}$$
$$= \frac{1}{a^7}$$

Example 10. Simplify $(x^{-5})^{-2}$.

Solution

$$(x^{-5})^{-2} = x^{-5(-2)}$$
$$= x^{10}$$

Example 13. Simplify $x^5 \cdot (x^{-3})^2$.

Solution

$$\begin{aligned} x^5 \cdot (x^{-3})^2 &= x^5 \cdot x^{-3(2)} \\ &= x^5 \cdot x^{-6} \\ &= x^{5+(-6)} \\ &= x^{-1} \\ &= \frac{1}{x} \end{aligned}$$

Example 16. Simplify $(3x^{-1})^4$.

Solution

$$\begin{aligned} (3x^{-1})^4 &= 3^4(x^{-1})^4 \\ &= 3^4x^{-4} \\ &= 81 \cdot \frac{1}{x^4} \\ &= \frac{81}{x^4} \end{aligned}$$

Example 19. Simplify $(2^{-1}x^2)^{-4}(2x^{-2})^{-3}$.

Solution

$$\begin{aligned} &(2^{-1}x^2)^{-4}(2x^{-2})^{-3} \\ &= 2^4x^{-8} \cdot 2^{-3}x^6 \\ &= 2^4 \cdot 2^{-3} \cdot x^{-8} \cdot x^6 \\ &= 2^1 \cdot x^{-2} \\ &= 2 \cdot \frac{1}{x^2} \\ &= \frac{2}{x^2} \end{aligned}$$

Example 21. Simplify $\frac{10x^{-8}}{2x^{-2}}$.

Solution

$$\begin{aligned} \frac{10x^{-8}}{2x^{-2}} &= \frac{10}{2} \cdot x^{-8-(-2)} \\ &= 5 \cdot x^{-8+2} \\ &= 5 \cdot x^{-6} \\ &= 5 \cdot \frac{1}{x^6} \\ &= \frac{5}{x^6} \end{aligned}$$

Example 23. Simplify $\frac{(-9x^{-3})(x^{-5})}{15x^{-10}}$.

Solution

$$\begin{aligned} &\frac{-9 \cdot x^{-3} \cdot x^{-5}}{15x^{-10}} \\ &= \frac{-9}{15} \cdot \frac{x^{-8}}{x^{-10}} \\ &= \frac{-3}{5} \cdot x^{-8-(-10)} \\ &= \frac{-3}{5} x^{-8+10} \\ &= \frac{-3}{5} \cdot \frac{x^2}{1} \\ &= \frac{-3x^2}{5} \end{aligned}$$

Answers to Trial Runs

page 269 **1.** $\frac{1}{25}$ **2.** $\frac{-1}{8}$ **3.** $\frac{4}{3}$ **4.** $\frac{1}{3a}$ **5.** $\frac{5}{x^2}$ **6.** $\frac{1}{25x^2}$

page 270 **1.** a^3 **2.** $\frac{1}{x^7}$ **3.** $\frac{3}{x^4}$ **4.** -5

page 272 **1.** $\frac{1}{x^8}$ **2.** a^{12} **3.** y **4.** $\frac{3}{x^5}$ **5.** $\frac{1}{9x^2}$ **6.** $\frac{8}{a^6}$ **7.** $\frac{27}{y^{12}}$ **8.** $\frac{4a^4}{b^6}$

page 274 **1.** x^{16} **2.** $\frac{1}{a^8}$ **3.** $\frac{2}{y^4}$ **4.** $\frac{b^3}{3a^5}$ **5.** $\frac{-x^6}{3}$ **6.** $\frac{x^4}{4}$

page 276 **1.** $\frac{y}{x}$ **2.** a^6b^9 **3.** $\frac{x^4}{y^2}$ **4.** $\frac{1}{9x^4y^6}$

Name ______________________ Date __________

EXERCISE SET 5.3

Simplify using the laws of exponents. Give answers with positive exponents.

______ **1.** $(2x^2)^3$

______ **2.** $(-3x^4)^2$

______ **3.** $\frac{a^5}{a^2}$

______ **4.** $\frac{a^9}{a^3}$

______ **5.** $\frac{5y^4}{y^7}$

______ **6.** $\frac{4y^5}{y^9}$

______ **7.** $\left(\frac{-3a^2}{b}\right)^3$

______ **8.** $\left(\frac{-2a^3}{b}\right)^5$

______ **9.** $(3x)^{-2}$

______ **10.** $(2x)^{-4}$

______ **11.** $(-5)^{-3}$

______ **12.** $(-4)^{-2}$

______ **13.** $\left(\frac{4}{5}\right)^{-2}$

______ **14.** $\left(\frac{2}{3}\right)^{-3}$

______ **15.** $(6x)^{-2}$

______ **16.** $(2x)^{-3}$

______ **17.** $6x^{-2}$

______ **18.** $2x^{-3}$

______ **19.** $4y^7 \cdot y^{-7}$

______ **20.** $5y^9y^{-9}$

______ **21.** $a^8 \cdot a^{-11}$

______ **22.** $a^{-9} \cdot a^6$

______ **23.** $(2x^{-3})(x^7)$

______ **24.** $(4x^{-2})(x^5)$

______ **25.** $(-5x^{-1})(x^{-2})$

______ **26.** $(-8x^{-2})(x^{-3})$

______ **27.** $(a^{-6})^2$

______ **28.** $(a^{-7})^3$

______ **29.** $(y^9)^{-1} \cdot (y^{-5})^{-2}$

______ **30.** $(y^{-3})^{-2} \cdot (y^5)^{-1}$

______ **31.** $9x^{-3}(x^{-7})^0$

______ **32.** $8x^{-7}(x^{-11})^0$

______ **33.** $(4x)^{-3}$

______ **34.** $(9x)^{-2}$

______ **35.** $(3a^{-4})^3$

______ **36.** $(6a^{-5})^2$

______ **37.** $(2^3y^{-5})^{-2}$

______ **38.** $(3y^{-7})^{-3}$

______ **39.** $(5a^3b^{-5})^2$

______ **40.** $(6a^{-3}b^7)^2$

______ **41.** $(9x^{-5}y^{-3})^{-2}$

______ **42.** $(11x^{-4}y^{-5})^{-2}$

______ **43.** $(7^{-1}x^3)^{-2} \cdot (7^{-1}x^2)^3$

______ **44.** $(5^{-3}x^4)^{-1} \cdot (5^{-1}x^2)^2$

______ **45.** $\frac{x^{13}}{x^{-2}}$

______ **46.** $\frac{x^9}{x^{-3}}$

_____ 47. $\frac{9a^{-5}}{15a^5}$

_____ 48. $\frac{8a^{-6}}{12a^6}$

_____ 49. $\frac{7a^{-4}b^{-5}}{21a^3b^7}$

_____ 50. $\frac{8a^{-9}b^{-2}}{32a^6b^9}$

_____ 51. $\frac{(-4x^{-3})(x^{-7})}{6x^{-12}}$

_____ 52. $\frac{(-9x^{-7})(x^{-4})}{-6x^{-8}}$

_____ 53. $\left(\frac{2x}{3}\right)^{-3}$

_____ 54. $\left(\frac{3x}{4}\right)^{-2}$

_____ 55. $\left(\frac{a^{-1}}{x^2}\right)^{-2}$

_____ 56. $\left(\frac{a^{-2}}{b^3}\right)^{-1}$

_____ 57. $\left(\frac{x^{-4}}{y^{-2}}\right)^{-3}$

_____ 58. $\left(\frac{x^{-5}}{y^{-3}}\right)^{2}$

_____ 59. $\left(\frac{-x^4}{3y^{-1}}\right)^{-2}$

_____ 60. $\left(\frac{-x^2}{2y^{-3}}\right)^{-3}$

☆ Stretching the Topics

Simplify.

_____ 1. $(3^{-1}x^5y^{-2}z^3)^{-2}(6x^{-2}y^{-4}z^{-6})^{-1}$

_____ 2. $\left(\frac{2a^{-2}b}{b^{-3}}\right)^2 \cdot \left(\frac{3a^4b^{-1}}{b^3}\right)^{-3}$

_____ 3. $\frac{(-x^{-3}y^{-2})^{-2}}{(x^{-2}y^2)^{-3}}$

Check your answers in the back of your book.

If you can simplify the expressions in **Checkup 5.3**, you are ready to do the **Review Exercises for Chapter 5.**

Name ______________________ Date __________

✓ CHECKUP 5.3

Simplify and give answers with positive exponents.

_______ **1.** $(2x)^{-5}$

_______ **2.** $a^7 \cdot a^{-3}$

_______ **3.** $3y^{-2}$

_______ **4.** $2x^{-8}(x^{-4})^{-2}$

_______ **5.** $\dfrac{-12x^5}{4x^{-2}}$

_______ **6.** $\dfrac{9a^{-3}}{18a^{-5}}$

_______ **7.** $\dfrac{(-2x^{-4})(x^6)}{6x^{-2}}$

_______ **8.** $\dfrac{9x^{-4}y^{-5}}{3x^4y^{-5}}$

_______ **9.** $\left(\dfrac{x^{-2}}{3}\right)^{-2}$

_______ **10.** $\left(\dfrac{3a^3}{b^{-2}}\right)^{-3}$

Check your answers in the back of your book.

If You Missed Problems:	You Should Review Examples:
1	3
2	5, 6
3	4
4	7–10
5–8	21–23
9, 10	24, 25

Summary

In this chapter we discussed integer exponents, beginning with whole-number exponents. We again state the definitions and laws that we used in working with exponents. Here a stands for any base, and m and n represent whole numbers.

	Symbols	Examples
Definition of a^n	$a^n = \underbrace{a \cdot a \cdot \ldots \cdot a}_{n \text{ factors}}$	$2^5 = 2 \cdot 2 \cdot 2 \cdot 2 \cdot 2 = 32$ $x^3 = x \cdot x \cdot x$
Definition of a^0	$a^0 = 1 \ (a \neq 0)$	$17^0 = 1$ $x^0 = 1$
First Law of Exponents	$a^m \cdot a^n = a^{m+n}$	$2^5 \cdot 2^3 = 2^{5+3} = 2^8$ $x^2 \cdot x = x^{2+1} = x^3$
Second Law of Exponents	$(a^m)^n = a^{m \cdot n}$	$(3^2)^4 = 3^{2 \cdot 4} = 3^8$ $(x^3)^6 = x^{3 \cdot 6} = x^{18}$
Third Law of Exponents	$(a \cdot b)^n = a^n \cdot b^n$	$(2 \cdot 5)^3 = 2^3 \cdot 5^3$ $(3xy^2)^4 = 3^4 \cdot x^4 \cdot (y^2)^4 = 81x^4y^8$
Fourth Law of Exponents	$\frac{a^m}{a^n} = a^{m-n}$ if $m \geq n$ $\frac{a^m}{a^n} = \frac{1}{a^{n-m}}$ if $n > m$	$\frac{2^9}{2^3} = 2^{9-3} = 2^6$ $\frac{y}{y^5} = \frac{1}{y^{5-1}} = \frac{1}{y^4}$
Fifth Law of Exponents	$\left(\frac{a}{b}\right)^n = \frac{a^n}{b^n}$	$\left(\frac{2}{3}\right)^4 = \frac{2^4}{3^4} = \frac{16}{81}$ $\left(\frac{x}{y}\right)^3 = \frac{x^3}{y^3}$ $\left(\frac{3x^3}{z}\right)^2 = \frac{(3x^3)^2}{z^2} = \frac{3^2(x^3)^2}{z^2} = \frac{9x^6}{z^2}$

We also learned that *negative integers* can be used as exponents and continue to obey all the laws of exponents.

	Symbols	Examples
Definition	$a^{-n} = \frac{1}{a^n}$	$x^{-3} = \frac{1}{x^3}$
First Law of Exponents	$a^m \cdot a^n = a^{m+n}$	$x^{-3} \cdot x^{-2} = x^{-5} = \frac{1}{x^5}$
		$x^2 \cdot x \cdot x^{-3} = x^0 = 1$
Second Law of Exponents	$(a^m)^n = a^{mn}$	$(x^{-3})^{-4} = x^{12}$
		$(x^{-1})^3 = x^{-3} = \frac{1}{x^3}$
Third Law of Exponents	$(ab)^n = a^n \cdot b^n$	$(2x)^{-3} = 2^{-3}x^{-3} = \frac{1}{8x^3}$
		$(x^{-1}y^2)^{-2} = x^2y^{-4} = \frac{x^2}{y^4}$
Fourth Law of Exponents	$\frac{a^m}{a^n} = a^{m-n}$	$\frac{x^3}{x^5} = x^{-2} = \frac{1}{x^2}$
	$(a \neq 0)$	$\frac{y^{-2}}{y^{-6}} = y^{-2-(-6)} = y^4$
Fifth Law of Exponents	$\left(\frac{a}{b}\right)^n = \frac{a^n}{b^n}$	$\left(\frac{2}{x}\right)^{-3} = \frac{2^{-3}}{x^{-3}} = \frac{x^3}{8}$
	$(b \neq 0)$	$\left(\frac{x^2}{2y^{-1}}\right)^{-2} = \frac{x^{-4}}{2^{-2}y^2} = \frac{2^2}{x^4y^2} = \frac{4}{x^4y^2}$

❑ Speaking the Language of Algebra

1. In the expression a^n, we call a the __________ and we call n the __________ . If n is a positive integer, the expression a^n tells us to use __________ as a factor in a product __________ times.
2. An exponent belongs only to the base to which it is attached unless __________ indicate otherwise.
3. To multiply powers of the same base, we keep the same base and __________ the exponents.
4. To raise a power to a power, we keep the same base and __________ the exponents.
5. To raise a product to a power, we raise each __________ to that power.
6. To divide powers of the same base, we keep the same base and __________ the exponents.
7. To raise a fraction to a power, we raise the __________ and the __________ to that power.
8. The expression a^{-1} tells us to find the __________ of a.
9. If a factor with a negative exponent appears in the denominator of a fraction, that factor should be moved to the __________ , where its exponent will be positive.
10. For any base a (except 0), $a^n \cdot a^{-n} =$ __________ .

Check your answers in the back of your book.

Name ______________________ Date __________

REVIEW EXERCISES for Chapter 5

Use exponents to rewrite each expression.

_______ 1. $-2 \cdot x \cdot x \cdot y \cdot y \cdot y$

_______ 2. $\left(\frac{1}{3}a\right)\left(\frac{1}{3}a\right)\left(\frac{1}{3}a\right)$

Write each expression without exponents.

_______ 3. $-3m^3n^2$

_______ 4. $(0.5a)^3$

Evaluate.

_______ 5. $(-3)^3$

_______ 6. $\left(\frac{-1}{2}\right)^4$

_______ 7. $-2a^2$ when $a = -1$

_______ 8. $(5m^2n)^0$ when $m = 3$ and $n = -1$

Simplify.

_______ 9. $x^7 \cdot x^3$

_______ 10. $-2a^3(a^2)(a^4)$

_______ 11. $9xy^6 \cdot x^4y^2$

_______ 12. $(m^2)^3$

_______ 13. $(-3y)^3$

_______ 14. $(x^3y^2z)^2$

_______ 15. $3a(a^4b^2)^3$

_______ 16. $\left(\frac{1}{3}x\right)^3$

_______ 17. $\frac{x^5}{x^8}$

_______ 18. $\frac{-40x^{12}}{8x^3}$

_______ 19. $\frac{-12x^6y^3}{4x^3y^9}$

_______ 20. $\frac{-x^9y^5z}{x^{12}y^3z}$

_______ 21. $\frac{(2x^3y)^3}{(-xy^2)^2}$

_______ 22. $\left(\frac{-3a^2}{2b}\right)^3$

_______ 23. $\left(\frac{22a^4b^2}{11ab^7}\right)^4$

_______ 24. $(5x)^{-2}$

_______ 25. $4y^{-3}$

_______ 26. $x^5 \cdot x^{-9}$

_______ 27. $4a^{-5} \cdot a^{-3} \cdot a^{10}$

_______ 28. $(2x^{-1})^3 \cdot (x^4)^{-2}$

_______ 29. $(-3x^{-1})^{-2} \cdot (x^{-2})^2$

_______ 30. $\frac{-35x^6}{5x^{-2}}$

_______ 31. $\frac{13a^{-2}}{52a^{-5}}$

_______ 32. $\left(\frac{3x}{2}\right)^{-4}$

_______ 33. $\left(\frac{-2y^{-3}}{y^{-4}}\right)^2$

_______ 34. $\frac{(-2x^{-3})^2(x^7)}{16x^{-3}}$

Check your answers in the back of your book.

If You Missed Exercises:	You Should Review Examples:	
1, 2	SECTION 5.1	1–3
3, 4		4–8
5–8		1, 9–11
9, 10		15–16
11		19
12		25
13		27
14–16		29, 30
17	SECTION 5.2	4
18–20		5–8
21		9
22		12
23		13
24, 25	SECTION 5.3	1–4
26, 27		5–8
28, 29		12–16
30, 31		21, 22
32, 33		24, 25
34		23

If you have completed the **Review Exercises** and corrected your errors, you are ready to take the **Practice Test for Chapter 5.**

Name ____________________ Date __________

PRACTICE TEST for Chapter 5

		SECTION	EXAMPLES
Evaluate each expression.			
________	**1.** $(-4)^3$	5.1	2
________	**2.** $\frac{1}{3}a^2b^3$ when $a = 3$ and $b = -1$	5.1	10
________	**3.** $4(2x - 3y)^0$ when $x = 1$ and $y = 1$	5.1	14
Simplify.			
________	**4.** $x^6 \cdot x^3$	5.1	15
________	**5.** $(-2y^3)(y^2)(y^3)$	5.1	16
________	**6.** $4x^5y^2z \cdot x^2yz^3$	5.1	19, 20
________	**7.** $(a^4)^3$	5.1	22
________	**8.** $(x^3)^4 \cdot (y^2)^3$	5.1	24
________	**9.** $(-3x)^3$	5.1	28
________	**10.** $(-2a^4b^3)^4$	5.1	29
________	**11.** $5a(a^2b)^3$	5.1	30
________	**12.** $\frac{x^7}{x^{15}}$	5.2	4
________	**13.** $\frac{-36x^{11}}{18x^6}$	5.2	5
________	**14.** $\frac{-24x^9y^3}{8x^4y^7}$	5.2	7
________	**15.** $\frac{-a^{12}b^3c^2}{a^{15}bc^2}$	5.2	8
________	**16.** $\frac{(-2x^2y^3)^5}{(x^2y)^4}$	5.2	9
________	**17.** $\left(\frac{-3a^5}{4b^2}\right)^3$	5.2	12
________	**18.** $\left(\frac{54x^4y^3}{9xy^7}\right)^4$	5.2	13
________	**19.** $7x^{-4}$	5.3	4

	Problem	Section	Example
________	**20.** $(-5w^{-3})(w^{-2})$	5.3	7
________	**21.** $(3a)^{-4}$	5.3	14
________	**22.** $(-5x^{-1}y^4)^{-2}(x^3y^{-2})^2$	5.3	19
________	**23.** $\dfrac{-12x^{-6}}{4x^{-2}}$	5.3	21
________	**24.** $\dfrac{28x^2y^{-4}}{7x^{-3}y}$	5.3	22
________	**25.** $\left(\dfrac{-4a^{-5}}{3b^5}\right)^{-1}$	5.3	26

Check your answers in the back of your book.

Name ______________________________ Date ____________

SHARPENING YOUR SKILLS after Chapters 1–5

Simplify.

SECTION

_______ **1.** $\dfrac{4(-9) + (-5)(-2)}{13(-1)}$ 2.3

_______ **2.** $-4[2(3a - 2b) - 3(2a - 3b)]$ 3.3

_______ **3.** $9x - 5y + 3z - 7x + 4y - 2z$ 3.2

_______ **4.** $19.24 - 3(6.13)$ 1.3

_______ **5.** $\dfrac{(-0.4)^2 - 2^2}{(-8)^2 - 5(6)(2)}$ 2.4

_______ **6.** $\dfrac{13}{28} - \dfrac{20}{21} + \dfrac{7}{12}$ 2.4

_______ **7.** Is -9 a solution for $7x - 4 = 6x + 5$? 4.1

_______ **8.** Write an algebraic expression for 8 times the sum of a variable and 13. 3.1

Solve.

_______ **9.** $\dfrac{3}{11}x = 6$ 4.1

_______ **10.** $5 - 0.3x = -7$ 4.2

_______ **11.** $4[2(x - 2) - (x + 4)] = 0$ 4.2

_______ **12.** $3[x - (2x - 1)] = -2(x - 3) - 6$ 4.2

_______ **13.** Solve $P = 2l + 2w$ for w. 4.2

_______ **14.** Sam and Ethel are collecting aluminum cans for recycling. Sam has collected 50 more cans than twice the number Ethel has collected. If the total number of cans both collected is 290, how many did each collect? 4.3

_______ **15.** The length of a rectangle is 4 times the width. If the perimeter is 40 feet, find its dimensions. 4.3

Check your answers in the back of your book.

Working with Polynomials 6

Mr. Abell's garden is in the shape of a square. Next year he plans to enlarge his garden by increasing the length by 6 feet and the width by 4 feet. Write an expression for the area of the garden after it has been enlarged.

Before you finish this chapter, you will be able to write an expression for the area of this garden. To do this, we must learn how to

1. Identify polynomials.
2. Add and subtract polynomials.
3. Multiply polynomials.
4. Divide polynomials.
5. Switch from word expressions to polynomials.

6.1 Simplifying Polynomials

Identifying Polynomials

Now that we have discussed the meaning of exponents, we can state the definition of a **polynomial**.

> **Definition of a Polynomial.** A polynomial in x is a sum of one or more terms of the form
>
> $$ax^n$$
>
> where a is a constant and n is a *whole number*.

It is important to understand that the exponent on the variable in each term of a polynomial must be a *whole number* and that a polynomial must be expressible as a *sum* of such terms.

Learning to recognize algebraic expressions that are *not* polynomials will help us understand what polynomials must look like.

Polynomials	Not Polynomials
x^9	$\frac{1}{x^9}$
$x^2 - 3x + 8$	$\frac{x^2 - 3x + 8}{x + 2}$
$\frac{1}{2}x^3 + 5x$	$\sqrt{x} + 5$

Mathematicians give different names to polynomials depending on how many *terms* the polynomial contains.

Number of Terms	Name of Polynomial	Examples
One	Monomial	$7x$ $-y^3$
Two	Binomial	$2x + 1$ $x^2 - 5y$
Three	Trinomial	$3a + 2b - 7$ $6x^2 + x - 2$

If a polynomial contains *more* than three terms, it is simply referred to as a polynomial of so-many terms.

$2x + 3y - 4z + 8$ is a polynomial of four terms

$a + b - c + d + e$ is a polynomial of five terms

Example 1. Name each polynomial.

$-16x$ is a monomial
$x^2 + x + 2$ is a trinomial.
$ab - c$ is a binomial.

Complete Example 2.

Example 2. Name each polynomial.

$x^2 - y^2$ is a ______.
xyz is a ______.
$6a^3 - a^2 + 3$ is a ______.

Check your work on page 295. ▶

We should combine like terms, if possible, before naming a polynomial. Remember that terms are called "like terms" if their variable parts contain exactly the same variable raised to the same power.

Example 3. $x^2 + x + 3x^2 + 2$.

Solution: We must combine like terms first.

$x^2 + x + 3x^2 + 2$ Look for like terms.
$= x^2 + 3x^2 + x + 2$ Rearrange terms.
$= 4x^2 + x + 2$ Combine like terms.

This is a trinomial.

Now you try Example 4.

Example 4. $ab + 2ab + ac$.

Solution: We must combine like terms first.

$ab + 2ab + ac$
$=$ ______ $+$ ______

This is a ______.

Check your work on page 295. ▶

Trial Run

Name each expression.

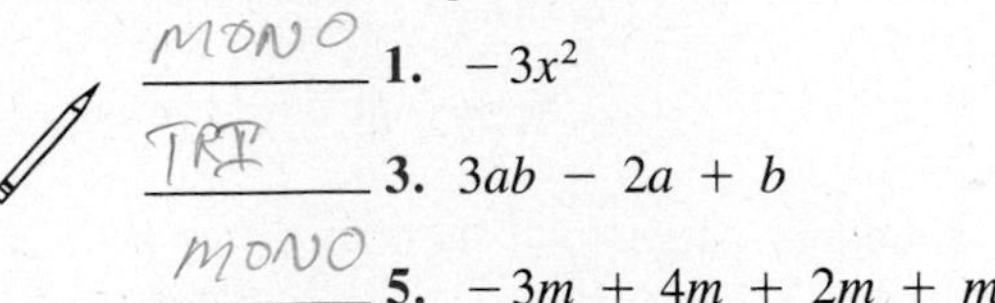

______ 1. $-3x^2$

______ 2. $-3x + 5$

______ 3. $3ab - 2a + b$

______ 4. $-2x + x - 2y^3$

______ 5. $-3m + 4m + 2m + m$

______ 6. $xy - 3xz + 4yz$

Answers are on page 295.

Simplifying Polynomials

Adding polynomials is simply a matter of **combining like terms**. Remember that *to combine like terms, we combine the numerical coefficients and keep the variable part.*

Example 5. Simplify

$(2x^2 + 7x - 3) + (x^2 - 5x - 1)$

Solution

$(2x^2 + 7x - 3) + (x^2 - 5x - 1)$
$= 2x^2 + 7x - 3 + x^2 - 5x - 1$
Remove parentheses.
$= 2x^2 + x^2 + 7x - 5x - 3 - 1$
Rearrange terms.
$= 3x^2 + 2x - 4$ Combine like terms.

Now you try Example 6.

Example 6. Simplify

$(-x^2 - 5x + 9) + (x^2 + x - 2)$

Solution

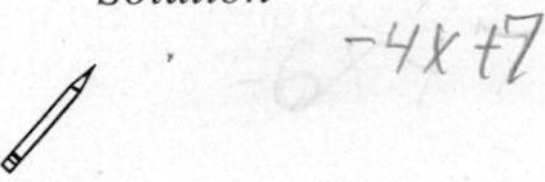

Check your work on page 295. ▶

To multiply a constant times a monomial, we use the method learned in Chapter 3. We multiply the constant times the numerical coefficient and keep the variable part.

Example 7. Simplify $5(-x)$.

Solution

$$5(-x)$$
$$= 5(-1x)$$
$$= 5(-1)x$$
$$= -5x$$

Now complete Example 8.

Example 8. Simplify $-\frac{3}{4}(4y^3)$.

Solution

$$\frac{-3}{4}(4y^3)$$
$$= \left(\frac{-3}{4} \cdot \frac{\quad}{\quad}\right)y^3$$
$$= \underline{\qquad}\, y^3$$

Check your work on page 295. ▶

The **distributive property** allows us to multiply constants times polynomials, just as we did in Chapter 3.

Example 9. Simplify $2(x^2 + 3x + 1)$.

Solution

$$2(x^2 + 3x + 1)$$
$$= 2x^2 + 6x + 2$$

You complete Example 10.

Example 10. Simplify $-(5a^2 - 3a + 2)$.

Solution

$$-(5a^2 - 3a + 2)$$
$$= -1(5a^2 - 3a + 2)$$
$$= \underline{\qquad\qquad}$$

Check your work on page 295. ▶

We must use all these ideas to simplify polynomial sums or differences.

Example 11. Simplify $3(2a - a^2) - (a^2 - a + 3)$.

Solution

$3(2a - a^2) - (a^2 - a + 3)$	
$= 3(2a - a^2) - 1(a^2 - a + 3)$	Insert 1 before second parentheses.
$= 6a - 3a^2 - a^2 + a - 3$	Use the distributive property.
$= -4a^2 + 7a - 3$	Combine like terms.

Now you complete Example 12.

Example 12. Simplify $2(x^2 - x) + 5(2x^2 - 1)$.

Solution

$2(x^2 - x) + 5(2x^2 - 1)$	
$= \underline{\quad}\, x^2 - \underline{\quad}\, x + \underline{\quad}\, x^2 - \underline{\quad}$	Use the distributive property.
$= \underline{\qquad\qquad}$	Combine like terms.

Check your work on page 295. ▶

Notice that it is customary to write polynomials with the exponents in decreasing order from left to right. This is called "writing a polynomial in descending powers of the variable."

Trial Run

Simplify.

______ 1. $5(-2xy)$

______ 2. $2(x + y) - 3(2x - y)$

______ 3. $5(x^2 - 1) + 2(2x^2 - 3x)$

______ 4. $(a^2 - 2a + 1) - (3a^2 + a + 5)$

______ 5. $-2(x^2 - 3) + (2x^2 - 5)$

______ 6. $(y^2 + 2y + 1) - (y^2 - 2y + 1)$

Answers are on page 295.

Examples You Completed

Example 2. Name each polynomial.

$x^2 - y^2$ is a binomial.
xyz is a monomial
$6a^3 - a^2 + 3$ is a trinomial.

Example 4. $ab + 2ab + ac$.

Solution: We must combine like terms first.

$$ab + 2ab + ac$$
$$= 3ab + ac$$

This is a binomial.

Example 6. Simplify $(-x^2 - 5x + 9) + (x^2 + x - 2)$.

Solution

$$(-x^2 - 5x + 9) + (x^2 + x - 2)$$
$$= -x^2 - 5x + 9 + x^2 + x - 2$$
$$= -x^2 + x^2 - 5x + x + 9 - 2$$
$$= -4x + 7$$

Example 8. Simplify $\frac{-3}{4}(4y^3)$.

Solution

$$\frac{-3}{4}(4y^3)$$
$$= \left(\frac{-3}{4} \cdot \frac{4}{1}\right)y^3$$
$$= -3y^3$$

Example 10. Simplify $-(5a^2 - 3a + 2)$.

Solution

$$-(5a^2 - 3a + 2)$$
$$= -1(5a^2 - 3a + 2)$$
$$= -5a^2 + 3a - 2$$

Example 12. Simplify $2(x^2 - x) + 5(2x^2 - 1)$.

Solution

$$2(x^2 - x) + 5(2x^2 - 1)$$
$$= 2x^2 - 2x + 10x^2 - 5$$
$$= 12x^2 - 2x - 5$$

Answers to Trial Runs

page 293 **1.** Monomial **2.** Binomial **3.** Trinomial **4.** Binomial **5.** Monomial **6.** Trinomial

page 295 **1.** $-10xy$ **2.** $-4x + 5y$ **3.** $9x^2 - 6x - 5$ **4.** $-2a^2 - 3a - 4$ **5.** 1 **6.** $4y$

Name ______________________ Date ____________

EXERCISE SET 6.1

Name each expression.

______ **1.** $-3x^2$

______ **2.** $-5y^4$

______ **3.** $7x - \frac{1}{4}y$

______ **4.** $-3x + \frac{2}{5}y$

______ **5.** $4ab - 2a + b$

______ **6.** $a - b + 3ab$

______ **7.** $-7m^2 + 2n^2 + 10m^2$

______ **8.** $8n^2 - 6m^2 + n^2$

______ **9.** $0.3a - 0.5b - 0.7a + 0.6b$

______ **10.** $0.9a - 0.6b - 0.4a + 0.9b$

______ **11.** $ab - ac + bc$

______ **12.** $3ab + 2ac - 5bc$

______ **13.** $4a^3 - 2a^2 + a - 5$

______ **14.** $-5a^3 - 4a^2 + 3a - 9$

Simplify.

______ **15.** $(x^2 - 3x + 5) + (3x^2 + x - 1)$

______ **16.** $(3x^2 - 4x + 2) + (x^2 + x - 5)$

______ **17.** $(a - 3b + 2c) + (4a + 3b - 5c)$

______ **18.** $(2a - 5b + c) + (7a + 5b - 2c)$

______ **19.** $(-5x^2 + x - 4) + (x^2 - 3x - 5)$

______ **20.** $(-4x^2 + 3x - 6) + (x^2 - x - 7)$

______ **21.** $-3(4x^2)$

______ **22.** $-2(5x^4)$

______ **23.** $4(-2xy)$

______ **24.** $7(-5xy)$

______ **25.** $\frac{2}{3}(9a)$

______ **26.** $\frac{3}{5}(15a)$

______ **27.** $0.4(-2ab^3)$

______ **28.** $0.5(-4ab^3)$

______ **29.** $-8\left(\frac{3}{4}x^5y^4\right)$

______ **30.** $-12\left(\frac{5}{6}x^4y^3\right)$

______ **31.** $5(x^2 - 3x + 2)$

______ **32.** $3(x^2 - 5x - 6)$

______ **33.** $-(6a - 4b + c)$

______ **34.** $-(3a - 2b + 5c)$

______ **35.** $-4(2a^2 - 2a + 1)$

______ **36.** $-3(5a^2 - a - 1)$

______ **37.** $\frac{1}{3}(9x^2 - 6x + 3)$

______ **38.** $\frac{1}{4}(8x^2 - 4x + 16)$

______ **39.** $(x + y) - (2x - y)$

______ **40.** $(2x + y) - (3x - y)$

______ **41.** $-2(x + y) + 3(2x - 5y)$

______ **42.** $-3(2x - y) + 4(x + y)$

______ **43.** $0.4(x^2 - 3x) - 0.5(3x - 2)$

______ **44.** $0.3(2x^2 - 5x) - 0.7(-2x + 5)$

______ **45.** $-3(y^2 + 3y - 5) - (2y^2 + 3y - 6)$

______ **46.** $-4(2y^2 - y - 4) - (y^2 - 2y + 2)$

______ **47.** $-3(a^2 + 2) + (3a^2 + 7)$

______ **48.** $(3x^2 - 6) - 3(x^2 - 2)$

______ **49.** $5(y^2 - 2y + 1) - 2(3y^2 - 5y + 4)$

______ **50.** $3(x^2 - 2x + 1) - 5(x^2 + x - 1)$

☆ Stretching the Topics

Simplify.

______ **1.** $2(h + 3k) - 4(2h - 5k) - (h + 2k) + 6k$

______ **2.** $2x(y + z) - 3x(2y - z) + x(y - 5z) - 3xz$

______ **3.** The length of a rectangle is $2a - b$ inches. The width is $2b$ inches. What is its perimeter?

Check your answers in the back of your book.

If you can complete **Checkup 6.1**, you are ready to go on to Section 6.2.

Name ______________________________ Date ______________

✓ CHECKUP 6.1

Name the following expressions.

________ **1.** $3x - y + 4xy$

________ **2.** $0.2a - 0.5b - 0.3a + 0.5b$

Simplify.

________ **3.** $(-x^2 + 3x - 5) + (4x^2 - 3x - 4)$

________ **4.** $-4(-3xy)$

________ **5.** $\frac{4}{7}(21a^3b^2)$

________ **6.** $0.5(4x^2 - 8x + 2)$

________ **7.** $-(3a - 4b - 7c)$

________ **8.** $4(3x - x^2) - (x^2 + x - 20)$

________ **9.** $5(2x^2 - x) - 3(4x^2 - 2)$

________ **10.** $(3y^2 - 2y + 5) - (y^2 + 2y - 5)$

Check your answers in the back of your book.

If You Missed Problems:	You Should Review Examples:
1, 2	1–4
3	5, 6
4, 5	7, 8
6, 7	9, 10
8–10	11, 12

6.2 Multiplying with Monomials

Multiplying Monomials

In Chapter 5 we learned to use the First Law of Exponents to multiply powers of the same base. Recall that $a^m \cdot a^n = a^{m+n}$. We use the commutative and associative properties of multiplication when multiplying monomials containing constants and variables.

Example 1. Simplify $2x^3(4x^2)$.

Solution

$$2x^3(4x^2)$$
$$= 2 \cdot x^3 \cdot 4 \cdot x^2$$
$$= 2 \cdot 4 \cdot x^3 \cdot x^2$$
$$= 8 \cdot x^{3+2}$$
$$= 8x^5$$

You try to complete Example 2.

Example 2. Simplify $-5y^8(6y^7)$.

Solution

$$-5y^8(6y^7)$$
$$= -5 \cdot y^8 \cdot 6 \cdot y^7$$
$$= -5 \cdot 6 \cdot y^8 \cdot y^7$$

Check your work on page 303 ▶

To multiply monomials, first multiply the numerical coefficients, then multiply the variables using the First Law of Exponents.

Example 3. Simplify $4a^5\left(\frac{1}{3}a^7\right)$.

Solution

$$4a^5\left(\frac{1}{3}a^7\right)$$
$$= 4 \cdot \frac{1}{3} \cdot a^5 \cdot a^7$$
$$= \frac{4}{3} \cdot a^{5+7}$$
$$= \frac{4}{3}a^{12}$$

You complete Example 4.

Example 4. Simplify $3x^2(-2x^4)(-7x)$.

Solution

$$3x^2(-2x^4)(-7x)$$
$$= 3(-2)(-7) \cdot x^2 \cdot x^4 \cdot x$$

Check your work on page 303. ▶

Example 5. Simplify $0.8x(3x^2y)$.

Solution

$$0.8x(3x^2y)$$
$$= 0.8 \cdot 3 \cdot x \cdot x^2 \cdot y$$
$$= 2.4x^3y$$

Try to complete Example 6.

Example 6. Simplify $(-2xy^2z)(-9x^2y^5z^7)$.

Solution

$$(-2xy^2z)(-9x^2y^5z^7)$$
$$= (-2)(-9) \cdot x \cdot x^2 \cdot y^2 \cdot y^5 \cdot z \cdot z^7$$
$$= \underline{\qquad}$$

Check your work on page 304. ▶

Trial Run

Find the products.

_______ 1. $(-4x^5)(7x^3)$

_______ 2. $(0.2a^3)(-5a^2)(6a^6)$

_______ 3. $(3x^2)(2x^5y^7)$

_______ 4. $(4x^2yz^5)\left(\frac{-1}{2}xy^3z^7\right)$

Answers are on page 304.

Multiplying Monomials Times Polynomials

The distributive property, the rules for signs, and the laws of exponents are all we need to multiply a monomial times a polynomial. The distributive property tells us to multiply each term inside the parentheses by the monomial outside the parentheses and the First Law of Exponents tells us how to simplify each of those multiplications.

Example 7. Simplify $x(x^2 + 3x - 1)$.

Solution

$$x(x^2 + 3x - 1)$$
$$= x \cdot x^2 + x \cdot 3x - x \cdot 1$$
$$= x^3 + 3x^2 - x$$

You complete Example 8.

Example 8. Simplify $2a(4a^3 - 5a - 3)$.

Solution

$$2a(4a^3 - 5a - 3)$$
$$= 2a \cdot 4a^3 - 2a \cdot 5a - 2a \cdot 3$$
$$= \_\_\_\_\_\_$$

Check your work on page 304. ▶

Remember that we learned to do such products without writing down each step, using arrows to keep our work straight.

Example 9. Simplify $-4xy(2x + y - 7)$.

Solution

$$-4xy(2x + y - 7)$$
$$= -8x^2y - 4xy^2 + 28xy$$

Complete Example 10.

Example 10. Simplify

$$x^3y(x^2y^2 - 4xy + 3xy^5).$$

Solution

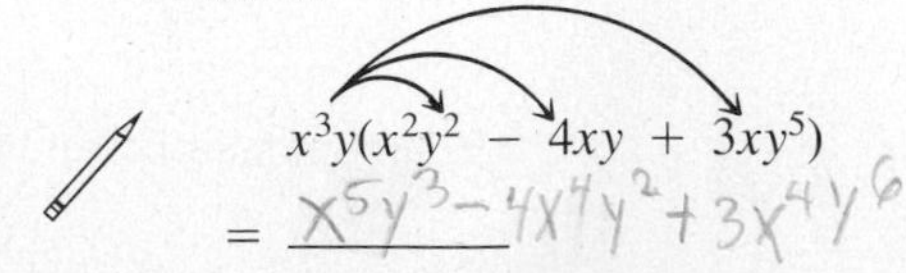

$$x^3y(x^2y^2 - 4xy + 3xy^5)$$
$$= \_\_\_\_\_\_$$

Check your work on page 304. ▶

Example 11. Simplify

$$x^2(2x - y) + x(xy - 3x^2)$$

Solution

$x^2(2x - y) + x(xy - 3x^2)$

$= 2x^3 - x^2y + x^2y - 3x^3$ Remove parentheses.

$= 2x^3 - 3x^3 - x^2y + x^2y$ Rearrange terms.

$= -x^3$ Combine like terms.

Now you try Example 12.

Example 12. Simplify

$$2a(x - 3xy) - 5x(a - 4ay)$$

Solution

Check your work on page 304. ▶

Trial Run

Simplify.

_______ **1.** $x(x^2 - 5x + 2)$

_______ **2.** $-2a^2(a - 6)$

_______ **3.** $3x(4x^2 - 2x + 7)$

_______ **4.** $a^2b(a^2b^2 - 3ab + 4ab^4)$

_______ **5.** $x^2(3x + y) - x(xy - 2x^2)$

_______ **6.** $2a(x - 2xy) - 3x(a + 3ay)$

Answers are on page 304.

▶ Examples You Completed

Example 2. Simplify $-5y^8(6y^7)$.

Solution

$-5y^8(6y^7)$

$= -5 \cdot y^8 \cdot 6 \cdot y^7$

$= -5 \cdot 6 \cdot y^8 \cdot y^7$

$= -30 \cdot y^{8+7}$

$= -30y^{15}$

Example 4. Simplify $3x^2(-2x^4)(-7x)$.

Solution

$3x^2(-2x^4)(-7x)$

$= 3(-2)(-7) \cdot x^2 \cdot x^4 \cdot x$

$= 3(14) \cdot x^{2+4+1}$

$= 42x^7$

Example 6. Simplify $(-2xy^2z)(-9x^2y^5z^7)$.

Solution

$$(-2xy^2z)(-9x^2y^5z^7)$$
$$= (-2)(-9) \cdot x \cdot x^2 \cdot y^2 \cdot y^5 \cdot z \cdot z^7$$
$$= 18x^3y^7z^8$$

Example 8. Simplify $2a(4a^3 - 5a - 3)$.

Solution

$$2a(4a^3 - 5a - 3)$$
$$= 2a \cdot 4a^3 - 2a \cdot 5a - 2a \cdot 3$$
$$= 8a^4 - 10a^2 - 6a$$

Example 10. Simplify $x^3y(x^2y^2 - 4xy + 3xy^2)$.

Solution

$$x^3y(x^2y^2 - 4xy + 3xy^5)$$
$$= x^5y^3 - 4x^4y^2 + 3x^4y^6$$

Example 12. Simplify $2a(x - 3xy) - 5x(a - 4ay)$.

Solution

$$2a(x - 3xy) - 5x(a - 4ay)$$
$$= 2ax - 6axy - 5ax + 20axy$$
$$= 2ax - 5ax - 6axy + 20axy$$
$$= -3ax + 14axy$$

Answers to Trial Runs

page 302 **1.** $-28x^8$ **2.** $-6a^{11}$ **3.** $6x^7y^7$ **4.** $-2x^3y^4z^{12}$

page 303 **1.** $x^3 - 5x^2 + 2x$ **2.** $-2a^3 + 12a^2$ **3.** $12x^3 - 6x^2 + 21x$ **4.** $a^4b^3 - 3a^3b^2 + 4a^3b^5$ **5.** $5x^3$ **6.** $-ax - 13axy$

Name ______________________ Date __________

EXERCISE SET 6.2

Simplify.

_____ **1.** $2x^2(5x^3)$

_____ **2.** $3x^3(7x^2)$

_____ **3.** $\frac{1}{2}x^7(4x)$

_____ **4.** $\frac{2}{3}x^5(3x)$

_____ **5.** $-2a^5(-5a^7)$

_____ **6.** $-3a^4(-5a^6)$

_____ **7.** $0.9y^3(-y)$

_____ **8.** $0.7y^4(-y)$

_____ **9.** $3m^2(-2m)m^5$

_____ **10.** $4m^3(-3m)(m^2)$

_____ **11.** $4x^2y^3(-3x^5y^2)$

_____ **12.** $8x^3y^4(-2x^2y^5)$

_____ **13.** $\frac{1}{4}x(-x^2y)(-8x^4y^3)$

_____ **14.** $\frac{-1}{5}x(-x^3y)(15x^4y^3)$

_____ **15.** $7xy(-2xy)(-3x^2y)$

_____ **16.** $-2xy^2(8x^2y^3)(-3xy)$

_____ **17.** $5x^2y^3z(7xyz^4)$

_____ **18.** $9x^2y^3z^5(3xyz^2)$

_____ **19.** $-3x^2y(8x^3z^2)$

_____ **20.** $-4x^3y^2(9x^5z^3)$

_____ **21.** $5x(2x^3)^2$

_____ **22.** $6x(3x^4)^2$

_____ **23.** $(5y^4)^2(-2y^5)^3$

_____ **24.** $(7y^2)^2(-y^4)^5$

_____ **25.** $(-6a^2b)(2ab^3)^2$

_____ **26.** $(2a^2b)(-3a^4b^2)^3$

_____ **27.** $5(2x + 3)$

_____ **28.** $4(x + 5)$

_____ **29.** $-7(2x^2 - 4)$

_____ **30.** $-3(4x^2 - 7)$

_____ **31.** $3x(2x - 4)$

_____ **32.** $5x(3x - 2)$

_____ **33.** $y^2(2y - 5)$

_____ **34.** $y^2(3y - 8)$

_____ **35.** $-a^2(a^2 - 2)$

_____ **36.** $-a^2(2a^2 - 3)$

_____ **37.** $4x^2(9x^2 - 4)$

_____ **38.** $5x^2(8x^2 - 1)$

_____ **39.** $-3a^3(7a^2 - 3a)$

_____ **40.** $-4a^3(6a^2 - 5a)$

_____ **41.** $2xy(5x + 2y)$

_____ **42.** $-5xy(3x + 5y)$

_____ **43.** $-x^2y^2(11x - 4y)$

_____ **44.** $-2x^2y^2(x - 7y)$

_____ **45.** $2a^3b^2(7a^2 - b^2)$.

_____ **46.** $6a^3b^2(a^2 - 2b^2)$

_____ **47.** $2m(m^2 - 5m + 6)$

_____ **48.** $9m(m^2 - m - 6)$

______ **49.** $5x^2y^2(3x^2y^2 - xy - 4)$

______ **50.** $4x^2y^2(5x^2y^2 - xy - 6)$

______ **51.** $-a^2b(3 - 2ab + 4a^2b^2)$

______ **52.** $-a^2b(5 - 3ab + 2a^2b^2)$

______ **53.** $4mn(8m^3 - 2m^2n + mn^2 - 3n^3)$

______ **54.** $5mn(6m^3 - m^2 + 2mn^2 - 7n^3)$

______ **55.** $x^2(2xy - 6) + 2x^2(xy + 3)$

______ **56.** $x(3xy - x^2) - x^2(y - x)$

______ **57.** $x^2(x^2 + 5) - 3(x^2 - 8)$

______ **58.** $x^2(x^2 - 4) + 2(x^2 - 5)$

______ **59.** $3m(3m + n) - n(3m + n)$

______ **60.** $7m(2m - n) - 2n(4m + n)$

______ **61.** $3ax(2 - 3y) - 2ax(1 - 2y)$

______ **62.** $5ax(3 - y) - 7ax(4 - 3y)$

______ **63.** $5xy(3x - 4y) - 2xy(4x - 5y)$

______ **64.** $8xy(2x - y) - xy(x - 2y)$

☆ Stretching the Topics

Simplify.

______ **1.** $-m^3n^2(4m^2 + 5mn + n^2) + m^3n^2(m^2 - 2mn + 3n^2)$

______ **2.** $xy^2(4x^2 - 3xy + 4xy^2) - x^2y(xy + 4y^2 - 3y^3)$

______ **3.** $2ab^2(-a^3b^{-2} + 3a^2b^{-1} + a - a^{-1}b) + 6a^2b(a^2b^{-1} - a - b + 4a^{-2}b^2)$

Check your answers in the back of your book.

If you can complete **Checkup 6.2**, you are ready to go on to Section 6.3.

Name ______________________ Date __________

✓ CHECKUP 6.2

Simplify.

______ **1.** $-3x^4(7x^7)$

______ **2.** $2a^3(-4a^5)(-a^4)$

______ **3.** $(0.5xy^3z)(-4x^3yz^2)$

______ **4.** $-2a^2(a^2 - 7a + 5)$

______ **5.** $3xy(5x - 2y + 4)$

______ **6.** $6x^2y^2(2x^2y^2 - 3xy + 4)$

______ **7.** $-2mn(4m^3 - m^2n + 2mn^2 - n^3)$

______ **8.** $2x^2(3xy - 2) - x^2(7xy - 4)$

______ **9.** $5x(3x + y) - y(4x - y)$

______ **10.** $-3ax(1 - 2y) + 2ax(5 - 3y)$

Check your answers in the back of your book.

If You Missed Problems:	You Should Review Examples:
1	1–3
2, 3	4–6
4–7	7–10
8–10	11, 12

6.3 Multiplying Binomials

Multiplying Binomials

Recall how we used the distributive property to multiply monomials times binomials.

$$2x(x + 3) = 2x^2 + 6x$$

$$5(x + 3) = 5x + 5$$

Now suppose that we wish to find the product

$$(2x + 5)(x + 3)$$

According to the distributive property, each term in the first binomial $2x + 5$ must be multiplied times the second binomial, $x + 3$. Then we combine like terms.

$$(2x + 5)(x + 3) = \underbrace{2x(x + 3)}_{\text{Multiply first term, } 2x, \text{ times } x + 3.} + \underbrace{5(x + 3)}_{\text{Multiply second term, } 5, \text{ times } x + 3.}$$

$$= 2x^2 + 6x + 5x + 15 \qquad \text{Use the distributive property again.}$$

$$= 2x^2 + 11x + 15 \qquad \text{Combine like terms.}$$

Example 1. Multiply $(x - 2)(3x + 1)$.

Solution

$$(x - 2)(3x + 1)$$

$$= x(3x + 1) - 2(3x + 1)$$

$$= 3x^2 + x - 6x - 2$$

$$= 3x^2 - 5x - 2$$

Now you try Example 2.

Example 2. Multiply $(5 - x)(4 + 7x)$.

Solution

$$(5 - x)(4 + 7x)$$

$$= 5(4 + 7x) - x(4 + 7x)$$

Check your work on page 314. ▶

Let us now consider some "special products."

Example 3. Find $(x + 5y)^2$.

Solution

$$(x + 5y)^2$$

$$= (x + 5y)(x + 5y)$$

$$= x(x + 5y) + 5y(x + 5y)$$

$$= x^2 + 5xy + 5xy + 25y^2$$

$$= x^2 + 10xy + 25y^2$$

You try Example 4.

Example 4. Find $(7 - a)^2$.

Solution

$$(7 - a)^2$$

$$= (7 - a)(7 - a)$$

$$= 7(7 - a) - a(7 - a)$$

Check your work on page 314. ▶

Examples 3 and 4 illustrate a very interesting "special product." In each case, we were asked to *square a binomial*. Notice how the terms in the answer are related to the terms in the original binomial.

first term ↓ $(x + 5y)^2$, ↑ second term

$$(x + 5y)^2 = x^2 + 10xy + 25y^2$$

square of first term ↓ x^2; twice the product of first and second terms ↑ $10xy$; square of second term ↓ $25y^2$

Indeed, this is always true.

Squaring a Binomial

$$(A + B)^2 = A^2 + 2AB + B^2$$

$$(A - B)^2 = A^2 - 2AB + B^2$$

The trinomial obtained by squaring a bonomial is sometimes called a **perfect square trinomial**.

Example 5. Multiply $(x + 3)(x - 3)$.

Solution

$$(x + 3)(x - 3)$$
$$= x(x - 3) + 3(x - 3)$$
$$= x^2 - 3x + 3x - 9$$
$$= x^2 - 9$$

Now try Example 6.

Example 6. Multiply $(8 - y)(8 + y)$.

Solution

$$(8 - y)(8 + y)$$
$$= 8(8 + y) - y(8 + y)$$

Check your work on page 314. ▶

Examples 5 and 6 illustrate another interesting "special product." In each case we multiplied the *sum* of two terms times the *difference* of the same two terms and our answer was just the **difference of the squares** of the original two terms. This will always occur.

Multiplying the Sum and Difference of Two Terms

$$(A + B)(A - B) = A^2 - B^2$$

Example 7. Use the formula for squaring a binomial to find $(3x - 4)^2$.

Solution

$$(3x - 4)^2$$
$$= (3x)^2 - 2(3x)(4) + 4^2$$
$$= 9x^2 - 24x + 16$$

Example 8. Use the formula for the difference of two squares to find $\left(y + \frac{6}{5}\right)\left(y - \frac{6}{5}\right)$.

Solution

$$\left(y + \frac{6}{5}\right)\left(y - \frac{6}{5}\right)$$
$$= y^2 - \left(\frac{6}{5}\right)^2$$
$$= y^2 - \frac{36}{25}$$

These formulas can come in handy, but remember that you can always multiply binomials using the distributive property.

Example 9. Multiply $(x^2 - 0.3)(x - 0.5)$.

Solution

$$\begin{aligned} &(x^2 - 0.3)(x - 0.5) \\ &= x^2(x - 0.5) - 0.3(x - 0.5) \\ &= x^3 - 0.5x^2 - 0.3x + 0.15 \end{aligned}$$

Now try Example 10.

Example 10. Multiply $(x^3 - 4)(7x^3 - 10)$.

Solution

$$\begin{aligned} &(x^3 - 4)(7x^3 - 10) \\ &= x^3(7x^3 - 10) - 4(7x^3 - 10) \end{aligned}$$

$7x^6 - 10x^3 - 28x^3 - 40$

$7x^6 - 38x^3 - 40$

Check your work on page 314. ▶

Trial Run

Multiply.

______ **1.** $(2x + 3)(x + 4)$

______ **2.** $(x - 5)(3x + 2)$

______ **3.** $(3 - x)(5 + 2x)$

______ **4.** $(-2a - 1)(3b + 2)$

______ **5.** $(2x - y)^2$

______ **6.** $\left(y + \frac{1}{5}\right)\left(y - \frac{1}{5}\right)$

Answers are on page 314.

Multiplying Binomials by FOIL

We shall now discover a shortcut that we may use in multiplying two binomials. This shortcut actually uses the distributive property with less writing. Notice what happens if we multiply our binomials using the following steps.

1. Multiply the **F**irst terms in the binomials.
2. Multiply the **O**uter terms in the binomials.
3. Multiply the **I**nner terms in the binomials.
4. Multiply the **L**ast terms in the binomials.
5. Combine the results of those multiplications.

We abbreviate First by **F**, Outer by **O**, Inner by **I**, and Last by **L**.

F L F O I L

$$(x + 7)(2x - 3) = 2x^2 - 3x + 14x - 21$$

I

O

$$= 2x^2 + 11x - 21$$

This is called the **FOIL method**.

Example 11. Multiply $(5x - 1)(3x + 3)$ using FOIL.

Solution

$$(5x - 1)(3x + 3)$$

F L

$$= (5x - 1)(3x + 3)$$

I

O

$$= 15x^2 + 15x - 3x - 3$$

$$= 15x^2 + 12x - 3$$

Now you use FOIL for Example 12.

Example 12. Multiply $(2x + 3y)(2x - 8y)$ using FOIL.

Solution

$$(2x + 3y)(2x - 8y)$$

F L

$$= (2x + 3y)(2x - 8y)$$

I

O

$= 4x^2 -$ ______ $+$ ______ $-$ ______

$= 4x^2 -$ ______ $-$ ______

Check your work on page 314. ▶

Remember that the FOIL method is merely a shortcut for writing down the terms of the product of two binomials. The process still uses the distributive property.

Trial Run

Multiply.

______ **1.** $(3x - 7)(x + 2)$

______ **2.** $(x^2 - 0.9)(x^2 + 0.7)$

______ **3.** $(7y - x)(2y + x)$

______ **4.** $(a - 2)(2b + 5)$

______ **5.** $(3x + 2)^2$

______ **6.** $(2x^3 - 7)(x^3 - 4)$

Answers are on page 314.

Multiplying More Polynomials

To find a product such as $(x + 2)(x^2 + 5x + 3)$ we must use the distributive property again. In this problem, we must multiply x times $x^2 + 5x + 3$ and then multiply 2 times $x^2 + 5x + 3$. Let's perform this multiplication.

$$\begin{aligned}(x + 2)(x^2 + 5x + 3) &= x(x^2 + 5x + 3) + 2(x^2 + 5x + 3)\\ &= x^3 + 5x^2 + 3x + 2x^2 + 10x + 6\\ &= x^3 + 7x^2 + 13x + 6\end{aligned}$$

> To multiply polynomials, we use the distributive property and then combine like terms.

Try to complete Example 13.

Example 13. Multiply

$$(x - 3)(x^2 + 2x - 4)$$

Solution

$$\begin{aligned}&(x - 3)(x^2 + 2x - 4)\\ &= x(x^2 + 2x - 4) - 3(x^2 + 2x - 4)\end{aligned}$$

Check your work on page 314. ▶

Example 14. Multiply

$$(2x^2 + 1)(x^2 - x - 1)$$

Solution

$$\begin{aligned}&(2x^2 + 1)(x^2 - x - 1)\\ &= 2x^2(x^2 - x - 1) + 1(x^2 - x - 1)\\ &= 2x^4 - 2x^3 - 2x^2 + x^2 - x - 1\\ &= 2x^4 - 2x^3 - x^2 - x - 1\end{aligned}$$

To multiply a monomial times two binomials it is easier to multiply the binomials together first and then multiply that product by the monomial.

Example 15. Simplify $2x(x - 3)(5x + 1)$.

Solution

$$\begin{aligned}&2x(x - 3)(5x + 1)\\ &= 2x(5x^2 + x - 15x - 3)\\ &= 2x(5x^2 - 14x - 3)\\ &= 10x^3 - 28x^2 - 6x\end{aligned}$$

You try Example 16.

Example 16. Simplify

$$-3y^2(6 - y)(5 + y)$$

Solution

$$-3y^2(6 - y)(5 + y)$$

Check your work on page 314. ▶

Unfortunately, there are no shortcuts to use here and lots of places to make errors. You should take your time and be careful when finding products of polynomials.

Trial Run

Simplify.

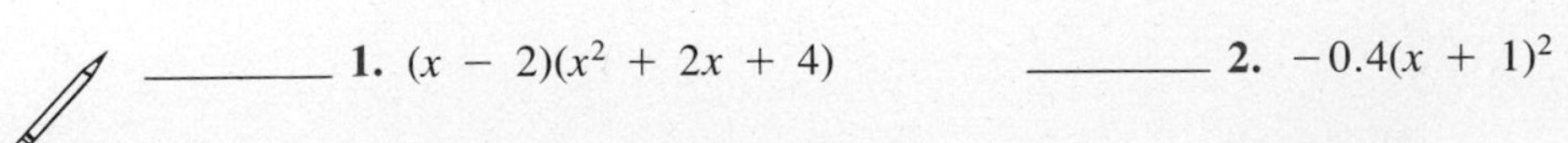

______ **1.** $(x - 2)(x^2 + 2x + 4)$

______ **2.** $-0.4(x + 1)^2$

_______ **3.** $3x(x - 2)(4x - 1)$

_______ **4.** $-y(y + 5)(y - 5)$

Answers are on page 314.

▶ Examples You Completed

Example 2. Multiply $(5 - x)(4 + 7x)$.

Solution

$$(5 - x)(4 + 7x)$$
$$= 5(4 + 7x) - x(4 + 7x)$$
$$= 20 + 35x - 4x - 7x^2$$
$$= 20 + 31x - 7x^2$$

Example 4. Find $(7 - a)^2$.

Solution

$$(7 - a)^2$$
$$= (7 - a)(7 - a)$$
$$= 7(7 - a) - a(7 - a)$$
$$= 49 - 7a - 7a + a^2$$
$$= 49 - 14a + a^2$$

Example 6. Multiply $(8 - y)(8 + y)$

Solution

$$(8 - y)(8 + y)$$
$$= 8(8 + y) - y(8 + y)$$
$$= 64 + 8y - 8y - y^2$$
$$= 64 - y^2$$

Example 10. Multiply $(x^3 - 4)(7x^3 - 10)$.

Solution

$$(x^3 - 4)(7x^3 - 10)$$
$$= x^3(7x^3 - 10) - 4(7x^3 - 10)$$
$$= 7x^6 - 10x^3 - 28x^3 + 40$$
$$= 7x^6 - 38x^3 + 40$$

Example 12. Multiply $(2x + 3y)(2x - 8y)$ using FOIL.

Solution

$$(2x + 3y)(2x - 8y)$$

F L

$$= (2x + 3y)(2x - 8y)$$

I

O

$$= 4x^2 - 16xy + 6xy - 24y^2$$
$$= 4x^2 - 10xy - 24y^2$$

Example 13. Multiply $(x - 3)(x^2 + 2x - 4)$.

Solution

$$(x - 3)(x^2 + 2x - 4)$$
$$= x(x^2 + 2x - 4) - 3(x^2 + 2x - 4)$$
$$= x^3 + 2x^2 - 4x - 3x^2 - 6x + 12$$
$$= x^3 - x^2 - 10x + 12$$

Example 16. Simplify $-3y^2(6 - y)(5 + y)$.

Solution

$$-3y^2(6 - y)(5 + y)$$
$$= -3y^2(30 + 6y - 5y - y^2)$$
$$= -3y^2(30 + y - y^2)$$
$$= -90y^2 - 3y^3 + 3y^4$$

Answers to Trial Runs

page 311 **1.** $2x^2 + 11x + 12$ **2.** $3x^2 - 13x - 10$ **3.** $15 + x - 2x^2$ **4.** $-6ab - 4a - 3b - 2$ **5.** $4x^2 - 4xy + y^2$ **6.** $y^2 - \frac{1}{25}$

page 312 **1.** $3x^2 - x - 14$ **2.** $x^4 - 0.2x^2 - 0.63$ **3.** $14y^2 + 5xy - x^2$ **4.** $2ab + 5a - 4b - 10$ **5.** $9x^2 + 12x + 4$ **6.** $2x^6 - 15x^3 + 28$

page 313 **1.** $x^3 - 8$ **2.** $0.4x^2 - 0.8x - 0.4$ **3.** $12x^3 - 27x^2 + 6x$ **4.** $-y^3 + 25y$

Name ______________________________ Date ____________

EXERCISE SET 6.3

Simplify.

______ **1.** $(x - 2)(x + 3)$

______ **2.** $(x + 4)(x - 1)$

______ **3.** $(x + 9)(x + 4)$

______ **4.** $(x + 5)(x + 7)$

______ **5.** $(x - 0.5)(x - 0.3)$

______ **6.** $(x - 0.2)(x - 0.8)$

______ **7.** $(y - 9)^2$

______ **8.** $(y - 6)^2$

______ **9.** $(a + 0.2)^2$

______ **10.** $(a + 0.3)^2$

______ **11.** $(x - 4)(x + 4)$

______ **12.** $(x - 5)(x + 5)$

______ **13.** $\left(y - \frac{1}{3}\right)\left(y + \frac{1}{3}\right)$

______ **14.** $\left(y - \frac{2}{5}\right)\left(y + \frac{2}{5}\right)$

______ **15.** $(4y - 1)(3y - 5)$

______ **16.** $(5y - 2)(3y - 1)$

______ **17.** $(2a + 7)(-3a + 5)$

$-6A^2+19A+7$ **18.** $(3a + 1)(-2a + 7)$

$-6A^2+21A-2A+7$

______ **19.** $(6x - 5)(6x + 5)$

______ **20.** $(5x - 3)(5x + 3)$

______ **21.** $(9 - 2y)(2 - y)$

______ **22.** $(8 - 3y)(3 - y)$

______ **23.** $(2a - b)(a - 5b)$

______ **24.** $(3a - b)(a - 4b)$

______ **25.** $(x - 3y)(2x + y)$

______ **26.** $(x - 2y)(3x + y)$

______ **27.** $(-x + y)(2x - 3y)$

______ **28.** $(-x + y)(3x - 5y)$

______ **29.** $(a + 3b)^2$

______ **30.** $(a + 5b)^2$

______ **31.** $(0.2 + 0.5x)(0.2 - 0.5x)$

______ **32.** $(0.3 - 0.2x)(0.3 + 0.2x)$

______ **33.** $\left(\frac{1}{2}x - y\right)^2$

______ **34.** $\left(\frac{1}{3}x - 6y\right)^2$

______ **35.** $(a - 2b)(c + d)$

______ **36.** $(x - y)(z + 2w)$

______ **37.** $(a^2 - 4)(a - 3)$

______ **38.** $(x^2 - 5)(x - 2)$

______ **39.** $(6a^2 - 1)(a^2 - 5)$

______ **40.** $(5a^2 - 2)(a^2 - 1)$

______ **41.** $(5a^2 - 2)^2$

______ **42.** $(4a^2 - 1)^2$

______ **43.** $(a^3 - 3)(a^3 + 1)$

______ **44.** $(x^3 - 4)(x^3 + 1)$

______ **45.** $(xy - 10)(xy + 5)$

______ **46.** $(xy - 7)(2xy + 1)$

______ **47.** $(ab - 2c)(ab + 2c)$

______ **48.** $(xy - 3z)(xy + 3z)$

______ **49.** $(3y - 2z)^2$

______ **50.** $(4a - 3b)^2$

______ **51.** $(6a^2 + b)(6a^2 - b)$

______ **52.** $(5x^2 - y)(5x^2 + y)$

______ **53.** $\left(10 + \frac{1}{2}x^2\right)^2$

______ **54.** $\left(6 + \frac{1}{3}x^2\right)^2$

______ **55.** $(x - 3)(x^2 - 2x + 1)$

______ **56.** $(x - 5)(x^2 - 6x + 9)$

______ **57.** $(b - 5)(b^2 + 5b + 25)$

______ **58.** $(x - 2)(x^2 + 2x + 4)$

______ **59.** $(x^2 - 4)(x^2 + 3x + 2)$

______ **60.** $(x^2 - 1)(x^2 - x - 2)$

______ **61.** $(1 - x + 2x^2)(2 + 3x)$

______ **62.** $(1 - 3x + x^2)(2 + x)$

______ **63.** $0.3x(x - 2)(x - 5)$

______ **64.** $0.5x(x - 3)(x + 1)$

______ **65.** $-y^2(7 - y)(7 + y)$

______ **66.** $-y^2(4 - y)(4 + y)$

☆ Stretching the Topics

Simplify.

______ **1.** $(x - 2y)(2x - y)(3x - 2y)$

______ **2.** $(x - 1)^3$

______ **3.** Find the area of a rectangle with a width of $2x - 3y$ inches and length of $x + y$ inches.

Check your answers in the back of your book.

If you can complete **Checkup 6.3**, you are ready to go on to Section 6.4.

Name ______________________ Date ____________

CHECKUP 6.3

Simplify.

______ 1. $(x - 2)(x + 5)$

______ 2. $(2y - 3)(y - 6)$

______ 3. $(5 - 2y)^2$

______ 4. $(2a + 5b)^2$

______ 5. $(5 - 2a)(5 + 2a)$

______ 6. $\left(y - \frac{3}{5}\right)\left(y + \frac{3}{5}\right)$

______ 7. $(a - 6)(a^2 + 7)$

______ 8. $(x^2 - 4)(3x^2 + 5)$

______ 9. $(x + 2)(x^2 - 2x + 4)$

______ 10. $-0.2x(3 - x)(5 + x)$

Check your answers in the back of your book.

If You Missed Problems:	You Should Review Examples:
1, 2	1
3, 4	3, 4, 7
5, 6	5, 6, 8
7	9
8	10
9	13
10	15

6.4 Dividing Polynomials

Now that we have learned to add, subtract, and multiply polynomials we shall consider the operation of division.

Dividing Polynomials by Monomials

To discover the method for dividing a polynomial by a monomial, we must use the definition of division $\left(\frac{A}{B} = A \cdot \frac{1}{B}\right)$ and the distributive property. Consider the following quotient.

$$\frac{3x^3 + 12x^2 + 3x}{3x}$$

$$= \frac{1}{3x}(3x^3 + 12x^2 + 3x)$$ Rewrite the quotient, using the definition of division.

$$= \frac{1}{3x} \cdot 3x^3 + \frac{1}{3x} \cdot 12x^2 + \frac{1}{3x} \cdot 3x$$ Use the distributive property.

$$= \frac{3x^3}{3x} + \frac{12x^2}{3x} + \frac{3x}{3x}$$ Rewrite each term, using the definition of division.

$$= x^2 + 4x + 1$$ Simplify each term.

We may skip two steps in this procedure provided we understand that the definition of division and the distributive property are being used mentally.

> To divide a polynomial by a monomial, we may divide each term of the polynomial by the monomial.

As usual, we can describe this method using symbols.

Dividing a Polynomial by a Monomial

$$\frac{A + B}{C} = \frac{A}{C} + \frac{B}{C} \qquad (C \neq 0)$$

Example 1. Find $\frac{2x^3 - 5x^2}{2x^2}$.

Solution

$$\frac{2x^3 - 5x^2}{2x^2}$$

$$= \frac{2x^3}{2x^2} - \frac{5x^2}{2x^2}$$ Divide each term by $2x^2$.

$$= x - \frac{5}{2}$$ Simplify each fraction.

Now try Example 2.

Example 2. Find $\frac{x^3y + x^2y^2 - 8xy^2}{xy}$.

Solution

$$\frac{x^3y + x^2y^2 - 8xy^2}{xy}$$

$$= \frac{x^3y}{xy} + \frac{x^2y^2}{xy} - \frac{8xy^2}{xy}$$

$$= x^2 + xy - 8y$$

Check your answer on page 323. ▶

Example 3. Find $\frac{x^2 + 3x - 1}{-x}$.

Solution

$$\frac{x^2 + 3x - 1}{-x}$$

$$= \frac{x^2}{-x} + \frac{3x}{-x} - \frac{1}{-x}$$ Divide each term by $-x$.

$$= -x - 3 + \frac{1}{x}$$ Simplify each fraction.

Example 4. Find $\frac{3a^2b - 6ab - 4b^2}{-3a^2b}$.

Solution

$$\frac{3a^2b - 6ab - 4b^2}{-3a^2b}$$

$$= \frac{3a^2b}{-3a^2b} - \frac{6ab}{-3a^2b} - \frac{4b^2}{-3a^2b}$$

$$= -1 + \frac{2}{a} + \frac{4b}{3a^2}$$

Trial Run

Find the quotients.

______ 1. $\frac{10x + 5}{5}$

______ 2. $\frac{x^2 + 3x + 1}{x}$

______ 3. $\frac{2 - 3y + y^2}{-y}$

______ 4. $\frac{x^3 - 5x^2 + x}{x^2}$

______ 5. $\frac{4x^2y - x^2y^2 + 3xy^2}{xy}$

______ 6. $\frac{14a^2 - 7ab + 3b^2}{7ab}$

Answers are on page 324.

Dividing Polynomials by Polynomials

To divide a polynomial by a polynomial, we use a method that is very similar to the method of **long division** in arithmetic.

$$\begin{array}{r} 23 \\ 17\overline{)391} \\ \underline{34} \\ 51 \\ \underline{51} \\ 0 \end{array}$$

CHECK:
$17(23) \stackrel{?}{=} 391$
$391 = 391$

Keeping the steps in this long division in mind, let's see how to divide polynomials, starting with the quotient $(x^2 + x - 12) \div (x - 3)$:

$$x - 3\overline{)x^2 + x - 12}$$

Set up quotient as long-division problem.

$$\begin{array}{r} x \\ x - 3\overline{)x^2 + x - 12} \end{array}$$

Divide first term (x) of divisor into first term (x^2) of dividend to get first term (x) of quotient.

$$\begin{array}{r} x \\ x - 3\overline{)x^2 + x - 12} \\ \underline{x^2 - 3x} \end{array}$$

Multiply quotient term (x) times *entire* divisor $(x - 3)$ and write product beneath like terms in dividend.

$$\begin{array}{r} x \\ x - 3\overline{)x^2 + x - 12} \\ \underline{x^2 - 3x } \\ 4x - 12 \end{array}$$

Subtract carefully, and bring down the next term of the dividend (-12).

$$\begin{array}{r} x + 4 \\ x - 3\overline{)x^2 + x - 12} \\ \underline{x^2 - 3x } \\ 4x - 12 \\ \underline{4x - 12} \\ 0 \end{array}$$

Divide first term (x) of divisor into first term ($4x$) of new dividend, then multiply new term (4) of dividend times divisor $(x - 3)$ and subtract again.

To check this problem, we use the same process as in arithmetic:

$$(x - 3)(x + 4) \stackrel{?}{=} x^2 + x - 12$$

$$x^2 + 4x - 3x - 12 \stackrel{?}{=} x^2 + x - 12$$

$$x^2 + x - 12 = x^2 + x - 12$$

Example 5. Find $(y^2 + 4y - 12) \div (y - 2)$ and check.

Solution

$$\begin{array}{r} y + 6 \\ y - 2\overline{)y^2 + 4y - 12} \\ \underline{y^2 - 2y } \\ 6y - 12 \\ \underline{6y - 12} \\ 0 \end{array}$$

Now you complete the check.

CHECK:

$$(y - 2)(y + 6) \stackrel{?}{=} y^2 + 4y - 12$$

Check your work on page 323. ▶

Example 6. Find $(x^3 - 2x^2 + 3x - 1) \div (x + 2)$.

Solution

$$\begin{array}{r} x^2 - 4x + 11 \\ x + 2\overline{)x^3 - 2x^2 + 3x - 1} \\ \underline{x^3 + 2x^2 } \\ -4x^2 + 3x \\ \underline{-4x^2 - 8x} \\ 11x - 1 \\ \underline{11x + 22} \\ -23 \end{array}$$

Here we have a *remainder* of -23, so we may write our answer in either of two ways.

$$(x^3 - 2x^2 + 3x - 1) \div (x + 2) = x^2 - 4x + 11 \quad \text{R } -23$$

or

$$(x^3 - 2x^2 + 3x - 1) \div (x + 2) = x^2 - 4x + 11 + \frac{-23}{x + 2}$$

In performing long division of polynomials, your work will be simplified if you remember to write the terms of the dividend and divisor in *descending powers* of the same variable and to leave space in the dividend for any missing terms.

Example 7. Find $(4x^4 + 11x^2 - 3) \div (2x - 1)$.

Solution: We arrange the terms in descending powers of x and note the missing x^3-term and x-term in the dividend.

$$\begin{array}{r|l}
 & 2x^3 + x^2 + 6x + 3 \\
\hline
2x - 1 & 4x^4 + 0x^3 + 11x^2 + 0x - 3 \\
 & \underline{4x^4 - 2x^3} \\
 & \phantom{4x^4 +{}} 2x^3 + 11x^2 \\
 & \phantom{4x^4 +{}} \underline{2x^3 - x^2} \\
 & \phantom{4x^4 + 0x^3 +{}} 12x^2 + 0x \\
 & \phantom{4x^4 + 0x^3 +{}} \underline{12x^2 - 6x} \\
 & \phantom{4x^4 + 0x^3 + 11x^2 +{}} 6x - 3 \\
 & \phantom{4x^4 + 0x^3 + 11x^2 +{}} \underline{6x - 3} \\
 & \phantom{4x^4 + 0x^3 + 11x^2 + 6x - {}} 0
\end{array}$$

$$(4x^4 + 11x^2 - 3) \div (2x - 1) = 2x^3 + x^2 + 6x + 3$$

Example 8. Find $(a^3 - 8) \div (a^2 + 2a + 4)$ and check.

Solution: We note the missing a^2-term and a-term in the dividend.

$$\begin{array}{r|l}
 & a - 2 \\
\hline
a^2 + 2a + 4 & a^3 + 0a^2 + 0a - 8 \\
 & \underline{a^3 + 2a^2 + 4a} \\
 & - 2a^2 - 4a - 8 \\
 & \underline{- 2a^2 - 4a - 8} \\
 & \phantom{a^3 - 2a^2 - 4a - {}} 0
\end{array}$$

CHECK:

$(a^2 + 2a + 4)(a - 2) \stackrel{?}{=} a^3 - 8$	
$a^2(a - 2) + 2a(a - 2) + 4(a - 2) \stackrel{?}{=} a^3 - 8$	Use distributive property.
$a^3 - 2a^2 + 2a^2 - 4a + 4a - 8 \stackrel{?}{=} a^3 - 8$	Remove parentheses.
$a^3 - 8 = a^3 - 8$	Combine like terms.

Trial Run

Divide.

_______ 1. $(x^2 + 3x + 2) \div (x + 1)$

_______ 2. $(3 + x^2 - x) \div (x - 2)$

_______ 3. $(9x^3 + 45x^2 - x - 5) \div (3x + 1)$

_______ 4. $(y^4 - 1) \div (y - 1)$

Answers are on page 324.

Examples You Completed

Example 2. Find $\dfrac{x^3y + x^2y^2 - 8xy^2}{xy}$.

Solution

$$\frac{x^3y + x^2y^2 - 8xy^2}{xy}$$
$$= \frac{x^3y}{xy} + \frac{x^2y^2}{xy} - \frac{8xy^2}{xy}$$
$$= x^2 + xy - 8y$$

Example 5. Find $(y^2 + 4y - 12) \div (y - 2)$ and check.

CHECK

$$(y - 2)(y + 6) \stackrel{?}{=} y^2 + 4y - 12$$
$$y^2 + 6y - 2y - 12 \stackrel{?}{=} y^2 + 4y - 12$$
$$y^2 + 4y - 12 = y^2 + 4y - 12$$

Answers to Trial Runs

page 320 **1.** $2x + 1$ **2.** $x + 3 + \frac{1}{x}$ **3.** $\frac{-2}{y} + 3 - y$ **4.** $x - 5 + \frac{1}{x}$ **5.** $4x - xy + 3y$
6. $\frac{2a}{b} - 1 + \frac{3b}{7a}$

page 323 **1.** $x + 2$ **2.** $x + 1 + \frac{5}{x - 2}$ **3.** $3x^2 + 14x - 5$ **4.** $y^3 + y^2 + y + 1$

Name ______________________________ Date ______________

EXERCISE SET 6.4

Find the quotients.

X+5 ______ 1. $\dfrac{4x + 20}{4}$

______ 2. $\dfrac{16x - 32}{16}$

8X-7 ______ 3. $\dfrac{8x^3 - 7x^2}{x^2}$

______ 4. $\dfrac{9x^5 - 5x^4}{x^3}$

4-3X ______ 5. $\dfrac{-8x + 6x^2}{-2x}$

______ 6. $\dfrac{-12x + 8x^2}{-4x}$

5X²Y²-2 ______ 7. $\dfrac{20x^3y^2 - 8xy}{4xy}$

______ 8. $\dfrac{15x^5y^3 - 9xy}{3xy}$

2D-3-1/4D ______ 9. $\dfrac{8d^3 - 12d^2 - d}{4d^2}$

______ 10. $\dfrac{16d^3 - 24d^2 - d}{8d^2}$

-Y²+A+B ______ 11. $\dfrac{y^3 - ay^2 - by}{-y}$

______ 12. $\dfrac{x^3 - mx^2 - nx}{-x}$

-M²-3MN+5N² ______ 13. $\dfrac{-m^4n - 3m^3n^2 + 5m^2n^3}{m^2n}$

______ 14. $\dfrac{-a^3b^2 - 5a^2b^3 + 4ab^4}{ab^2}$

3B+4C-D/2 ______ 15. $\dfrac{6ab + 8ac - ad}{2a}$

______ 16. $\dfrac{15xw + 9xy - xz}{3x}$

X+7 ______ 17. $(x^2 + 10x + 21) \div (x + 3)$

______ 18. $(x^2 - 6x - 16) \div (x + 2)$

A-4 ______ 19. $(2a^2 - 5a - 12) \div (2a + 3)$

______ 20. $(10a^2 + 11a - 6) \div (5a - 2)$

Z²-3Z+2 ______ 21. $(z^3 - 8 + 14z - 7z^2) \div (z - 4)$

______ 22. $(x^3 + 6 - 11x + 2x^2) \div (x - 2)$

? ______ 23. $(x^3 - 27) \div (x + 3)$

______ 24. $(x^3 - 64) \div (x + 4)$

______ 25. $(4x^3 - 39x^2 + 92x - 63) \div (x^2 - 8x + 9)$

______ 26. $(3x^3 - 22x^2 + 47x - 28) \div (x^2 - 5x + 4)$

______ 27. $(6x^4 - 7x^3 + 5x^2 - 10x - 3) \div (2x - 3)$

______ 28. $(8x^4 + 2x^3 + 3x^2 + 12x + 5) \div (2x + 1)$

______ 29. $(3x^3 - 4x + 2) \div (x - 1)$

______ 30. $(4x^3 - 5x + 11) \div (x + 2)$

☆ Stretching the Topics

Find the quotients.

________ **1.** $(x^5 - 32y^5) \div (x - 2y)$

________ **2.** $(6a^4 - a^3b - 2a^2b^2 - 11ab^3 - 6b^4) \div (2a - 3b)$

________ **3.** The area of a rectangle is $12x^2y^3$. The length of the rectangle is $3xy^3$. Write expressions for the width and the perimeter of the rectangle.

Check your answers in the back of your book.

If you can complete **Checkup 6.4**, you are ready to go on to Section 6.5.

Name ______________________ Date ____________

CHECKUP 6.4

Find the quotients.

_______ **1.** $\dfrac{6x^4 - x^2}{3x^2}$

_______ **2.** $\dfrac{a^3b - 4a^2b^2 - 3ab^3}{-a^2b}$

_______ **3.** $(2y^2 + 11y + 9) \div (y + 4)$

_______ **4.** $(x^3 - x^2 - 7x + 3) \div (x - 3)$

_______ **5.** $(9x^4 + 8x^2 - 1) \div (3x - 1)$

_______ **6.** $(a^3 - 27) \div (a^2 + 3a + 9)$

Check your answers in the back of your book.

If You Missed Problems:	You Should Review Examples:
1, 2	1–4
3, 4	5, 6
5	7
6	8

6.5 Switching from Word Expressions to Polynomials

Let's continue our practice of switching word expressions to variable expressions, making use of some of the skills learned in this chapter. As before, we always begin by deciding what the variable represents. Then we translate the word expressions in the problem into variable expressions. An illustration will sometimes help us get started.

Example 1. If one side of a square is s, write an expression for the perimeter. Write an expression for the area.

Solution

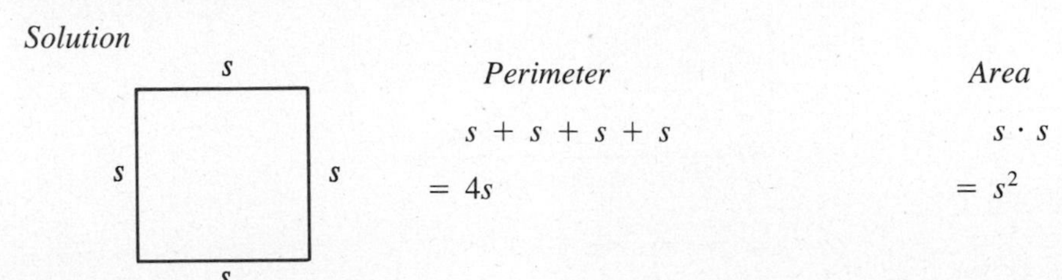

Perimeter	*Area*
$s + s + s + s$	$s \cdot s$
$= 4s$	$= s^2$

Now you complete Example 2.

Example 2. If one side of a square is $2x$, write an expression for the perimeter. Write an expression for the area.

Solution

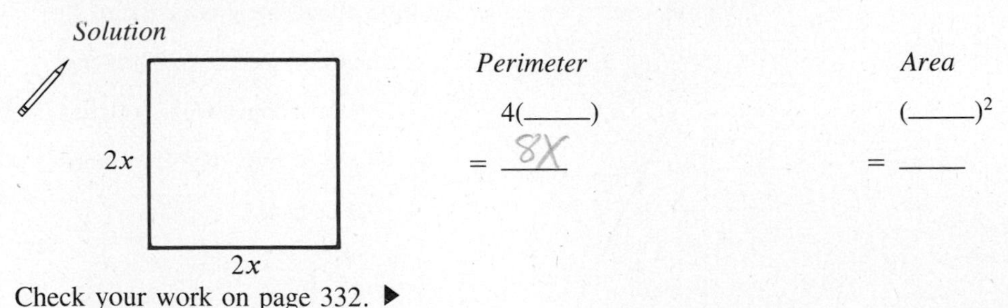

Check your work on page 332. ▶

Example 3. If the length of a rectangle is l and its width is w, write an expression for the perimeter. Write an expression for the area.

Solution

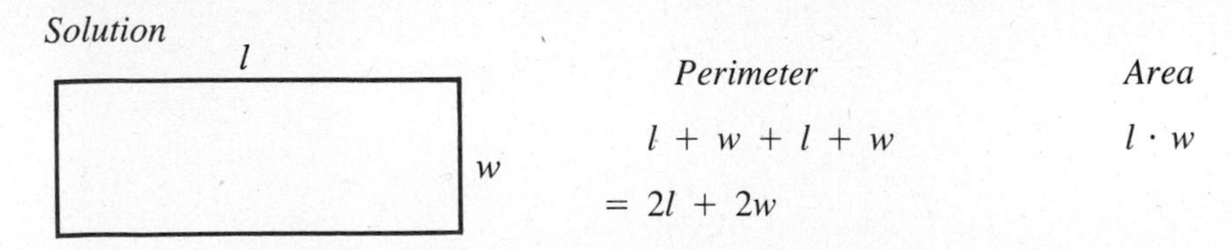

Perimeter	*Area*
$l + w + l + w$	$l \cdot w$
$= 2l + 2w$	

Example 4. If the length of a rectangular room is 5 feet more than its width, write an expression for the perimeter. Write an expression for the area.

Solution: We know the least about the width, so we let

x = width (feet)

$x + 5$ = length (feet)

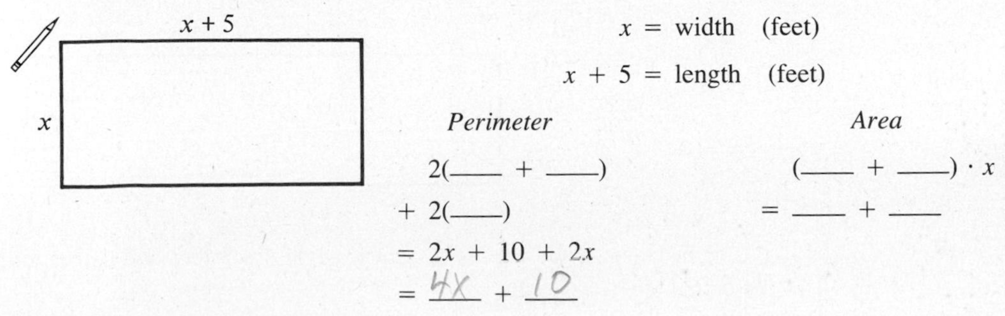

Check your work on page 332. ▶

A triangle in which one side is perpendicular to another side is called a **right triangle**. The perpendicular sides form a **right angle** (90°). The side opposite the right angle is called the **hypotenuse** of the right triangle. The other sides are called **legs**.

Example 5. If a and b are legs of a right triangle with hypotenuse c, write an expression for the perimeter of the triangle. Write an expression for the area.

Solution

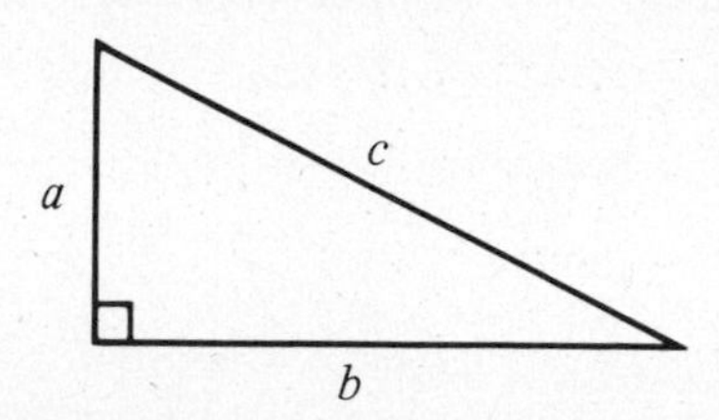

Perimeter

$a + b + c$

Area

$\frac{a \cdot b}{2}$

Now you try Example 6.

Example 6. One leg of a right triangle is twice the other leg and the hypotenuse is 3 inches more than the longer leg. Write an expression for the perimeter. Write an expression for the area.

Solution: Since we know the least about the shorter leg of the right triangle, we let

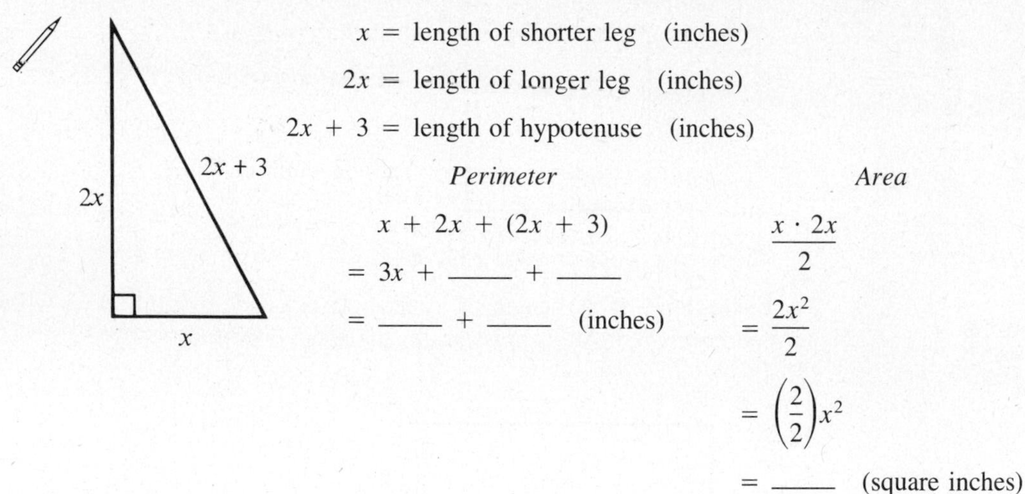

x = length of shorter leg (inches)

$2x$ = length of longer leg (inches)

$2x + 3$ = length of hypotenuse (inches)

Perimeter

$x + 2x + (2x + 3)$

$= 3x + \_\_\_\_ + \_\_\_\_$

$= \_\_\_\_ + \_\_\_\_$ (inches)

Area

$\frac{x \cdot 2x}{2}$

$= \frac{2x^2}{2}$

$= \left(\frac{2}{2}\right)x^2$

$= \_\_\_\_$ (square inches)

Check your work on page 332. ▶

A Greek mathematician named **Pythagoras** discovered that if the lengths of the legs of a right triangle are squared and then added together, their sum is equal to the square of the length of the hypotenuse.

Pythagorean Theorem

$$a^2 + b^2 = c^2$$

For instance, if the legs of a right triangle are 3 feet and 4 feet long, the hypotenuse must be 5 feet long, because

$$3^2 + 4^2 = 5^2$$
$$9 + 16 = 25$$
$$25 = 25$$

Example 7. If one leg of a right triangle is 5 inches longer than the other leg, write an algebraic expression for the square of the hypotenuse.

Solution: We know the least about the shorter leg, so we let

$$a = \text{length of first leg (inches)}$$
$$a + 5 = \text{length of second leg (inches)}$$
$$a^2 = \text{square of first leg}$$
$$(a + 5)^2 = \text{square of second leg}$$

We may write the square of the hypotenuse as

$$a^2 + (a + 5)^2$$
$$= a^2 + a^2 + 10a + 25$$
$$= 2a^2 + 10a + 25$$

Consecutive integers are integers that are next to each other when the list of integers is written in order. For instance

22, 23, 24	are consecutive integers
$-2, -1, 0, 1$	are consecutive integers

To get from one integer to the next consecutive integer, do you see that we must add 1? So if a first integer is x, the next integer must be $x + 1$. Sometimes we must work with sums or products of two consecutive integers. Let

$$x = \text{first integer}$$
$$x + 1 = \text{next consecutive integer}$$

Sum	*Product*
$x + (x + 1)$	$x(x + 1)$
$= 2x + 1$	$= x^2 + x$

Try to complete Example 8.

Example 8. Write an expression for the square of the sum of two consecutive integers.

Solution

Let x = first integer

_____ = next consecutive integer

_____ = sum of two consecutive integers

The square of the sum of two consecutive integers is

$$(\underline{2X} + \underline{1})^2 = \underline{4X^2} + \underline{4X} + \underline{1}$$

Check your work on page 333. ▶

Now that we have practiced switching from words to algebraic expressions, let's return to the problem stated at the beginning of this chapter.

Example 9. Mr. Abell's garden is in the shape of a square. Next year he plans to enlarge his garden by increasing the length by 6 feet and the width by 4 feet. Write an expression for the area of the garden after it has been enlarged.

Solution

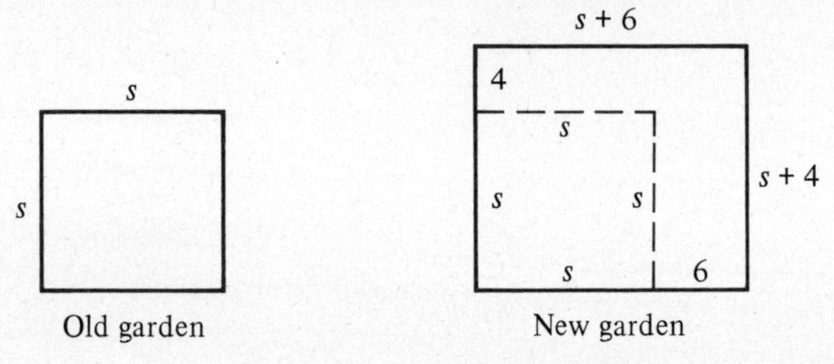

Old garden New garden

The area of the new garden is found by multiplying the length times the width.

$$\text{Area: } (s + 6)(s + 4)$$
$$= s^2 + 6s + 4s + 24$$
$$= s^2 + 10s + 24$$

▶ Examples You Completed

Example 2. If one side of a square is $2x$, write an expression for the perimeter. Write an expression for the area.

Solution

Perimeter	*Area*
$4(2x)$	$(2x)^2$
$= 8x$	$= 4x^2$

Example 4

Solution

Perimeter	*Area*
$2(x + 5) + 2(x)$	$(x + 5) \cdot x$
$= 2x + 10 + 2x$	$= x^2 + 5x$
$= 4x + 10$	

Example 6

Solution

Perimeter	*Area*
$x + 2x + (2x + 3)$	$\frac{x \cdot 2x}{2}$
$= 3x + 2x + 3$	
$= 5x + 3$	$= \frac{2x^2}{2}$
	$= x^2$

Example 8. Write an expression for the square of the sum of two consecutive integers.

Solution

$$\text{Let } x = \text{first integer}$$

$$x + 1 = \text{next consecutive integer}$$

$$2x + 1 = \text{sum of two consecutive integers}$$

The square of the sum of two consecutive integers is

$$(2x + 1)^2 = 4x^2 + 4x + 1$$

Name ______________________ Date ____________

EXERCISE SET 6.5

Write and simplify the expression indicated.

_______ **1.** If the width of a rectangle is 7 less than the length, write an expression for the perimeter. Write an expression for the area.

_______ **2.** If the length of a rectangle is 10 more than the width, write an expression for the perimeter. Write an expression for the area.

_______ **3.** If one side of a square is $3x$, write an expression for the perimeter. Write an expression for the area.

_______ **4.** If one side of a square is $5x$, write an expression for the perimeter. Write an expression for the area.

_______ **5.** One leg of a right triangle is 7 more than the other leg and the hypotenuse is 8 more than the shorter leg. Write an expression for the perimeter. Write an expression for the area.

_______ **6.** One leg of a right triangle is 2 more than the other leg and the hypotenuse is 4 more than the shorter leg. Write an expression for the perimeter. Write an expression for the area.

_______ **7.** If the length of one leg of a right triangle is twice the length of the other leg, write an expression for the square of the hypotenuse.

_______ **8.** If the length of one leg of a right triangle is 4 times the length of the other leg, write an expression for the square of the hypotenuse.

_______ **9.** Write an expression for the product of two consecutive integers.

_______ **10.** Write an expression for the sum of the squares of two consecutive integers.

_______ **11.** One number is 12 more than the other. Write an expression for their product.

_______ **12.** One integer is 3 less than twice another. Write an expression for their product.

_______ **13.** One number is 9 less than another. Write an expression for the square of the larger added to 5 times the smaller.

_______ **14.** One number is 2 more than another. Write an expression for the square of the smaller added to 7 times the larger.

_______ **15.** Write an expression for the product of two consecutive integers increased by 3 times the smaller.

_______ **16.** Write an expression for the product of two consecutive integers decreased by twice the larger.

_______ **17.** Wick plans to plow a rectangular field that is 15 rods longer than it is wide. Write an expression for the area.

_______ **18.** Kaaryn is building a room on her house that is 5 feet longer than it is wide. Write an expression for the area of the room.

_______ **19.** Lamar is building a rectangular pen for his pony that is 15 feet longer than it is wide. Write an expression for the number of feet of fence he will need.

_______ **20.** Carol is buying lace to border a table cloth that is 20 inches longer than it is wide. Write an expression for the number of inches of lace she will need.

_______ **21.** Maxine has a vegetable garden that is a square. She plans to increase the size of the garden by increasing the length of one side by 10 feet and decreasing the other by 2 feet. Write an expression for the area of the larger garden.

_______ **22.** Robert has a square family room that he plans to enlarge by increasing the length of one side by 6 feet and increasing the other by 2 feet. Write an expression for the area of the enlarged family room.

_______ **23.** The length of a room is 6 feet longer than twice its width. Write an expression for the number of square feet of carpet needed to cover the floor.

_______ **24.** A yard is 6 feet longer than it is wide. Write an expression for the number of square feet of sodding it would take to cover the yard.

_______ **25.** A flower bed is in the shape of a right triangle with one side 9 feet shorter than the other. Write an expression for the area.

_______ **26.** Chris is building a corner table for her kitchen. The top will be a right triangle wih equal sides. Write an expression for the area.

_______ **27.** One leg of a right triangle is 5 more than twice the other leg. Write an expression for the square of the hypotenuse.

_______ **28.** One leg of a right triangle is 3 more than 4 times the other leg. Write an expression for the square of the hypotenuse.

_______ **29.** If one number is 5 times the other, write an expression for the product of the numbers increased by twice the sum.

_______ **30.** If one number is 2 smaller than twice the other, write an expression for the product of the two numbers decreased by 3 times their sum.

☆ Stretching the Topics

_______ **1.** The width of a rectangle is 3 less than one-half its length. Write an expression for the area of a square with a side equal to the rectangle's width.

_______ **2.** If a train travels $2t^2 - 1$ miles in t hours, write an expression for the rate of speed of the train.

_______ **3.** If n represents an odd number, write an expression for the product of the next two consecutive higher even numbers.

Check your answers in the back of your book.

If you can complete **Checkup 6.5**, you are ready to do the **Review Exercises for Chapter 6.**

Name ____ Date ____________

✓ CHECKUP 6.5

______ **1.** If one side of a square is $3x$, write an expression for the perimeter. Write an expression for the area.

______ **2.** If the length of a rectangle is 3 less than 7 times the width, write an expression for the perimeter. Write an expression for the area.

______ **3.** One leg of a right triangle is 7 more than the other and the hypotenuse is 3 more than twice the shorter leg. Write an expression for the perimeter. Write an expression for the area.

______ **4.** If one leg of a right triangle is 7 more than the other, write an expression for the square of the hypotenuse.

______ **5.** Write an expression for the square of the sum of two consecutive integers decreased by twice the smaller.

Check your answers in the back of your book.

If You Missed Problem:	You Should Review Example:
1	2
2	4
3	6
4	7
5	8

Summary

In this chapter we learned to work with **polynomials**, which are sums of terms that are constants times whole number powers of variables. We identified polynomials according to the number of terms they contain.

Name	Number of Terms	Examples
Monomial	One	$3x$ $-x^2y$
Binomial	Two	$3x + 1$ $x^2 - y^2$
Trinomial	Three	$x^2 + x - 2$ $a + b + c$

Then we discussed the methods for performing the basic operations of addition, subtraction, multiplication, and division with polynomials.

In Order to:	We Must:	Example
Add polynomials	Combine like terms.	$(x^2 + 2x - 1) + (2x^2 - 5)$ $= 3x^2 + 2x - 6$
Subtract polynomials	Remove parentheses and combine like terms.	$(3x^2 - 5x + 2) - (x^2 - x + 1)$ $= 3x^2 - 5x + 2 - x^2 + x - 1$ $= 2x^2 - 4x + 1$
Multiply monomials	Multiply coefficients, multiply variables using the First Law of Exponents.	$-3x^2y(2x^3y)$ $= -6x^5y^2$
Multiply a monomial times a polynomial	Use the distributive property.	$-3x^2(x^2 - 5x + 1)$ $= -3x^4 + 15x^3 - 3x^2$
Multiply binomials	Use FOIL as a shortcut for the distributive property.	$(2x + 3)(x - 1)$ $= 2x^2 - 2x + 3x - 3$ $= 2x^2 + x - 3$
Multiply polynomials	Use the distributive property.	$(x + 3)(x^2 - 5x - 2)$ $= x(x^2 - 5x - 2) + 3(x^2 - 5x - 2)$ $= x^3 - 5x^2 - 2x + 3x^2 - 15x - 6$ $= x^3 - 2x^2 - 17x - 6$
Divide a polynomial by a monomial	Divide each term of the polynomial by the monomial.	$\frac{5x^2 - 3x - 6}{3x}$ $= \frac{5x^2}{3x} - \frac{3x}{3x} - \frac{6}{3x}$ $= \frac{5x}{3} - 1 - \frac{2}{x}$
Divide polynomials	Use long division.	$x + 2 \overline{)x^2 + 5x + 6}$ quotient $x + 3$ $\underline{x^2 + 2x}$ $3x + 6$ $\underline{3x + 6}$ 0

In multiplying binomials we discovered two *special products* which can be found quickly using formulas.

Special Product	Formula	Examples
Square of a binomial	$(A + B)^2 = A^2 + 2AB + B^2$ $(A - B)^2 = A^2 - 2AB + B^2$	$(x + 5)^2 = x^2 + 10x + 25$ $(2y - 3)^2 = 4y^2 - 12y + 9$
Product of sum and difference of two terms	$(A + B)(A - B) = A^2 - B^2$	$(2x + 7)(2x - 7)$ $= 4x^2 - 49$

Finally, we practiced switching from word expressions to variable expressions that were polynomials.

❑ Speaking the Language of Algebra

1. The polynomial $7y$ is called a __________ because it contains __________ term. The polynomial $5x - y$ is a __________ because it contains __________ terms. The polynomial $6 - x + 3x^2$ is a __________ because it contains __________ terms.

2. To add polynomials, we must __________ __________ __________.

3. To multiply a monomial times a polynomial, we must use the __________ property.

4. The FOIL method can be used only when multiplying two __________.

5. To square a binomial, we use the special product formula $(A + B)^2 =$ __________

6. To multiply the sum and difference of the same two terms, we use the special product formula $(A + B)(A - B) =$ __________.

7. To divide a polynomial by a monomial, we divide each term of the polynomial by the __________.

8. To divide two polynomials, we use the method of __________ __________.

Check your answers in the back of your book.

Name ______________________ Date ____________

REVIEW EXERCISES for Chapter 6

Name each expression.

_______ **1.** $3a + 2b$

_______ **2.** $7x - 2y + 5x$

_______ **3.** $-3m^2$

_______ **4.** $x^2 - 3x - 1$

Simplify.

_______ **5.** $(3x^2 - 2x) + (5x - 3)$

_______ **6.** $(-3a^2 + 2a - 4) + (a^2 - 5a + 4)$

_______ **7.** $-2(7xy)$

_______ **8.** $\frac{2}{3}(9x^2)$

_______ **9.** $-4(0.8a^2b^3)$

_______ **10.** $5(-x^2 - 3x + 2)$

_______ **11.** $-(x^2 - 2x + 1)$

_______ **12.** $0.2(8x^2 - 3x + 2)$

_______ **13.** $(4x - 3y) - (2x + 3y)$

_______ **14.** $-5(2a + b) + 3(a + b)$

_______ **15.** $2(x^2 - x - 9) - (5x^2 + 2x - 4)$

_______ **16.** $x^7 \cdot x^3$

_______ **17.** $2a^3(-3a)(a^4)$

_______ **18.** $9xy^6(-2x^4y^2)$

_______ **19.** $(m^2)^3$

_______ **20.** $(-3y)^4$

_______ **21.** $(x^3y^2z)^2$

_______ **22.** $3a(a^4b^2)^3$

_______ **23.** $\left(\frac{1}{2}x\right)^3$

_______ **24.** $0.8(5x - 3)$

_______ **25.** $7y(3y + 2)$

_______ **26.** $3m^2(m^2 - 4mn - 5n^2)$

_______ **27.** $5xy(7x - 11y)$

_______ **28.** $-a^2b(6 - 3ab + 4a^2b^2)$

_______ **29.** $4xy(3x - y) - 5xy(2x - y)$

_______ **30.** $(x + 9)(x - 8)$

_______ **31.** $(3y - 2)(4y + 1)$

_______ **32.** $(2a - 3b)(2a + 3b)$

_______ **33.** $(x - 5y)^2$

_______ **34.** $(2xy - 1)^2$

_______ **35.** $(m^2 - 3)(m^2 + 5)$

_______ **36.** $(3x^2 - 5y^2)(2x^2 + 4y^2)$

_______ **37.** $3x(3x - 2)^2$

_______ **38.** $2x(x - 3)(x + 3)$

_______ **39.** $-4a^2(3 - 2a)(3 + 2a)$

_______ **40.** $(y - 1)(y^2 - 3y - 4)$

Find the quotients.

_______ **41.** $\dfrac{3x^3 - 6x^2 - 2x}{3x}$

_______ **42.** $\dfrac{x^3y + 15x^2y^2 - 5xy^2}{-xy}$

_______ **43.** $(5x^2 + 33x - 14) \div (5x - 2)$

_______ **44.** $(x^2 - 2x + 5) \div (x + 4)$

_______ **45.** $(x^3 - 3x^2 + x + 4) \div (x + 1)$

_______ **46.** $(8x^4 - 2x^2 + 1) \div (2x - 1)$

_______ **47.** $(y^3 + 1) \div (y^2 + y + 1)$

_______ **48.** Joyce is adding a family room to her house. If the width of the room is to be 10 feet less than the length, write an expression for the area of the room, and simplify.

_______ **49.** If one leg of a right triangle is 2 inches longer than twice the other leg, write an expression for the square of the hypotenuse, and simplify.

_______ **50.** Write an expression for the product of two consecutive integers, decreased by 3 times the square of the smaller one.

Check your answers in the back of your book.

If You Missed Exercises:	You Should Review Examples:	
1–4	SECTION 6.1	1–4
5, 6		5, 6
7–9		7, 8
12–15		9–12
16–23	SECTION 6.2	1–6
24–28		7–10
29		11, 12
30, 31	SECTION 6.3	1, 2
32		8
33, 34		7
35, 36		9, 10
37–39		13–16
41, 42	SECTION 6.4	1–4
43, 44		5
45		6
46		7
47		8
48	SECTION 6.5	4
49		7
50		8

If you have completed the **Review Exercises** and corrected your errors, you are ready to take the **Practice Test for Chapter 6**.

Name ____________________ Date __________

PRACTICE TEST for Chapter 6

Name each expression.	SECTION	EXAMPLES
______ **1.** $-4a - 3b + 4a - b$	6.1	1
______ **2.** $3x^2 - 5x + 2$	6.1	3
Simplify.		
______ **3.** $(2x + 7y - 3z) + (4x - y + 2z)$	6.1	6
______ **4.** $(-0.3)(7a^2)$	6.1	8
______ **5.** $5(2x^2 - x + 3) - (x^2 + 3x + 1)$	6.1	11
______ **6.** $y^5(-3y)(5y^4)$	6.2	4
______ **7.** $(-4x^3y^2)(5x^2y)$	6.2	5, 6
______ **8.** $-5c(a^2 - 3ac - c^2)$	6.2	8
______ **9.** $8xy(2x - 3y) - 4xy(x - 6xy)$	6.2	11
______ **10.** $(x + 11)(x - 10)$	6.3	1
______ **11.** $(4x - 5)(3x - 2)$	6.3	2
______ **12.** $(a - 7)^2$	6.3	3, 4, 7
______ **13.** $\left(x - \frac{2}{3}\right)\left(x + \frac{2}{3}\right)$	6.3	5, 6, 8
______ **14.** $(5a + 3c)(4a - c)$	6.3	11, 12
______ **15.** $(y^2 - 0.5)(y^2 + 0.7)$	6.3	9, 10
______ **16.** $(4x - z)^2$	6.3	7
______ **17.** $(x + 2)(x^2 - 2x + 4)$	6.3	13
______ **18.** $-2y(y + 9)(y - 9)$	6.3	15, 16
______ **19.** $-3(4x - y)^2$	6.3	15, 16
______ **20.** $\dfrac{9x^3 - 15x^2 - 12x}{3x}$	6.4	1, 2
______ **21.** $\dfrac{6x^2 - 12xy - 5y}{-6x^2y}$	6.4	3, 4
______ **22.** $(6x^2 - 17x - 14) \div (3x + 2)$	6.4	5

________	**23.** $(x^3 - 5x - 9) \div (x - 3)$	6.4	6
________	**24.** If one leg of a right triangle is 2 inches longer than 3 times the other leg, write an expression for the square of the hypotenuse, and simplify.	6.5	7
________	**25.** If all four sides of a square patio are increased by 3 feet, write an expression for the area of the new patio, and simplify.	6.5	9

Check your answers in the back of your book.

Name ________________________________ Date ______________

SHARPENING YOUR SKILLS after Chapters 1–6

		SECTION
______	**1.** Change $\frac{5}{8}$ to a decimal number.	1.3
______	**2.** Use the associative property for addition to rewrite $(-3 + 4) + (-10)$.	2.3
______	**3.** Evaluate $\frac{2a - 5b}{a - b}$ when $a = 4$ and $b = -2$.	3.1
______	**4.** Is -2 a solution for $3x - 2(x + 5) = -12$?	4.1
______	**5.** Evaluate $-2(x^2 - xy)$ when $x = -1$ and $y = 2$.	5.1

Simplify.

______	**6.** $\left(\frac{2}{3}\right)^2 - \frac{5}{12} + \frac{5}{6}$	1.2
______	**7.** $\frac{-3(8 - 4) - 7(-4)}{-6 + [(-2)^2 \cdot 5]}$	2.4
______	**8.** $-3a + 7b - 4c + 9a - 3b + 5c$	3.2
______	**9.** $6[-4(x - y) - (2x - 3y)]$	3.3
______	**10.** $(-3x)^2(xy)(x^3y^4)$	5.1
______	**11.** $\frac{-72x^3y^2z}{81xy^5z^4}$	5.2
______	**12.** $(-7y^{-5})(y^3)(-y)^2$	5.3

Solve.

______	**13.** $4(x - 5) - 7 = 3(2x - 4) - 5x$	4.2
______	**14.** $4[2(x - 2) - (x + 4)] = 0$	4.2
______	**15.** The length of a rectangle is 3 less than twice the width. If the perimeter of the rectangle is 30 feet, find the dimensions of the rectangle.	4.3

Check your answers in the back of your book.

Factoring Polynomials and Solving Quadratic Equations

7

After a tree was broken over by a storm, the part that was bending over was 26 feet long. The part left standing was 14 feet less than the distance from the foot of the tree to the point where the broken part touched the ground. Find the height of the tree before the storm.

Before you finish this chapter, you will be able to solve this problem using a **second-degree equation**. A second-degree equation is a polynomial equation in which the largest exponent on the variable is 2.

One way to solve a second-degree equation is to rewrite the polynomial as a product of **factors**. Our purpose in this chapter is to learn procedures for figuring out what factors were multiplied together to give a certain polynomial.

In this chapter we learn to

1. Find the common factor of a polynomial.
2. Factor a difference of two squares.
3. Factor a general trinomial.
4. Solve quadratic equations by factoring.
5. Switch from word statements to quadratic equations.

7.1 Looking for Common Factors

Factoring is a process by which we try to rewrite an expression as a **product**. In order to figure out what quantities would multiply together to give a certain product, you must be very familiar with the multiplication of polynomials.

Looking for Common Monomial Factors

Recall that when we multiplied monomials times polynomials the products were found using the distributive property. For example,

$$-7(2x - y) = -14x + 7y$$

$$8(x^2 + 3x - 5) = 8x^2 + 24x - 40$$

Can you write $2x + 2y$ as a product of factors?

$$2x + 2y = 2(x + y)$$

In factoring, you can *always* decide whether your "guess" is correct. All you must do is multiply your factors together and see if your answer is the same as the original polynomial.

Let's try factoring

$$9y^2 + 27y - 18$$

We notice that 3 is a factor of each term, but 9 is also a factor of each term. Which do we choose? We must agree to choose the *largest possible* common factor. Here we choose 9 and write

$$9y^2 + 27y - 18 = 9(y^2 + 3y - 2)$$

Example 1. Factor $-5x^2 + 10$.

Solution

$$-5x^2 + 10 = -5(x^2 - 2)$$

Notice that if the first numerical coefficient in a polynomial is negative, we usually agree to factor out a negative common factor.

You try Examples 2 and 3.

Example 2. Factor $14x^2 - 70x - 7$.

Solution

$$14x^2 - 70x - 7 = 7(2x^2 - 10x - 1)$$

Example 3. Factor $-19x + 19$.

Solution

$$-19x + 19 = -19(x - 1)$$

Did you remember to check your factors by multiplying? Check your work on page 352. ▶

If you always remember to check your factors by multiplying, you will not make factoring errors such as these.

Polynomial	Wrong	Right
$-7x + 7$	$-7(x + 1)$	$-7(x - 1)$
$2x^2 + 2$	$2(x^2)$	$2(x^2 + 1)$
$3x + 6y$	$3(x + 6y)$	$3(x + 2y)$

The process used to factor a polynomial such as $3x^2 + 6x$ is similar to the process used in factoring out common numerical factors. Once again we are looking for factors common to every term in the polynomial. Those factors may contain a numerical part and/or a variable part.

In the product $3x^2 + 6x$, we notice that both terms contain a factor of 3 *and* a factor of x. Would you agree that we may write

$$3x^2 + 6x = 3x(x + 2)?$$

Let's factor $98a^2 - 28a$.

$98a^2 - 28a$	Notice that 2 and a are factors of both terms.
$= 2a(49a - 14)$	Notice that 7 is a factor of $49a - 14$.
$= 2a \cdot 7(7a - 2)$	It is not too late to factor out the 7.
$= 14a(7a - 2)$	Simplify the monomial factor.

Indeed, $14a$ could have been factored out at the start, but you were not wrong if you did not spot it right away. *It is never too late to factor out a common factor.*

Example 4. Factor $10xy - 2x$.

Solution

$$10xy - 2x = 2x(5y - 1)$$

Now try Example 5.

Example 5. Factor $3x^2 - 15x$.

Solution

$$3x^2 - 15x = \underline{\qquad\qquad}$$

Check your factors on page 352. ▶

In the polynomial $x^5 + 2x^3 + x^2$ we see that each term contains a power of x. The highest power of x that can be factored out is actually the lowest power contained in any one of the terms. In this case, the lowest power is x^2, so

$$x^5 + 2x^3 + x^2 = x^2(x^3 + 2x + 1)$$

Example 6. Factor $8x^7 - 2x^6 + 10x^5 - 2x^3$.

Solution

$$8x^7 - 2x^6 + 10x^5 - 2x^3$$
$$= 2x^3(4x^4 - x^3 + 5x^2 - 1)$$

Try Example 7.

Example 7. Factor $-x^3 + 3x^2 - 4x$.

Solution

$$-x^3 + 3x^2 - 4x$$
$$= -x(\underline{\qquad\qquad})$$

Remember to check by multiplying. Check your factors on page 352. ▶

Example 8. Factor $3a^2x^2 - 5a^2x$.

Solution

$$3a^2x^2 - 5a^2x = a^2x(3x - 5)$$

Now try Example 9.

Example 9. Factor $x^7y^8 - 2x^5y^6$.

Solution

$$x^7y^8 - 2x^5y^6$$
$$= \underline{\qquad\qquad}$$

Check your factors on page 352. ▶

To factor out the largest common factor of a polynomial, we follow these steps.

1. Look for the largest monomial factor common to all terms.
2. Use the distributive property to rewrite the polynomial as a product of that monomial factor and a new polynomial factor.
3. Check by multiplying your factors.

Trial Run

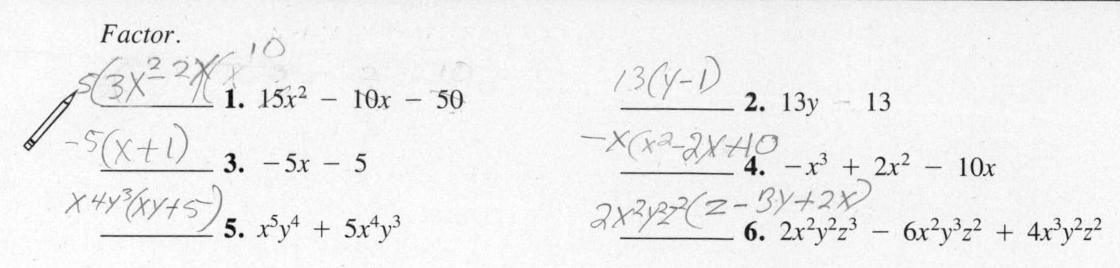

Factor.

_____ 1. $15x^2 - 10x - 50$

_____ 2. $13y - 13$

_____ 3. $-5x - 5$

_____ 4. $-x^3 + 2x^2 - 10x$

_____ 5. $x^5y^4 + 5x^4y^3$

_____ 6. $2x^2y^2z^3 - 6x^2y^3z^2 + 4x^3y^2z^2$

Answers are on page 353.

Looking for Common Binomial Factors (Factoring by Grouping)

In an expression such as $3(x + 1) + y(x + 1)$ notice that there are just *two terms* to consider. The first term is $3(x + 1)$ and the second term is $y(x + 1)$. You should see that both terms contain the common factor $(x + 1)$. The distributive property still allows us to write

$$3(x + 1) + y(x + 1) = (x + 1)(3 + y)$$

Let's try factoring another expression of this type. Factor $a(x + y) + (x + y)$.

$a(x + y) + (x + y)$	Note that there are two terms.
$= a(x + y) + 1(x + y)$	Rewrite the expression.
$= (x + y)(a + 1)$	The binomial factor $(x + y)$ is common to both terms.

Example 10. Factor $(x + 5) + 2a(x + 5)$.

Solution

$$(x + 5) + 2a(x + 5)$$
$$= 1(x + 5) + 2a(x + 5)$$
$$= (x + 5)(1 + 2a)$$

Now finish Example 11.

Example 11. Factor $x(a - b) + (a - b)$.

Solution

$$x(a - b) + (a - b)$$
$$= x(a - b) + 1(a - b)$$
$$= (a - b)(\_\_\_\_)$$

Check your work on page 352. ▶

Now try Example 12 and 13.

Example 12. Factor $y(x^2 + 3) - 2(x^2 + 3)$.

Solution

$$y(x^2 + 3) - 2(x^2 + 3)$$
$$= (\_\_\_\_)(\_\_\_\_)$$

Check your factors on page 352. ▶

Example 13. Factor $3x(x^2 + 1) - (x^2 + 1)$.

Solution

Check your factors on page 352. ▶

Suppose that we wish to factor the four-term polynomial $5x + 5y + bx + by$. We must try to *group* the terms into *two pairs of terms*, each pair containing a common factor. We notice that the first two terms contain the common factor 5 and the second two terms contain the common factor b. We can rewrite our four terms as

$$\begin{aligned} 5x + 5y + bx + by &= 5(x + y) + b(x + y) \\ &= (x + y)(5 + b) \end{aligned}$$

and we have written our four-term polynomial as a product of factors. This technique is called **factoring by grouping**.

Let's try factoring another polynomial of four terms by grouping.

$$6y^2 + 3y - 2ay - a$$

The first two terms contain the common factor $3y$ and the second two terms contain a common factor a or $-a$. Which should we remove? Look at the choices.

$$\begin{aligned} &6y^2 + 3y - 2ay - a \\ &= 3y(2y + 1) + a(-2y - 1) \end{aligned} \qquad \text{or} \qquad \begin{aligned} &6y^2 + 3y - 2ay - a \\ &= 3y(2y + 1) - a(2y + 1) \end{aligned}$$

Both versions are correct, but which one will allow us to continue factoring? The second version leaves us with the common binomial factor $(2y + 1)$ in both terms, so we choose that version.

$$\begin{aligned} 6y^2 + 3y - 2ay - a &= 3y(2y + 1) - a(2y + 1) \\ &= (2y + 1)(3y - a) \end{aligned}$$

Example 14. Factor $4ax - 12ay + x - 3y$.

Solution

$$\begin{aligned} &4ax - 12ay + x - 3y \\ &= 4a(x - 3y) + 1(x - 3y) \\ &= (x - 3y)(4a + 1) \end{aligned}$$

Now complete Example 15.

Example 15. Factor $8rx - 4r - 2x + 1$.

Solution

$8rx - 4r - 2x + 1$

$= 4r(\_\_\_\_\_) - 1(\_\_\_\_\_)$

$= (\_\_\_\_\_)(\_\_\_\_\_)$

Check your work on page 352. ▶

Example 16. Factor $x^3 + 5x^2 + 2x + 10$.

Solution

$$\begin{aligned} &x^3 + 5x^2 + 2x + 10 \\ &= x^2(x + 5) + 2(x + 5) \\ &= (x + 5)(x^2 + 2) \end{aligned}$$

You complete Example 17.

Example 17. Factor $-2x^2 - 2y^2 + 9x^2y + 9y^3$.

Solution

$-2x^2 - 2y^2 + 9x^2y + 9y^3$

$= -2(\_\_\_\_\_) + 9y(\_\_\_\_\_)$

$= (\_\_\_\_\_)(\_\_\_\_\_)$

Check your work on page 352. ▶

Trial Run

Factor.

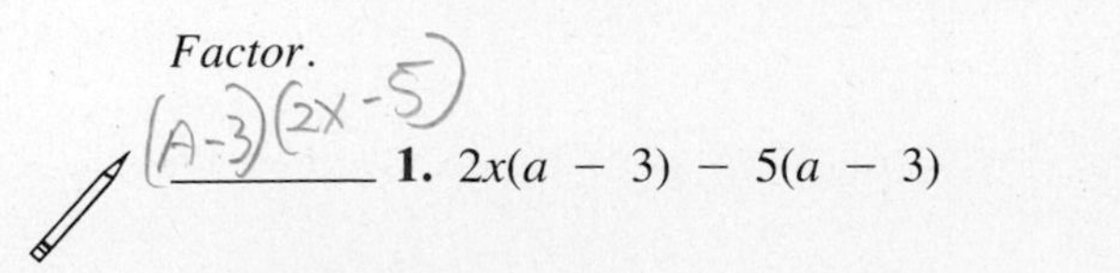

_____ 1. $2x(a - 3) - 5(a - 3)$

_____ 2. $y(a - b) + (a - b)$

________ **3.** $4x + 4y + bx + by$

________ **4.** $x^3 - 7x^2 + 3x - 21$

________ **5.** $3ax - 15ay + 2x - 10y$

________ **6.** $8a^2 + 4a - 2ab - b$

Answers are on page 353.

▶ Examples You Completed

Example 2. Factor $14x^2 - 70x - 7$.

Solution

$$14x^2 - 70x - 7 = 7(2x^2 - 10x - 1)$$

Example 3. Factor $-19x + 19$.

Solution

$$-19x + 19 = -19(x - 1)$$

Example 5. Factor $3x^2 - 15x$.

Solution

$$3x^2 - 15x = 3x(x - 5).$$

Example 7. Factor $-x^3 + 3x^2 - 4x$.

Solution

$$-x^3 + 3x^2 - 4x$$
$$= -x(x^2 - 3x + 4)$$

Example 9. Factor $x^7y^8 - 2x^5y^6$.

Solution

$$x^7y^8 - 2x^5y^6 = x^5y^6(x^2y^2 - 2)$$

Example 11. Factor $x(a - b) + (a - b)$.

Solution

$$x(a - b) + (a - b)$$
$$= x(a - b) + 1\,(a - b)$$
$$= (a - b)(x + 1)$$

Example 12. Factor $y(x^2 + 3) - 2(x^2 + 3)$.

Solution

$$y(x^2 + 3) - 2(x^2 + 3)$$
$$= (x^2 + 3)(y - 2)$$

Example 13. Factor $3x(x^2 + 1) - (x^2 + 1)$.

Solution

$$3x(x^2 + 1) - (x^2 + 1)$$
$$= 3x(x^2 + 1) - 1(x^2 + 1)$$
$$= (x^2 + 1)(3x - 1)$$

Example 15. Factor $8rx - 4r - 2x + 1$.

Solution

$$8rx - 4r - 2x + 1$$
$$= 4r(2x - 1) - 1(2x - 1)$$
$$= (2x - 1)(4r - 1)$$

Example 17. Factor $-2x^2 - 2y^2 + 9x^2y + 9y^3$.

Solution

$$-2x^2 - 2y^2 + 9x^2y + 9y^3$$
$$= -2(x^2 + y^2) + 9y(x^2 + y^2)$$
$$= (x^2 + y^2)(-2 + 9y)$$

Answers to Trial Runs

page 350 **1.** $5(3x^2 - 2x - 10)$ **2.** $13(y - 1)$ **3.** $-5(x + 1)$ **4.** $-x(x^2 - 2x + 10)$
5. $x^4y^3(xy + 5)$ **6.** $2x^2y^2z^2(z - 3y + 2x)$

page 351 **1.** $(a - 3)(2x - 5)$ **2.** $(a - b)(y + 1)$ **3.** $(x + y)(4 + b)$ **4.** $(x - 7)(x^2 + 3)$
5. $(x - 5y)(3a + 2)$ **6.** $(2a + 1)(4a - b)$

Name ____________________ Date __________

EXERCISE SET 7.1

Factor.

____ **1.** $7x + 14$

____ **2.** $5x - 15$

____ **3.** $21x - 28$

____ **4.** $9x + 21$

____ **5.** $42 - 6y$

____ **6.** $45 - 9y$

____ **7.** $6a^2 - 36$

____ **8.** $8a^2 - 40$

____ **9.** $-12x + 18y$

____ **10.** $-15x + 20y$

____ **11.** $30x^2 + 15y^2$

____ **12.** $60x^2 - 12y^2$

____ **13.** $7x^2 - 21x - 7$

____ **14.** $9x^2 + 27x + 9$

____ **15.** $10y^2 + 15y - 35$

____ **16.** $12y^2 - 18y - 33$

____ **17.** $28 - 7a + 35a^2$

____ **18.** $33 - 22a + 55a^2$

____ **19.** $50x^2 - 10xy + 20y^2$

____ **20.** $36x^2 - 24xy + 48y^2$

____ **21.** $2x^3 - 6x^2 + 4x + 8$

____ **22.** $8x^3 - 16x^2 + 24x + 8$

____ **23.** $x^2 + x$

____ **24.** $2x^2 + x$

____ **25.** $-y^3 + y^2 - 2y$

____ **26.** $-y^3 - y^2 + 4y$

____ **27.** $2b^3 - 6b^2 - 4b$

____ **28.** $5b^3 - 10b^2 + 25b$

____ **29.** $7b^3 + 2b^5 - 3b^7$

____ **30.** $9b^4 + 4b^6 - 2b^8$

____ **31.** $3x^3 - 6x^2y + 12xy^2$

____ **32.** $5x^3 + 15x^2y + 55xy^2$

____ **33.** $-a^4b + 6a^3b^2 - 5a^2b^3$

____ **34.** $-a^5b - 4a^4b^2 - 5a^3b^3$

____ **35.** $4x^3y^3 - 20x^2y^2 + 2xy$

____ **36.** $9x^4y^4 + 18x^3y^3 - 3x^2y^2$

____ **37.** $x^3y^2z - 2x^4y^3z - x^5y^4z$

____ **38.** $x^6y^4z^2 + 5x^4y^2z^2 - x^2z^2$

____ **39.** $63x^2 + 72x$

____ **40.** $36x^3 + 45x^2$

____ **41.** $100a^3 - 60b^3$

____ **42.** $75a^3 - 100b^3$

____ **43.** $16b^3 + 64b^4$

____ **44.** $12b^5 + 60b^7$

____ **45.** $6a^3 - 42a^2 - 78a$

____ **46.** $9a^3 - 72a^2 + 81a$

____ **47.** $5x^3y + 80xy^3$

____ **48.** $7x^4y^2 - 56x^3y^3$

____ **49.** $-7b^2 - 42b^3 + 35b^4$

____ **50.** $-12b^4 + 36b^5 - 48b^6$

_______ **51.** $9x^3y - 12x^2y^2 + 20xy^3$

_______ **52.** $27x^3y + 45x^2y^2 - 63xy^3$

_______ **53.** $72a^4 - 48a^3 - 8a^2$

_______ **54.** $72a^5 - 54a^4 - 6a^3$

_______ **55.** $12x^3y - 8x^2y^2 + 4xy^3$

_______ **56.** $21x^3y - 15x^2y^2 - 3xy^3$

_______ **57.** $9x^4yz + 18x^3y^2z - 27x^2y^3z - 45xy^4z$

_______ **58.** $35xy^4a - 28xy^3a - 14xy^2a + 7xya$

_______ **59.** $5a + ac + 5b + bc$

_______ **60.** $7x + 7y + xz + yz$

_______ **61.** $x^3 - 5x^2 + 6x - 30$

_______ **62.** $x^3 - 9x^2 + 9x - 81$

_______ **63.** $70a^2 - 40ab + 21a - 12b$

_______ **64.** $30a^2 - 75ab + 4a - 10b$

_______ **65.** $3xy - 6x - y + 2$

_______ **66.** $2ab - 10a - b + 5$

_______ **67.** $2xy^2 + 8x + 3y^2 + 12$

_______ **68.** $3ab^2 - 2a + 12b^3 - 8b$

_______ **69.** $3x^3 - 6x^2 + x - 2$

_______ **70.** $2y^3 - 6y^2 + y - 3$

☆ Stretching the Topics

Factor.

_______ **1.** $2(m - n)^2 - (m - n)$

_______ **2.** $3x^{2m+5} + 9x^{2m}$

_______ **3.** $\frac{1}{6}Bh + \frac{1}{6}bh + \frac{2}{3}Mh$

Check your answers in the back of your book.

If you can complete **Checkup 7.1**, you are ready to go on to Section 7.2.

Name ______________________ Date ____________

CHECKUP 7.1

Factor.

_____ 1. $3x + 15$

_____ 2. $35 - 14y$

_____ 3. $-x^2 + x$

_____ 4. $15x^2 - 10xy + 25y^2$

_____ 5. $8a^3 - 16a^2 - 40a$

_____ 6. $9a^3b^3 - 27a^2b^2 + 3ab$

_____ 7. $-8m^3 - 24m$

_____ 8. $x^4yz - x^3y^2z + x^2y^3z$

_____ 9. $5m + 5n + pm + pn$

_____ 10. $3x^2 + 6x - x - 2$

Check your answers in the back of your book.

If You Missed Problems:	You Should Review Examples:
1–3	1–5
4–8	6–9
9–10	14–17

7.2 Factoring the Difference of Two Squares

In Chapter 5 we noted that whenever we multiplied the sum of two terms times the difference of those same two terms, the product was the **difference of the squares** of the two terms. For example, remember that

$$(x + 7)(x - 7) = x^2 - 49$$

In this chapter we are concerned with *reversing* the operation of multiplication. Factoring is a process by which we rewrite a polynomial as a product of factors.

Let's try to factor $x^2 - 16$. Is this a difference of two squares? Is x^2 a square? Of course. Is 16 a square? Yes, 16 is 4^2. You should agree that

$$x^2 - 16 = (x + 4)(x - 4)$$

As usual, we can check our factors by multiplication.

We see that the difference of two squares can be factored in the following way.

Difference of Two Squares

$$A^2 - B^2 = (A + B)(A - B)$$

Let's factor $25y^2 - 121$.

Is this a difference of two terms?	Yes.
Is the first term a square?	Yes, $25y^2$ is the square of $5y$.
What is A?	$A = 5y$.
Is the second term a square?	Yes, 121 is the square of 11.
What is B?	$B = 11$.

$$25y^2 - 121 = (5y + 11)(5y - 11)$$

The difference of two squares is not difficult to spot. You must look for two terms with a minus sign in between and decide whether each term is the square of some expression. If so, one factor will be the *sum* of those expressions and the other factor will be the *difference* of those expressions. Remember to remove any common monomial factors first.

Example 1. Factor $36 - x^2$.

Solution

$$36 - x^2$$
$$= (6 + x)(6 - x)$$

You try Example 2.

Example 2. Factor $\frac{1}{9}a^2 - y^2$.

Solution

$$\frac{1}{9}a^2 - y^2$$
$$= (\_\_\_\_\_\_)(\_\_\_\_\_\_)$$

Check your factors on page 361. ▶

Example 3. Factor $5x^2 - 20$.

Solution

$$5x^2 - 20$$
$$= 5(x^2 - 4)$$
$$= 5(x + 2)(x - 2)$$

Try Example 4 and watch for a *common factor*.

Example 4. Factor $3x^3 - 3x$.

Solution

$$3x^3 - 3x$$
$$= 3x(\underline{\qquad\qquad})$$
$$= 3x(\underline{\qquad\qquad})(\underline{\qquad\qquad})$$

Check your factors on page 361. ▶

Note that we removed the common monomial factors in Examples 3 and 4. Do not think that you should skip the step in which you factor out the common monomial factor. Even mathematicians who have been factoring for years would not dream of jumping from the original polynomial to its completely factored form.

Now try Example 5.

Example 5. Factor $99x^2y^2 - 44x^2z^2$.

Solution

Check your factors on page 362. ▶

Example 6. Factor $x^2 + 4y^2$.

Solution: $x^2 + 4y^2$ is a *sum* of two squares. It is *not* a *difference* of two squares. It cannot be factored.

We note that a *sum of two squares* cannot be factored in the context of this course. When a polynomial cannot be factored we say that the polynomial is **prime** (or **irreducible**).

Example 7. Factor $x^4 - 1$.

Solution

$$x^4 - 1$$
$$= (x^2 + 1)(x^2 - 1)$$
$$= (x^2 + 1)(x + 1)(x - 1)$$

Now try Example 8.

Example 8. Factor $16y^4 - 81z^4$.

Solution

$$16y^4 - 81z^4$$
$$= (4y^2 + 9z^2)(\underline{\qquad\qquad})$$
$$= (4y^2 + 9z^2)(\underline{\qquad\qquad})(\underline{\qquad\qquad})$$

Check your factors on page 362. ▶

Notice in Examples 7 and 8 that after factoring the first time, one of our factors was the difference of two squares. Therefore, we continued factoring because an expression should always be factored completely.

Since it is important that we be able to recognize numbers that are squares, let's complete the following table of squares of the first 20 natural numbers.

Table of Squares

n	n^2	n	n^2
1	1	11	121
2	____	12	____
3	____	13	____
4	____	14	196
5	____	15	____
6	____	16	256
7	____	17	289
8	64	18	____
9	81	19	361
10	____	20	____

Check your table on page 362.

Trial Run

Factor.

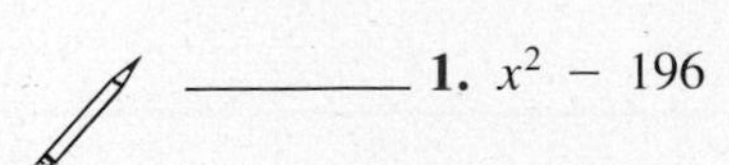

______ **1.** $x^2 - 196$

______ **2.** $9x^2 - 49$

______ **3.** $32 - 2x^2$

______ **4.** $x^2 + y^2$

______ **5.** $12a^2 - 75b^2$

______ **6.** $45x^3 - 5x$

Answers are on page 362.

Examples You Completed

Example 2. Factor $\frac{1}{9}a^2 - y^2$.

Solution

$$\frac{1}{9}a^2 - y^2$$
$$= \left(\frac{1}{3}a + y\right)\left(\frac{1}{3}a - y\right)$$

Example 4. Factor $3x^3 - 3x$.

Solution

$$3x^3 - 3x$$
$$= 3x(x^2 - 1)$$
$$= 3x(x + 1)(x - 1)$$

Example 5. Factor $99x^2y^2 - 44x^2z^2$.

Solution

$$99x^2y^2 - 44x^2z^2$$
$$= 11x^2(9y^2 - 4z^2)$$
$$= 11x^2(3y + 2z)(3y - 2z)$$

Example 8. Factor $16y^4 - 81z^4$.

Solution

$$16y^4 - 81z^4$$
$$= (4y^2 + 9z^2)(4y^2 - 9z^2)$$
$$= (4y^2 + 9z^2)(2y + 3z)(2y - 3z)$$

Table of Squares

n	n^2	n	n^2
1	1	11	121
2	4	12	144
3	9	13	169
4	16	14	196
5	25	15	225
6	36	16	256
7	49	17	289
8	64	18	324
9	81	19	361
10	100	20	400

Answers to Trial Run

page 361 **1.** $(x - 14)(x + 14)$ **2.** $(3x + 7)(3x - 7)$ **3.** $2(4 + x)(4 - x)$ **4.** $x^2 + y^2$
5. $3(2a - 5b)(2a + 5b)$ **6.** $5x(3x + 1)(3x - 1)$

Name ______________________ Date __________

EXERCISE SET 7.2

Factor.

_______ **1.** $x^2 - 25$

_______ **2.** $x^2 - 16$

_______ **3.** $9y^2 - 1$

_______ **4.** $144y^2 - 1$

_______ **5.** $49 - a^2$

_______ **6.** $36 - a^2$

_______ **7.** $4a^2 - 121$

_______ **8.** $9a^2 - 100$

_______ **9.** $25b^2 - \frac{4}{9}$

_______ **10.** $49b^2 - \frac{25}{64}$

_______ **11.** $x^2 - y^2$

_______ **12.** $a^2 - b^2$

_______ **13.** $a^2 - 0.36b^2$

_______ **14.** $a^2 - 0.81b^2$

_______ **15.** $4x^2 - 225y^2$

_______ **16.** $9x^2 - 25y^2$

_______ **17.** $m^2n^2 - 144$

_______ **18.** $x^2y^2 - 169$

_______ **19.** $9a^2b^2 - c^2$

_______ **20.** $16a^2b^2 - c^2$

_______ **21.** $0.64x^2y^2 - 0.25z^2$

_______ **22.** $0.16x^2y^2 - 0.81z^2$

_______ **23.** $x^4 - 81$

_______ **24.** $x^4 - 16$

_______ **25.** $25x^4 - 121$

_______ **26.** $9x^4 - 64$

_______ **27.** $225 - a^4b^4$

_______ **28.** $289 - a^4b^4$

_______ **29.** $3x^2 - 75$

_______ **30.** $2x^2 - 32$

_______ **31.** $11x^2 - 44y^2$

_______ **32.** $5x^2 - 45y^2$

_______ **33.** $16y - 9x^2y$

_______ **34.** $25y - 4x^2y$

_______ **35.** $6m^3 - 150m$

_______ **36.** $7m^3 - 112m$

_______ **37.** $49a^2b - \frac{1}{4}b^3$

_______ **38.** $64a^2b - \frac{1}{9}b^3$

_______ **39.** $2x^2y^2 - 72y^4$

_______ **40.** $3x^2y^2 - 48y^4$

_______ **41.** $3m^4 - 243$

_______ **42.** $5m^4 - 80$

_______ **43.** $2x^3y - 8xy^3$

_______ **44.** $3x^3y - 27xy^3$

_______ **45.** $27x^3y^3 - 12xy$

_______ **46.** $50x^3y^3 - 2xy$

_______ 47. $a^6 - 25a^2$

_______ 48. $a^6 - 81a^2$

_______ 49. $289x^2y^2z^2 - z^2$

_______ 50. $256x^2y^2z^2 - z^2$

☆ Stretching the Topics

Factor.

_______ 1. $(a + b)^2 - 9c^2$

_______ 2. $(2x - y)^2 - (y + 3z)^2$

_______ 3. $x^{2a} - 25y^{2b}$

Check your answers in the back of your book.

If you can do **Checkup 7.2,** you are ready to go to Section 7.3.

Name ______________________ Date ____________

✓ CHECKUP 7.2

Factor.

______ 1. $x^2 - 64$

______ 2. $25 - y^2$

______ 3. $9a^2 - 100$

______ 4. $x^2 - \frac{25}{49}y^2$

______ 5. $x^4 - 16$

______ 6. $3x^2 - 0.75y^2$

______ 7. $7x^2y^2 - 28y^4$

______ 8. $5m^4 - 405n^4$

______ 9. $4x^3y - 400xy^3$

______ 10. $3a - 243a^3$

Check your answers in the back of the book.

If You Missed Problems:	You Should Review Examples:
1–4	1–2
5	7
6–7	3, 4
8	3, 8
9	5
10	4, 1

7.3 Factoring Trinomials

Recall from Chapter 5 the FOIL method that we used to multiply trinomials.

$$\overset{F\qquad L}{(x + 5)(x - 2)} = \overset{F}{x^2} \overset{O}{- 2x} \overset{I}{+ 5x} \overset{L}{- 10} = x^2 + 3x - 10$$

(I = inner terms, O = outer terms)

It seems that whenever we multiply two binomials, our product often turns out to be a trinomial. In this section we must learn how to decide what binomials were multiplied together to give trinomials; we must learn how to *factor* trinomials.

Factoring Trinomials (Leading Coefficient of 1)

Notice in our FOIL example that the first term in our trinomial (the F term) came from multiplying the first terms of the binomials. The last term in the trinomial (the L term) came from multiplying the second terms of the binomials. The middle term in the trinomial was the sum of the O and I terms.

Keeping that in mind, let's try to factor. $x^2 + 4x + 3$

We are looking for two binomials. $(\_\_\_\_\_\_)(\_\_\_\_\_\_)$

Since the first term of our trinomial is x^2, we can write: $(x \quad)(x \quad)$

Since the last term of our trinomial is 3, we know the second terms of our binomials must be 1 and 3 or -1 and -3.

$(x + 1)(x + 3)$

or

$(x - 1)(x - 3)$

Do both of these give the correct product?

$(x + 1)(x + 3)$	$(x - 1)(x - 3)$
$= x^2 + 3x + x + 3$	$= x^2 - 3x - x + 3$
$= x^2 + 4x + 3$	$= x^2 - 4x + 3$
Yes	No

Thus in factored form we write

$$x^2 + 4x + 3 = (x + 1)(x + 3)$$

Checking your factors by multiplying is a *must*. Suppose that we consider another problem where a choice must be made from more possibilities.

Let's factor the trinomial. $x^2 - 7x + 12$

We are looking for two binomials. $(x \quad)(x \quad)$

Since the last term of our trinomial is 12, there are several possibilities for the second terms of the binomials.

$(x + 1)(x + 12)$

$(x + 6)(x + 2)$

$(x + 3)(x + 4)$

$(x - 1)(x - 12)$

$(x - 6)(x - 2)$

$(x - 3)(x - 4)$

Since the middle term of the trinomial is negative ($-7x$), we may reject the first three possibilities because they will yield a *positive* middle term. We check the last three possibilities.

$$(x - 1)(x - 12) = x^2 - 12x - x + 12$$
$$= x^2 - 13x + 12$$
$$(x - 6)(x - 2) = x^2 - 2x - 6x + 12$$
$$= x^2 - 8x + 12$$
$$(x - 3)(x - 4) = x^2 - 4x - 3x + 12$$
$$= x^2 - 7x + 12$$

We find that only the last pair of factors gives the correct middle term and we conclude that

$$x^2 - 7x + 12 = (x - 3)(x - 4)$$

After some practice you will learn to leave out the possibilities that do not seem likely. You need not list *all* the possibilities, but be sure to check your factors by multiplying before deciding that they are correct.

Example 1. Factor $x^2 - 9xy + 20y^2$.

Solution

$$x^2 - 9xy + 20y^2$$
$$= (x - 5y)(x - 4y)$$

You try Example 2.

Example 2. Factor $x^2 + 12xy + 20y^2$.

Solution

$$x^2 + 12xy + 20y^2$$
$$= (\_\_\_\_\_\_)(\_\_\_\_\_\_)$$

Remember to check your factors by multiplying. Check your factors on page 376. ▶

Example 3. Factor $8 + 6x + x^2$.

Solution

$$8 + 6x + x^2$$
$$= (4 + x)(2 + x)$$

You factor Example 4.

Example 4. Factor $15 - 8y + y^2$.

Solution

$$15 - 8y + y^2$$
$$= (\_\_\_\_\_\_)(\_\_\_\_\_\_)$$

Check your factors on page 376. ▶

Trial Run

Factor.

______ **1.** $x^2 + 8x + 12$

______ **2.** $12 + 13a + a^2$

______ **3.** $y^2 - 9y + 18$

______ **4.** $x^2 - 6xy + 9y^2$

Answers are on page 377.

What happens when the last term of our trinomial is *negative*?

Let's try to factor. $\quad x^2 - 5x - 6$

We are looking for two binomials. $\quad (x \quad)(x \quad)$

Since the last term of the trinomial is -6, there are several possibilities for the second terms of the binomials.

$$(x + 1)(x - 6)$$
$$(x - 1)(x + 6)$$
$$(x + 2)(x - 3)$$
$$(x - 2)(x + 3)$$

Which of these possibilities gives the correct product?

$(x + 1)(x - 6)$	$(x - 1)(x + 6)$	$(x + 2)(x - 3)$	$(x - 2)(x + 3)$
$= x^2 - 6x + x - 6$	$= x^2 + 6x - x - 6$	$= x^2 - 3x + 2x - 6$	$= x^2 + 3x - 2x - 6$
$= x^2 - 5x - 6$	$= x^2 + 5x - 6$	$= x^2 - x - 6$	$= x^2 + x - 6$

Our trinomial can be factored as

$$x^2 - 5x - 6 = (x + 1)(x - 6)$$

Example 5. Factor $x^2 - x - 12$.

Solution

$$x^2 - x - 12$$
$$= (x - 4)(x + 3)$$

Try Example 6.

Example 6. Factor $x^2 + 4x - 12$.

Solution

$$x^2 + 4x - 12$$
$$= (\_\_\_\_\_\_)(\_\_\_\_\_\_)$$

Check your factors on page 376. ▶

Example 7. Factor $3x^2 - 33xy - 36y^2$.

Solution

$$3x^2 - 33xy - 36y^2$$
$$= 3(x^2 - 11xy - 12y^2)$$
$$= 3(x - 12y)(x + y)$$

Notice that we remembered to look first for a *common monomial factor*.

You try Example 8.

Example 8. Factor $x^3 - 13x^2 - 30x$.

Solution

$$x^3 - 13x^2 - 30x$$
$$= x(\_\_\_\_\_\_)$$
$$= x(\_\_\_\_\_\_)(\_\_\_\_\_\_)$$

Check your factors on page 376. ▶

The order in which we write the factors is not important, but the factors themselves are **unique**, which means that if a trinomial can be factored, there is *one and only one* correct pair of factors. After some practice you will become skillful at finding the correct pair.

Trial Run

Factor.

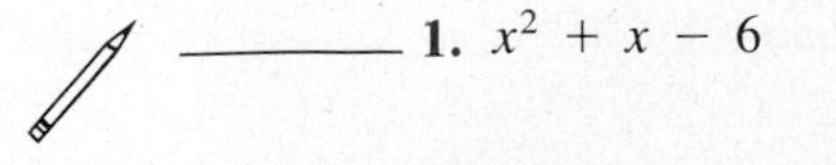

______ 1. $x^2 + x - 6$

______ 2. $10 - 3a - a^2$

______ 3. $2m^2 - 10m - 48$

______ 4. $a^2 - 14ab - 15b^2$

Answers are on page 377.

Factoring Trinomials (Leading Coefficient Not 1)

We have learned to factor trinomials in which the coefficient of the squared term is 1. There are many trinomials in which the coefficient of the squared term is *not* 1. The techniques used to factor such trinomials are exactly the same as the techniques already learned. We continue to look for a common monomial factor first and then for two binomials that multiply together to give the trinomial.

Let's try to factor. $3x^2 + 7x + 2$

Since there is no common monomial factor, we know the first terms in the binomials must multiply to give $3x^2$. $(3x \quad)(x \quad)$

Since the last term of the trinomial is 2, there are two possibilities for the second terms of the binomials.

$$(3x + 2)(x + 1)$$
$$(3x + 1)(x + 2)$$

We check both possibilities by multiplying.

$$\begin{aligned}(3x + 2)(x + 1) &= 3x^2 + 3x + 2x + 2\\ &= 3x^2 + 5x + 2\\ (3x + 1)(x + 2) &= 3x^2 + 6x + x + 2\\ &= 3x^2 + 7x + 2\end{aligned}$$

Therefore, our trinomial factors as

$$3x^2 + 7x + 2 = (3x + 1)(x + 2)$$

If the first term and/or the third term of the trinomial can be obtained in several ways, there may be many possible pairs of binomial factors to try.

Let's look at: $2x^2 - x - 10$

Noting that the terms contain *no common monomial factor*, we list the possible pairs of binomials factors.

$$(2x + 2)(x - 5)$$
$$(2x - 2)(x + 5)$$
$$(2x + 10)(x - 1)$$
$$(2x - 10)(x + 1)$$
$$(2x + 1)(x - 10)$$
$$(2x - 1)(x + 10)$$
$$(2x + 5)(x - 2)$$
$$(2x - 5)(x + 2)$$

We can reject the first four pairs of factors because in each pair, the first factor contains a common monomial factor of 2. If these were the correct binomials, a common factor of 2 could have been removed from the *original* trinomial. Now we can check the remaining four pairs of factors.

$$\begin{aligned}&(2x + 1)(x - 10)\\ &= 2x^2 - 10x + x - 10\\ &= 2x^2 + 9x - 10\end{aligned}$$

$$\begin{aligned}&(2x - 1)(x + 10)\\ &= 2x^2 + 10x - x - 10\\ &= 2x^2 + 9x - 10\end{aligned}$$

$$\begin{aligned}&(2x + 5)(x - 2)\\ &= 2x^2 - 4x + 5x - 10\\ &= 2x^2 + x - 10\end{aligned}$$

$$\begin{aligned}&(2x - 5)(x + 2)\\ &= 2x^2 + 4x - 5x - 10\\ &= 2x^2 - x - 10\end{aligned}$$

Since the last pair gives the product we were seeking,

$$2x^2 - x - 10 = (2x - 5)(x + 2)$$

Writing all the possible pairs of factors can sometimes be a major task. It is almost easier to jump in and try some, hoping to hit the correct pair before too long. You should *not* erase pairs of factors after rejecting them or you are likely to lose track of which pairs you have tried.

Example 9. Factor $8x^2 + 2x - 15$.

Solution

$$8x^2 + 2x - 15$$
$$= (2x + 3)(4x - 5)$$

You try Example 10.

Example 10. Factor $7x^2 - x - 6$.

Solution

$$7x^2 - x - 6$$
$$= (\_\_\_\_\_\_)(\_\_\_\_\_\_)$$

Check your factors on page 376. ▶

Example 11. Factor $-21y - 5y^2 + 6y^3$

Solution

$$-21y - 5y^2 + 6y^3$$
$$= -y(21 + 5y - 6y^2)$$
$$= -y(3 + 2y)(7 - 3y)$$

Now try Example 12.

Example 12. Factor $3x^3y + 6x^2y^2 + 3xy^3$.

Solution

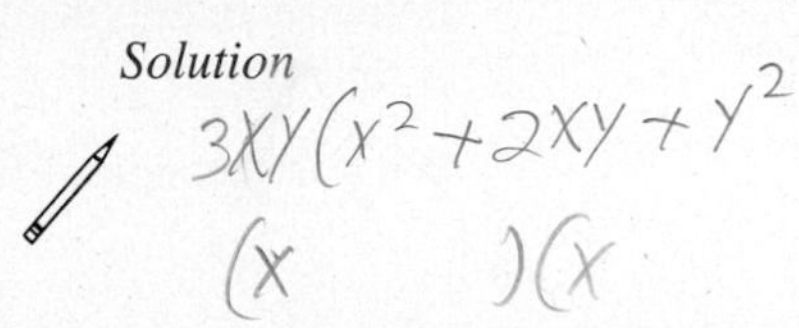

Check your factors on page 376. ▶

Trial Run

Factor.

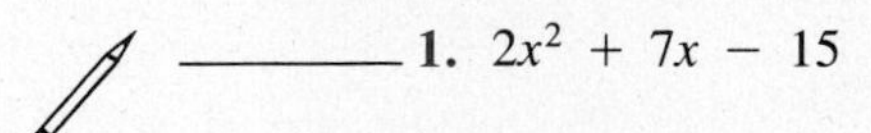

______ **1.** $2x^2 + 7x - 15$

______ **2.** $2 - 11m + 12m^2$

______ **3.** $2x^2 - 9xy + 10y^2$

______ **4.** $4ax^3 - 12ax^2 + 9ax$

Answers are on page 377.

Factoring Perfect Square Trinomials

In Chapter 5 we learned that the square of a binomial yielded a perfect square trinomial found by the following formulas.

$$(A + B)^2 = A^2 + 2AB + B^2$$
$$(A - B)^2 = A^2 - 2AB + B^2$$

If we recognize that a trinomial that we wish to factor is indeed a perfect square trinomial, we may reverse these formulas to find the factors. In a perfect square trinomial, the first term must be the square of some quantity A, the last term must be the square of some quantity B, and the middle term must be twice the product of A times B. Then we can use one of our formulas.

Perfect Square Trinomials

$$A^2 + 2AB + B^2 = (A + B)^2$$
$$A^2 - 2AB + B^2 = (A - B)^2$$

Notice that the sign of the middle term tells us which formula to use.

Consider the trinomial $x^2 + 6x + 9$.

Is the first term a square?	Yes, x^2 is the square of x.
What is A?	$A = x$.
Is the third term a square?	Yes, 9 is the square of 3.
What is B?	$B = 3$.
Is the second term twice the product of A times B?	Yes, $6x = 2 \cdot 3 \cdot x$.
What is the sign of the middle term?	The middle term is positive.

$$x^2 + 6x + 9 = (x + 3)^2$$

Example 13. Factor $x^2 - 10xy + 25y^2$.

Solution: Is $x^2 - 10xy + 25y^2$ a perfect square trinomial? Yes, because

$$A = x$$

$$B = 5y$$

$$2AB = 2 \cdot x \cdot 5y = 10xy$$

The middle term is negative.

$$x^2 - 10xy + 25y^2 = (x - 5y)^2$$

You try Example 14.

Example 14. Factor $4a^2 + 28a + 49$.

Solution: Is $4a^2 + 28a + 49$ a perfect square trinomial? Yes, because

$A =$ _____

$B =$ _____

$2AB = 2 \cdot$ _____ $\cdot$ _____ $=$ _____

The middle term is _______ .

$4a^2 + 28a + 49 = ($_____ $+$ _____$)^2$

Check your work on page 376. ▶

Example 15. Factor $y^2 + \frac{2}{3}y + \frac{1}{9}$.

Solution: Is $y^2 + \frac{2}{3}y + \frac{1}{9}$ a perfect square trinomial? Yes, because

$$A = y$$

$$B = \frac{1}{3}$$

$$2AB = 2 \cdot y \cdot \frac{1}{3} = \frac{2}{3}y$$

The middle term is positive.

$$y^2 + \frac{2}{3}y + \frac{1}{9} = \left(y + \frac{1}{3}\right)^2$$

You try Example 16.

Example 16. Factor $x^2 - 0.8x + 0.16$.

Solution: Is $x^2 - 0.8x + 0.16$ a perfect square trinomial? Yes, because

$A =$ _____

$B = 0.4$

$2AB = 2 \cdot$ _____ $(0.4) =$ _____

The middle term is _______ .

$x^2 - 0.8x + 0.16 = (x -$ _____$)^2$

Check your work on page 376. ▶

Although perfect square trinomials can always be factored by the trial-and-error process discussed earlier, you should try to recognize them and use the formulas whenever possible.

_______ **3.** $8x^2 - 26x + 15$

_______ **4.** $6x^2 + 5x - 6$

Answers are on page 377.

Factoring Trinomials of the Form $ax^4 + bx^2 + c$ (Optional)

Let's try to factor $x^4 - 5x^2 - 6$. From our work with multiplication, we can perhaps see that this trinomial is the product of two binomials.

We consider: $x^4 - 5x^2 - 6$

We look for two binomials like: $(x^2 \quad)(x^2 \quad)$

Since the constant term is -6, we know we need one positive number and one negative number as our second terms in the binomials. $(x^2 - \_\_\_\_)(x^2 + \_\_\_\_)$

By our usual methods, we decide on the correct numbers. $(x^2 - 6)(x^2 + 1)$

Therefore, we see that

$$x^4 - 5x^2 - 6 = (x^2 - 6)(x^2 + 1)$$

Example 19. Factor $x^4 - 8x^2 - 9$.

Solution

$$x^4 - 8x^2 - 9$$
$$= (x^2 + 1)(x^2 - 9)$$
$$= (x^2 + 1)(x + 3)(x - 3)$$

Notice that we continue factoring when possible.

You try Example 20.

Example 20. Factor $x^4 - 5x^2 + 4$.

Solution

$$x^4 - 5x^2 + 4$$
$$= (x^2 - 4)(x^2 - \_\_\_\_)$$
$$=$$

Check your factors on page 377. ▶

Example 21. Factor $5 - 10x^2 + 5x^4$.

Solution

$$5 - 10x^2 + 5x^4$$
$$= 5(1 - 2x^2 + x^4)$$
$$= 5(1 - x^2)(1 - x^2)$$
$$= 5(1 + x)(1 - x)(1 + x)(1 - x)$$
$$= 5(1 + x)^2(1 - x)^2$$

Now try Example 22.

Example 22. Factor $3x^4 + 3x^2 - 60$.

Solution

Check your factors on page 377. ▶

Trial Run

Factor.

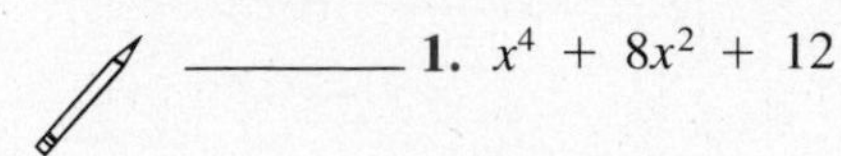

______ 1. $x^4 + 8x^2 + 12$

______ 2. $2x^4 + 8x^2 + 6$

______ 3. $x^4 + 2x^2 - 3$

______ 4. $x^4 - 13x^2 + 36$

Answers are on page 377.

Examples You Completed

Example 2. Factor $x^2 + 12xy + 20y^2$.

Solution

$$x^2 + 12xy + 20y^2$$
$$= (x + 10y)(x + 2y)$$

Example 4. Factor $15 - 8y + y^2$.

Solution

$$15 - 8y + y^2$$
$$= (5 - y)(3 - y)$$

Example 6. Factor $x^2 + 4x - 12$.

Solution

$$x^2 + 4x - 12$$
$$= (x + 6)(x - 2)$$

Example 8. Factor $x^3 - 13x^2 - 30x$.

Solution

$$x^3 - 13x^2 - 30x$$
$$= x(x^2 - 13x - 30)$$
$$= x(x - 15)(x + 2)$$

Example 10. Factor $7x^2 - x - 6$.

Solution

$$7x^2 - x - 6$$
$$= (7x + 6)(x - 1)$$

Example 12. Factor $3x^3y + 6x^2y^2 + 3xy^3$.

Solution

$$3x^3y + 6x^2y^2 + 3xy^3$$
$$= 3xy(x^2 + 2xy + y^2)$$
$$= 3xy(x + y)(x + y)$$

Example 14. Factor $4a^2 + 28a + 49$.

Solution: Is $4a^2 + 28a + 49$ a perfect square trinomial? Yes, because

$$A = 2a$$
$$B = 7$$
$$2AB = 2 \cdot 2a \cdot 7 = 28a$$

The middle term is positive.

$$4a^2 + 28a + 49 = (2a + 7)^2$$

Example 16. Factor $x^2 - 0.8x + 0.16$.

Solution: Is $x^2 - 0.8x + 0.16$ a perfect square trinomial? Yes, because

$$A = x$$
$$B = 0.4$$
$$2AB = 2 \cdot x\,(0.4) = 0.8x$$

The middle term is negative.

$$x^2 - 0.8x + 0.16 = (x - 0.4)^2$$

Example 17. Use the *ac*-method to factor $2x^2 - x - 15$.

Solution

$$a \cdot c = (2)(-15) = -30$$

Pairs of factors:

$(1)(-30)$	$(-1)(30)$
$(2)(-15)$	$(-2)(15)$
$(3)(-10)$	$(-3)(10)$
$(5)(-6)$	$(-5)(6)$

Correct pair: 5 and -6.

$$\begin{aligned} &2x^2 - x - 15 \\ &= 2x^2 + 5x - 6x - 15 \\ &= x(2x + 5) - 3(2x + 5) \\ &= (2x + 5)(x - 3) \end{aligned}$$

Example 20. Factor $x^4 - 5x^2 + 4$.

Solution

$$\begin{aligned} &x^4 - 5x^2 + 4 \\ &= (x^2 - 4)(x^2 - 1) \\ &= (x + 2)(x - 2)(x + 1)(x - 1) \end{aligned}$$

Example 22. Factor $3x^4 + 3x^2 - 60$.

Solution

$$\begin{aligned} &3x^4 + 3x^2 - 60 \\ &= 3(x^4 + x^2 - 20) \\ &= 3(x^2 + 5)(x^2 - 4) \\ &= 3(x^2 + 5)(x + 2)(x - 2) \end{aligned}$$

Answers to Trial Run

page 368 **1.** $(x + 6)(x + 2)$ **2.** $(12 + a)(1 + a)$ **3.** $(y - 6)(y - 3)$ **4.** $(x - 3y)(x - 3y)$

page 369 **1.** $(x + 3)(x - 2)$ **2.** $(5 + a)(2 - a)$ **3.** $2(m - 8)(m + 3)$ **4.** $(a - 15b)(a + b)$

page 371 **1.** $(2x - 3)(x + 5)$ **2.** $(2 - 3m)(1 - 4m)$ **3.** $(2x - 5y)(x - 2y)$
4. $ax(2x - 3)(2x - 3)$

page 373 **1.** $(x + 6)^2$ **2.** $(10y - x)^2$ **3.** $\left(a - \frac{1}{7}\right)^2$ **4.** $(0.1 + x)^2$

page 374 **1.** $(3x - 1)(2x - 3)$ **2.** $(5x + 2)(x - 3)$ **3.** $(4x - 3)(2x - 5)$ **4.** $(3x - 2)(2x + 3)$

page 376 **1.** $(x^2 + 6)(x^2 + 2)$ **2.** $2(x^2 + 1)(x^2 + 3)$ **3.** $(x^2 + 3)(x + 1)(x - 1)$
4. $(x + 3)(x - 3)(x + 2)(x - 2)$

Name ______________________ Date ____________

EXERCISE SET 7.3

Factor completely.

$(x-3)(x-4)$ **1.** $x^2 - 7x + 12$

________ **2.** $x^2 - 7x + 10$

$(x+0.5)(x+0.4)$ **3.** $x^2 + 0.9x + 0.20$

________ **4.** $x^2 + 0.8x + 0.15$

$(x-12)(x-12)$ **5.** $x^2 - 24x + 144$

________ **6.** $x^2 - 18x + 81$

$(A+2)(x-7)$ **7.** $a^2 - 5a - 14$

________ **8.** $a^2 + 6a - 27$

$(x+4)(x+4)$ **9.** $x^2 + 8x + 16$

________ **10.** $x^2 + 14x + 49$

$(x+0.5)(x+0.7)$ **11.** $x^2 - 0.2x - 0.35$

________ **12.** $x^2 + 1.1x - 0.26$

$(x+9)(x-8)$ **13.** $x^2 + x - 72$

________ **14.** $x^2 + x - 30$

$(x-2y)(x-2y)$ **15.** $x^2 - 4xy + 4y^2$

________ **16.** $x^2 - 8xy + 16y^2$

$(x-\frac{1}{5}y)(x-\frac{1}{5}y)$ **17.** $x^2 - \frac{2}{5}xy + \frac{1}{25}y^2$

________ **18.** $x^2 - \frac{2}{7}xy + \frac{1}{49}y^2$

$(M+3N)(M-2N)$ **19.** $m^2 + mn - 6n^2$

________ **20.** $m^2 - 4mn - 45n^2$

$(M-13)(M-3)$ **21.** $39 - 16m + m^2$

________ **22.** $72 - 17m + m^2$

$(xy+0.5)(xy-0.3)$ **23.** $x^2y^2 + 0.2xy - 0.15$

________ **24.** $x^2y^2 + 0.5xy - 0.24$

________ **25.** $x^4 - 5x^2 + 6$

________ **26.** $x^4 - 11x^2 + 24$

________ **27.** $3x^2 - 18x + 27$

________ **28.** $5x^2 - 40x + 80$

________ **29.** $-2y^3 + 6y^2 + 20y$

________ **30.** $-3y^3 - 24y^2 + 27y$

________ **31.** $4m^4 + 4m^2n^2 - 24n^4$

________ **32.** $5m^4 - 25m^2n^2 - 30n^4$

________ **33.** $x^3y - 18x^2y^2 + 80xy^3$

________ **34.** $x^3y - 15x^2y^2 + 26xy^3$

________ **35.** $2a^4b - 28a^3b^2 - 64a^2b^3$

________ **36.** $3a^3b^2 - 15a^2b^3 - 450ab^4$

________ **37.** $2x^2 + 7x + 3$

________ **38.** $3x^2 + 16x + 5$

$(3A-5)(3A-5)$ **39.** $9a^2 - 30a + 25$

________ **40.** $25a^2 - 20a + 4$

________ **41.** $10y^2 - 7y - 12$

________ **42.** $14y^2 + 17y - 6$

________ **43.** $8m^2 + 14m - 15$

________ **44.** $8m^2 + 18m - 35$

________ **45.** $6x^2 - 5xy - 56y^2$

________ **46.** $16x^2 + 6xy - 27y^2$

________ **47.** $81x^2 - 36xy + 4y^2$

________ **48.** $25x^2 - 30xy + 9y^2$

________ **49.** $121 + 132a + 36a^2$

$36A^2 + 132A + 121$

$(6A+11)(6A+11)$

________ **50.** $144 + 168a + 49a^2$

_______ **51.** $30c^2 - 145cd + 45d^2$

_______ **52.** $20c^2 - 68cd + 24d^2$

_______ **53.** $-8y^2 + 56y - 98$

_______ **54.** $-48y^2 + 72y - 27$

_______ **55.** $2x^2 - 7x - 3$

_______ **56.** $5x^2 - x - 8$

_______ **57.** $4a^2b^4 - 25a^2b^3 + 6a^2b^2$

_______ **58.** $4a^4b^2 - 23a^3b^2 - 6a^2b^2$

_______ **59.** $24x^3y + 26x^2y - 70xy$

_______ **60.** $105x^3y + 57x^2y - 72xy$

_______ **61.** $0.09k^2 - 3hk + 25h^2$

_______ **62.** $0.04h^2 - 1.2hk + 9k^2$

_______ **63.** $x^4 - 9x^2 + 20$

_______ **64.** $x^4 - 8x^2 - 9$

☆ Stretching the Topics

Factor completely.

_______ **1.** $(a + b)^2 - 5(a + b) - 14$

_______ **2.** $x^5y - 53x^3y^3 + 196xy^5$

_______ **3.** $5a^{2x} - 8a^xb^y - 21b^{2y}$

Check your answers in the back of your book.

If you can complete **Checkup 7.3**, you are ready to go on to Section 7.4.

Name ______________________ Date ____________

CHECKUP 7.3

Factor.

______ 1. $x^2 - 3xy - 40y^2$

______ 2. $21 - 10x + x^2$

______ 3. $x^2 + 0.9x - 0.22$

______ 4. $5x^2 - 10xy - 15y^2$

______ 5. $x^2 - 12x + 36$

______ 6. $x^3 - \frac{2}{3}x^2 + \frac{1}{9}x$

______ 7. $2x^3y + 14x^2y^2 - 16xy^3$

______ 8. $5x^2 - 17x + 6$

______ 9. $-8y + 2y^2 + 3y^3$

______ 10. $12a^3b + 26a^2b^2 - 10ab^3$

Check your answers in the back of your book.

If You Missed Problems:	You Should Review Examples:
1–4	1–6
5, 6	13–16
7	7, 8
8	9, 10
9, 10	11, 12

7.4 Using All Types of Factoring

Factoring is not a guessing game. If you are systematic in your approach, you will become successful after lots of practice. Get into the habit of asking yourself these questions.

1. Do all terms contain a common factor? If so, use the distributive property to rewrite the expression.

$$AB + AC = A(B + C)$$

2. Does the expression contain exactly *two* terms that are perfect squares separated by a minus sign? If so, use the formula for factoring the difference of two squares.

$$A^2 - B^2 = (A + B)(A - B)$$

3. Does the expression contain exactly *four* terms? If so, try to group the terms two by two.
4. Does the expression contain exactly *three* terms? If so, is it the square of a binomial?

$$A^2 + 2AB + B^2 = (A + B)^2 \quad \text{or} \quad A^2 - 2AB + B^2 = (A - B)^2$$

Or is it the product of two different binomials?

In some cases you will use more than one factoring technique. In other cases you will discover that the expression cannot be factored.

Example 1. Factor completely $3x^3 - 48x$.

Solution

$3x^3 - 48x$

$= 3x(x^2 - 16)$ Remove common factor $3x$.

$= 3x(x + 4)(x - 4)$ Factor the difference of two squares.

Example 2. Factor completely $x^2 + xy + 6x + 6y$.

Solution

$x^2 + xy + 6x + 6y$

$= x^2 + xy + 6x + 6y$ Mentally group the terms two by two.

$= x(x + y) + 6(x + y)$ Remove the common factor from each pair.

$= (x + y)(x + 6)$ Remove the common binomial factor, $(x + y)$.

Example 3. Factor completely $x^4 + 8x^2 - 9$.

Solution

$x^4 + 8x^2 - 9$

$= (x^2 + 9)(x^2 - 1)$ Factor the trinomial as a product of two binomials.

$= (x^2 + 9)(x + 1)(x - 1)$ Factor the difference of two squares.
The sum of two squares will not factor.

Trial Run

Factor completely.

_______ 1. $10x^2 - 90y^2$

_______ 2. $-45x^2 + 30x - 5$

_______ 3. $x^4 - 3x^2 - 4$

_______ 4. $ac + bc - ad - bd$

Answers are given below.

Answers to Trial Run

1. $10(x + 3y)(x - 3y)$ **2.** $-5(3x - 1)^2$ **3.** $(x + 2)(x - 2)(x^2 + 1)$ **4.** $(a + b)(c - d)$

Name ______________________ Date __________

EXERCISE SET 7.4

Factor completely.

_______ **1.** $x^2 - 11x + 30$

_______ **2.** $x^2 - 14x + 48$

_______ **3.** $42c^2 - 66cd + 24d^2$

_______ **4.** $35c^2 - 91cd + 56d^2$

_______ **5.** $20ax + 4ay - 15x - 3y$

_______ **6.** $10ax - 15ay - 8x + 12y$

_______ **7.** $49x^2 - 42xy + 9y^2$

_______ **8.** $81x^2 - 36xy + 4y^2$

_______ **9.** $16b^2 - 121$

_______ **10.** $49b^2 - 169$

_______ **11.** $12a^3b - 16a^2 + 20ab^3$

_______ **12.** $35a^3b - 25a^2b^2 + 20b^2$

_______ **13.** $2a^3 - 200a$

_______ **14.** $3a^3 - 75a$

_______ **15.** $a^2b^2 - 11abc - 12c^2$

_______ **16.** $a^2b^2 - 13abc - 14c^2$

_______ **17.** $9x^6 + 30x^4y^2 + 25x^2y^4$

_______ **18.** $16x^6 + 24x^4y^2 + 9x^2y^4$

_______ **19.** $169 + 52a + 4a^2$

_______ **20.** $144 - 120a + 25a^2$

_______ **21.** $-24x^2 + 8x - 56$

_______ **22.** $-28x^2 - 21x + 35$

_______ **23.** $a^3 - 3a^2 + 9a - 27$

_______ **24.** $a^3 - 4a^2 + 16a - 64$

_______ **25.** $x^4 - 2x^2 - 63$

_______ **26.** $x^4 - 22x^2 - 75$

_______ **27.** $14a^3 - 28a^2 + 14a$

_______ **28.** $6a^3 - 36a^2 + 54a$

_______ **29.** $x^3yz - 8x^2y^2z + 15xy^3z$

_______ **30.** $x^3yz - 4x^2y^2z - 21xy^3z$

_______ **31.** $-x^4 + 19x^3 - 34x^2$

_______ **32.** $-x^4 + 21x^3 - 38x^2$

_______ **33.** $21x^2 - 9x + 3$

_______ **34.** $18x^2 - 7x + 5$

_______ **35.** $x^2 - 18xy + 81y^2$

_______ **36.** $x^2 - 14xy + 49y^2$

_______ **37.** $10x^2 - 3xy - 27y^2$

_______ **38.** $28x^2 + xy - 15y^2$

_______ **39.** $-30y^2 - 225y + 40$

_______ **40.** $-28y^2 - 56y + 105$

_______ **41.** $4a^2b^4 - 28a^2b^3 + 49a^2b^2$

_______ **42.** $9a^2b^4 - 30a^2b^3 + 25a^2b^2$

_______ **43.** $x^3 - 4x^2 - 9x + 36$

_______ **44.** $2x^3 - 3x^2 - 8x + 12$

_______ **45.** $4x^3 - 20x^2y + 16xy^2$

_______ **46.** $6x^3 - 42x^2y - 48xy^2$

_______ **47.** $a^6 - 16a^2$

_______ **48.** $a^6 - 81a^2$

_______ **49.** $-121y^2 + 16x^2y^2$

_______ **50.** $-144y^2 + 25x^2y^2$

☆ Stretching the Topics

Factor completely.

_______ **1.** $-12a^5b^2 + 19a^3b^2 + 21ab^2$

_______ **2.** $(a + b)^2 - 10c(a + b) + 25c^2$

_______ **3.** $4(x + y)^2 + 12(x + y)(x - y) + 9(x - y)^2$

Check your answers in the back of your book.

If you can complete **Checkup 7.4**, you are ready to go on to Section 7.5.

Name ______________________ Date ____________

CHECKUP 7.4

Factor completely.

______ 1. $9x^2 - 42x + 49$

______ 2. $18x^2 - 63x - 9$

______ 3. $4x^2 - 36$

______ 4. $81x^4 - y^4$

______ 5. $x^3 - x^2 - 9x + 9$

______ 6. $y^2 - 3y - 54$

______ 7. $9m^2 - 18m - 16$

______ 8. $64a^2b - 25b^3$

______ 9. $30x^3y - 18x^2y - 99xy$

______ 10. $10x^4y^2 + 80x^2y^4 - 480y^6$

Check your answers in the back of your book.

If You Missed Problem:	You Should Review Example:	
1	SECTION 7.3	14
2	SECTION 7.1	2
3	SECTION 7.2	4
4		8
5	SECTION 7.1	15
6	SECTION 7.3	5
7		10
8	SECTION 7.2	4
9	SECTION 7.3	7
10		21

7.5 Using Factoring to Solve Quadratic Equations

In Chapter 4 we learned to solve equations such as

$$3x + 1 = 6 \qquad 5 - 2x = 25 \qquad 2(y + 3) = y - 1$$

by finding the value of the variable that satisfied the equation. In each of these equations, the variable appeared with an exponent of 1. Such equations are called **first-degree equations** or **linear equations**.

A polynomial equation containing one variable in which the largest exponent on the variable is a 2 is called a **second-degree equation** or a **quadratic equation**. Before we learn to solve quadratic equations, we must make a very important observation about *zero* as the result of multiplication.

The Zero Product Rule

If we are told that two quantities are multiplied together and the product is zero, what can we conclude? If the product of two factors is zero, then one or the other or both factors *must* be zero. We state this **zero product rule** in general as follows.

> **Zero Product Rule.** If $A \cdot B = 0$, then $A = 0$ or $B = 0$ or both.

This rule is very important in learning to solve quadratic equations.

Example 1. Solve $(x + 3)(x - 5) = 0$.

Solution: We observe that this product has two factors and we know that either one of them could be zero.

$$x + 3 = 0 \quad \textbf{or} \quad x - 5 = 0$$

Solving each of these equations, we have

$$x + 3 - 3 = 0 - 3 \quad \text{or} \quad x - 5 + 5 = 0 + 5$$

$$x = -3 \quad \text{or} \quad x = 5$$

Now we check our solutions in the original equation.

CHECK: $x = -3$

$$(x + 3)(x - 5) = 0$$
$$(-3 + 3)(-3 - 5) \stackrel{?}{=} 0$$
$$0(-8) \stackrel{?}{=} 0$$
$$0 = 0$$

CHECK: $x = 5$

$$(x + 3)(x - 5) = 0$$
$$(5 + 3)(5 - 5) \stackrel{?}{=} 0$$
$$8(0) \stackrel{?}{=} 0$$
$$0 = 0$$

Example 2. Solve $(x - 7)(x - 7) = 0$ and check.

Solution

$$(x - 7)(x - 7) = 0$$

$$x - 7 = 0 \quad \text{or} \quad x - 7 = 0$$

$$x = 7 \quad \text{or} \quad x = 7$$

Both solutions are the same, so we say that the solution is $x = 7$.

CHECK: $x = 7$

$$(x - 7)(x - 7) = 0$$

$$(7 - 7)(7 - 7) \stackrel{?}{=} 0$$

$$(0)(0) \stackrel{?}{=} 0$$

$$0 = 0$$

You check the solutions for Example 3.

Example 3. Solve $x(5 + x) = 0$ and check.

Solution

$$x(5 + x) = 0$$

$$x = 0 \quad \text{or} \quad 5 + x = 0$$

$$x = 0 \quad \text{or} \quad x = -5$$

CHECK: $x = 0$ CHECK: $x = -5$

Check your work on page 396. ▶

Trial Run

Solve and check.

_______ 1. $(x - 11)(x - 2) = 0$

_______ 2. $(y - 3)(y + 5) = 0$

_______ 3. $(a + 12)(a + 5) = 0$

_______ 4. $x(x - 1) = 0$

_______ 5. $(1 - x)(2 + x) = 0$

_______ 6. $(9 - x)(9 - x) = 0$

Answers are on page 397.

Solving Quadratic Equations by Factoring

We have just learned to solve an equation in which a product of factors is zero. Now we shall see how the zero product rule can help us solve a quadratic equation. Let's solve

$$x^2 + 3x + 2 = 0$$

If we could rewrite the polynomial on the left-hand side as a *product*, we could use the zero product rule to solve the equation.

Example 4. Solve $x^2 + 3x + 2 = 0$ and check.

Solution

$$x^2 + 3x + 2 = 0$$

$$(x + 2)(x + 1) = 0 \qquad \text{Factor the polynomial.}$$

$$x + 2 = 0 \quad \text{or} \quad x + 1 = 0 \qquad \text{Zero product rule.}$$

$$x = -2 \quad \text{or} \quad x = -1 \qquad \text{Solve for } x.$$

CHECK: $x = -2$

$$x^2 + 3x + 2 = 0 \qquad \text{Original equation.}$$

$$(-2)^2 + 3(-2) + 2 \stackrel{?}{=} 0 \qquad \text{Substitute } -2 \text{ for } x.$$

$$4 - 6 + 2 \stackrel{?}{=} 0 \qquad \text{Simplify.}$$

$$0 = 0 \qquad \text{Simplify.}$$

CHECK: $x = -1$

$$x^2 + 3x + 2 = 0 \qquad \text{Original equation.}$$

$$(-1)^2 + 3(-1) + 2 \stackrel{?}{=} 0 \qquad \text{Substitute } -1 \text{ for } x.$$

$$1 - 3 + 2 \stackrel{?}{=} 0 \qquad \text{Simplify.}$$

$$0 = 0 \qquad \text{Simplify.}$$

The solutions are $x = -2$ or $x = -1$.

Example 5. Solve $3y^2 - 12 = 0$ and check.

Solution

$$3y^2 - 12 = 0$$

$$3(y^2 - 4) = 0$$

Notice that we removed the *common factor* 3.

$$3(y + 2)(y - 2) = 0$$

Since the constant factor 3 can *never* be zero, we know

$$y + 2 = 0 \quad \text{or} \quad y - 2 = 0$$

$$y = -2 \quad \text{or} \quad y = 2$$

Now you check these solutions.

CHECK: $y = -2$

$$3y^2 - 12 = 0$$

$$3(\_\_\_)^2 - 12 \stackrel{?}{=} 0$$

$$3(\_\_\_) - 12 \stackrel{?}{=} 0$$

$$\_\_\_ - 12 \stackrel{?}{=} 0$$

$$\_\_\_ = 0$$

CHECK: $y = 2$

$$3y^2 - 12 = 0$$

$$3(\_\_\_)^2 - 12 \stackrel{?}{=} 0$$

$$3(\_\_\_) - 12 \stackrel{?}{=} 0$$

$$\_\_\_ - 12 \stackrel{?}{=} 0$$

$$\_\_\_ = 0$$

Check your work on page 396. ▶

Now try Example 6.

Example 6. Solve $5x^2 + 10x = 0$.

Solution

$$5x^2 + 10x = 0$$

$$5x(\underline{x+2}) = 0$$

$5x = 0$ or $\underline{\qquad} = 0$

$x = 0$ or $x = \underline{-2}$

Check your work on page 396. ▶

Try Example 7.

Example 7. Solve $0 = x^2 - 12x + 36$.

Solution $(x-6)(x-6)$

The solution is $x = \underline{6}$.

Check your work on page 396. ▶

Trial Run

Solve.

______ 1. $x^2 - 4x - 5 = 0$

______ 2. $y^2 - 36 = 0$

______ 3. $x^2 - 7x = 0$

______ 4. $a^2 + 10a + 25 = 0$

______ 5. $3x^2 - 15x + 18 = 0$

______ 6. $27 + 6m - m^2 = 0$

$-m^2 + 6m + 27$

Answers are on page 397.

More Quadratic Equations

Remember that solving quadratic equations depends very much on the zero product rule. To solve quadratic equations, *we must be sure that our polynomial equals zero*. Then we may factor, set each factor equal to zero, and solve.

Example 8. Solve $x^2 - x = 20$ and check.

Solution

$x^2 - x = 20$	
$x^2 - x - 20 = 20 - 20$	Subtract 20 from both sides.
$x^2 - x - 20 = 0$	Simplify.
$(x - 5)(x + 4) = 0$	Factor the polynomial.
$x - 5 = 0$ or $x + 4 = 0$	Zero product rule.
$x = 5$ or $x = -4$	Solve for x.

CHECK: $x = 5$

$$x^2 - x = 20$$
$$(5)^2 - 5 \stackrel{?}{=} 20$$
$$25 - 5 \stackrel{?}{=} 20$$
$$20 = 20$$

CHECK: $x = -4$

$$x^2 - x = 20$$
$$(-4)^2 - (-4) \stackrel{?}{=} 20$$
$$16 + 4 \stackrel{?}{=} 20$$
$$20 = 20$$

The solutions are $x = 5$ or $x = -4$.

Complete the missing steps in Example 9.

Example 9. Solve $x^2 - 10 = 3x$.

Solution

$x^2 - 10 = 3x$	
$x^2 - 10 - 3x = 3x - 3x$	Subtract $3x$ from both sides.
$x^2 - 10 - 3x = 0$	Simplify.
$x^2 - 3x - 10 = 0$	Rearrange terms.
$(\_\_\_\_\_\_)(\_\_\_\_\_\_) = 0$	Factor.
$\_\_\_\_ = 0$ or $\_\_\_\_ = 0$	Zero product rule.
$x = $ 6 or $x = $ 13	Solve for x.

Check your work on page 396. ▶

You finish Example 10.

Example 10. Solve $y^2 = 16$.

Solution

$$y^2 = 16$$
$$y^2 - 16 = 0$$

Check your work on page 396. ▶

Example 11. Solve $8x^2 + 2x = 3$.

Solution

$$8x^2 + 2x = 3$$
$$8x^2 + 2x - 3 = 0$$
$$(4x + 3)(2x - 1) = 0$$
$$4x + 3 = 0 \quad \text{or} \quad 2x - 1 = 0$$
$$4x = -3 \qquad 2x = 1$$
$$x = \frac{-3}{4} \quad \text{or} \quad x = \frac{1}{2}$$

Trial Run

Solve.

______ **1.** $x^2 + 2x = 35$

______ **2.** $y^2 = 64$

________ **3.** $x^2 = 3x$

________ **4.** $y^2 = 7y - 12$

________ **5.** $3 = 10x^2 - 13x$

________ **6.** $15 - 2a = a^2$

Answers are on page 397.

The method we have just learned will also work to solve equations containing parentheses or like terms that must be combined. Remember, to solve a quadratic equation, our aim is to have zero on one side of the equation and all other terms on the other side.

Let's work an example.

Example 12. Solve $3(x^2 - 1) - 9 = x(x + 10)$ and check.

Solution

$3(x^2 - 1) - 9 = x(x + 10)$	
$3x^2 - 3 - 9 = x^2 + 10x$	Remove parentheses.
$3x^2 - 12 = x^2 + 10x$	Combine like terms.
$3x^2 - 12 - x^2 = x^2 + 10x - x^2$	Subtract x^2 from both sides.
$2x^2 - 12 = 10x$	Combine like terms.
$2x^2 - 12 - 10x = 10x - 10x$	Subtract $10x$ from both sides.
$2x^2 - 10x - 12 = 0$	Combine and rearrange terms.
$2(x^2 - 5x - 6) = 0$	Factor out common factor.
$2(x - 6)(x + 1) = 0$	Factor the polynomial
$x - 6 = 0 \quad \text{or} \quad x + 1 = 0$	Use zero product rule.
$x = 6 \quad \text{or} \quad x = -1$	Solve for x.

CHECK: $x = 6$	CHECK: $x = -1$
$3(x^2 - 1) - 9 = x(x + 10)$	$3(x^2 - 1) - 9 = x(x + 10)$
$3[(6)^2 - 1] - 9 \stackrel{?}{=} 6(6 + 10)$	$3[(-1)^2 - 1] - 9 \stackrel{?}{=} -1(-1 + 10)$
$3[36 - 1] - 9 \stackrel{?}{=} 6(16)$	$3[1 - 1] - 9 \stackrel{?}{=} -1(9)$
$3[35] - 9 \stackrel{?}{=} 96$	$3[0] - 9 \stackrel{?}{=} -9$
$105 - 9 \stackrel{?}{=} 96$	$0 - 9 \stackrel{?}{=} -9$
$96 = 96$	$-9 = -9$

Now you complete Examples 13 and 14.

Example 13. Solve $4x^2 + x = 3(x^2 - 4x)$.

Solution

$$4x^2 + x = 3(x^2 - 4x)$$
$$4x^2 + x = 3x^2 - 12x$$
$$x^2 + x = -12x$$
$$x^2 + 13x = 0$$

Check your work on page 397. ▶

Example 14. $5x(x - 2) - 6 = 4x^2 - 2(5x + 1)$.

Solution

$$5x(x - 2) - 6 = 4x^2 - 2(5x + 1)$$
$$5x^2 - 10x - 6 = 4x^2 - 10x - 2$$
$$x^2 - 10x - 6 = -10x - 2$$
$$x^2 - 6 = -2$$

Check your work on page 397. ▶

Trial Run

Solve.

_______ **1.** $4(x^2 - 2x) = 3x^2 - 15$

_______ **2.** $7x^2 - 6x = 3x(2x + 3)$

_______ **3.** $3y(y - 6) - 21 = 2y^2 - 6(3y + 2)$

_______ **4.** $3(x^2 - 2) + 16 = x(x + 12)$

_______ **5.** $3y(y - 5) = 2y(y - 3)$

_______ **6.** $5a(a - 1) + 36 = 3(2a^2 - 15) - 5a$

Answers are on page 397.

To solve quadratic equations we have learned to use the following steps.

1. Work with parentheses if necessary.
2. Get zero on one side of the equation and all other terms on the other side (combining like terms when possible).
3. Rewrite the equation with the polynomial in factored form.
4. Set each factor containing the variable equal to zero.
5. Solve for the variable and check.

▶ Examples You Completed

Example 3. Solve $x(5 + x) = 0$ and check.

CHECK: $x = 0$

$$x(5 + x) = 0$$
$$0(5 + 0) \stackrel{?}{=} 0$$
$$0(5) \stackrel{?}{=} 0$$
$$0 = 0$$

CHECK: $x = -5$

$$x(5 + x) = 0$$
$$-5(5 + -5) \stackrel{?}{=} 0$$
$$-5(0) \stackrel{?}{=} 0$$
$$0 = 0$$

Example 5. Solve $3y^2 - 12 = 0$ and check.

CHECK: $y = -2$

$$3y^2 - 12 = 0$$
$$3(-2)^2 - 12 \stackrel{?}{=} 0$$
$$3(4) - 12 \stackrel{?}{=} 0$$
$$12 - 12 \stackrel{?}{=} 0$$
$$0 = 0$$

CHECK: $y = 2$

$$3y^2 - 12 = 0$$
$$3(2)^2 - 12 \stackrel{?}{=} 0$$
$$3(4) - 12 \stackrel{?}{=} 0$$
$$12 - 12 \stackrel{?}{=} 0$$
$$0 = 0$$

Example 6. Solve $5x^2 + 10x = 0$.

Solution

$$5x^2 + 10x = 0$$
$$5x(x + 2) = 0$$
$$5x = 0 \quad \text{or} \quad x + 2 = 0$$
$$x = 0 \quad \text{or} \quad x = -2$$

Example 7. Solve $0 = x^2 - 12x + 36$.

Solution

$$0 = x^2 - 12x + 36$$
$$0 = (x - 6)(x - 6)$$
$$0 = x - 6 \quad \text{or} \quad 0 = x - 6$$
$$6 = x \quad \text{or} \quad 6 = x$$

The solution is $x = 6$.

Example 9. Solve $x^2 - 10 = 3x$.

Solution

$$x^2 - 10 = 3x$$
$$x^2 - 10 - 3x = 3x - 3x$$
$$x^2 - 10 - 3x = 0$$
$$x^2 - 3x - 10 = 0$$
$$(x - 5)(x + 2) = 0$$
$$x - 5 = 0 \quad \text{or} \quad x + 2 = 0$$
$$x = 5 \quad \text{or} \quad x = -2$$

Example 10. Solve $y^2 = 16$.

Solution

$$y^2 = 16$$
$$y^2 - 16 = 0$$
$$(y + 4)(y - 4) = 0$$
$$y + 4 = 0 \quad \text{or} \quad y - 4 = 0$$
$$y = -4 \quad \text{or} \quad y = 4$$

Example 13. Solve $4x^2 + x = 3(x^2 - 4x)$.

Solution

$$4x^2 + x = 3(x^2 - 4x)$$
$$4x^2 + x = 3x^2 - 12x$$
$$x^2 + x = -12x$$
$$x^2 + 13x = 0$$
$$x(x + 13) = 0$$
$$x = 0 \quad \text{or} \quad x + 13 = 0$$
$$x = 0 \quad \text{or} \quad x = -13$$

Example 14. $5x(x - 2) - 6 = 4x^2 - 2(5x + 1)$.

Solution

$$5x(x - 2) - 6 = 4x^2 - 2(5x + 1)$$
$$5x^2 - 10x - 6 = 4x^2 - 10x - 2$$
$$x^2 - 10x - 6 = -10x - 2$$
$$x^2 - 6 = -2$$
$$x^2 - 4 = 0$$
$$(x + 2)(x - 2) = 0$$
$$x + 2 = 0 \quad \text{or} \quad x - 2 = 0$$
$$x = -2 \quad \text{or} \quad x = 2$$

Answers to Trial Runs

page 390 **1.** $x = 11$ or $x = 2$ **2.** $y = 3$ or $y = -5$ **3.** $a = -12$ or $a = -5$ **4.** $x = 0$ or $x = 1$ **5.** $x = 1$ or $x = -2$ **6.** $x = 9$

page 392 **1.** $x = 5$ or $x = -1$ **2.** $y = 6$ or $y = -6$ **3.** $x = 0$ or $x = 7$ **4.** $a = -5$ **5.** $x = 3$ or $x = 2$ **6.** $m = 9$ or $m = -3$

page 393 **1.** $x = -7$ or $x = 5$ **2.** $y = 8$ or $y = -8$ **3.** $x = 0$ or $x = 3$ **4.** $y = 4$ or $y = 3$ **5.** $x = \frac{3}{2}$ or $x = \frac{-1}{5}$ **6.** $a = -5$ or $a = 3$

page 395 **1.** $x = 3$ or $x = 5$ **2.** $x = 0$ or $x = 15$ **3.** $y = 3$ or $y = -3$ **4.** $x = 5$ or $x = 1$ **5.** $y = 0$ or $y = 9$ **6.** $a = 9$ or $a = -9$

Name ______________________ Date __________

EXERCISE SET 7.5

Solve.

_______ 1. $(x - 3)(x - 2) = 0$

_______ 2. $(x - 7)(x - 5) = 0$

_______ 3. $(x + 0.2)(x - 0.6) = 0$

_______ 4. $(x + 0.5)(x - 1.1) = 0$

_______ 5. $(a + 13)(a + 5) = 0$

_______ 6. $(a + 1)(a + 9) = 0$

_______ 7. $x\left(x - \frac{7}{8}\right) = 0$

_______ 8. $x\left(x + \frac{9}{10}\right) = 0$

_______ 9. $(7 - x)(9 + x) = 0$

_______ 10. $(2 - x)(8 + x) = 0$

_______ 11. $2(y - 3)(y + 1) = 0$

_______ 12. $-3(y + 6)(y - 1) = 0$

_______ 13. $x^2 - 6x + 8 = 0$

_______ 14. $x^2 - 9x + 8 = 0$

_______ 15. $x^2 - \frac{25}{16} = 0$

_______ 16. $x^2 - \frac{25}{49} = 0$

_______ 17. $y^2 + 9y + 18 = 0$

_______ 18. $y^2 + 8y + 15 = 0$

_______ 19. $m^2 + 1.5m = 0$

_______ 20. $m^2 - 0.9m = 0$

_______ 21. $z^2 + 3z - 54 = 0$

_______ 22. $z^2 + z - 42 = 0$

_______ 23. $2x^2 - 10x = 0$

_______ 24. $3x^2 - 9x = 0$

_______ 25. $z^2 - 30z + 225 = 0$

_______ 26. $z^2 - 26z + 169 = 0$

_______ 27. $5y^2 - 500 = 0$

_______ 28. $3y^2 - 192 = 0$

_______ 29. $18n - 9n^2 = 0$

_______ 30. $20n - 5n^2 = 0$

_______ 31. $33 - 8y - y^2 = 0$

_______ 32. $80 - 2y - y^2 = 0$

_______ 33. $5x^2 + 13x - 6 = 0$

_______ 34. $6x^2 + 19x - 7 = 0$

_______ 35. $x^2 + 0.3x = 0.18$

_______ 36. $x^2 - 0.7x = 0.18$

_______ 37. $y^2 = 81$

_______ 38. $y^2 = 121$

_______ 39. $2x^2 = 5x$

_______ 40. $5x^2 - 3x = 0$

_______ 41. $72 = x^2 - x$

_______ 42. $54 = x^2 - 3x$

_______ 43. $10 - 3a = a^2$

_______ 44. $33 - 8a = a^2$

_______ 45. $x^2 - 16x = -64$

_______ 46. $x^2 - 20x = -100$

_______ 47. $10x^2 + 4 = 22x$

_______ 48. $12x^2 + 15 = 27x$

_______ 49. $144 = a^2$

_______ 50. $225 = a^2$

_______ 51. $y^2 = -17y$

_______ 52. $y^2 = -15y$

_______ 53. $100 = y^2 - 15y$

_______ 54. $34 = y^2 + 15y$

_______ 55. $x^2 - 1.4x = -0.45$

_______ 56. $x^2 - 1.1x = -0.28$

_______ 57. $x^2 + 6 = -5x$

_______ 58. $x^2 + 24 = -11x$

_______ 59. $18y^2 = 8$

_______ 60. $20y^2 = 5$

_______ 61. $2x^2 - 12x = -18$

_______ 62. $3x^2 - 24x = -48$

_______ 63. $5y^2 = 5y$

_______ 64. $11y^2 = 22y$

_______ 65. $54 + 12x = 2x^2$

_______ 66. $45 - 40x = 5x^2$

_______ 67. $10y^2 = 360$

_______ 68. $9y^2 = 9$

_______ 69. $5(x^2 - 1) = 4(x^2 + 5)$

_______ 70. $3(x^2 - 2) = 2(x^2 + 15)$

_______ 71. $7a(3a - 1) = 2(3a - 1)$

_______ 72. $5a(4a - 3) = 2a - 3$

_______ 73. $2x(x - 4) - 28 = x^2 - 2(3x - 10)$

_______ 74. $3x(2x - 1) + 8 = 5x^2 - 3(3x - 5)$

_______ 75. $4(x^2 - 2x) + 25 = x(3x + 2)$

_______ 76. $5(x^2 - 4x) + 9 = 2x(2x - 7)$

_______ 77. $5x(x - 1) + 8 = 3(x^2 + 2) - 9x$

_______ 78. $7x(x - 1) - 60 = 2(2x^2 + 15) - 4x$

_______ 79. $5(y^2 + 2) = 2(2y + 5)$

_______ 80. $4(y^2 - 3) = 3(3y - 4)$

☆ Stretching the Topics

Solve.

_______ 1. $0.05m^2 + 0.13m = -0.06$

_______ 2. $(2x + 1)(x - 2) + x^2 = -2(5x + 2)$

_______ 3. $(x - 1)^2 + (x + 1)^2 + (x - 1)(x + 1) - 28 = 0$

Check your answers in the back of your book.

If you can solve the equations in **Checkup 7.5**, you are ready to go on to Section 7.6.

Name ______________________ Date ____________

✓ CHECKUP 7.5

Solve.

_______ 1. $2x(x - 13) = 0$

_______ 2. $a^2 - 2a - 63 = 0$

_______ 3. $y^2 - \frac{16}{25} = 0$

_______ 4. $3x^2 + 30x + 63 = 0$

_______ 5. $7x^2 - 2.8x = 0$

_______ 6. $x^2 - 3x = 40$

_______ 7. $3x^2 + 18 = 15x$

_______ 8. $a^2 = 6a - 9$

_______ 9. $5(x^2 - 4) = 4(x^2 + 11)$

_______ 10. $4y(y + 1) + 7 = 3(y^2 - 1) - 3y$

Check your answers in the back of your book.

If You Missed Problems:	You Should Review Examples:
1	1–3
3–5	4–7
6–8	8–11
9, 10	12–14

7.6 Switching from Word Statements to Quadratic Equations

In Chapter 4 we learned to switch from word statements to first degree equations. Now we shall put those same skills to work in switching from word statements to quadratic equations.

Example 1. The Watsons wish to build a rectangular concrete patio which is 6 feet longer than it is wide, but they only have enough concrete mix to cover 280 square feet. *What dimensions* should their patio have?

Solution: We have italicized what we are being asked to find.

Let $w = $ width (feet)

Then $w + 6 = $ length (feet)

w w

$w + 6$

$(w + 6)w = $ area of patio (square feet) — Remember, area of a rectangle is the *product* of length and width.

Here $(w + 6)w = 280$ — Patio area must be covered by 280 square feet of concrete.

$w^2 + 6w = 280$ — Remove parentheses.

$w^2 + 6w - 280 = 0$ — Set the polynomial equal to zero.

$(w + 20)(w - 14) = 0$ — Factor.

$w + 20 = 0$ or $w - 14 = 0$ — Zero product rule.

$w = -20$ or $w = 14$ — Solve for w.

The width: $w = 14$ feet — The measure of a side cannot be a negative number.

The length: $w + 6 = 14 + 6$

$= 20$ feet

The patio's dimensions are 14 feet by 20 feet.

CHECK: If the patio's dimensions are 14 feet by 20 feet, the area will be

$$14 \cdot 20 = 280 \text{ square feet}$$

Now you try to complete Example 2.

Example 2. *Find the length* of one side of a square checkerboard with area 64 square inches.

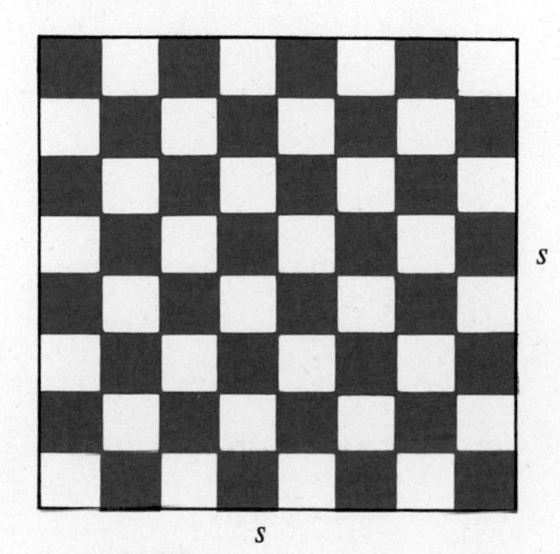

Solution:

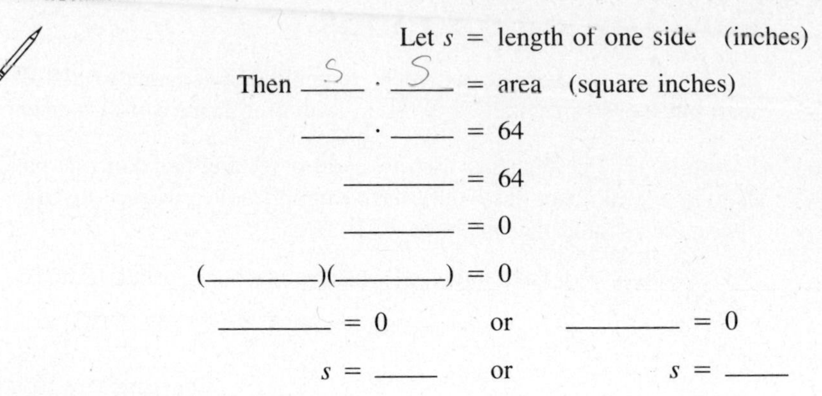

Let s = length of one side (inches)

Then __S__ · __S__ = area (square inches)

____ · ____ = 64

______ = 64

______ = 0

(______)(______) = 0

______ = 0 or ______ = 0

s = ____ or s = ____

The length of one side is ____ inches.

Check your work on page 407. ▶

Now let's return to the problem stated at the beginning of this chapter and see if we can solve it.

Example 3. After a tree was broken over by a storm, the part that was bending over was 26 feet long. The part left standing was 14 feet less than the distance from the foot of the tree to the point where the broken part touched the ground. *Find the height of the tree* before the storm.

Solution: Let's illustrate this problem.

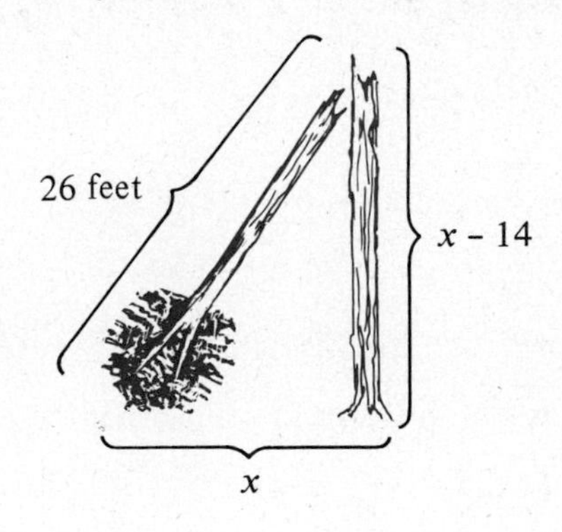

Let x = distance on ground from foot of tree to top of tree (feet)

$x - 14$ = height of part of tree left standing (feet)

26 = height of broken part of tree (feet)

We must use the Pythagorean Theorem that we learned in Chapter 4. Here the legs are x and $x - 14$, and the hypotenuse is 26.

$x^2 + (x - 14)^2 = 26^2$	Pythagorean Theorem: $a^2 + b^2 = c^2$; here $a = x$, $b = x - 14$, and $c = 26$.
$x^2 + x^2 - 28x + 196 = 676$	Square the binomial.
$2x^2 - 28x + 196 = 676$	Combine like terms.
$2x^2 - 28x - 480 = 0$	Subtract 676 from both sides.
$2(x^2 - 14x - 240) = 0$	Remove common monomial factor, 2.
$2(x - 24)(x +$ __10__$) = 0$	Factor the trinomial.

$x - 24 = 0$ or $x + \_\_\_\_ = 0$ — Use the zero product rule.

$x = \_\_\_\_$ or $x = \_\_\_\_$ — Solve for x.

The distance from the foot of the tree is

$$x = \_\_\_\_ \text{ feet}$$

The height of the part left standing is

$$x - 14 = \_\_\_\_ \text{ feet}$$

The height of the tree before the storm was

$$26 + (x - 14)$$
$$= 26 + \_\_\_\_\_\_$$
$$= \_\_\_\_\_\_ \text{ feet}$$

Check your work on page 407. ▶

Let's work together on Example 4.

Example 4. If the product of two consecutive integers is 90, *find the integers.*

Solution

Let $x =$ first integer

$x + 1 =$ next consecutive integer — Recall that the next consecutive integer is found by adding 1.

$x(x + 1) =$ product of the consecutive integers

Here $x(x + 1) = 90$

$x^2 + x = 90$ — Remove parentheses.

$x^2 + x - 90 = 0$ — Subtract 90 from both sides.

$(x + 10)(x - 9) = 0$ — Factor the trinomial.

$x + 10 = 0$ or $x - 9 = 0$ — Use the zero product rule.

$x = -10$ or $x = 9$ — Solve for x.

We have two perfectly good solutions.

If first integer: $x = -10$	If first integer: $x = 9$
then next integer: $x + 1 = -10 + 1 = -9$	then next integer: $x + 1 = 9 + 1 = 10$
CHECK: $(-10)(-9) = 90$	CHECK: $(9)(10) = 90$

The two pairs of consecutive integers are

$$-10 \text{ and } -9 \quad \text{or} \quad 9 \text{ and } 10$$

Example 5. The sum of the squares of two consecutive odd integers is 2 more than 5 times their sum. *Find the two odd integers.*

Solution: In the set of all integers, consecutive *odd* integers are *two* units apart. Let

$x =$ first odd integer

$x + 2 =$ next consecutive odd integer

Now we must identify each expression used in our problem.

sum of squares of two odd integers: $x^2 + (x + 2)^2$

sum of two odd integers: $x + (x + 2) = 2x + 2$

5 times sum of two odd integers: $5(2x + 2)$

2 more than 5 times sum of odd integers: $5(2x + 2) + 2$

Finally, we translate the entire statement into an equation.

$$x^2 + (x + 2)^2 = 5(2x + 2) + 2$$

$$x^2 + x^2 + 4x + 4 = 10x + 10 + 2 \qquad \text{Remove parentheses.}$$

$$2x^2 + 4x + 4 = 10x + 12 \qquad \text{Combine like terms.}$$

$$2x^2 - 6x - 8 = 0 \qquad \text{Get all terms to one side.}$$

$$2(x^2 - 3x - 4) = 0 \qquad \text{Factor out common factor.}$$

$$2(x - 4)(x + 1) = 0 \qquad \text{Factor the trinomial.}$$

$$x - 4 = 0 \quad \text{or} \quad x + 1 = 0 \qquad \text{Use zero product rule.}$$

$$x = 4 \quad \text{or} \quad x = -1$$

We reject the solution $x = 4$ because 4 is even rather than odd. Our only solution is

$$x = -1 \qquad \text{First odd integer.}$$

$$x + 2 = -1 + 2 \qquad \text{Next consecutive odd integer.}$$

$$= 1$$

The consecutive odd integers are -1 and 1. Let's check this solution using the words of the problem.

$$\text{sum of squares} \stackrel{?}{=} 2 + 5 \text{ times sum}$$

$$(-1)^2 + 1^2 \stackrel{?}{=} 2 + 5(-1 + 1)$$

$$1 + 1 \stackrel{?}{=} 2 + 5(0)$$

$$2 \stackrel{?}{=} 2 + 0$$

$$2 = 2$$

▶ Examples You Completed

Example 2. Find the length of one side of a square checkerboard with area 64 square inches.

Solution

Let s = length of one side

$$\text{Then } s \times s = \text{area}$$

$$s \times s = 64$$

$$s^2 = 64$$

$$s^2 - 64 = 0$$

$$(s + 8)(s - 8) = 0$$

$$s + 8 = 0 \quad \text{or} \quad s - 8 = 0$$

$$s = -8 \quad \text{or} \quad s = 8$$

The length of one side is 8 inches.

Example 3. Finish solving

$$2x^2 - 28x - 480 = 0$$

Solution

$$2x^2 - 28x - 480 = 0$$

$$2(x^2 - 14x - 240) = 0$$

$$2(x - 24)(x + 10) = 0$$

$$x - 24 = 0 \quad \text{or} \quad x + 10 = 0$$

$$x = 24 \quad \text{or} \quad x = -10$$

Distance from foot of tree:

$$x = 24 \text{ feet}$$

Height of standing part:

$$x - 14 = 10 \text{ feet}$$

Height of tree before storm:

$$26 + (x - 14)$$

$$= 26 + 10$$

$$= 36 \text{ feet}$$

Name ______________________ Date ____________

EXERCISE SET 7.6

Solve.

________ **1.** One number is 5 more than another. The square of the smaller added to 3 times the larger is 43. Find the numbers.

________ **2.** One number is 3 more than another. The square of the smaller added to 5 times the larger is 65. Find the numbers.

________ **3.** The width of a rectangle is 1 less than the length. The area is 72 square feet. Find the length and the width.

________ **4.** The width of a rectangle is 2 less than the length. The area is 63 square feet. Find the length and the width.

________ **5.** Three times the square of an integer added to 6 times the integer is 105. Find the integer.

________ **6.** If 4 times an integer is subtracted from twice the square of the integer, the result is 48. Find the integer.

________ **7.** If the area of a square is 81 square centimeters, find the length of a side of the square.

________ **8.** The area of a square is 36 square centimeters. Find the length of a side of the square.

________ **9.** One leg of a right triangle is 7 centimeters less than the other leg and the hypotenuse is 17 centimeters. Find the lengths of the two legs.

________ **10.** One leg of a right triangle is 4 centimeters more than the other leg and the hypotenuse is 20 centimeters. Find the lengths of the two legs.

________ **11.** The length of a rectangle is twice its width. The area is 128 square feet. Find the dimensions of the rectangle.

________ **12.** The length of a rectangle is 3 times its width. The area is 75 square feet. Find the dimensions of the rectangle.

________ **13.** The Haskins plan to buy a rectangular lot that is 50 feet longer than it is wide. If the area of the lot is 30,000 square feet, find its dimensions.

________ **14.** The Perrys plan to fence a dog pen that is to be 8 feet longer than it is wide. They have enough fence to enclose an area of 240 square feet. Find the dimensions of the pen.

________ **15.** A sail is to be manufactured in the shape of a right triangle. If the hypotenuse must be 13 feet long and one leg is 7 feet more than the other leg, what must be the dimensions of the sail?

________ **16.** A flower bed is in the shape of a right triangle. One side is 2 feet shorter than the other side and the hypotenuse is 10 feet. Find how many feet of picket fence will be needed to border the flower bed.

_______ **17.** Find two consecutive integers whose product is 56.

_______ **18.** Find two consecutive integers whose product is 72.

_______ **19.** The sum of the squares of two consecutive integers is 61. Find the integers.

_______ **20.** The sum of the squares of two consecutive integers is 113. Find the integers.

_______ **21.** The product of two consecutive even integers is 4 more than five times the smaller. Find the integers.

_______ **22.** The product of two consecutive odd integers is 7 more than their sum. Find the integers.

☆ Stretching the Topics

Solve.

_______ **1.** A 2-inch square is to be cut out of each corner of a square piece of tin. The sides are then turned up to form a open box with a volume of 288 cubic inches. What must be the length of a side of the original square of tin?

_______ **2.** The total area of a circular cylinder is represented by $S = 2\pi r^2 + 2\pi rh$. Factor this expression and use the factored form to find the total area of a cylinder whose radius is 6 inches and whose height is 16 inches. Use $\pi = 3.14$.

_______ **3.** Find two consecutive even integers such that the square of the larger subtracted from 12 times the smaller is equal to their product.

Check your answers in the back of your book.

If you can do the problems in **Checkup 7.6**, you are ready to do the **Review Exercises for Chapter 7**.

Name ______________________ Date ____________

CHECKUP 7.6

Solve.

______ **1.** The length of a rectangle is 2 centimeters more than the width and the area is 35 centimeters. Find the dimensions.

______ **2.** The length of the Abell's living room is the same as the width. If it takes 225 square feet of carpet to cover the room, find the length of a side of the room.

______ **3.** The support beams of a carport form a right triangle. The hypotenuse is 25 feet and the one leg is 5 feet less than the other. Find the length of the piece of lumber that would be needed to cut all three.

______ **4.** One number is four more than another. The square of the smaller added to twice the larger is 32. Find the numbers.

______ **5.** The sum of the squares of two consecutive integers is 13. Find the integers.

Check your answers in the back of your book.

If You Missed Problems:	You Should Review Example:
1	1
2	2
3	3
4, 5	4

Summary

In this chapter we learned to rewrite a polynomial as a product of factors. Different types of **factoring** were used in different situations, and we agreed to look for certain patterns in our search for factors.

Type of Factoring	How to Recognize	Examples
Common factor	Look for a factor common to all terms. $AB + AC = A(B + C)$	$3x^2 - 6x^2 = 3x^2(x - 2)$ $x(a + b) - (a + b) = (a + b)(x - 1)$
By grouping	Look for four terms that can be rewritten as two pairs of terms, each pair containing a common factor.	$x^3 + x^2 + 2x + 2$ $= x^2(x + 1) + 2(x + 1)$ $= (x + 1)(x^2 + 2)$
Difference of two squares	Look for two square terms with a minus sign between them. $A^2 - B^2 = (A + B)(A - B)$	$4x^2 - 9 = (2x + 3)(2x - 3)$ $\frac{1}{100}x^2 - \frac{1}{9}y^2 = \left(\frac{1}{10}x + \frac{1}{3}y\right)\left(\frac{1}{10}x - \frac{1}{3}y\right)$
Perfect square trinomial	Look for trinomial with first term A^2, third term B^2, and middle term $2AB$. $A^2 + 2AB + B^2 = (A + B)^2$ $A^2 - 2AB + B^2 = (A - B)^2$	$x^2 - 16x + 64 = (x - 8)^2$ $\frac{1}{121}y^2 + \frac{2}{11}y + 1 = \left(\frac{1}{11}y + 1\right)^2$
General trinomial	Look for two binomials, by trial and error, that multiply to give the trinomial.	$x^2 - 4x - 21 = (x - 7)(x + 3)$ $6y^2 + 11xy - 10x^2 = (2y + 5x)(3y - 2x)$

Remember that the order in which factors are written is not important, but the factors themselves are unique. A polynomial can be factored completely in only one way, and you can always check your factors by multiplying. If a polynomial cannot be written as a product of factors, we say that it is **prime**.

In this chapter we also learned to solve **quadratic** or second-degree equations using the process of factoring and the **zero product rule**. Solving quadratic equations required the following procedure.

1. Work with parentheses if necessary.
2. Get zero on one side of the equation and all other terms on the other side, combining like terms if possible.
3. Factor the polynomial.
4. Set each factor equal to zero.
5. Solve for the variable and check solutions in the original equation.

❑ Speaking the Language of Algebra

1. The process of rewriting an expression as a product is called __________ .

2. In factoring, the first thing to look for is a __________ __________ .

3. A polynomial of the form $A^2 - B^2$ is called a __________ __________ __________ __________ .

4. A polynomial of the form $A^2 + 2AB + B^2$ or $A^2 - 2AB + B^2$ is called a __________ __________ __________ .

5. If a polynomial cannot be written as a product of factors, we say that it is __________ .

6. An equation in which the variable appears with an exponent of 2 is called a __________ equation or a __________ – __________ equation.

7. The rule which states that if a product of factors is zero, then at least one of the factors must be zero is called the __________ __________ rule.

Check your answers in the back of your book.

Name ______________________ Date __________

REVIEW EXERCISES for Chapter 7

Completely factor the following expressions.

_______ **1.** $10x - 15$

_______ **2.** $2y^3 - \frac{3}{5}y^2$

_______ **3.** $10x^2y + 4xy^2$

_______ **4.** $-3a^2b + 2a^3b^2 - 4a^4b^3$

_______ **5.** $4x + 4y + xz + yz$

_______ **6.** $25 - y^2$

_______ **7.** $3m^2 - 48n^2$

_______ **8.** $18x^4 - 8x^2y^2$

_______ **9.** $a^2 - 16ab + 64b^2$

_______ **10.** $x^2 - 8x + 15$

_______ **11.** $x^2y^2 - 3xy - 10$

_______ **12.** $4a^2 - 20ab + 2b^2$

_______ **13.** $6x^2 - 11x - 7$

_______ **14.** $-2x^2 + 4xy + 6y^2$

_______ **15.** $10m^2 + 15m - 220$

_______ **16.** $169 - 4a^2$

_______ **17.** $8x^2 - 11x + 3$

_______ **18.** $3y^4 + 63y^2 - 300$

_______ **19.** $x^3 - 4x^2 - x + 4$

Solve for the variable.

_______ **20.** $(x - 7)(x + 8) = 0$

_______ **21.** $x^2 - 169 = 0$

_______ **22.** $x^2 + 14x + 49 = 0$

_______ **23.** $m^2 + \frac{2}{3}m = 0$

_______ **24.** $162 - 2a^2 = 0$

_______ **25.** $z^2 + 17z + 52 = 0$

_______ **26.** $y^2 + 11y = -24$

_______ **27.** $y^2 = -2.3y$

_______ **28.** $x^2 + 63 = 24x$

_______ **29.** $5x^2 = 405$

_______ **30.** $5x(x - 2) - 18 = 4x^2 - 3(x - 4)$

_______ **31.** The length of a rectangular bathroom is 4 feet more than the width. If 60 square feet of linoleum will cover the floor, what are the dimensions of the bathroom?

_______ **32.** If the area of a square is 225 square centimeters, find the length of a side of the square.

_______ **33.** One integer is 4 less than another and the product of the two is 117. Find the integers.

_______ **34.** Hannah wants to border her antique shawl with lace. The shawl is in the shape of a right triangle with one side 1 foot longer than the other side, and the hypotenuse is 5 feet. How many feet of lace should she buy?

________ **35.** The product of two consecutive odd integers is 7 more than 4 times the larger. Find the integers.

Check your answers in the back of your book.

If You Missed Exercises:	You Should Review Examples:	
1–4	SECTION 7.1	1–9
5		10–17
6–8	SECTION 7.2	1–8
9	SECTION 7.3	13–16
10–12		1–8
13–19		9–12
20	SECTION 7.5	1–3
21–25		4–7
26–29		8–11
30		12–14
31–35	SECTION 7.6	1–5

If you have completed the **Review Exercises** and corrected your errors, you are ready to take the **Practice Test for Chapter 7**.

Name ______________________________ Date ______________

PRACTICE TEST for Chapter 7

Factor each expression completely.	Section	Examples
________ **1.** $-10x + 25$	7.1	1
________ **2.** $33x^2 - 77x$	7.1	4
________ **3.** $6y^5 + 3y^4 - 15y^2$	7.1	6
________ **4.** $-8a^3b + 14a^2b^2 - 8ab^3$	7.1	8
________ **5.** $ay - 3y + 5a - 15$	7.1	14
________ **6.** $16x^2 - 25a^2$	7.2	2
________ **7.** $0.36 - a^2b^2$	7.2	1
________ **8.** $7x^2 - 7$	7.2	3
________ **9.** $a^2 + 9ab + 20b^2$	7.3	1
________ **10.** $x^2 - 13x + 22$	7.3	2
________ **11.** $20 + 8x - x^2$	7.3	4
________ **12.** $20x^2 - 28x - 3$	7.3	10
________ **13.** $-3a^2b + 24ab^2 - 48b^3$	7.3	11, 12
________ **14.** $x^4 - 10x^2 + 9$	7.3	19
Solve for the variable.		
________ **15.** $8x(11 + x) = 0$	7.5	3
________ **16.** $x^2 + 1.4x = 0$	7.5	6
________ **17.** $3n^2 - 75 = 0$	7.5	5
________ **18.** $a^2 + 17a + 70 = 0$	7.5	7
________ **19.** $y^2 = \frac{25}{121}$	7.5	10
________ **20.** $x^2 = 8x - 16$	7.5	9, 7
________ **21.** $5y^2 - 10y = 40$	7.5	8
________ **22.** $5x(x - 2) - 12 = 4x^2 - 2(x - 4)$	7.5	15

________	**23.** The width of a rectangle is 6 meters less than the length. The area of the rectangle is 40 square meters. Find the length and the width.	7.6	1
________	**24.** The hypotenuse of a right triangle is 3 yards longer than twice one of the legs. If the remaining leg is 12 yards long, find the length of the hypotenuse.	7.6	3

Check your answers in the back of your book.

Name ______________________ Date __________

SHARPENING YOUR SKILLS after Chapters 1–7

Simplify. SECTION

	Problem	Section
______	**1.** $\left(\frac{-4}{5}\right)^3$	2.4
______	**2.** $(21.3)(0.72)$	1.3
______	**3.** $\frac{5}{9} - \left(\frac{2}{5} - \frac{11}{15}\right)$	2.4
______	**4.** $-\lvert 3(-7)\rvert + \lvert -15\rvert$	2.1
______	**5.** $(-6)^2(0.4)\left(\frac{-1}{2}k\right)$	3.3
______	**6.** $3[3x - (x - 4)] - [2(x + 2) - 5x]$	3.3
______	**7.** $(-2x^2)^3 \cdot (-xy^2)^2$	5.1
______	**8.** $(y^2 - 2yz + 2z^2) - 2(-y^2 + 3yz - z^2)$	6.1
______	**9.** $(5x^2y)(-3xy)(3x^4y^5)$	6.2
______	**10.** $-a^2b(5 - 2ab + 7a^2b^2)$	6.2
______	**11.** $(2a - 3b)(4a + 5b)$	6.3
______	**12.** $-y^2(y - 3)(y + 5)$	6.3

Find the quotients.

	Problem	Section
______	**13.** $\frac{15x^3y - 12x^2y^2 + 4xy^3}{-3xy}$	6.4
______	**14.** $(4x^2 - 17x - 14) \div (x - 5)$	6.4

Solve.

	Problem	Section
______	**15.** $\frac{2}{3}x - 15 = 9$	4.2
______	**16.** $11 = 5(x + 3) + 6$	4.2
______	**17.** $-0.5x + 3 = 0.1x - 21$	4.2
______	**18.** Evaluate $-4a^2b^3$ when $a = 2$ and $b = -3$.	5.1
______	**19.** If the width of a rectangle is 10 less than 3 times the length, write an expression for the area. Find the area if the length is 15 meters.	3.1
______	**20.** Solve $A = p + prt$ for t.	4.2

Check your answers in the back of your book.

Working with Rational Expressions 8

At the beginning of the semester the number of males in Harry's math class was 3 more than twice the number of females. After 5 males and 5 females had dropped the class, the ratio of females to males was 1 to 4. How many males and females were in Harry's math class at the beginning of the semester?

Before you complete this chapter, you will be able to solve this problem using a fractional equation. In Chapters 1 and 2 we learned to work with numerical fractions. Now that we have discussed the algebra of polynomials, we turn our attention to the algebra of fractions containing variables. In this chapter we learn to

1. Simplify rational algebraic expressions.
2. Multiply and divide rational algebraic expressions.
3. Add and subtract rational algebraic expressions.
4. Solve equations containing rational algebraic expressions.
5. Switch from word statements to equations containing rational algebraic expressions.

8.1 Simplifying Rational Algebraic Expressions

Recognizing Rational Algebraic Expressions

An algebraic expression that can be written as a fraction with a polynomial numerator and a polynomial denominator is called a **rational algebraic expression**. For example,

$$\frac{5}{2x} \qquad \frac{x^2}{7x} \qquad \frac{y+3}{y+5} \qquad \frac{a^2+5}{2a} \qquad \frac{x^2+7x+6}{x^2-1}$$

are all examples of rational algebraic expressions.

We can *evaluate* rational expressions by substituting some numerical value for the variable.

Example 1. Evaluate $\frac{y+3}{y-5}$ when $y = 0$.

Solution

$$\frac{y+3}{y-5}$$

$$= \frac{0+3}{0-5}$$

$$= \frac{3}{-5}$$

$$= \frac{-3}{5}$$

You complete Example 2.

Example 2. Evalaute $\frac{x+3}{5x}$ when $x = 2$.

Solution

$$\frac{x+3}{5x}$$

$$= \frac{2+3}{5 \cdot 2}$$

$$= \frac{\quad}{10}$$

$$= \underline{\qquad}$$

Did you remember to reduce your fraction?
Check your work on page 429. ▶

Notice that we chose to write our final fraction in Example 1 with the negative sign in the numerator rather than in the denominator. In mathematics it is customary to "let the numerator carry the sign" for the fraction.

Look at the following three fractions:

$$\frac{-6}{2} \qquad \frac{6}{-2} \qquad -\frac{6}{2}$$

$$= -3 \qquad = -3 \qquad = -3$$

Each of these fractions has the same value, but we prefer the first form.

$$\frac{-a}{b} = \frac{a}{-b} = -\frac{a}{b}$$

all represent the same fraction, but

$$\frac{-a}{b}$$

is the preferred form in this text.

Now you try to evaluate the fraction in Example 3.

Example 3. Evaluate $\frac{y+3}{y-5}$ when $y = -3$.

Solution

$$\frac{y+3}{y-5}$$

$$= \frac{-3+3}{-3-5}$$

$$= \frac{\quad}{-8}$$

$$= \underline{\quad\quad}$$

Check your work on page 429. ▶

Example 4. Evaluate $\frac{y+3}{y-5}$ when $y = 5$.

Solution

$$\frac{y+3}{y-5}$$

$$= \frac{5+3}{5-5}$$

$$= \frac{8}{0} \quad \text{is undefined}$$

Notice that our answer in Example 4 was undefined, because we learned in Chapters 1 and 2 that **division by zero is always undefined**. In working with rational algebraic expressions, we must always avoid substituting for the variable any number that will make the denominator equal zero. The process of finding the value of the variable that will make the denominator zero is called **finding the restrictions on the variable**.

Let's find the restrictions on the variable for the fraction $\frac{5x}{x-7}$. Since the denominator is zero when

$$x - 7 = 0$$

$$x = 7$$

the restriction is

$$x \neq 7$$

Example 5. Find the restrictions on the variable for $\frac{5x}{3x-2}$.

Solution: The denominator is zero when

$$3x - 2 = 0$$

$$3x = 2$$

$$x = \frac{2}{3}$$

so the restriction is

$$x \neq \frac{2}{3}$$

You complete Example 6.

Example 6. Find the restrictions on the variable for $\frac{x+3}{4x}$.

Solution: The denominator is zero when

$$4x = 0$$

$$x = \underline{\quad\quad}$$

so the restriction is

$$x \neq \underline{\quad\quad}$$

Check your work on page 429. ▶

Trial Run

Evaluate and reduce if possible.

________ **1.** $\frac{x+5}{6x}$ when $x = 3$

________ **2.** $\dfrac{y-5}{y+2}$ when $y = 0$

________ **3.** $\dfrac{a-4}{a^2-4}$ when $a = 4$

Find the restrictions on the variable.

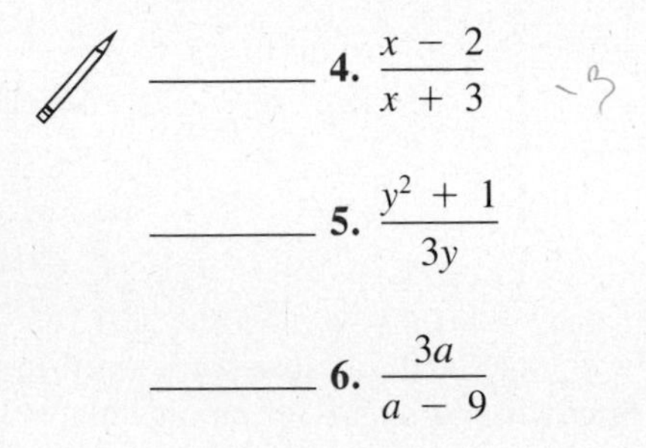

________ **4.** $\dfrac{x-2}{x+3}$

________ **5.** $\dfrac{y^2+1}{3y}$

________ **6.** $\dfrac{3a}{a-9}$

Answers are on page 430.

Reducing Rational Algebraic Expressions Containing Monomials

In arithmetic you learned to reduce rational numbers (or fractions) by dividing numerator and denominator by common factors. For example,

$$\frac{15}{21} = \frac{\overset{1}{\cancel{3}} \cdot 5}{\underset{1}{\cancel{3}} \cdot 7} = \frac{5}{7}$$

To reduce a rational algebraic expression with a monomial numerator and a monomial denominator, we may again look for common factors *or* we may use the Fourth Law of Exponents.

Original Fraction	Reducing by Common Factors	Reducing by Fourth Law of Exponents
$\dfrac{x^3}{x}$	$\dfrac{x^3}{x} = \dfrac{\overset{1}{\cancel{x}} \cdot x \cdot x}{\underset{1}{\cancel{x}}} = x^2$	$\dfrac{x^3}{x} = x^{3-1} = x^2$
$\dfrac{y^2}{y^5}$	$\dfrac{y^2}{y^5} = \dfrac{\overset{1}{\cancel{y}} \cdot \overset{1}{\cancel{y}}}{\underset{1}{\cancel{y}} \cdot \underset{1}{\cancel{y}} \cdot y \cdot y \cdot y} = \dfrac{1}{y^3}$	$\dfrac{y^2}{y^5} = \dfrac{1}{y^{5-2}} = \dfrac{1}{y^3}$
$\dfrac{6a^2x}{3ax}$	$\dfrac{6a^2x}{3ax} = \dfrac{\overset{1}{\cancel{3}} \cdot 2 \cdot \overset{1}{\cancel{a}} \cdot a \cdot \overset{1}{\cancel{x}}}{\underset{1}{\cancel{3}} \cdot \underset{1}{\cancel{a}} \cdot \underset{1}{\cancel{x}}} = 2a$	$\dfrac{6a^2x}{3ax} = \dfrac{6}{3} \cdot \dfrac{a^2}{a} \cdot \dfrac{x}{x}$ $= 2a^{2-1}x^{1-1} = 2a$

Either approach to reducing monomial fractions yields the correct result. In this chapter we shall most often use the method of reducing by dividing numerator and denominator by common factors.

Reducing Fractions

$$\frac{AC}{BC} = \frac{A}{B} \quad (B \neq 0,\ C \neq 0)$$

Example 7. Reduce $\frac{-3x^3}{12x^7}$.

Solution

$$\frac{-3x^3}{12x^7}$$

$$= \frac{\overset{-1}{\cancel{-3}}\overset{1}{\cancel{x^3}}}{\underset{4}{\cancel{12}}\underset{x^4}{\cancel{x^7}}}$$ Divide numerator and denominator by 3. Divide numerator and denominator by x^3.

$$= \frac{-1}{4x^4}$$ Simplify numerator and denominator.

Example 8. Reduce $\frac{22a^2bc^3}{55abc^5}$.

Solution

$$\frac{22a^2bc^3}{55abc^5}$$

$$= \frac{\overset{2}{\cancel{22}}\overset{a}{\cancel{a^2}}\overset{1}{\cancel{b}}\overset{1}{\cancel{c^3}}}{\underset{5}{\cancel{55}}\underset{1}{\cancel{a}}\underset{1}{\cancel{b}}\underset{c^2}{\cancel{c^5}}}$$ Divide numerator and denominator by common factors: 11, a, b, and c^3.

$$= \frac{2a}{5c^2}$$ Simplify numerator and denominator.

Trial Run

Reduce the fractions.

_______ 1. $\frac{-15x^7}{3x^2}$

_______ 2. $\frac{2x^4}{10x^5}$

_______ 3. $\frac{-3x^5y^3}{12x^4y^7}$

_______ 4. $\frac{64a^2b^5}{72a^3b^5}$

_______ 5. $\frac{-x^2y^3z}{x^9y^3z^5}$

_______ 6. $\frac{24a^7b^2c^3}{8a^2b^2c}$

Answers are on page 430.

Reducing Rational Algebraic Expressions Containing Polynomials

Factoring will continue to provide the key to reducing fractions as we consider fractions that contain polynomials in the numerator and denominator. Remember, we reduce fractions by dividing the numerator and denominator by *common factors*.

Let's try some examples, remembering to *factor the numerator and denominator completely before looking for common factors*.

Example 9. Reduce $\frac{2x + 2y}{2a - 4b}$.

Solution

$$\frac{2x + 2y}{2a - 4b}$$

$$= \frac{\overset{1}{\cancel{2}}(x + y)}{\underset{1}{\cancel{2}}(a - 2b)}$$

$$= \frac{x + y}{a - 2b}$$

You try Example 10.

Example 10. Reduce $\frac{-2x - 2y}{3x + 3y}$.

Solution

$$\frac{-2x - 2y}{3x + 3y}$$

$$= \frac{-2(x + y)}{3(x + y)}$$

$$=$$

Check your work on page 429. ▶

Example 11. Reduce $\frac{z^2 - 9}{2z + 6}$.

Solution

$$\frac{z^2 - 9}{2z + 6}$$

$$= \frac{\overset{1}{\cancel{(z + 3)}}(z - 3)}{2\underset{1}{\cancel{(z + 3)}}}$$

$$= \frac{z - 3}{2}$$

Now complete Example 12.

Example 12. Reduce $\frac{3x}{9x^2 - 15xy}$.

Solution

Check your work on page 429. ▶

Example 13. Reduce $\frac{3x^2 - 3xy - 6y^2}{x^3 - xy^2}$.

Solution

$$\frac{3x^2 - 3xy - 6y^2}{x^3 - xy^2}$$

$$= \frac{3(x^2 - xy - 2y^2)}{x(x^2 - y^2)}$$

$$= \frac{3(x - 2y)\overset{1}{\cancel{(x + y)}}}{x(x - y)\underset{1}{\cancel{(x + y)}}}$$

$$= \frac{3(x - 2y)}{x(x - y)}$$

You complete Example 14.

Example 14. Reduce $\frac{x^2 - 5x - 14}{x^2 - 14x + 49}$.

Solution

Check your work on page 429. ▶

Notice that we generally write our final fraction with the numerator and denominator in factored form to be sure that we have not overlooked a common factor.

Let's take a close look at the steps we could use to reduce thc fraction $\frac{x-5}{5-x}$.

$\frac{x-5}{5-x}$	Original fraction.
$=\frac{x-5}{-x+5}$	Rewrite numerator and denominator with terms in the same order.
$=\frac{x-5}{-1(x-5)}$	Factor -1 from the denominator.
$=\frac{\overset{1}{\cancel{x-5}}}{-1\underset{1}{\cancel{(x-5)}}}$	Divide numerator and denominator by the common factor.
$=\frac{1}{-1}$	
$=-1$	Simplify.

Similar reasoning can be used to reduce any fraction of the form $\frac{A-B}{B-A}$.

$\frac{A-B}{B-A}$	Original fraction.
$=\frac{A-B}{-A+B}$	Rewrite terms in same order.
$=\frac{A-B}{-1(A-B)}$	Factor -1 from denominator.
$=\frac{\overset{1}{\cancel{A-B}}}{-1\underset{1}{\cancel{(A-B)}}}$	Divide numerator and denominator by common factor $(A-B)$.
$=-1$	Simplify.

From now on, whenever you see the factors $A-B$ and $B-A$ in the numerator and denominator of a fraction, you will know that their quotient is -1.

$$\frac{A-B}{B-A}=-1 \qquad \text{provided that } B \neq A$$

Try to complete Example 15.

Example 15. Reduce $\frac{9 - 3x}{x - 3}$.

Solution

$$\frac{9 - 3x}{x - 3}$$

$$= \frac{3(3 - x)}{x - 3}$$

Check your work on page 430. ▶

Example 16. Reduce $\frac{36 - x^2}{x^2 - 3x - 18}$.

Solution

$$\frac{36 - x^2}{x^2 - 3x - 18}$$

$$= \frac{(6 + x)(6 - x)}{(x - 6)(x + 3)}$$

$$= \frac{(6 + x)\overset{-1}{\cancel{(6 - x)}}}{\underset{1}{\cancel{(x - 6)}}(x + 3)}$$

$$= \frac{-1(6 + x)}{x + 3}$$

$$\text{or} \quad \frac{-6 - x}{x + 3}$$

Trial Run

Reduce each fraction.

_______ 1. $\frac{4x + 4y}{8}$

_______ 2. $\frac{3x + 3y}{3x - 3y}$

_______ 3. $\frac{4x + 4y}{8x + 8y}$

_______ 4. $\frac{x^2 + 3x}{2x + 6}$

_______ 5. $\frac{x^2 - 5x + 6}{x^2 - 4}$

_______ 6. $\frac{a^2 - 16}{12 - 3a}$

Answers are on page 430.

By now you should agree that factoring is an important tool in our work with fractions. Once the numerators and denominators are in factored form, we may reduce or multiply or divide with ease. It is important that you recognize the difference between factors and terms.

Remember that *terms* are quantities that are being added together, but *factors* are quantities that are being multiplied together. We reduce fractions only by dividing numerators and denominators by common *factors*.

Cannot Be Reduced	Can Be Reduced
$\frac{x+5}{5}$	$\frac{5x}{5} = x$
$\frac{4+xy}{x}$	$\frac{4xy}{x} = 4y$
$\frac{x+2}{x+3}$	$\frac{2x}{3x} = \frac{2}{3}$

▶ Examples You Completed

Example 2. Evaluate $\frac{x+3}{5x}$ when $x = 2$.

Solution

$$\frac{x+3}{5x}$$

$$= \frac{2+3}{5(2)}$$

$$= \frac{5}{10}$$

$$= \frac{1}{2}$$

Example 3. Evaluate $\frac{y+3}{y-5}$ when $y = -3$.

Solution

$$\frac{y+3}{y-5}$$

$$= \frac{-3+3}{-3-5}$$

$$= \frac{0}{-8}$$

$$= 0$$

Example 6. Find the restrictions on the variable for $\frac{x+3}{4x}$.

Solution: The denominator is zero when

$$4x = 0$$

$$x = 0$$

so the restriction is

$$x \neq 0$$

Example 10. Reduce $\frac{-2x-2y}{3x+3y}$.

Solution

$$\frac{-2x-2y}{3x+3y}$$

$$= \frac{-2\overset{1}{\cancel{(x+y)}}}{3\underset{1}{\cancel{(x+y)}}}$$

$$= \frac{-2}{3}$$

Example 12. Reduce $\frac{3x}{9x^2-15xy}$.

Solution

$$\frac{3x}{9x^2-15xy}$$

$$= \frac{\overset{1}{\cancel{3x}}}{\underset{1}{\cancel{3x}}(3x-5y)}$$

$$= \frac{1}{3x-5y}$$

Example 14. Reduce $\frac{x^2-5x-14}{x^2-14x+49}$.

Solution

$$\frac{x^2-5x-14}{x^2-14x+49}$$

$$= \frac{\overset{1}{\cancel{(x-7)}}(x+2)}{\underset{1}{\cancel{(x-7)}}(x-7)}$$

$$= \frac{x+2}{x-7}$$

Example 15. Reduce $\frac{9 - 3x}{x - 3}$.

Solution

$$\frac{9 - 3x}{x - 3}$$

$$= \frac{3\overset{-1}{\cancel{(3 - x)}}}{\underset{1}{\cancel{x - 3}}}$$

$$= 3(-1)$$

$$= -3$$

Answers to Trial Runs

page 423 **1.** $\frac{4}{9}$ **2.** $\frac{-5}{2}$ **3.** 0 **4.** $x \neq -3$ **5.** $y \neq 0$ **6.** $a \neq 9$

page 425 **1.** $-5x^5$ **2.** $\frac{1}{5x}$ **3.** $\frac{-x}{4y^4}$ **4.** $\frac{8}{9a}$ **5.** $\frac{-1}{x^7z^4}$ **6.** $3a^5c^2$

page 428 **1.** $\frac{x + y}{2}$ **2.** $\frac{x + y}{x - y}$ **3.** $\frac{1}{2}$ **4.** $\frac{x}{2}$ **5.** $\frac{x - 3}{x + 2}$ **6.** $\frac{-(a + 4)}{3}$

Name ______________________ Date __________

EXERCISE SET 8.1

Find the restrictions on the variable.

_______ 1. $\frac{6}{x^2}$

_______ 2. $\frac{-4}{x^2}$

_______ 3. $\frac{x^2}{x+1}$

_______ 4. $\frac{x^2}{x+2}$

_______ 5. $\frac{2(x+2)}{3x}$

_______ 6. $\frac{5(x-1)}{4x}$

_______ 7. $\frac{y-1}{8-y}$

_______ 8. $\frac{y-3}{5-y}$

_______ 9. $\frac{4x}{2x-1}$

_______ 10. $\frac{3x}{3x-2}$

Reduce the fractions.

_______ 11. $\frac{8x^5}{4x^2}$

_______ 12. $\frac{9x^7}{3x^3}$

_______ 13. $\frac{-5a^6}{20a^5}$

_______ 14. $\frac{-4a^7}{12a^6}$

_______ 15. $\frac{6x^5y^2}{2x^3y^4}$

_______ 16. $\frac{21x^3y}{7x^2y^6}$

_______ 17. $\frac{-15ab^4}{20a^3b^2}$

_______ 18. $\frac{-18a^2b}{27a^8b^2}$

_______ 19. $\frac{25x^2yz}{20x^4yz^2}$

_______ 20. $\frac{28x^3yz^2}{35x^5y^4z^2}$

_______ 21. $\frac{2x+2y}{6}$

_______ 22. $\frac{4x-4y}{12}$

_______ 23. $\frac{3a+15b}{3}$

_______ 24. $\frac{7a+21b}{7}$

_______ 25. $\frac{18xy-14x}{2x}$

_______ 26. $\frac{12xy-6x}{3x}$

_______ 27. $\frac{8x}{64x+4x^2}$

_______ 28. $\frac{15x}{27x+3x^2}$

_______ 29. $\frac{36a^2+36b^2}{30}$

_______ 30. $\frac{21a^2+21b^2}{49}$

________ 31. $\dfrac{-25x^2 - 25y^2}{-5}$

________ 32. $\dfrac{-24x^2 - 24y^2}{-8}$

________ 33. $\dfrac{49x^3}{7x^2 + 7x}$

________ 34. $\dfrac{81x^4}{9x^2 + 9x}$

________ 35. $\dfrac{-12a^2b - 15abc}{18ab}$

________ 36. $\dfrac{-10a^2b - 18abc}{22ab}$

________ 37. $\dfrac{8x^2 + 12x + 4}{40}$

________ 38. $\dfrac{15x^2 + 12x + 9}{39}$

________ 39. $\dfrac{4a + 4b}{4x + 4y}$

________ 40. $\dfrac{3a + 3b}{3x + 3y}$

________ 41. $\dfrac{3x - 3y}{6x + 6y}$

________ 42. $\dfrac{4x - 4y}{12x + 12y}$

________ 43. $\dfrac{9x + 9y}{18x + 18y}$

________ 44. $\dfrac{5x + 5y}{15x + 15y}$

________ 45. $\dfrac{-4a - 4b}{12a + 12b}$

________ 46. $\dfrac{-3a - 3b}{15a + 15b}$

________ 47. $\dfrac{x^2 - y^2}{3x + 3y}$

________ 48. $\dfrac{x^2 - y^2}{8x - 8y}$

________ 49. $\dfrac{x^2 - 49}{4x + 28}$

________ 50. $\dfrac{x^2 - 25}{4x + 20}$

________ 51. $\dfrac{15x - 15y}{5x^2 - 5y^2}$

________ 52. $\dfrac{12x - 12y}{3x^2 - 3y^2}$

________ 53. $\dfrac{-7a - 7b}{3a^2 - 3b^2}$

________ 54. $\dfrac{-5a - 5b}{2a^2 - 2b^2}$

________ 55. $\dfrac{x^2 - 4y^2}{3x + 6y}$

________ 56. $\dfrac{x^2 - 9y^2}{2x + 6y}$

________ 57. $\dfrac{x^2 + 9x + 20}{x^2 - 16}$

________ 58. $\dfrac{x^2 + 7x + 12}{x^2 - 9}$

________ 59. $\dfrac{x^2 + 6x + 8}{x^2 + 7x + 10}$

________ 60. $\dfrac{x^2 + 8x + 15}{x^2 + 5x + 6}$

________ 61. $\dfrac{25x^2 - 225}{5x^2 - 50x + 105}$

________ 62. $\dfrac{16x^2 - 144}{4x^2 - 20x + 24}$

________ 63. $\dfrac{3x^2 + 14xy - 5y^2}{3x^2 + 2xy - y^2}$

________ 64. $\dfrac{6x^2 + xy - y^2}{2x^2 - xy - y^2}$

______ **65.** $\dfrac{4a^2 - 8ab - 12b^2}{2a^2 - 12ab + 18b^2}$

______ **66.** $\dfrac{5a^2 - 4ab - b^2}{3a^2 - 6ab + 3b^2}$

______ **67.** $\dfrac{x^2 - 25}{x^2 + 25}$

______ **68.** $\dfrac{4x^2 - 9}{4x^2 + 9}$

______ **69.** $\dfrac{2b - 1}{4b^2 - 4b + 1}$

______ **70.** $\dfrac{3b + 1}{9b^2 + 6b + 1}$

______ **71.** $\dfrac{x^4 - 16y^4}{x^2 + 2xy - 8y^2}$

______ **72.** $\dfrac{x^4 - 81y^4}{x^2 + 6xy - 27y^2}$

______ **73.** $\dfrac{a - b}{9b - 9a}$

______ **74.** $\dfrac{2a - b}{5b - 10a}$

______ **75.** $\dfrac{x^2 - 25}{5 - x}$

______ **76.** $\dfrac{x^2 - 49}{7 - x}$

☆ Stretching the Topics

______ **1.** Find the restrictions on the variable for the expression $\dfrac{2m^2 - 5m - 12}{2m^3 + 11m^2 + 12m}$.

______ **2.** Reduce $\dfrac{6a^2b - 12abc + 6bc^2}{9c^2 - 9a^2}$.

______ **3.** Reduce $\dfrac{6x^n(b + c)^n}{3x^{n-1}(b + c)}$.

Check your answers in the back of your book.

If you can do **Checkup 8.1,** you are ready to go on to Section 8.2.

Name ______________________________ Date ____________

✓ CHECKUP 8.1

Find the restrictions on the variable.

______ 1. $\frac{x + 5}{3x}$

______ 2. $\frac{x + 4}{2x - 1}$

Reduce the fractions.

______ 3. $\frac{x^{12}y^5}{x^9y^7}$

______ 4. $\frac{-5x^4}{20x^5}$

______ 5. $\frac{16xy - 20x}{4x}$

______ 6. $\frac{6xy}{30x^2y - 24xy^2}$

______ 7. $\frac{3a + 3b}{12a - 12b}$

______ 8. $\frac{5y - 5x}{15x^2 - 15y^2}$

______ 9. $\frac{m^2 - 4n^2}{2m - 4n}$

______ 10. $\frac{x^2 + 6x + 8}{x^2 + 7x + 10}$

Check your answers in the back of your book.

If You Missed Problems:	You Should Review Examples:
1, 2	5, 6
3, 4	7, 8
5, 6	12
7	9, 10
8	15, 16
9, 10	14

8.2 Multiplying and Dividing Rational Algebraic Expressions

Multiplying and Dividing Monomial Fractions

Recall that the product of fractions is found by multiplying their numerators and multiplying their denominators and then reducing. We continue to use this method to multiply monomial fractions and then reduce, first reducing the numerical part and then reducing the variable parts.

Example 1. Find $\frac{-6x^2y}{5z} \cdot \frac{15xz^2}{2y^4}$.

Solution

$$\frac{-6x^2y}{5z} \cdot \frac{15xz^2}{2y^4}$$

$$= \frac{-6 \cdot 15 \cdot x^2 \cdot x \cdot y \cdot z^2}{5 \cdot 2 \cdot y^4 \cdot z}$$ Multiply numerators. Multiply denominators.

$$= \frac{-\overset{-3}{\cancel{6}} \cdot \overset{3}{\cancel{15}} \cdot x^3 \cdot \overset{1}{\cancel{y}} \cdot \overset{z}{\cancel{z^2}}}{\underset{1}{\cancel{5}} \cdot \underset{1}{\cancel{2}} \cdot \underset{y^3}{\cancel{y^4}} \cdot \underset{1}{\cancel{z}}}$$ Divide numerator and denominator by common factors.

$$= \frac{-9x^3z}{y^3}$$

As was true with numerical fractions, we may reduce *before* multiplying rational algebraic expressions. This is done by dividing any one numerator and any one denominator by the same factor. Then we multiply the factors remaining in the numerators and denominators.

Example 2. Find $\frac{7a^2x}{3b} \cdot \frac{9b^3}{14ax}$.

Solution

$$\frac{7a^2x}{3b} \cdot \frac{9b^3}{14ax}$$

$$= \frac{\overset{1}{\cancel{7}}\overset{a}{\cancel{a^2}}\overset{1}{\cancel{x}}}{\underset{1}{\cancel{3}}\underset{1}{\cancel{b}}} \cdot \frac{\overset{3}{\cancel{9}}\overset{b^2}{\cancel{b^3}}}{\underset{2}{\cancel{14}}\underset{1}{\cancel{a}}\underset{1}{\cancel{x}}}$$ Divide numerator and denominator by common factors.

$$= \frac{3ab^2}{2}$$ Multiply remaining factors.

Since we know that division of fractions requires multiplication by the reciprocal of the divisor, we may use our multiplication skills to divide monomial fractions. Just remember to *invert the divisor and multiply*.

Dividing Fractions

$$\frac{A}{B} \div \frac{C}{D} = \frac{A}{B} \cdot \frac{D}{C} \qquad (B, C, D \neq 0)$$

Example 3. Find $\dfrac{6x^4}{5y} \div \dfrac{9x^3}{10y^4}$.

Solution

$$\frac{6x^4}{5y} \div \frac{9x^3}{10y^4}$$

$$= \frac{6x^4}{5y} \cdot \frac{10y^4}{9x^3}$$ Invert divisor and multiply.

$$= \frac{\overset{2\,x}{\cancel{6}\cancel{x^4}}}{\underset{1\,1}{\cancel{5}\cancel{y}}} \cdot \frac{\overset{2\,y^3}{\cancel{10}\cancel{y^4}}}{\underset{3\,1}{\cancel{9}\cancel{x^3}}}$$ Reduce.

$$= \frac{4xy^3}{3}$$ Multiply remaining factors.

Now you try Example 4.

Example 4. Find $\dfrac{-7x^2y^4}{8z} \div \dfrac{5x^5y}{12z}$.

Solution

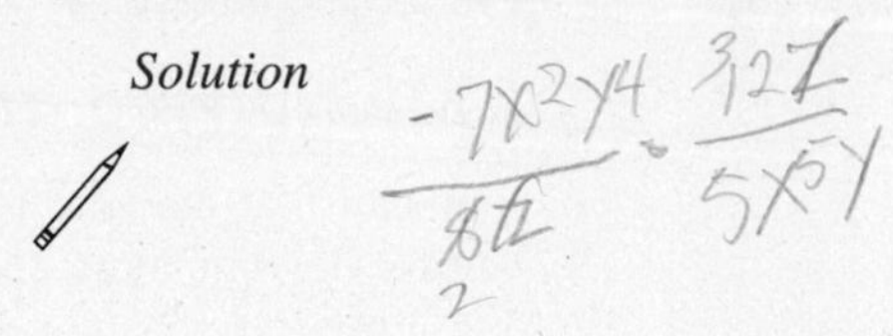

Check your work on page 441. ▶

Trial Run

Simplify.

______ 1. $\dfrac{-8x^3y}{3z} \cdot \dfrac{9xz^3}{4y^4}$

______ 2. $\dfrac{a^2bc^3}{7} \cdot \dfrac{14}{a^4c^5}$

______ 3. $\dfrac{9x^2y}{5xz^3} \cdot \dfrac{25xz^2}{3xy} \cdot \dfrac{xy^2}{15}$

______ 4. $\dfrac{4x}{y} \div \dfrac{3y}{-x}$

______ 5. $\dfrac{12a^3x^5}{5y^2} \div \dfrac{16x^5}{10y^2}$

Answers are on page 441.

Multiplying and Dividing Polynomial Fractions

The methods we have already learned are all that we need to multiply fractions with polynomial numerators and denominators.

> **Multiplying Fractions**
>
> 1. Factor all numerators and denominators.
> 2. Reduce, by dividing out any factor common to a numerator and a denominator.
> 3. Multiply remaining numerator factors and multiply remaining denominator factors.

Example 5. Find $\dfrac{x^2 - 4}{5x + 5} \cdot \dfrac{x + 1}{x + 2}$.

Solution

$$\frac{x^2 - 4}{5x + 5} \cdot \frac{x + 1}{x + 2}$$

$$= \frac{\overset{1}{\cancel{(x + 2)}}(x - 2)}{5\underset{1}{\cancel{(x + 1)}}} \cdot \frac{\overset{1}{\cancel{x + 1}}}{\underset{1}{\cancel{x + 2}}}$$

$$= \frac{x - 2}{5}$$

Now you try to complete Example 6.

Example 6. Find $\dfrac{x^2 + 7x + 10}{x^2 + 2x} \cdot \dfrac{x^3}{4x + 20}$.

Solution

$$\frac{x^2 + 7x + 10}{x^2 + 2x} \cdot \frac{x^3}{4x + 20}$$

$$= \frac{(x + 5)(x + 2)}{x(x + 2)} \cdot \frac{x^3}{4(x + 5)}$$

Check your work on page 441. ▶

Example 7. Find $\dfrac{5x + 5y}{6a - 3x} \cdot \dfrac{6x - 12a}{10x + 10y}$.

Solution

$$\frac{5x + 5y}{6a - 3x} \cdot \frac{6x - 12a}{10x + 10y}$$

$$= \frac{\overset{1}{\cancel{5}}\overset{1}{\cancel{(x + y)}}}{\underset{1}{\cancel{3}}\underset{1}{\cancel{(2a - x)}}} \cdot \frac{\overset{2}{\cancel{6}}\overset{-1}{\cancel{(x - 2a)}}}{\underset{2}{\cancel{10}}\underset{1}{\cancel{(x + y)}}}$$

$$= \frac{-2}{2}$$

$$= -1$$

Now try Example 8.

Example 8. Find $\dfrac{x^2 - 2x - 3}{x^2 - 1} \cdot \dfrac{2x^2 + 5x - 3}{x^2 - 9}$.

Solution

Check your work on page 441. ▶

Once again, we declare that if we can multiply polynomial fractions, we can surely divide them. We simply must remember to *invert the divisor and multiply.*

Example 9. Find $\frac{x^2 - 25}{5} \div (x - 5)$.

Solution

$$\frac{x^2 - 25}{5} \div (x - 5).$$

$$= \frac{x^2 - 25}{5} \div \frac{x - 5}{1}$$

$$= \frac{x^2 - 25}{5} \cdot \frac{1}{x - 5}$$

$$= \frac{(x + 5)\overset{1}{\cancel{(x - 5)}}}{5} \cdot \frac{1}{\underset{1}{\cancel{x - 5}}}$$

$$= \frac{x + 5}{5}$$

Now complete Example 10.

Example 10. Find $\frac{x^2 + 12x + 36}{5x^3 + 30x^2} \div \frac{7x + 42}{10x^4}$.

Solution

$$\frac{x^2 + 12x + 36}{5x^3 + 30x^2} \div \frac{7x + 42}{10x^4}$$

$$= \frac{x^2 + 12x + 36}{5x^3 + 30x^2} \cdot \frac{10x^4}{7x + 42}$$

$$= \frac{(x + 6)(\quad\quad)}{5x^2(\quad\quad)} \cdot \frac{10x^4}{7(\quad\quad)}$$

Check your work on page 441. ▶

Trial Run

Simplify.

_______ 1. $\frac{2x + 2}{x + 3} \cdot \frac{1}{x + 1}$

_______ 2. $\frac{x^2 + 5x + 6}{x^2 + 3x} \cdot \frac{x^3}{3x + 6}$

_______ 3. $\frac{3x + 3y}{4a - 8b} \cdot \frac{8a - 16b}{12x + 12y}$

_______ 4. $\frac{x^2 - 3x + 2}{x^2 - 1} \cdot \frac{2x^2 + x - 1}{x^2 - 4x + 4}$

_______ 5. $\frac{16 - x^2}{3} \div \frac{x^2 - 8x + 16}{3x - 12}$

_______ 6. $\frac{x^2 - 2x - 35}{3x^2 + 15x} \div \frac{4x - 28}{9x^3}$

Answers are on page 441.

▶ Examples You Completed

Example 4. $\frac{-7x^2y^4}{8z} \div \frac{5x^5y}{12z}$.

Solution

$$\frac{-7x^2y^4}{8z} \div \frac{5x^5y}{12z}$$

$$= \frac{-7x^2y^4}{8z} \cdot \frac{12z}{5x^5y}$$

$$= \frac{-7\overset{1}{\cancel{x^2}}\overset{y^3}{\cancel{y^4}}}{\underset{2}{\cancel{8}}\underset{1}{\cancel{z}}} \cdot \frac{\overset{3}{\cancel{12}}\overset{1}{\cancel{z}}}{5\underset{x^3}{\cancel{x^5}}\underset{1}{\cancel{y}}}$$

$$= \frac{-21y^3}{10x^3}$$

Example 6. Find $\frac{x^2 + 7x + 10}{x^3 + 2x} \cdot \frac{x^3}{4x + 20}$.

Solution

$$\frac{x^2 + 7x + 10}{x^2 + 2x} \cdot \frac{x^3}{4x + 20}$$

$$= \frac{\overset{1}{\cancel{(x + 5)}}\overset{1}{\cancel{(x + 2)}}}{\underset{1}{\cancel{x}}\underset{1}{\cancel{(x + 2)}}} \cdot \frac{\overset{x^2}{\cancel{x^3}}}{4\underset{1}{\cancel{(x + 5)}}}$$

$$= \frac{x^2}{4}$$

Example 8. Find $\frac{x^2 - 2x - 3}{x^2 - 1} \cdot \frac{2x^2 + 5x - 3}{x^2 - 9}$.

Solution

$$\frac{x^2 - 2x - 3}{x^2 - 1} \cdot \frac{2x^2 + 5x - 3}{x^2 - 9}$$

$$= \frac{\overset{1}{\cancel{(x - 3)}}\overset{1}{\cancel{(x + 1)}}}{\underset{1}{\cancel{(x + 1)}}(x - 1)} \cdot \frac{(2x - 1)\overset{1}{\cancel{(x + 3)}}}{\underset{1}{\cancel{(x - 3)}}\underset{1}{\cancel{(x + 3)}}}$$

$$= \frac{2x - 1}{x - 1}$$

Example 10. Find $\frac{x^2 + 12x + 36}{5x^3 + 30x^2} \div \frac{7x + 42}{10x^4}$.

Solution

$$\frac{x^2 + 12x + 36}{5x^3 + 30x^2} \div \frac{7x + 42}{10x^4}$$

$$= \frac{x^2 + 12x + 36}{5x^3 + 30x^2} \cdot \frac{10x^4}{7x + 42}$$

$$= \frac{\overset{1}{\cancel{(x + 6)}}\overset{1}{\cancel{(x + 6)}}}{\underset{1}{\cancel{5x^2}}\underset{1}{\cancel{(x + 6)}}} \cdot \frac{\overset{2\,x^2}{\cancel{10x^4}}}{7\underset{1}{\cancel{(x + 6)}}}$$

$$= \frac{2x^2}{7}$$

Answers to Trial Runs

page 438 **1.** $\frac{-6x^4z^2}{y^3}$ **2.** $\frac{2b}{a^2c^2}$ **3.** $\frac{x^2y^2}{z}$ **4.** $\frac{-4x^2}{3y^2}$ **5.** $\frac{3a^3}{2}$

page 440 **1.** $\frac{2}{x + 3}$ **2.** $\frac{x^2}{3}$ **3.** $\frac{1}{2}$ **4.** $\frac{2x - 1}{x - 2}$ **5.** $-(4 + x)$ or $-4 - x$ **6.** $\frac{3x^2}{4}$

Name ______________________ Date ____________

EXERCISE SET 8.2

Simplify.

______ 1. $\frac{x}{y} \cdot \frac{1}{y}$

______ 2. $\frac{-x}{y} \cdot \frac{1}{3}$

______ 3. $\frac{3xy}{4z} \cdot \frac{16z}{12y}$

______ 4. $\frac{5xy}{9z} \cdot \frac{27z}{15y}$

______ 5. $\frac{3x}{2y} \cdot \frac{10y^2}{x^2} \cdot \frac{7}{x}$

______ 6. $\frac{5x}{3y} \cdot \frac{21y^2}{x^2} \cdot \frac{2}{x}$

______ 7. $\frac{1}{a} \div \frac{1}{ab}$

______ 8. $\frac{1}{a} \div \frac{1}{a^2}$

______ 9. $\frac{x}{3} \div -3$

______ 10. $\frac{x}{5} \div -15$

______ 11. $\frac{x^2y}{2z} \div \frac{xy}{8z^2}$

______ 12. $\frac{x^3y}{3z^2} \div \frac{x^2y}{18z}$

______ 13 $\frac{4ab^3}{c} \div \frac{4a^3}{5b^2c}$

______ 14. $\frac{9a^3b^2}{c} \div \frac{9a}{7b^3c}$

______ 15. $\frac{4y^3}{5x} \div \frac{2xy}{45}$

______ 16. $\frac{9y^3}{7x} \div \frac{3xy}{28}$

______ 17. $\frac{34b^3}{81a^3} \div \frac{17a^2}{18a^4b}$

______ 18. $\frac{38b^3}{25a^4} \div \frac{19a}{10a^3b}$

______ 19. $\frac{15x^2}{16y^3} \div \frac{9y^8}{8x^4}$

______ 20. $\frac{12x^3}{49y^3} \div \frac{16y^9}{21x^3}$

______ 21. $\frac{2x + 2}{x + 3} \cdot \frac{1}{x + 1}$

______ 22. $\frac{3x + 3}{x + 4} \cdot \frac{1}{x + 1}$

______ 23. $\frac{3x + 6}{x^2} \cdot \frac{x}{8x + 16}$

______ 24. $\frac{5x + 10}{x^3} \cdot \frac{x^4}{15x + 30}$

______ 25. $\frac{2a + 6b}{a^2b} \cdot \frac{ab^2}{a + 3b}$

______ 26. $\frac{7a + 14b}{a^3b^2} \cdot \frac{a^2b^3}{a + 2b}$

______ 27. $\frac{m^2 + n^2}{6} \cdot \frac{42}{3m^2 + 3n^2}$

______ 28. $\frac{m^2 + n^2}{19} \cdot \frac{38}{5m^2 + 5n^2}$

______ 29. $\frac{x^2 - 9y^2}{x^2 + 9y^2} \cdot \frac{x^2 + 9y^2}{x + 3y}$

______ 30. $\frac{x^2 - 25y^2}{x^2 + 25y^2} \cdot \frac{x^2 + 25y^2}{x + 5y}$

______ 31. $\frac{x^2 - 4}{3x - 9} \cdot \frac{6 - 2x}{x^2 + 4x + 4}$

______ 32. $\frac{x^2 - 9}{2x - 4} \cdot \frac{6 - 3x}{x^2 - 6x + 9}$

_______ 33. $\frac{(x+3)^2}{64yz} \cdot \frac{40y^2z}{x^2-9}$

_______ 34. $\frac{(x+5)^2}{81yz^3} \cdot \frac{72y^5z}{x^2-25}$

_______ 35. $\frac{2x^2+3x+1}{2x^2+5x+3} \cdot \frac{2x^2+11x+12}{2x^2+13x+6}$

_______ 36. $\frac{20x^2-7x-3}{4x^2-7x-2} \cdot \frac{3x^2-5x-2}{15x^2-4x-3}$

_______ 37. $\frac{6x^2+xy-y^2}{2x^2-9xy-5y^2} \cdot \frac{x^2-25y^2}{15x-5y}$

_______ 38. $\frac{8x^2+18xy-5y^2}{16x^2-8xy+y^2} \cdot \frac{4x^2+3xy-y^2}{12x+30y}$

_______ 39. $\frac{x^2-49}{y} \div \frac{x+7}{y^2}$

_______ 40. $\frac{x^2-64}{y^4} \div \frac{x+8}{y^3}$

_______ 41. $\frac{x^2+7x+10}{63x} \div \frac{7x+14}{9x}$

_______ 42. $\frac{x^2-12x+32}{60x} \div \frac{5x-20}{12x}$

_______ 43. $\frac{x+2}{4x-8} \div \frac{x^2-7x-18}{2x-4}$

_______ 44. $\frac{x+7}{8x-56} \div \frac{x^2+4x-21}{4x-28}$

_______ 45. $\frac{2-x}{3} \div \frac{(x-2)^2}{9}$

_______ 46. $\frac{5-x}{4} \div \frac{(x-5)^2}{16}$

_______ 47. $\frac{x^2-8x+15}{x^2-25} \div \frac{x^2-9}{x+5}$

_______ 48. $\frac{x^2-7x-18}{x^2-81} \div \frac{x^2+4x+4}{x+9}$

_______ 49. $\frac{a^2-2ab+b^2}{10a^5b^5} \div \frac{a-b}{50a^3b^3}$

_______ 50. $\frac{a^2-6ab+9b^2}{14a^4b^7} \div \frac{a-3b}{70a^3b^2}$

_______ 51. $\frac{xy-y^2}{x^2-2xy+y^2} \div \frac{5xy-3y^2}{5x^2-8xy+3y^2}$

_______ 52. $\frac{x^2-7xy}{x^2-14xy+49y^2} \div \frac{3x^2+2xy}{3x^2-19xy-14y^2}$

☆ Stretching the Topics

Simplify.

_______ 1. $\frac{-(3x^3y)^2}{10x} \cdot \frac{(-3x^2)^3}{(2x^2y)^2}$

_______ 2. $\frac{a^2-b^2}{a^2} \cdot \frac{a^2b-b^3+a^2c-cb^2}{ab+ac}$

_______ 3. $\frac{5x^2+10xy}{3xy-3y^2} \cdot \frac{2x-3y}{x^2+4xy+4y^2} \div \frac{4x^2-9y^2}{4x^2-4y^2}$

Check your answers in the back of your book.

If you can simplify the products and quotients in **Checkup 8.2**, you are ready to go on to Section 8.3.

Name ______________________ Date ____________

CHECKUP 8.2

Simplify.

______ 1. $\dfrac{-8x^3}{3y} \cdot \dfrac{21xy^2}{4yz}$

______ 2. $\dfrac{15a^3b}{4c} \div \dfrac{21b^2}{8c^3}$

______ 3. $\dfrac{-9x^3y}{16z} \div \dfrac{7x^5y^2}{12z^3}$

______ 4. $\dfrac{2x + 8y}{xy^2} \cdot \dfrac{x^2y}{x^2 + 4xy}$

______ 5. $\dfrac{x^2 - 9}{3x - 12} \cdot \dfrac{2x - 8}{x^2 + 6x + 9}$

______ 6. $\dfrac{3}{y - x} \cdot \dfrac{4x - 4y}{12}$

______ 7. $\dfrac{x^2 - 64}{y^2} \div \dfrac{x + 8}{y}$

______ 8. $\dfrac{6x + 18}{54x} \div \dfrac{x^2 + 9x + 18}{3x}$

______ 9. $\dfrac{x^2 - 121}{x^2 - 5x - 24} \cdot \dfrac{x^2 - 16x + 64}{x^2 + 14x + 33}$

______ 10. $\dfrac{n^2 + 3n - 18}{n^2 - 36} \div \dfrac{2n^2 - 5n - 3}{4n^2 - 1}$

Check your answers in the back of your book.

If You Missed Problems:	You Should Review Examples:
1	1, 2
2, 3	3, 4
4–6	5–7
7, 8	9, 10
9, 10	8, 10

8.3 Building Rational Algebraic Expressions

We spent time in Section 7.1 reducing algebraic fractions to simplest form by dividing numerator and denominator by common factors. For the work that lies ahead, we must review the methods for **building fractions**.

Building Monomial Fractions

We learned to build numerical fractions in Chapter 1. The same method works when we wish to build fractions containing monomials.

$$\frac{5x}{3y} = \frac{?}{12y^4}$$ We must find a numerator for the fraction on the right.

$$\frac{5x}{3y} = \frac{?}{3y(4y^3)}$$ By what factor was the old denominator ($3y$) multiplied to arrive at $12y^4$? The factor must be $4y^3$.

$$\frac{5x}{3y} = \frac{5x(4y^3)}{3y(4y^3)}$$ If we multiplied $3y$ by $4y^3$, we must *also* multiply $5x$ by $4y^3$ (because $\frac{4y^3}{4y^3} = 1$).

$$\frac{5x}{3y} = \frac{20xy^3}{12y^4}$$ We have found our new fraction.

To build an old fraction to a new fraction with a certain denominator, we see that we must use these steps.

Building Fractions

1. Decide by what factor the old denominator was multiplied to arrive at the new denominator.
2. Then multiply the old numerator by the same factor to arrive at the new numerator.

Example 1. Find the new fraction.

$$\frac{5}{xy} = \frac{?}{3x^2y^3}$$

Solution

$$\frac{5}{xy} = \frac{?}{3x^2y^3}$$

$$\frac{5}{xy} = \frac{?}{(xy)(3xy^2)}$$

$$\frac{5}{xy} = \frac{5(3xy^2)}{(xy)(3xy^2)}$$

$$\frac{5}{xy} = \frac{15xy^2}{3x^2y^3}$$

Now complete Example 2.

Example 2. Find the new fraction.

$$\frac{-7a}{2bc} = \frac{?}{6b^2c^2}$$

Solution

$$\frac{-7a}{2bc} = \frac{?}{6b^2c^2}$$

$$\frac{-7a}{2bc} = \frac{?}{2bc(3bc)}$$

$$\frac{-7a}{2bc} = \frac{-7a(\_\_\_)}{2bc(3bc)}$$

$$= \frac{\_\_\_}{6b^2c^2}$$

Check your work on page 450. ▶

Trial Run

Write the new fraction.

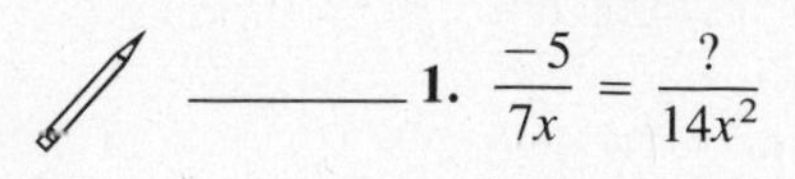

_______ 1. $\dfrac{-5}{7x} = \dfrac{?}{14x^2}$

_______ 2. $\dfrac{8x}{5y} = \dfrac{?}{15y^4}$

_______ 3. $\dfrac{2a}{9bc} = \dfrac{?}{27abc}$

_______ 4. $\dfrac{ax}{by} = \dfrac{?}{a^2b^2y}$

Answers are on page 450.

Building Polynomial Fractions

Building fractions containing polynomials requires a bit more care because the building factor may not be obvious. If the old and new denominators are written in factored form, however, the procedure used is exactly the same as we have already discussed.

$$\frac{3}{x+2} = \frac{?}{x(x+2)}$$

By what factor was the old denominator $(x + 2)$ multiplied to arrive at $x(x + 2)$? The factor must be x.

$$\frac{3}{x+2} = \frac{x(3)}{x(x+2)}$$

If we multiplied the old denominator by x, we must *also* multiply the old numerator by x.

$$\frac{3}{x+2} = \frac{3x}{x(x+2)}$$

We have found our new fraction. Notice that we leave the denominator in factored form because that form will be more useful in adding fractions.

Example 3. Find the new fraction.

$$\frac{2x-5}{x-1} = \frac{?}{(x-1)(x+1)}$$

Solution

$$\frac{2x-5}{x-1} = \frac{?}{(x-1)(x+1)}$$

$$\frac{2x-5}{x-1} = \frac{(2x-5)(x+1)}{(x-1)(x+1)}$$

$$\frac{2x-5}{x-1} = \frac{2x^2-3x-5}{(x-1)(x+1)}$$

Example 4. Find the new fraction.

$$\frac{x+2}{x+3} = \frac{?}{x^2(x+3)}$$

Solution

$$\frac{x+2}{x+3} = \frac{?}{x^2(x+3)}$$

$$\frac{x+2}{x+3} = \frac{x^2(x+2)}{x^2(x+3)}$$

$$\frac{x+2}{x+3} = \frac{x^3+2x^2}{x^2(x+3)}$$

If the new denominator is *not* in factored form, we must factor it before we can decide what the building factors must be.

$$\frac{7x}{x-3} = \frac{?}{x^2-9}$$

We must find our new numerator.

$$\frac{7x}{x-3} = \frac{?}{(x-3)(x+3)}$$

First we factor the new denominator. By what factor was the old denominator $(x - 3)$ multiplied to arrive at the new denominator $(x - 3)(x + 3)$? The factor must be $(x + 3)$.

$$\frac{7x}{x-3} = \frac{7x(x+3)}{(x-3)(x+3)}$$ We must multiply the old numerator by $x + 3$.

$$\frac{7x}{x-3} = \frac{7x^2+21x}{(x-3)(x+3)}$$ We have found our new fraction.

You try to complete Example 5.

Example 5. Find the new fraction.

$$\frac{x+6}{2x+3} = \frac{?}{2x^2-7x-15}$$

Solution

$$\frac{x+6}{2x+3} = \frac{?}{2x^2-7x-15}$$

$$\frac{x+6}{2x+3} = \frac{?}{(2x+3)(x-5)}$$

$$\frac{x+6}{2x+3} = \frac{(x+6)(\_\_\_\_\_\_)}{(2x+3)(x-5)}$$

$$\frac{x+6}{2x+3} = \frac{\_\_\_\_\_\_}{(2x+3)(x-5)}$$

Check your work on page 450. ▶

Example 6. Find the new fraction.

$$\frac{a-8}{a+2} = \frac{?}{3a^3-12a}$$

Solution

$$\frac{a-8}{a+2} = \frac{?}{3a^3-12a}$$

$$\frac{a-8}{a+2} = \frac{?}{3a(a^2-4)}$$

$$\frac{a-8}{a+2} = \frac{?}{3a(a-2)(a+2)}$$

$$\frac{a-8}{a+2} = \frac{3a(a-2)(a-8)}{3a(a-2)(a+2)}$$

$$= \frac{3a(a^2-10a+16)}{3a(a-2)(a+2)}$$

$$\frac{a-8}{a+2} = \frac{3a^3-30a^2+48a}{3a(a-2)(a+2)}$$

Trial Run

Find the new fraction.

______ **1.** $\frac{1}{3} = \frac{?}{3(a-2)}$

______ **2.** $\frac{-4}{x+5} = \frac{?}{x(x+5)}$

______ **3.** $\frac{x}{2x-y} = \frac{?}{(2x-y)(2x+y)}$

______ **4.** $\frac{5x}{x-4} = \frac{?}{x^2-16}$

______ **5.** $\frac{x+1}{x+4} = \frac{?}{x^3+4x^2}$

______ **6.** $\frac{x-3}{2x-1} = \frac{?}{2x^2+9x-5}$

Answers are on page 450.

▶ Examples You Completed

Example 2. Find the new fraction.

$$\frac{-7a}{2bc} = \frac{?}{6b^2c^2}$$

Solution

$$\frac{-7a}{2bc} = \frac{?}{6b^2c^2}$$

$$\frac{-7a}{2bc} = \frac{?}{2bc(3bc)}$$

$$\frac{-7a}{2bc} = \frac{-7a(3bc)}{2bc(3bc)}$$

$$= \frac{-21abc}{6b^2c^2}$$

Example 5. Find the new fraction.

$$\frac{x + 6}{2x + 3} = \frac{?}{2x^2 - 7x - 15}$$

Solution

$$\frac{x + 6}{2x + 3} = \frac{?}{2x^2 - 7x - 15}$$

$$\frac{x + 6}{2x + 3} = \frac{?}{(2x + 3)(x - 5)}$$

$$\frac{x + 6}{2x + 3} = \frac{(x + 6)(x - 5)}{(2x + 3)(x - 5)}$$

$$\frac{x + 6}{2x + 3} = \frac{x^2 + x - 30}{(2x + 3)(x - 5)}$$

Answers to Trial Runs

page 448 **1.** $\frac{-10x}{14x^2}$ **2.** $\frac{24xy^3}{15y^4}$ **3.** $\frac{6a^2}{27abc}$ **4.** $\frac{a^3bx}{a^2b^2y}$

page 449 **1.** $\frac{a - 2}{3(a - 2)}$ **2.** $\frac{-4x}{x(x + 5)}$ **3.** $\frac{2x^2 + xy}{(2x + y)(2x - y)}$ **4.** $\frac{5x^2 + 20x}{(x + 4)(x - 4)}$ **5.** $\frac{x^3 + x^2}{x^2(x + 4)}$ **6.** $\frac{x^2 + 2x - 15}{(2x - 1)(x + 5)}$

Name ______________________ Date __________

EXERCISE SET 8.3

Find the new fraction.

_______ **1.** $\frac{2}{3} = \frac{?}{3x}$

_______ **2.** $\frac{4}{5} = \frac{?}{5x}$

_______ **3.** $\frac{-1}{5x} = \frac{?}{15x^2}$

_______ **4.** $\frac{-1}{6x} = \frac{?}{24x^2}$

_______ **5.** $\frac{4}{xy} = \frac{?}{2x^2y^2}$

_______ **6.** $\frac{5}{xy} = \frac{?}{4x^2y^2}$

_______ **7.** $\frac{3x}{7y} = \frac{?}{21y^3}$

_______ **8.** $\frac{4x}{9y} = \frac{?}{36y^3}$

_______ **9.** $\frac{2a}{-bc} = \frac{?}{15abc}$

_______ **10.** $\frac{5a}{-xy} = \frac{?}{10axy}$

_______ **11.** $x = \frac{?}{3y}$

_______ **12.** $a = \frac{?}{2y}$

_______ **13.** $\frac{-x^2}{y^2} = \frac{?}{3y^3}$

_______ **14.** $\frac{-a^2}{b^2} = \frac{?}{12b^4}$

_______ **15.** $-2b^2 = \frac{?}{4a^2b^2}$

_______ **16.** $-3x^2 = \frac{?}{9x^2y^2}$

_______ **17.** $\frac{2}{3} = \frac{?}{12(x - y)}$

_______ **18.** $\frac{3}{4} = \frac{?}{20(a - b)}$

_______ **19.** $\frac{5}{2x} = \frac{?}{10x^3}$

_______ **20.** $\frac{3}{5x} = \frac{?}{15x^4}$

_______ **21.** $\frac{-4}{a} = \frac{?}{a(a + b)}$

_______ **22.** $\frac{-7}{x} = \frac{?}{x(x + y)}$

_______ **23.** $\frac{3x}{x + y} = \frac{?}{x^2 + xy}$

_______ **24.** $\frac{2y}{x - y} = \frac{?}{xy - y^2}$

_______ **25.** $\frac{2a}{b + 1} = \frac{?}{b^2 - 1}$

_______ **26.** $\frac{3a}{b - 2} = \frac{?}{b^2 - 4}$

_______ **27.** $\frac{6x^2}{x - 5} = \frac{?}{x^2 - 3x - 10}$

_______ **28.** $\frac{5x}{x - 7} = \frac{?}{x^2 - 9x + 14}$

_______ **29.** $\frac{x + 5}{x + 3} = \frac{?}{x^3 + 3x^2}$

_______ **30.** $\frac{a + 3}{a + 2} = \frac{?}{a^4 + 2a^3}$

_______ **31.** $\frac{a - 1}{a - 3} = \frac{?}{a^2 - 7a + 12}$

_______ **32.** $\frac{x - 2}{x - 5} = \frac{?}{x^2 - 8x + 15}$

______ 33. $\dfrac{-3y}{x^2 - x} = \dfrac{?}{x^3 - 2x^2 + x}$

______ 34. $\dfrac{-2x}{y^3 - 2y^2} = \dfrac{?}{y^4 - 4y^2}$

______ 35. $\dfrac{x - y}{2x + 2y} = \dfrac{?}{2x^2 - 4xy - 6y^2}$

______ 36. $\dfrac{x + 2y}{3x + 3y} = \dfrac{?}{3x^2 + 9xy - 12y^2}$

☆ Stretching the Topics

Find the new fraction.

______ 1. $\dfrac{x - 2}{2x^2 - 3x} = \dfrac{?}{6x^4 + x^3 - 15x^2}$

______ 2. $\dfrac{5}{a + b} = \dfrac{?}{ax - ay + bx - by}$

______ 3. $\dfrac{-3}{x - 2} = \dfrac{?}{x^4 - 3x^2 - 4}$

Check your answers in the back of your book.

If you can complete **Checkup 8.3**, you are ready to go to Section 8.4.

Name ________________________________ Date ____________

✓ CHECKUP 8.3

Find the new fraction.

______ 1. $\frac{6}{ab} = \frac{?}{9a^3b^2}$

______ 2. $\frac{-5x}{2yz} = \frac{?}{10y^2z^3}$

______ 3. $\frac{-x}{6y^2} = \frac{?}{18y^4}$

______ 4. $a^2 = \frac{?}{16a^2b^2}$

______ 5. $\frac{x - 5}{x + 2} = \frac{?}{(x + 2)(x + 2)}$

______ 6. $\frac{2a - b}{a - 3b} = \frac{?}{a^3(a - 3b)}$

______ 7. $\frac{2a}{a + b} = \frac{?}{a^2b + ab^2}$

______ 8. $\frac{5x}{x + 6} = \frac{?}{x^2 - 36}$

______ 9. $\frac{x - 2}{x - 1} = \frac{?}{x^2 + 3x - 4}$

______ 10. $\frac{a - b}{5a + 5b} = \frac{?}{5a^2 + 10ab + 5b^2}$

Check your answers in the back of your book.

If You Missed Problems:	You Should Review Examples:
1–4	1, 2
5, 6	3, 4
7–10	5, 6

8.4 Adding and Subtracting Rational Algebraic Expressions

Adding and Subtracting Fractions with the Same Denominator

The procedure for combining algebraic fractions having the same denominator is exactly the same as the procedure for combining numerical fractions.

Combining Fractions with the Same Denominator

1. Combine the numerators.
2. Keep the same denominator.
3. Factor the numerator and denominator and reduce the fraction if possible.

Example 1. Find $\frac{a}{5} + \frac{2a}{5}$.

Solution

$$\frac{a}{5} + \frac{2a}{5}$$

$$= \frac{a + 2a}{5}$$

$$= \frac{3a}{5}$$

You complete Example 2.

Example 2. Find $\frac{5x}{4y} - \frac{x}{4y} + \frac{3}{4y} + \frac{5}{4y}$.

Solution

$$\frac{5x}{4y} - \frac{x}{4y} + \frac{3}{4y} + \frac{5}{4y}$$

$$= \frac{5x - x + 3 + 5}{4y}$$

$$= \frac{4x + 8}{4y}$$

$$= \frac{4(x+2)}{4y}$$

$$= \frac{x+2}{y}$$

Check your work on page 460. ▶

Example 3. Find $\frac{2a - b}{4a} - \frac{2a - 3b}{4a}$.

Solution: Be careful with the minus sign before the second fraction. It shows that the *entire* second numerator is being subtracted. Parentheses will help.

$$\frac{2a - b}{4a} - \frac{2a - 3b}{4a}$$

$$= \frac{2a - b - (2a - 3b)}{4a}$$ Combine the numerators and keep the same denominator.

$$= \frac{2a - b - 2a + 3b}{4a}$$ Remove parentheses.

$$= \frac{2b}{4a}$$ Combine like terms.

$$= \frac{b}{2a}$$ Reduce the fraction.

Let's try adding fractions with more than one term in the denominators. As long as the denominators all match, our method will be the same. Try Example 4.

Example 4. Find

$$\frac{x^2}{x+1} - \frac{2x}{x+1} - \frac{3}{x+1}.$$

Solution

$$\frac{x^2}{x+1} - \frac{2x}{x+1} - \frac{3}{x+1}$$

$$= \frac{x^2 - 2x - 3}{x+1}$$

$$= \frac{(x-3)(\_\_\_\_\_\_)}{x+1}$$

$$= \_\_\_\_\_\_$$

Check your work on page 460. ▶

Example 5. Find $\frac{2x-3}{x^2-1} - \frac{x-4}{x^2-1}$.

Solution

$$\frac{2x-3}{x^2-1} - \frac{x-4}{x^2-1}$$

$$= \frac{2x - 3 - 1(x-4)}{x^2-1}$$

Check your work on page 460. ▶

Complete Example 5.

Trial Run

Combine the fractions and reduce if possible.

______ 1. $\frac{4x}{5y} + \frac{6x}{5y}$

______ 2. $\frac{7x}{3y} + \frac{2}{3y} - \frac{x}{3y} + \frac{1}{3y}$

______ 3. $\frac{7a-3b}{10b} - \frac{a-3b}{10b}$

______ 4. $\frac{3}{x+7} + \frac{4}{x+7}$

______ 5. $\frac{4x}{x-2y} - \frac{3x}{x-2y} + \frac{2y}{x-2y}$

______ 6. $\frac{x^2}{x+2} - \frac{13x}{x+2} - \frac{30}{x+2}$

______ 7. $\frac{2a^2}{a^2-8a} - \frac{17a}{a^2-8a} + \frac{8}{a^2-8a}$

______ 8. $\frac{5x-3}{x^2-4} - \frac{4x-1}{x^2-4}$

Answers are on page 461.

Adding and Subtracting Fractions with Different Denominators

The method for adding and subtracting numerical fractions with different denominators works for algebraic fractions also.

Adding and Subtracting Fractions

1. Factor the denominators of the original fractions.
2. Choose the LCD. (It is the product of all the different factors appearing in the original denominators, with each factor used the most times that it appears in any single denominator.)
3. Build new fractions from the original fractions, with each new fraction having the LCD as its denominator.
4. Simplify each numerator.
5. Combine the numerators to form one fraction with the LCD as its denominator.
6. Simplify the numerator and reduce the fraction if possible.

You try to complete Example 6.

Example 6. Simplify $\frac{x}{3} + \frac{2x}{5} - \frac{x}{9}$.

Solution

$$\frac{x}{3} + \frac{2x}{5} - \frac{x}{9}$$

$$= \frac{x}{3} + \frac{2x}{5} - \frac{x}{3 \cdot 3} \qquad \text{LCD: } 3 \cdot 3 \cdot 5$$

$$= \frac{x \cdot 3 \cdot 5}{3 \cdot 3 \cdot 5} + \frac{2x \cdot 3 \cdot 3}{3 \cdot 3 \cdot 5} - \frac{x \cdot 5}{3 \cdot 3 \cdot 5}$$

$$= \frac{\quad}{45} + \frac{\quad}{45} - \frac{\quad}{45}$$

$$= \frac{\quad + \quad - \quad}{45}$$

$$= \frac{\quad}{45}$$

Check your work on page 460. ▶

Example 7. Simplify $\frac{3}{y} + \frac{7}{x^2} + \frac{5}{3x}$.

Solution

$$\frac{3}{y} + \frac{7}{x^2} + \frac{5}{3x} \qquad \text{LCD: } 3x^2y.$$

$$= \frac{3 \cdot 3x^2}{3x^2y} + \frac{7 \cdot 3 \cdot y}{3x^2y} + \frac{5 \cdot x \cdot y}{3x^2y}$$

$$= \frac{9x^2}{3x^2y} + \frac{21y}{3x^2y} + \frac{5xy}{3x^2y}$$

$$= \frac{9x^2 + 21y + 5xy}{3x^2y}$$

Example 8. Find $\frac{1}{x+2} + \frac{1}{5x}$.

Solution

$$\frac{1}{x+2} + \frac{1}{5x} \qquad \text{LCD: } 5x(x+2).$$

$$= \frac{1 \cdot 5x}{5x(x+2)} + \frac{1(x+2)}{5x(x+2)}$$

$$= \frac{5x}{5x(x+2)} + \frac{x+2}{5x(x+2)}$$

$$= \frac{5x+x+2}{5x(x+2)}$$

$$= \frac{6x+2}{5x(x+2)}$$

$$= \frac{2(3x+1)}{5x(x+2)}$$

You complete Example 9.

Example 9. Find $\frac{4}{2x-1} - \frac{2}{2x^2-x}$.

Solution

$$\frac{4}{2x-1} - \frac{2}{2x^2-x}$$

$$= \frac{4}{2x-1} - \frac{2}{x(2x-1)}$$

LCD: $x(2x-1)$.

$$= \frac{4 \cdot x}{x(2x-1)} - \frac{2}{x(2x-1)}$$

Check your work on page 461. ▶

Notice that we factored the numerator in the final step of Example 8 to see whether a factor common to the numerator and denominator might appear.

Example 10. Find $\frac{x}{(x+3)^2} + \frac{2x}{x^2-9}$.

Solution

$$\frac{x}{(x+3)^2} + \frac{2x}{x^2-9}$$

$$= \frac{x}{(x+3)^2} + \frac{2x}{(x+3)(x-3)}$$

Notice that $(x+3)$ appears twice in the first denominator. LCD: $(x+3)^2(x-3)$.

$$= \frac{x(x-3)}{(x+3)^2(x-3)} + \frac{2x(x+3)}{(x+3)^2(x-3)}$$

$$= \frac{x^2-3x}{(x+3)^2(x-3)} + \frac{2x^2+6x}{(x+3)^2(x-3)}$$

$$= \frac{x^2-3x+2x^2+6x}{(x+3)^2(x-3)}$$

$$= \frac{3x^2+3x}{(x+3)^2(x-3)}$$

$$= \frac{3x(x+1)}{(x+3)^2(x-3)}$$

Example 11. Find $\frac{7x}{x^2 - x - 2} + \frac{2}{3x - 6}$.

Solution

$$\frac{7x}{x^2 - x - 2} + \frac{2}{3x - 6}$$

$$= \frac{7x}{(x + 1)(x - 2)} + \frac{2}{3(x - 2)} \qquad \text{LCD: } 3(x + 1)(x - 2).$$

$$= \frac{7x \cdot 3}{3(x + 1)(x - 2)} + \frac{2(x + 1)}{3(x + 1)(x - 2)}$$

$$= \frac{21x}{3(x + 1)(x - 2)} + \frac{2x + 2}{3(x + 1)(x - 2)}$$

$$= \frac{21x + 2x + 2}{3(x + 1)(x - 2)}$$

$$= \frac{23x + 2}{3(x + 1)(x - 2)}$$

Adding fractions is not a quick procedure, but each step we have used is necessary for obtaining the correct result.

Trial Run

Simplify.

5x+12x

_______ **1.** $\frac{x}{4} + \frac{3x}{5}$

_______ **2.** $\frac{5x}{9} - \frac{x}{6} + \frac{1}{4}$

_______ **3.** $\frac{3}{x} + \frac{2}{x^2} - \frac{1}{3x}$

_______ **4.** $\frac{5}{2x} - \frac{4}{x + 3}$

_______ **5.** $\frac{1}{x - 1} + \frac{3}{x + 2}$

_______ **6.** $\frac{x}{x - 9} - \frac{2x - 4}{x^2 - 11x + 18}$

Answers are on page 461.

▶ Examples You Completed

Example 2. Find $\frac{5x}{4y} - \frac{x}{4y} + \frac{3}{4y} + \frac{5}{4y}$.

Solution

$$\frac{5x}{4y} - \frac{x}{4y} + \frac{3}{4y} + \frac{5}{4y}$$

$$= \frac{5x - x + 3 + 5}{4y}$$

$$= \frac{4x + 8}{4y}$$

$$= \frac{\overset{1}{\cancel{4}}(x + 2)}{\underset{1}{\cancel{4}}y}$$

$$= \frac{x + 2}{y}$$

Example 4. Find $\frac{x^2}{x + 1} - \frac{2x}{x + 1} - \frac{3}{x + 1}$.

Solution

$$\frac{x^2}{x + 1} - \frac{2x}{x + 1} - \frac{3}{x + 1}$$

$$= \frac{x^2 - 2x - 3}{x + 1}$$

$$= \frac{(x - 3)\overset{1}{\cancel{(x + 1)}}}{\underset{1}{\cancel{x + 1}}}$$

$$= x - 3$$

Example 5 Find $\frac{2x - 3}{x^2 - 1} - \frac{x - 4}{x^2 - 1}$.

Solution

$$\frac{2x - 3}{x^2 - 1} - \frac{x - 4}{x^2 - 1}$$

$$= \frac{2x - 3 - (x - 4)}{x^2 - 1}$$

$$= \frac{2x - 3 - x + 4}{x^2 - 1}$$

$$= \frac{x + 1}{x^2 - 1}$$

$$= \frac{\overset{1}{\cancel{x + 1}}}{\underset{1}{\cancel{(x + 1)}}(x - 1)}$$

$$= \frac{1}{x - 1}$$

Example 6. Simplify $\frac{x}{3} + \frac{2x}{5} - \frac{x}{9}$.

Solution

$$\frac{x}{3} + \frac{2x}{5} - \frac{x}{9}$$

$$= \frac{x}{3} + \frac{2x}{5} - \frac{x}{3 \cdot 3} \quad \text{LCD: } 3 \cdot 3 \cdot 5.$$

$$= \frac{x \cdot 3 \cdot 5}{3 \cdot 3 \cdot 5} + \frac{2x \cdot 3 \cdot 3}{3 \cdot 3 \cdot 5} - \frac{x \cdot 5}{3 \cdot 3 \cdot 5}$$

$$= \frac{15x}{45} + \frac{18x}{45} - \frac{5x}{45}$$

$$= \frac{15x + 18x - 5x}{45}$$

$$= \frac{28x}{45}$$

Example 9. Find $\dfrac{4}{2x-1} - \dfrac{2}{2x^2 - x}$.

Solution

$$\frac{4}{2x-1} - \frac{2}{2x^2-x}$$

$$= \frac{4}{2x-1} - \frac{2}{x(2x-1)} \qquad \text{LCD: } x(2x-1).$$

$$= \frac{4 \cdot x}{x(2x-1)} - \frac{2}{x(2x-1)}$$

$$= \frac{4x-2}{x(2x-1)}$$

$$= \frac{2\overset{1}{\cancel{(2x-1)}}}{x\underset{1}{\cancel{(2x-1)}}}$$

$$= \frac{2}{x}$$

Answers to Trial Runs

page 456 1. $\dfrac{2x}{y}$ 2. $\dfrac{2x+1}{y}$ 3. $\dfrac{3a}{5b}$ 4. $\dfrac{7}{x+7}$ 5. $\dfrac{x+2y}{x-2y}$ 6. $x - 15$ 7. $\dfrac{2a-1}{a}$ 8. $\dfrac{1}{x+2}$

page 459 1. $\dfrac{17x}{20}$ 2. $\dfrac{14x+9}{36}$ 3. $\dfrac{2(4x+3)}{3x^2}$ 4. $\dfrac{-3(x-5)}{2x(x+3)}$ 5. $\dfrac{4x-1}{(x-1)(x+2)}$ 6. $\dfrac{x-2}{x-9}$

Name ______________________ Date ____________

EXERCISE SET 8.4

Combine the following fractions. Write the answers in reduced form.

_____ **1.** $\frac{5x}{3} + \frac{4x}{3}$

_____ **2.** $\frac{9x}{7} + \frac{5x}{7}$

_____ **3.** $\frac{2a}{7} + \frac{3a}{7} - \frac{6}{7}$

_____ **4.** $\frac{5a}{9} + \frac{2a}{9} - \frac{4}{9}$

_____ **5.** $\frac{7}{5x} - \frac{2}{5x}$

_____ **6.** $\frac{15}{11x} - \frac{4}{11x}$

_____ **7.** $\frac{5x}{2y} - \frac{3}{2y} - \frac{x}{2y} + \frac{1}{2y}$

_____ **8.** $\frac{8x}{5y} - \frac{3}{5y} + \frac{2x}{5y} - \frac{2}{5y}$

_____ **9.** $\frac{2a - 3}{3b} - \frac{2a + 3}{3b}$

_____ **10.** $\frac{5a - 2}{7b} - \frac{a - 2}{7b}$

_____ **11.** $\frac{4y}{x + 3y} + \frac{5y}{x + 3y}$

_____ **12.** $\frac{6x}{x + 2y} - \frac{4y}{x + 2y}$

_____ **13.** $\frac{3x}{x - 1} - \frac{3}{x - 1}$

_____ **14.** $\frac{5x}{x - 2} - \frac{10}{x - 2}$

_____ **15.** $\frac{3x}{2x + 5} - \frac{x}{2x + 5}$

_____ **16.** $\frac{7x}{3x - 1} - \frac{4x}{3x - 1}$

_____ **17.** $\frac{x^2}{x^2 - 4} + \frac{2x}{x^2 - 4}$

_____ **18.** $\frac{x^2}{x^2 - 9} + \frac{3x}{x^2 - 9}$

_____ **19.** $\frac{3x^2}{x + 5} - \frac{75}{x + 5}$

_____ **20.** $\frac{2x^2}{x + 4} - \frac{32}{x + 4}$

_____ **21.** $\frac{2a^2}{a^2 + 7a} + \frac{11a}{a^2 + 7a} - \frac{21}{a^2 + 7a}$

_____ **22.** $\frac{2a^2}{a^2 + 3a} + \frac{a}{a^2 + 3a} - \frac{15}{a^2 + 3a}$

_____ **23.** $\frac{7x - y}{3x^2 - 6xy} - \frac{4x - y}{3x^2 - 6xy}$

_____ **24.** $\frac{6x - y}{4x^2 - 8xy} - \frac{2x - y}{4x^2 - 8xy}$

_____ **25.** $\frac{x}{6} - \frac{y}{10}$

_____ **26.** $\frac{x}{9} - \frac{y}{6}$

_____ **27.** $\frac{3}{x} + \frac{5}{x^2} - \frac{2}{5x^2}$

_____ **28.** $\frac{5}{x} + \frac{6}{x} - \frac{7}{3x^2}$

_____ **29.** $7 - \frac{1}{y}$

_____ **30.** $9 - \frac{1}{y}$

_____ **31.** $\frac{4}{x} - 2x$

_____ **32.** $\frac{6}{x} - 3x$

_______ 33. $\frac{5}{x} + \frac{2}{y}$

_______ 34. $\frac{7}{x} - \frac{11}{y}$

_______ 35. $\frac{2}{x} - \frac{3}{y} + \frac{2}{5x} - \frac{1}{3y}$

_______ 36. $\frac{3}{x} - \frac{2}{y} + \frac{1}{4x} - \frac{4}{3y}$

_______ 37. $\frac{4}{x^2y} - \frac{5}{xy^2}$

_______ 38. $\frac{3}{x^2y} - \frac{7}{xy^2}$

_______ 39. $\frac{x - 2y}{3x^2y} - \frac{x + 2y}{4xy^2} + \frac{1}{6xy}$

_______ 40. $\frac{2x - y}{2x^2y} - \frac{2x + y}{5xy^2} - \frac{8}{5xy}$

_______ 41. $\frac{7}{x + 2} - \frac{5}{(x + 2)^2}$

_______ 42. $\frac{11}{x + 3} - \frac{4}{(x + 3)^2}$

_______ 43. $\frac{3}{x + 4} - \frac{5}{x - 4}$

_______ 44. $\frac{4}{x + 5} - \frac{3}{x - 5}$

_______ 45. $\frac{x}{3} - \frac{2}{x + 1}$

_______ 46. $\frac{x}{4} - \frac{3}{x - 1}$

_______ 47. $\frac{4}{3x + 3} - \frac{5}{x + 1}$

_______ 48. $\frac{7}{2x + 2} - \frac{4}{x + 1}$

_______ 49. $\frac{7}{x - 2} - \frac{5}{x + 2}$

_______ 50. $\frac{4}{x - 3} - \frac{5}{x + 3}$

_______ 51. $\frac{x}{x^2 - 25} - \frac{1}{x^2 - 5x}$

_______ 52. $\frac{x}{x^2 - 49} - \frac{3}{x^2 - 7x}$

_______ 53. $\frac{1}{x^2 - 2x} - \frac{2}{x^3 - 2x^2}$

_______ 54. $\frac{1}{x^2 - 3x} - \frac{3}{x^3 - 3x^2}$

_______ 55. $\frac{4}{x^2 - 49} - \frac{1}{x^2 + 14x + 49}$

_______ 56. $\frac{5}{x^2 - 36} - \frac{1}{x^2 + 12x + 36}$

_______ 57. $\frac{x + 6}{x^2 + 2x - 8} + \frac{3}{x + 4}$

_______ 58. $\frac{x - 8}{x^2 - 11x + 18} - \frac{4}{x - 9}$

_______ 59. $\frac{2x - 6}{2x^2 - 7x + 3} - \frac{5}{2x^2 - x}$

_______ 60. $\frac{3x - 21}{3x^2 - 22x + 7} - \frac{4}{3x^2 - x}$

_______ 61. $\frac{3}{x^2 - 81} - \frac{1}{x^2 - 18x + 81}$

_______ 62. $\frac{4}{x^2 - 64} - \frac{2}{x^2 - 16x + 64}$

_______ 63. $\frac{x}{2x^2 + x - 6} - \frac{1}{2x^2 - x - 3}$

_______ 64. $\frac{x}{2x^2 - 11x + 15} + \frac{2}{2x^2 - 9x + 10}$

☆ Stretching the Topics

Combine the fractions and reduce.

_______ **1.** $\dfrac{5}{n^2 - 8n + 15} - \dfrac{3}{n - 5} + \dfrac{2}{3 - n}$

_______ **2.** $\dfrac{x}{x^2 + 4x - 21} + \dfrac{4}{x^2 + 7x} + \dfrac{2x}{x^2 - 9}$

_______ **3.** $\dfrac{3a + 3b}{xa + 3ya + xb - 3yb} + \dfrac{5}{3y - x}$

Check your answers in the back of your book.

If you can complete **Checkup 8.4,** you are ready to go on to Section 8.5.

Name ______________________ Date __________

✓ CHECKUP 8.4

Combine the fractions and, if possible, reduce.

______ 1. $\frac{7x}{4} - \frac{9x}{4}$

______ 2. $\frac{6a}{5b} - \frac{4}{5b} + \frac{4a}{5b} - \frac{1}{5b}$

______ 3. $\frac{x^2}{x - 1} - \frac{9x}{x - 1} + \frac{8}{x - 1}$

______ 4. $\frac{7x - 10}{2x^2 - 6x} - \frac{2x + 5}{2x^2 - 6x}$

______ 5. $\frac{3}{xy} + \frac{2}{y^2}$

______ 6. $\frac{4}{x - 3} + \frac{7}{x}$

______ 7. $\frac{7}{2x + 2} + \frac{4}{x + 1}$

______ 8. $\frac{4}{x^2 - 25} - \frac{1}{x^2 + 10x + 25}$

______ 9. $\frac{3}{3x^2 - 2x} - \frac{2}{3x^3 - 2x^2}$

______ 10. $\frac{x + 8}{x^2 - 2x - 63} - \frac{1}{x + 7}$

Check your answers in the back of your book.

If You Missed Problems:	You Should Review Examples:
1, 2	1, 2
3, 4	4, 5
5, 6	7, 8
7	9
8, 9	10
10	11

8.5 Working with Complex Fractions

A **complex fraction** is a fraction that contains fractions in its numerator and/or its denominator. Simplifying complex fractions requires that you be able to add, subtract, multiply, and divide fractions.

Let's simplify the complex fraction

$$\frac{\frac{x}{y^2}}{\frac{x^2}{5y^6}}$$

$$\frac{\frac{x}{y^2}}{\frac{x^2}{5y^6}} = \frac{x}{y^2} \div \frac{x^2}{5y^6}$$ Rewrite the complex fraction as a division problem.

$$= \frac{x}{y^2} \cdot \frac{5y^6}{x^2}$$ Invert the divisor and change the operation to multiplication.

$$= \frac{\overset{1}{\cancel{x}}}{\underset{1}{\cancel{y^2}}} \cdot \frac{5\overset{y^4}{\cancel{y^6}}}{\underset{x}{\cancel{x^2}}}$$ Reduce before multiplying.

$$= \frac{5y^4}{x}$$ Multiply.

Example 1. Simplify $\dfrac{\frac{8a^3}{3b^2}}{\frac{16a^5}{15b^8}}$.

Solution

$$\frac{\frac{8a^3}{3b^2}}{\frac{16a^5}{15b^8}} = \frac{8a^3}{3b^2} \div \frac{16a^5}{15b^8}$$

$$= \frac{\overset{1}{\cancel{8}}\cancel{a^3}}{\underset{1}{\cancel{3}}\underset{1}{\cancel{b^2}}} \cdot \frac{\overset{5}{\cancel{15}}\overset{b^6}{\cancel{b^8}}}{\underset{2}{\cancel{16}}\underset{a^2}{\cancel{a^5}}}$$

$$= \frac{5b^6}{2a^2}$$

Now you try Example 2.

Example 2. Simplify $\dfrac{\frac{x^2 - 4}{x^3}}{\frac{5x + 10}{x^2 + x}}$.

Solution

$$\frac{\frac{x^2 - 4}{x^3}}{\frac{5x + 10}{x^2 + x}} = \frac{x^2 - 4}{x^3} \div \frac{5x + 10}{x^2 + x}$$

$$= \frac{x^2 - 4}{x^3} \cdot \frac{x^2 + x}{5x + 10}$$

$$= \frac{(x + 2)(\underline{\qquad})}{x^3} \cdot \frac{x(x + 1)}{5(\underline{\qquad})}$$

$$= \frac{(\underline{\qquad})(\underline{\qquad})}{5x^2}$$

Check your work on page 474. ▶

Trial Run

Simplify.

______ 1. $\dfrac{\frac{5a^2}{2b^3}}{\frac{25a^3}{12b^7}}$

______ 2. $\dfrac{\frac{1}{x + 6}}{\frac{1}{2x + 12}}$

______ 3. $\dfrac{2x - 1}{\frac{6x - 3}{5}}$

______ 4. $\dfrac{\frac{2x + 2}{x^2}}{\frac{x^2 - 1}{3x^3}}$

Answers are on page 474.

Now we must develop a procedure for simplifying a complex fraction in which the numerator and/or denominator is a sum or difference containing fractional expressions. Let's consider the complex fraction

$$\frac{y - \frac{1}{y}}{y - 1}$$

$$\frac{y - \frac{1}{y}}{y - 1} = \left[\frac{y}{1} - \frac{1}{y}\right] \div [y - 1]$$

Rewrite the complex fraction as a division problem.

$$= \frac{x^3\left(x + \frac{2}{x}\right)}{x^3\left(x - \frac{4}{x^3}\right)}$$ Multiply numerator and denominator by LCD: x^3.

$$= \frac{x^3(x) + \frac{x^3}{1}\left(\frac{2}{x}\right)}{x^3(x) - \frac{x^3}{1}\left(\frac{4}{x^3}\right)}$$ Use the distributive property.

$$= \frac{x^4 + 2x^2}{x^4 - 4}$$ Simplify each term.

$$= \frac{x^2\overset{1}{\cancel{(x^2 + 2)}}}{\underset{1}{\cancel{(x^2 + 2)}}(x^2 - 2)}$$ Factor numerator and denominator.

$$= \frac{x^2}{x^2 - 2}$$ Reduce.

Since either of the two methods for simplifying complex fractions is acceptable, you should decide which method you prefer and learn to use it successfully.

Trial Run

Simplify.

_______ 1. $\dfrac{x - \dfrac{9}{x}}{x + 3}$

_______ 2. $\dfrac{1 + \dfrac{1}{x}}{1 - \dfrac{1}{x^2}}$

_______ 3. $\dfrac{2x + \dfrac{5x - 3}{x}}{\dfrac{x}{3} - \dfrac{3}{x}}$

_______ 4. $\dfrac{\dfrac{1}{x} + \dfrac{1}{y}}{\dfrac{1}{x} - \dfrac{1}{y}}$

Answers are on page 474.

▶ Examples You Completed

Example 2. Simplify $\dfrac{\dfrac{x^2 - 4}{x^3}}{\dfrac{5x + 10}{x^2 + x}}$.

Solution

$$\frac{\dfrac{x^2 - 4}{x^3}}{\dfrac{5x + 10}{x^2 + x}} = \frac{x^2 - 4}{x^3} \div \frac{5x + 10}{x^2 + x}$$

$$= \frac{x^2 - 4}{x^3} \cdot \frac{x^2 + x}{5x + 10}$$

$$= \frac{\overset{1}{\cancel{(x + 2)}}(x - 2)}{\underset{x^2}{\cancel{x^3}}} \cdot \frac{\overset{1}{\cancel{x}}(x + 1)}{5\underset{1}{\cancel{(x + 2)}}}$$

$$= \frac{(x - 2)(x + 1)}{5x^2}$$

Answers to Trial Runs

page 470 **1.** $\dfrac{6b^4}{5a}$ **2.** 2 **3.** $\dfrac{5}{3}$ **4.** $\dfrac{6x}{x - 1}$

page 473 **1.** $\dfrac{x - 3}{x}$ **2.** $\dfrac{x}{x - 1}$ **3.** $\dfrac{3(2x - 1)}{x - 3}$ **4.** $\dfrac{y + x}{y - x}$

Name ________________________ Date __________

EXERCISE SET 8.5

Simplify the fractions.

_____ 1. $\dfrac{\frac{5}{12}}{\frac{35}{48}}$

_____ 2. $\dfrac{\frac{8}{15}}{\frac{21}{56}}$

_____ 3. $\dfrac{\frac{x^9}{y^3}}{\frac{4x^4}{y^2}}$

_____ 4. $\dfrac{\frac{x^{12}}{2y^5}}{\frac{3x^7}{10y^3}}$

_____ 5. $\dfrac{\frac{x - 3y}{4x^2}}{\frac{x^2 - 9y^2}{12x^3}}$

_____ 6. $\dfrac{\frac{x + 5y}{3y^5}}{\frac{x^2 - 25y^2}{9y^6}}$

_____ 7. $\dfrac{\frac{1}{x^2 - 16}}{\frac{4}{5x + 20}}$

_____ 8. $\dfrac{\frac{1}{x^2 - 36}}{\frac{9}{2x - 12}}$

_____ 9. $\dfrac{x^2 + 7x}{\frac{x^2 - 49}{x}}$

_____ 10. $\dfrac{x^2 + 9x}{\frac{x^2 - 81}{x}}$

_____ 11. $\dfrac{\frac{9x^2 - 9x}{3}}{x^4 - x^3}$

_____ 12. $\dfrac{\frac{4x^2 + 4x}{2}}{x^3 + x^2}$

_____ 13. $\dfrac{\frac{25x^2 - 4}{9x^3}}{\frac{5x + 2}{18x}}$

_____ 14. $\dfrac{\frac{9x^2 - 16}{12x^4}}{\frac{3x - 4}{9x}}$

_____ 15. $\dfrac{3 + \frac{3}{4}}{1 - \frac{3}{8}}$

_____ 16. $\dfrac{5 - \frac{1}{3}}{2 - \frac{5}{6}}$

_____ 17. $\dfrac{x - \frac{49}{x}}{x + 7}$

_____ 18. $\dfrac{x - \frac{25}{x}}{x - 5}$

_____ 19. $\dfrac{\frac{1}{x}}{1 - \frac{1}{x}}$

_____ 20. $\dfrac{\frac{2}{x^2}}{1 + \frac{2}{x^2}}$

_____ 21. $\dfrac{5 - \dfrac{1}{x}}{25 - \dfrac{1}{x^2}}$

_____ 22. $\dfrac{9 - \dfrac{1}{x}}{81 - \dfrac{1}{x^2}}$

_____ 23. $\dfrac{\dfrac{1}{2} - \dfrac{1}{5}}{\dfrac{2}{3} - \dfrac{4}{15}}$

_____ 24. $\dfrac{\dfrac{1}{3} - \dfrac{1}{4}}{\dfrac{1}{2} + \dfrac{1}{8}}$

_____ 25. $\dfrac{\dfrac{1}{x^2} - \dfrac{1}{9}}{\dfrac{1}{x} - \dfrac{1}{3}}$

_____ 26. $\dfrac{\dfrac{1}{x^2} - \dfrac{1}{16}}{\dfrac{1}{x} - \dfrac{1}{4}}$

_____ 27. $\dfrac{6 - \dfrac{5x + 25}{x^2}}{\dfrac{3}{5} + \dfrac{1}{x}}$

_____ 28. $\dfrac{10 - \dfrac{x + 3}{x^2}}{\dfrac{x}{3} - \dfrac{1}{5}}$

_____ 29. $\dfrac{1 - \dfrac{4}{x} - \dfrac{45}{x^2}}{\dfrac{x}{9} - \dfrac{9}{x}}$

_____ 30. $\dfrac{1 + \dfrac{1}{x} - \dfrac{72}{x^2}}{\dfrac{x}{8} - \dfrac{8}{x}}$

☆ Stretching the Topics

Simplify the fractions.

_____ 1. $\dfrac{\dfrac{4x}{y} - 5 + \dfrac{x}{y}}{\dfrac{x}{y} - \dfrac{8y}{x} - 2}$

_____ 2. $\dfrac{\dfrac{3y - 1}{y - 3} + 5}{\dfrac{y + 3}{y - 2}}$

_____ 3. $\dfrac{\dfrac{2}{x^2 - 4x - 21} - \dfrac{1}{x - 7}}{\dfrac{-1}{x - 7} + \dfrac{3}{x + 3}}$

Check your answers in the back of your book.

If you can complete **Checkup 8.5**, you are ready to go on to Section 8.6.

Name ______________________ Date ____________

✓ CHECKUP 8.5

Simplify the fractions.

______ 1. $\dfrac{\dfrac{x^9}{y^3}}{\dfrac{x^{12}}{3y^2}}$

______ 2. $\dfrac{\dfrac{15x^8}{4}}{\dfrac{5x^5}{6y^4}}$

______ 3. $\dfrac{\dfrac{3x - 5}{5x}}{\dfrac{9x^2 - 25}{20x^3}}$

______ 4. $\dfrac{\dfrac{x^2 - 9}{x}}{\dfrac{3x - 9}{x^2 + x}}$

______ 5. $\dfrac{\dfrac{x^2 + 8x}{x^2 - 64}}{3}$

______ 6. $\dfrac{x - \dfrac{16}{x}}{x - 4}$

______ 7. $\dfrac{25 - \dfrac{16}{a^2}}{5 - \dfrac{4}{a}}$

______ 8. $\dfrac{\dfrac{1}{x^2} - \dfrac{1}{81}}{\dfrac{1}{x} - \dfrac{1}{9}}$

______ 9. $\dfrac{5 - \dfrac{9x - 4}{x^2}}{\dfrac{5}{4} - \dfrac{1}{x}}$

______ 10. $\dfrac{1 - \dfrac{2}{x} - \dfrac{24}{x^2}}{\dfrac{x}{4} - \dfrac{4}{x}}$

Check your answers in the back of your book.

If You Missed Problems:	You Should Review Examples:
1, 2	1
3–5	2
6–8	4
9	3
10	3, 4

8.6 Solving Fractional Equations

Fractional equations are equations that contain rational expressions. From our earlier work with equations we recall that we can

1. Add (or subtract) the same quantity to (or from) both sides of an equation.
2. Multiply (or divide) both sides of an equation by the same nonzero quantity.

To solve a fractional equation, our first goal is to *get rid of all the denominators* in the equation. We can accomplish this by multiplying both sides of the equation by the LCD for all the fractions in the equation.

Solve $\frac{x}{3} + \frac{5}{6} = \frac{9x}{2}$ LCD: $6 \left(\text{or } \frac{6}{1}\right)$.

$\frac{6}{1}\left(\frac{x}{3} + \frac{5}{6}\right) = \frac{6}{1} \cdot \frac{9x}{2}$ Multiply both sides by LCD.

$\frac{6}{1}\left(\frac{x}{3}\right) + \frac{6}{1}\left(\frac{5}{6}\right) = \frac{6}{1} \cdot \frac{9x}{2}$ Use the distributive property on the left side.

$\frac{\overset{2}{\cancel{6}}}{1} \cdot \frac{x}{\underset{1}{\cancel{3}}} + \frac{\overset{1}{\cancel{6}}}{1} \cdot \frac{5}{\underset{1}{\cancel{6}}} = \frac{\overset{3}{\cancel{6}}}{1} \cdot \frac{9x}{\underset{1}{\cancel{2}}}$ Multiply fractions in each term by the usual method.

$2x + 5 = 27x$ Simplify.

$5 = 25x$ Subtract $2x$ from both sides.

$\frac{5}{25} = \frac{25x}{25}$ Divide both sides by 25.

$\frac{1}{5} = x$ Simplify.

Example 1. Solve $\frac{3a}{4} = 10 + \frac{a}{3}$ and check.

Solution: LCD is 12.

$$\frac{\overset{3}{\cancel{12}}}{1}\left(\frac{3a}{\underset{1}{\cancel{4}}}\right) = 12(10) + \frac{\overset{4}{\cancel{12}}}{1}\left(\frac{a}{\underset{1}{\cancel{3}}}\right)$$

$$3(3a) = 120 + 4a$$

$$9a = 120 + 4a$$

$$5a = 120$$

$$\frac{5a}{5} = \frac{120}{5}$$

$$a = 24$$

CHECK:

$$\frac{3a}{4} = 10 + \frac{a}{3}$$

$$\frac{3(24)}{4} \stackrel{?}{=} 10 + \frac{24}{3}$$

$$3(6) \stackrel{?}{=} 10 + 8$$

$$18 = 18$$

Example 2. Solve $\frac{7}{x^2} - \frac{3}{x} = \frac{5}{2x}$.

Solution

$$\frac{7}{x^2} - \frac{3}{x} = \frac{5}{2x}$$

The LCD is $2x^2$, but remember that we agreed to multiply both sides of equations by *non-zero* quantities only. We must find the *restrictions on the variable* so that the LCD is never zero. Here we must find the value of x for which

$$2x^2 = 0$$

$$x^2 = 0$$

$$x = 0$$

Therefore, we agree that $x \neq 0$ and multiply both sides of the equations by $2x^2$.

$$\frac{\overset{2}{\cancel{2x^2}}}{1}\left(\frac{7}{\underset{1}{\cancel{x^2}}}\right) - \frac{\overset{2x}{\cancel{2x^2}}}{1}\left(\frac{3}{\underset{1}{\cancel{x}}}\right) = \frac{\overset{x}{\cancel{2x^2}}}{1}\left(\frac{5}{\underset{1}{\cancel{2x}}}\right)$$

$$14 - 6x = 5x$$

$$14 = 11x$$

$$\frac{14}{11} = x$$

From now on we shall find the *restrictions on the variable* as soon as we have chosen the LCD for a fractional equation.

Trial Run

Solve.

$\frac{7}{13} = x$ 1. $\frac{x}{5} + \frac{7}{10} = \frac{3x}{2}$

140 2. $\frac{2a}{7} = 5 + \frac{a}{4}$

$\frac{8}{7} = x$ 3. $\frac{4}{x^2} - \frac{2}{x} = \frac{3}{2x}$

$\frac{10}{3}$ 4. $\frac{y}{5} = 4 - y$

Trial Run

Solve.

_______ 1. $\frac{x}{x-6} + \frac{3}{2} = \frac{4}{x-6}$

_______ 2. $\frac{4}{x-5} - \frac{5}{x+1} = \frac{3}{2x+2}$

_______ 3. $\frac{10}{3x} = \frac{2}{x+9}$

_______ 4. $\frac{x}{x-2} + \frac{7}{3} = \frac{2}{x-2}$

_______ 5. $\frac{x}{x-4} - \frac{8}{x-4} = 2$

_______ 6. $\frac{5}{x+3} = \frac{7}{x^2-9}$

Answers are on page 483.

The Example You Completed

Example 3. Solve $\frac{5}{2x} = \frac{2}{x+7}$.

Solution: LCD is $2x(x + 7)$. Restrictions are $x \neq 0$, $x \neq -7$.

$$\frac{\overset{1}{\cancel{2x}}(x+7)}{1}\left(\frac{5}{\underset{1}{\cancel{2x}}}\right) = \frac{2x\overset{1}{\cancel{(x+7)}}}{1}\left(\frac{2}{\underset{1}{\cancel{x+7}}}\right)$$

$$5(x + 7) = 2x \cdot 2$$

$$5x + 35 = 4x$$

$$x + 35 = 0$$

$$x = -35$$

Answers to Trial Runs

page 480 **1.** $x = \frac{7}{13}$ **2.** $a = 140$ **3.** $x = \frac{8}{7}$ **4.** $y = \frac{10}{3}$ **5.** $x = \frac{-4}{7}$ **6.** $x = \frac{6}{5}$

page 483 **1.** $x = \frac{26}{5}$ **2.** $x = \frac{73}{5}$ **3.** $x = \frac{-45}{2}$ **4.** No solution **5.** $x = 0$ **6.** $x = \frac{22}{5}$

Name ______________________ Date ____________

EXERCISE SET 8.6

Solve.

______ 1. $\frac{x}{5} = \frac{7}{10}$

______ 2. $\frac{x}{8} = \frac{9}{20}$

______ 3. $\frac{x}{4} = \frac{x - 2}{2}$

______ 4. $\frac{y}{10} = \frac{y - 10}{5}$

______ 5. $\frac{2x}{7} = \frac{x - 2}{6}$

______ 6. $\frac{3x}{5} = \frac{x - 1}{4}$

______ 7. $\frac{y + 5}{5} = \frac{y - 5}{10}$

______ 8. $\frac{y + 6}{6} = \frac{y - 6}{18}$

______ 9. $\frac{x}{9} + \frac{x}{2} = 5$

______ 10. $\frac{x}{3} - \frac{x}{9} = 2$

______ 11. $\frac{x + 1}{3} - \frac{x}{9} = 4$

______ 12. $\frac{x + 3}{5} - \frac{x}{25} = 1$

______ 13. $\frac{3}{y} - \frac{7}{y} = 4$

______ 14. $\frac{5}{y} - \frac{11}{y} = 6$

______ 15. $\frac{7}{x} - \frac{3}{2} = \frac{5}{2x}$

______ 16. $\frac{11}{x} - \frac{4}{3} = \frac{1}{3x}$

______ 17. $\frac{3}{x - 4} = \frac{2}{x - 1}$

______ 18. $\frac{-1}{x + 2} = \frac{2}{x + 5}$

______ 19. $\frac{z - 1}{z + 1} = 2$

______ 20. $\frac{z + 4}{z - 4} = 3$

______ 21. $\frac{2x - 1}{x^2} - \frac{5}{3x} = 0$

______ 22. $\frac{3x + 1}{x^2} - \frac{7}{5x} = 0$

______ 23. $\frac{x}{x - 7} + \frac{5}{4} = \frac{1}{x - 7}$

______ 24. $\frac{x}{x - 9} + \frac{1}{2} = \frac{5}{x - 9}$

______ 25. $\frac{8}{x - 3} - \frac{2}{x + 4} = \frac{5}{2x - 6}$

______ 26. $\frac{7}{x - 5} - \frac{3}{x + 1} = \frac{2}{3x + 3}$

______ 27. $\frac{11}{4x} = \frac{5}{x + 4}$

______ 28. $\frac{3}{7x} = \frac{2}{x + 7}$

______ 29. $\frac{a}{a - 4} + \frac{a - 8}{a - 4} = 3$

______ 30. $\frac{2a}{a - 3} + \frac{a - 9}{a - 3} = 1$

______ 31. $\frac{4}{x - 7} = \frac{5}{x^2 - 49}$

______ 32. $\frac{3}{x - 5} = \frac{2}{x^2 - 25}$

_______ 33. $\frac{4}{2x - 1} + \frac{1}{2x + 1} = \frac{2}{4x^2 - 1}$

_______ 34. $\frac{5}{3x + 1} - \frac{2}{3x - 1} = \frac{1}{9x^2 - 1}$

_______ 35. $\frac{7}{a - 2} = \frac{3}{a^2 - 7a + 10}$

_______ 36. $\frac{3}{a - 9} = \frac{2}{a^2 - 6a - 27}$

_______ 37. $\frac{3b}{4b^2 - 1} = \frac{-1}{2b - 1}$

_______ 38. $\frac{4b}{9b^2 - 4} = \frac{-2}{3b + 2}$

_______ 39. $\frac{3}{x^2 - x - 12} - \frac{2}{x - 4} = \frac{1}{2x + 6}$

_______ 40. $\frac{1}{x^2 - 6x - 7} - \frac{3}{x - 7} = \frac{2}{3x + 3}$

_______ 41. $\frac{5}{x - 1} = \frac{3x - 2}{x^2 + 15x - 16}$

_______ 42. $\frac{3}{x - 2} = \frac{7x - 4}{x^2 - 8x + 12}$

_______ 43. $\frac{5x + 3}{x - 7} = 2$

_______ 44. $\frac{6x + 5}{x - 9} = 3$

_______ 45. $\frac{a - 2}{a + 5} = \frac{a - 1}{a + 1}$

_______ 46. $\frac{a - 7}{a - 2} = \frac{a + 3}{a + 2}$

_______ 47. $\frac{x}{x - 5} + \frac{2}{5} = \frac{5}{x - 5}$

_______ 48. $\frac{x}{x - 4} + \frac{3}{2} = \frac{4}{x - 4}$

_______ 49. $\frac{2x}{2x - 3} - \frac{6}{2x - 3} = 2$

_______ 50. $\frac{3x}{3x - 1} - \frac{4}{3x - 1} = 4$

_______ 51. $x + \frac{40}{x} = 13$

_______ 52. $x - \frac{28}{x} = 3$

_______ 53. $2 - \frac{15}{x^2} = \frac{7}{x}$

_______ 54. $3 + \frac{5}{x} = \frac{2}{x^2}$

_______ 55. $\frac{y}{y + 12} = \frac{1}{y}$

_______ 56. $\frac{y}{y + 6} = \frac{1}{y}$

_______ 57. $\frac{3x}{2x - 1} = \frac{2}{x - 1}$

_______ 58. $\frac{2x}{x + 6} = \frac{1}{x - 5}$

_______ 59. $\frac{x}{x + 1} = \frac{x + 3}{x^2 - 1}$

_______ 60. $\frac{2x}{x - 3} = \frac{x - 2}{x^2 - 9}$

☆ Stretching the Topics

Solve.

_______ **1.** $\dfrac{6}{2x^2 + 5x - 12} - \dfrac{2}{4x^2 - 9} = \dfrac{4}{2x^2 + 11x + 12}$

_______ **2.** $\dfrac{x}{a} - \dfrac{2x}{b} + \dfrac{3x}{c} = k$ for x.

_______ **3.** $I = \dfrac{E}{\dfrac{Rn + r}{n}}$ for n.

Check your answers in the back of your book.

If you can complete **Checkup 8.6**, you are ready to go on to Section 8.7.

Name ______________________ Date __________

CHECKUP 8.6

Solve.

______ 1. $\frac{14}{y} = \frac{7}{9}$

______ 2. $\frac{x}{2} = \frac{x-3}{5}$

______ 3. $\frac{x-4}{5} = x + 8$

______ 4. $\frac{y+5}{2} = \frac{y-4}{3}$

______ 5. $\frac{6}{a} - \frac{4}{9} = \frac{8}{9a}$

______ 6. $\frac{5}{11x} = \frac{2}{x+11}$

______ 7. $\frac{6x}{x-3} - \frac{x+15}{x-3} = 6$

______ 8. $\frac{2x-1}{x+2} - \frac{x-3}{x-2} = 1$

______ 9. $\frac{x-1}{2x+4} - \frac{x+3}{2x-4} = \frac{2}{x^2-4}$

______ 10. $\frac{2x-1}{x+4} - \frac{x}{x+3} = \frac{-1}{x^2+7x+12}$

Check your answers in the back of your book.

If You Missed Problems:	You Should Review Examples:
1–5	1, 2
6, 7	3, 4
8–10	5, 6

8.7 Switching from Word Statements to Fractional Equations

Suppose that we again practice switching from word statements to equations. Remember that in each problem we must identify the variable and translate the word phrases into variable expressions before coming up with an equation to solve.

Example 1. Maria and Tony earn the same monthly salary. Maria saves $\frac{1}{5}$ of her salary and Tony saves $\frac{1}{6}$ of his salary. Together they save \$550 each month. What is the monthly salary?

Solution: We are asked to find the unknown monthly salary, so we let

$$x = \text{monthly salary} \quad \text{(dollars)}$$

Then

$$\frac{1}{5}x = \text{Maria's savings} \quad \text{(dollars)}$$

and

$$\frac{1}{6}x = \text{Tony's savings} \quad \text{(dollars)}$$

$$\frac{1}{5}x + \frac{1}{6}x = \text{total savings} \quad \text{(dollars)}$$

So

$$\frac{1}{5}x + \frac{1}{6}x = 550$$

$$\frac{x}{5} + \frac{x}{6} = \frac{550}{1} \qquad \text{LCD: } 5 \cdot 6 = 30.$$

$$\frac{30}{1}\left(\frac{x}{5}\right) + \frac{30}{1}\left(\frac{x}{6}\right) = 30(550)$$

$$6x + 5x = 16{,}500$$

$$11x = 16{,}500$$

$$\frac{11x}{11} = \frac{16{,}500}{11}$$

$$x = 1500$$

The monthly salary is \$1500.

Example 2. On long trips, Mary drives her car 5 miles per hour faster than her sister Jo. If Mary can drive 312 miles in the same amount of time that Jo drives 282 miles, how fast does each sister drive?

Solution: The formulas relating distance to rate and time are

distance = rate × time	$\text{rate} = \dfrac{\text{distance}}{\text{time}}$	$\text{time} = \dfrac{\text{distance}}{\text{rate}}$

We know each sister's distance, and we know that they travel those distances in the same amount of time, but we do not know their rates. The rates are unknown quantities.

For Jo	*For Mary*
$r = \text{rate}$	$r + 5 = \text{rate}$
$282 = \text{distance}$	$312 = \text{distance}$
$\frac{282}{r} = \text{time}$	$\frac{312}{r+5} = \text{time}$

Since the sister's driving times are the *same*, we know

$$\text{Jo's time} = \text{Mary's time}$$

$$\frac{282}{r} = \frac{312}{r+5} \qquad \text{LCD: } r(r+5).$$

$$\frac{r(r+5)}{1}\left(\frac{282}{r}\right) = \frac{r(r+5)}{1}\left(\frac{312}{r+5}\right)$$

$$282(r+5) = 312r$$

$$282r + 1410 = 312r$$

$$1410 = 30r$$

$$\frac{1410}{30} = \frac{30r}{30}$$

$$47 = r$$

Jo's rate	*Mary's rate*
$r = 47$ mph	$r + 5 = 52$ mph

Jo drives at a rate of 47 mph and Mary drives at a rate of 52 mph.

Ratios are handy tools to use in solving everyday mathematical problems.

A **ratio** is a way of comparing two quantities by means of a fraction.

First Quantity	Second Quantity	Ratio	Interpretation
23 males	11 females	$\frac{\text{males}}{\text{females}} = \frac{23}{11}$	For every 23 males, there are 11 females.
\$250 saved	\$750 spent	$\frac{\text{spending}}{\text{saving}} = \frac{750}{250} = \frac{3}{1}$	For every \$3 spent, \$1 was saved.
7 games won	5 games lost	$\frac{\text{wins}}{\text{losses}} = \frac{7}{5}$	For every 7 games won, there were 5 games lost.

You try to set up the ratio in Example 3.

Example 3. Martin earns \$1500 per month and Clarence earns \$1200 per month. Find the ratio of Martin's salary to Clarence's salary.

Solution

$$\frac{\text{Martin's salary}}{\text{Clarence's salary}} = \frac{1500}{1200}$$

$$= \frac{5}{4}$$

Check your work on page 497. ▶

Example 4. If Martin's salary is raised to \$1800, what must be Clarence's new salary so that the ratio will continue to be $\frac{5}{4}$?

Solution: Let C be Clarence's new salary. Then

$$\frac{1800}{C} = \frac{5}{4} \qquad \text{LCD: } 4C.$$

$$\frac{4C}{1}\left(\frac{1800}{C}\right) = \frac{4C}{1}\left(\frac{5}{4}\right)$$

$$4(1800) = 5C$$

$$7200 = 5C$$

$$\frac{7200}{5} = \frac{5C}{5}$$

$$1440 = C$$

Clarence's new salary must be \$1440.

In Example 4, we wrote an equation that said that two ratios were equal. Such an equation is called a **proportion**.

A *proportion* is an equation that states that two ratios are equal.

It is easier to solve fractional equations that are proportions by using the definition of equal fractions.

Equality of Fractions

If $\frac{A}{B} = \frac{C}{D}$ then $AD = BC$

(provided that $B \neq 0$, $D \neq 0$).

A numerical example should convince you that this definition makes sense.

$$\frac{9}{12} = \frac{3}{4}$$

$$9(4) = 12(3)$$

$$36 = 36$$

We shall use the definition of equality of fractions (sometimes called the method of **cross products**) to solve all our proportion problems. You must realize, however, that we can use this method only when solving a fractional equation that states that *one* fraction is equal to *one* other fraction. We cannot use this method if either side of a fractional equation contains a sum of fractions. In those situations, we *must* multiply both sides by the LCD.

Example 5. In an orchard of 95 trees, the ratio of pear trees to apple trees is 2 to 3. How many pear trees and how many apple trees are in the orchard?

Solution: If we let

$$x = \text{number of pear trees}$$

the number of apple trees can be found by subtracting the number of pear trees (x) from the total number of trees (95).

$$95 - x = \text{number of apple trees}$$

$$\frac{x}{95 - x} = \text{ratio of pears to apples}$$

and we write the proportion

$$\frac{x}{95 - x} = \frac{2}{3}$$

$3x = 2(95 - x)$	Use equality of fractions.
$3x = 190 - 2x$	Remove parentheses.
$5x = 190$	Isolate the variable term.
$\frac{5x}{5} = \frac{190}{5}$	Isolate the variable.
$x = 38$	Simplify.

There are 38 pear trees in the orchard.
There are $95 - 38 = 57$ apple trees.

Now we are ready to solve the problem stated at the beginning of the chapter.

Example 6. At the beginning of the semester the number of males in Harry's math class was 3 more than twice the number of females. After 5 males and 5 females had dropped the class, the ratio of females to males was 1 to 4. How many males and females were in Harry's math class at the beginning of the semester?

Solution: Since the original number of males is described in terms of the number of females, we let

$$x = \text{original number of females}$$

	Beginning of Semester	End of Semester
Number of females	x	$x - 5$
Number of males	$2x + 3$	$(2x + 3) - 5$ $= 2x - 2$

Since we know the ratio of females to males is $\frac{1}{4}$ at the end of the semester,

$$\frac{x-5}{2x-2} = \frac{1}{4}$$

$4(x-5) = 2x-2$	Use equality of fractions.
$4x-20 = 2x-2$	Remove parentheses.
$2x =$ ____	Isolate the variable term.
$x =$ ____	Isolate the variable.

	Beginning of Semester	End of Semester
Number of females	$x =$ ____	$x-5 =$ ____ -5 $=$ ____
Number of males	$2x+3 = 2($____$)+3$ $=$ ____	$2x-2 = 2($____$)-2$ $=$ ____

There were ____ females and ____ males in Harry's math class at the beginning of the semester. Check your work on page 497. ▶

If two or more people or machines work together to complete a job, we expect that less time will be needed than if one person or machine were to complete the same job alone. Fractional equations can be used to solve **work problems**.

Suppose that it takes Linda 4 hours to rake the lawn, but it takes her older sister Nancy only 2 hours to rake the same lawn. If they rake the lawn *together*, how long will it take them?

Linda's rate	*Nancy's rate*
$\frac{1}{4}$ lawn per hour	$\frac{1}{2}$ lawn per hour

The amount of lawn raked by each sister is found by multiplying her hourly raking rate times the number of hours she rakes.

> Amount of work = (rate of work) · (time worked)

Since Nancy and Linda work together for the *same* number of hours, we can let

$$t = \text{hours worked}$$

Linda's work	*Nancy's work*
$\frac{1}{4} \cdot t$	$\frac{1}{2} \cdot t$

The total work to be done is 1 lawn, so

$$\text{Linda's work} + \text{Nancy's work} = 1 \text{ lawn}$$

$$\frac{1}{4}t + \frac{1}{2}t = 1$$

$$\frac{1}{4}\cdot\frac{t}{1} + \frac{1}{2}\cdot\frac{t}{1} = 1$$

$$\frac{t}{4} + \frac{t}{2} = 1 \qquad \text{LCD: 4.}$$

$$\frac{4}{1}\cdot\frac{t}{4} + \frac{4}{1}\cdot\frac{t}{2} = 4\cdot 1$$

$$t + 2t = 4$$

$$3t = 4$$

$$t = \frac{4}{3} \text{ hours}$$

Together the sisters can rake the lawn in $\frac{4}{3}$ or $1\frac{1}{3}$ hours (or 1 hour 20 minutes). Notice that the amount of time needed is *less* than either sister's time alone. This will always be true.

Example 7 A tub can be filled with hot water in 15 minutes and with cold water in 10 minutes. How long will it take to fill the tub if both faucets are allowed to run at the given rate?

Solution: We let t = time for both faucets.

	Rate	Time	Work
Hot water	$\frac{1}{15}$	t	$\frac{1}{15}\cdot t$
Cold water	$\frac{1}{10}$	t	$\frac{1}{10}\cdot t$

$$\frac{1}{15}t + \frac{1}{10}t = 1 \text{ tub}$$

$$\frac{t}{15} + \frac{t}{10} = 1 \qquad \text{LCD: 30.}$$

$$\frac{30}{1}\cdot\frac{t}{15} + \frac{30}{1}\cdot\frac{t}{10} = 30(1)$$

$$2t + 3t = 30$$

$$5t = 30$$

$$t = 6 \text{ minutes}$$

The tub will fill in 6 minutes.

▶ Examples You Completed

Example 3

Solution

$$\frac{\text{Martin's salary}}{\text{Clarence's salary}} = \frac{1500}{1200}$$
$$= \frac{5}{4}$$

Example 6

Solution

$$\frac{x - 5}{2x - 2} = \frac{1}{4}$$
$$4(x - 5) = 2x - 2$$
$$4x - 20 = 2x - 2$$
$$2x = 18$$
$$x = 9$$

	Beginning of Semester	End of Semester
Number of females	$x = 9$	$x - 5 = 9 - 5$ $= 4$
Number of males	$2x + 3 = 2(9) + 3$ $= 21$	$2x - 2 = 2(9) - 2$ $= 16$

There were 9 females and 21 males in Harry's math class at the beginning of the semester.

Name ______________________________ Date ____________

EXERCISE SET 8.7

For each problem, write an equation and then solve it.

_______ **1.** Eric and Kate each own interest in the Specialty Shop. Each month Eric receives $\frac{1}{10}$ of the profits and Kate receives $\frac{1}{8}$ of the profits. Together one month they received $630. How much profit did the Specialty shop make that month?

_______ **2.** Each month Marco saves $\frac{1}{10}$ of his paycheck in the credit union and $\frac{1}{12}$ of his paycheck in his savings account. If he saves a total of $154 each month, what is Marco's monthly paycheck?

_______ **3.** Louise's average driving speed is 6 miles per hour faster than her husband's average speed. If Louise can drive 280 miles in the same length of time that her husband drives 250 miles, how fast does each drive?

_______ **4.** Driving 3 miles per hour faster than his normal rate, Henry can travel 638 miles in the same time it takes him to travel 605 miles at his normal rate. What is Henry's normal rate of speed?

_______ **5.** José and Rick assembled 36 parts of a carburetor in an hour. The ratio of the number Rick did to the number José did is 4 to 5. Find how many parts each assembled during the hour.

_______ **6.** The ratio of seniors to graduate students in an upper division course is 7 to 3. If there are 16 more seniors than graduate students, how many graduate students are in the class?

_______ **7.** On a road map, 2 inches represents 30 miles. If the distance between Birmingham and Chattanooga is 11 inches on the map, how far apart are the two cities?

_______ **8.** The ratio of students to teachers at Banton High School is 3 teachers for every 70 students. If the school presently has 24 teachers, what is the enrollment of the student body?

_______ **9.** At State University the ratio of in-state students to out-of-state students is 5 to 2. If there are 3750 more in-state students than out-of-state students, how many out-of-state students are enrolled?

_______ **10.** In the freshmen class at P.J.C, there are 50 more computer science majors than there are math majors. If the ratio of computer science majors to math majors is 7 to 2, find how many freshmen are computer science majors.

_______ **11.** The width of a rectangle is $\frac{2}{3}$ of its length. If the perimeter is 60 feet, find the dimensions. (*Remember*: $P = 2l + 2w$.)

_______ **12.** The width of a rectangular lot is $\frac{2}{5}$ of its length. If the perimeter of the rectangle is 140 feet, find the dimensions of the lot.

________ **13.** A number is added to the numerator, and twice the same number is subtracted from the denominator of the fraction $\frac{4}{15}$. The resulting fraction is equal to $\frac{7}{9}$. Find the number.

________ **14.** If the denominator of $\frac{x}{12}$ is decreased by x, the new fraction is equal to $\frac{1}{2}$. Find the value of x.

________ **15.** It takes 5 hours to fill a watering tank for the elephants at the zoo using a pump. The tank can be filled in 10 hours using a hose. How long will it take to fill the tank using both the pump and the hose at the same time?

________ **16.** Sam can paint Sherry's house in 5 days. Sherry can paint it by herself in 7 days. How long will it take them to paint the house if they work together?

________ **17.** Louis can clean the dog pens at the Humane Shelter in 8 hours and Margaret can do it in 6 hours. How long will it take them to clean the pens if they work together?

________ **18.** Charlie can clean the theater after a play in 6 hours and Ricky can clean the theater in 4 hours. If they work together, how long will it take them to clean the theater?

________ **19.** If an $8000 investment earns $480 in 6 months, how much will $15,000 earn invested at the same rate for the same period of time?

________ **20.** The interest owed on a loan of $5000 borrowed for 6 months is $300. If $12,000 is borrowed for the same period of time at the same rate of interest, how much interest will be owed?

☆ Stretching the Topics

Write an equation and solve.

________ **1.** In a fraction the denominator is 4 greater than the numerator. If 24 is added to the numerator, the fraction will be equal to the reciprocal of the original fraction. What is the original numerator of the fraction?

________ **2.** Two pipes are pouring water into an irrigation reservoir. The first pipe can fill it in 20 days but if both pipes are used the reservoir can be filled in 12 days. How many days would it take the second pipe to fill the reservoir if the first pipe is not open?

________ **3.** Carlita's scout troop hiked 10 miles to Horseshoe Cave for a picnic. They walked 3 miles per hour faster going than they did returning. The total time spent hiking was 7 hours. How fast did the troop hike on the return trip?

Check your answers in the back of your book.

If you can solve the problems in **Checkup 8.7**, you are ready to go on to the **Review Exercises for Chapter 8**.

Name ______________________ Date ____________

✓ CHECKUP 8.7

For each problem, write an equation and then solve it.

______ **1.** Sidney owns $\frac{3}{8}$ interest in a farm and his brother, Wallis, owns $\frac{1}{4}$ interest. When the farm was sold, their combined share was $150,000. What was the total selling price of the farm?

______ **2.** In a certain math course the ratio of success to failure is 20 to 3. If there are 115 students enrolled in the course, how many can be expected to succeed?

______ **3.** Faye's average jogging speed is 2 miles per hour faster than her daughter's average speed. If Faye can jog 9 miles in the same length of time that her daughter can jog 6 miles, how fast does each jog?

______ **4.** If the denominator of the fraction $\frac{x}{7}$ is decreased by x, the new fraction is equal to $\frac{2}{5}$. Find the value of x.

______ **5.** It takes Daphne 4 hours to mow the playground with her riding lawn mower, but it takes Sara 6 hours with her push mower. If they both begin mowing together, how long will it take them to finish?

Check your answers in the back of your book.

If You Missed Problem:	You Should Review Example:
1	1
2	5
3	2
4	6
5	7

Summary

In this chapter we learned to work with **rational algebraic expressions**, which are fractions with polynomial numerators and polynomial denominators.

In Order to:	We Must:	Examples
Restrict the variable in a rational expression	Exclude any value of the variable that makes the denominator zero.	For $\frac{3x}{y^2}$, $y \neq 0$ For $\frac{3}{x^2 + 2x} = \frac{3}{x(x + 2)}$, $x \neq 0, x \neq -2$
Reduce a rational expression	Factor numerator and denominator. Divide numerator and denominator by any common factors.	$\frac{x^9y}{x^2y^3} = \frac{x^7}{y^2}$ $\frac{3x + 6}{x^2 - 4} = \frac{3(x + 2)}{(x + 2)(x - 2)}$ $= \frac{3}{x - 2}$
Multiply rational expression	Factor each numerator and denominator. Reduce if possible. Then multiply remaining numerators and denominators.	$\frac{x^2y}{3z} \cdot \frac{2z^2}{x} = \frac{2xyz}{3}$ $\frac{x^2 + x}{10} \cdot \frac{5x - 5}{x^2 - 1}$ $= \frac{x(x + 1)}{10} \cdot \frac{5(x - 1)}{(x + 1)(x - 1)}$ $= \frac{x}{2}$
Divide rational expressions	Invert the divisor and multiply.	$\frac{2x}{3} \div \frac{x^2}{6} = \frac{2x}{3} \cdot \frac{6}{x^2}$ $= \frac{4}{x}$ $\frac{a - 4}{a^2} \div \frac{a^2 - 3a - 4}{7a}$ $= \frac{a - 4}{a^2} \cdot \frac{7a}{(a - 4)(a + 1)}$ $= \frac{7}{a(a + 1)}$

In Order to:	We Must:	Examples
Build a rational expression	Multiply numerator and denominator by the same factor.	$\frac{3}{x+3} = \frac{?}{x^2+5x+6}$ $= \frac{?}{(x+3)(x+2)}$ $= \frac{3(x+2)}{(x+3)(x+2)}$ $= \frac{3x+6}{x^2+5x+6}$
Add (or subtract) rational expressions	Choose the LCD and build each fraction to a new fraction having the LCD. Combine numerators over LCD. Simplify and reduce if possible.	$\frac{x}{x+1} - \frac{3}{x+2}$ LCD: $(x+1)(x+2)$ $= \frac{x(x+2)}{(x+1)(x+2)} + \frac{-3(x+1)}{(x+1)(x+2)}$ $= \frac{x^2+2x-3x-3}{(x+1)(x+2)}$ $= \frac{x^2-x-3}{(x+1)(x+2)}$
Simplify a complex fraction (Method 1)	Rewrite the complex fraction as a division problem. Write the dividend and the divisor as single fractions. Invert the divisor and multiply.	$\dfrac{\frac{2}{x}+\frac{1}{3}}{\frac{2x+12}{x^2}}$ $= \left[\frac{2}{x}+\frac{1}{3}\right] \div \left[\frac{2x+12}{x^2}\right]$ $= \left[\frac{6}{3x}+\frac{x}{3x}\right] \div \left[\frac{2x+12}{x^2}\right]$ $= \frac{6+x}{3x} \cdot \frac{x^2}{2x+12}$ $= \frac{\overset{1}{\cancel{6+x}}}{3\underset{1}{\cancel{x}}} \cdot \frac{\overset{x}{\cancel{x^2}}}{2\underset{1}{\cancel{(x+6)}}} = \frac{x}{6}$
Simplify a complex fraction (Method 2)	Multiply numerator and denominator by LCD for *both* numerator and denominator.	$\dfrac{\frac{2}{x}+\frac{1}{3}}{\frac{2x+12}{x^2}}$ LCD: $3x^2$ $= \dfrac{3x^2\left[\frac{2}{x}+\frac{1}{3}\right]}{3x^2\left[\frac{2x+12}{x^2}\right]}$ $= \frac{6x+x^2}{3(2x+12)} = \frac{x\overset{1}{\cancel{(6+x)}}}{6\underset{1}{\cancel{(x+6)}}}$ $= \frac{x}{6}$

In Order to:	We Must:	Examples
Solve a fractional equation	Multiply both sides of the equation by the LCD. Solve and check solutions against restrictions on the variable.	$\frac{3}{x} + \frac{5}{x+1} = \frac{1}{2x}$ LCD: $2x(x+1)$; $x \neq 0$, $x \neq -1$ $\frac{2x(x+1)}{1}\left[\frac{3}{x}\right] + \frac{2x(x+1)}{1}\left[\frac{5}{x+1}\right] = \frac{2x(x+1)}{1}\left[\frac{1}{2x}\right]$ $6(x+1) + 10x = x + 1$ $6x + 6 + 10x = x + 1$ $15x = -5$ $x = \frac{-5}{15}$ $x = \frac{-1}{3}$
Solve a proportion	Use the definition of equality of fractions (cross products).	$\frac{x+3}{x} = \frac{6}{5}$ $5(x+3) = 6x$ $5x + 15 = 6x$ $15 = x$

Having mastered the skills for working with rational algebraic expressions, we practiced switching from word statements to fractional equations.

❑ Speaking the Language of Algebra

1. A fraction with a polynomial numerator and a polynomial denominator is called a ________ ________ ________ .
2. We must sometimes find restrictions on the variable in a rational expression because division by ________ is ________ .
3. To multiply rational algebraic expressions, we multiply ________ and multiply ________ .
4. In addition and subtraction of rational algebraic expressions, we must be sure the fractions have the same ________ .
5. After a complex fraction has been rewritten as a division problem, the dividend and divisor must each be expressed as a ________ ________ .
6. To solve a fractional equation, we multiply both sides of the equation by the ________ ________ ________ for all the fractions in the equation.
7. A solution for a new version of an equation that does *not* satisfy the original equation is called an ________ solution.
8. An equation that states that two ratios are equal is called a ________ .

Check your answers in the back of your book.

Name ______________________________ Date ______________

REVIEW EXERCISES for Chapter 8

Evaluate and reduce if possible.

______ 1. $\frac{a - 3}{a + 6}$ when $a = 12$

______ 2. $\frac{x - 5}{x^2 - 25}$ when $x = -10$

Find the restrictions on the variable for each expression.

______ 3. $\frac{x^2}{x + 4}$

______ 4. $\frac{5x}{2x - 1}$

Reduce the fractions.

______ 5. $\frac{12x^7}{6x^2}$

______ 6. $\frac{45x^4y^3}{36x^2y}$

______ 7. $\frac{5a + 15b}{5}$

______ 8. $\frac{64x^5}{8x^3 + 8x}$

______ 9. $\frac{-7a - 7b}{35a + 35b}$

______ 10. $\frac{x^2 - 16}{5x^2 + 20x}$

______ 11. $\frac{5a + 6b}{25a^2 - 36b^2}$

______ 12. $\frac{9x - 9y}{3y^2 - 3x^2}$

______ 13. $\frac{x^2 - 10x + 25}{2x^2 - 13x + 15}$

______ 14. $\frac{12x^2 + 8xy - 4y^2}{12x - 4y}$

Simplify.

______ 15. $\frac{3x}{7y} \cdot \frac{49y^2}{x^3} \cdot \frac{x^2y}{18}$

______ 16. $\frac{a^3b^2}{3c} \div \frac{a^2b}{27c^2}$

______ 17. $\frac{10a + 20b}{a^3b} \cdot \frac{a^4b^5}{a^2 + 2ab}$

______ 18. $\frac{(x + 4)^2}{64yz^2} \div \frac{3x^2 - 48}{24y^3z}$

______ 19. $\frac{10x^2 - 11x - 6}{2x^2 + 5x - 12} \cdot \frac{3x^2 + 11x - 4}{15x^2 - 11x + 2}$

______ 20. $\frac{4x^2 - 8xy}{x^2 - 4y^2} \div \frac{16x^2 - 24xy}{2x^2 + xy - 6y^2}$

Find the new fraction.

______ 21. $\frac{3x}{7y} = \frac{?}{35y^2}$

______ 22. $\frac{3y}{y - 2} = \frac{?}{y^2 - 4}$

Perform the operations and simplify.

______ 23. $\frac{7x}{2y} + \frac{5}{2y} - \frac{3x}{2y} + \frac{1}{2y}$

______ 24. $\frac{2x^2 - 8x}{6x^2 + 7x - 3} + \frac{x - 15}{6x^2 + 7x - 3}$

________ 25. $\frac{8}{x} + \frac{1}{x^2} - \frac{3}{4x^2}$

________ 26. $\frac{x + 6}{3x} - \frac{3x - 4}{2x}$

________ 27. $\frac{x + y}{4x^2y} - \frac{x - 2y}{3xy^2} - \frac{21}{12xy}$

________ 28. $\frac{5x}{6x - 18} - \frac{3x}{2x - 6}$

________ 29. $\frac{2a}{a^2 - 4} - \frac{3}{a^2 - a - 2}$

________ 30. $\frac{x + 5}{x^2 + 3x - 18} - \frac{4}{x + 6}$

________ 31. $\dfrac{\frac{a^3b^2}{6c^3}}{\frac{a^4b^7}{15c}}$

________ 32. $\dfrac{\frac{1}{x^2} - \frac{1}{49}}{\frac{1}{x} - \frac{1}{7}}$

Solve.

________ 33. $\frac{x - 3}{5} - \frac{x}{10} = 1$

________ 34. $\frac{2x - 5}{x^2} - \frac{7}{3x} = 0$

________ 35. $\frac{y}{y - 3} + \frac{7}{3} = \frac{3}{y - 3}$

________ 36. $\frac{a + 3}{a + 1} = \frac{a - 3}{a - 2}$

________ 37. $\frac{a}{a - 4} + \frac{2}{a} = \frac{16}{a^2 - 4a}$

________ 38. $\frac{x + 3}{x + 5} + \frac{2}{x - 9} = \frac{-20}{x^2 - 4x - 45}$

________ 39. If 4 gallons of paint will cover 600 square feet, how many gallons will be needed to paint 1800 square feet?

________ 40. The number of games the Braves lost last year was 5 less than the number they won. If the ratio of the games won to the games lost is 3 to 2, find how many games the Braves played last year.

________ 41. It takes Megan 3 hours to prepare the medicine for the patients in the north wing of the hospital, but Fred can do it in 2 hours. If they work together, how long will it take them to prepare the medicine?

________ 42. On a vacation it took Maurice the same time to drive 275 miles as it took Denise to drive 325 miles. Maurice's rate of speed was 10 miles per hour less than Denise's. How fast was Denise driving?

Check your answers in the back of your book.

If You Missed Exercises:	You Should Review Examples:	
1, 2	SECTION 8.1	1–4
3, 4		5, 6
5, 6		7, 8
7–9		9–12
10–14		13–16
15, 16	SECTION 8.2	1–5
17–20		6–11
21	SECTION 8.3	1, 2
22		3–6
23, 24	SECTION 8.4	2–5
25–27		6–8
28–30		9–11
31	SECTION 8.5	1, 2
32		3, 4
33, 34	SECTION 8.6	1, 2
35–38		3–6
39–42	SECTION 8.7	1–7

If you have completed the **Review Exercises** and corrected your errors, you are ready to take the **Practice Test for Chapter 8**.

Name ______________________ Date ____________

PRACTICE TEST for Chapter 8

		Section	Examples
______	1. Evaluate $\frac{3x - 2}{x + 5}$ when $x = 3$.	8.1	1, 2
______	2. Find the restrictions on the variable for the expression $\frac{y + 2}{3y + 1}$.	8.1	5

Reduce the fractions.

		Section	Examples
______	3. $\frac{-11x^3y^4}{44x^6y^3}$	8.1	7, 8
______	4. $\frac{36x^4}{6x^3 + 12x^2}$	8.1	12
______	5. $\frac{2x^2 - 10x}{x^2 - 25}$	8.1	13, 14
______	6. $\frac{3x^2 - 3y^2}{-9x - 9y}$	8.1	15, 16

Simplify.

		Section	Examples
______	7. $\frac{2a^2}{3b} \cdot \frac{6b^3}{a^5} \cdot \frac{ab}{10}$	8.2	2
______	8. $\frac{27y}{20x^4} \div \frac{15y^5}{8x^2}$	8.2	4
______	9. $\frac{x^2 - 5x - 6}{3x + 3} \div \frac{5x^2 - 30x}{6x^2}$	8.2	10
______	10. $\frac{6x^2 - 7x - 5}{2x^2 - 9x - 5} \cdot \frac{3x^2 - 14x - 5}{9x^2 - 12x - 5}$	8.2	8
______	11. $\dfrac{\frac{4x^2 - 1}{x^5}}{\frac{6x - 3}{x^2 + x}}$	8.5	2

Combine the fractions.

		Section	Examples
______	12. $\frac{4x}{5a} + \frac{3}{5a} + \frac{x}{5a} - \frac{8}{5a}$	8.4	2
______	13. $\frac{2x^2 - x}{3x^2 + 5x - 2} + \frac{x^2 - 3x + 1}{3x^2 + 5x - 2}$	8.4	5

________ **14.** $\dfrac{4}{a^2b} - \dfrac{3}{5ab} + \dfrac{1}{3ab^2}$ 8.4 7

________ **15.** $\dfrac{6}{y + 5} - \dfrac{6}{y - 5}$ 8.4 8, 9

________ **16.** $\dfrac{3}{a^2 + a - 12} - \dfrac{2}{a^2 - 3a}$ 8.4 11

Solve.

________ **17.** $\dfrac{x + 2}{9} - \dfrac{x}{27} = 1$ 8.6 1

________ **18.** $\dfrac{5}{3x} - \dfrac{4}{x + 7} = 0$ 8.6 3

________ **19.** $\dfrac{x - 5}{5x^2 - 3x - 2} = \dfrac{2}{x - 1}$ 8.6 5, 6

________ **20.** On a recent plane flight, the ratio of first class passengers to tourist class passengers was 2 to 11. If the number of tourist passengers was 6 more than 5 times the number of first class passengers, how many first class passengers were on the flight? 8.6 6

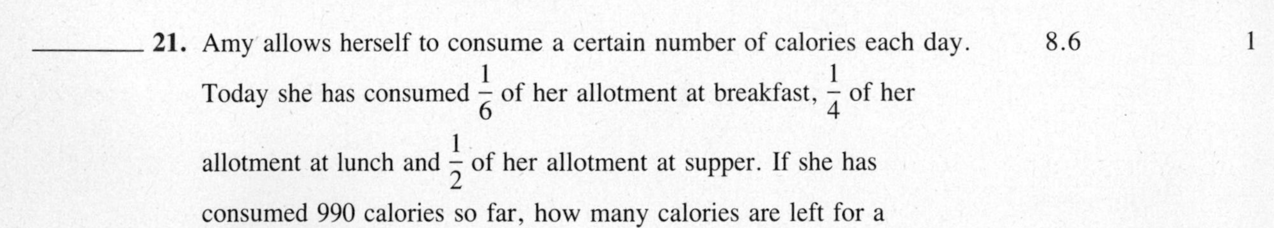

________ **21.** Amy allows herself to consume a certain number of calories each day. Today she has consumed $\frac{1}{6}$ of her allotment at breakfast, $\frac{1}{4}$ of her allotment at lunch and $\frac{1}{2}$ of her allotment at supper. If she has consumed 990 calories so far, how many calories are left for a midnight snack? 8.6 1

________ **22.** Mel can clean the oven at the pizza parlor in 5 hours, but Luigi can do the same job in 4 hours. How long will it take them if they work together? 8.6 7

Check your answers in the back of your book.

Name ______________________ Date ____________

SHARPENING YOUR SKILLS after Chapters 1–8

Simplify. SECTION

_______ **1.** $\dfrac{5(-2)^2 - 3 \cdot 4}{(-6)(4) + 3 \cdot 2^3}$ 2.3

_______ **2.** $\left(\dfrac{2}{3}\right)^2 \div \left(\dfrac{3}{5} \cdot \dfrac{10}{27}\right)$ 2.4

_______ **3.** $-5[3(2x - y) - 7(x + 2y)]$ 3.3

_______ **4.** $(-2x^2y)^3(xy^3)^2$ 5.1

_______ **5.** $\left(\dfrac{-27a^2b^5}{28a^7}\right)^2$ 5.2

_______ **6.** $(-4a^3b)(3ab^2)^2$ 6.2

_______ **7.** $(2a - 3b)^2$ 6.3

_______ **8.** $(x - 0.5y)(x + 0.5y)$ 6.3

_______ **9.** $-2x(x - 5y)(x + 2y)$ 6.3

Factor.

_______ **10.** $15x^3y - 65x^2y^2 + 20xy^3$ 7.1

_______ **11.** $2x^2 - 98$ 7.2

_______ **12.** $x^2 - 10xy + 25y^2$ 7.3

_______ **13.** $-15x^2 - 80x + 60$ 7.3

Solve.

_______ **14.** $(4x - 3) - (2x + 6) = -9$ 4.2

_______ **15.** $2(5y - 1) - 5 = 3(y + 2)$ 4.2

_______ **16.** $4y(3y - 4) = 5y(y + 1)$ 7.5

_______ **17.** One leg of a right triangle is 7 inches less than the other leg. The hypotenuse is 13 inches. Find the two legs. 7.6

_______ **18.** In a basketball game, the winning team's score was 17 less than twice the points scored by the losing team. If the winner's score was 59, what was the loser's score? 4.3

Check your answers in the back of your book.

Working with Inequalities 9

To qualify for federal funding, a special program must enroll between 150 and 210 students. Moreover, the number of rural students must be twice the number of nonrural students. How many of each type of student must be enrolled in the special program?

In Chapter 1 we learned that numbers can be compared easily by describing their locations on the number line. Statements that compare two quantities using phrases such as

"is less than"
"is greater than"
"is between"

are called **inequalities**.

In this chapter we learn how to

1. Write and graph numerical inequalities and variable inequalities.
2. Simplify and solve first-degree inequalities.
3. Combine inequalities containing variables.
4. Switch from word statements to variable inequalities.

9.1 Writing and Graphing Inequalities

Mathematicians have invented some handy symbols to be used in writing inequalities. If we are given any two real numbers, a and b, it is always possible to compare them.

Symbols	Words	Number Line
$a < b$	a is less than b	a lies to the left of b
$a > b$	a is greater than b	a lies to the right of b
$a = b$	a is equal to b	a and b are at the same point

Let's practice comparing some numbers using symbols.

Example 1. Fill in the blanks with <, >, or =.

$$2 \_\_\_ 10$$

$$-5 \_\_\_ 0$$

$$\frac{1}{2} \_\_\_ \frac{1}{3}$$

Solution

$$2 < 10$$

$$-5 < 0$$

$$\frac{1}{2} > \frac{1}{3}$$

You complete Example 2.

Example 2. Fill in the blanks with <, >, or =.

$$1 \_\_\_ 0$$

$$-3 \_\_\_ -7$$

$$\frac{-5}{2} \_\_\_ \frac{-10}{4}$$

Solution

$$1 \_\_\_ 0$$

$$-3 \_\_\_ -7$$

$$\frac{-5}{2} \_\_\_ -\frac{10}{4}$$

Check your work on page 518. ▶

When one number lies between two other numbers, we may use a three-part inequality to express that fact. For example, to say that 5 is between 2 and 9, we may write

$$2 < 5 < 9$$

This inequality says that 5 is greater than 2 *and* less than 9.

Example 3. Write an inequality to say that 0 is between -1 and 3.

Solution

$$-1 < 0 < 3$$

Now you try Example 4.

Example 4. Write an inequality to say that $\frac{-1}{2}$ is between $\frac{-3}{4}$ and $\frac{-1}{4}$.

Solution

$$\frac{-3}{4} < \frac{\_\_\_}{\_\_\_} < \frac{\_\_\_}{\_\_\_}$$

Check your work on page 518. ▶

Suppose that we are asked to find a number that is less than 7. One student might respond with the answer 6. Another might suggest 0. Still others might offer $\frac{5}{2}$ or -3 or -659. Which answer is correct? All of them are correct, of course. There is, in fact, an endless list of numbers less than 7. We might say there is an *infinite* number of correct answers to such a problem.

In the problem described, we were being asked to find all the values of x that would satisfy the inequality

$$x < 7$$

We discovered that there is no practical way to *list* all the solutions to such a **variable inequality**. One way to get a better idea of what numbers are in this infinite list is to use the number line.

On the number line, all the numbers that satisfy the inequality $x < 7$ are the numbers that lie to the *left* of 7. We can picture those numbers by

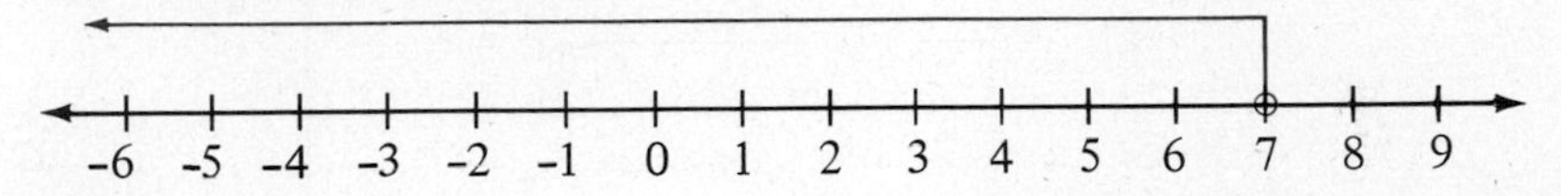

Such a picture is called a **graph** of the inequality. Notice the *open dot* at the point 7 which shows that 7 itself is *not* a solution to the inequality but that any number up to 7 (such as 6.9 or 6.999 and so on) *does* satisfy the inequality. Notice also that the arrow pointing to the left on the graph shows that it extends forever in that direction.

Let's practice graphing solutions to inequalities.

Example 5. Graph the solutions for

$$a > -2$$

Solution: Our solutions are all numbers to the ________ of -2.

$^-4$ $^-3$ $^-2$ $^-1$ 0 1 2 3

Check your work on page 519. ▶

Example 6. Graph the solutions for

$$-1 < x < 3$$

Solution: This inequality says that the middle quantity, x, lies *between* the other quantities.

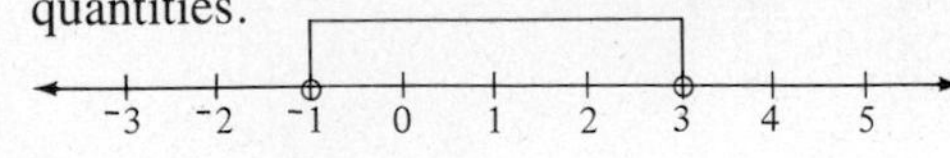

Suppose that we are asked to find all numbers that are less than 5 *or* equal to 5. In symbols, this is written $x \leq 5$. On the number line, we need all the numbers that lie to the *left* of 5 or *at* the point 5.

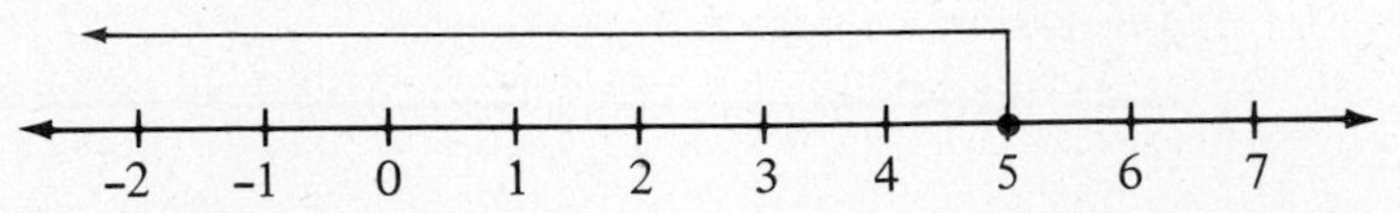

Note the *solid dot* at 5 to show that 5 itself is a solution to the inequality.

Symbols	Words	Number Line
$x \leq a$	x is less than *or* equal to a	a
$x \geq a$	x is greater than *or* equal to a	a

Example 7. Graph the solutions for $y \geq 0$.

Solution: We need numbers to the right of 0 or at 0.

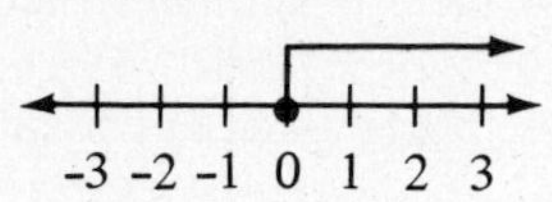

You graph Example 8.

Example 8. Graph the solutions for

$$x \leq -3$$

Solution: We need numbers to the ________ of -3 or at -3.

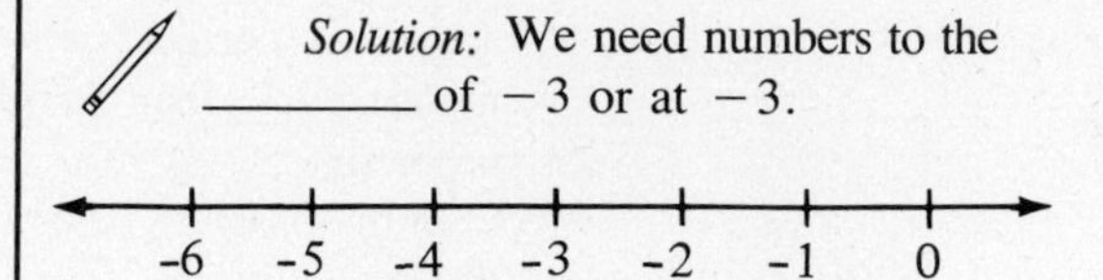

Check your work on page 519. ▶

Example 9. Graph the solutions for

$$\frac{-1}{2} \leq x < 4$$

Solution

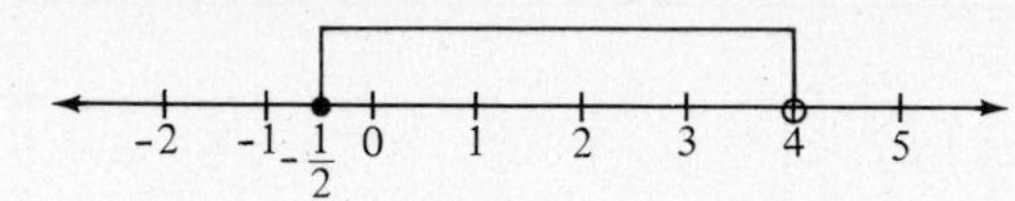

Now try Example 10.

Example 10. Graph the solutions for

$$0 < x \leq \frac{3}{2}$$

Solution

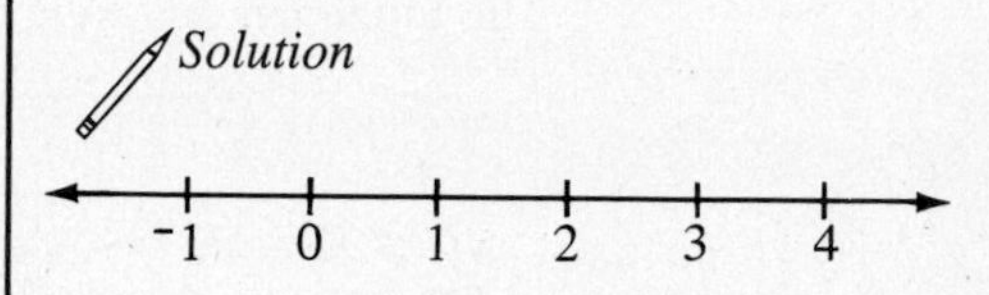

Check your work on page 519. ▶

Trial Run

Graph the solutions on number lines.

1. $a > 5$

2. $x \leq -2$

3. $0 \leq x \leq \frac{5}{2}$

4. $-2 < x \leq 1$

Answers are on page 519.

▶ Examples You Completed

Example 2

Solution

$$1 > 0$$

$$-3 > -7$$

$$\frac{-5}{2} = \frac{-10}{4}$$

Example 4. Write an inequality to say that $\frac{-1}{2}$ is between $\frac{-3}{4}$ and $\frac{-1}{4}$.

Solution

$$\frac{-3}{4} < -\frac{1}{2} < \frac{-1}{4}$$

Example 5. Graph the solutions for $a > -2$.

Solution: Our solutions are all numbers to the right of -2.

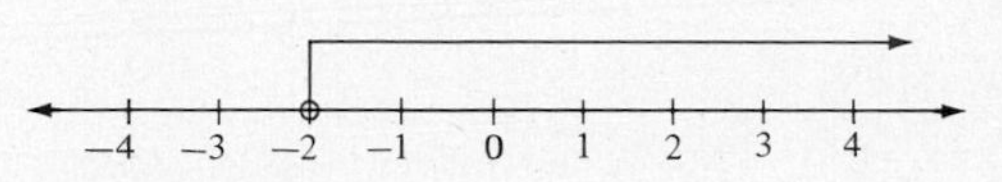

Example 8. Graph the solutions for $x \leq -3$.

Solution: We need numbers to the left of -3 or at -3.

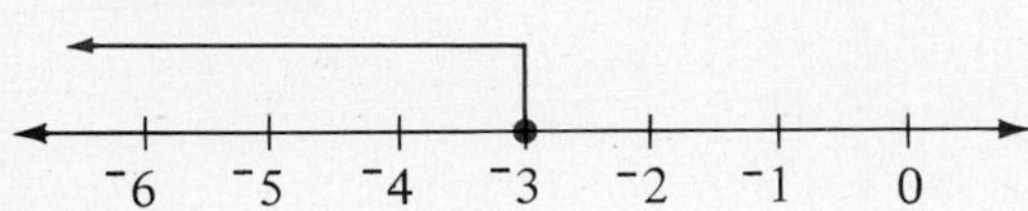

Example 10. Graph the solutions for

$$0 < x \leq \frac{3}{2}$$

Solution

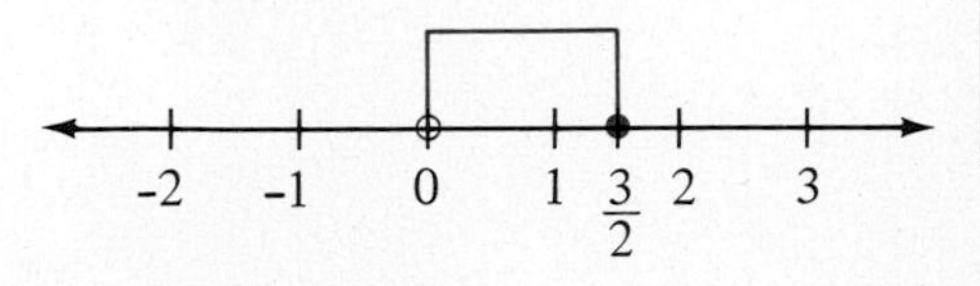

Answers to Trial Run

page 518 **1.**

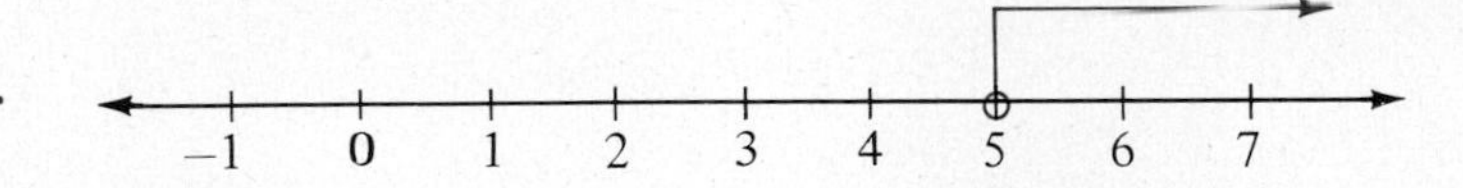

2.

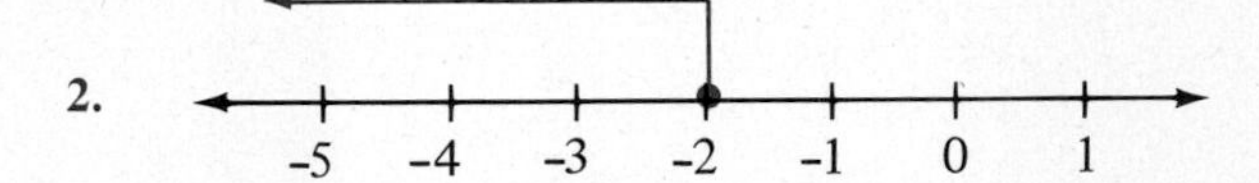

3.

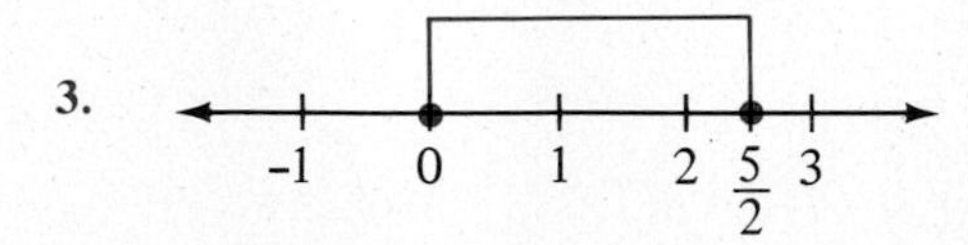

4.

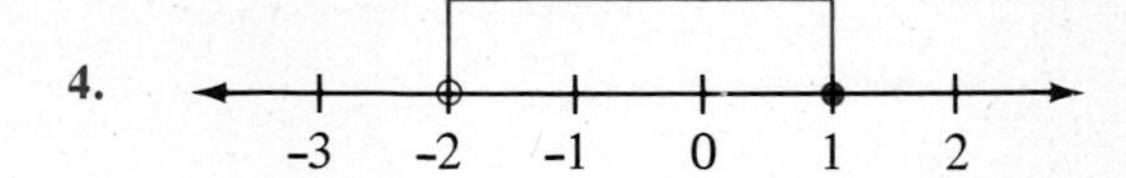

Name ______________________________ Date ______________

EXERCISE SET 9.1

Fill each blank with the correct symbol (<, >, =).

1. -11 ____ 0
2. -9 ____ -1
3. 3 ____ -19
4. 5 ____ -23
5. $\frac{1}{3}$ ____ $\frac{4}{12}$
6. $\frac{10}{15}$ ____ $\frac{2}{3}$
7. -23 ____ -2
8. -45 ____ 0
9. $\frac{-800}{50}$ ____ -15
10. $\frac{-600}{30}$ ____ -19
11. -2 ____ -10
12. 5 ____ -35
13. -1.5 ____ -5.9
14. -1.7 ____ -3.8

Write an inequality for each of the following.

________ **15.** -5 is between -11 and 0.

________ **16.** -6 is between -9 and -1.

________ **17.** 3 is greater than -2 and less than 7.

________ **18.** 4 is greater than 0 and less than 10.

________ **19.** $\frac{-2}{5}$ is between 0 and $\frac{-4}{5}$.

________ **20.** $\frac{-3}{8}$ is between $\frac{-7}{8}$ and $\frac{-1}{8}$.

Graph the solutions on a number line.

21. $x > 1$
22. $x > 3$
23. $a \leq 3$
24. $a \leq 0$
25. $y \geq -2$
26. $y \geq -3$
27. $b < 0$
28. $b < -6$
29. $0 < x < 6$
30. $-2 < x < 3$
31. $-1 \leq x < \frac{7}{2}$
32. $0 \leq x < \frac{5}{2}$
33. $-6 \leq a \leq -2$
34. $-5 \leq a \leq -3$
35. $\frac{-3}{2} < y \leq 0$
36. $-1 < y \leq \frac{1}{2}$
37. $3 < x < 4$
38. $-1 < x < 0$
39. $-5 \leq x \leq 1$
40. $-4 \leq x \leq 2$

☆ Stretching the Topics

Graph.

1. $\frac{-1}{4} > x$

2. $x \neq \frac{-1}{2}$

3. $\frac{1}{2}x < -3.2$

Check your answers in the back of your book.

If you complete **Checkup 9.1**, you are ready to go on to Section 9.2.

Name ______________________ Date ____________

✓ CHECKUP 9.1

Fill each blank with the correct symbol (<, >, =).

1. -7 ____ 1

2. $\frac{6}{10}$ ____ $\frac{15}{25}$

3. -18 ____ -22

4. 0 ____ $\frac{-4}{5}$

Write an inequality for each of the following.

____ **5.** -2 is between -5 and 0.

____ **6.** 3 is between $\frac{3}{2}$ and $\frac{7}{2}$.

Graph the solution on a number line.

7. $a > -1$

8. $x \leq 0$

9. $-3 \leq y \leq \frac{7}{2}$

10. $-2 < x \leq 2$

Check your answers in the back of your book.

If You Missed Problems:	You Should Review Examples:
1–4	1, 2
5, 6	3, 4
7, 8	5, 7, 8
9, 10	6, 9, 10

9.2 Solving Inequalities

We have already learned how to use the operations of addition, subtraction, multiplication, and division on both sides of an equation. Is an *inequality* still true after we perform those same operations on both sides? We must investigate before we can decide.

Operating on Numerical Inequalities

Let's operate on both sides of the inequality

$$6 < 10$$

first with some positive numbers:

Add 3	Subtract 15	Multiply by 3	Divide by 2
$6 + 3 \overset{?}{<} 10 + 3$	$6 - 15 \overset{?}{<} 10 - 15$	$6 \cdot 3 \overset{?}{<} 10 \cdot 3$	$\frac{6}{2} \overset{?}{<} \frac{10}{2}$
$9 \overset{?}{<} 13$	$-9 \overset{?}{<} -4$	$18 \overset{?}{<} 30$	$3 \overset{?}{<} 5$
True	True	True	True
$9 < 13$	$-9 < -4$	$18 < 30$	$3 < 5$

and then with some negative numbers:

Add -7	Subtract -4	Multiply by -5	Divide by -2
$6 + (-7) \overset{?}{<} 10 + (-7)$	$6 - (-4) \overset{?}{<} 10 - (-4)$	$6(-5) \overset{?}{<} 10(-5)$	$\frac{6}{-2} \overset{?}{<} \frac{10}{-2}$
$-1 \overset{?}{<} 3$	$10 \overset{?}{<} 14$	$-30 \overset{?}{<} -50$	$-3 \overset{?}{<} -5$
True	True	False	False
$-1 < 3$	$10 < 14$	$-30 > -50$	$-3 > -5$

These examples may help you see that:

> If we add (or subtract) the same positive or negative quantity to (or from) both sides of an inequality, the *direction* of the resulting inequality will remain the *same*.
>
> If we multiply (or divide) both sides of an inequality by the same *positive* number, the *direction* of the resulting inequality will remain the *same*.
>
> If we multiply (or divide) both sides of an inequality by the same *negative* number, the *direction* of the resulting inequality must be *reversed*.

Example 1. Perform the indicated operation on both sides of the given inequality.

Inequality	Operation	Result
$-1 < 5$	add 3	$-1 + 3 < 5 + 3$ ____ $<$ ____
$12 > 7$	subtract 5	$12 - 5 > 7 - 5$ ____ $>$ ____
$-2 < -1$	multiply by 5	$5(-2) < 5(-1)$ ____ $<$ ____
$12 < 21$	divide by 3	$\frac{12}{3} < \frac{21}{3}$ ____ $<$ ____
$18 > 9$	divide by -3	$\frac{18}{-3} < \frac{9}{-3}$ ____ $<$ ____
$16 > -7$	multiply by -1	$-1(16) < -1(-7)$ ____ $<$ ____

Check your work on page 530. ▶

Solving Variable Inequalities

Now that we know how to operate on both sides of an inequality, we may solve variable inequalities such as

$$3x - 4 \geq 17$$

This inequality is similar to a first-degree equation because the variable term has an exponent of 1. As with a first-degree equation, our goal is to *isolate the variable*.

$3x - 4 \geq 17$	The original inequality.
$3x - 4 + 4 \geq 17 + 4$	Add 4 to both sides.
$3x \geq 21$	Simplify.
$\frac{3x}{3} \geq \frac{21}{3}$	Divide both sides by 3.
$x \geq 7$	Simplify.

We may graph the solutions on a number line.

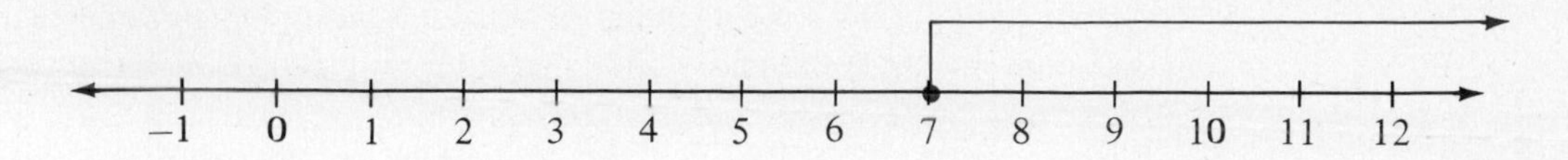

Solving a First-Degree Inequality

1. Perform any necessary additions and/or subtractions on both sides of the inequality to isolate the variable term.
2. Perform any necessary multiplications and/or divisions on both sides of the inequality, remembering to *reverse* the direction of the inequality whenever multiplying or dividing by a *negative* number.

Example 2. Solve and graph the solutions for

$$1 - y < -3$$

Solution

$$1 - y < -3$$

$-y < -4$ Subtract 1.

$\frac{-1y}{-1} > \frac{-4}{-1}$ Divide by -1.

$y > 4$ Simplify.

Notice that we reversed the direction of the inequality when we divided both sides by -1.

You try Example 3.

Example 3. Solve and graph the solutions for

$$\frac{x}{2} - 4 \leq 1$$

Solution

$$\frac{x}{2} - 4 \leq 1$$

$\frac{x}{2} \leq 5$ Add 4.

$\frac{2}{1}\left(\frac{x}{2}\right) \leq 2(\_\_\_\_)$ Multiply by 2.

$x \leq \_\_\_\_$ Simplify.

Check your work on page 530. ▶

Example 4. Solve $\frac{x+3}{2} > \frac{-1}{3}$

Solution

$\frac{x+3}{2} > \frac{-1}{3}$ LCD is $2 \cdot 3$ or 6.

$\frac{6}{1}\left(\frac{x+3}{2}\right) > \frac{6}{1}\left(\frac{-1}{3}\right)$ Multiply by LCD.

$3(x + 3) > 2(-1)$ Simplify.

$3x + 9 > -2$ Remove parentheses.

$3x > -11$ Subtract 9.

$x > \frac{-11}{3}$ Divide by 3.

Now you complete Example 5.

Example 5. Solve $\frac{-3x}{2} + 5 \geq 2$.

Solution

$$\frac{-3x}{2} + 5 \geq 2$$

$\frac{-3x}{2} \geq -3$ Subtract 5.

$-\frac{2}{3}\left(\frac{-3x}{2}\right) \leq \frac{-2}{3}\left(\frac{\quad}{1}\right)$

Multiply by reciprocal of $\frac{-3}{2}$

$x \leq \_\_\_\_$ Simplify.

Check your work on page 531. ▶

To solve inequalities containing parentheses or variables on both sides, we use the same methods we used for equations. We must remove parentheses and then get all variable terms on one side and constant terms on the other side.

Example 6. Solve $5(x + 1) - 3 \leq 12$.

Solution

$5(x + 1) - 3 \leq 12$	
$5x + 5 - 3 \leq 12$	Remove parentheses
$5x + 2 \leq 12$	Combine like terms
$5x \leq 10$	Subtract 2.
$\frac{5x}{5} \leq \frac{10}{5}$	Divide by 5.
$x \leq 2$	Simplify.

Now you complete Example 7.

Example 7. Solve

$$4(x - 1) - 10 \geq 7x + 1$$

Solution

$$4(x - 1) - 10 \geq 7x + 1$$

$$4x - 4 - 10 \geq 7x + 1$$

$$4x - 14 \geq 7x + 1$$

Check your work on page 531. ▶

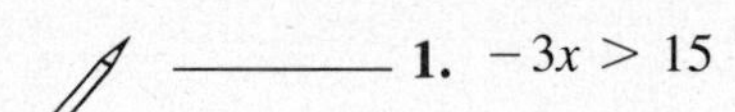

Trial Run

Solve each inequality.

_______ **1.** $-3x > 15$

_______ **2.** $a + 2.1 < -7.1$

_______ **3.** $\frac{2a}{5} - 3 \geq 1$

_______ **4.** $9 - 6y \leq -6$

_______ **5.** $2y - 3(y + 1) > 5$

_______ **6.** $5y - 7 \leq 2y + 5$

Answers are on page 531.

Solving Three-Part Inequalities

Suppose that we consider a three-part inequality in which x is not isolated in the middle. For example, let's solve

$$-1 < 3x + 5 < 8$$

Remember that this inequality says that

$$3x + 5 > -1 \quad \text{and} \quad 3x + 5 < 8$$

Let's solve both inequalities.

$3x > -6$	Subtract 5	$3x < 3$
$x > -2$	Divide by 3	$x < 1$

We conclude that x must be greater than -2 and less than 1, and we write the solution as: $-2 < x < 1$.

A shorter way of solving the original inequality allows us to operate on all three parts at the same time. We decide what operations must be performed to isolate the variable in the middle, and then perform those operations on all three parts of the inequality.

$-1 < 3x + 5 < 8$	The original inequality.
$-1 - 5 < 3x + 5 - 5 < 8 - 5$	Subtract 5.
$-6 < 3x < 3$	Simplify.
$\frac{-6}{3} < \frac{3x}{3} < \frac{3}{3}$	Divide by 3.
$-2 < x < 1$	Simplify.

Notice that this solution matches the earlier solution.

Example 8. Solve $-3 \le -5x - 3 < 2$.

Solution

$$-3 \le -5x - 3 < 2$$

$$-3 + 3 \le -5x - 3 + 3 < 2 + 3$$

$$0 \le -5x < 5$$

$$\frac{0}{-5} \ge \frac{-5x}{-5} > \frac{5}{-5}$$

$$0 \ge x > -1$$

It is customary to write three-part inequalities with the smaller number on the left. Here we write

$$-1 < x \le 0$$

Now you try Example 9.

Example 9. Solve $-5 < \frac{1}{2}x + 3 \le 0$.

Solution:

$-3 \quad -3 \quad -3$

$-8 < \frac{1}{2}x \le -3$

$-16 < x \le -6$

Check your work on page 531. ▶

Trial Run

Solve each inequality.

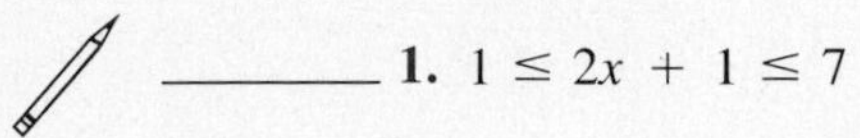

______ 1. $1 \le 2x + 1 \le 7$

______ 2. $-3 < \frac{1}{2}x - 4 < 1$

______ 3. $1 \le 2 - y < 4$

______ 4. $-1 \le 2 - \frac{3}{4}x \le 8$

Answers are on page 531.

Examples You Completed

Example 1

Results

$$-1 + 3 < 5 + 3$$
$$2 < 8$$
$$12 - 5 > 7 - 5$$
$$7 > 2$$
$$5(-2) < 5(-1)$$
$$-10 < -5$$
$$\frac{12}{3} < \frac{21}{3}$$
$$4 < 7$$
$$\frac{18}{-3} < \frac{9}{-3}$$
$$-6 < -3$$
$$-1(16) < -1(-7)$$
$$-16 < 7$$

Example 3. Solve and graph the solutions for

$$\frac{x}{2} - 4 \le 1$$

Solution

$$\frac{x}{2} - 4 \le 1$$
$$\frac{x}{2} \le 5$$
$$2\left(\frac{x}{2}\right) \le 2(5)$$
$$x \le 10$$

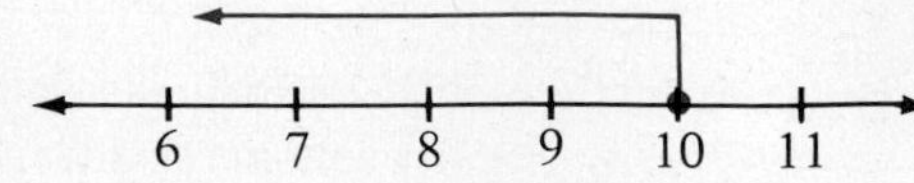

Example 5. Solve $\frac{-3x}{2} + 5 \geq 2$.

Solution

$$\frac{-3x}{2} + 5 \geq 2$$

$$\frac{-3x}{2} \geq -3$$

$$-\frac{2}{3}\left(\frac{-3x}{2}\right) \leq \frac{-2}{3} \cdot \frac{(-3)}{1}$$

$$x \leq 2$$

Example 7. Solve $4(x - 1) - 10 \geq 7x + 1$.

Solution

$$4(x - 1) - 10 \geq 7x + 1$$

$$4x - 4 - 10 \geq 7x + 1$$

$$4x - 14 \geq 7x + 1$$

$$-3x - 14 \geq 1$$

$$-3x \geq 15$$

$$\frac{-3x}{-3} \leq \frac{15}{-3}$$

$$x \leq -5$$

Example 9. Solve $-5 < \frac{1}{2}x + 3 \leq 0$.

Solution

$$-5 < \frac{1}{2}x + 3 \leq 0$$

$$-5 - 3 < \frac{1}{2}x + 3 - 3 \leq 0 - 3$$

$$-8 < \frac{1}{2}x \leq -3$$

$$2(-8) < \frac{2}{1}\left(\frac{1}{2}x\right) \leq 2(-3)$$

$$-16 < x \leq -6$$

Answers to Trial Runs

page 528 **1.** $x < -5$ **2.** $a < -9.2$ **3.** $a \geq 10$ **4.** $y \geq \frac{5}{2}$ **5.** $y < -8$ **6.** $y \leq 4$

page 530 **1.** $0 \leq x \leq 3$ **2.** $2 < x < 10$ **3.** $-2 < y \leq 1$ **4.** $-8 \leq x \leq 4$

Name ______________________ Date ____________

EXERCISE SET 9.2

Perform the indicated operation on both sides of the inequality.

_______ **1.** $-3 < 5$, add 4

_______ **2.** $2 < 7$, add -2

_______ **3.** $13 > 2$, add -7

_______ **4.** $3 > -2$, add 6

_______ **5.** $2 > -3$, multiply by 4

_______ **6.** $4 > -5$, multiply by 3

_______ **7.** $-18 < -6$, divide by 3

_______ **8.** $-21 < -14$, divide by 7

_______ **9.** $13 > 6$, subtract 20

_______ **10.** $12 > 3$, subtract 15

_______ **11.** $10 > -5$, divide by -5

_______ **12.** $12 > -16$, divide by -4

_______ **13.** $-13 < -1$, multiply by -1

_______ **14.** $-19 < -3$, divide by -1

_______ **15.** $-8 < 16$, divide by -2

_______ **16.** $16 > 10$, multiply by -3

Solve and graph each inequality.

_______ **17.** $x + 3 \le 7$

_______ **18.** $x + 5 \le 7$

_______ **19.** $3x > -15$

_______ **20.** $4x > -16$

_______ **21.** $-2a > 7$

_______ **22.** $-3a > 9$

_______ **23.** $3a - 3 \ge 12$

_______ **24.** $4a - 5 \ge 15$

_______ **25.** $\frac{x}{3} - 1 < -2$

_______ **26.** $\frac{x}{5} - 3 < -4$

_______ **27.** $-5z + 12 \le 17$

_______ **28.** $-6z + 14 \le 26$

_______ **29.** $3 - y > 1$

_______ **30.** $4 - y > 1$

_______ **31.** $\frac{5x}{2} + 1 \ge 6$

_______ **32.** $\frac{4x}{3} + 5 \ge 9$

Solve each inequality.

_______ **33.** $4(x - 5) \ge -2$

_______ **34.** $3(x - 6) \ge -4$

_______ **35.** $4y - 2(y - 5) < 10$

_______ **36.** $7y - 3(y - 4) < 12$

_______ **37.** $\frac{2x - 9}{3} > 5$

_______ **38.** $\frac{5x + 1}{4} > 9$

_______ **39.** $2a - 4 > 5a - 1$

_______ **40.** $5a - 8 > 7a - 12$

_______ **41.** $5x - (x - 4) \le 2x + 5$

_______ **42.** $3x - (x - 5) \le x + 7$

________ 43. $\frac{x-4}{3} > \frac{2x-1}{2}$

________ 44. $\frac{3x-1}{5} > \frac{4x-3}{2}$

________ 45. $2(z-2) - 3 < 5(z+5) - 23$

________ 46. $3(z+1) + 5 < 9(2z+3) + 1$

________ 47. $\frac{-x}{4} + 2 \geq \frac{x}{3} - 5$

________ 48. $\frac{-x}{3} + 2 \geq \frac{x}{5} - 6$

________ 49. $\frac{x}{4} - \frac{2x}{3} \geq \frac{1}{12}$

________ 50. $\frac{x}{5} - \frac{5x}{3} \geq \frac{4}{15}$

________ 51. $4 < x + 6 < 9$

________ 52. $-5 < x - 3 < 1$

________ 53. $-1 \leq -2x \leq 8$

________ 54. $-6 \leq -4x \leq 4$

________ 55. $-5 \leq 3x + 4 \leq 16$

________ 56. $-7 \leq 5x + 3 \leq 28$

________ 57. $7 \leq \frac{1}{2}x + 9 < 17$

________ 58. $-7 \leq \frac{1}{3}x - 6 < -5$

________ 59. $4 \leq 6 - 3x \leq 9$

________ 60. $-5 \leq 7 - 2x \leq 5$

☆ Stretching the Topics

Solve.

________ 1. $7x - 3[4 - (5 - x)] \leq 7x - (5 - 2x)$

________ 2. $\frac{4(y-6)}{-3} > \frac{6-2y}{5}$

________ 3. $-1 \leq -3x - 5[2(x-1) - (3x+4)] < 7$

Check your answers in the back of your book.

If you can solve the inequalities in **Checkup 9.2**, you are ready to go on to Section 9.3.

Name ______________________________ Date ______________

✓ CHECKUP 9.2

Perform the indicated operation on both sides of the inequality.

_______ **1.** $-2 > -9$, add 5

_______ **2.** $4 > -6$, divide by -2

Solve and graph each inequality.

_______ **3.** $x - 5 \geq -7$

_______ **4.** $-5a \leq 15$

_______ **5.** $\frac{-4x}{3} + 7 < 3$

_______ **6.** $-6 \leq 5y - 1 \leq 9$

Solve each inequality.

_______ **7.** $9(a - 2) \leq -3$

_______ **8.** $\frac{4x - 1}{2} > -3$

_______ **9.** $5y - 2 < 9y$

_______ **10.** $-2 \leq \frac{3x}{2} - 5 \leq 4$

Check your answers in the back of your book.

If You Missed Problems:	You Should Review Examples:
1, 2	1
3–5	2–5
6	8
7, 8	6
9	7
10	9

9.3 Combining Variable Inequalities (Optional)

Sometimes it is necessary to find the values of a variable that satisfy more than one inequality at the same time. The graphs of the solutions for each inequality will help us decide on the correct solutions for more than one inequality.

Satisfying One Inequality and Another Inequality

To find solutions satisfying one inequality *and* another, let's graph both inequalities on the *same* number line and see where the graphs *overlap*.

Example 1. Solve $x < 3$ and $x > -1$.

Solution

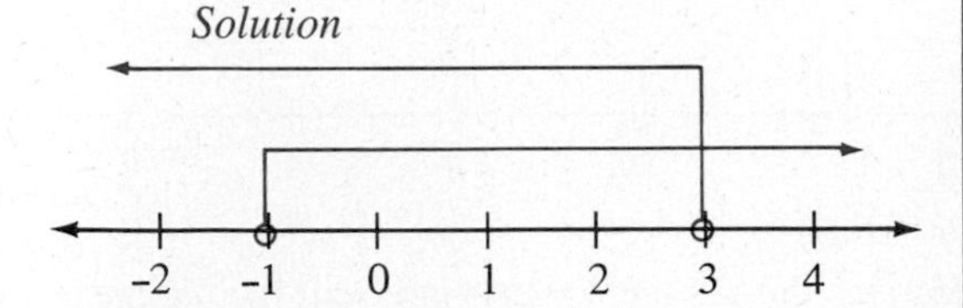

The graphs overlap *between* -1 and 3. The inequality $-1 < x < 3$ describes all numbers satisfying *both* of the original inequalities.

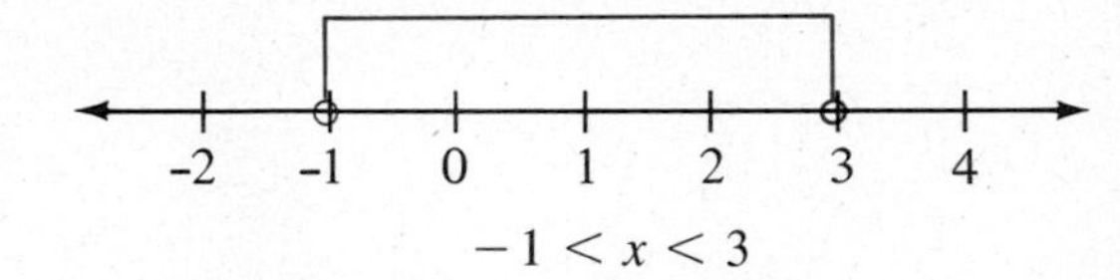

$-1 < x < 3$

Example 2. Solve $x > 2$ and $x > 5$.

Solution

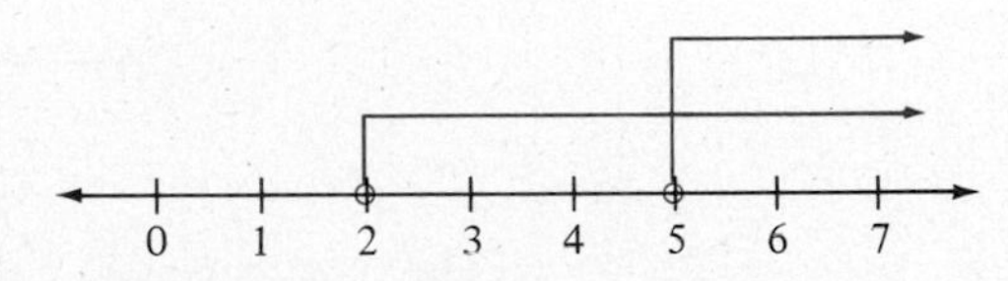

Both graphs appear at the same time to the *right* of 5. The inequality $x > 5$ describes all numbers satisfying *both* of the original inequalities.

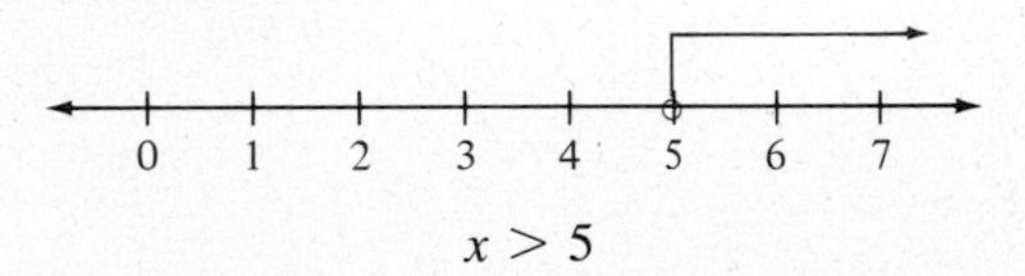

$x > 5$

Example 3. Solve $x < 7$ and $x \leq -1$.

Solution

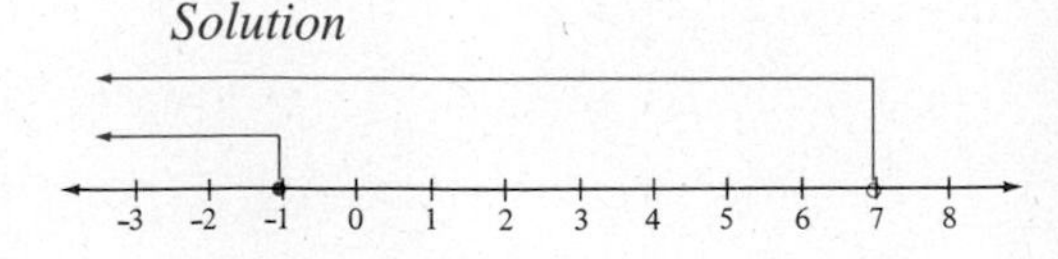

Both graphs appear at the same time to the left of -1 or at -1. The inequality $x \leq -1$ describes all numbers satisfying *both* of the original inequalities.

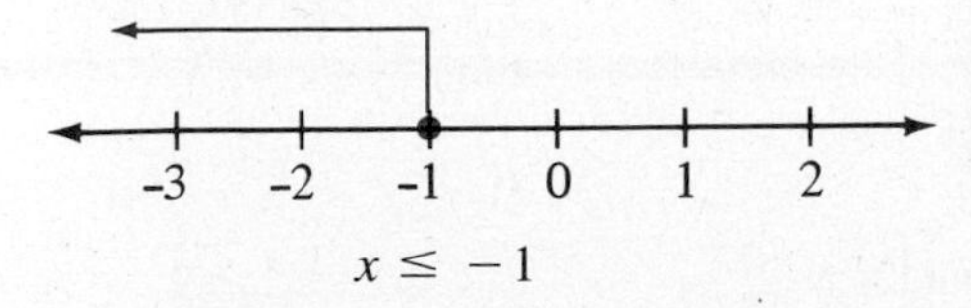

$x \leq -1$

Example 4. Solve $x > 2$ and $x \leq 0$.

Solution

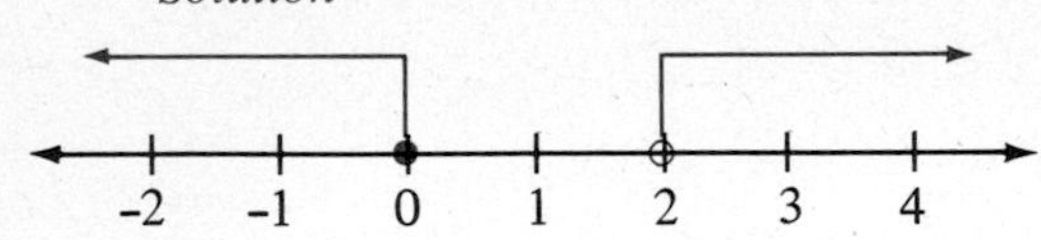

The graphs *never* overlap here. There are *no* numbers that satisfy *both* inequalities at the same time. There is *no* solution.

Satisfying One Inequality and Another Inequality by Graphing

1. Isolate the variable in each inequality.
2. Graph the solutions for the inequalities on the same number line.
3. Find the numbers that are included in *both* graphs. Look for the overlap.
4. Write a single inequality that describes the numbers satisfying both of the original inequalities.

Example 5. Solve $2 + 3x \geq 11$ and $5x - 1 > 2$.

Solution

$$2 + 3x \geq 11 \quad \text{and} \quad 5x - 1 > 2$$

$$3x \geq 9 \qquad 5x > 3 \qquad \text{Isolate variable terms.}$$

$$\frac{3x}{3} \geq \frac{9}{3} \qquad \frac{5x}{5} > \frac{3}{5} \qquad \text{Isolate the variables.}$$

$$x \geq 3 \quad \text{and} \quad x > \frac{3}{5} \qquad \text{Simplify.}$$

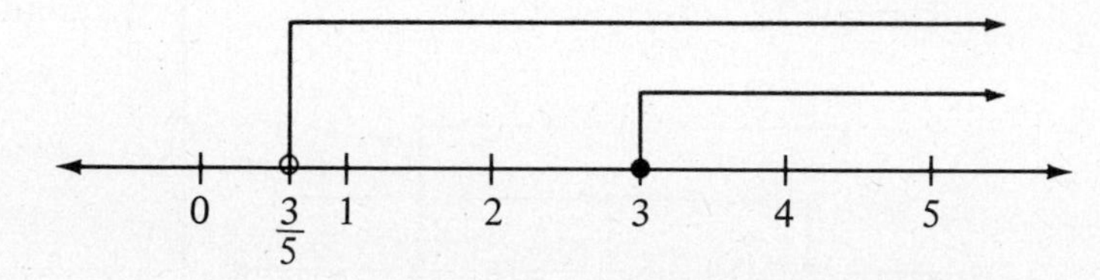

The inequality $x \geq 3$ describes the numbers satisfying both inequalities.

Try to complete Example 6.

Example 6. Solve $1 - x < 8$ and $\frac{2}{3}x \leq 0$.

Solution

$$1 - x < 8 \quad \text{and} \quad \frac{2}{3}x \leq 0$$

$$-x < 7 \qquad \text{Isolate variable terms.}$$

$$\frac{-1x}{-1} > \frac{7}{-1} \qquad \frac{3}{2}\left(\frac{2}{3}x\right) \leq \frac{3}{2}(0) \qquad \text{Isolate the variables.}$$

$$x > -7 \quad \text{and} \quad x \leq 0 \qquad \text{Simplify.}$$

$^{-}9 \quad ^{-}8 \quad ^{-}7 \quad ^{-}6 \quad ^{-}5 \quad ^{-}4 \quad ^{-}3 \quad ^{-}2 \quad ^{-}1 \quad 0 \quad 1 \quad 2$

The inequality ________ describes the numbers satisfying both inequalities. This inequality states that x is ________ than -7 and ________ than or equal to 0.
Check your work on page 541. ▶

Trial Run

Solve the inequalities.

________ **1.** $x < -3$ and $x \leq 2$

________ **2.** $x \leq 7$ and $x > -1$

________ **3.** $5x + 6 > -4$ and $4 - x \leq 8$

________ **4.** $2x - 1 < 0$ and $2x - 3 \geq 1$

Answers are on page 541.

Satisfying One Inequality or Another Inequality

Even in everyday English, the word *and* is much more demanding than the word *or*. If we are told that we must meet one condition *and* another, we shall be successful only when we have met *both* of those conditions. On the other hand, if we are told that we must meet one condition *or* another, we shall be successful if we meet *either* the first condition *or* the second condition or perhaps both.

The same is true for satisfying one inequality *or* another; any numbers that satisfy *either* of the inequalities (or perhaps both) will be acceptable solutions. To find those solutions, we can graph both inequalities on the same number line and see where *either* one or the other (or both) of the inequalities is satisfied.

Example 7. Solve $x > 1$ or $x \geq 5$.

Solution

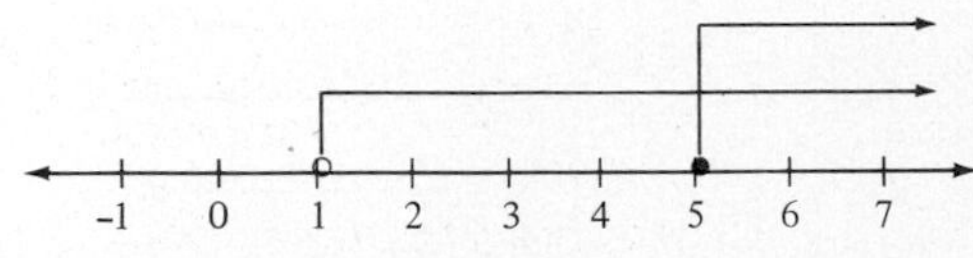

All the numbers that satisfy *either* (or both) of the inequalities lie to the right of 1. We describe those numbers with the inequality $x > 1$.

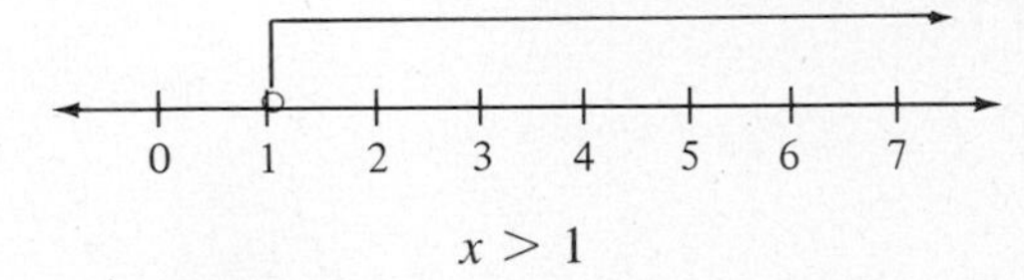

$x > 1$

Example 8. Solve $x \leq 0$ or $x \leq 2$.

Solution

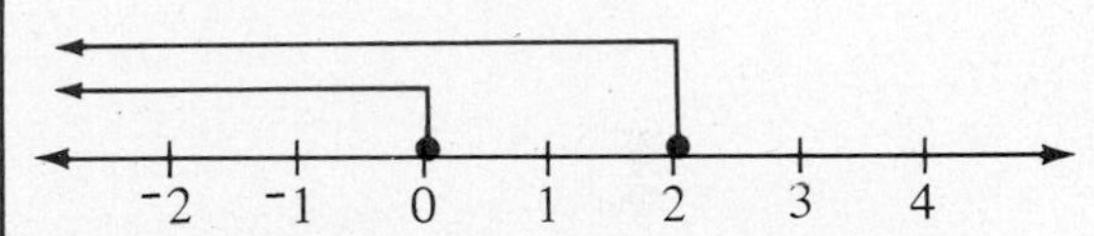

All the numbers that satisfy *either* (or both) of the inequalities lie to the left of 2 or at 2. We describe those numbers with the inequality $x \leq 2$.

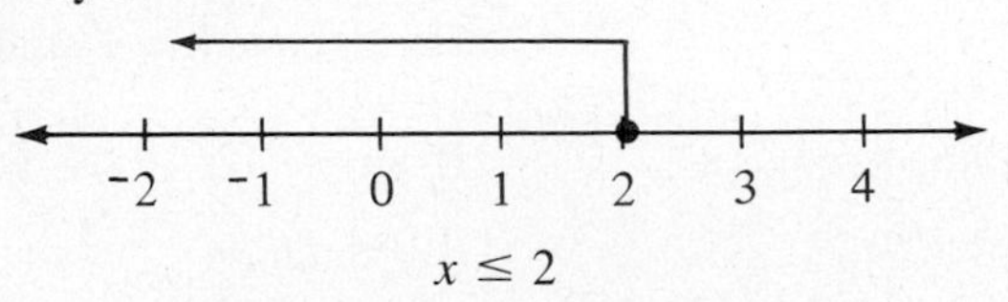

$x \leq 2$

Example 9. Solve $x < -1$ or $x > \frac{3}{2}$.

Solution

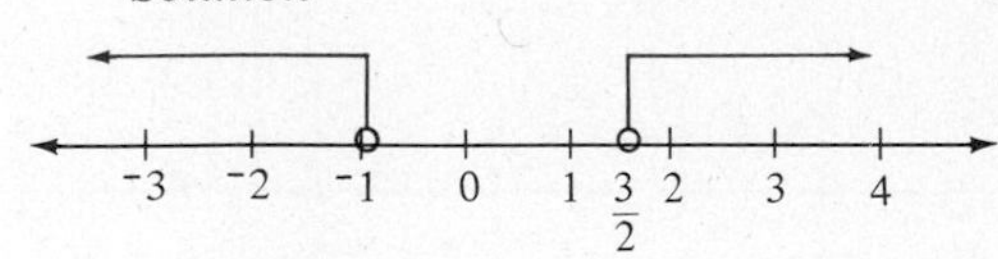

All the numbers that satisfy *either* of the inequalities lie to the left of -1 *or* to the right of $\frac{3}{2}$. Our solution cannot be written as a single inequality. We describe the solution by

$$x < -1 \quad \text{or} \quad x > \frac{3}{2}$$

Example 10. Solve $x < 5$ or $x > 3$.

Solution

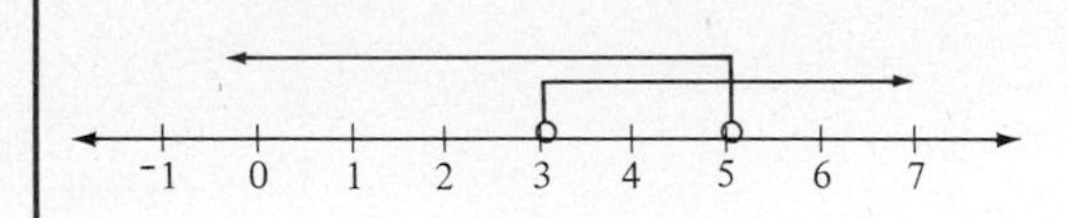

All the numbers on the number line satisfy either (or both) of the inequalities. Our solutions are all real numbers.

Satisfying One Inequality or Another Inequality by Graphing

1. Isolate the variable in each inequality.
2. Graph the solutions for the inequalities on the same number line.
3. Find the numbers that are included in *either* graph.
4. Describe the numbers satisfying either of the original inequalities by writing a single inequality or a pair of inequalities (using the word *or*).

Example 11. Solve $-3 + 7x > 4$ or $\frac{-5}{2}x \leq 0$.

Solution

$$-3 + 7x > 4 \quad \text{or} \quad \frac{-5}{2}x \leq 0$$

$$7x > 7 \qquad \frac{2}{1}\left(\frac{-5}{2}x\right) \leq 2(0)$$

$$\frac{7x}{7} > \frac{7}{7} \qquad -5x \leq 0$$

$$\frac{-5x}{-5} \geq \frac{0}{-5}$$

$$x > 1 \quad \text{or} \quad x \geq 0$$

Our solution is $x \geq 0$.

Try to complete Example 12.

Example 12. Solve $5x - 1 > 14$ or $-2x + 5 \geq 0$.

Solution

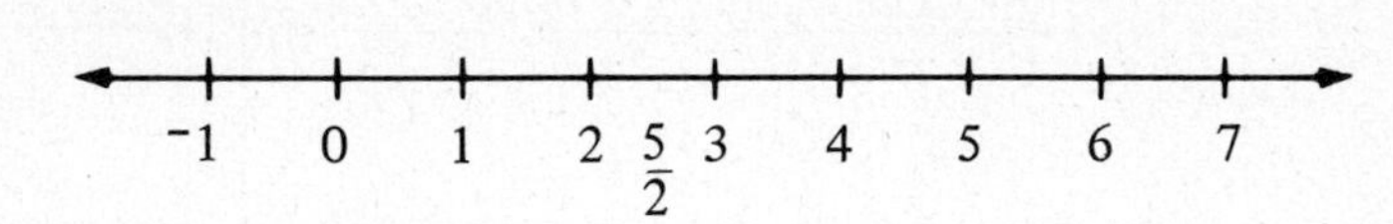

The solution is ______________________ .

Check your work on page 541. ▶

Trial Run

Solve the inequalities.

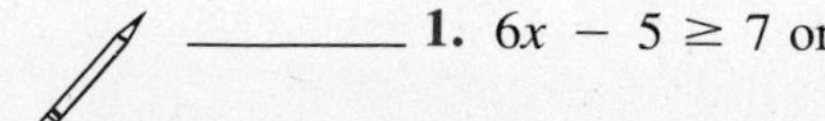

________ **1.** $6x - 5 \geq 7$ or $3 - x < 8$ ________ **2.** $3x - 1 < 2$ or $x - 9 \leq -6$

________ **3.** $2x + 13 < 9$ or $7x + 9 > 16$ ________ **4.** $\frac{x}{5} + 6 \leq 7$ or $\frac{2}{3}x < 0$

Answers are on page 541.

Remember, to satisfy one inequality *and* another inequality, we must find those numbers that satisfy *both* inequalities at the same time. To satisfy one inequality *or* another inequality, we must find those numbers that satisfy either the first inequality *or* the second inequality (or perhaps both).

▶ Examples You Completed

Example 6

Solution

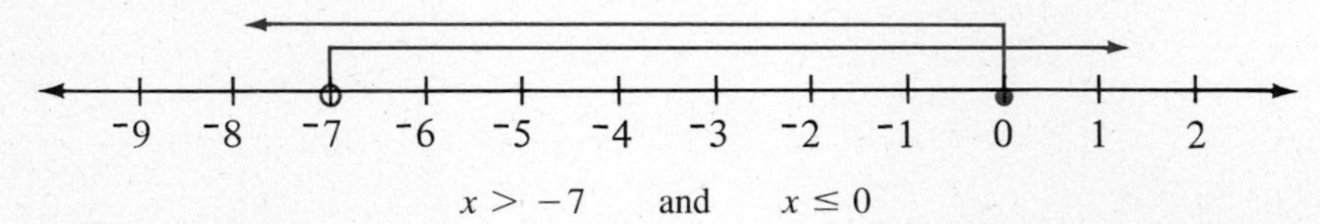

The inequality $-7 < x \leq 0$ describes the numbers satisfying both inequalities. This inequality states that x is greater than -7 and less than or equal to 0.

Example 12. Solve $5x - 1 > 14$ or $-2x + 5 \geq 0$.

Solution

$$5x - 1 > 14 \quad \text{or} \quad -2x + 5 \geq 0$$

$$5x > 15 \qquad -2x \geq -5$$

$$\frac{5x}{5} > \frac{15}{5} \qquad \frac{-2x}{-2} \leq \frac{-5}{-2}$$

$$x > 3 \quad \text{or} \quad x \leq \frac{5}{2}$$

The solution is $x \leq \frac{5}{2}$ or $x > 3$.

Answers to Trial Runs

page 538 **1.** $x < -3$ **2.** $-1 < x \leq 7$ **3.** $x > -2$ **4.** No solution

page 540 **1.** $x > -5$ **2.** $x \leq 3$ **3.** $x < -2$ or $x > 1$ **4.** $x \leq 5$

Name ______________________ Date __________

EXERCISE SET 9.3

Find the solution for the pairs of inequalities.

______ **1.** $x \geq -2$ and $x \leq 3$

______ **2.** $x \geq -6$ and $x \leq 2$

______ **3.** $x > -6$ and $x > 0$

______ **4.** $x \geq -5$ and $x > -1$

______ **5.** $x < 5$ or $x < -2$

______ **6.** $x < -1$ or $x < 3$

______ **7.** $x > 1$ or $x < -2$

______ **8.** $x > 0$ or $x < -3$

______ **9.** $x > 5$ and $x < -1$

______ **10.** $x > -1$ and $x < -3$

______ **11.** $x > \frac{1}{2}$ and $x < 4$

______ **12.** $x > -2$ and $x < 3$

______ **13.** $x > 0$ or $x > 4$

______ **14.** $x > -3$ or $x > 2$

______ **15.** $x + 6 < 8$ and $x + 2 > 0$

______ **16.** $x + 5 < 6$ and $x + 5 > 1$

______ **17.** $3x + 15 \geq 21$ and $5x - 2 < 13$

______ **18.** $2x + 9 \geq 11$ and $3x - 27 < -15$

______ **19.** $7x + 10 > 10$ or $9 - 2x \leq 19$

______ **20.** $11x - 3 > 8$ or $5x - 7 \geq 3$

______ **21.** $7 - 3x > 10$ or $\frac{2}{3}x > 0$

______ **22.** $12 - 5x > 22$ or $\frac{4}{5}x > 0$

______ **23.** $\frac{x}{2} + 3 > 5$ and $8 - \frac{x}{3} \leq 7$

______ **24.** $\frac{x}{3} - 1 \geq 0$ and $5 - \frac{x}{2} < -6$

______ **25.** $5x - 3 < 12$ or $2x - 3 < -5$

______ **26.** $2x - 13 < -7$ or $3x + 11 < 5$

______ **27.** $3x + 1 < 7$ or $2x - 3 \leq 3$

______ **28.** $7x + 2 \leq 9$ or $3x - 1 < -4$

______ **29.** $-x - 6 < 3$ and $\frac{2x}{3} - 1 > 1$

______ **30.** $-x + 9 < 5$ and $\frac{5x}{2} - 4 > 1$

______ **31.** $3x - 5 < 2x + 6$ and $4x - 1 \geq 2x + 9$

______ **32.** $2x + 5 > -4x - 7$ and $2x - 11 \leq -6x + 13$

______ **33.** $2(x - 5) > 3(x - 2)$ or $3(x - 1) > 2(x + 5)$

______ **34.** $4(x - 3) > 5(x - 1)$ or $7(x - 5) > 2(3x - 5)$

______ **35.** $\frac{5}{6}x - \frac{2}{3}x \geq \frac{1}{2}$ and $\frac{1}{5}x - \frac{3}{10}x < \frac{1}{2}$

______ **36.** $\frac{2}{7}x - \frac{1}{2}x > \frac{3}{14}$ and $\frac{1}{5}x - \frac{2}{3}x \leq \frac{14}{15}$

______ **37.** $\frac{-x}{2} + \frac{x}{8} \leq 3$ or $\frac{2x}{3} - \frac{x}{4} \geq 5$

______ **38.** $\frac{-x}{5} + \frac{x}{10} \geq 3$ or $\frac{3x}{4} - \frac{x}{2} \geq 2$

☆ Stretching the Topics

Solve.

______ **1.** $14 - 3(2x - 1) < 5[x - (2x + 4)]$ and $-[8 - 2(x - 5) < 6 - 2(x + 5)$

______ **2.** $\frac{3x - 2}{5} + \frac{x}{3} \leq 2x - 1$ or $\frac{x - 5}{2} - \frac{2x}{7} \geq x - 3$

______ **3.** $2x - (4 - x) \geq 11$ or $9 < 2x + 1 < 15$

Check your answers in the back of your book.

If you can solve the inequalities in **Checkup 9.3**, you are ready to go on to Section 9.4.

Name ______________________ Date __________

CHECKUP 9.3

Find the solution for the pairs of inequalities.

_______ **1.** $x \geq -1$ and $x > 3$

_______ **2.** $x > -2$ or $x \geq 3$

_______ **3.** $a \leq -2$ or $a \geq 4$

_______ **4.** $a > -2$ and $a < -5$

_______ **5.** $x < 0$ or $x \geq -6$

_______ **6.** $x < 4$ and $x > 0$

_______ **7.** $4 - 3x > -8$ and $2x - 5 > -13$

_______ **8.** $2x + 2 < 5$ and $5x - 7 > -12$

_______ **9.** $\frac{x}{5} - 2 > -1$ or $7 - \frac{x}{3} < 1$

_______ **10.** $4x - 3 > 13$ or $-2x + 3 > 0$

Check your answers in the back of your book.

If You Missed Problems:	You Should Review Examples:
1	2
2	7
3	9
4	4
5	10
6	1
7, 8	5, 6
9, 10	11, 12

9.4 Switching from Word Statements to Inequalities

Many everyday math problems involve inequalities rather than equations.

Example 1. Write the inequality that describes each word statement.

Words	Inequality
To qualify as a sophomore, a student must have completed at least 30 hours of coursework.	h ≥ 30
The age of a Miss America contestant must be at least 18 but not more than 25 years.	18 ≤ a ≤ 25

Check your answers on page 549. ▶

Suppose that we try solving some word problems using inequalities. We must continue with our usual methods: Identify the variable, write any expression containing the variable, write an inequality from the information in the problem, and solve.

Example 2. Mr. Angelo, a retiree, will lose some retirement benefits if he earns over $8000 per year as a night watchman. If he has already earned $7200, how many more hours may he work at $5 per hour without losing any retirement benefits?

Solution

$$\text{Let } h = \text{hours to be worked}$$
$$5h = \text{dollars to be earned}$$
$$7200 + 5h = \text{total earnings} \quad \text{(dollars)}$$

If Mr. Angelo does not wish to lose benefits, he must earn $8000 or less. So

$$7200 + 5h \leq 8000$$
$$5h \leq 8000 - 7200$$
$$5h \leq 800$$
$$\frac{5h}{5} \leq \frac{800}{5}$$
$$h \leq 160$$

Mr. Angelo may work 160 or fewer hours.

Example 3. In order to receive a B in her math course, Henrietta must have an average of at least 80 on six tests. Her grades on the first five tests were 73, 85, 80, 79, and 72. What grade must she receive on the sixth test in order to earn a B in the course?

Solution

$$\text{Let} \quad G = \text{grade on sixth test}$$
$$\frac{73 + 85 + 80 + 79 + 72 + G}{6} = \text{average grade for six tests}$$

Since Henrietta must have at least an 80 average, we write

$$\frac{73 + 85 + 80 + 79 + 72 + G}{6} \geq 80$$

$$\frac{389 + G}{6} \geq 80$$

$$\frac{6}{1}\left(\frac{389 + G}{6}\right) \geq 6(80)$$

$$389 + G \geq 480$$

$$G \geq 480 - 389$$

$$G \geq 91$$

Henrietta must receive a 91 or better on her last test.

Let's solve the problem stated at the beginning of this chapter.

Example 4. To qualify for federal funding, a special program must enroll between 150 and 210 students. Moreover, the number of rural students enrolled must be twice the number of nonrural students. How many of each type of student must be enrolled in the special program?

Solution

Let x = number of nonrural students

$2x$ = number of rural students

$x + 2x$ = total number of students

Since we are working with a "between" statement, we must use a three-part inequality.

$$150 \leq x + 2x \leq 210$$

$$150 \leq 3x \leq 210$$

$$\frac{150}{3} \leq \frac{3x}{3} \leq \frac{210}{3}$$

$$50 \leq x \leq 70$$

Between 50 and 70 nonrural students must be enrolled. To find the number of rural students, we must look at $2x$. Since

$$50 \leq x \leq 70$$

we can multiply all parts of the inequality by 2 to obtain $2x$ in the middle

$$2(50) \leq 2 \cdot x \leq 2(70)$$

$$100 \leq 2x \leq 140$$

Between 100 and 140 rural students must be enrolled in the program.

Example 5. A total of 200 spaces is available for a guided tour of Cape Canaveral. If an adult ticket sells for \$5 and a child's ticket for \$3, how many of each must be sold in order for the money from ticket sales to total at least \$900?

Solution: We shall let a represent the number of adult tickets and use a chart to organize our information.

	Adults	Children
Number of tickets	a	$200 - a$
Dollars from tickets	$5 \cdot a$	$3(200 - a)$

The total dollars from ticket sales must be at least $900.

$$5a + 3(200 - a) \geq 900$$
$$5a + 600 - 3a \geq 900$$
$$2a + 600 \geq 900$$
$$2a \geq 300$$
$$a \geq 150$$

At least 150 adult tickets must be sold, but what about children's tickets?

$$\begin{aligned} 200 - a &= 200 - 150 \\ &= 50 \end{aligned}$$

Therefore, 50 *or fewer* children's tickets must be sold. Remember that the total number of tickets available is just 200.

▶ Example You Completed

Example 1. Write the inequality that describes each word statement.

Words	Inequality
To qualify as a sophomore, a student must have completed at least 30 hours of coursework.	$h \geq 30$
The age of a Miss America contestant must be at least 18 but not more than 25 years.	$18 \leq a \leq 25$

Name ______________________________ Date ______________

EXERCISE SET 9.4

Solve.

______ **1.** In order for the Barminskis to have good television reception, the top of their antenna must be at least 24 feet above the ground. If the height of their house is 16 feet, what antenna lengths would assure good reception?

______ **2.** The total passenger weight of the persons riding the elevator at Holt's Department Store must not exceed 2175 pounds. If the average weight of an adult is 145 pounds, how many adults are allowed in the elevator?

______ **3.** In addition to her yearly salary of $12,380, Ms. Greenwell sells vacuum cleaners to supplement her income. She receives a commission of $48.75 on each vacuum cleaner she sells. If she wants to keep her total yearly income below $20,960 for tax purposes, how many cleaners should she sell each year?

______ **4.** Ella's diet requires her to have no more than 1100 calories per day. For lunch she can have 100 more calories than she has for breakfast, and for dinner she can have twice the number of calories she has for breakfast. Find the possible number of calories that she can have for each meal.

______ **5.** The perimeter of a triangle can be no more than 36 feet. One side is 7 feet and the other two sides are the same. Find the possible lengths for the two equal sides.

______ **6.** If 5 times a number added to 9 is less than or equal to 4 times the number added to 12, find the values the number can have.

______ **7.** The sum of 7 times a number and 13 is greater than 27 and less than 48. Find the possible values for the number.

______ **8.** The Ovell Memorial Theater has only 400 seats. If the theatre company charges admissions of $4.25 for adults and $3 for children, how many adult tickets, can they sell and still cover their production expenses of $1600?

______ **9.** The Quarterback Club is selling candy bars to raise money. Large bars sell for $2 and small bars for $1. They decide that they can sell 650 candy bars. Find the number of $2 bars they can sell so that their total sales will be greater than or equal to $1000.

______ **10.** If the product of 8 and a number is added to 15, the result is greater than or equal to 11 times the number. Find the possible values for the number.

______ **11.** The length of a cattle-loading pen is to be 6 feet longer than the width. The perimeter must be less than or equal to 108 feet, which is the amount of fence available. Find the possible widths of the pen.

______ **12.** The perimeter of a triangle must be less than 75 centimeters. One side is 22 cm and the second side is to be 3 cm more than the third side. Find the possible lengths of the third side.

_______ **13.** In order for the Associated Student Government to break even on a concert, they must take in at least \$34,800. The price of a ticket sold in advance is \$6 and the price is \$8 at the door. If ASG sells 5000 tickets altogether, how many tickets must be sold at the door so that they do not lose money?

_______ **14.** The length of a rectangle is to be 2 meters less than 5 times the width. If the difference between the two dimensions is not more than 4 meters, find the possible values for the width.

_______ **15.** Mr. Geiger wishes to use no more than 90 feet of fencing for a dog pen. If he wishes the length to be 7 more feet than the width, what are the possible dimensions of the pen?

_______ **16.** In order to receive an A in her math course, Irene must have an average of at least 90 on five tests. If her grades on the first four tests were 96, 82, 98, and 80, what grades can she receive on the fifth test in order to earn an A in the course?

_______ **17.** The length of a rectangle is to be 7 meters more than twice the width. If the perimeter must be no more than 48 meters, find the possible values for the width.

_______ **18.** If Lamar works for an hourly wage of \$4.25, what number of hours can he work so that his gross pay will be at least \$153?

_______ **19.** The constitution of Associated Student Government states that the entertainment committee may have no fewer than 12 members and no more than 30. It also states that the number of senior members shall be twice the number of junior members. Find how many juniors must be appointed to the committee.

_______ **20.** The Press Club is having a dinner for its 150 members. It will cost them \$4.25 to serve a chicken plate and \$5.50 to serve a ham plate. If they have only \$765, how many ham plates could be served?

☆ Stretching the Topics

_______ **1.** Juan earns a salary of \$15,000 a year plus a 12% commission on all sales he makes. If Juan's income last year was more than \$16,200 but at most \$18,000, what was the total amount of his sales?

_______ **2.** Frank is enclosing a rectangular garden by using his garage for one length and fence for the other three sides. The length is to be 20 feet more than $\frac{2}{3}$ the width. Find the possible widths if the maximum amount of fencing available is 164 feet.

Check your answers in the back of your book.

If you can solve the problems in **Checkup 9.4**, you are ready to do the **Review Exercises for Chapter 9**.

6.

Name ______________________________ Date ______________

Solve.

________ **1.** A rental agency charges a flat fee of \$15, plus \$2.75 an hour for use of a garden tiller. If Irma rents the tiller, for how many hours can she use it and spend no more than \$26?

________ **2.** Weight Worrier Bob has lost 3, 2, 4, and 2 pounds during the first 4 weeks of his diet. How many pounds will he have to lose during the fifth week so that his average weight loss will be at least 3 pounds per week?

________ **3.** The length of a rectangular playground is to be twice its width. The dimensions must be such that the playground can be enclosed with no more than 327 feet of available fence. Find the possible dimensions the playground can have.

________ **4.** The Shanklin Theater has a seating capacity of 1200. If the drama group charges admissions of \$6.00 for adults and \$4.00 for children, find the number of adult tickets they can sell and still cover their production costs of \$6700.

________ **5.** If George earns \$37.50, \$58.25, \$61.15 and \$42.35 during the first 4 days of his yard sale, how much will he have to sell during the fifth day so that his sales average is at least \$50 per day?

Check your answers in the back of your book.

If You Missed Problem:	You Should Review Example:
1	2
2	3
3	2
4	5
5	3

Summary

In this chapter we learned to solve and graph first-degree variable inequalities.

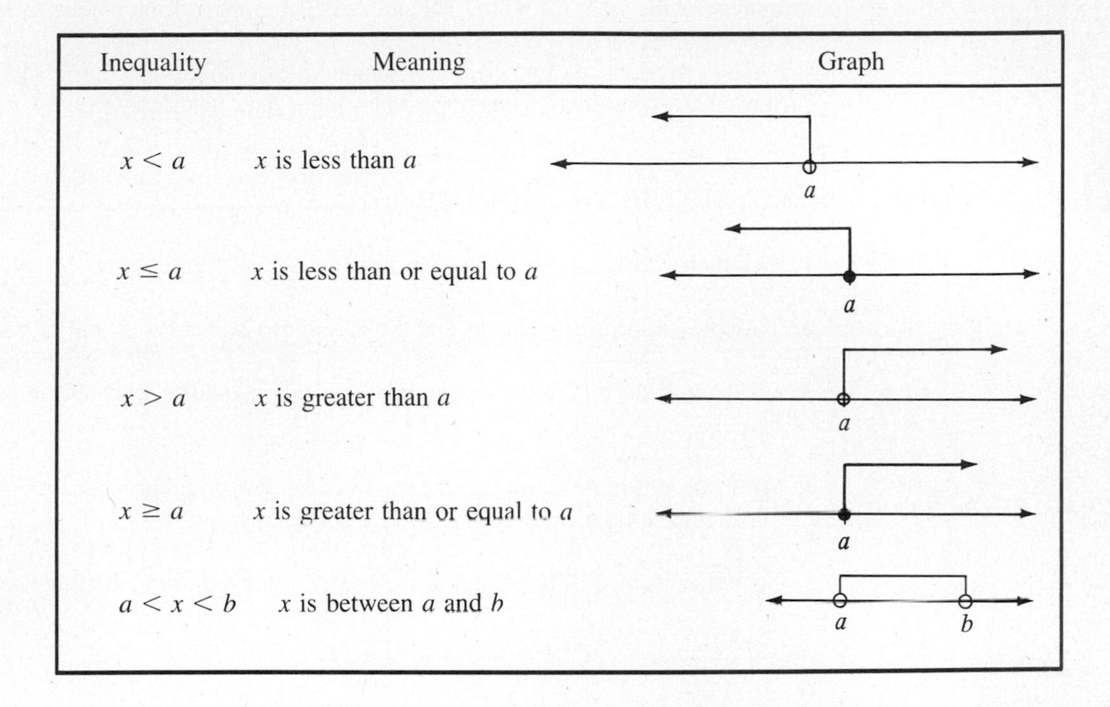

Inequality	Meaning	Graph
$x < a$	x is less than a	
$x \leq a$	x is less than or equal to a	
$x > a$	x is greater than a	
$x \geq a$	x is greater than or equal to a	
$a < x < b$	x is between a and b	

We discovered that first-degree inequalities can be solved by methods similar to those used for solving first-degree equations.

Operation	Statement	Examples
Addition (or subtraction)	We may add (or subtract) the same quantity to (or from) both sides of an inequality. The direction of the inequality remains the same.	$x - 2 < 8$ $x < 10$ $x + 3 \geq -1$ $x \geq -4$
Multiplication (or division) by a *postive* quantity	We may multiply (or divide) both sides of an inequality by a *positive* quantity. The direction of the inequality remains the same.	$\frac{x}{2} \leq 5$ $x \leq 10$ $3x > -1$ $x > \frac{-1}{3}$
Multiplication (or division) by a *negative* quantity	We may multiply (or divide) both sides of an inequality by a *negative* quantity. The direction of the inequality is *reversed*.	$-4x > 8$ $x < -2$ $\frac{-x}{3} \leq -1$ $x \geq 3$

In an optional section we discussed solutions to pairs of inequalities. We learned that the numbers satisfying one inequality *and* another inequality must satisfy *both* inequalities at the same time. Numbers satisfying one inequality *or* another inequality may satisfy either one inequality *or* the other (or both). The number-line graphs of the inequalities helped us find the correct solutions.

Finally, we used our knowledge of inequalities to solve word problems involving phrases such as at least, no more than, fewer than, more than, and less than.

Speaking the Language of Algebra

1. Statements such as $x < 2$ and $y \geq 3$ are called __________ .

2. To write an inequality stating that the variable x is between 2 and 5, we write ________ .

3. On the number line, all numbers satisfying the inequality $x < -7$ lie to the __________ of -7.

4. We must remember to reverse the direction of an inequality whenever we __________ or __________ both sides by a __________ quantity.

5. To write an inequality stating that 3 times some number, n, is at least 18.7, we write __________ .

Check your answers in the back of your book.

Name ____________________ Date ____________

REVIEW EXERCISES for Chapter 9

Fill each blank with the correct symbols ($<$, $>$, $-$).

1. -27 __$<$__ -3

2. 0.3 __$<$__ -1.1

Write an inequality for each of the following.

__$-10 < -7 < 3$__ **3.** -7 is between -10 and 3.

__$0 < \frac{5}{3} < 4$__ **4.** $\frac{5}{3}$ is between 0 and 4.

Graph the solutions on a number line.

______ **5.** $y \geq -5$

______ **6.** $a > 0$

______ **7.** $-3 < x \leq 7$

______ **8.** $-2 \leq y \leq \frac{5}{2}$

Perform the operation indicated, on both sides of the inequality.

__$4 < 13$__ **9.** $-5 < 4$, add 9

__$21 < 30$__ **10.** $7 < 10$, multiply by -3

__$-5 < 2$__ **11.** $10 > -4$, divide by -2

__$-9 > -15$__ **12.** $-3 > -9$, subtract 6

Solve and graph each inequality.

__$x \leq 4$__ **13.** $3x + 9 \leq 21$

__$-y > -5$__ **14.** $8 - y > 3$

__$A \geq -8$__ **15.** $\frac{a}{2} + 2 \geq -2$

______ **16.** $\frac{-3x}{2} + 6 < -3$

Solve each inequality.

__$x < 5$__ **17.** $2x - 5(x + 5) < -10$

__$2 \geq x$__ **18.** $8x - 9 \geq 3x + 1$

__$x > \frac{-9}{5}$__ **19.** $\frac{x - 7}{4} < \frac{3x + 1}{2}$

______ **20.** $\frac{-x}{3} + 2 \geq \frac{x}{2} - 1$

______ **21.** $3(y - 1) - y \geq 4(y + 2) + 1$

______ **22.** $-3 \leq 3y - 1 \leq 5$

______ **23.** $-8 \leq 4 - 2x < 0$

______ **24.** $-3 \leq \frac{5x}{3} + 2 \leq 7$

Find the solution for the pairs of inequalities.

______ **25.** $x + 4 > 7$ or $x - 6 \geq -1$

______ **26.** $\frac{-x}{3} \leq -2$ and $\frac{3x}{2} - 5 > 1$

______ **27.** $4(x - 2) > 2(x - 1)$ or $5(3 - x) < -2(3x - 7)$

Solve.

________ **28.** Carl's salary for a 40-hour week is \$140, but he finds that he needs to earn at least \$218 a week to pay his bills. If he can earn \$5.20 per hour working overtime, how many hours could he work overtime per week and still pay all his bills?

________ **29.** The perimeter of a rectangle can be no more than 72 feet and the length must be twice the width. Find the possible dimensions of the rectangle.

________ **30.** The Band Boosters Club is having a dinner to raise money for the purchase of uniforms. They decide that they can serve 500 people. If the tickets are \$5 for adults and \$3 for children, find the number of adult tickets the club should sell so that total sales will be greater than or equal to \$2100.

Check your answers in the back of your book.

If You Missed Exercises:	You Should Review	Examples:
1, 2	SECTION 9.1	1, 2
3, 4		3, 4
5–8		5–10
9–12	SECTION 9.2	1
13–16		2–5
17–21		6, 7
22–24		8, 9
25, 27	SECTION 9.3	7–12
26		1–6
28–30	SECTION 9.4	1–5

If you have completed the **Review Exercises** and corrected your errors, you are ready to take the **Practice Test for Chapter 9**.

Name ______________________________ Date ____________

PRACTICE TEST for Chapter 9

Problem	Section	Example
______ 1. Write an inequality that says that $\frac{-1}{2}$ is between -1 and $\frac{1}{4}$.	9.1	3
2. Graph the solutions for $x \leq -1$ on the number line.	9.1	8

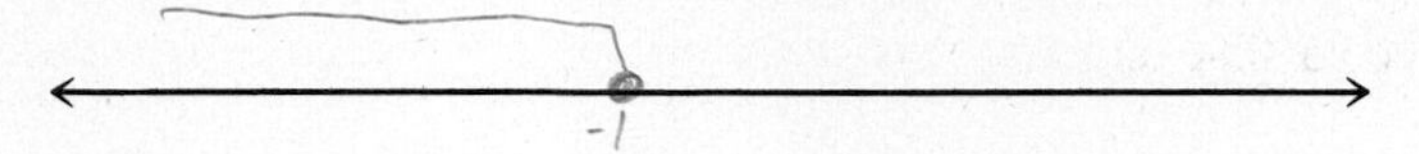

Problem	Section	Example
3. Graph the solutions for $-2 < a \leq \frac{5}{3}$ on the number line.	9.1	9

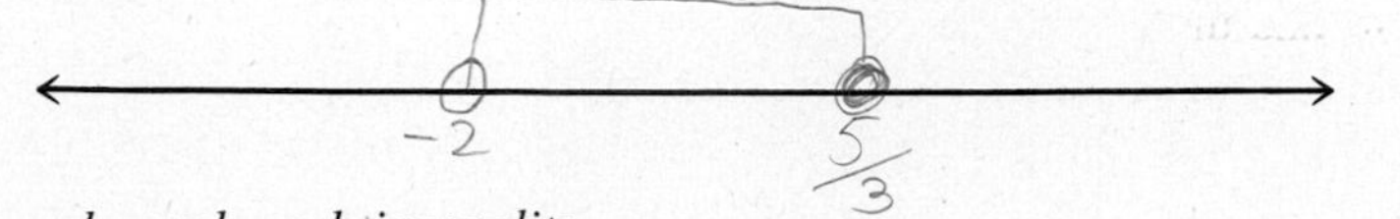

Solve and graph each inequality.

Problem	Section	Example
______ 4. $7 - x \geq 5$	9.2	2
______ 5. $\frac{2x}{5} - 1 > 1$	9.2	5

Solve each inequality.

Problem	Section	Example
______ 6. $5x - 3(2x + 4) < -3$	9.2	6
______ 7. $8x + 7 > 5x + 7$	9.2	7
______ 8. $\frac{3x + 1}{5} < \frac{2x + 1}{4}$	9.2	4
______ 9. $6(a - 3) - 4a \geq 3(a - 2) + 5$	9.2	7
______ 10. $\frac{3y}{2} + 5 \leq \frac{-y}{9} + 1$	9.2	4
______ 11. $-5 \leq 4x + 3 \leq -1$	9.2	8
______ 12. $-3 < 5 - 2x < 3$	9.2	8
______ 13. $-8 \leq \frac{5}{4}x + 2 \leq 2$	9.2	9
______ 14. $2 - x \geq 1$ *and* $\frac{x}{5} + 3 < 2$	9.3	5

NO 15. $4(x + 2) < 3(x - 1)$ *or* $7(2 - x) < 11 - 4(x + 3)$ 9.3 12

______ 16. For income tax purposes, typewriter salesman Thomas wishes his yearly gross income to remain below \$30,000. If he is paid a monthly salary of \$2000 plus a bonus of \$75 for each typewriter he sells, how many typewriters should he sell this year? 9.4 1

Check your answers in the back of your book.

2400
75|6000

Name ______________________________ Date ____________

SHARPENING YOUR SKILLS after Chapters 1–9

		SECTION
______	**1.** Change $\frac{5}{8}$ to a percent.	1.3
______	**2.** Compare $\frac{4}{5}$ and $\frac{5}{7}$ using $<$ or $>$.	2.1
______	**3.** Find $12\frac{1}{2}\%$ of \$3540.	1.3
______	**4.** Evaluate $(x - 2y)^2 - 4xy$ when $x = 1$ and $y = -2$.	5.1
______	**5.** Find the restrictions on the variable for $\frac{x - 5}{x^2 - 16}$.	8.1

Simplify.

______	**6.** $\frac{2}{3}[2(6x - 2y) - (9x - y)]$	3.3
______	**7.** $(9xy^3)(-x^2y)(-2xy^2)$	6.2
______	**8.** $(x - 1)(x^2 + x + 1)$	6.3
______	**9.** $-4x(2 - x)(3 + x)$	6.3
______	**10.** $\frac{m^2 - 9n^2}{2m + 6n}$	8.1
______	**11.** $\frac{x^2 - 49}{4x - 8} \cdot \frac{3x - 6}{x^2 + 14x + 49}$	8.2
______	**12.** $\frac{5}{4x + 4} + \frac{2}{x + 1}$	8.4
______	**13.** $\dfrac{\frac{x - 2}{x^3y^4}}{\frac{x^2 - 4}{x^2y}}$	8.5

Factor.

______	**14.** $16a^3b^3 - 24a^2b^2 + 8ab$	7.3
______	**15.** $6m^4 - 96$	7.2
______	**16.** $-x^3y - 8x^2y^2 - 15xy^3$	7.3

Solve.

_______ **17.** $\frac{1}{2}[3x - (x + 2)] = 8$ 4.2

_______ **18.** $0.2(5y - 1) = 0.3(7 + y) + 0.5$ 4.2

_______ **19.** $x^2 + 10x - 39 = 0$ 7.5

_______ **20.** $3x^2 = 27x$ 7.5

_______ **21.** $\frac{x}{3} = \frac{x - 2}{6}$ 8.6

_______ **22.** $\frac{4}{a - 3} = \frac{-2a}{a^2 + 5a - 24}$ 8.6

_______ **23.** Earl has 56 feet of border to use for enclosing his tulip bed. If he wishes the bed to be 3 times as long as it is wide, what should be the dimensions of the bed? 4.3

_______ **24.** The sum of the squares of two consecutive integers is 181. Find the integers. 7.6

_______ **25.** Driving her car Louise's average speed is 6 miles per hour faster than her husband's average speed. If Louise can drive 280 miles in the same length of time her husband drives 250 miles, how fast does each drive? 8.6

Check your answers in the back of your book.

Working with Equations in Two Variables 10

At a local restaurant, chicken sandwiches are sold for $1.50 each. Use an equation to help a waitress make a chart showing the cost, y, of x chicken sandwiches.

So far in our study of algebra we have concentrated on expressions and equations containing *one* variable. We have learned to solve first-degree and second-degree (quadratic) equations by finding the values of the variable that make the equations true statements.

In this chapter we deal with first-degree equations that contain *two* variables, and we learn how to

1. Write equations containing two variables.
2. Find solutions of equations containing two variables.
3. Graph solutions of equations containing two variables.
4. Graph solutions of constant equations.

10.1 Switching from Word Statements to Equations Containing Two Variables

Suppose that a student wishes to earn \$72 in a week, working at two part-time jobs. One job pays \$3 per hour and one pays \$4 per hour. Let's write an expression for the amount of money the student can earn in 1 week. Since we cannot assume that the number of hours spent at one job is the same as the number of hours spent at the other job, we must use **two variables**.

Let

$$x = \text{hours per week spent at first job}$$

$$y = \text{hours per week spent at second job}$$

$$3x = \text{dollars per week earned at first job}$$

$$4y = \text{dollars per week earned at second job}$$

$$3x + 4y = \text{total earnings in 1 week}$$

If the student wishes to earn \$72 per week, we may write

$$3x + 4y = 72$$

There are many possible values of x and y that will satisfy this equation.

Example 1. If the student works 4 hours at the first job, how many hours must she spend at the second job?

Solution: Let $x = 4$ and find y.

$$3x + 4y = 72$$

$$3(4) + 4y = 72$$

$$12 + 4y = 72$$

$$4y = 60$$

$$y = 15$$

She can work 4 hours at the first job and 15 hours at the second job.

Now you complete Example 2.

Example 2. If she works 12 hours at the first job, how many hours must she spend at the second job?

Solution: Let $x =$ ____ and find y.

$$3x + 4y = 72$$

$$3(\_\_\_\_) + 4y = 72$$

She can work ____ hours at the first job and ____ hours at the second job.

Check your work on page 569. ▶

Finding Ordered Pairs

In each part of the student's work problem, we found an x-value *and* a y-value that would satisfy the equation $3x + 4y = 72$. Sometimes we choose to write these corresponding x-values and y-values in the form of **ordered pairs** enclosed in parentheses with a comma between.

Corresponding Values	Ordered Pair
$x = 4, \quad y = 15$	(4, 15)
$x = 12, \quad y = 9$	(12, 9)
$x = 0, \quad y = 18$	(0, 18)

The first number in the ordered pair is always the x-value and the second number in the ordered pair is always the corresponding y-value. An ordered pair **satisfies** an equation if we can make the equation *true* by substituting the x-value and y-value from the ordered pair into the equation. In other words, solutions for equations containing two variables are ordered pairs of numbers.

Example 3. Does (3, 1) satisfy

$$x + 2y = 5?$$

Solution: From (3, 1) we know that $x = 3$ and $y = 1$.

$$x + 2y = 5$$

$$3 + 2(1) \stackrel{?}{=} 5$$

$$3 + 2 \stackrel{?}{=} 5$$

$$5 = 5$$

Yes, (3, 1) satisfies the equation $x + 2y = 5$.

You complete Example 4.

Example 4. Does $(0, -5)$ satisfy

$$9x - 2y = 6?$$

Solution: From $(0, -5)$ we know that $x =$ _____ and $y =$ _____.

$$9x - 2y = 6$$

$$9(\_\_\_) - 2(\_\_\_) \stackrel{?}{=} 6$$

Check your work on page 569. ▶

Suppose that we wish to find values of x and y that satisfy a particular equation. Let's find some ordered pairs that satisfy

$$y = 2x + 3$$

We may *choose any x-value* that we wish, substitute that value for x in the equation, and then find the corresponding y-value. We can use a table to organize our work.

x	$y = 2x + 3$	(x, y)
-2	$y = 2(-2) + 3 = -4 + 3 = -1$	$(-2, -1)$
0	$y = 2(0) + 3 = 0 + 3 = 3$	$(0, 3)$
$\frac{1}{2}$	$y = 2\left(\frac{1}{2}\right) + 3 = 1 + 3 = 4$	$\left(\frac{1}{2}, 4\right)$
0.4	$y = 2(0.4) + 3 = 0.8 + 3 = 3.8$	$(0.4, 3.8)$

Since we could continue in this way forever, we agree that there is an *infinite number* of ordered pairs that will satisfy the equation $y = 2x + 3$.

Finding ordered pairs to satisfy the equation $5x + y = 16$ looks a bit more difficult. First we can *isolate* y.

$5x + y = 16$	Original equation.
$y = 16 - 5x$	Subtract $5x$ from both sides to isolate y.

Now we see that to find a y-value corresponding to some x-value, we must multiply the x-value by 5 and subtract that quantity from 16. For variety we often choose a negative x-value, a positive x-value, and zero.

Example 5. Complete the table of ordered pairs that satisfy $y = 16 - 5x$.

Solution

x	$y = 16 - 5x$	(x, y)
1	$y = 16 - 5(1) = 16 - \underline{5} = \underline{11}$	$(1, \underline{11})$
0	$y = 16 - 5(0) = 16 - \underline{0} = \underline{16}$	$(0, \underline{16})$
-3	$y = 16 - 5(-3) = 16 + \underline{15} = \underline{31}$	$(-3, \underline{31})$
$\frac{2}{5}$	$y = 16 - 5\left(\frac{2}{5}\right) = 16 - \underline{2} = \underline{14}$	$\left(\frac{2}{5}, \underline{14}\right)$

Check your work on page 570. ▶

Example 6. Complete a table of three ordered pairs that satisfy $2x + 3y = 0$.

Solution: First we isolate y.

$$2x + 3y = 0 \quad \text{Original equation.}$$

$$3y = -2x \quad \text{Subtract } 2x \text{ to isolate } y\text{-term.}$$

$$y = \frac{-2x}{3} \quad \text{Divide by 3 to isolate } y.$$

x	$y = \frac{-2x}{3}$	(x, y)
-3	$y = \frac{-2(-3)}{3} = \frac{6}{3} = 2$	$(-3, 2)$
0	$y = \frac{-2(0)}{3} = \frac{0}{3} = 0$	$(0, 0)$
1	$y = \frac{-2(1)}{3} = \frac{-2}{3}$	$\left(1, \frac{-2}{3}\right)$

You may have noticed that first-degree equations in two variables can always be written in the form "numerical coefficient times x, plus numerical coefficient times y, equals constant." Using a, b, and c to represent constants, we note:

General First-Degree Equation in Two Variables

$$ax + by = c$$

where a and b are not both zero.

To find several ordered pairs that satisfy a first-degree equation in two variables, we have used the following procedure.

Finding Ordered Pairs

1. Isolate y in the original equation.
2. Choose any x-values and find the corresponding y-values.
3. Write each solution as an ordered pair (x, y).

Trial Run

Decide if the given ordered pair satisfies the equation.

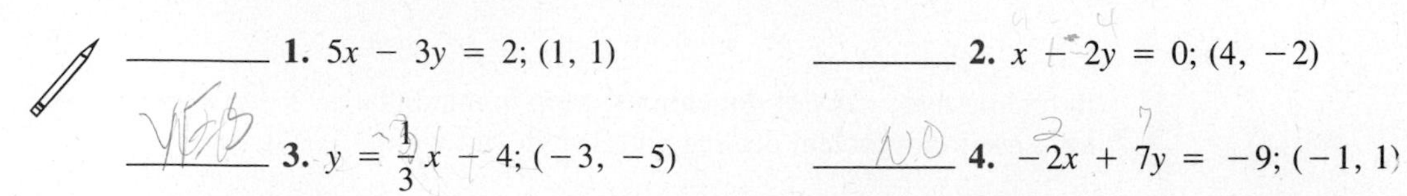

_______ **1.** $5x - 3y = 2$; $(1, 1)$

_______ **2.** $x + 2y = 0$; $(4, -2)$

_______ **3.** $y = \frac{1}{3}x - 4$; $(-3, -5)$

_______ **4.** $-2x + 7y = -9$; $(-1, 1)$

Find three ordered pairs that satisfy each equation. There are lots of correct answers.

_______ **5.** $y = 2x + 5$

_______ **6.** $x - y = -2$

_______ **7.** $3x + y = 4$

_______ **8.** $5x - 2y = 1$

Answers are on page 570.

Writing Equations from Word Statements

The student's work hours problem pointed out the everyday usefulness of equations in two variables. The methods we have just learned will help us find ordered pairs that satisfy such equations. As always when dealing with word problems, we must be sure to identify the variables in each situation.

Example 7. Aaron Gray wishes to fence in a rectangular portion of his yard for a dog run. If he has 60 feet of fencing to use, what are three possible pairs of dimensions for the run?

Solution: Let's use an illustration.

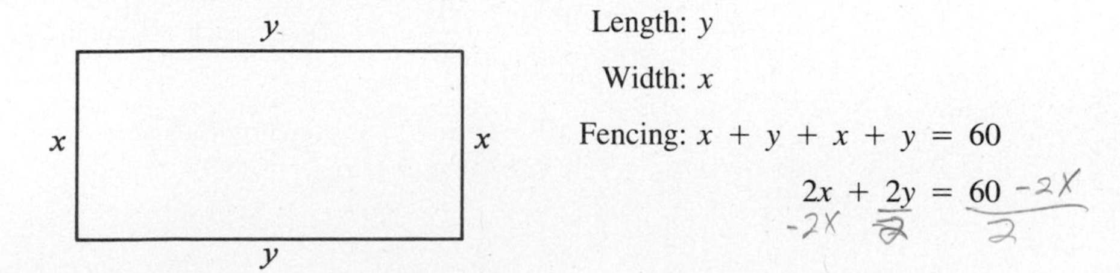

Length: y

Width: x

Fencing: $x + y + x + y = 60$

$$2x + 2y = 60$$

Now we isolate y to make it easier to find ordered pairs.

$$2x + 2y = 60$$

$$2y = 60 - 2x \qquad \text{Subtract } 2x \text{ to isolate } y\text{-term.}$$

$$\frac{2y}{2} = \frac{60 - 2x}{2} \qquad \text{Divide by 2 to isolate } y.$$

$$y = \frac{\overset{1}{\cancel{2}}(30 - x)}{\underset{1}{\cancel{2}}} \qquad \text{Factor the numerator.}$$

$$y = 30 - x \qquad \text{Reduce.}$$

In finding three pairs of dimensions, keep in mind that *zero* and *negative* x-values would make no sense in a problem dealing with length and width.

x	$y = 30 - x$	(x, y)	Width	Length
10	$y = 30 - 10 = 20$	(10, 20)	10 feet	20 feet
15	$y = 30 - 15 = 15$	(15, 15)	15 feet	15 feet
$8\frac{1}{2}$	$y = 30 - 8\frac{1}{2} = 21\frac{1}{2}$	$\left(8\frac{1}{2}, 21\frac{1}{2}\right)$	$8\frac{1}{2}$ feet	$21\frac{1}{2}$ feet

Keep in mind that there are infinitely many other possibilities.

Example 8. Doug and Bob plan to take turns driving on a 600-mile trip. If Doug drives at a rate of 50 mph and Bob drives at 60 mph, write three ordered pairs which describe the number of hours that each might drive.

Solution: Remember that distance = rate × time.

Let x = hours Doug drives Then $50x$ = miles Doug drives

y = hours Bob drives $60y$ = miles Bob drives

$50x + 60y$ = total miles driven

$$50x + 60y = 600$$

We isolate y.

$$60y = 600 - 50x \qquad \text{Subtract } 50x \text{ to isolate } y\text{-term.}$$

$$y = \frac{600 - 50x}{60} \qquad \text{Divide by 60 to isolate } y.$$

$$y = \frac{600}{60} - \frac{50x}{60} \qquad \text{Divide each polynomial term by 60.}$$

$$y = 10 - \frac{5}{6}x \qquad \text{Simplify each term.}$$

x	$y = 10 - \frac{5}{6}x$	(x, y)	Doug's Time	Bob's Time
12	$y = 10 - \frac{5}{6}(12) = 10 - 10 = 0$	(12, 0)	12 hours	0 hours
6	$y = 10 - \frac{5}{6}(6) = 10 - 5 = 5$	(6, 5)	6 hours	5 hours
0	$y = 10 - \frac{5}{6}(0) = 10 - 0 = 10$	(0, 10)	0 hours	10 hours

Now let's solve the problem stated at the beginning of the chapter.

Example 9. At a local restaurant, chicken sandwiches sell for $1.50. Use an equation to help a waitress make a chart showing the cost, y, of x chicken sandwiches.

Solution: We can let

$$x = \text{number of sandwiches}$$

$$1.5\,x = \text{cost of sandwiches}$$

Then

$$y = 1.5x$$

x	$y = 1.5x$	(x, y)	Number of Sandwiches	Cost
1	$y = 1.5(1) = 1.5$	(1, 1.5)	1	$1.50
2	$y = 1.5(2) = 3$	(2, 3)	2	3.00
3	$y = 1.5(3) = 4.5$	(3, 4.5)	3	4.50
4	$y = 1.5(4) = 6$	(4, 6)	4	6.00
5	$y = 1.5(5) = 7.5$	(5, 7.5)	5	7.50
6	$y = 1.5(6) = 9$	(6, 9)	6	9.00
7	$y = 1.5(7) = 10.5$	(7, 10.5)	7	10.50
8	$y = 1.5(8) = 12$	(8, 12)	8	12.00
9	$y = 1.5(9) = 13.5$	(9, 13.5)	9	13.50
10	$y = 1.5(10) = 15$	(10, 15)	10	15.00

▶ Examples You Completed

Example 2. If she works 12 hours at the first job, how many hours must she spend at the second job.?

Solution: Let $x = 12$ and find y.

$$3x + 4y = 72$$

$$3(12) + 4y = 72$$

$$36 + 4y = 72$$

$$4y = 36$$

$$y = 9$$

She can work 12 hours at the first job and 9 hours at the second job.

Example 4. Does $(0, -5)$ satisfy $9x - 2y = 6$?

Solution: From $(0, -5)$ we know that

$$x = 0 \text{ and } y = -5$$

$$9x - 2y = 6$$

$$9(0) - 2(-5) \stackrel{?}{=} 6$$

$$0 + 10 \stackrel{?}{=} 6$$

$$10 \neq 6$$

No, $(0, -5)$ does *not* satisfy the equation $9x - 2y = 6$.

Example 5. Complete the table of ordered pairs that satisfy $y = 16 - 5x$.

Solution

x	$y = 16 - 5x$	(x, y)
1	$y = 16 - 5(1) = 16 - 5 = 11$	$(1, 11)$
0	$y = 16 - 5(0) = 16 - 0 = 16$	$(0, 16)$
-3	$y = 16 - 5(-3) = 16 + 15 = 31$	$(-3, 31)$
$\frac{2}{5}$	$y = 16 - 5\left(\frac{2}{5}\right) = 16 - 2 = 14$	$\left(\frac{2}{5}, 14\right)$

Answers to Trial Run

page 567 **1.** Yes **2.** No **3.** Yes **4.** No

Name ________________________________ Date ______________

EXERCISE SET 10.1

Decide if the given ordered pair satisfies the equation.

________ **1.** $6x - 5y = 2$; $(2, 2)$

________ **2.** $4x - 3y = 1$; $(1, 1)$

________ **3.** $2x - 4y = 0$; $(2, -1)$

________ **4.** $3x - 9y = 0$; $(3, -1)$

________ **5.** $y = \frac{1}{5}x - 6$; $(5, -5)$

________ **6.** $y = \frac{1}{2}x - 5$; $(2, -4)$

________ **7.** $-3x + 2y = 7$; $(1, 2)$

________ **8.** $-4x + 5y = 11$; $(-4, 5)$

________ **9.** $5x + 3y = 8$; $(0, 0)$

________ **10.** $7x + 11y = 18$; $(0, 0)$

________ **11.** $2x + 3y = -6$; $(3, -4)$

________ **12.** $3x + 8y = -21$; $(1, -3)$

Find three ordered pairs that satisfy each equation.

________ **13.** $x + y = 2$

________ **14.** $x - y = 3$

________ **15.** $2x + y = 4$

________ **16.** $x + 2y = 6$

________ **17.** $2x - 3y = 1$

________ **18.** $4x - 5y = 1$

________ **19.** $x - 2y = 0$

________ **20.** $2x + y = 0$

________ **21.** $x + 3y = 3$

________ **22.** $4x + y = 5$

________ **23.** $y = x + 6$

________ **24.** $y = x - 5$

________ **25.** $y = 3x - 5$

________ **26.** $y = 4x - 7$

________ **27.** $4x + 3y = -6$

________ **28.** $3x + 2y = -5$

Write each of the following word statements as an equation containing two variables.

X+8=4Y **29.** One number is 8 more than 4 times another.

X-3=6Y **30.** One number is 3 less than 6 times another.

X+Y=28 **31.** The sum of two numbers is 28.

X+10=2Y **32.** The sum of one number and 10 is twice another number.

________ **33.** The length of a rectangle is 5 less than twice its width.

________ **34.** The length of a rectangle is 2 more than three times the width.

________ **35.** The Math Club treasurer has a stack of both \$1 and \$5 bills. Their total value is \$100.

________ **36.** Steve bought a stack of books at a yard sale. Some cost \$2 each and some cost \$3 each. The total cost was \$38.

________ **37.** The perimeter of a rectangle is 48 feet.

________ **38.** The perimeter of a rectangle is 64 feet.

________ **39.** Rick Yunkus has savings accounts in two different banks. The total amount of his savings is \$12,500.

________ **40.** The value of the nickels and dimes Peggy has in her change purse is \$1.25.

________ **41.** Henry earned \$147.20 last week, receiving \$3.20 per hour for regular time and \$4.80 per hour for overtime.

________ **42.** For the basketball game, some student and some adult tickets were sold. Student tickets sell for \$1.50 and adult tickets for \$3.00. The total gate receipts were \$1500.

________ **43.** In the Humane Society animal shelter, the number of cats is 2 less than 5 times the number of dogs.

________ **44.** In the Chemistry Club, the number of boys is 7 less than 3 times the number of girls.

________ **45.** Kelly walked 20 miles in the Walk-a-Thon to raise money for the Special Olympics. Part of the time she averaged 3 miles per hour and the remainder she averaged 2 miles per hour.

________ **46.** When Duane went on vacation, he made part of the trip by bus and the other part by train. He averaged 50 miles per hour on the bus and 60 miles per hour by train. The total trip was 600 miles.

☆ Stretching the Topics

________ **1.** Does the ordered pair $\left(\frac{1}{2}, \frac{-5}{6}\right)$ satisfy the equation $\frac{1}{3}x - \frac{2}{5}y = \frac{1}{2}$?

________ **2.** Complete the following table and write the equation showing the relation that exists between x and y.

x	1	2	3	4	5	
y	4	7	10		16	19

________ **3.** Write an equation containing two variables that describes the following situation. Carlos has a certain amount of money to invest. He wishes to invest part of the money at $8\frac{1}{2}\%$ and part at 12% so that he can obtain an annual income of 10% on the entire amount.

Check your answers in the back of your book.

If you can complete **Checkup 10.1**, you are ready to go on to Section 10.2.

Name ______________________ Date __________

CHECKUP 10.1

Decide if the given ordered pair satisfies the equation.

_______ **1.** $2x - 6y = 4$; $(1, -1)$

_______ **2.** $y = \frac{1}{3}x - 2$; $(6, -1)$

_______ **3.** $-4x + y = 0$; $(0, 0)$

_______ **4.** $5x - y = -4$; $(-2, 6)$

Find three ordered pairs that satisfy each equation.

_______ **5.** $x - y = -2$

_______ **6.** $2x - y = 0$

_______ **7.** $y = 3x - 2$

Write each word statement as an equation containing two variables.

_______ **8.** The amount of money Mr. Brinkley has invested at Union Bank is \$500 less than twice the amount invested at Farmers State Bank.

_______ **9.** Janyce bought a box of antique bottles. Some cost \$2 each and some cost \$3 each. The total cost was \$17.

_______ **10.** The perimeter of a rectangular lot is 196 feet.

Check your answers in the back of your book.

If You Missed Problems:	You Should Review Examples:
1–4	3, 4
5–7	5, 6
8, 9	8
10	7

10.2 Graphing First-Degree Equations in Two Variables

Earlier we learned to use the **number line** to graph solutions for first-degree equations and inequalities containing one variable. We would like to graph solutions for first-degree equations in two variables, but one number line will not work very well, since every solution is an ordered *pair* of numbers. Every solution pairs an x-value with a corresponding y-value.

Over three hundred years ago, a mathematician and philosopher named René Descartes thought of a way to graph ordered pairs using *two* number lines. He placed the two number lines at right angles (perpendicular) to each other and labeled the horizontal line the ***x*-axis** and the vertical line the ***y*-axis**. Naming the point at which the two lines intersect the **origin**, he marked off the units in the usual way. Descartes came up with the following diagram, called the **Cartesian coordinate plane** in his honor.

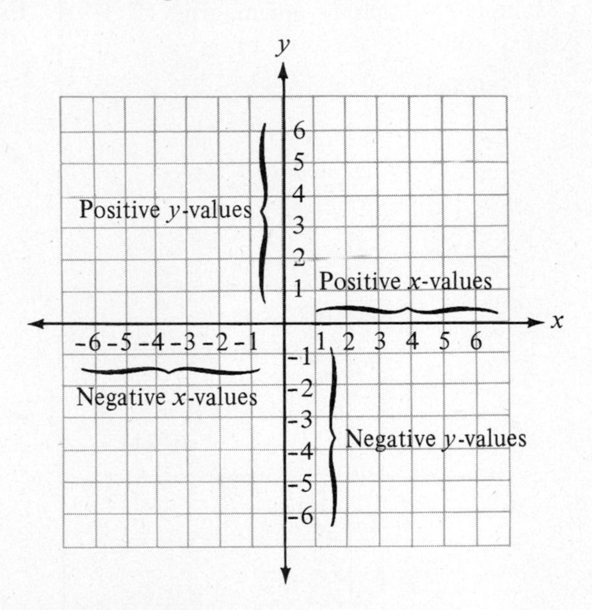

Graphing Ordered Pairs

Now we must discover how to use the Cartesian coordinate plane to graph ordered pairs. To graph an ordered pair such as (2, 3) we must keep in mind that

(2, 3) means "when $x = 2$, then $y = 3$"

On the Cartesian plane, we may use a dotted vertical line to show where x is 2 and a dashed horizontal line to show where y is 3.

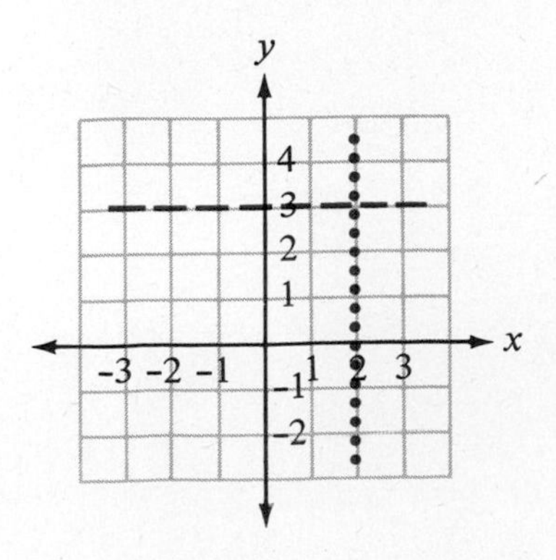

Since we wish x to be 2 and y to be 3 *at the same time*, the point where the dotted line crosses the dashed line must represent the graph of the ordered pair (2, 3).

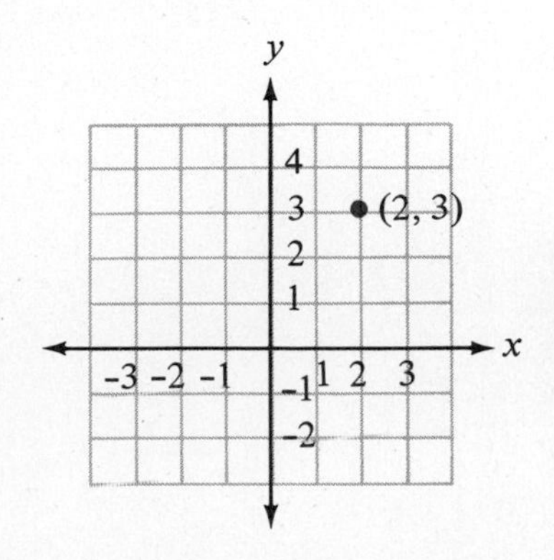

When plotting points in the Cartesian plane, we shall not sketch the dotted and dashed lines. You should do that mentally and then locate the point in the plane. All points should be labeled with an ordered pair. The point corresponding to an ordered pair contains two numbers, called **coordinates**.

Coordinates of Ordered Pairs

x-coordinate: the x-value in an ordered pair
y-coordinate: the y-value in an ordered pair

Example 1. Graph the ordered pairs (5, 1) and (1, 5).

Solution

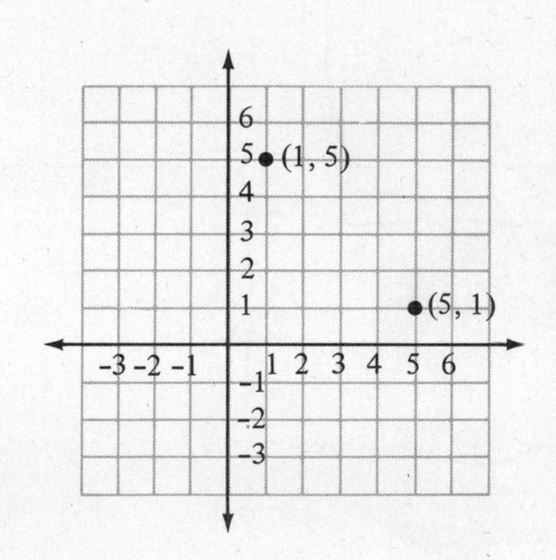

Notice that (5, 1) and (1, 5) do *not* represent the same point.

Now you graph the ordered pairs in Example 2.

Example 2. Graph the ordered pairs (1, −2) (−3, 4), (−5, −5), and $\left(2, \frac{7}{2}\right)$.

Solution

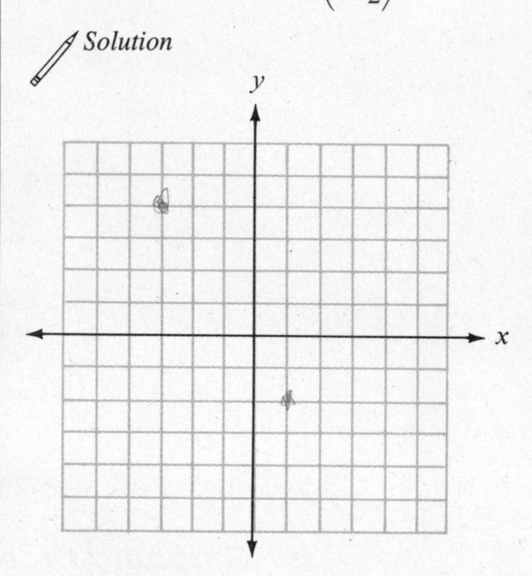

Check your work on page 583. ▶

Suppose that we plot some points for which *zero* is the x-coordinate and/or the y-coordinate.

Example 3. Graph the ordered pairs (0, 0), (3, 0), and (−5, 0).

Solution

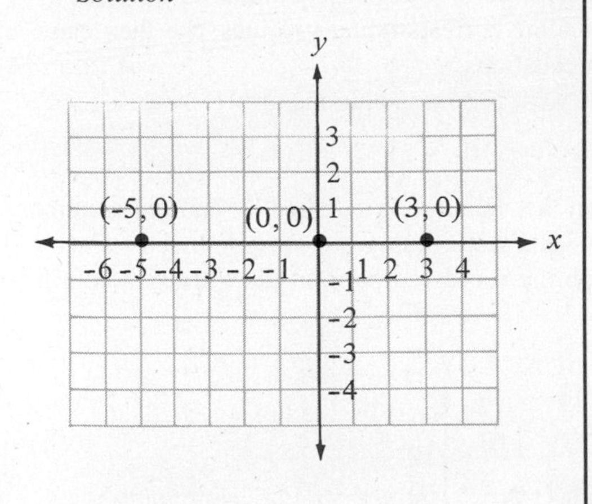

Example 4. Graph the ordered pairs (0, 0), (0, 2), and (0, −3).

Solution

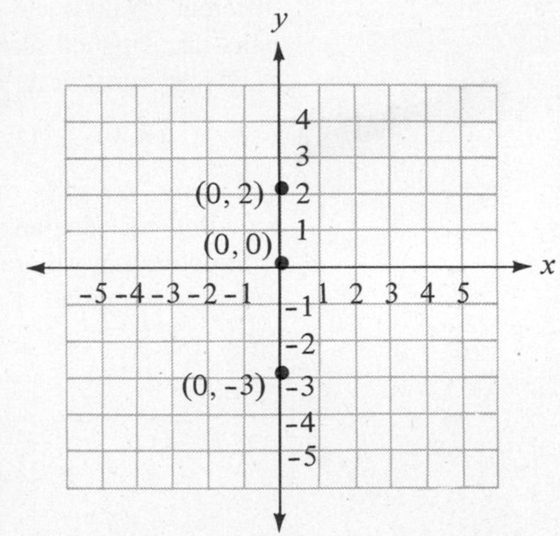

If a point has a y-coordinate of 0, that point lies on the x-axis.

If a point has an x-coordinate of 0, that point lies on the y-axis.

From our work in this section, we observe that each ordered pair of numbers corresponds to *one and only one* point in the Cartesian coordinate plane. Moreover, each point in the plane corresponds to *one and only one* ordered pair of numbers.

Trial Run

1. *Graph the ordered pairs* (0, −2), (−1, −3), (−1, 2), and $\left(\frac{3}{2}, 2\right)$.

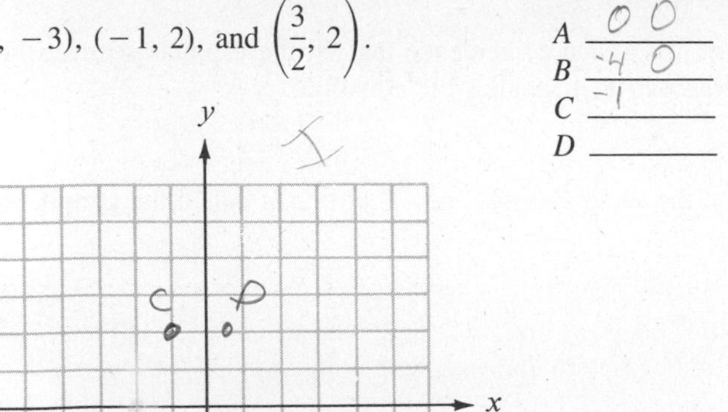

2. *Give the coordinates of each point.*

A 0 0 ______

B -4 0 ______

C -1 ______

D ______

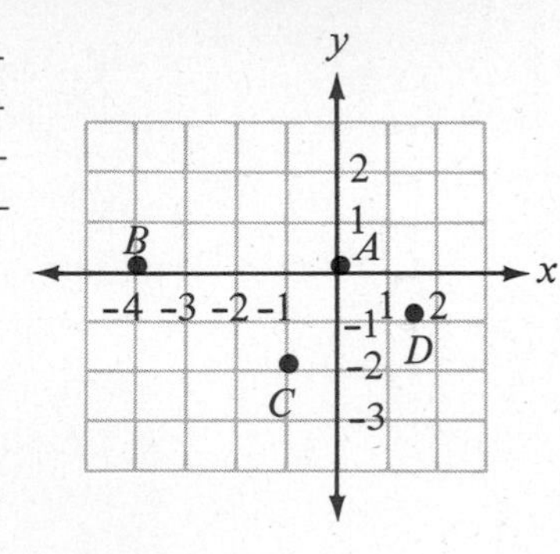

Answers are on page 583.

Graphing Equations by "Choose an x, Find a y" Method

In Section 10.1 we learned how to find ordered pairs of numbers that would satisfy equations in two variables. We considered first-degree equations of the form $ax + by = c$. By choosing different x-values and finding corresponding y-values, we then came up with several ordered pairs that satisfied such equations.

For the equation $x + y = 7$, some solutions are

$$(7, 0) \qquad (5, 2) \qquad (10, -3) \qquad (3, 4) \qquad (0, 7) \qquad (-2, 9) \qquad (1, 6)$$

and so on. We agree that this equation is satisfied by an *infinite* number of ordered pairs, each found by choosing an x-value and finding a corresponding y-value.

Suppose that we graph these ordered pairs in the Cartesian coordinate plane, remembering to label each point.

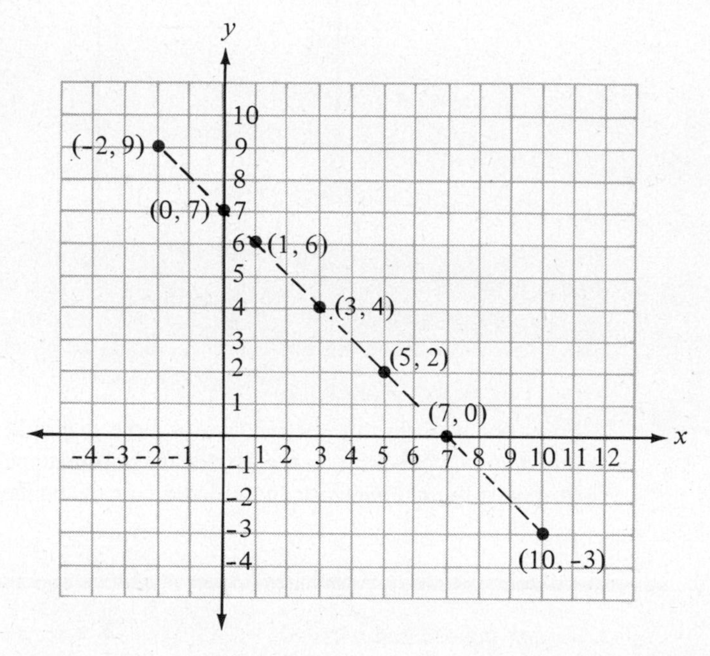

It is not just a nice coincidence that all these points seem to lie in a **straight line**. Indeed, *any* ordered pair that satisfies the equation

$$x + y = 7$$

will lie on the same straight line. That line is called the **graph of the equation**.

Graph of an Equation. The graph of an equation is the set of all points corresponding to the ordered pairs that satisfy the equation.

Since it is impossible to find *all* points that satisfy an equation, we shall settle for a reasonable number of ordered pairs, say 3 or 4. Although it is true that two points are all that you need to construct a straight line, we shall use at least three points just to check our arithmetic. If three points that satisfy a first-degree equation do *not* lie in a straight line, that is a signal that you have made an error in substitution, and you should check each of your points again.

Example 5. Graph $2x - y = 6$.

Solution: In order to "choose an x and find a y," we discovered that it is helpful to isolate y right away.

$2x - y = 6$	Original equation.
$-y = -2x + 6$	Subtract $2x$ from both sides.
$y = 2x - 6$	Divide both sides by -1.

Now we find ordered pairs, organizing our work in a table.

x	$y = 2x - 6$	(x, y)
-2	$y = 2(-2) - 6 = -4 - 6 = -10$	$(-2, -10)$
0	$y = 2(0) - 6 = 0 - 6 = -6$	$(0, -6)$
3	$y = 2(3) - 6 = 6 - 6 = 0$	$(3, 0)$
5	$y = 2(5) - 6 = 10 - 6 = 4$	$(5, 4)$

Referring to the last column, we plot the four points.

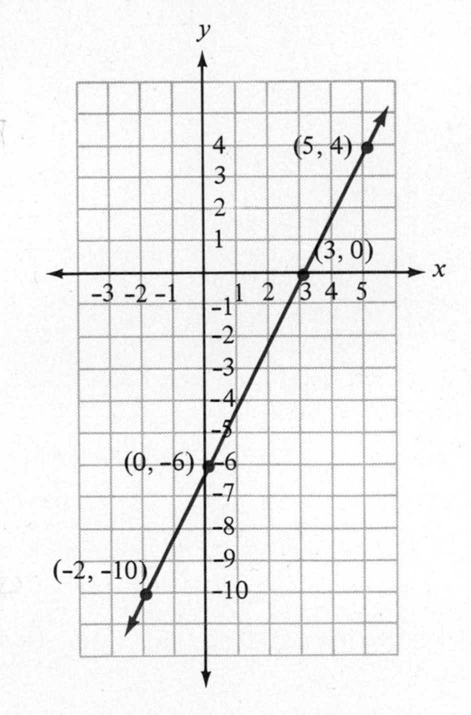

Notice that the ordered pairs satisfying the equation $2x - y = 6$ all lie on a straight line.

Graph of a First-Degree Equation. The graph of an equation of the type

$$ax + by = c$$

is a straight line.

You complete Example 6.

Example 6. Graph $y = 5x$.

Solution: Since y is isolated already, we shall "choose an x and find a y" to obtain points. Complete the last column of the table.

x	$y = 5x$	(x, y)
-1	$y = 5(-1) = -5$	$(-1, -5)$
0	$y = 5(0) = 0$	(____, ____)
$\frac{1}{2}$	$y = 5\left(\frac{1}{2}\right) = \frac{5}{2}$	(____, ____)
1	$y = 5(1) = 5$	(____, ____)

Now plot these points.

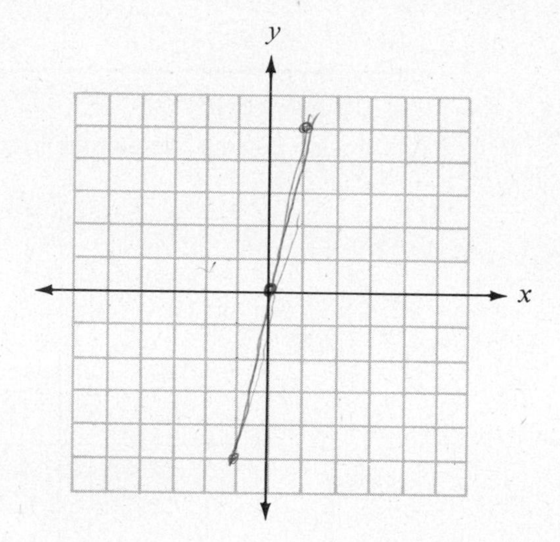

Check your work on page 583. ▶

The equation $y = 5x$ may not appear to be of the general first-degree form

$$ax + by = c$$

but by subtracting $5x$ from both sides of the equation we may rewrite $y = 5x$ as

$$-5x + y = 0 \qquad \text{or} \qquad 5x - y = 0$$

which *is* general first-degree form (with $c = 0$).

Because of their graphs, we call such equations **linear equations**.

> **Linear Equation.** An equation of the type
>
> $$ax + by = c$$
>
> is called a linear equation.

The method we have developed for graphing linear equations is sometimes called the **arbitrary point** method.

> **Graphing Linear Equations ($ax + by = c$)**
>
> 1. Isolate y in the equation.
> 2. Choose at least three x-values and find the corresponding y-values by substituting into the equation with y isolated.
> 3. Write each solution as an ordered pair (x, y).
> 4. In the Cartesian plane plot the point corresponding to each ordered pair.
> 5. Join the points with a straight line.

Example 7. Graph the linear equation $3x + 6y = 4$.

Solution: We first isolate y.

$$3x + 6y = 4$$

$$6y = 4 - 3x \qquad \text{Subtract } 3x \text{ to isolate } y\text{-term.}$$

$$\frac{6y}{6} = \frac{4 - 3x}{6} \qquad \text{Divide by 6 to isolate } y.$$

$$y = \frac{4}{6} - \frac{3x}{6} \qquad \text{Divide each polynomial term by 6.}$$

$$y = \frac{2}{3} - \frac{1}{2}x \qquad \text{Simplify each term.}$$

x	$y = \frac{2}{3} - \frac{1}{2}x$	(x, y)
-1	$y = \frac{2}{3} - \frac{1}{2}(-1) = \frac{2}{3} + \frac{1}{2} = \frac{4}{6} + \frac{3}{6} = \frac{7}{6}$	$\left(-1, \frac{7}{6}\right)$
0	$y = \frac{2}{3} - \frac{1}{2}(0) = \frac{2}{3}$	$\left(0, \frac{2}{3}\right)$
2	$y = \frac{2}{3} - \frac{1}{2}(2) = \frac{2}{3} - 1 = \frac{2}{3} - \frac{3}{5} = \frac{-1}{3}$	$\left(2, \frac{-1}{3}\right)$

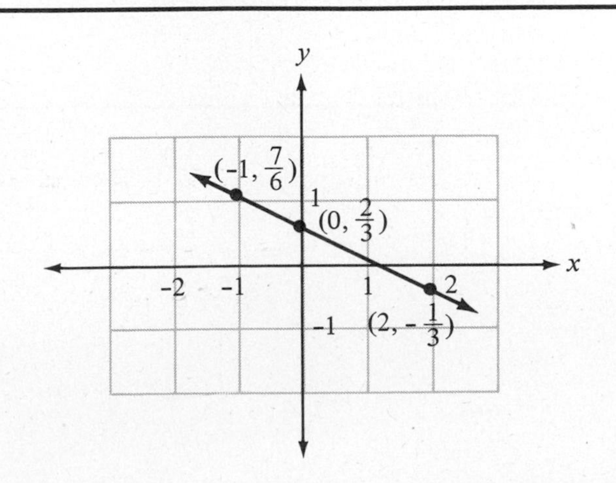

Trial Run

Graph each linear equation by the "choose an x, find a y" method.

1. $x + y = -3$

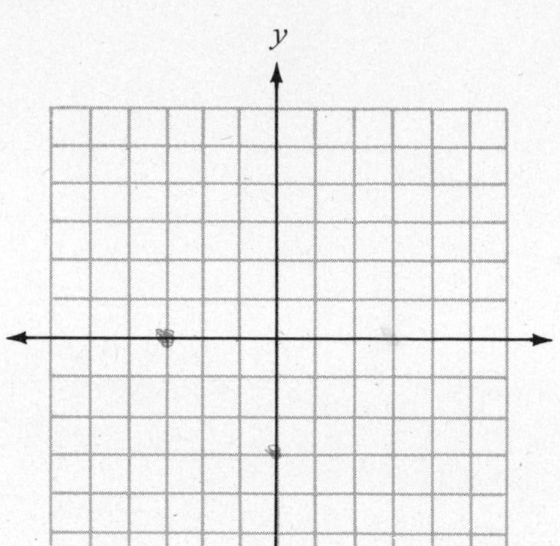

2. $y = 2x - 1$

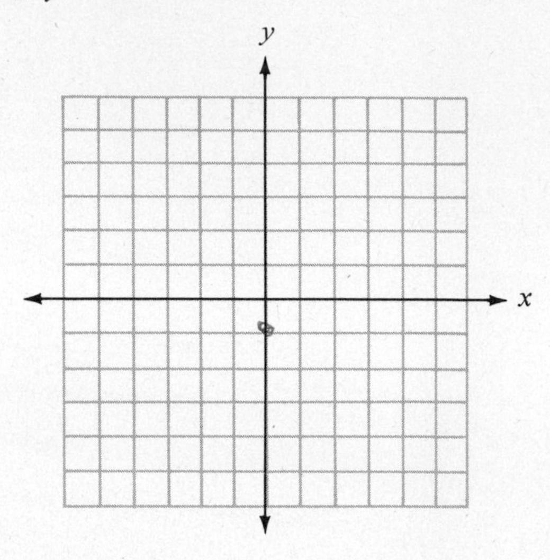

3. $2x - y = 3$

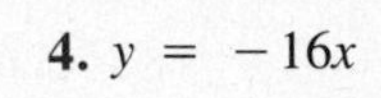

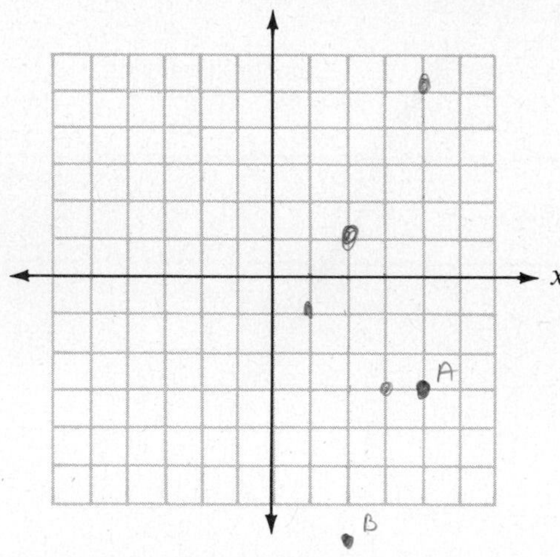

4. $y = -16x$

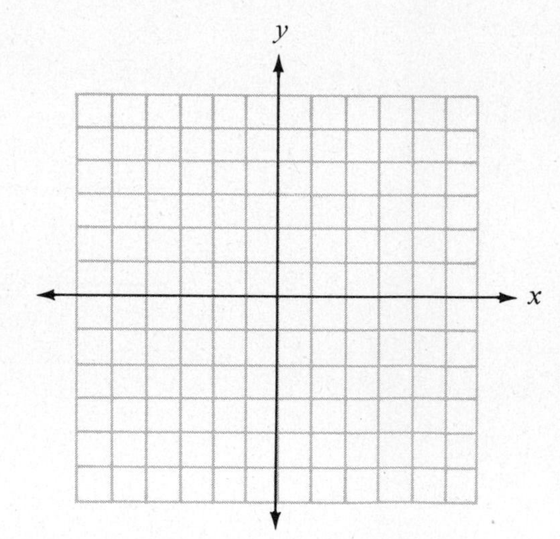

Answers are on page 583.

▶ Examples You Completed

Example 2. Graph the ordered pairs $(1, -2)$, $(-3, 4)$, $(-5, -5)$, and $\left(2, \frac{7}{2}\right)$.

Solution

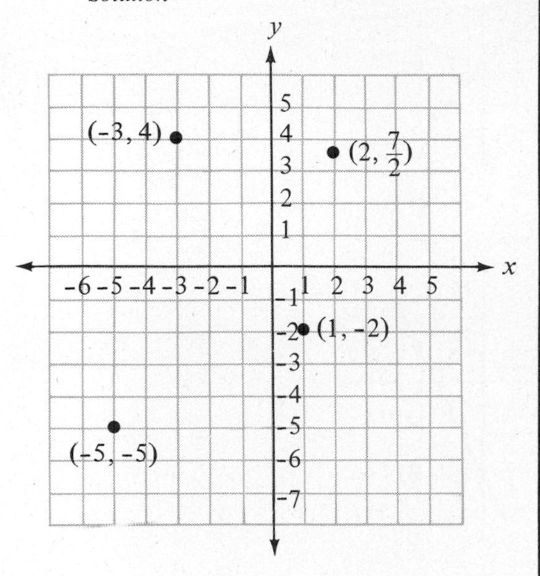

Example 6

Solution

x	$y = 5x$	(x, y)
-1	$y = 5(-1) = -5$	$(-1, -5)$
0	$y = 5(0) = 0$	$(0, 0)$
$\frac{1}{2}$	$y = 5\left(\frac{1}{2}\right) = \frac{5}{2}$	$\left(\frac{1}{2}, \frac{5}{2}\right)$
1	$y = 5(1) = 5$	$(1, 5)$

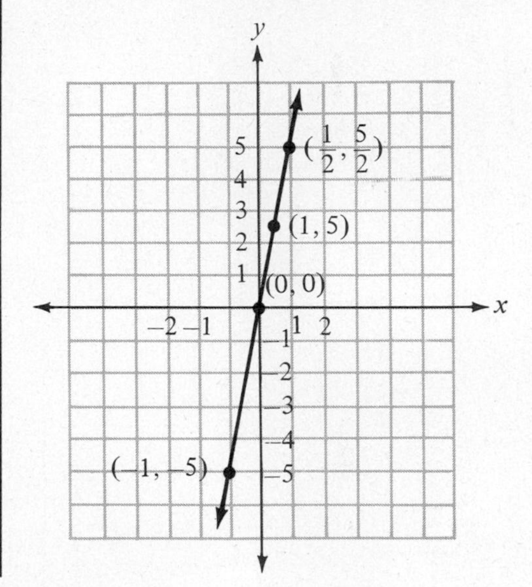

Answers to Trial Runs

page 577 **1.**

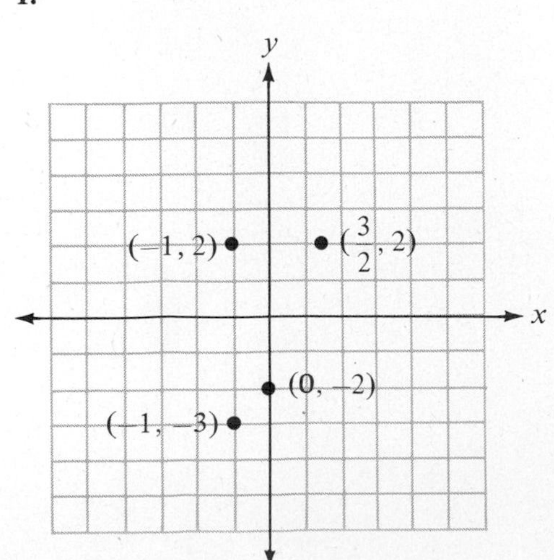

2. $A = (0, 0)$; $B = (-4, 0)$; $C = (-1, -2)$; $D = \left(\frac{3}{2}, -1\right)$

page 582 **1.**

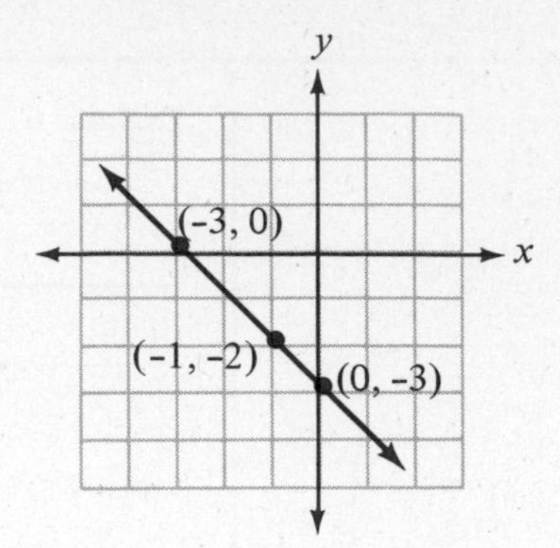

2.

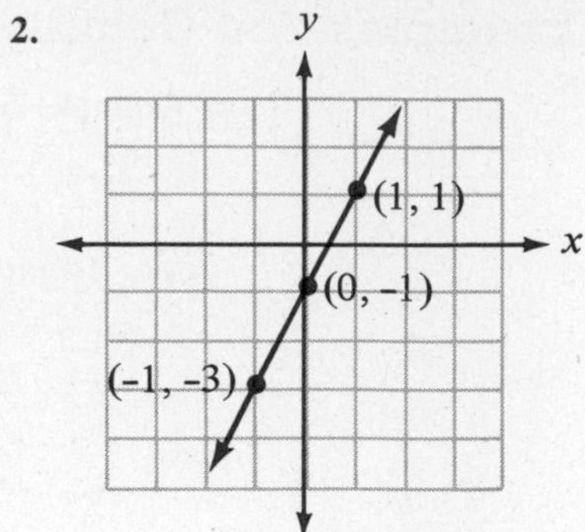

3.

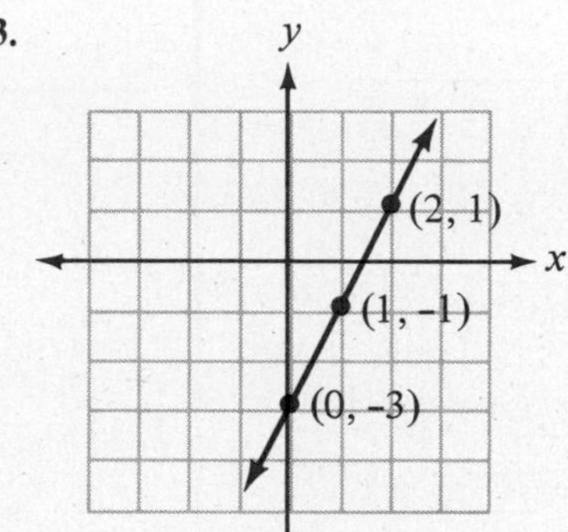

4.

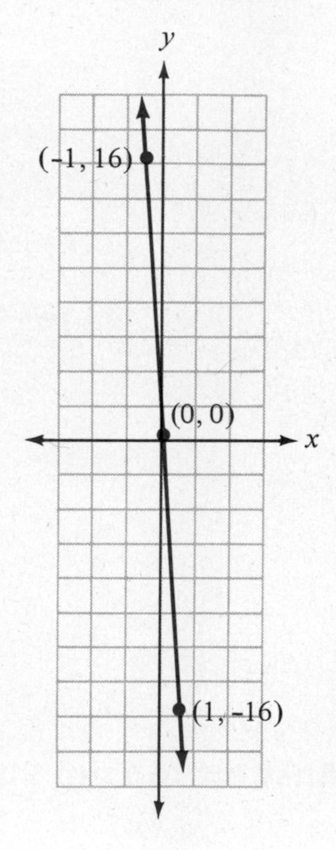

Name ____________________________ Date ____________

EXERCISE SET 10.2

Graph the following points.

1. A: (2, 0)

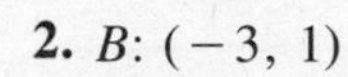

2. B: $(-3, 1)$

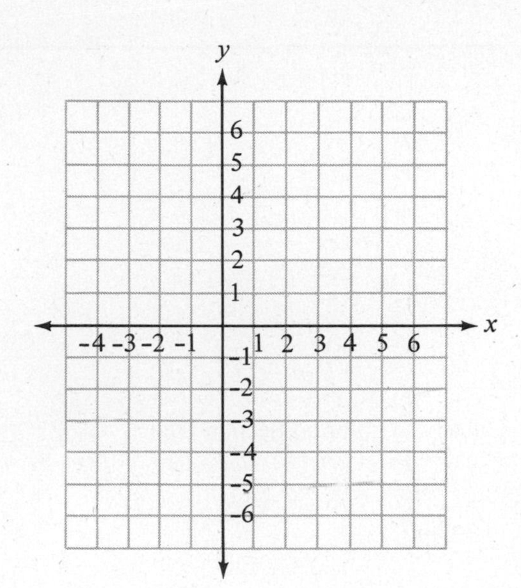

3. C: $(0, -5)$

4. D: $(-3, -4)$

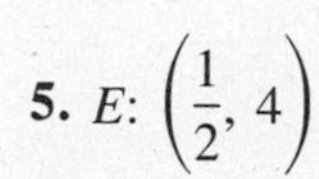

5. E: $\left(\frac{1}{2}, 4\right)$

6. F: (0, 0)

7. G: (6, 5)

8. H: $(3, -2)$

Give the coordinates of each point.

________ **9.** A

________ **10.** B

________ **11.** C

________ **12.** D

________ **13.** E

________ **14.** F

________ **15.** G

________ **16.** H

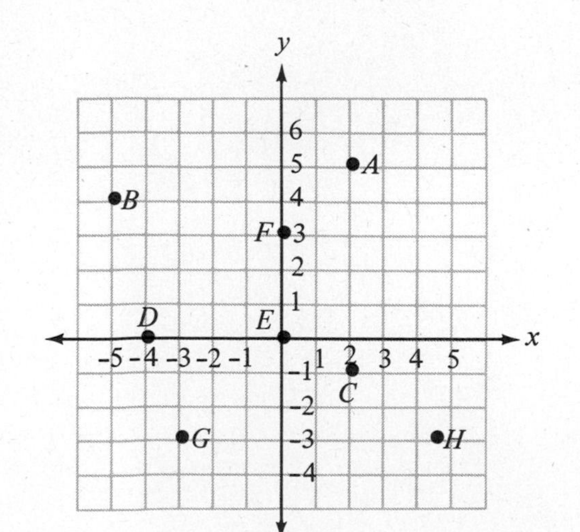

Graph each equation.

17. $x + y = 2$

18. $x + y = 5$

19. $y = 25x$

20. $3y = 15x$

21. $y = 4x - 2$

22. $y = 3x - 1$

23. $2x + y = 1$

24. $3x + y = 2$

25. $x + 6y = 8$

26. $x + 4y = 5$

27. $2x - y = -5$

28. $3x - y = -6$

☆ Stretching the Topics

Graph each equation.

_______ **1.** $7 - (3 - 3y) = -5 + 2x$

_______ **2.** $0.04x + 0.2y = 0.7$

_______ **3.** $\frac{2}{3}x + \frac{1}{6}y = 4$

Check your answers in the back of your book.

If you can complete **Checkup 10.2**, you are aready to go on to Section 10.3.

Name ________________________________ Date ______________

✓ CHECKUP 10.2

Graph each of the following points.

1. A: (3, 0)

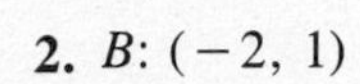

2. B: $(-2, 1)$

3. C: $(2, -3)$

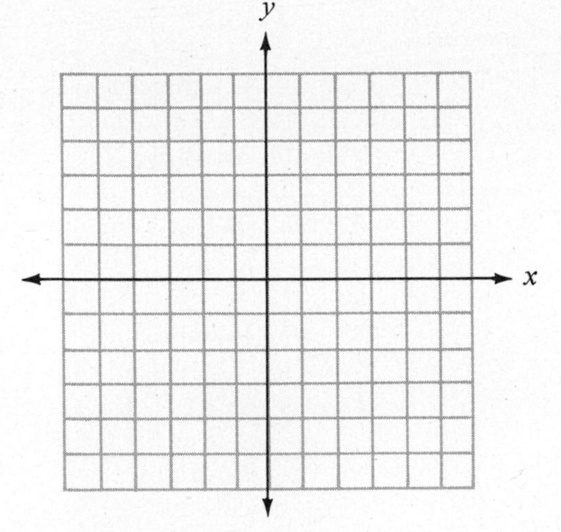

Give the coordinates of each point.

_______ 4. A

_______ 5. B

_______ 6. C

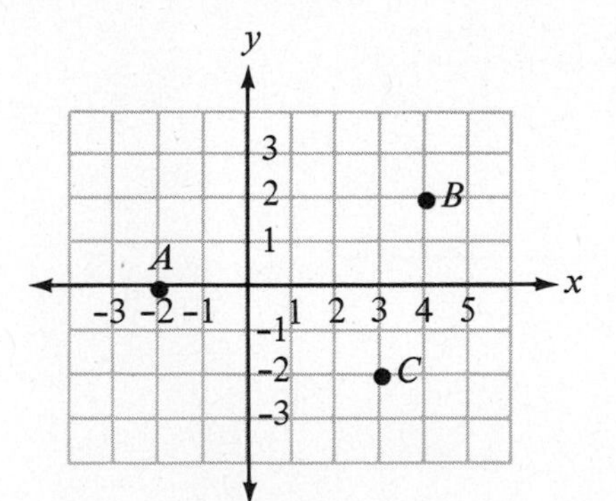

Graph each equation.

7. $y = 2x$

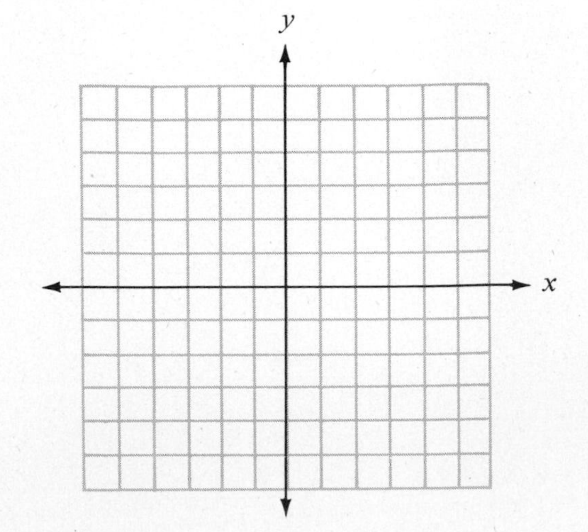

8. $x + y = 1$

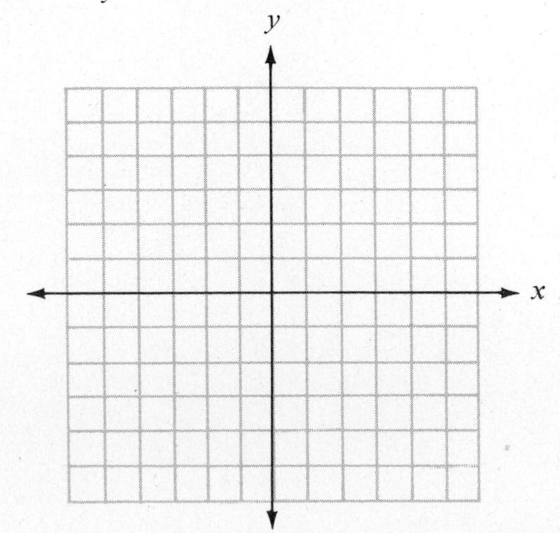

9. $x - 2y = 4$

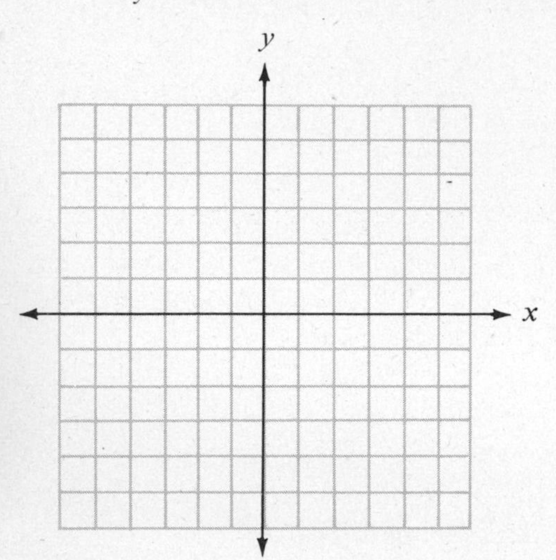

10. $3x - y = 0$

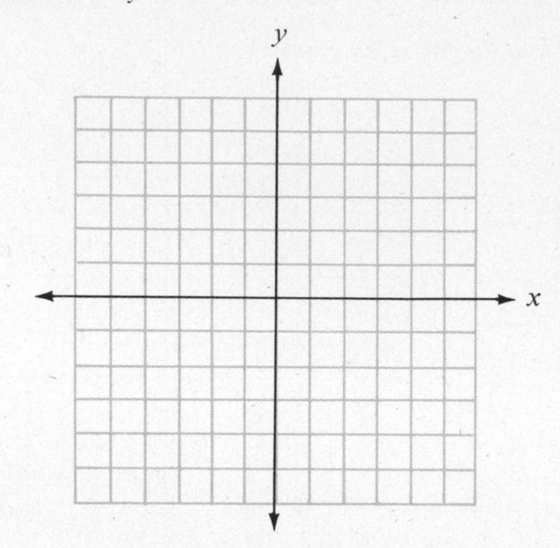

Check your answers in the back of your book.

If You Missed Problems:	You Should Review Examples:
1–3	1–4
4–6	1–4
7	6
8–10	5, 7

10.3 Graphing Linear Equations by Other Methods

Sometimes when we isolate y in an equation of the form $ax + by = c$, substituting into the final form becomes awkward. Perhaps we can find a less complicated method for graphing linear equations in which y is not already isolated.

Graphing Linear Equations by the Intercepts Method

If you glance back at the linear equations we have already graphed, you will notice that each straight line crosses the x-axis at some point and also crosses the y-axis at some point. These points are especially interesting to us and are called the **intercepts** for a graph.

Intercepts
x-intercept: the point at which a graph crosses the x-axis
y-intercept: the point at which a graph crosses the y-axis

Earlier we observed that *each point on the x-axis has a y-coordinate of 0* and that *each point on the y-axis has an x-coordinate of 0.* Putting this observation together with the idea of intercepts, we may say that

The x-intercept for a graph is the point that has a y-coordinate of 0.

The y-intercept for a graph is the point that has an x-coordinate of 0.

To find the x-intercept for a graph, we must let $y = 0$ in the given equation and find the corresponding x-value. To find the y-intercept for a graph, we must let $x = 0$ in the given equation and find the corresponding y-value.

Example 1. Use intercepts to graph $3x + 2y = 6$.

Solution

$$3x + 2y = 6$$

To find the x-intercept, we substitute 0 for y.

$$3x + 2(0) = 6$$
$$3x + 0 = 6$$
$$3x = 6$$
$$x = 2$$

The x-intercept is $(2, 0)$.

To find the y-intercept, we substitute 0 for x.

$$3(0) + 2y = 6$$
$$0 + 2y = 6$$
$$2y = 6$$
$$y = 3$$

The y-intercept is $(0, 3)$.

Because there is less chance of error when we substitute 0 for a variable, we trust these two points to be accurate and use them to graph our line.

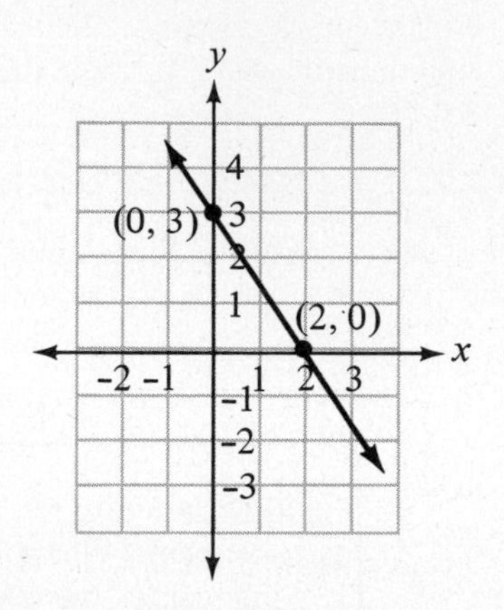

Now you help complete Example 2.

Example 2. Use intercepts to graph $5x - 2y = 15$.

Solution

$$5x - 2y = 15$$

To find the x-intercept, we substitute 0 for _____ .

$$5x - 2(0) = 15$$

The x-intercept is (__3__ , 0)

To find the y-intercept, we substitute 0 for _____ .

$$5(0) - 2y = 15$$

The y-intercept is (0, __-7.5__)

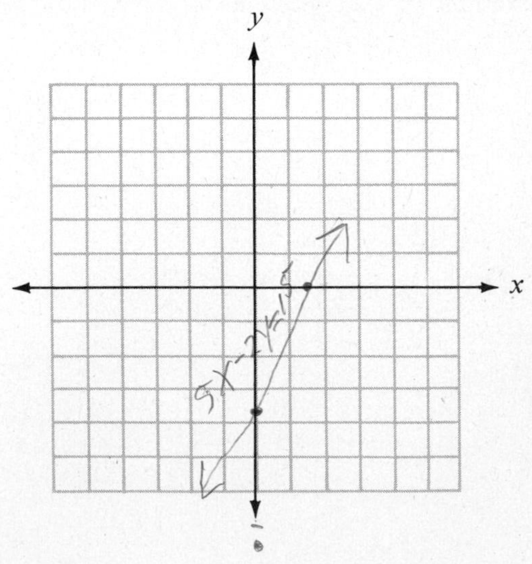

Check your work on page 596. ▶

Example 3. Graph $2x + 4y = 0$.

Solution

$$2x + 4y = 0$$

Find x-intercept: let $y = 0$.

$$2x + 4(0) = 0$$
$$2x = 0$$
$$x = 0$$

The x-intercept is $(0, 0)$.

Find y-intercept: let $x = 0$.

$$2(0) + 4y = 0$$
$$4y = 0$$
$$y = 0$$

The y-intercept is $(0, 0)$.

Here the x-intercept and y-intercept are both $(0, 0)$. Since one point will not give us a straight line, we must choose another x-value and find the corresponding y-value. Suppose that we let $x = 2$. If $x = 2$, then

$$2(2) + 4y = 0$$
$$4 + 4y = 0$$
$$4y = -4$$
$$y = -1$$

Our second point is $(2, -1)$.

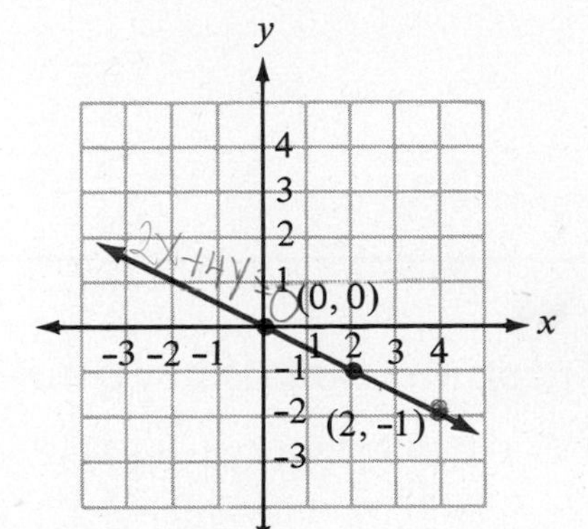

From Example 3 we see that whenever the x-intercept and y-intercept are both $(0, 0)$ we must find a second point on the graph by choosing an x and finding the corresponding y.

Trial Run

Use intercepts to graph each linear equation.

1. $5x - 3y = -15$

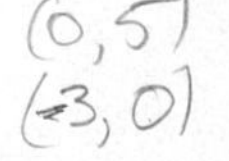

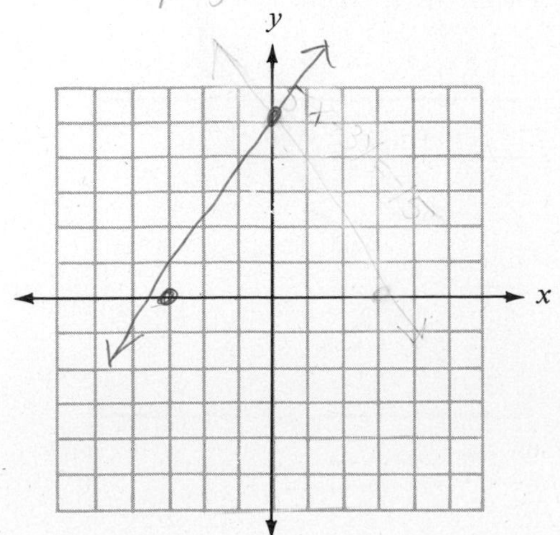

2. $-2x - 3y = 6$

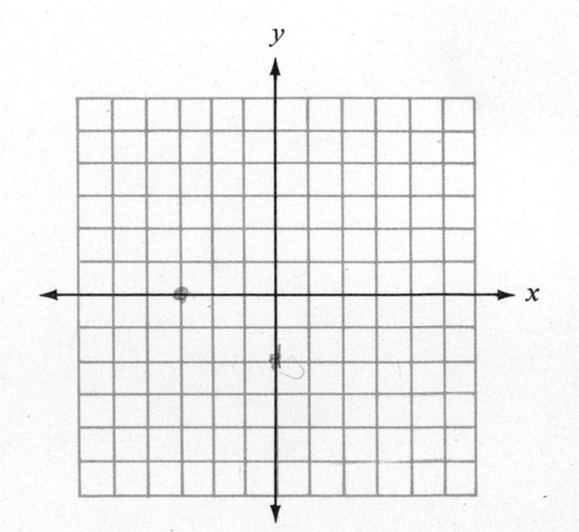

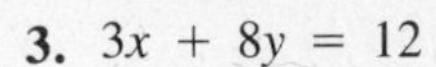

3. $3x + 8y = 12$

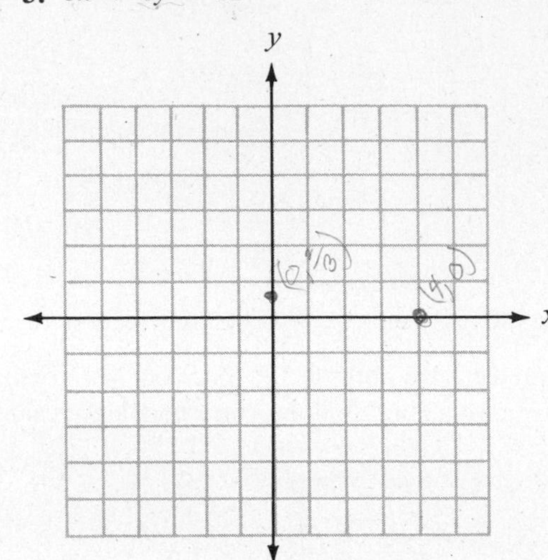

4. $4x - y = 0$

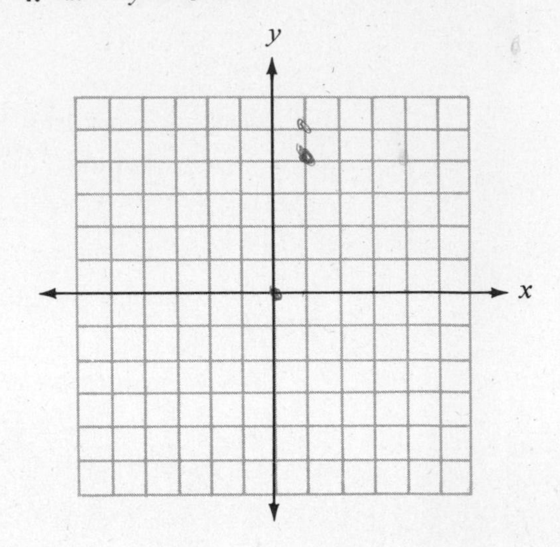

Answers are on page 597.

Graphing Constant Equations

In the general first-degree equation for a straight line,

$$ax + by = c$$

we agree that a, b, and c represent constants. We have graphed several equations in which c was 0. Until now, however, we have not considered any equations in which a is zero or b is zero. In other words, we have not discussed equations in which the x-term or the y-term is missing.

We must investigate what happens when we use the Cartesian coordinate plane to graph equations such as

$$y = 4 \qquad \text{or} \qquad x = 1$$

Example 4. Graph $y = 4$.

Solution: The equation $y = 4$ states that y is always 4. What about x? This equation says that "no matter what x is, y is *always* 4." Let's look at some ordered pairs.

x	$y = 4$	(x, y)
-2	$y = 4$	$(-2, 4)$
0	$y = 4$	$(0, 4)$
$\frac{7}{8}$	$y = 4$	$\left(\frac{7}{8}, 4\right)$
3	$y = 4$	$(3, 4)$

Plotting these points, we have

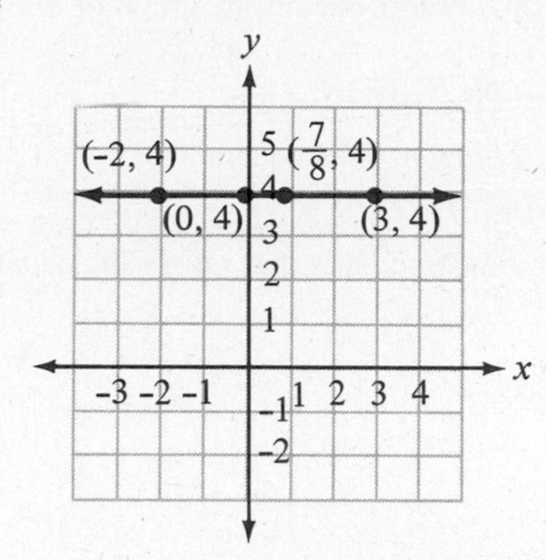

The graph of $y = 4$ in the Cartesian plane is a **horizontal line** (parallel to the x-axis) crossing the y-axis at (0, 4).

Example 5. Graph $y + 3 = 0$.

Solution: First we isolate y.

$$y + 3 = 0$$

$$y = -3$$

No matter what x is, y is *always* -3. Our graph should be a horizontal line, crossing the y-axis at $(0, -3)$.

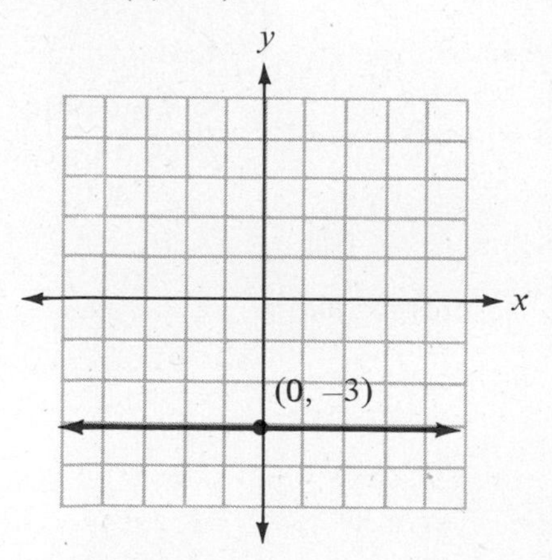

You complete Example 6.

Example 6. Graph $2y - 1 = 0$.

Solution: First we isolate y.

$$2y - 1 = 0$$

$$2y = 1$$

$$y = \frac{1}{2}$$

No matter what x is, y is *always* _____ . Our graph is a ________ line, crossing the y-axis at (0, _____).

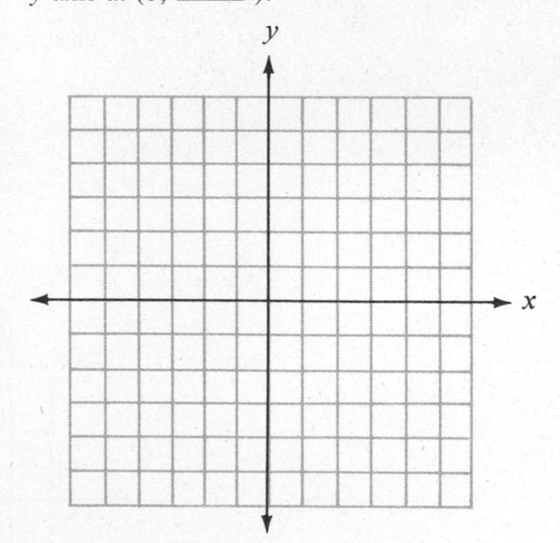

Check your graph on page 596. ▶

> The graph of an equation of the form
>
> $$y = k$$
>
> where k is a constant, is a **horizontal line** (parallel to the x-axis) crossing the y-axis at $(0, k)$.

Now let us consider the graph of a linear equation in which the y-term is missing.

Example 7. Graph $x = -3$.

Solution: The equation $x = -3$ states that x is always -3. No matter what y is, x is *always* -3.

Some ordered pairs that satisfy the equation are

(x, y)

$(-3, -1)$

$(-3, 0)$

$\left(-3, \frac{1}{2}\right)$

$(-3, 2)$

The graph of the equation is

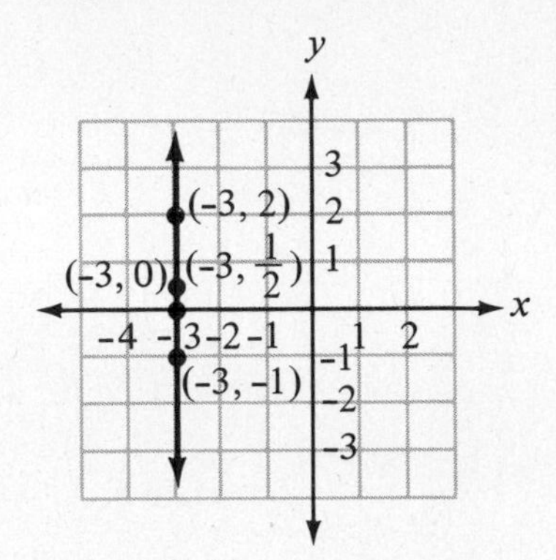

The graph of $x = -3$ is a **vertical line** (parallel to the y-axis) crossing the x-axis at $(-3, 0)$.

Example 8. Graph $3x + 4 = 0$.

Solution: First we isolate x.

$$3x + 4 = 0$$

$$3x = -4$$

$$x = \frac{-4}{3}$$

No matter what y is, x is *always* $\frac{-4}{3}$. Our graph should be a vertical line, crossing the x-axis at $\left(\frac{-4}{3}, 0\right)$

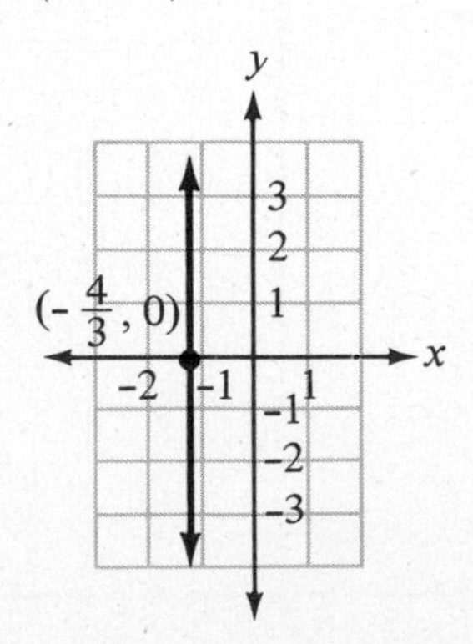

Now you try Example 9.

Example 9. Graph $x - 6 = 0$.

Solution: First we isolate x.

$$x - 6 = 0$$

$$x = 6$$

No matter what y is, x is *always* ______. Our graph is a ________ line, crossing the x-axis at (__6__ , 0).

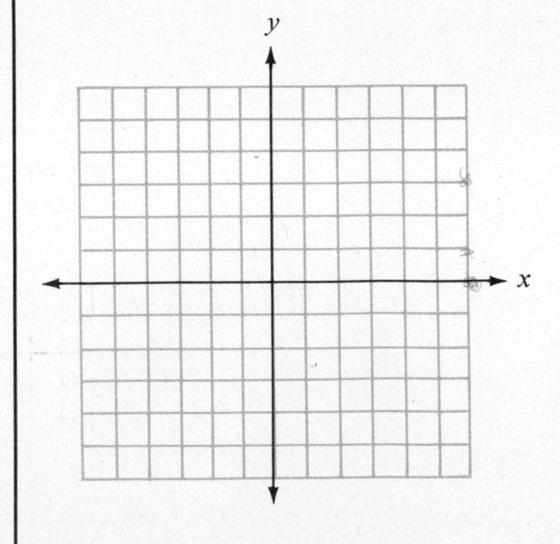

Check your work on page 596. ▶

> The graph of an equation of the form
>
> $$x = k$$
>
> where k is a constant, is a **vertical line** (parallel to the y-axis) crossing the x-axis at the point $(k, 0)$.

Trial Run

Graph each equation.

1. $y = 3$

2. $3y + 8 = 6$

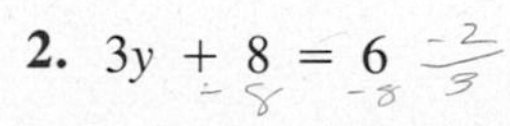

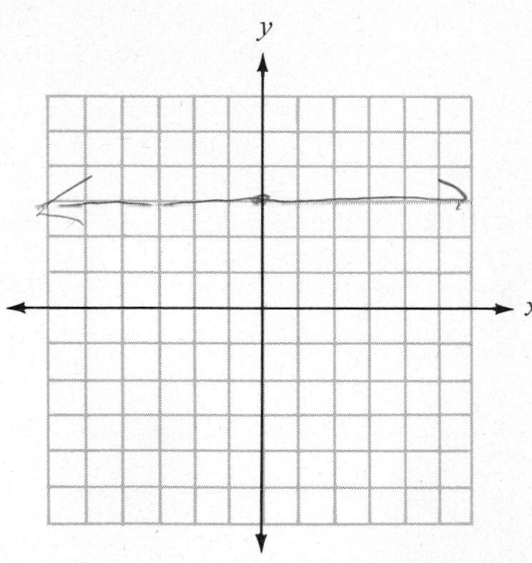

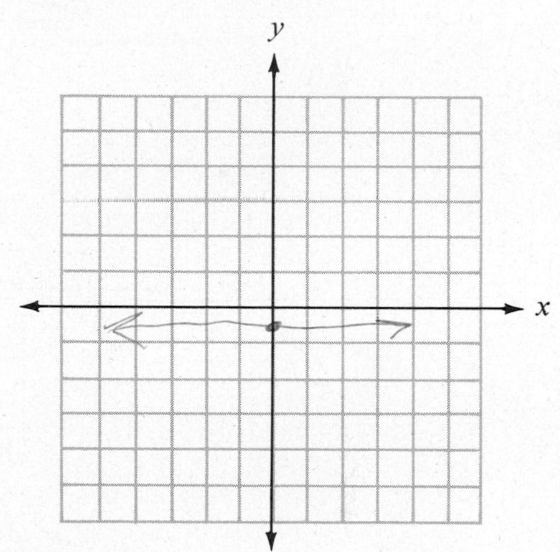

3. $x + 5 = 0$

4. $4x - 8 = 0$

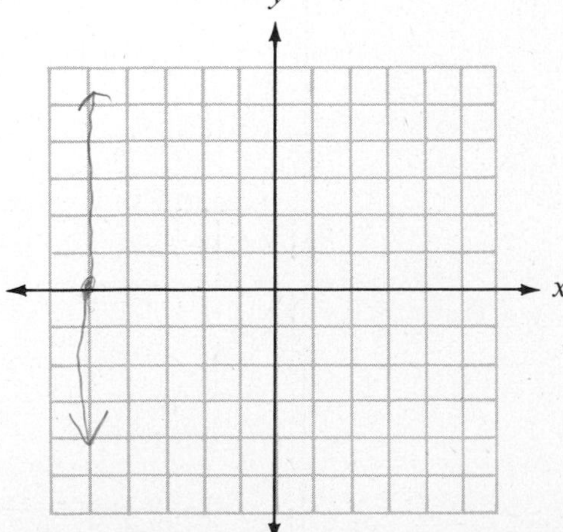

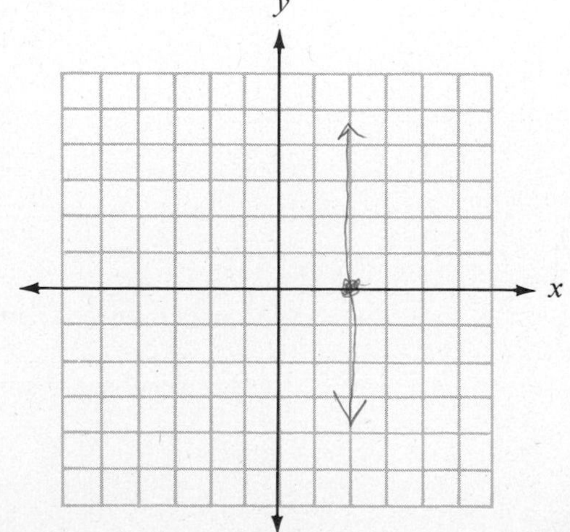

Answers are on page 597.

▶ Examples You Completed

Example 2. Use intercepts to graph $5x - 2y = 15$.

Solution

$$5x - 2y = 15$$

To find the x-intercept, we substitute 0 for y.

$$5x - 2(0) = 15$$
$$5x = 15$$
$$x = 3$$

The x-intercept is $(3, 0)$.

To find the y-intercept, we substitute 0 for x.

$$5(0) - 2y = 15$$
$$-2y = 15$$
$$y = \frac{15}{-2}$$
$$y = \frac{-15}{2}$$

The y-intercept is $\left(0, \frac{-15}{2}\right)$.

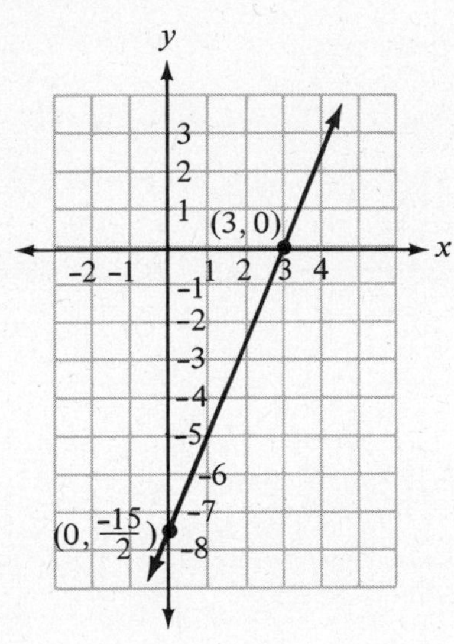

Example 6. Graph $2y - 1 = 0$.

Solution: First we isolate y.

$$2y - 1 = 0$$
$$2y = 1$$
$$y = \frac{1}{2}$$

No matter what x is, y is always $\frac{1}{2}$. Our graph is a horizontal line, crossing the y-axis at $\left(0, \frac{1}{2}\right)$.

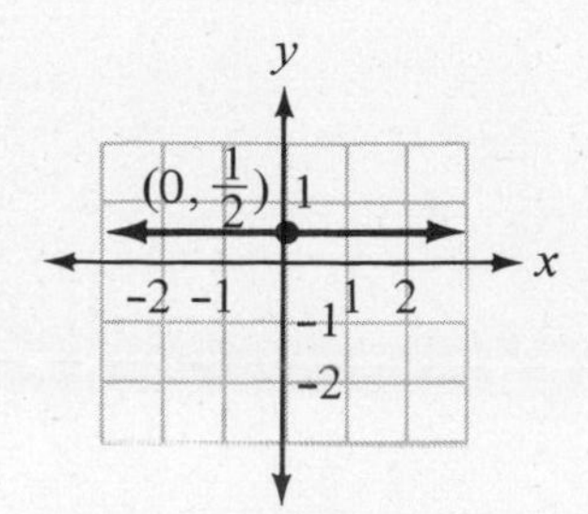

Example 9. Graph $x - 6 = 0$.

Solution: First we isolate x.

$$x - 6 = 0$$
$$x = 6$$

No matter what y is, x is always 6. Our graph is a vertical line, crossing the x-axis at $(6, 0)$.

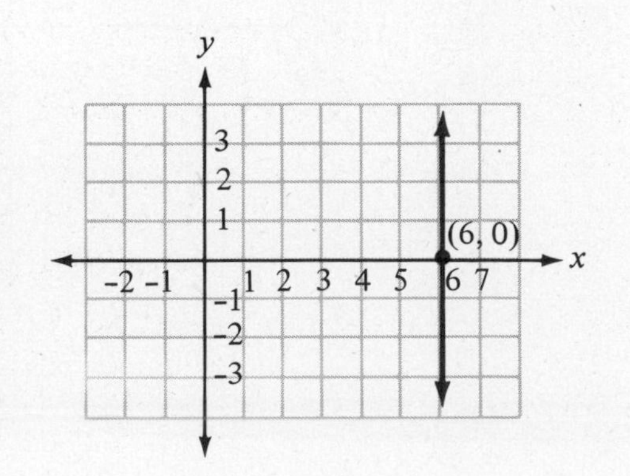

Answers to Trial Runs

page 591 **1.**

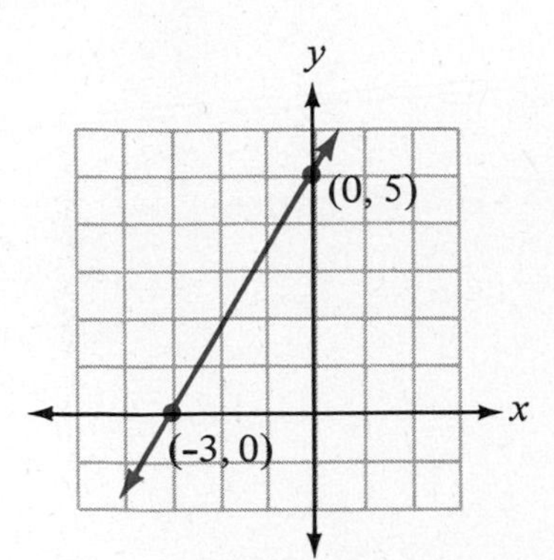

2.

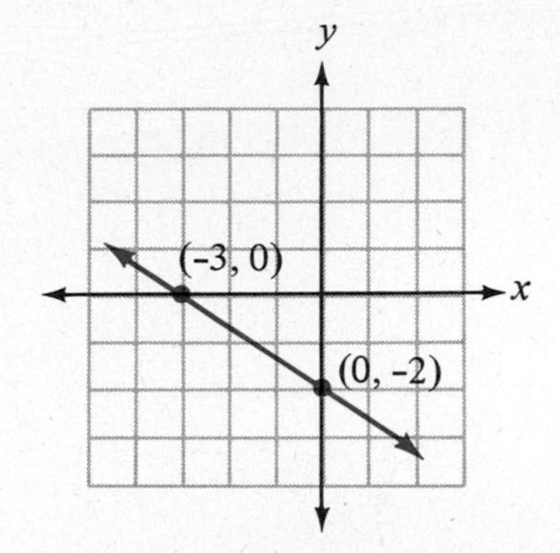

3.

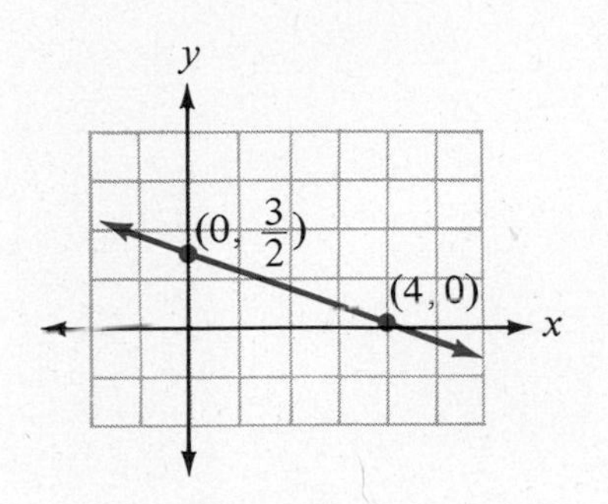

4.

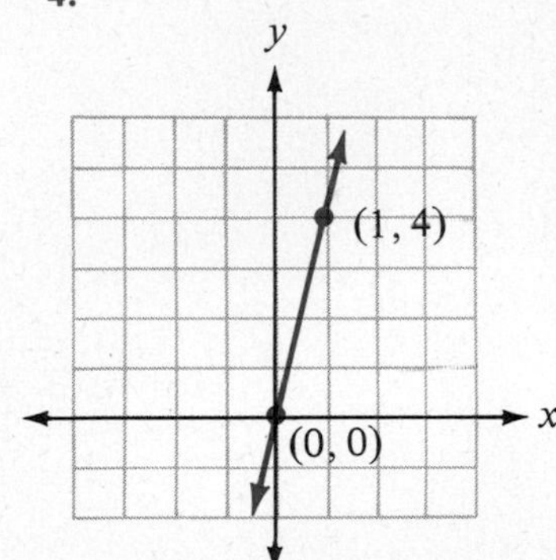

page 595 **1.**

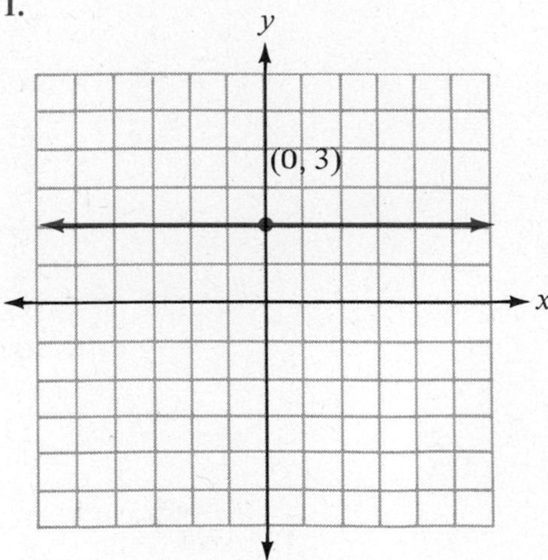

2.

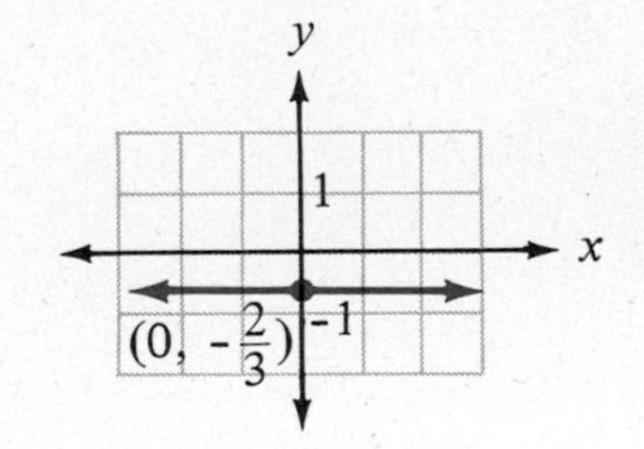

3.

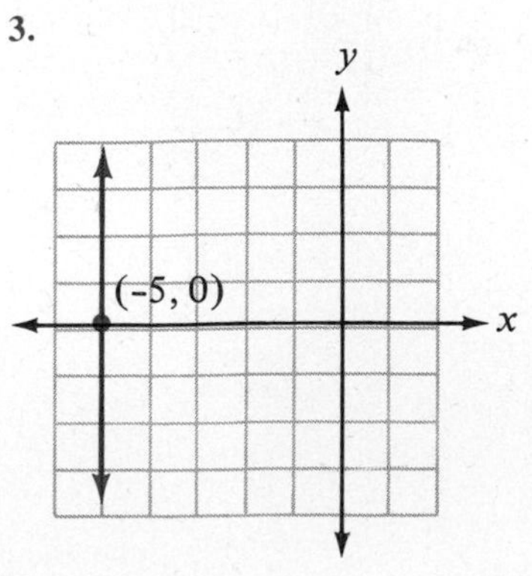

4.

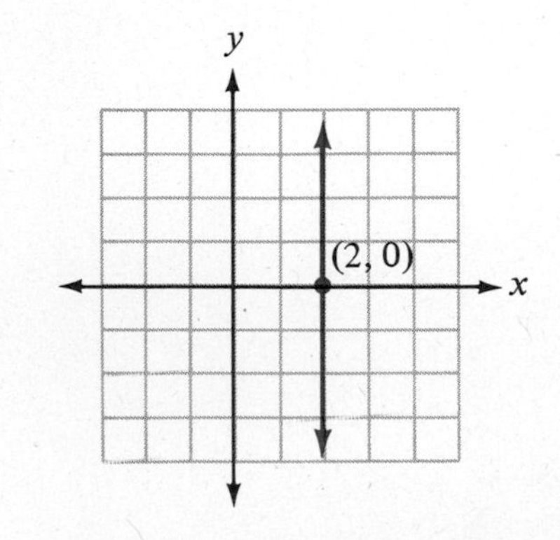

Name ________________________________ Date ____________

EXERCISE SET 10.3

Graph each equation by the intercept method.

1. $x + y = 3$
2. $x + y = -2$
3. $x - y = -5$
4. $x - y = 4$
5. $3x + y = 6$
6. $5x + y = 5$
7. $x - 2y = 8$
8. $x - 3y = -6$
9. $5x + 3y = -15$
10. $3x + 4y = 12$
11. $x + 2y = 5$
12. $x + 4y = 6$
13. $3x - y = 7$
14. $2x - y = -3$
15. $3x - 4y = -14$
16. $2x - 3y = 10$
17. $-5x + 2y = 15$
18. $-7x + 3y = 14$
19. $-x - y = 3$
20. $-x - y = 1$

Graph the lines represented by the following equations.

21. $y = 4$
22. $y = -3$
23. $x - 2 = 0$
24. $x - 5 = 0$
25. $y + 4 = 0$
26. $y + 1 = 0$
27. $3y - 9 = 0$
28. $5y - 10 = 0$
29. $2x = 7$
30. $3x = 5$
31. $2x + 5 = 0$
32. $4x + 18 = 0$
33. $6y - 16 = 0$
34. $8y - 28 = 0$
35. $9x - 27 = 0$
36. $7x + 35 = 0$

☆ Stretching the Topics

1. Graph $-1.2x + 3.6y = 2.4$ using the intercept method.

2. Graph $\frac{-1}{5}x + \frac{2}{3}y = 2$ using the intercept method.

3. Graph $\frac{x - 5}{2} = \frac{x - 3}{5}$.

Check your answers in the back of your book.

If you can graph each equation in **Checkup 10.3**, you are ready to go on to Section 10.4.

Name ______________________ Date ____________

✓ CHECKUP 10.3

Graph each of the following equations.

1. $2x - y = 4$

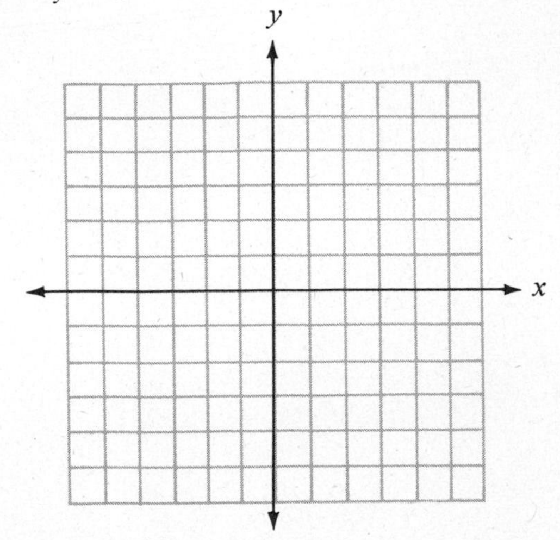

2. $x + 2y = 5$

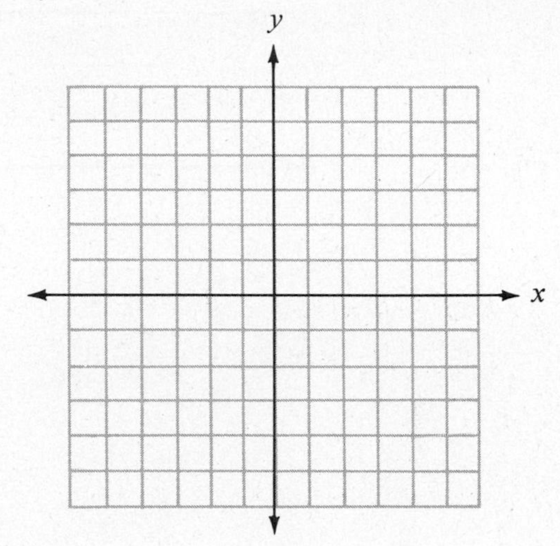

3. $-2x + 3y = -12$

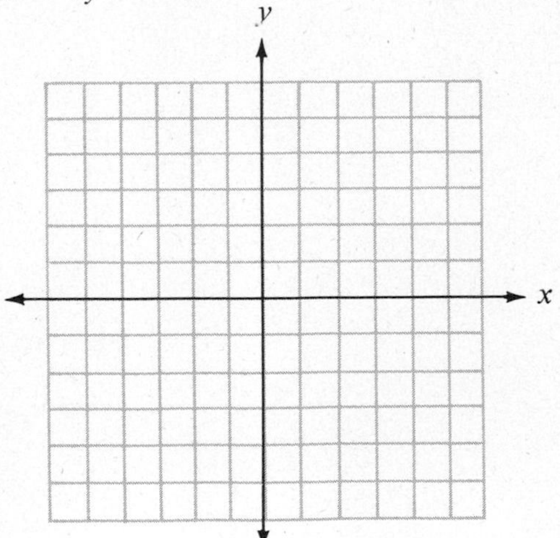

4. $y - 4 = 0$

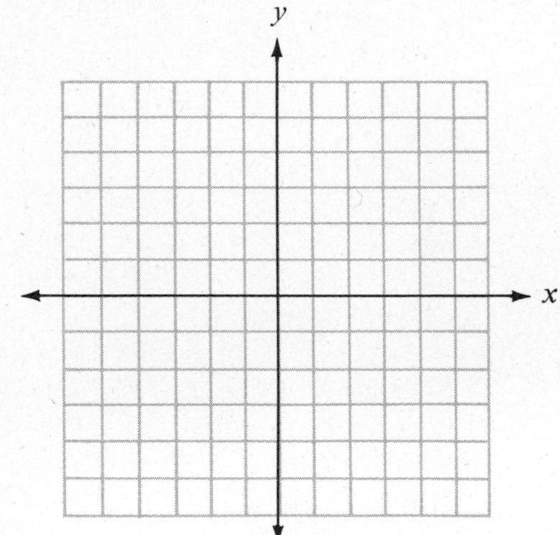

5. $3x = 8$

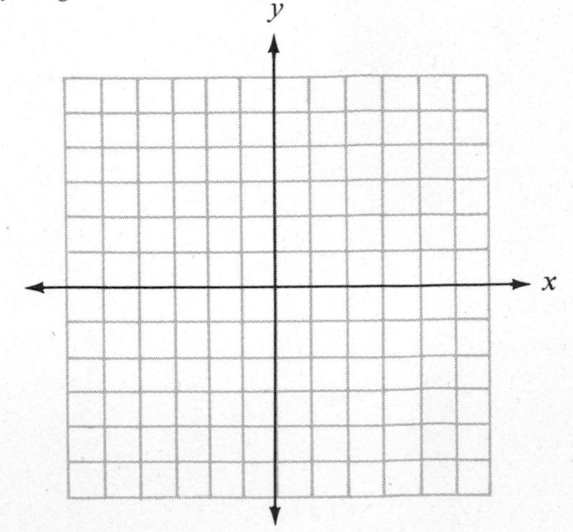

6. $3y + 6 = 0$

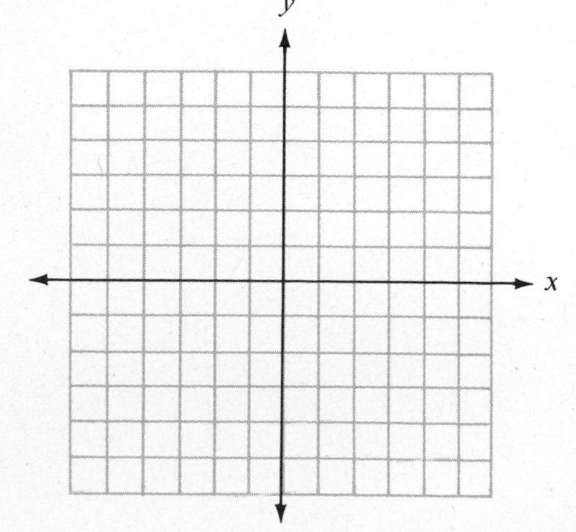

Check your answers in the back of your book.

If You Missed Problems:	You Should Review Examples:
1–3	1–3
4	5
5	8
6	6

10.4 Graphing Lines Using the Slope

Let's take a look at the graphs of several linear equations.

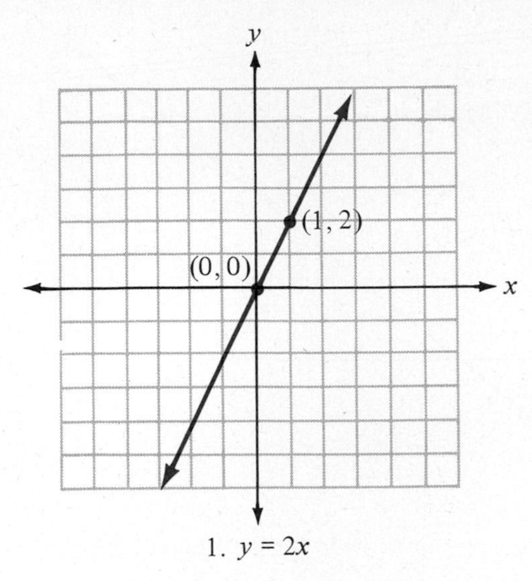

1. $y = 2x$

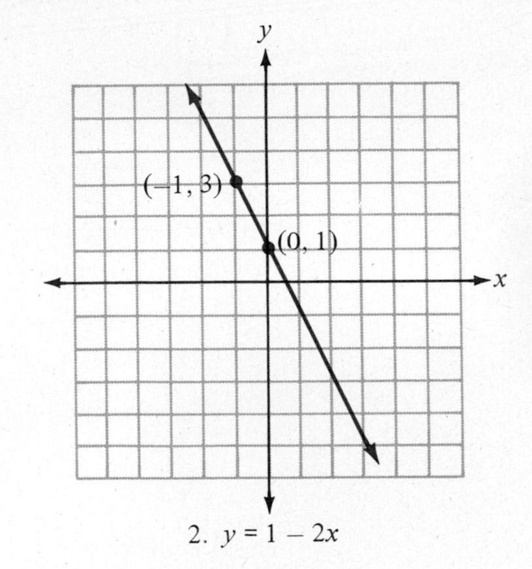

2. $y = 1 - 2x$

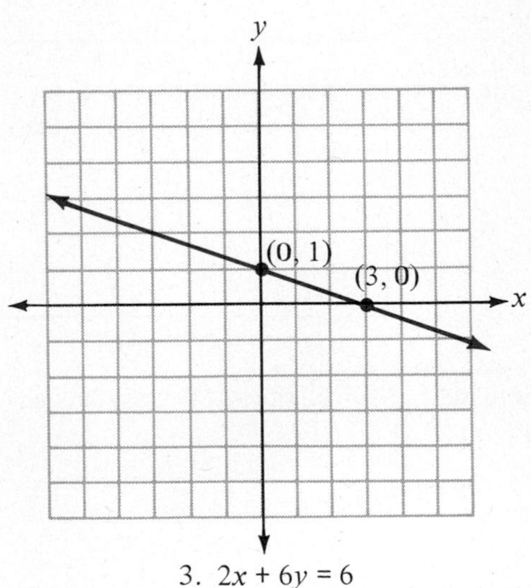

3. $2x + 6y = 6$

If we wish to describe the **steepness** of these lines, we might say that lines (1) and (2) seem to be equally steep, but in opposite directions. Line (3) seems to be less steep than the other two lines.

Defining the Slope of a Line

We can measure the steepness or **slope** of a line by locating any two points on that line and then comparing the units of change in the y-values (sometimes called the **rise**) to the units of change in the x-values (sometimes called the **run**).

$$\text{Slope} = \frac{\text{rise}}{\text{run}}$$

$$= \frac{\text{vertical change}}{\text{horizontal change}}$$

$$= \frac{\text{change in } y}{\text{change } x}$$

To find the numerical value of the slope of a line joining two points, we can let the first point be (x_1, y_1) and the second point be (x_2, y_2).

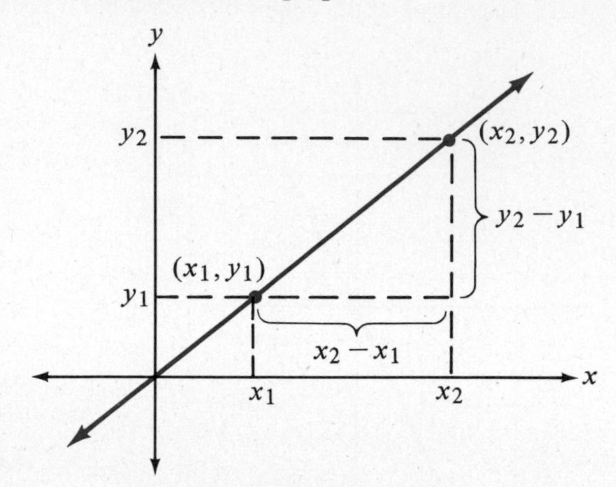

It is customary to use the letter ***m*** to stand for the slope of the line.

Definition of Slope. If (x_1, y_1) and (x_2, y_2) are two points on a line then the slope of that line is defined by

$$m = \frac{y_2 - y_1}{x_2 - x_1}$$

provided that $x_2 \neq x_1$.

Example 1. Find the slope of the line through (1, 3) and (4, 7).

Solution

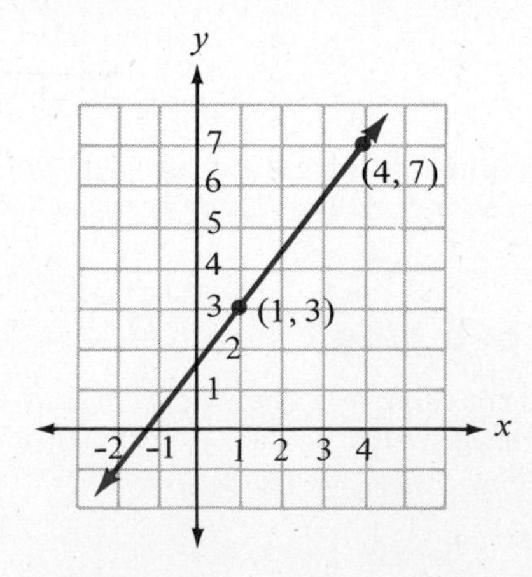

Let $(x_1, y_1) = (1, 3)$

and $(x_2, y_2) = (4, 7)$

Then $m = \dfrac{y_2 - y_1}{x_2 - x_1}$

$= \dfrac{7 - 3}{4 - 1}$

$m = \dfrac{4}{3}$

Notice that this line slopes *upward* from left to right. To move from the first point to the second point, we must go to the *right* 3 units and *up* 4 units.

You complete Example 2.

Example 2. Find the slope of the line through $(-4, 1)$ and $(3, -2)$.

Solution

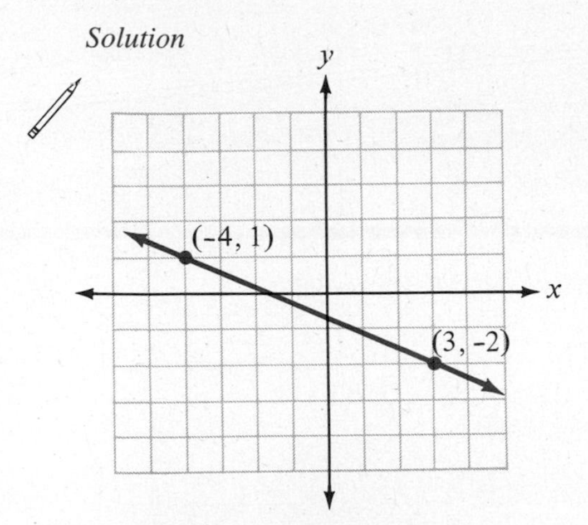

Let $(x_1, y_1) = (-4, 1)$

and $(x_2, y_2) = (3, -2)$

Then $m = \dfrac{y_2 - y_1}{x_2 - x_1}$

$= \dfrac{-2 - 1}{3 - (-4)}$

$m =$ ______

Notice that this line slopes ________ from left to right. To move from the first point to the second point, we must go to the ________ 7 units and ________ 3 units.

Check your work on page 613. ▶

From Examples 1 and 2 we make an important observation.

> If a line slopes *upward* from left to right, its slope is *positive*.
>
> If a line slopes *downward* from left to right, its slope is *negative*.

Example 3. Find the slope of the line through $(1, 4)$ and $(-2, 4)$.

Solution

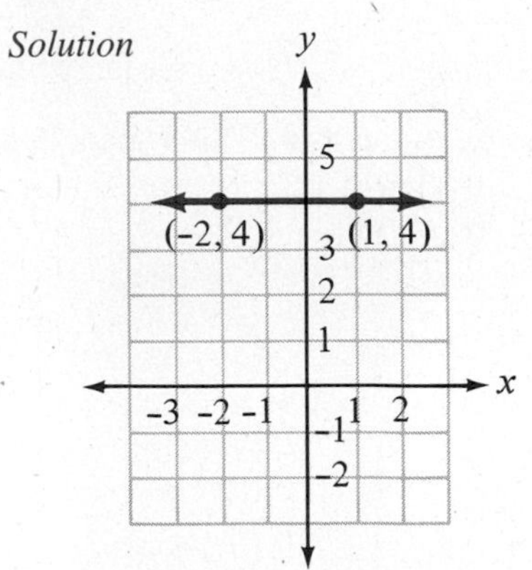

Let $(x_1, y_1) = (1, 4)$

and $(x_2, y_2) = (-2, 4)$

Then $m = \dfrac{y_2 - y_1}{x_2 - x_1}$

$= \dfrac{4 - 4}{-2 - 1}$

$= \dfrac{0}{-3}$

$m = 0$

Example 4. Find the slope of the line through $(2, 3)$ and $(2, -1)$.

Solution

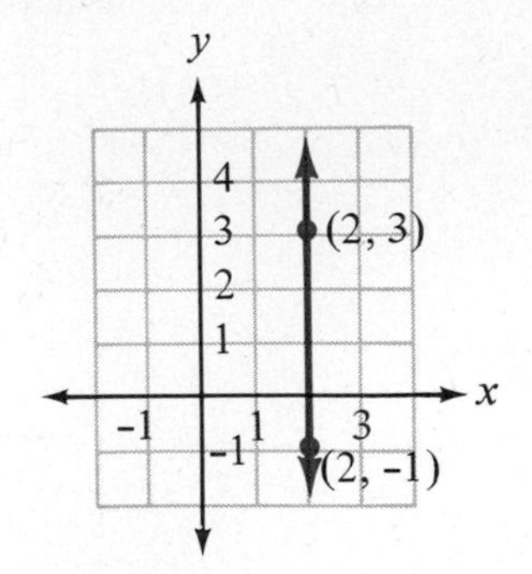

Let $(x_1, y_1) = (2, 3)$

and $(x_2, y_2) = (2, -1)$

Then $m = \dfrac{y_2 - y_1}{x_2 - x_1}$

$= \dfrac{-1 - 3}{2 - 2}$

$= \dfrac{-4}{0}$

m is undefined.

From Examples 3 and 4 we observe that:

> The slope of a horizontal line is zero.
>
> The slope of a vertical line is undefined.

Trial Run

Find the slope of the line through the given points.

______ **1.** (2, 1) and (−3, 0)

______ **2.** (−4, 3) and (−5, 8)

______ **3.** (−3, 5) and (4, 5)

______ **4.** (2, 2) and (−7, −7)

______ **5.** (9, −3) and (3, 1)

______ **6.** (7, −1) and (7, 5)

Answers are on page 614.

Using the Slope for Graphing

If we know one point on a line and the slope of that line, we can find another point on the same line using the fact that

$$\text{slope} = \frac{\text{change in } y}{\text{change in } x}$$

For instance, if we know that the point (2, 1) is on a line with slope $\frac{3}{4}$, we know we must start at (2, 1) and then go to the *right* 4 units and *up* 3 units to arrive at another point on the same line.

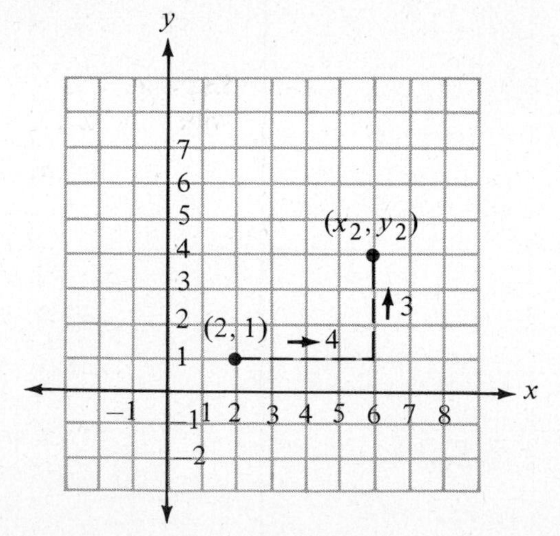

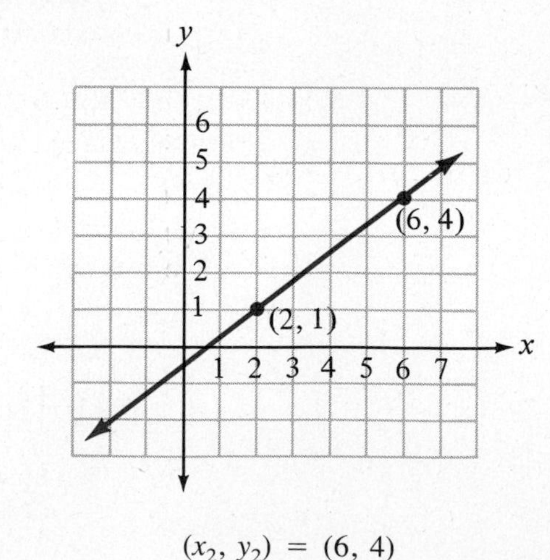

$(x_2, y_2) = (6, 4)$

$(x_1, y_1) = (2, 1)$

$$m = \frac{3 \uparrow}{4 \rightarrow}$$

The coordinates of the second point (x_2, y_2) are found by noting the changes in x and y that are described by the slope.

Old Coordinate	Change	New Coordinate
$x_1 = 2$	$+4$	$x_2 = 2 + 4 = 6$
$y_1 = 1$	$+3$	$y_2 = 1 + 3 = 4$

Example 5. Graph the line through the origin with slope 3.

Solution: Our first point is the origin (0, 0) and we write our slope 3 as a fraction.

$$m = 3 = \frac{3 \uparrow}{1 \rightarrow}$$

$$(x_1, y_1) = (0, 0)$$

Starting at the point (0, 0) our slope says we must move to the right 1 unit and up 3 units.

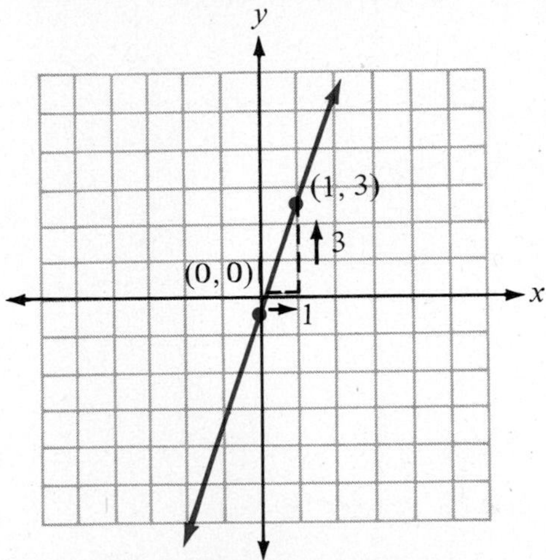

Now you try Example 6.

Example 6. Graph the line through (3, −2) with slope $\frac{-1}{4}$.

Solution

$(x_1, y_1) = (\_\_\_\_, \_\_\_\_)$

$$m = \frac{-1 \downarrow}{4 \rightarrow}$$

Starting at the point (____, ____) our slope says we must move to the ______ 4 units and ______ 1 unit.

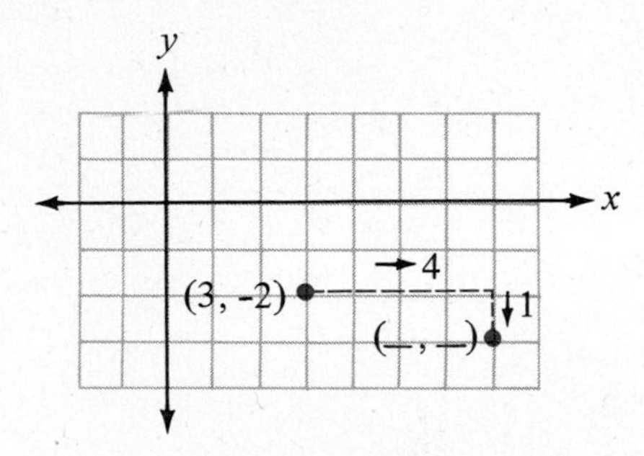

Check your work on page 613. ▶

Example 7. Graph the line through (5, −2) with undefined slope.

Solution: Our first point is (5, −2) and the slope is undefined. Recall that if the slope of a line is undefined, the line is *vertical*.

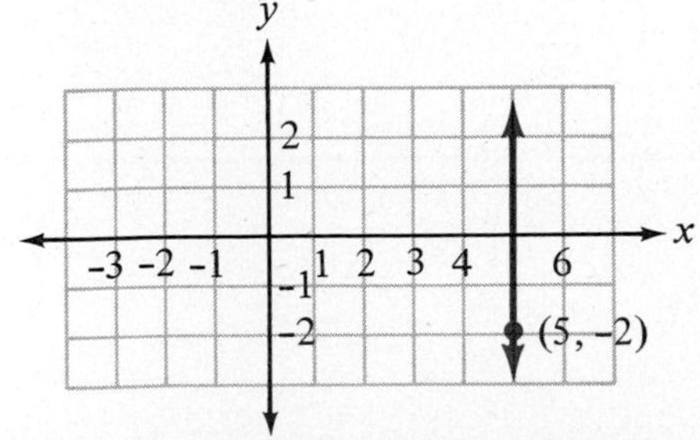

What is the equation for this line? The equation for a vertical line is always $x = k$, a constant. Here the equation must be

$$x = 5$$

Now you try Example 8.

Example 8. Graph the line through (5, −2) with slope zero.

Solution: Our first point is (____, ____) and the slope is ____. Recall that if the slope if a line is zero, the line is ______ .

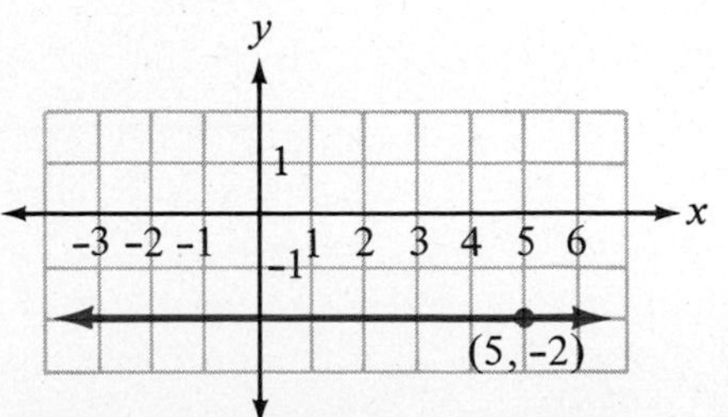

What is the equation for this line? The equation for a horizontal line is always ______ . Here the equation must be

____ = ____

Check your work on page 613. ▶

Trial Run

Graph the line through the given point with the given slope.

1. (0, 0); $m = \frac{5}{2}$

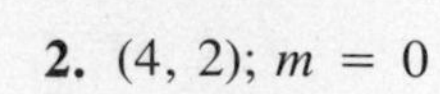

2. (4, 2); $m = 0$

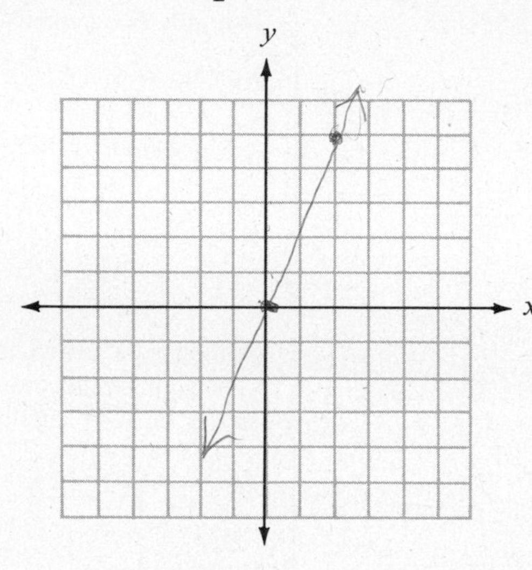

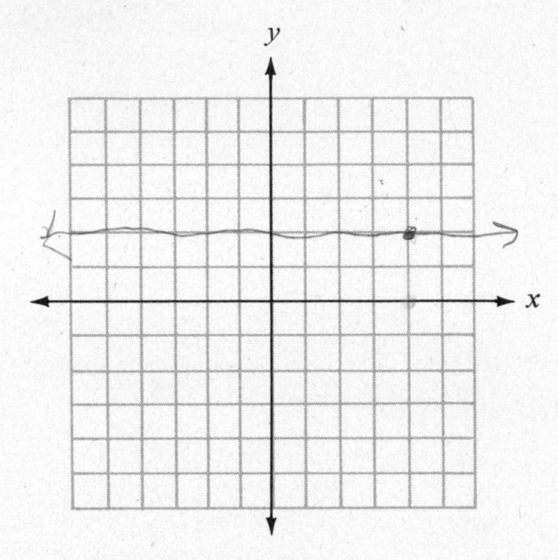

3. $(-2, -4)$; $m = -1$

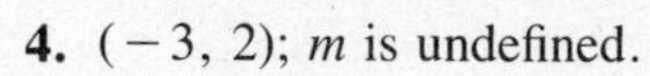

4. $(-3, 2)$; m is undefined.

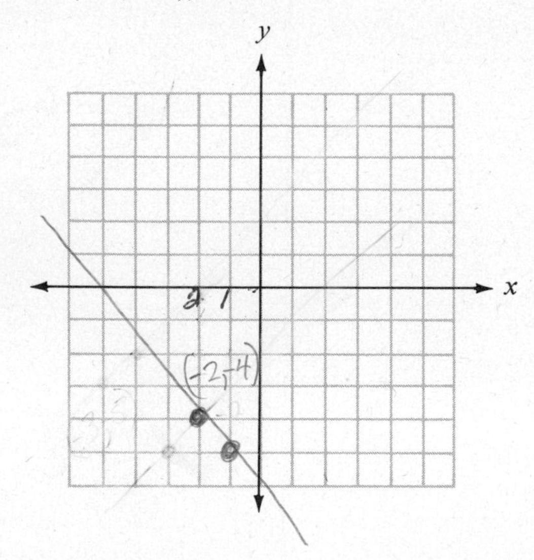

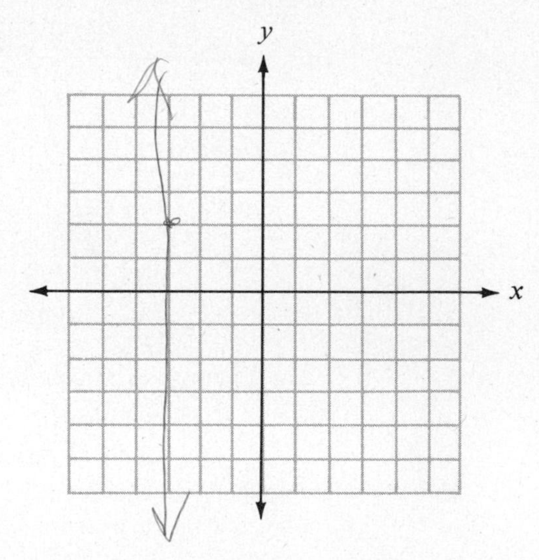

Answers are on page 614.

Graphing Lines by the Slope-Intercept Method

Let's take another look at the graphs at the beginning of Section 10.4. For each graph the y-intercept has been found by the usual method. Let's find the slope of each line using the points labeled on each graph.

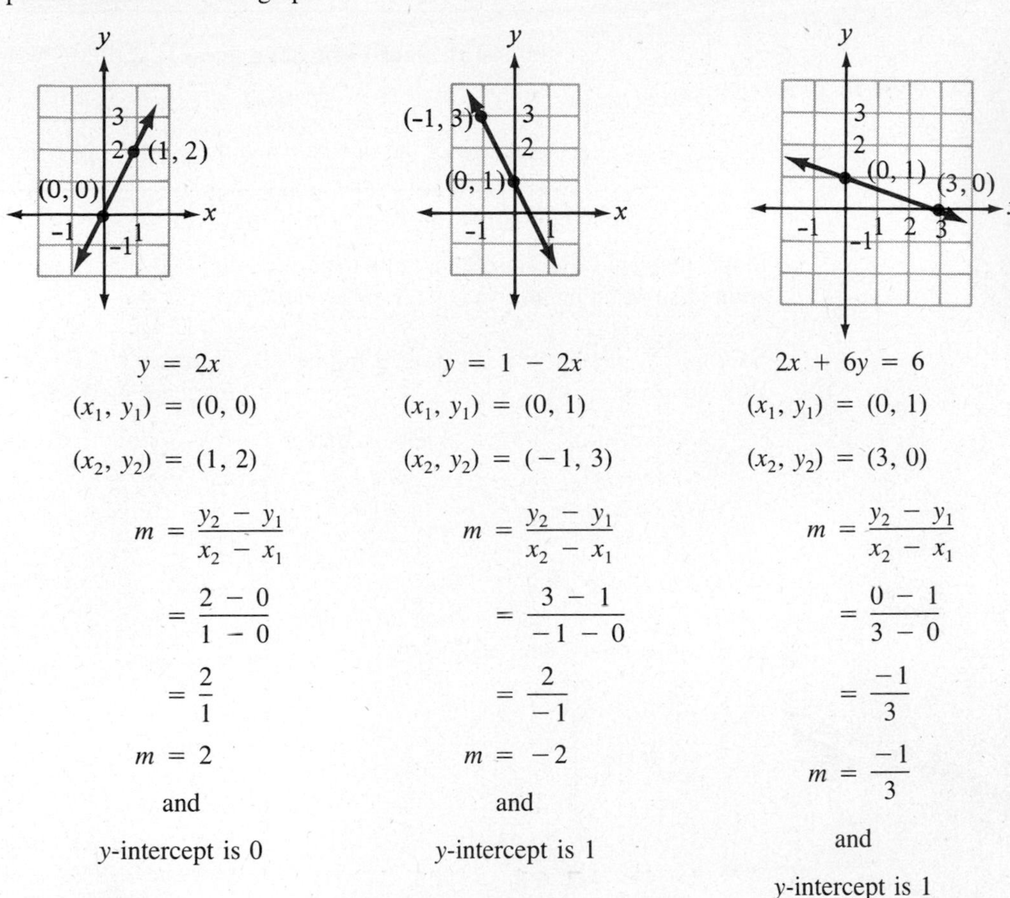

$y = 2x$	$y = 1 - 2x$	$2x + 6y = 6$
$(x_1, y_1) = (0, 0)$	$(x_1, y_1) = (0, 1)$	$(x_1, y_1) = (0, 1)$
$(x_2, y_2) = (1, 2)$	$(x_2, y_2) = (-1, 3)$	$(x_2, y_2) = (3, 0)$
$m = \dfrac{y_2 - y_1}{x_2 - x_1}$	$m = \dfrac{y_2 - y_1}{x_2 - x_1}$	$m = \dfrac{y_2 - y_1}{x_2 - x_1}$
$= \dfrac{2 - 0}{1 - 0}$	$= \dfrac{3 - 1}{-1 - 0}$	$= \dfrac{0 - 1}{3 - 0}$
$= \dfrac{2}{1}$	$= \dfrac{2}{-1}$	$= \dfrac{-1}{3}$
$m = 2$	$m = -2$	$m = \dfrac{-1}{3}$
and	and	and
y-intercept is 0	y-intercept is 1	y-intercept is 1

Let's rewrite each line's equation in the same form, with y isolated on the left-hand side and the x-term and constant term on the right-hand side.

$y = 2x$ becomes

$$y = 2x + 0$$

$y = 1 - 2x$ becomes

$$y = -2x + 1$$

$2x + 6y = 6$ becomes

$$6y = -2x + 6$$

$$y = \frac{-2x}{6} + \frac{6}{6}$$

$$y = \frac{-1}{3}x + 1$$

Now we summarize the information about these equations and their slopes and y-intercepts.

Equation	Slope	y-Intercept
$y = 2x + 0$	2	0
$y = -2x + 1$	-2	1
$y = \dfrac{-1}{3}x + 1$	$\dfrac{-1}{3}$	1

It is not just a coincidence that when y is isolated in a linear equation, the slope of its graph matches the coefficient of the x-term and the y-intercept matches the constant term. In fact (unless the coefficient of y is zero) every linear equation can be put into this form.

> **Slope-Intercept Form.** If a linear equation is written
>
> $$y = mx + b$$
>
> the slope of the line is m and the y-intercept is b.

Slope-intercept form is useful because it allows us to graph a line just by looking at the equation without substituting any values for either variable.

Example 9. Graph $y = -3x + 2$ by the slope-intercept method.

Solution: $y = -3x + 2$ tells us that

$$m = -3$$

$$= \frac{-3}{1}$$

and $b = 2$. We start at the point (0, 2) and move to the right 1 unit and down 3 units.

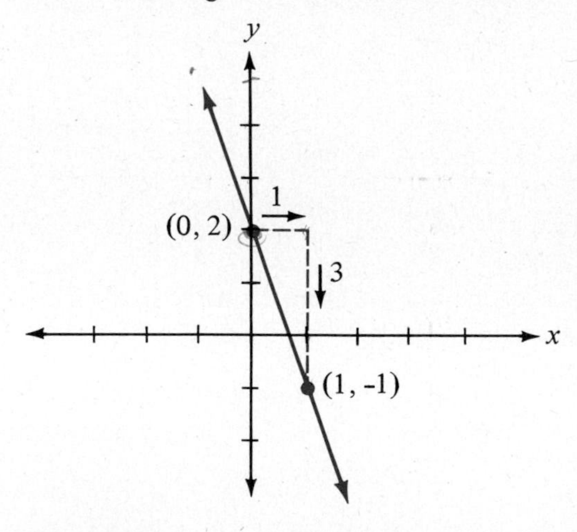

Remember that y must be isolated in order to use the slope-intercept method.

Try to complete Example 10.

Example 10. Graph $x + y = 0$ by the slope-intercept method.

Solution: First we must isolate y in the equation.

$$x + y = 0$$

$$y = -x$$

$$y = -1x + 0$$

Now we know that

$$m = \underline{-1} = \frac{-1}{1}$$

$$b = \underline{\quad\quad}$$

We start at (0, 0) and move to the right 1 unit and ___Down___ 1 unit.

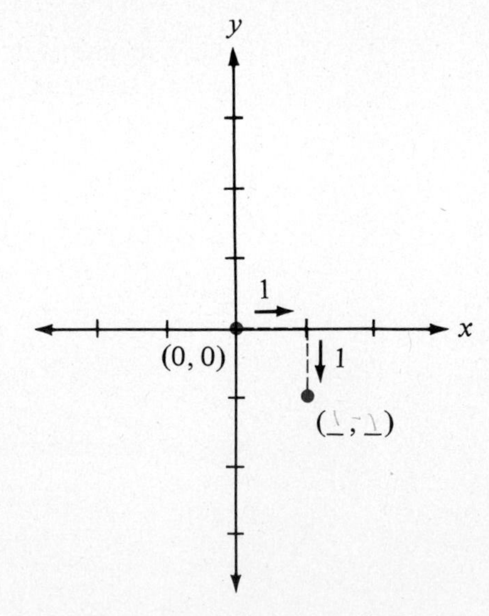

Check your work on page 613. ▶

Example 11. Graph $2x - 3y = 9$ by the slope intercept method.

Solution: First we must isolate y.

$$2x - 3y = 9$$

$$-3y = -2x + 9 \quad \text{Subtract } 2x.$$

$$y = \frac{-2x + 9}{-3} \quad \text{Divide by } -3.$$

$$y = \frac{-2x}{-3} + \frac{9}{-3} \quad \text{Rewrite the quotient on the right.}$$

$$y = \frac{2}{3}x - 3 \quad \text{Slope-intercept form.}$$

Now we know that

$$m = \frac{2 \uparrow}{3 \rightarrow}$$

$$b = -3$$

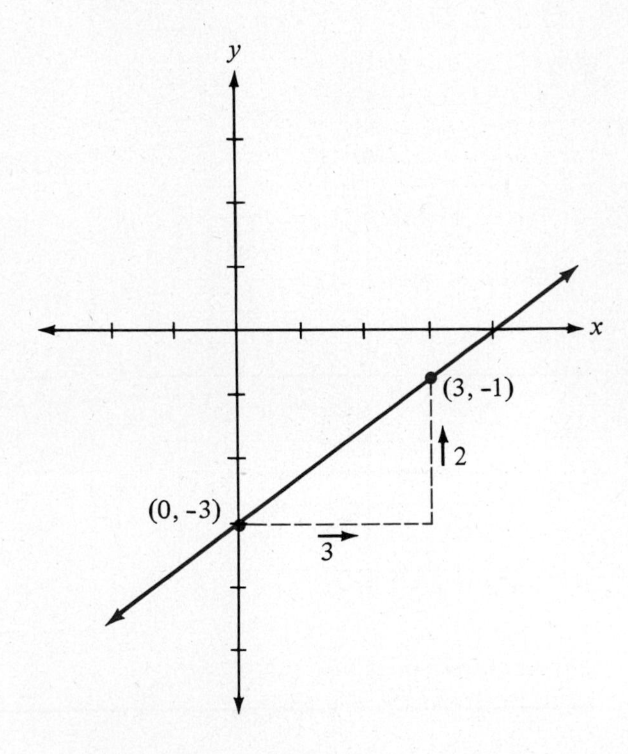

Trial Run

Graph each equation by the slope-intercept method.

1. $y = -3x + 6$

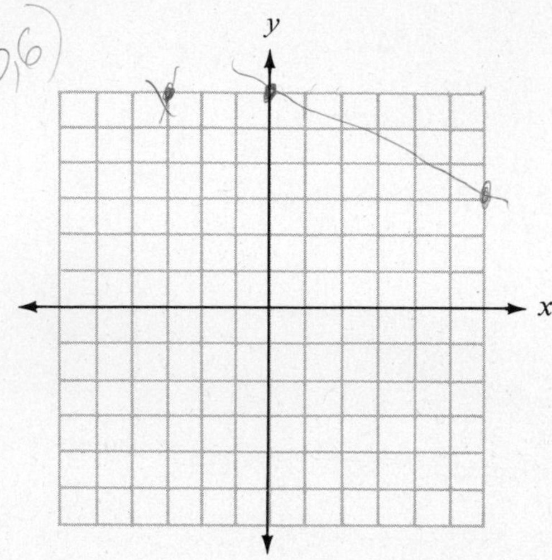

2. $x - y = 4$

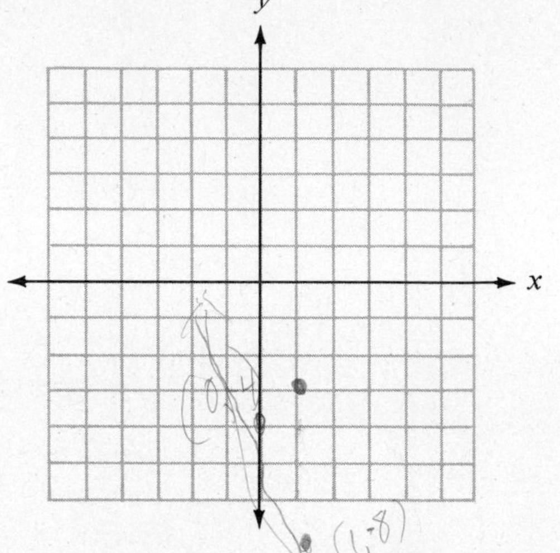

3. $x + 4y = -8$

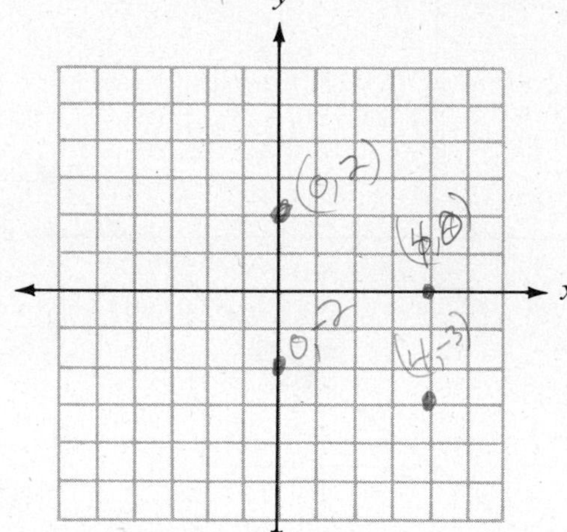

4. $3x - 4y = 12$

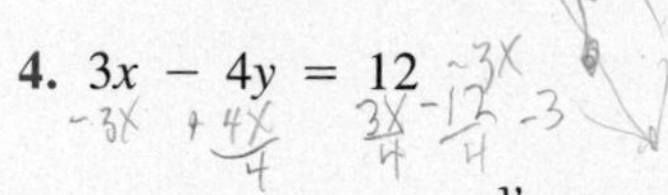

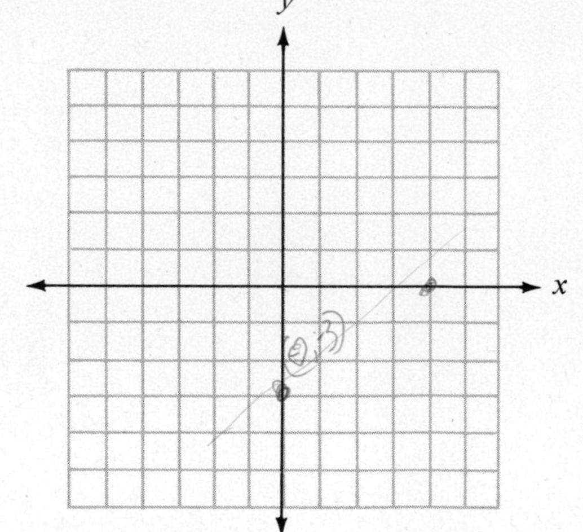

Answers are on page 614.

▶ Examples You Completed

Example 2. Find the slope of the line through $(-4, 1)$ and $(3, -2)$.

Solution

$$\text{Let } (x_1, y_1) = (-4, 1)$$

$$\text{and } (x_2, y_2) = (3, -2)$$

$$\text{Then } m = \frac{y_2 - y_1}{x_2 - x_1}$$

$$= \frac{-2 - 1}{3 - (-4)}$$

$$m = \frac{-3}{7}$$

Notice that this line slopes downward from left to right. To move from the first point to the second point, we must go to the right 7 units and down 3 units.

Example 6. Graph the line through $(3, -2)$ with slope $\frac{-1}{4}$.

Solution

$$(x_1, y_1) = (3, -2)$$

$$m = \frac{-1}{4}$$

Starting at the point $(3, -2)$, our slope says that we must move to the right 4 units and down 1 unit.

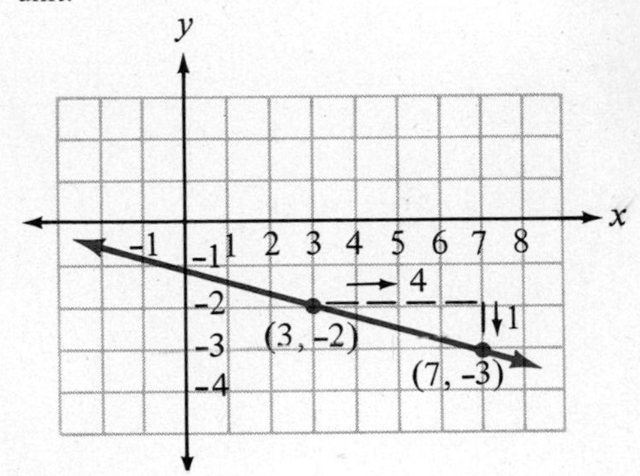

Example 8. Graph the line through $(5, -2)$ with slope zero.

Solution: Our first point is $(5, -2)$ and the slope is 0. Recall that if the slope of a line is zero, the line is horizontal.

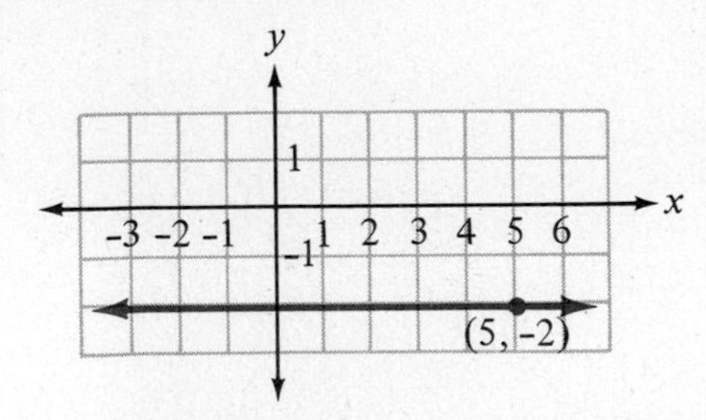

What is the equation for this line? The equation for a horizontal line is always y equals a constant. Here the equation must be

$$y = -2$$

Example 10. Graph $x + y = 0$ by the slope-intercept method.

Solution: First we must isolate y in the equation.

$$x + y = 0$$

$$y = -x$$

$$y = -1x + 0$$

Now we know that

$$m = -1 = \frac{-1}{1}$$

$$b = 0$$

We start at $(0, 0)$ and move to the right 1 unit and down 1 unit.

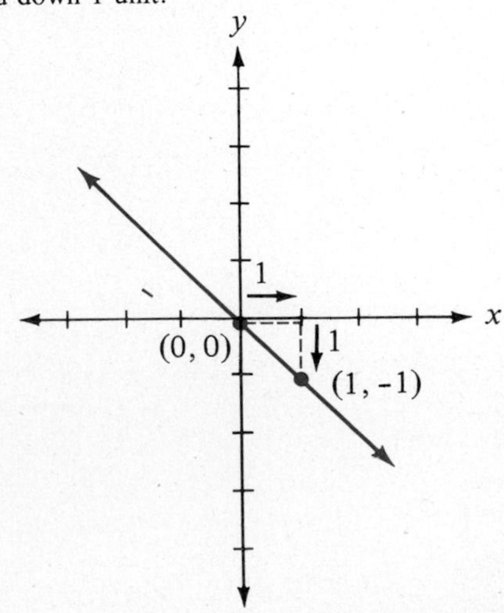

Answers to Trial Runs

page 606 **1.** $\frac{1}{5}$ **2.** -5 **3.** 0 **4.** 1 **5.** $\frac{-2}{3}$ **6.** Undefined

page 608 **1.**

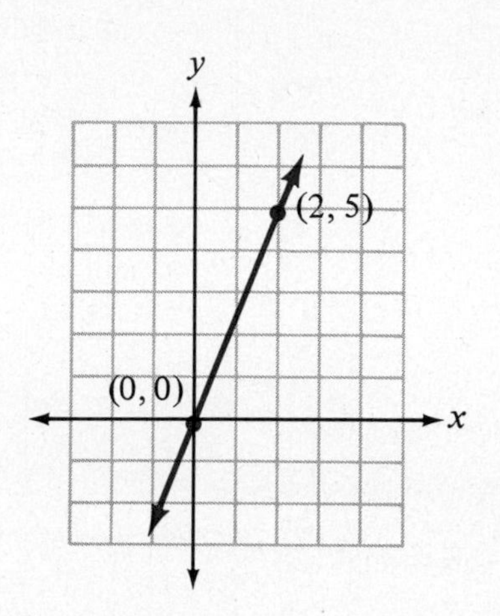

2.

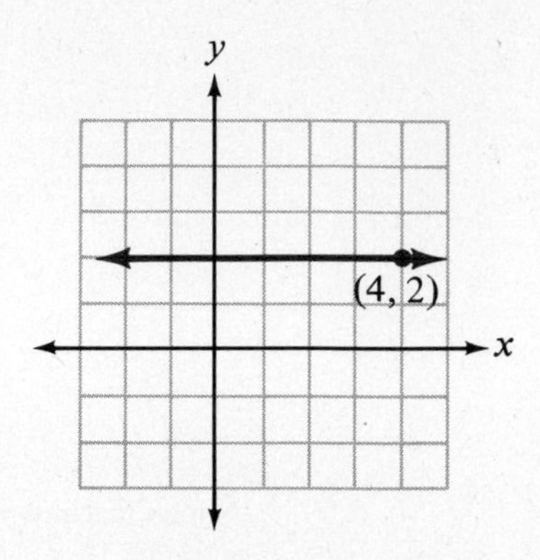

3.

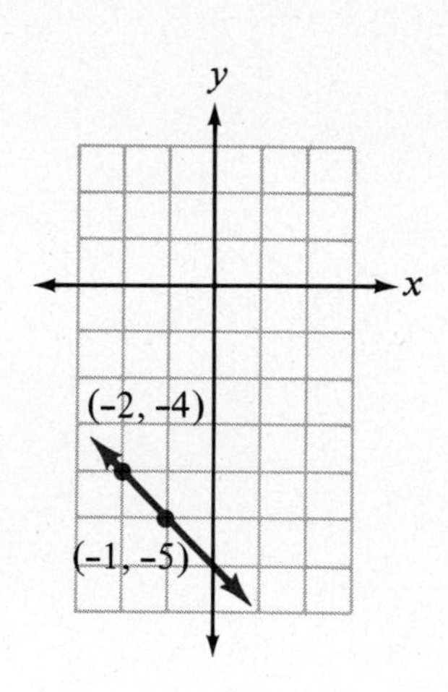

4.

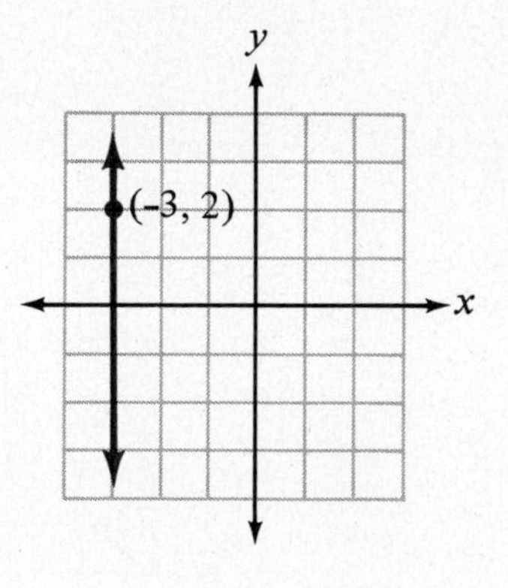

page 612 **1.**

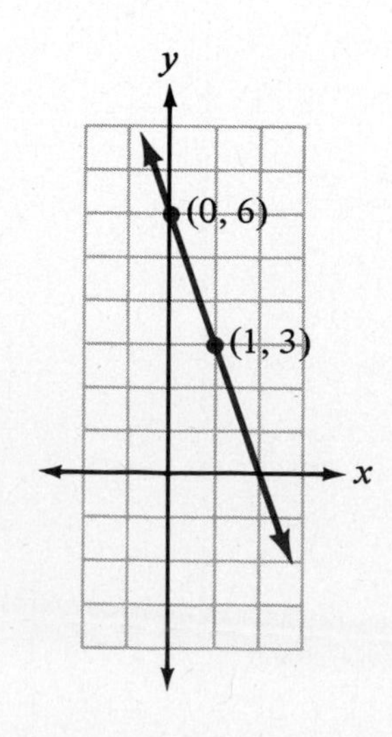

2.

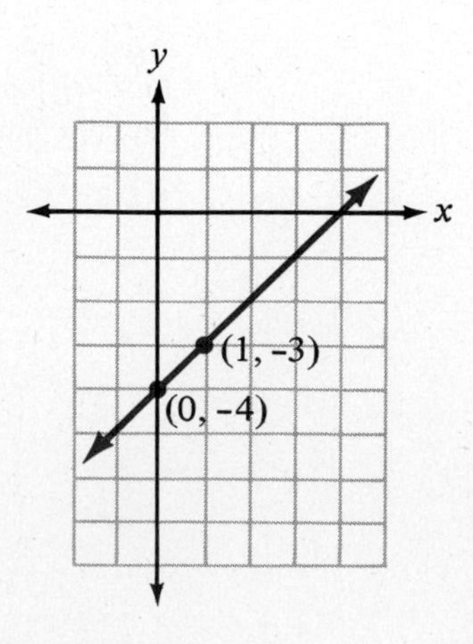

3.

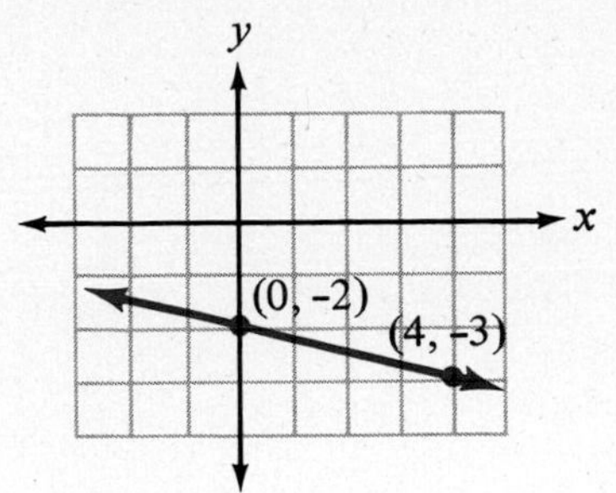

4.

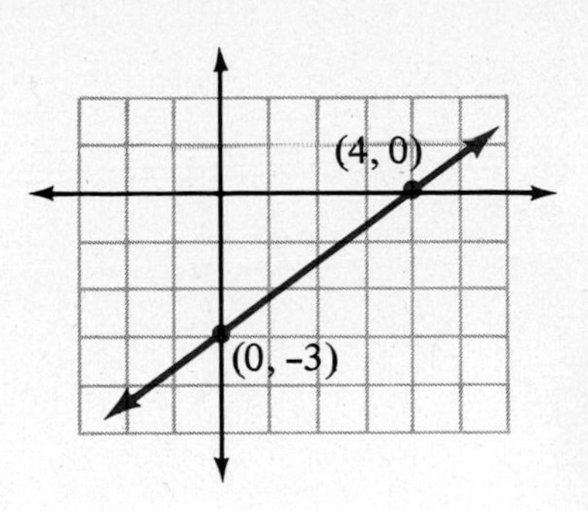

Name ______________________ Date __________

EXERCISE SET 10.4

Find the slope of the line through the given points.

_______ **1.** $(0, 5)$ and $(-3, 0)$

_______ **2.** $(0, -7)$ and $(4, 0)$

_______ **3.** $(5, -1)$ and $(5, 4)$

_______ **4.** $(2, -7)$ and $(2, 3)$

_______ **5.** $(6, -3)$ and $(8, -5)$

_______ **6.** $(9, -7)$ and $(5, -3)$

_______ **7.** $(3, 5)$ and $(-2, 5)$

_______ **8.** $(-4, -3)$ and $(5, -3)$

_______ **9.** $(2, -3)$ and $(5, 6)$

_______ **10.** $(1, -2)$ and $(-3, 6)$

Graph the line through the given point with the given slope.

11. $(1, 3)$; $m = 2$

12. $(2, 5)$; $m = 3$

13. $(5, -1)$; $m = -3$

14. $(4, -3)$; $m = -2$

15. $(3, 5)$; $m = 0$

16. $(1, -4)$; $m = 0$

17. $(-2, 3)$; $m = \frac{-1}{2}$

18. $(-1, -1)$; $m = \frac{-1}{4}$

19. $(5, -4)$; m is undefined

20. $(3, -6)$; m is undefined

Use the slope-intercept method to graph each equation.

21. $y = -2x + 1$

22. $y = -3x + 4$

23. $x - y = 9$

24. $x - y = 6$

25. $x + 3y = 15$

26. $x + 6y = 12$

27. $3x - 2y = 6$

28. $5x - 2y = 10$

29. $3x + 5y = 0$

30. $4x + 3y = 0$

☆ Stretching the Topics

_______ **1.** Find the slope of the line through $\left(\frac{1}{2}, \frac{-2}{3}\right)$ and $\left(\frac{-3}{4}, \frac{1}{6}\right)$.

_______ **2.** If the coordinates of two points are represented by $(3k, -2c)$ and $(7k, 5c)$, find an expression for the slope of a line through the two points.

_______ **3.** Find the missing coordinate if a line through $(-5, 3)$ and $(x, -2)$ has a slope of $\frac{-1}{2}$.

Check your answers in the back of your book.

If you can graph the equations in **Checkup 10.4**, you are ready to do the **Review Exercises for Chapter 10**.

Name ______________________________ Date ____________

✓ CHECKUP 10.4

Find the slope for the line through the given points.

______ **1.** (16, −4) and (9, 3) ______ **2.** (1, 5) and (−3, 5)

Graph the line through the given point with the given slope.

3. $(2, -7);\ m = \frac{-3}{2}$

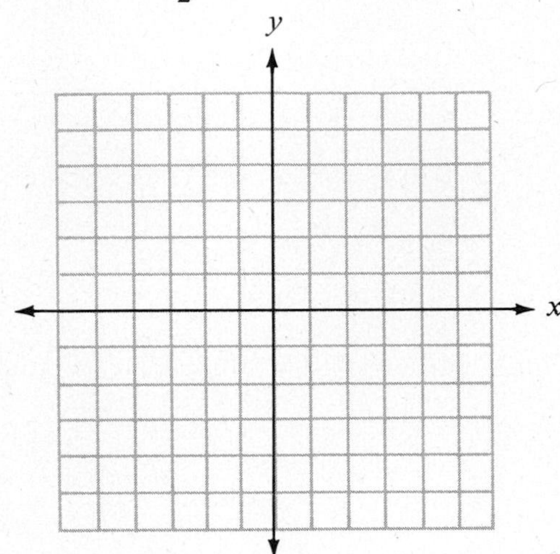

4. $(5, -1);\ m = 0$

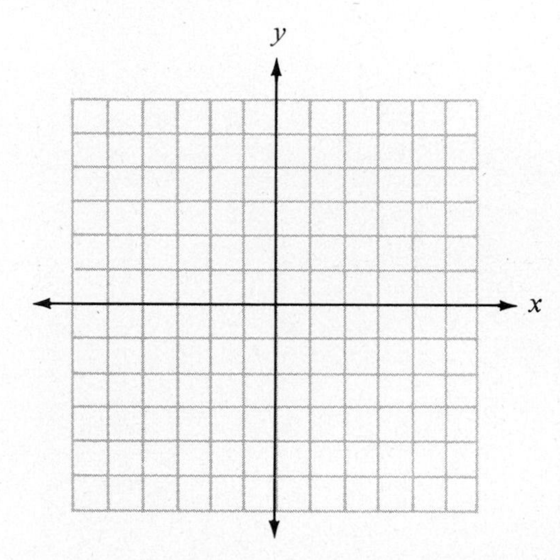

Use the slope-intercept method to graph the equations.

5. $2x - y = 3$

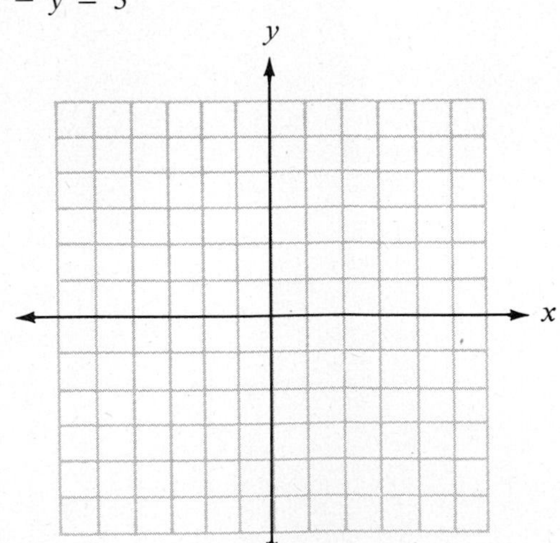

6. $3x - 4y = 0$

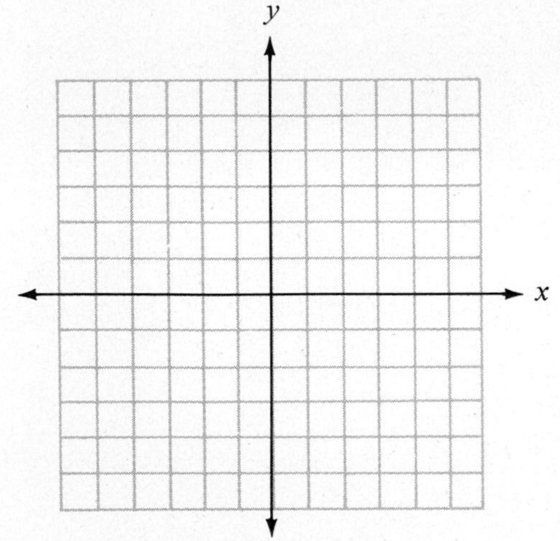

Check your answers in the back of your book.

If You Missed Problems:	You Should Review Examples:
1, 2	1–4
3	6
4	8
5	11
6	10

Summary

In this chapter we have considered equations containing two variables. In particular, we investigated first-degree equations of the form $ax + by = c$, where a and b are not both zero.

We discovered that solutions to such equations are ordered pairs of numbers (x, y) that can be graphed as points in the Cartesian coordinate plane. Every first-degree equation will yield a graph that is a straight line. The *slope* m of a straight line through the points (x_1, y_1) and (x_2, y_2) can be found by $m = \dfrac{y_2 - y_1}{x_2 - x_1}$.

A straight line can be graphed by three different methods.

Method	Procedure
Arbitrary points	Find three points by choosing x-values and finding corresponding y-values.
Intercepts	Find the x-intercept (by letting $y = 0$). Find the y-intercept (by letting $x = 0$).
Slope-intercept	Write equation in the form $y = mx + b$, where m is slope and b is y-intercept.

We also learned to recognize and graph constant equations.

Equation	Graph
$y = k$, a constant	Horizontal line, crossing the y-axis at $(0, k)$
$x = k$, a constant	Vertical line, crossing the x-axis at $(k, 0)$

❑ Speaking the Language of Algebra

1. For the point $(3, -2)$, the x-coordinate is ________ and the y-coordinate is ________ .

2. The point at which a graph crosses the x-axis is called its ________ . The point at which a graph crosses the y-axis is called its ________ .

3. The point $(0, 0)$ is called the ________ .

4. The graph of all points (x, y) that satisfy an equation $ax + by = c$ is a ________ ________ .

5. If (x_1, y_1) and (x_2, y_2) are two points on a line, the slope m of the line is found by $m =$ ________ .

6. The slope of a horizontal line is ________ . Every horizontal line corresponds to an equation of the form ________ .

7. The slope of a vertical line is ________ . Every vertical line corresponds to an equation of the form ________ .

8. If a first-degree equation is written in the form $y = mx + b$, then m is the ________ of the line and b is its ________ .

Check your answers in the back of your book.

Name ________________________________ Date ____________

REVIEW EXERCISES for Chapter 10

Decide if the given point satisfies the equation.

________ **1.** $y = \frac{1}{4}x - 3$, $(-8, -5)$

________ **2.** $4x - y = 5$, $(0, -5)$

Find three ordered pairs that satisfy each equation.

________ **3.** $2x - y = 6$

________ **4.** $x - 3y = 0$

________ **5.** $y = -2x + 5$

________ **6.** $x - y = -2$

Write each of the following word statements as an equation containing two variables, then list three ordered pairs that will satisfy the equation.

________ **7.** The length of a rectangle is 5 less than 3 times the width.

________ **8.** Regina sold some of her blouses at a yard sale. For some she got \$5 and for others \$3. The total amount she made from the sale of her blouses was \$45.

9. Graph the following points.

A: $(-4, 2)$ B: $(0, -5)$ C: $(-2, -4)$

10. Give the coordinates of each point.

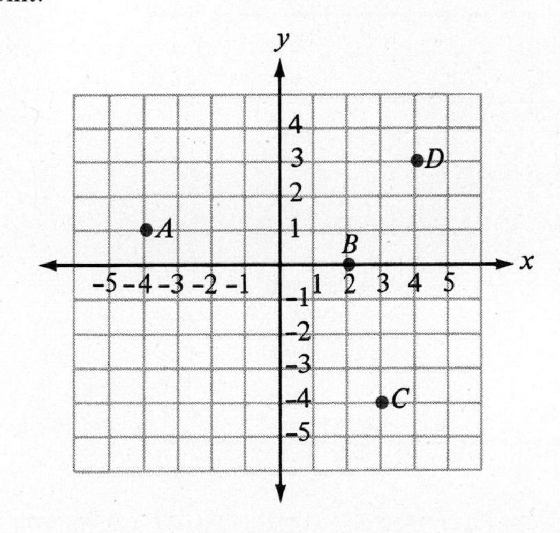

Graph each equation.

11. $x + y = 5$

12. $y = -3x$

13. $4x - y = 0$

14. $3x - y = 6$

Graph each equation by the intercept method.

15. $x + y = 6$

16. $x - 3y = -6$

17. $2x - y = 7$

18. $5x - 3y = 15$

19. $x = 4$

20. $2y + 12 = 0$

Find the slope of the line through the given points.

_______ **21.** $(3, 2)$ and $(0, -1)$

_______ **22.** $(-7, 3)$ and $(4, 3)$

_______ **23.** $(-1, -3)$ and $(-5, 4)$

_______ **24.** $(-5, 3)$ and $(-5, 10)$

Graph the line through the given point with the given slope.

25. $(0, 3)$; $m = -4$

26. $(-1, 2)$; $m = 0$

27. $(3, -3)$; m is undefined

28. $(4, -2)$; $m = \frac{3}{2}$

Use the slope-intercept method to graph each of the following.

29. $y = -5x + 4$

30. $x + y = 7$

31. $x + 4y = 20$

32. $2x - 7y = 14$

33. $3y = 2x$

34. $4x - 5y = 0$

Check your answers in the back of your book.

If You Missed Exercises:	You Should Review	Examples:
1, 2	SECTION 10.1	3, 4
3–6		5, 6
7, 8		7, 9
9, 10	SECTION 10.2	1–4
11–14		5–7
15–18	SECTION 10.3	1–3
19, 20		4–7
21–24	SECTION 10.4	1–4
25–28		5–8
29–34		9–11

If you have completed the **Review Exercises** and corrected your errors, you are ready to take the **Practice Test for Chapter 10**.

Name ______________________________ Date ____________

PRACTICE TEST for Chapter 10

	Section	Examples
________ **1.** Does the ordered pair $(-4, 3)$ satisfy the equation $y = \frac{1}{2}x + 1$?	10.1	3, 4
________ **2.** Does the ordered pair $(2, -3)$ satisfy the equation $5x + y = 7$?	10.1	3, 4

Write each word statement as an equation containing two variables.

________ **3.** Mary's weight is 15 pounds more than half her father's weight.	10.1	7–9
________ **4.** Clay spent \$15.40 at the post office on 22-cent stamps and 14-cent stamps.	10.1	7–9

Graph each equation using three ordered pairs. Label each point.

5. $x - y = 4$ — 10.2 — 5, 7

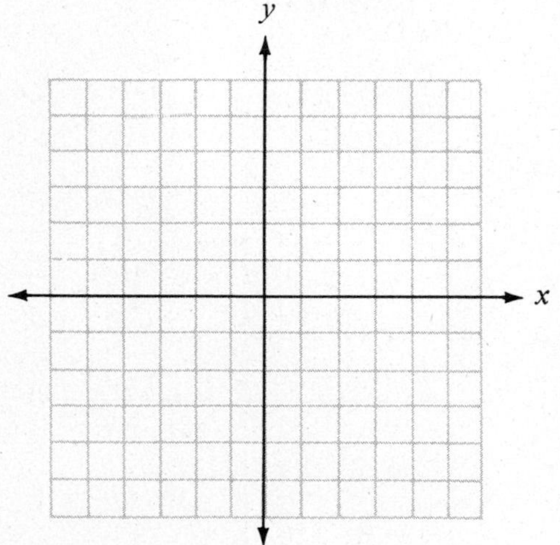

6. $y = 6 - 2x$ — 10.2 — 6

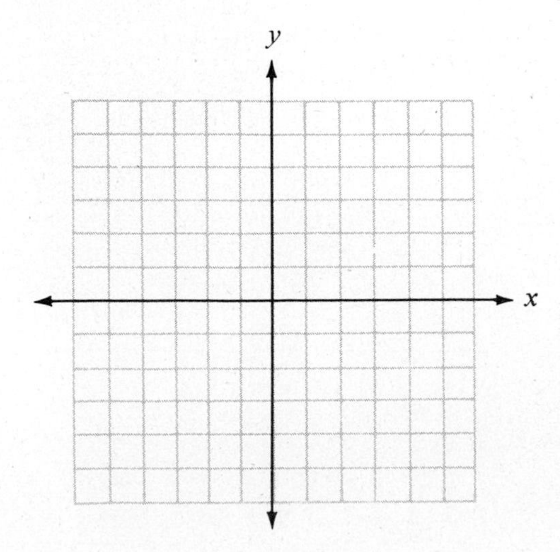

7. $2x + y = 0$ 10.2 5, 7

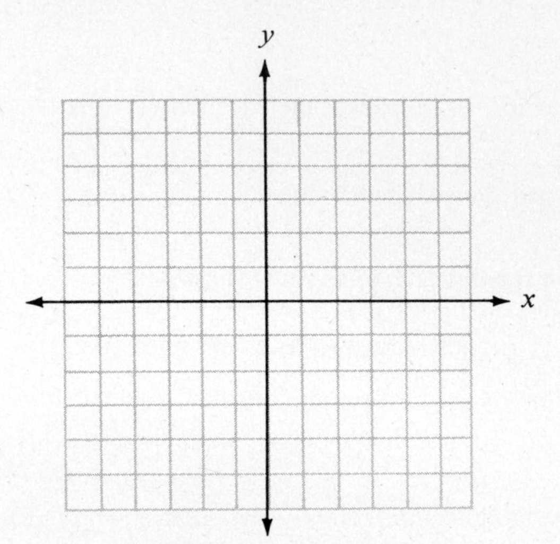

Graph each equation by the intercept method. Label the intercepts.

8. $x - 5y = -10$ 10.3 2

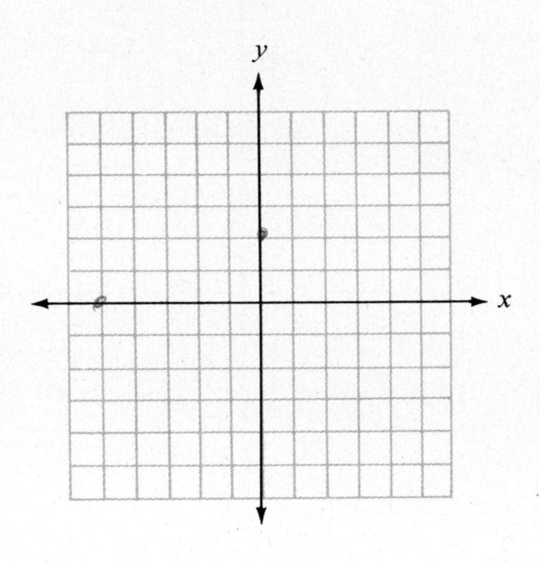

9. $4x + 3y = 8$ 10.3 1

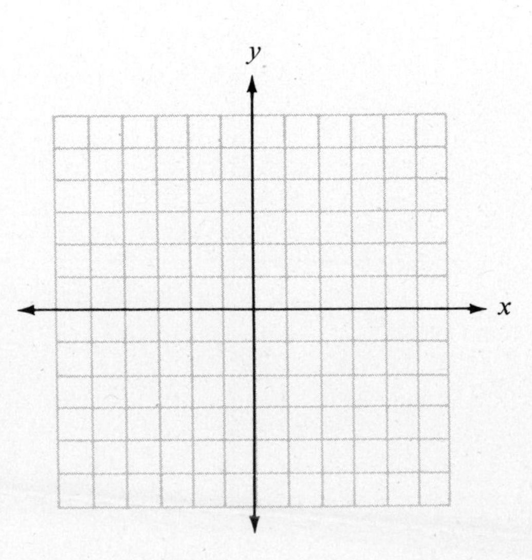

10. $-3x - 2y = 0$ 10.3 3

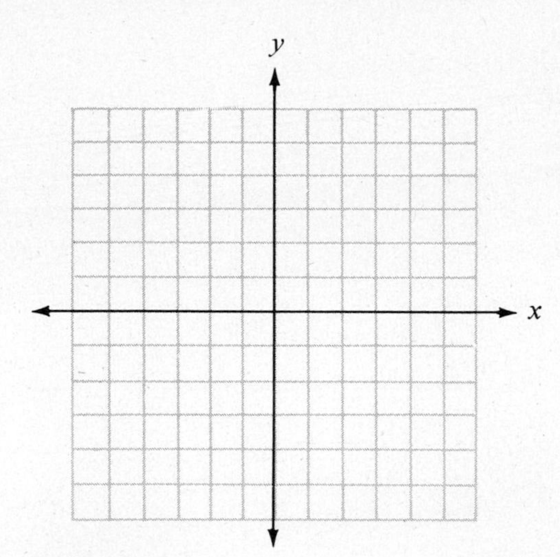

Graph each equation by inspection. Label the intercept.

11. $y - 4 = 0$ 10.3 5

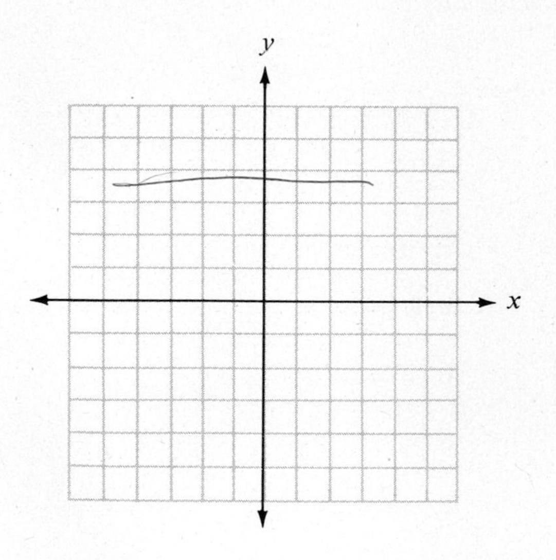

12. $3x + 2 = 0$ 10.3 8

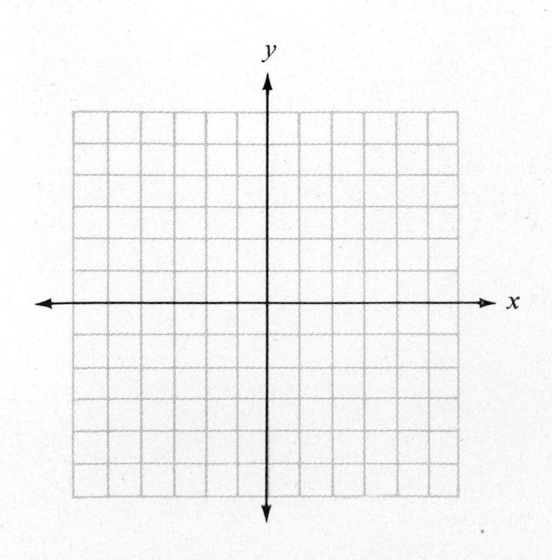

_______ **13.** Find the slope of the line through (5, 2) and (−3, 1). 10.4 1, 2

14. Graph the line through the point (2, −3) with slope $\frac{4}{3}$. 10.4 6

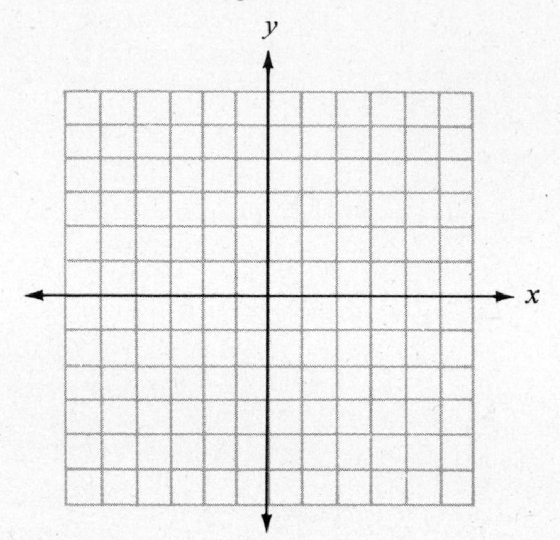

Graph each equation by the slope-intercept method. Label points.

15. $y = x + 4$ 10.4 9

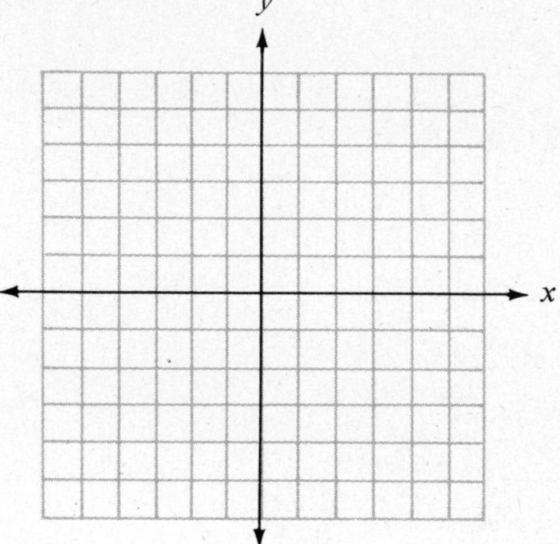

16. $6x + 4y = -12$ 10.4 11

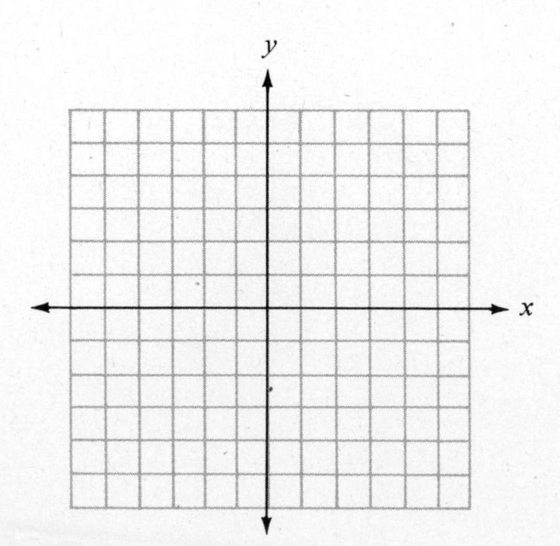

Check your answers in the back of your book.

Name ________________________ Date ____________

SHARPENING YOUR SKILLS after Chapters 1–10

		Section
______	**1.** Change 19.3% to a decimal number.	1.3
______	**2.** $0.1107 \div 1.23$	1.3
______	**3.** $\frac{2}{3} - \frac{7}{6} + \frac{4}{9}$	2.4
______	**4.** $0.3[-2.1 - (3 - 5.6)]$	2.4

Simplify.

		Section
______	**5.** $\frac{4}{5}(20x - 15y)$	3.3
______	**6.** $(-6)\left(\frac{-1}{2}\right)^2(y)$	3.3
______	**7.** $\frac{(-2a^2b^6)^4}{(a^3b^2)^5}$	5.2
______	**8.** $(-2x^{-1}y^2)^3(x^2y^3)^{-1}$	5.3
______	**9.** $5x^2(2x^3y)^3$	6.2
______	**10.** $-5ax(2 - 3y) + 10ax(1 - 2y)$	6.2
______	**11.** $(x^2 - 3)(x^2 + 3)$	6.3
______	**12.** $\frac{9x^2 - 45x - 36}{9x}$	6.4
______	**13.** $(x^3 - x + 5) \div (x + 2)$	6.4
______	**14.** $\frac{2x + 3y}{4x^2 + 12xy + 9y^2}$	8.1
______	**15.** $\frac{(3y - 1)^2}{14x^2y^2} \cdot \frac{21xy}{9y^2 - 1}$	8.2
______	**16.** $\frac{\frac{-36x^2y^3}{z^4}}{\frac{15x^4y}{z^7}}$	8.5
______	**17.** $\frac{3}{x + 4} - \frac{2}{5x}$	8.4

Solve.

_______ **18.** $2(5y - 1) - 5 = 3(7 + y)$ 4.2

_______ **19.** $63 + 12x - 3x^2 = 0$ 7.5

_______ **20.** $3y(y + 2) + 10 = 2(y^2 - 5) - 3y$ 7.5

_______ **21.** $\dfrac{a + 3}{a + 1} = \dfrac{a - 3}{a - 2}$ 8.6

_______ **22.** $\dfrac{2}{3b - 1} + \dfrac{3}{b + 2} = 0$ 8.6

_______ **23.** $\dfrac{5x - 2}{3} > -2$ 9.2

_______ **24.** $-2 \leq 3y - 5 \leq 7$ 9.2

_______ **25.** $5 - 2x > 1$ or $3x - 6 < 3$ 9.3

_______ **26.** Solve $A = P + Prt$ for P. 4.2

_______ **27.** Solve and graph the inequality $\dfrac{-3x}{2} + 1 \leq 4$. 9.2

_______ **28.** If Abe earns $270 per week, for how many weeks must he work to earn at least $1890? 9.4

_______ **29.** The width of a rectangle is 4 centimeters less than the length and the area is 45 square centimeters. Find the dimensions. 7.6

_______ **30.** If the denominator of the fraction $\dfrac{x}{7}$ is decreased by x, the new fraction is equal to $\dfrac{2}{5}$. Find the value of x. 8.6

Check your answers in the back of your book.

Solving Systems of Linear Equations 11

José works part-time mowing lawns and life-guarding at the swimming pool. When he mows for 5 hours and guards for 6 hours, he earns $42. When he mows for 11 hours and guards for 4 hours, he earns $51. What is his hourly wage for each job?

In this problem there are two unknown quantities and two conditions that must be satisfied. To solve this problem, we would like to find out what ordered pair of values (if any) will satisfy two equations at the same time. This is called **solving a system of linear equations** and in this chapter we learn how to

1. Solve a system of linear equations by graphing.
2. Solve a system of linear equations by substitution.
3. Solve a system of linear equations by elimination.
4. Switch from words to systems of linear equations.

11.1 Solving a System of Equations by Graphing

One way in which we can try to locate common solutions for a system of two equations is by looking at the *graphs* of the two equations. We shall graph two equations in the same Cartesian plane and see what conclusions we can draw.

Consider the following system of two linear equations and graph each line by the intercept method.

(1) $x + y = 7$ and (2) $2x - y = 5$

(1) $x + y = 7$	(2) $2x - y = 5$
if $x = 0, y = 7$	if $x = 0, y = -5$
$(0, 7)$	$(0, -5)$
if $y = 0, x = 7$	if $y = 0, x = \frac{5}{2}$
$(7, 0)$	$\left(\frac{5}{2}, 0\right)$

Now we graph both equations in the same plane, labeling the lines (1) and (2).

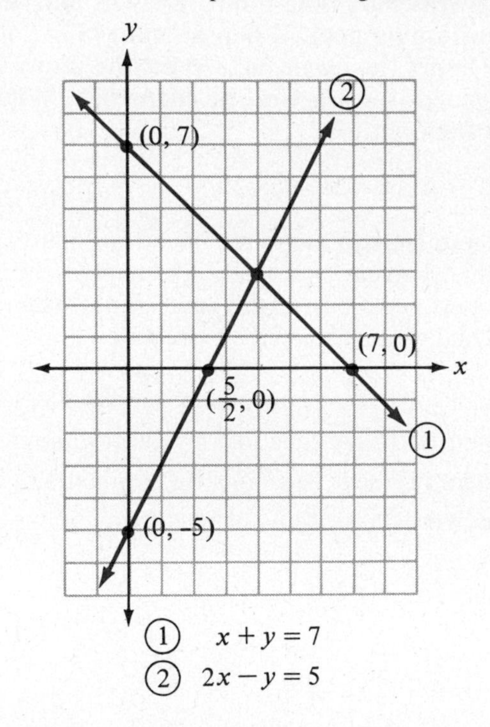

These lines have just one point in common—the point at which the lines cross. That point appears to be (4, 3). We can see if that point is shared by both equations by checking the ordered pair (4, 3) in each one.

$$\begin{array}{rl} (1)\ x + y &= 7 \\ 4 + 3 &\stackrel{?}{=} 7 \\ 7 &= 7 \end{array} \qquad \begin{array}{rl} (2)\ 2x - y &= 5 \\ 2(4) - 3 &\stackrel{?}{=} 5 \\ 8 - 3 &\stackrel{?}{=} 5 \\ 5 &= 5 \end{array}$$

Indeed, the ordered pair (4, 3) does satisfy both equations. The ordered pair (4, 3) is the only solution for the system of equations.

Try Example 1.

Example 1. Use graphing to solve the system. Then check your solution.

$$(1)\ 2x + 3y = 12$$
$$(2)\ x - y = 1$$

Solution: Use intercepts.

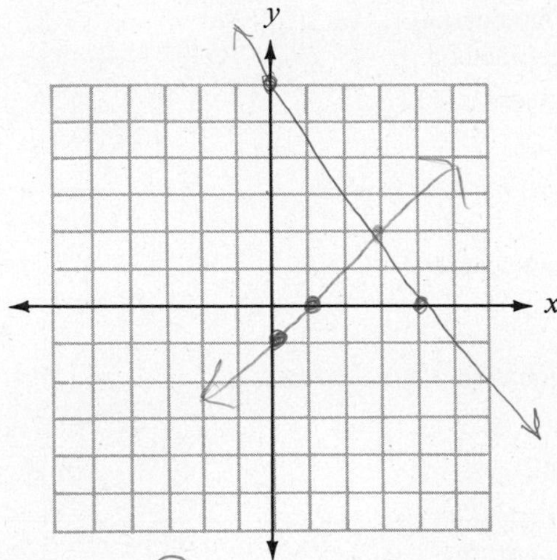

(1) $2x + 3y = 12$

if $x = 0$, $y =$ 4

(0, 4)

if $y = 0$, $x =$ 6

(6, 0)

(2) $x - y = 1$

if $x = 0$, $y =$ −1

(0, −1)

if $y = 0$, $x =$ 1

(1, 0)

The point (3, 2) seems to be the solution for the system.

CHECK: The solution is (____ , ____).

(1) $2x + 3y = 12$

$2(\_\_\_\_) + 3(\_\_\_\_) \stackrel{?}{=} 12$

$\_\_\_\_ + \_\_\_\_ \stackrel{?}{=} 12$

$\_\_\_\_ = 12$

(2) $x - y = 1$

$\_\_\_\_ - \_\_\_\_ \stackrel{?}{=} 1$

$\_\_\_\_ = 1$

Check your work on page 635. ▶

Example 2. Use graphing to solve the system.

$$(1)\ x + y = 3$$
$$(2)\ x + y = -1$$

Solution: Use intercepts.

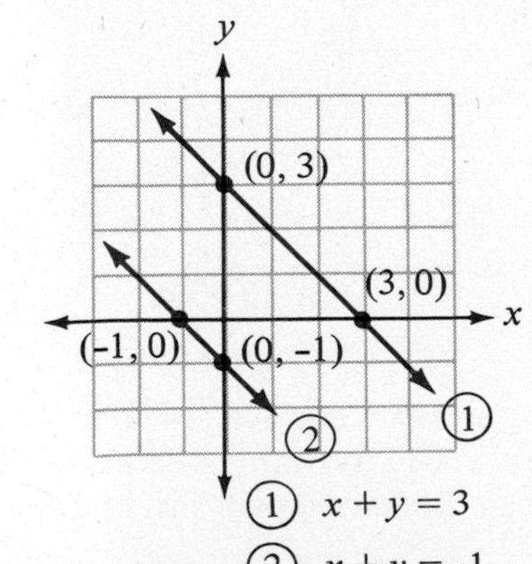

(1) $x + y = 3$

if $x = 0$, $y = 3$

(0, 3)

if $y = 0$, $x = 3$

(3, 0)

(2) $x + y = -1$

if $x = 0$, $y = -1$

(0, −1)

if $y = 0$, $x = -1$

(−1, 0)

Our graphs do not cross; the lines are *parallel*. A system such as this has *no solution* and is called an **inconsistent system**.

Example 3. Use graphing to solve the system.

$$(1)\ y = 5x - 2$$

$$(2)\ y = 3$$

Solution: For equation (1) we shall use the slope-intercept method.

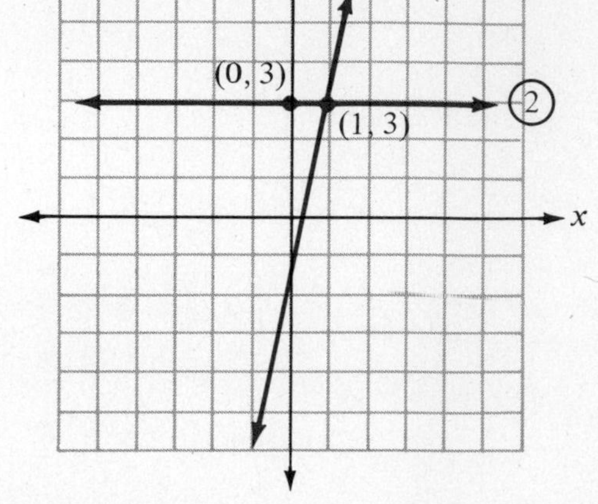

$$(1)\quad y = 5x - 2$$

$$\text{slope: } m = 5 = \frac{5}{1}$$

$$y\text{-intercept: } b = -2$$

$$(0,\ -2)$$

We recognize equation (2) as a constant equation.

$$(2)\ y = 3$$

The graph is a horizontal line crossing the y-axis at the point (0, 3).

The point (1, 3) appears to be the common solution. You should check it by substitution in each equation.

Example 4. Use graphing to solve the system.

$$(1)\quad x + \ y = 3$$

$$(2)\ 2x + 2y = 6$$

Solution: Use intercepts.

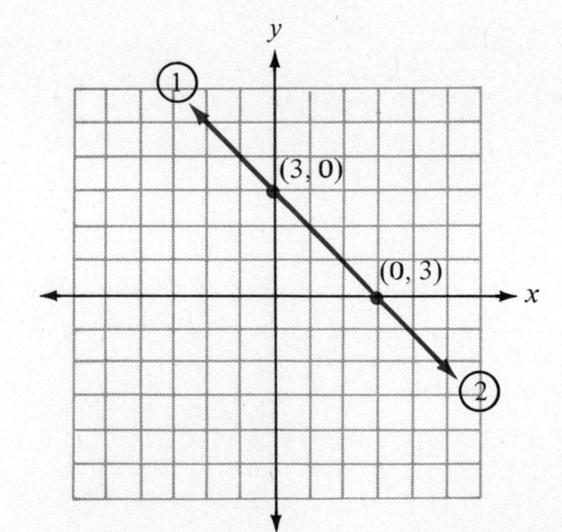

(1) $x + y = 3$

if $x = 0,\ y = 3$

(0, 3)

if $y = 0,\ x = 3$

(3, 0)

(2) $2x + 2y = 6$

if $x = 0,\ y = 3$

(0, 3)

if $y = 0,\ x = 3$

(3, 0)

The graph for both equations is the *same line*! Any solution for one of the equations will be a solution for the other equation. Such a system is said to be a **dependent system**. Here we say the solutions for the system are all ordered pairs that satisfy $x + y = 3$ (or $2x + 2y = 6$).

Notice that equation (2) is a constant multiple of equation (1) in Example 4. This will always be true if the equations in a *dependent* system are both in standard form.

Trial Run

Use graphing to solve each system.

1. $x - y = 3$
$y = 2x - 7$

2. $y = 2x$
$x + 2y = 10$

3. $x - y = 5$
$y = x + 2$

4. $2x + 4y = 2$
$3x + 6y = 3$

Answers are on page 636.

Example You Completed

Example 1. Use graphing to solve the system.

$$(1)\quad 2x + 3y = 12$$
$$(2)\quad x - y = 1$$

Solution: Use intercepts.

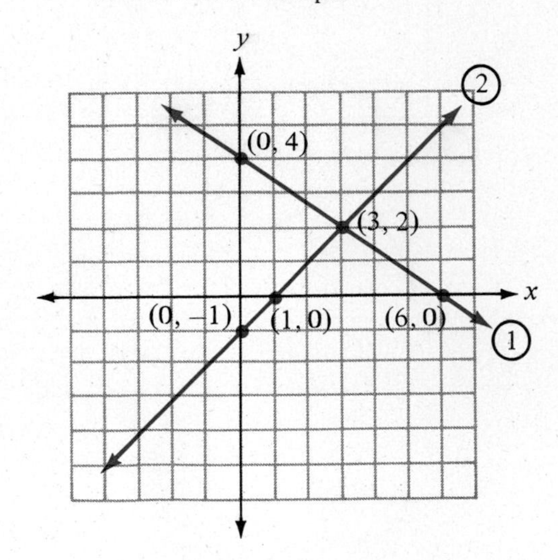

(1) $2x + 3y = 12$

if $x = 0$, $y = 4$

$(0, 4)$

if $y = 0$, $x = 6$

$(6, 0)$

(2) $x - y = 1$

if $x = 0$, $y = -1$

$(0, -1)$

if $y = 0$, $x = 1$

$(1, 0)$

The point (3, 2) seems to be the solution for the system.

CHECK: The solution is (3, 2).

(1) $2x + 3y = 12$

$2(3) + 3(2) \stackrel{?}{=} 12$

$6 + 6 \stackrel{?}{=} 12$

$12 = 12$

(2) $x - y = 1$

$3 - 2 \stackrel{?}{=} 1$

$1 = 1$

Answers to Trial Run

page 635 **1.**

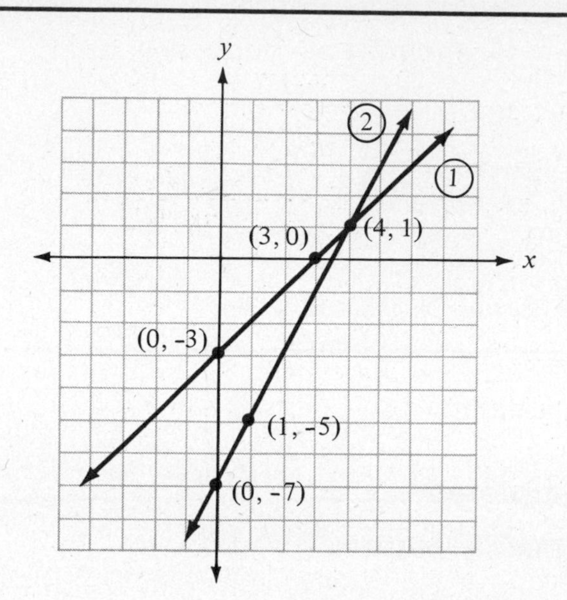

Solution: (4, 1)

2.

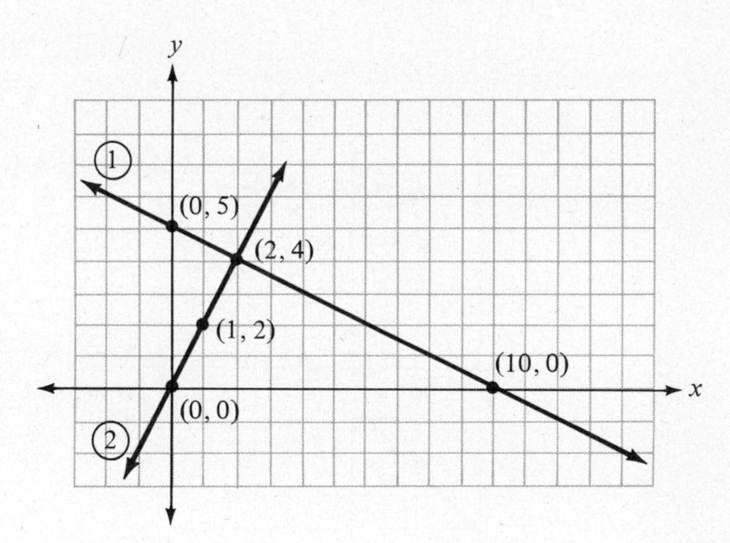

Solution: (2, 4)

3.

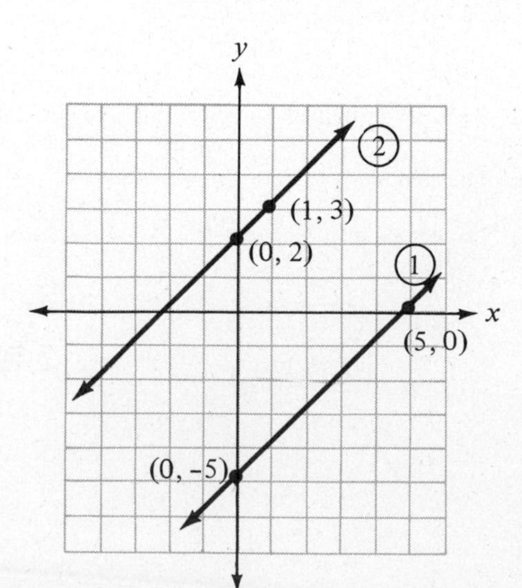

No solution.

4.

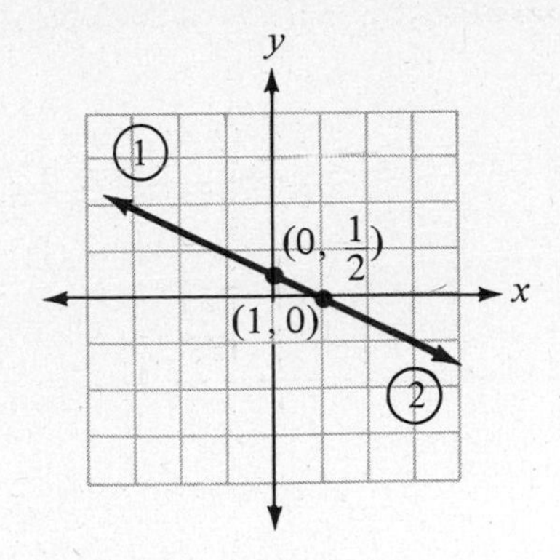

Solution: All ordered pairs that satisfy $2x + 4y = 2$ (or $3x + 6y = 3$).

Name ______________________ Date __________

EXERCISE SET 11.1

Solve each system by graphing.

1. $x - y = -2$
   $x + y = 4$

2. $x + y = 6$
   $x - y = -4$

3. $3x + 2y = 6$
   $2x + y = 2$

4. $5x - 4y = -60$
   $x + y = 6$

5. $3x - y = 6$
   $y = 6x$

6. $-2x + y = 8$
   $y = 4x$

7. $x + y = -5$
   $-2x - 2y = 8$

8. $x - y = 3$
   $3x - 3y = -6$

9. $4x - y = 8$
   $2x + y = 4$

10. $-3x + y = 9$
    $-2x - y = 6$

11. $2x - y = 2$
    $-6x + 3y = -6$

12. $-3x + 3y = 6$
    $2x - 2y = -4$

13. $3x - 2y = 0$
    $x - y = -1$

14. $4x + 3y = 0$
    $x + y = -1$

15. $2x - y = 10$
    $y = 4$

16. $3x + y = 9$
    $y = -3$

17. $2x + y = 9$
    $x - y = 3$

18. $6x + y = -3$
    $x + y = 7$

19. $4x - 2y = 16$
    $-2x + y = -4$

20. $-3x + 2y = -6$
    $6x - 4y = 8$

☆ Stretching the Topics

Solve each system by graphing.

1. $x + 2(y + 5) = 3(x + y + 1)$
   $0 = 5y - 3(x - 10)$

2. $\dfrac{2x - y}{4} = x - y$
   $2x - 3y = 0$

Check your answers in the back of your book.

If you can solve the systems in **Checkup 11.1**, you are ready to go on to Section 11.2.

Name ______________________ Date __________

✓ CHECKUP 11.1

Solve each system by graphing.

1. $x - y = 2$
$x + y = 0$

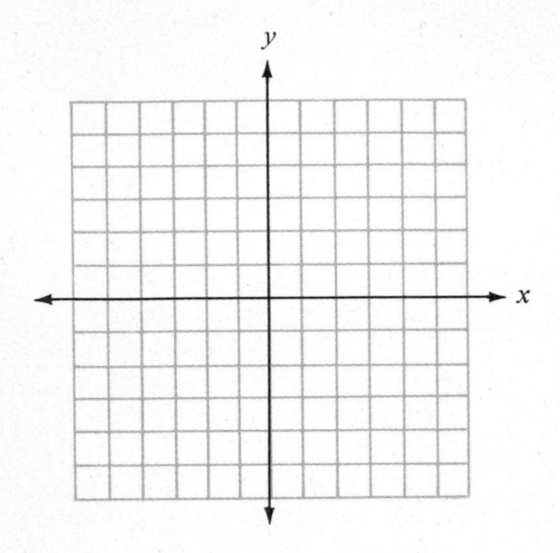

2. $-4x + 2y = 8$
$2x - y = 2$

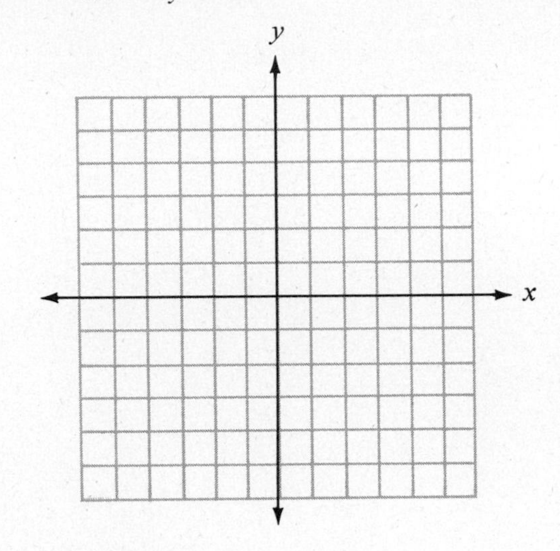

3. $2x + 3y = 6$
$x = -3$

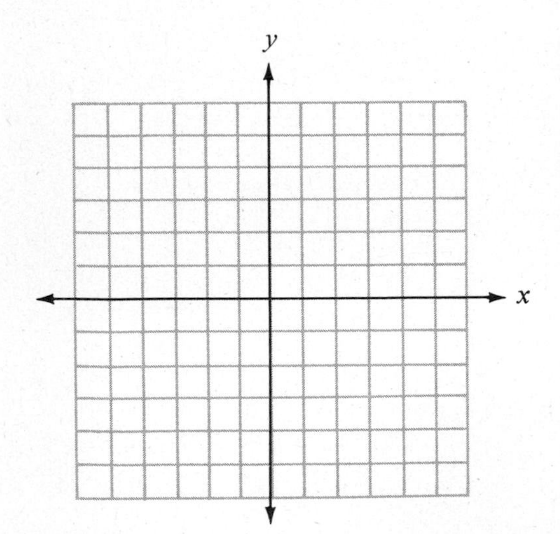

4. $-x + 2y = 4$
$5x - 10y = -20$

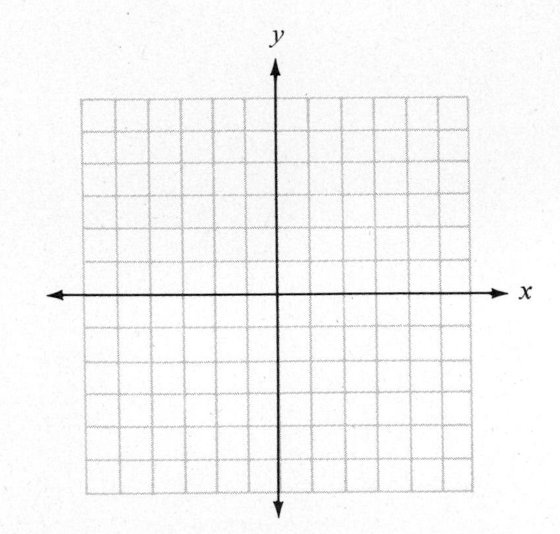

Check your answers in the back of your book.

If You Missed Problem:	You Should Review Example:
1	1
2	2
3	3
4	4

11.2 Solving a System of Equations by Substitution

We have seen that the solution for a system of equations can often by found by graphing each equation and finding the point common to both graphs. Sometimes, however, it is difficult (if not impossible) to find the exact coordinates of the point where two lines cross. We need a method to find a solution without graphing.

In finding the solution for a system of equations, remember that we are looking for a common x-value and a common y-value that satisfy *both* equations. The method we shall use is called **solving a system by substitution**. The method of substitution works especially well when one variable is already isolated in one of the original equations or when one of the variables has a coefficient of 1.

Example 1. Use substitution to solve the system.

$$(1)\ y = 2x - 1$$
$$(2)\ 2x + 3y = 5$$

Solution: Equation (1) states that y and $2x - 1$ are the same. Wherever y is used, we may substitute $2x - 1$. In particular, we can substitute the expression $2x - 1$ for y in equation (2).

$$(2)\ 2x + 3y = 5$$

becomes

$2x + 3(2x - 1) = 5$	Substitute $2x - 1$ for y in equation (2).
$2x + 6x - 3 = 5$	Remove parentheses.
$8x - 3 = 5$	Combine like terms.
$8x = 8$	Add 3 to isolate x-term.
$x = 1$	Divide by 8 to isolate x.

Now we know the x-value for the common solution. To find the corresponding y-value, we can substitute 1 for x in either of the original equations.

(1) $y = 2x - 1$	or	(2) $2x + 3y = 5$
$y = 2(1) - 1$		$2(1) + 3y = 5$
$y = 2 - 1$		$2 + 3y = 5$
$y = 1$		$3y = 3$
		$y = 1$

The solution for the system is the ordered pair (1, 1). Graphically, this means that the two lines intersect at the point (1, 1).

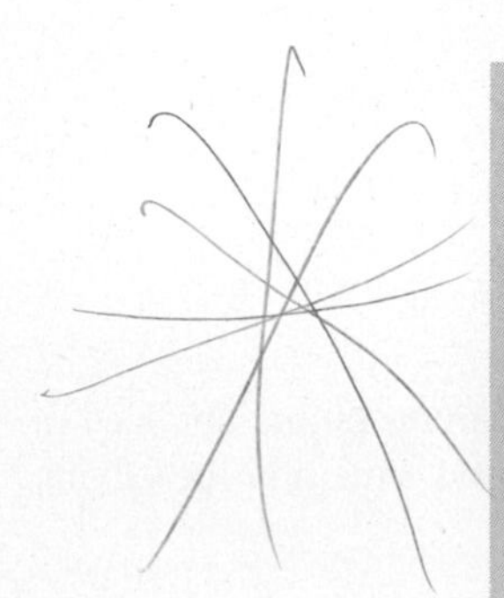

Solving a System by Substitution

1. Look for a variable appearing in an equation with a coefficient of 1, and isolate that variable in its equation.
2. Substitute the expression obtained for the isolated variable into the other equation.
3. Solve for the value of the remaining variable.
4. Substitute that value into either of the original equations and solve for the other variable.
5. Write the solution as an ordered pair.

Example 2. Use substitution to solve the system.

$$(1)\ 5x + 2y = 9$$
$$(2)\ x + 3y = 4$$

Solution: We can isolate x in equation (2).

$$(2)\ x + 3y = 4$$
$$x = 4 - 3y$$

Now we substitute $4 - 3y$ for x in equation (1).

$$(1)\ 5x + 2y = 9$$

becomes

$$5(4 - 3y) + 2y = 9$$
$$20 - 15y + 2y = 9$$
$$20 - 13y = 9$$
$$-13y = -11$$
$$y = \frac{-11}{-13}$$
$$y = \frac{11}{13}$$

To find the corresponding x-value, return to equation (2) with x isolated and substitute $\frac{11}{13}$ for y.

$$x = 4 - 3y$$
$$x = 4 - 3\left(\frac{11}{13}\right)$$
$$= 4 - \frac{33}{13}$$
$$= \frac{52}{13} - \frac{33}{13}$$
$$x = \frac{19}{13}$$

The solution is $\left(\frac{19}{13}, \frac{11}{13}\right)$.

Now try Example 3.

Example 3. Use substitution to solve the system.

$$(1)\ 2x + 3y = 4$$
$$(2)\ 3x + y = 7$$

Solution: We can isolate y in equation (2).

$$(2)\ 3x + y = 7$$
$$y = 7 - 3x$$

Now we substitute ________ for y in equation (1).

$$(1)\ 2x + 3y = 4$$

becomes

$$2x + 3(\underline{\qquad\qquad}) = 4$$

To find the corresponding y-value, return to equation (2) with y isolated and substitute ________ for x.

$$y = 7 - 3x$$
$$y = 7 - 3(\underline{\qquad\qquad})$$

The solution is (________, ________)

Check your work on page 647. ▶

Can you imagine how difficult it would have been to locate the common points obtained in Examples 2 and 3 using the method of *graphing*? That is why we needed a method for solving a system *algebraically* without relying on a graph.

Example 4. Use substitution to solve the system.

$$(1)\ y = 3x + 2$$

$$(2)\ 3x - y = 3$$

Solution: Since y is isolated in equation (1), we can substitute $3x + 2$ for y in equation (2).

$$(2)\ 3x - y = 3$$

becomes

$$3x - (3x + 2) = 3$$

$$3x - 3x - 2 = 3$$

$$-2 = 3$$

This result is disturbing, since we know that

$$-2 \neq 3$$

To determine what this result means for our system, let's graph the original equations in the same plane.

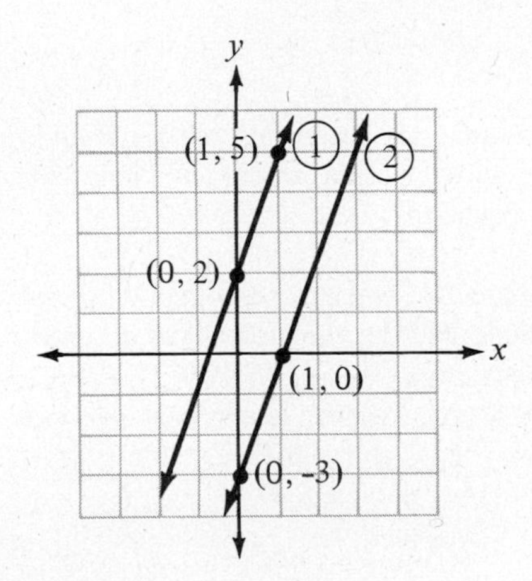

Our two lines are *parallel*; the equations have no ordered pair in common. There is *no solution* for this system.

In general, if the method of substitution yields a *false* statement, we must conclude that there is *no* ordered pair that satisfies the system of equations. The system is *inconsistent* and the graphs of the equations are parallel lines.

Example 5. Use substitution to solve the system.

$$(1)\ 4x + 2y = 10$$

$$(2)\ y = 5 - 2x$$

Solution: Since y is isolated in equation (2), we can substitute $5 - 2x$ for y in equation (1).

$$(1)\ 4x + 2y = 10$$

becomes

$$4x + 2(5 - 2x) = 10$$

$$4x + 10 - 4x = 10$$

$$10 = 10$$

This is certainly a true statement, but it contains no variable, so we have not found either coordinate of a common point. To determine what this result means for our system, let's graph the original equations in the same plane.

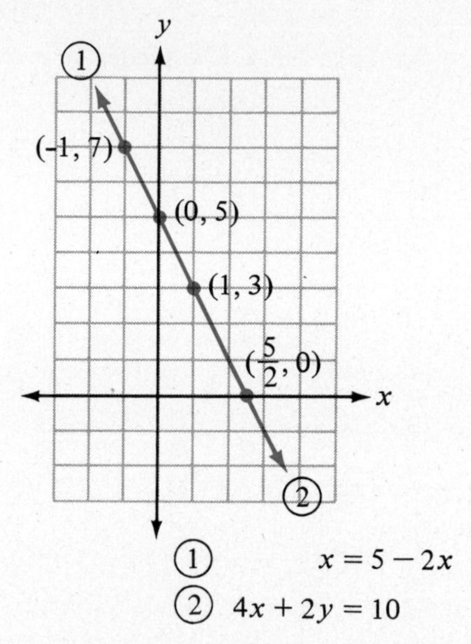

Both equations graph into the *same line*; any solution for one equation is a solution for the other equation. Here any ordered pair that satisfies $y = 5 - 2x$ (or $4x + 2y = 10$) will satisfy both equations.

In general, whenever the method of substitution yields a statement that is *always true*, we conclude that the equations have *all* ordered pairs in common, and the system is *dependent*.

Let's note here the results that can occur when we solve a system by substitution.

If substitution yields *one* ordered pair, the graphs intersect in *one point* and the system is said to be **independent and consistent**.

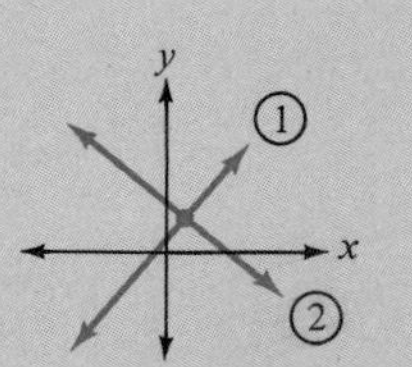

If substitution yields a *false* statement, the system has *no solution*. The graphs are **parallel lines** and the system is **inconsistent**.

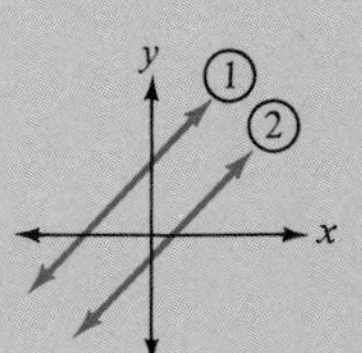

If substitution yields a statement that is *always true*, the system has an *infinite number* of solutions. Any solution for one equation is a solution for both equations. The graphs are the **same line** and the system is **dependent**.

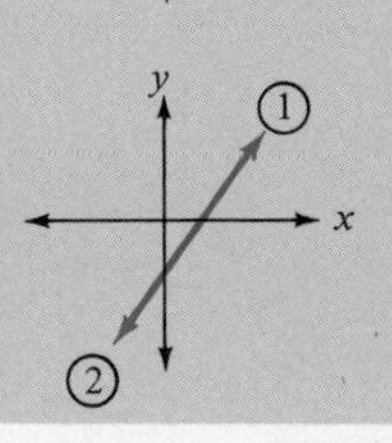

Trial Run

Use substitution to solve each system.

_______ 1. $y = 7x - 3$
$y = 6 - 2x$

_______ 2. $y = 3x$
$3x - 2y = 6$

_______ 3. $x + 4y = 5$
$2y + x = 7$

_______ 4. $y + 2 = 4x$
$3y + 2x = 1$

_______ 5. $y = 3x + 7$
$2y - 6x = 5$

_______ 6. $x = 4y + 1$
$8y = 2x - 2$

Answers are on page 648.

The Example You Completed

Example 3. Use substitution to solve the system.

$$(1)\ 2x + 3y = 4$$
$$(2)\ 3x + y = 7$$

Solution: We can isolate y in equation (2).

$$(2)\ 3x + y = 7$$
$$y = 7 - 3x$$

Now we substitute $7 - 3x$ for y in equation (1).

$$(1)\ 2x + 3y = 4$$

becomes

$$2x + 3(7 - 3x) = 4$$
$$2x + 21 - 9x = 4$$
$$21 - 7x = 4$$
$$-7x = -17$$
$$x = \frac{-17}{-7}$$
$$x = \frac{17}{7}$$

To find the corresponding y-value, return to equation (2) with y isolated and substitute $\frac{17}{7}$ for x.

$$y = 7 - 3x$$

$$y = 7 - 3\left(\frac{17}{7}\right)$$

$$= 7 - \frac{51}{7}$$

$$= \frac{49}{7} - \frac{51}{7}$$

$$y = -\frac{2}{7}$$

The solution is $\left(\frac{17}{7}, \frac{-2}{7}\right)$.

Answers to Trial Run

page 647 **1.** (1, 4) **2.** (−2, −6) **3.** (−3, 2) **4.** $\left(\frac{1}{2}, 0\right)$ **5.** No solution

6. All ordered pairs that satisfy $x = 4y + 1$ (or $8y = 2x - 2$)

60

Name ______________________________ Date ______________

EXERCISE SET 11.2

Solve each system by the substitution method.

______ **1.** $8x - 3y = 7$
$x = 2$

______ **2.** $7x - 11y = -1$
$x = 3$

______ **3.** $3x - 8y = 13$
$y = 2x$

______ **4.** $5x - 3y = 28$
$y = -3x$

______ **5.** $y = 4x - 5$
$5x - y = 11$

______ **6.** $y = 3x + 8$
$7x - y = -4$

______ **7.** $2x - 5y = 9$
$x - 3y = 2$

______ **8.** $3x - 2y = 15$
$x - 2y = 1$

______ **9.** $10x - 5y = 20$
$y = 2x - 4$

______ **10.** $3x - 15y = 15$
$x = 5y + 5$

______ **11.** $4x - 3y = 7$
$2x - y = -1$

______ **12.** $5x - 2y = 12$
$4x - y = 3$

______ **13.** $x + 2y = -1$
$4x + 6y = 5$

______ **14.** $x - 3y = 5$
$3x - 7y = 4$

______ **15.** $3x - 2y = 12$
$x + y = 6$

______ **16.** $2x - 5y = 11$
$x + y = 3$

______ **17.** $x + 3y = 10$
$x - 3 = 0$

______ **18.** $x + 9y = 1$
$x + 5 = 0$

______ **19.** $14x - 2y = 10$
$7x - y = 2$

______ **20.** $10x + 2y = 3$
$5x + y = -1$

______ **21.** $x - y = -2$
$11x - 3y = 2$

______ **22.** $x - y = -4$
$7x + 5y = -4$

______ **23.** $x - 2y = -3$
$3x - 7y = 1$

______ **24.** $x + 5y = 1$
$2x + 8y = -6$

______ **25.** $3x + y = 2$
$4x - 2y = -3$

______ **26.** $5x + y = -1$
$13x + 3y = 2$

☆ Stretching the Topics

Solve by the substitution method.

______ **1.** $8x - 5y = 0.9$
$6x + y = 0.2$

______ **2.** $3x + 9y = 21$
$\frac{2}{3}x + y = \frac{1}{3}$

________ **3.** Solve for x and y in terms of a.

$$\begin{aligned} ax + 3ay &= 5a^2 \\ x - \quad y &= a \end{aligned}$$

Check your answers in the back of your book.

If you can solve the systems in **Checkup 11.2,** you are ready to to go on to Section 11.3.

Name ______________________________ Date ____________

✓ CHECKUP 11.2

Solve each system by the substitution method.

_______ 1. $5x - 2y = 3$
$y = 2x$

_______ 2. $x - y = -3$
$-4x + 4y = 1$

_______ 3. $x - 6 = 0$
$3x - 4y = 8$

_______ 4. $9x - 4y = 10$
$2x + y = 6$

_______ 5. $4x + 7y = 16$
$x + 4y = 7$

Check your answers in the back of your book.

If You Missed Problems:	You Should Review Examples:
1	1
2	4
3	1
4, 5	2, 3

11.3 Solving a System of Equations by Elimination (or Addition)

Besides graphing and substitution, there is another method for solving systems of equations. This method uses two properties of equations with which we are already familiar.

1. We may multiply both sides of an equation by any constant (except zero).
2. We may add equal quantities to both sides of an equation without changing the solutions of the original equation.

Suppose that we wish to solve the system

$$(1)\ 3x + 2y = 9$$
$$(2)\ x - 2y = 7$$

Notice what happens when we add the left-hand side of equation (1) to the left-hand side of equation (2) *and* add the right-hand side of equation (1) to the right-hand side of equation (2). (We know that this is permissible because of the second property stated above.)

$$\begin{array}{rl} (1)\ 3x + 2y = & 9 \\ (2)\ \underline{x - 2y =} & \underline{7} \\ 4x + 0 = & 16 \end{array}$$

Notice that we have completely eliminated the variable y; our new equation contains the single variable x. Let's solve for x.

$$4x = 16$$
$$x = \frac{16}{4}$$
$$x = 4$$

We may find the corresponding y-value by substituting 4 for x into either equation (1) or equation (2). Let's use equation (2) and then check our solution in equation (1).

$$\begin{aligned} (2)\ x - 2y &= 7 \\ 4 - 2y &= 7 \\ -2y &= 3 \\ y &= \frac{3}{-2} \\ y &= \frac{-3}{2} \\ (x, y) &= \left(4, \frac{-3}{2}\right) \end{aligned}$$

CHECK:

$$\begin{aligned} (1)\qquad 3x + 2y &= 9 \\ 3(4) + 2\left(\frac{-3}{2}\right) &\stackrel{?}{=} 9 \\ 12 - 3 &\stackrel{?}{=} 9 \\ 9 &= 9 \end{aligned}$$

The ordered pair $\left(4, \frac{-3}{2}\right)$ satisfies both equations.

This method, called the **method of elimination** (or the **method of addition**), works nicely if one of the variables appears in the equations with *coefficients that are opposites*. This is an important observation.

Now you complete Example 1.

Example 1. Use elimination to solve the system.

$$(1) \quad x + 3y = 5$$
$$(2) \quad -x - 7y = 3$$

Solution: Since the coefficients of x are *opposites*, we shall eliminate the variable x and solve for y.

$$\begin{aligned} (1) \quad x + 3y &= 5 \\ (2) \quad \underline{-x - 7y} &= \underline{3} \\ 0 - 4y &= 8 \\ y &= \frac{8}{-4} \\ y &= \_\_\_\_ \end{aligned}$$

Now we can substitute ____ for y into equation (1) and check our solution in equation (2).

$$\begin{aligned} (1) \quad x + 3y &= 5 \\ x + 3(\_\_\_\_) &= 5 \\ x - \_\_\_\_ &= 5 \\ x &= \_\_\_\_ \\ (x, y) &= (\_\_\_\_, \_\_\_\_) \end{aligned}$$

CHECK:

$$\begin{aligned} (2) \quad -x - 7y &= 3 \\ \_\_\_\_ - 7(\_\_\_\_) &\stackrel{?}{=} 3 \\ \_\_\_\_ + \_\_\_\_ &\stackrel{?}{=} 3 \\ \_\_\_\_ &= 3 \end{aligned}$$

The ordered pair (____ , ____) satisfies both equations.

Check your work on page 659. ▶

Elimination works well when one of the variables has opposite coefficients in the two equations. But can the elimination method be used when the original coefficients are not opposites? Yes, provided that we *make the coefficients opposites*.

Example 2. Use elimination to solve the system.

$$(1) \quad 2x + 3y = 7$$
$$(2) \quad x + 5y = 4$$

Solution: Here neither variable seems ready for elimination. Suppose that we wish to eliminate the variable x. To do that, the x-term in equation (2) would have to be $-2x$. If we multiply equation (2) by -2, we will have what we want. (Remember that it is permissible to multiply both sides of an equation by any constant except zero.)

$$(2) \quad x + 5y = 4$$

becomes

$$-2[x + 5y = 4]$$
$$-2x - 10y = -8$$

and our new system is

$$\begin{aligned}
(1)\quad 2x + 3y &= 7 \\
(2)\quad -2x - 10y &= -8 \\
\hline
-7y &= -1 \\
y &= \frac{-1}{-7} \\
y &= \frac{1}{7}
\end{aligned}$$

Substituting $\frac{1}{7}$ for y in the original equation (2), we have

$$\begin{aligned}
x + 5y &= 4 \\
x + 5\left(\frac{1}{7}\right) &= 4 \\
x + \frac{5}{7} &= 4 \\
x &= 4 - \frac{5}{7} \\
x &= \frac{28}{7} - \frac{5}{7} \\
x &= \frac{23}{7}
\end{aligned}$$

The solution for the system is $\left(\frac{23}{7}, \frac{1}{7}\right)$.

Example 3. Use elimination to solve the system.

$$\begin{aligned}
(1)\quad 4x - 7y &= 20 \\
(2)\quad 5x + 2y &= 3
\end{aligned}$$

Solution: To eliminate either variable in these equations, we must multiply *both* equations by constants. We may choose to work with either variable. Since the y coefficients are already opposite in sign, let's work to eliminate the variable y.

$$\begin{aligned}
(1)\quad 4x - 7y &= 20 \\
(2)\quad 5x + 2y &= 3
\end{aligned}$$

To eliminate y, the coefficients must be opposite multiples of 7 and 2. The smallest multiple of 7 and 2 is 14, so we must make the coefficients -14 and 14.

Multiply equation (1) by 2.

$$\begin{aligned}
2[4x - 7y &= 20] \\
8x - 14y &= 40
\end{aligned}$$

Multiply equation (2) by 7.

$$\begin{aligned}
7[5x + 2y &= 3] \\
35x + 14y &= 21
\end{aligned}$$

$$\begin{aligned}
(1)\quad 8x - 14y &= 40 \\
(2)\quad 35x + 14y &= 21 \\
\hline
43x + 0 &= 61 \\
43x &= 61 \\
x &= \frac{61}{43}
\end{aligned}$$

The thought of substituting this value for x into either equation is not very appealing. It might be easier to start all over again and eliminate the variable x, so that we are left with the variable y.

$$\begin{aligned}
(1)\quad 4x - 7y &= 20 \\
(2)\quad 5x + 2y &= 3
\end{aligned}$$

To eliminate x, the coefficients must be opposite multiples of 4 and 5. The smallest multiple of 4 and 5 is 20, so we must make the coefficients 20 and -20.

Multiply equation (1) by 5.

$$5[4x - 7y = 20]$$
$$20x - 35y = 100$$

Multiply equation (2) by -4.

$$-4[5x + 2y = 3]$$
$$-20x - 8y = -12$$

$$\begin{array}{lrcl} (1) & 20x - 35y & = & 100 \\ (2) & -20x - 8y & = & -12 \\ \hline & 0 - 43y & = & 88 \end{array}$$

$$y = \frac{88}{-43}$$

$$y = \frac{-88}{43}$$

The solution is the ordered pair $\left(\frac{61}{43}, \frac{-88}{43}\right)$.

Example 4. Use elimination to solve the system.

$$7x + 6y = 8$$
$$5x + 4y = 6$$

Solution: Let's eliminate y. To do so, the coefficients of y must be opposite multiples of 6 and 4. The smallest multiple of 6 and 4 is 12, so we must make the coefficients 12 and -12.

Multiply equation (1) by 2.

$$2[7x + 6y = 8]$$
$$14x + 12y = 16$$

Multiply equation (2) by -3.

$$-3[5x + 4y = 6]$$
$$-15x - 12y = -18$$

$$\begin{array}{lrcl} (1) & 14x + 12y & = & 16 \\ (2) & -15x - 12y & = & -18 \\ \hline & -x & = & -2 \end{array}$$

$$x = 2$$

Now we can substitute 2 for x in either of the original equations. Let's use equation (2).

$$(2) \quad 5x + 4y = 6$$
$$5(2) + 4y = 6$$
$$10 + 4y = 6$$
$$4y = -4$$
$$y = -1$$

The solution is the ordered pair $(2, -1)$.

In general, to solve a system of equations by the method of elimination, we should keep these steps in mind.

Solving a System by Elimination (Addition)

1. Write both equations in the form $ax + by = c$.
2. Look to see if the coefficient of one of the variables in one equation is the opposite of the coefficient of the same variable in the other equation.
3. If the coefficients are not opposites, multiply one or both equations by constants that will make the coefficients opposites.
4. Add the equations together to eliminate one variable.
5. Solve for the value of the remaining variable.
6. Substitute that value into either of the original equations and solve for the other variable; *or* repeat the addition process, eliminating the other variable.
7. Write the solution as an ordered pair.

Trial Run

Use elimination to solve each system.

_______ **1.** $\begin{aligned} 3x - y &= 2 \\ 2x + y &= 8 \end{aligned}$

_______ **2.** $\begin{aligned} 11x - 8y &= 4 \\ -5x + 8y &= -4 \end{aligned}$

_______ **3.** $\begin{aligned} -2x + y &= 1 \\ 4x + 3y &= 8 \end{aligned}$

_______ **4.** $\begin{aligned} 7x - 3y &= -9 \\ 2x + 5y &= 4 \end{aligned}$

Answers are on page 660.

Let's consider some more systems of equations.

Example 5. Use elimination to solve

$$(1)\quad 2x + 2y = 5$$

$$(2)\quad 3x + 3y = 1$$

Solution: To eliminate the variable x, we can multiply equation (1) by 3 and multiply equation (2) by -2.

$$3[2x + 2y = 5]$$

$$-2[3x + 3y = 1]$$

$$\begin{aligned} (1)\quad 6x + 6y &= 15 \\ (2)\quad \underline{-6x - 6y} &= \underline{-2} \\ 0 + 0 &= 13 \\ 0 &= 13 \end{aligned}$$

The elimination process has yielded a *false statement*. As before, we must conclude that there is *no ordered pair* common to both equations. Indeed, if we were to graph these equations, the two lines would be *parallel*. There is *no solution* for this system.

Example 6. Use elimination to solve the system.

$$(1)\ y = 3x + 1$$

$$(2)\ 6x - 2y + 2 = 0$$

Solution: First we line up like variables beneath each other.

$$(1)\quad -3x + y = 1$$

$$(2)\quad 6x - 2y = -2$$

We can eliminate the variable y by multiplying equation (1) by 2.

$$2[-3x + y = 1]$$

$$-6x + 2y = 2$$

$$\begin{array}{lrcr} (1) & -6x + 2y & = & 2 \\ (2) & 6x - 2y & = & -2 \\ \hline & 0 + 0 & = & 0 \\ & 0 & = & 0 \end{array}$$

We seem to have eliminated both variables! What does this mean? The statement $0 = 0$ is *always true*, but what is the solution for our system? If we were to graph these equations, they would both be the *same line*. Our conclusion is that these equations have *all* ordered pairs in common. All ordered pairs that satisfy $y = 3x + 1$ (or $6x - 2y + 2 = 0$) will satisfy this system.

Let's review the results that can occur when elimination is used to solve a system of equations.

If elimination yields *one* ordered pair, the graphs intersect in *one point*, and the system is **independent and consistent**.

If elimination yields a *false* statement, the system has *no solution*. The graphs are *parallel lines* and the system is **inconsistent**.

If elimination yields a statement that is *always true*, the system has an infinite number of solutions. Any solution for one equation is a solution for both equations. The graphs are the *same line* and the system is **dependent**.

Trial Run

Use elimination to solve each system.

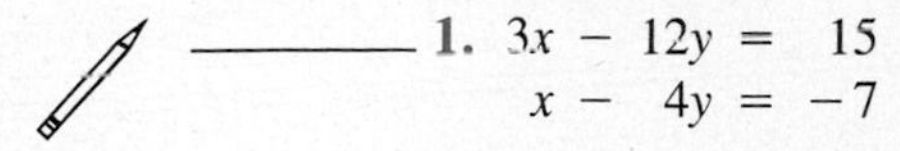

______ 1. $\begin{aligned} 3x - 12y &= 15 \\ x - 4y &= -7 \end{aligned}$

______ 2. $\begin{aligned} 5x + 6y &= 4 \\ 4x - 3y &= -2 \end{aligned}$

______ 3. $\begin{aligned} 5x - 10y &= -15 \\ -4x + 8y &= 12 \end{aligned}$

______ 4. $\begin{aligned} -9x + 12y &= -6 \\ 15x - 20y &= 1 \end{aligned}$

Answers are on page 660.

The Example You Completed

Example 1. Use elimination to solve the system.

$$\begin{aligned} (1) \quad x + 3y &= 5 \\ -x - 7y &= 3 \end{aligned}$$

Solution: Since the coefficients of x are *opposites*, we shall eliminate the variable x and solve for y.

$$\begin{aligned} (1) \quad x + 3y &= 5 \\ (2) \quad -x - 7y &= 3 \\ \hline 0 - 4y &= 8 \\ y &= \frac{8}{-4} \\ y &= -2 \end{aligned}$$

Now we can substitute -2 for y into equation (1) and check our solution in equation (2).

$$\begin{aligned} (1) \quad x + 3y &= 5 \\ x + 3(-2) &= 5 \\ x - 6 &= 5 \\ x &= 11 \\ (x, y) &= (11, -2) \end{aligned}$$

CHECK:

$$\begin{aligned} (2) \quad -x - 7y &= 3 \\ -11 - 7(-2) &\stackrel{?}{=} 3 \\ -11 + 14 &\stackrel{?}{=} 3 \\ 3 &= 3 \end{aligned}$$

The ordered pair $(11, -2)$ satisfies both equations.

Answers to Trial Runs

page 657 **1.** (2, 4) **2.** $\left(0, \frac{-1}{2}\right)$ **3.** $\left(\frac{1}{2}, 2\right)$ **4.** $\left(\frac{33}{41}, \frac{-46}{41}\right)$

page 659 **1.** No solution **2.** $\left(0, \frac{2}{3}\right)$ **3.** All ordered pairs that satisfy $5x - 10y = -15$
4. No solution

Name ________________________________ Date ____________

EXERCISE SET 11.3

Solve each system by the elimination method.

_______ **1.** $3x - y = 4$
$2x + y = 1$

_______ **2.** $4x - y = 5$
$3x + y = 9$

_______ **3.** $-2x - y = 3$
$2x + 7y = -9$

_______ **4.** $4x - 5y = 6$
$-4x - y = 6$

_______ **5.** $2x - 4y = 1$
$4y = 7$

_______ **6.** $4x - 9y = -9$
$9y = 1$

_______ **7.** $3x + 4y = -2$
$2x - 4y = 12$

_______ **8.** $5x + 3y = -1$
$x - 3y = 7$

_______ **9.** $6x - 4y = 1$
$-6x - 4y = 7$

_______ **10.** $-5x - 3y = 2$
$5x - 3y = 10$

_______ **11.** $3x - 2y = 0$
$5x + 2y = 4$

_______ **12.** $4x + 5y = 0$
$4x + y = 2$

_______ **13.** $7x - 10y = 4$
$x + 5y = 7$

_______ **14.** $8x - 4y = 6$
$3x + 2y = 4$

_______ **15.** $3x + 4y = 5$
$-9x - 7y = 5$

_______ **16.** $5x + 3y = 7$
$-10x - 13y = 7$

_______ **17.** $2x - 8y = 9$
$-x + 4y = -6$

_______ **18.** $-3x + 3y = -1$
$x - y = 8$

_______ **19.** $5x - 2y = 5$
$4x - 6y = 15$

_______ **20.** $10x - 9y = 12$
$2x - 3y = 4$

_______ **21.** $3x + 4y = 9$
$9x + 4y = 5$

_______ **22.** $7x + 6y = 10$
$7x + 12y = 9$

_______ **23.** $5x - 4y = 1$
$2x + 3y = 5$

_______ **24.** $7x - 6y = 20$
$3x + 5y = 1$

_______ **25.** $14x + 7y = 7$
$-6x - 3y = -3$

_______ **26.** $5x - 15y = 10$
$-6x + 18y = -12$

_______ **27.** $11x - 9y = 14$
$7x + 3y = 6$

_______ **28.** $3x - 5y = 15$
$11x + 15y = -5$

_______ **29.** $7x - 5y = 2$
$2x - 3y = -1$

_______ **30.** $2x - 3y = 5$
$5x - 2y = -4$

☆ Stretching the Topics

_______ **1.** $0.8b = 0.2a + 6.4$
$0.7a - 0.2b = 1$

_______ **2.** $\frac{3}{5}x - \frac{2}{3}y = 3$
$\frac{2}{5}x + \frac{1}{3}y = -2$

_______ **3.** Find a value of k so that the following system is dependent.

$4x - 2y = 7$

$12x - ky = 21$

Check your answers in the back of your book.

If you can solve the systems in **Checkup 11.3**, you are ready to go on to Section 11.4.

Name ______________________________ Date ____________

CHECKUP 11.3

Solve each system by the elimination method.

______ **1.** $3x + y = 6$
$5x - y = 10$

______ **2.** $3a + 5b = -1$
$2a - b = 8$

______ **3.** $2x - 3y = 5$
$-4x + 6y = -7$

______ **4.** $8x + 5y = 9$
$6x + 9y = 5$

______ **5.** $9x + 5y = 53$
$5x - 6y = -32$

Check your answers in the back of your book.

If You Missed Problem:	You Should Review Example:
1	1
2	2
3	5
4	4
5	3

11.4 Switching from Word Statements to Systems of Equations

Throughout our study of algebra, we have practiced switching word statements into variable equations. Sometimes we may be asked to solve a problem containing more than one unknown quantity; in such a situation we may wish to use *two* variables. In dealing with two variables, we must translate the given information into *two* equations and solve the resulting system.

Example 1. For a recent chili supper a total of 191 tickets were sold. Adult tickets sold for \$2 each and children's tickets sold for \$1 each. If \$304 worth of tickets were sold, how many adult tickets and how many children's tickets were sold?

Solution: We are asked to find two unknown quantities here, so we may use two variables. Our problem gives us information of two kinds: information about the *number* of tickets and information about *money* from the tickets.

	Number of Tickets	Money from Tickets
Adult	x	$2 \cdot x$
Children's	y	$1 \cdot y$
Total	$x + y$	$2x + y$

(1) $x + y = 191$ (2) $2x + y = 304$

Now we can use elimination to solve the system. Let's eliminate y.

$$\begin{aligned} (1)\quad x + y &= 191 \\ (2)\quad 2x + y &= 304 \end{aligned}$$

$$\begin{aligned} (1)\quad -x - y &= -191 && \text{Multiply equation (1) by } -1. \\ (2)\quad 2x + y &= 304 && \text{Equation (2).} \\ x + 0 &= 113 && \text{Add the equations.} \\ x &= 113 && \text{Solve for } x. \end{aligned}$$

We know that 113 adult tickets were sold and we can substitute 113 for x in equation (1) to find y. Then we can check our solution in equation (2).

$$\begin{aligned} (1)\quad x + y &= 191 \\ 113 + y &= 191 \\ y &= 78 \end{aligned}$$

CHECK:

$$\begin{aligned} (2)\quad 2x + y &= 304 \\ 2(113) + 78 &\stackrel{?}{=} 304 \\ 226 + 78 &\stackrel{?}{=} 304 \\ 304 &= 304 \end{aligned}$$

There were 113 adult tickets and 78 children's tickets sold.

Let's return to the problem stated at the beginning of the chapter and use a system of equations to find José's hourly wages.

Example 2. José works part-time mowing lawns and lifeguarding at the swimming pool. When he mows for 5 hours and guards for 6 hours, he earns \$42. When he mows for 11 hours and guards for 4 hours, he earns \$51. What is his hourly wage for each job?

Solution: Since we are asked to find two unknown quantities, we may use two variables. Our problem gives us information about two different situations. Let x = hourly mowing rate (in dollars) and let y = hourly guarding rate (in dollars).

	Situation 1	Situation 2
Mowing rate	x	x
Mowing time	5	11
Mowing pay	$5 \cdot x$	$11 \cdot x$
Guarding rate	y	y
Guarding time	6	4
Guarding pay	$6 \cdot y$	$4 \cdot y$
Total pay	$5x + 6y$	$11x + 4y$
	(1) $5x + 6y = 42$	(2) $11x + 4y = 51$

We can use elimination to solve the system. To eliminate y, we note that the smallest multiple of 6 and 4 is 12.

$$(1)\quad 5x + 6y = 42$$

$$(2)\quad 11x + 4y = 51$$

(1)	$2[5x + 6y = 42]$	Multiply equation (1) by 2.
(2)	$-3[11x + 4y = 51]$	Multiply equation (2) by -3.
(1)	$10x + 12y = 84$	Remove parentheses in equation (1).
(2)	$-33x - 12y = -153$	Remove parentheses in equation (2).
	$-23x + 0 = -69$	Add the equations.
	$-23x = -69$	Simplify.
	$x = \dfrac{-69}{-23}$	Divide by -23 to isolate x.
	$x = 3$	Simplify.

Now we know that José earns $3 per hour mowing lawns. To find y, we can substitute 3 for x in equation (2) and check our solution in equation (1).

$$(2)\quad 11x + 4y = 51$$
$$11(3) + 4y = 51$$
$$33 + 4y = 51$$
$$4y = 18$$
$$y = \frac{18}{4}$$
$$y = 4.5$$

CHECK:

$$(1)\quad 5x + 6y = 42$$
$$5(3) + 6(4 \cdot 5) \stackrel{?}{=} 42$$
$$15 + 27 \stackrel{?}{=} 42$$
$$42 = 42$$

José earns $3 per hour mowing lawns and $4.50 per hour lifeguarding.

Now you try completing Example 3.

Example 3. In Mario's psychology class, the male students outnumber the female students by 10. If there are 46 students altogether in the class, how many are male and how many are female?

Solution: We are asked to find two unknown quantities so we may use _____ variables. We can let x = number of females and let y = number of males. Our problem gives us information about male students outnumbering female students and about the total number of students in the class.

"Outnumbering" Information		"Total Number" Information	
Number of females	x	Number of females	x
Number of males	y	Number of males	y
Number of males	$x + 10$	Total number	$x + y$

(1) $y = x + 10$ (2) $x + y =$ _____

We can use substitution to solve this system.

$$(1)\ y = x + 10$$

$$(2)\ x + y = 46$$

We substitute _______ for y in equation (2) and solve for x.

$$(2)\ x + (\_\_\_\_\_\_\_) = 46$$

Now we can substitute _____ for x in equation (1) and find y.

$$(1)\ y = x + 10$$

There are _____ females and _____ males in Mario's psychology class.

Check your work on page 669. ▶

Example 4. When Marty jogs around the perimeter of a certain rectangular field, she covers a distance of 540 yards. When she jogs around another rectangular field that is twice as wide and 60 yards longer, she covers a distance of 860 yards. Find the dimensions of the smaller field.

Solution: We must find the length and width of the smaller field, so we may use two variables. Let l = length of smaller field (in yards) and let w = width of smaller field (in yards). Information about the smaller field and larger field will provide us with two equations.

Remember that the perimeter of a rectangle can be found by the formula $p = 2l + 2w$.

	Smaller Field	Larger Field
Length	l	$l + 60$
Width	w	$2 \cdot w$
Perimeter	$2l + 2w$	$2(l + 60) + 2(2w)$

(1) $2l + 2w = 540$ (2) $2(l + 60) + 4w = 860$

Our system becomes

$$(1)\ 2l + 2w = 540$$
$$(2)\ 2(l + 60) + 4w = 860$$

Simplifying equation (2) we have

$$\begin{aligned}(2)\ 2(l + 60) + 4w &= 860 \\ 2l + 120 + 4w &= 860 \\ 2l + 4w &= 740\end{aligned}$$

Now we may use elimination to solve the system. Let's eliminate l.

$$(1)\ 2l + 2w = 540$$
$$(2)\ 2l + 4w = 740$$

$$\begin{aligned}(1)\ -2l - 2w &= -540 && \text{Multiply equation (1) by } -1. \\ (2)\ 2l + 4w &= 740 && \text{Equation (2).} \\ 0 + 2w &= 200 && \text{Add the equations.} \\ w &= 100 && \text{Solve for } w.\end{aligned}$$

The width of the smaller field is 100 yards, and we can substitute 100 for w in equation (1) to find the length, l. Then you should check the solution in equation (2).

$$\begin{aligned}(1)\ 2l + 2w &= 540 \\ 2l + 2(100) &= 540 \\ 2l + 200 &= 540 \\ 2l &= 340 \\ l &= 170 \\ (l, w) &= (170,100)\end{aligned}$$

CHECK:

$$(2)\ 2(l + 60) + 4w = 860$$
$$2(\_\_\_\_ + 60) + 4(\_\_\_\_) \stackrel{?}{=} 860$$

You complete the check. Check your work on page 669.

The smaller field is 170 yards long and 100 yards wide.

To use a system of equations to solve a problem stated in words, we must follow these steps.

1. Decide what the two unknown quantities are and identify them, using a different variable to represent each quantity.
2. From the information in the problem, write *two* different equations containing the two variables.
3. Solve the resulting system by the method of your choice (substitution or elimination).
4. Check your solution in the original word problem.

▶ Examples You Completed

Example 3

Solution: We are asked to find two unknown quantities, so we need two equations. Our problem gives us information about male students outnumbering female students and about the total number of students in the class.

"Outnumbering" Information		"Total Number" Information	
Number of females	x	Number of females	x
Number of males	y	Number of males	y
Number of males	$x + 10$	Total number	$x + y$

(1) $y = x + 10$ (2) $x + y = 46$

We can use substitution to solve this system.

$$(1)\ y = x + 10$$
$$(2)\ x + y = 46$$

We substitute $x + 10$ for y in equation (2).

$$(2)\ x + (x + 10) = 46$$
$$2x + 10 = 46$$
$$2x = 36$$
$$x = 18$$

Now we can substitute 18 for x in equation (1).

$$(1)\ y = x + 10$$
$$y = 18 + 10$$
$$y = 28$$

There are 18 females and 28 males in Mario's psychology class.

Example 4

CHECK:

$$(2)\quad 2(l + 60) + 4w = 860$$
$$2(170 + 60) + 4(100) \stackrel{?}{=} 860$$
$$2(230) + 400 \stackrel{?}{=} 860$$
$$460 + 400 \stackrel{?}{=} 860$$
$$860 = 860$$

Name ______________________ Date ____________

EXERCISE SET 11.4

Change each word problem to a system of equations using two variables. Solve each system by the better method (substitution or elimination).

________ **1.** One number is 9 more than twice the other. The sum of the two numbers is 6. Find the numbers.

________ **2.** The sum of two numbers is 10. One number is 14 less than 3 times the other. Find the numbers.

________ **3.** The length of a rectangle is 5 feet more than twice the width. If the perimeter of the rectangle is 58 feet, find the dimensions.

________ **4.** For her flower bed, Nancy needed 3 times as many boxes of petunias as she did marigolds. If she bought 12 boxes of flowers, how many boxes of each type did she buy?

________ **5.** The cashier at Mary's Market has a stack of 25 bills. Some are 5-dollar bills and some are 1-dollar bills. The total value of the stack of bills is \$77. How many bills of each denomination does the cashier have?

________ **6.** The length of a carpet is 20 feet less than 3 times the width. The difference between the length and the width is 4 feet. Find the dimensions of the carpet.

________ **7.** Jay sold a box of books at a yard sale for \$51. He sold some of the books for \$2 each and some for \$3 each. If there were 20 books in the box, how many did he sell at each price?

________ **8.** When Hector went to see his father he made part of the trip by bus and the other part by taxi. The time he traveled by taxi was twice the time he rode in the bus. He averaged 40 miles per hour on the bus and 30 miles per hour in the taxi. If the total trip was 25 miles, how much time did he spend traveling by bus and how much time by taxi?

________ **9.** A.J. has money invested in two different companies. The sum of twice the amount invested in Abell's Alarm Service and the amount invested in Betti's Boutique is \$11,000. The amount invested in Betti's Boutique is \$2000 more than the amount invested in Abell's Alarm Service. Find the amount of each investment.

________ **10.** The sum of 3 times one number and 4 times a second number is -17. The difference between twice the first number and the second number is 18. Find the two numbers.

________ **11.** At a sidewalk sale Kris bought three pairs of jeans and two tops at a cost of \$72. Kelly bought five pairs of jeans and three tops from the same racks for \$116. What was the price of each pair of jeans and each top?

________ **12.** For a concert, the sum of the 3 times the number of tickets sold in advance and the number of tickets sold at the door was 2825. The advance tickets sold for \$8 and tickets at the door cost \$10. If the total ticket receipts were \$16,700, how many tickets of each type were sold?

_______ **13.** The sum of the measures of the three angles of any triangle is 180°. One angle measures 90°, one of the remaining angles measures 6° less than 3 times the other. Find the measure of each angle.

_______ **14.** The length of a rectangular bulletin board is 7 inches more than twice the width. If the perimeter is 224 inches, find the dimensions of the bulletin board.

_______ **15.** At the Record Mart going-out-of-business sale, all records were sold at the same price and all tapes were sold at the same price. Cynthia bought 6 records and 4 tapes for $58. Netta bought 5 records and 3 tapes for $46. What is the sale price of each record and each tape?

_______ **16.** The treasurer of the Lion's Club has a stack of 50 bills. Some are 5-dollar bills and some are 10-dollar bills. The total value of the stack of bills is $390. How many bills of each type does he have?

_______ **17.** A blouse and skirt cost $49.00. The skirt cost $7 more than the blouse. What is the cost of each?

_______ **18.** The perimeter of a room is 54 feet. Twice the length increased by three times the width is 66 feet. What are the dimensions of the room?

_______ **19.** A druggist wishes to put 500 grains of a certain blood pressure medicine into 3-grain and 2-grain capsules. If he fills 200 capsules, how many capsules of each kind does he fill?

_______ **20.** In making a trip of 180 miles, Ronald traveled part of the way by train and the rest of the distance by bus. Two-thirds of the distance traveled by train equals $\frac{5}{6}$ of the distance by bus. How far did he travel by train and how far by bus?

☆ Stretching the Topics

_______ **1.** Coletta has invested a total of $16,000, one part at 8% per year and the rest at 10% per year. How much does she have invested at each rate if her simple interest at the end of 1 year is $8\frac{3}{4}\%$ of the total amount?

_______ **2.** The formula for finding the cost y of producing x telephones is $y = \$25x + \$2000$. The formula for finding the revenue y from selling x telephones is $y = \$40x - \$550$. The company will break even when the cost and the revenue are the same. Find the number of telephones that must be sold for the company to break even.

Check your answers in the back of your book.

If you can solve the problems in **Checkup 11.4**, you are ready to do the **Review Exercises for Chapter 11**.

Name ______________________ Date __________

✓ CHECKUP 11.4

Solve.

________ **1.** Lenora works part-time as a waitress and part-time as a window washer. When she works as a waitress for 5 hours and washes windows for 9 hours, she earns \$61. When she washes windows 3 hours and waits tables for 12 hours, she earns \$63. How much does she earn per hour at each job?

________ **2.** The Fashion Outlet sells all its shirts at one price and all its ties at another price. Pat bought three shirts and two ties for \$49 and Jon bought four shirts and three ties for \$67. What was the price of each shirt and each tie?

________ **3.** The width of a table is 20 inches less than the length. If the perimeter is 200 inches, find the dimensions of the table.

________ **4.** The sum of 5 times one number and twice the second is 1. The difference between the first and the second is 17. Find the numbers.

________ **5.** A total of 2825 tickets were sold for the concert for total ticket receipts of \$25,898. If advance tickets sold for \$8 and tickets at the door sold for \$10, how many tickets of each type were sold?

Check your answers in the back of your book.

If You Missed Problems:	You Should Review Example:
1, 2	2
3, 4	3
5	1

Summary

When we find all ordered pairs that satisfy two equations containing two variables, we are *solving a system* of two equations. In this chapter we discussed three methods for solving systems.

Method	Useful When:
Graphing	Common point is easy to identify from the graphs of the equations.
Substitution	One of the variables is easily isolated in one of the equations.
Elimination (addition)	We can make the coefficient of one of the variables in one of the equations the opposite of the coefficient of the same variable in the other equation.

No matter which method we used, we discovered three possible situations that could occur in solving a system of two first-degree equations.

Type of System	Solution	Graphs	Signal
Independent and consistent	One ordered pair	Lines cross at only one point	Solving algebraically yields one x-value and one y-value
Inconsistent	None	Lines are parallel	Solving algebraically yields a false statement
Dependent	Any solution for one equation is a solution for the other	Lines are the same	Solving algebraically yields a statement that is always true

Having learned to solve systems of equations algebraically, we put our methods to use in solving problems stated in words.

❑ Speaking the Language of Algebra

1. In solving a system of two first-degree equations, we are seeking ordered pairs that __________ both equations.

2. When one variable is easily isolated in one of the equations in a system, we can solve the system by the method of __________ .

3. If the method of elimination yields a false statement, we know the system has __________ solution(s).

4. If the method of graphing yields two straight lines crossing at only one point, we know the system has __________ solution(s).

Check your answers in the back of your book.

Name ______________________ Date ____________

REVIEW EXERCISES for Chapter 11

Solve each system by graphing.

______ **1.** $2x - y = 1$
$x + 2y = 8$

______ **2.** $5x - 6y = 4$
$x = 2$

______ **3.** $x - 3y = 7$
$x - 2y = 0$

______ **4.** $3x + y = -1$
$2x + y = 0$

Solve each system by the substitution method.

______ **5.** $6x + y = -3$
$y = 3$

______ **6.** $9x - 5y = 12$
$y = -3x$

______ **7.** $4x - y = 0$
$3x - 2y = 10$

______ **8.** $x - 3y = 8$
$-2x + 6y = 20$

______ **9.** $x = 2y + 7$
$2x - y = 23$

______ **10.** $x - y = -5$
$x - 9y = -13$

Solve each system by the elimination method.

______ **11.** $2x + y = 12$
$3x - y = 13$

______ **12.** $x + 3y = 9$
$4x + 5y = 15$

______ **13.** $2x - y = -6$
$-4x + 2y = -5$

______ **14.** $3x + 11y = 0$
$-2x + 3y = -5$

______ **15.** $-6x + 5y = 9$
$24x - 20y = -36$

______ **16.** $5x + 6y = 17$
$8x - 9y = -10$

Solve by any method you choose.

______ **17.** $3x - y = -8$
$-5x + 4y = 32$

______ **18.** $x - 2y = 0$
$6x - 13y = 5$

______ **19.** $4x + 15y = -18$
$6x - 5y = 28$

______ **20.** $2x - 3y = 4$
$5x + 4y = 13$

Change each word problem to a system of equations using two variables. Solve each system by the better method (substitution or elimination).

______ **21.** The sum of two numbers is 22. The difference between twice the first number and the second number is 8. Find both numbers.

______ **22.** The length of a rectangular field is 10 meters more than the width. The perimeter of the field is 116 meters. Find the length and width of the field.

______ **23.** Chico has savings accounts at two different banks. The amount at the Union Bank is $100 less than 3 times the amount at the National Bank. The sum of the two accounts is $1700. Find the amount in each savings account.

_______ 24. Marian bought some collectible bottles at the flea market for $76. She paid $5 each for some of the bottles and $3 each for the others. If she bought 20 bottles altogether, how many did she buy at each price?

Check your answers in the back of your book.

If You Missed Exercises:	You Should Review Examples:	
1, 3, 4	SECTION 11.1	1
2		3
5, 6, 9	SECTION 11.2	1
7, 10		2, 3
8		5
11, 12	SECTION 11.3	1, 2
13, 15		5, 6
14, 16		3, 4
17–20	SECTION 11.2	1–5
	SECTION 11.3	1–4
21–24	SECTION 11.4	1–3

If you have completed the **Review Exercises** and corrected your errors, you are ready to take the **Practice Test for Chapter 11**.

Name ______________________________ Date ______________

PRACTICE TEST for Chapter 11

Solve each system by graphing. — SECTION | EXAMPLE

________ **1.** $2x + y = 4$
$x - 3y = -5$ — 11.1 | 1

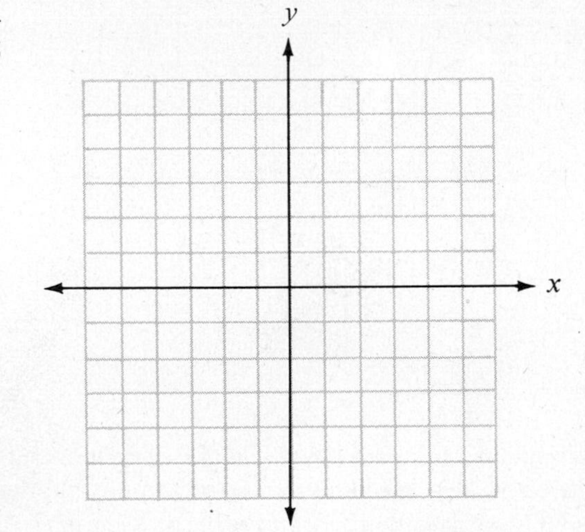

________ **2.** $x + 5 = 0$
$2x + 3y = -1$ — 11.1 | 3

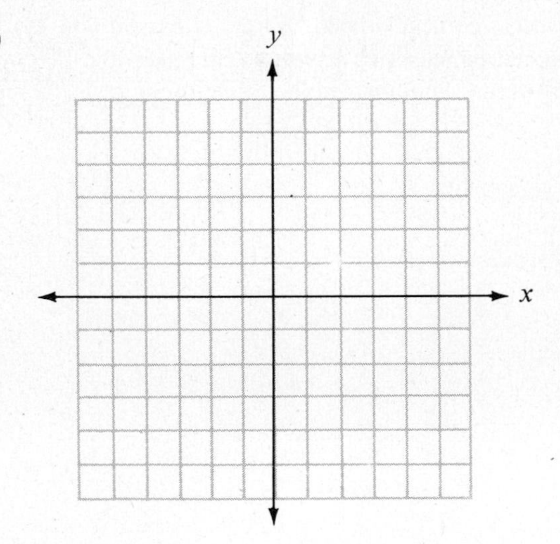

Solve each system by the substitution method.

________ **3.** $-4x + y = -3$
$7x - 2y = 4$ — 11.2 | 3

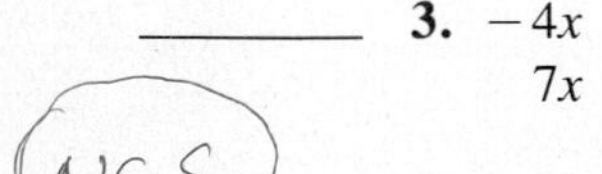

________ **4.** $x = 3y - 1$
$2x - 6y = 3$ — 11.2 | 4

Solve each system by the elimination method.

________ **5.** $6x + 5y = -7$
$-2x + 3y = -7$ — 11.3 | 2

ACS DEPENDANT

________ **6.** $3x - 3y = 6$
$5y = 5x - 10$ — 11.3 | 6

Solve each system by any method you choose.

______	**7.** $x = 3y - 2$ $3x - 5y = 10$	11.2 11.3	2 2
______	**8.** $4x - 3y = -1$ $-5x + 2y = -4$	11.3	3
______	**9.** $6x - 7y = 8$ $7x - 6y = 5$	11.3	3
______	**10.** $8x + 5y = 2$ $3x - 7y = 1$	11.3	3

Use a system of equations to solve each problem.

______	**11.** When Ben mows the Smith's yard for 3 hours and digs the Harmons' garden for 2 hours, he earns $18. When he mows for 5 hours and digs for 4 hours, he earns $33. What is Ben's hourly wage at each job?	11.4	1, 2
______	**12.** The perimeter of Martin's corn field is 320 yards. If Martin doubles the width of the field and increases its length by 30 yards, the perimeter will be 500 yards. Find the original dimensions of the corn field.	11.4	4

Check your answers in the back of your book.

Name ______________________ Date ____________

SHARPENING YOUR SKILLS after Chapters 1–11

		SECTION
______	**1.** Compare -2.39 and -2.184 using $<$ or $>$.	2.1
______	**2.** Evaluate $\frac{(m - 7)^2}{2n}$ when $m = 3$ and $n = 5$.	5.1
______	**3.** Find the restrictions on the variable for $\frac{y - 3}{2y^2 - y - 15}$.	8.1
______	**4.** Write an inequality that says $\frac{-2}{3}$ is between -1 and $\frac{-1}{5}$.	9.1

Simplify.

______	**5.** $\lvert -9 - 3\rvert - \lvert -15\rvert$	2.1
______	**6.** $\left(\frac{-2}{3}\right)^2 \div \frac{45}{14}$	2.4
______	**7.** $-4[2a - (5a - 4)] + 3[(2a - 3) + 7]$	3.3
______	**8.** $(0.8x^4y^3z^2)(-5xy^2z^3)$	6.2
______	**9.** $2a(a - 5)(a^2 + 2)$	6.3
______	**10.** $\frac{3x - 6}{x - 6} \cdot \frac{2}{x - 2}$	8.2
______	**11.** $\frac{3x}{4x - 12} - \frac{2x}{5x - 15}$	8.4

Factor.

______	**12.** $81x^4 - 16$	7.2
______	**13.** $7x^3y - 175xy^3$	7.2
______	**14.** $2ax + 2bx - 3ay - 3by$	7.1
______	**15.** $x^2y^2 - 8xyz - 9z^2$	7.3

Solve.

______	**16.** $5 = 3(x - 2) - 5(x + 1)$	4.2
______	**17.** $3x^3 - 27x^2 = 0$	7.5
______	**18.** $4y(3y - 4) = 5y(y + 1)$	7.5

________ 19. $\frac{18}{7x} = \frac{3}{x + 7}$ 8.6

________ 20. $5y - 6 < 11y - 6$ 9.2

________ 21. $\frac{x}{2} - 4 > -1$ and $6 - \frac{x}{3} \leq 5$ 9.3

22. Graph $x - 2y = 2$ by the intercept method. Label the intercepts. 10.3

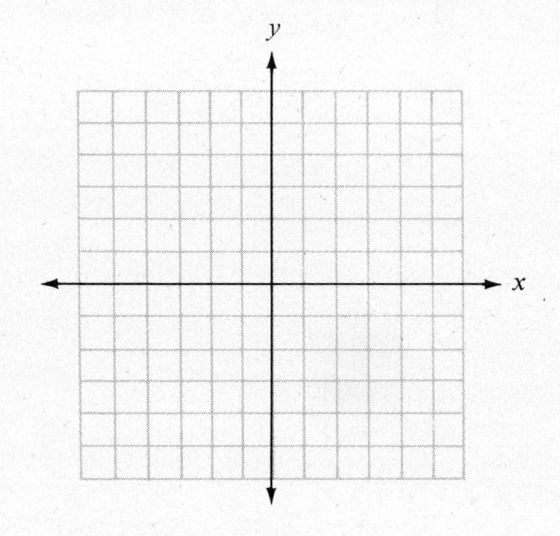

________ 23. Find the slope of the line through $(-3, 2)$ and $(2, 5)$. 10.4

________ 24. A plumber charges a flat rate of \$25 per visit plus \$15 per hour for labor. If materials cost \$58 and Ms. Carrell's total bill was \$128, how many hours did the plumber work? 4.3

________ 25. The number of rows of seats in the Little Theater is 5 more than the number of seats per row. If there are 300 seats in the theater, how many rows are there? 7.6

Check your answers in the back of your book.

Working with Square Roots 12

A ladder that is 20 feet long is leaning against the wall of a building. If the foot of the ladder is 10 feet from the bottom of the wall, how far up the wall is the top of the ladder?

It appears that this problem might be solved using the Pythagorean Theorem, but the solution is not in the set of numbers that we have been using. After working in this chapter you will be able to find the solution.

In earlier chapters we learned to recognize numbers that were perfect squares of rational numbers. Now we turn our attention to numbers that are squares of numbers that are not rational. In this chapter we

1. Define and simplify principal square roots.
2. Add and subtract radical expressions.
3. Multiply and divide radical expressions.
4. Rationalize denominators.
5. Solve quadratic equations with irrational solutions.
6. Switch from words to quadratic equations.

12.1 Working with Radical Expressions

We have talked often about *squares* of numbers and variables in our work with polynomials and factoring.

Finding Square Roots

In determining whether a certain number was a square, we searched for a number that could be multiplied times itself to give the original number. Such a number is called the **square root** of the original number. We may say that

$$3 \text{ is a square root of } 9 \quad \text{because} \quad 3^2 = 9$$

$$-3 \text{ is a square root of } 9 \quad \text{because} \quad (-3)^2 = 9$$

$$4 \text{ is a square root of } 16 \quad \text{because} \quad 4^2 = 16$$

$$-4 \text{ is a square root of } 16 \quad \text{because} \quad (-4)^2 = 16$$

$$1 \text{ is a square root of } 1 \quad \text{because} \quad 1^2 = 1$$

$$-1 \text{ is a square root of } 1 \quad \text{because} \quad (-1)^2 = 1$$

To avoid confusion, mathematicians have agreed that

> The **principal square root** of a positive number is always the *positive* square root.

From now on, we shall use the phrase *square root* to mean the *principal* (positive) square root. Mathematicians have invented a symbol to represent the principal square root of a number. That symbol is called a **radical** and it looks like this: $\sqrt{}$. The quantity under the radical is called the **radicand**.

The expression $\sqrt{9}$ stands for the principal square root of 9. We may write

$$\sqrt{9} = 3 \quad \text{because} \quad 3^2 = 9$$

$$\sqrt{25} = 5 \quad \text{because} \quad 5^2 = 25$$

$$\sqrt{81} = 9 \quad \text{because} \quad 9^2 = 81$$

$$\sqrt{1} = 1 \quad \text{because} \quad 1^2 = 1$$

$$\sqrt{0} = 0 \quad \text{because} \quad 0^2 = 0$$

In general, we agree that if b is not negative, then

$$\boxed{\sqrt{a} = b \quad \text{means} \quad b^2 = a}$$

In this radical statement, a is called the *radicand* and b is called the *root*.

What happens if we try to find the square root of a *negative* number? What is the square root of -9? We must find a number that will multiply times itself to give -9. Will 3 work? No, because $3 \cdot 3 = 9$. Will -3 work? No, because $(-3)(-3) = 9$. Indeed, there is no possible real number square root for -9. As a matter of fact, *it is impossible to find the square root of any negative number among the real numbers.*

The square root of a negative number is *not* a real number.

Let's practice finding some square roots. You complete Example 1.

Example 1. Find $\sqrt{36}$.

Solution

$$\sqrt{36} = \_\_\_\_$$

because $(\_\_\_\_)^2 = 36$.

Check your work on page 691. ▶

Example 2. Find $\sqrt{\frac{4}{9}}$.

Solution

$$\sqrt{\frac{4}{9}} = \frac{2}{3}$$

because $\left(\frac{2}{3}\right)^2 = \frac{2^2}{3^2} = \frac{4}{9}$

Now complete Example 3.

Example 3. Find $\sqrt{1}$.

Solution

$$\sqrt{1} = \_\_\_\_$$

because $(\_\_\_\_)^2 = 1$.

Check your work on page 691. ▶

Example 4. Find $\sqrt{-16}$.

Solution: We can find no real number whose square is -16.

It is worthwhile to note what happens when we square a square root.

Example 5. Find $(\sqrt{9})^2$.

Solution

$$(\sqrt{9})^2 = (\sqrt{9})(\sqrt{9})$$
$$= (3)(3)$$
$$(\sqrt{9})^2 = 9$$

You complete Example 6.

Example 6. Find $(\sqrt{36})^2$.

Solution

$$(\sqrt{36})^2 = (\sqrt{36})(\sqrt{36})$$
$$= (\_\_\_\_)(\_\_\_\_)$$
$$(\sqrt{36})^2 = \_\_\_\_$$

Check your work on page 691. ▶

From these examples, we agree that

$$(\sqrt{a})^2 = (\sqrt{a})(\sqrt{a}) = a \quad \text{provided that } a \text{ is not negative}$$

This rule allows us to compute squares of square roots without finding the square root first.

Example 7. Find $(\sqrt{17})^2$.

Solution

$$(\sqrt{17})^2 = 17$$

You complete Example 8.

Example 8. Find $(\sqrt{53})^2$.

Solution

$$(\sqrt{53})^2 = \_\_\_\_$$

Check your work on page 691. ▶

To express the square root of a variable expression, it is absolutely necessary that the variable expression represent a positive quantity or zero. Why? Because we have no way to find the square root of a negative quantity.

> Throughout this chapter, we assume that variable expressions under square root radicals do *not* represent negative quantities.

Once we have agreed on this very important condition, we may simplify some radicals containing variables.

Example 9. Find $\sqrt{x^2}$, $\sqrt{y^4}$, and $\sqrt{\dfrac{a^2}{4}}$.

Solution

$\sqrt{x^2} = x$ because $(x)^2 = x^2$

$\sqrt{y^4} = y^2$ because $(y^2)^2 = y^4$

$\sqrt{\dfrac{a^2}{4}} = \dfrac{a}{2}$ because $\left(\dfrac{a}{2}\right)^2 = \dfrac{a^2}{4}$

Now complete Example 10.

Example 10. Find $(\sqrt{x^2})^2$, $(\sqrt{y})^2$, and $(\sqrt{5x})^2$.

Solution

$(\sqrt{x^2})^2 = x^2$

$(\sqrt{y})^2 =$ ____

$(\sqrt{5x})^2 =$ ____

Check your work on page 691. ▶

Trial Run

Simplify the radical expressions.

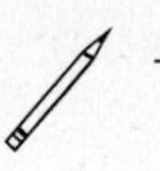

_______ 1. $\sqrt{64}$

_______ 2. $\sqrt{100}$

_______ 3. $(\sqrt{5})^2$

_______ 4. $\sqrt{y^2}$

_______ 5. $\sqrt{\dfrac{9}{x^2}}$

_______ 6. $(\sqrt{2y})^2$

Answers are on page 692.

So far we have practiced finding square roots that were **rational numbers**. Recall that rational numbers are numbers that can be expressed as fractions of *integers*.

$$\sqrt{9} = 3 \quad \text{is rational because} \quad 3 = \frac{3}{1}$$

$$\sqrt{\frac{4}{25}} = \frac{2}{5} \quad \text{is rational because} \quad \frac{2}{5} = \frac{2}{5}$$

We do not have to look very far in the set of natural numbers $\{1, 2, 3, 4, \ldots\}$ before we find a number whose square root is *not* rational. For example,

$$\sqrt{2} \text{ is not rational}$$

You could hunt forever and never find a rational number that can be multiplied times itself to give an answer of 2. This does *not* mean that $\sqrt{2}$ does not exist. It merely means that $\sqrt{2}$ is *not rational*; $\sqrt{2}$ cannot be expressed as a quotient of integers.

A number that is *not* rational is called an **irrational number**. Its value can be approximated, but it cannot be expressed as a fraction of integers. When we approximate irrational square roots, we use the symbol $\doteq$ to mean "is approximately equal to." (A table of square roots appears inside the back cover of your book.)

Irrational Number	Approximate Value	Check
$\sqrt{2}$	$\sqrt{2} \doteq 1.414$	$(1.414)^2 = 1.999396 \doteq 2$
$\sqrt{3}$	$\sqrt{3} \doteq 1.732$	$(1.732)^2 = 2.999824 \doteq 3$
$\sqrt{5}$	$\sqrt{5} \doteq 2.236$	$(2.236)^2 = 4.999696 \doteq 5$

In order to represent *exact* values of square roots, we must leave them in radical form. Let's look at some square roots and decide whether they are rational or irrational numbers. We shall simplify those square roots that are rational.

Square Root	Simplest Form	Rational or Irrational?
$\sqrt{0}$	0	Rational
$\sqrt{1}$	1	Rational
$\sqrt{2}$	$\sqrt{2}$	Irrational
$\sqrt{3}$	$\sqrt{3}$	Irrational
$\sqrt{4}$	2	Rational
$\sqrt{7}$	$\sqrt{7}$	Irrational
$\sqrt{9}$	3	Rational
$\sqrt{19}$	$\sqrt{19}$	Irrational
$\sqrt{100}$	10	Rational

Trial Run

Tell whether the radical expressions are rational or irrational numbers.

_______ 1. $\sqrt{49}$

_______ 2. $\sqrt{13}$

_______ 3. $\sqrt{0}$

_______ 4. $\sqrt{\frac{9}{16}}$

Answers are on page 692.

Simplifying Radical Expressions

Let's consider a method for simplifying radical expressions such as $\sqrt{36}$. We know that

$$\sqrt{36} = 6 \qquad \text{and} \qquad \sqrt{4} \cdot \sqrt{9} = 2 \cdot 3 = 6$$

so here we conclude that

$$\sqrt{36} = \sqrt{4} \cdot \sqrt{9}$$

Do you suppose that the square root of a product is *always* equal to the product of the square roots of the factors? Look at another example.

$$\sqrt{144} = 12 \qquad \text{and} \qquad \sqrt{9} \cdot \sqrt{16} = 3 \cdot 4 = 12$$

so

$$\sqrt{144} = \sqrt{9} \cdot \sqrt{16}$$

Although we have verified this rule for only two examples, it is indeed true that

> The square root of a product of factors is equal to the product of the square roots of the factors.
>
> $$\sqrt{a \cdot b} = \sqrt{a} \cdot \sqrt{b} \qquad \text{provided that } a \text{ and } b \text{ are not negative}$$

We make use of this rule in simplifying radical expressions.

Example 11. Simplify $\sqrt{4x^2}$.

Solution

$$\sqrt{4x^2} = \sqrt{4 \cdot x^2} = \sqrt{4} \cdot \sqrt{x^2} = 2 \cdot x = 2x$$

You complete Example 12.

Example 12. Simplify $\sqrt{9a^2b^4}$.

Solution

$\sqrt{9a^2b^4}$

$= \sqrt{9 \cdot a^2 \cdot b^2 \cdot b^2}$

$= \sqrt{\_\_} \cdot \sqrt{\_\_} \cdot \sqrt{\_\_} \cdot \sqrt{\_\_}$

$= \_\_\_\_$

Check your work on page 691. ▶

Sometimes a radical will not be removed completely in the simplification process. The best that we can do is to rewrite the radicand as a product of factors, some of which are perfect squares. The factors whose roots are *not* rational must then be left under the radical.

Example 13. Simplify $\sqrt{12}$.

Solution: Remember that we wish to write 12 as a product of factors, where at least one factor is a perfect square.

$$\sqrt{12} = \sqrt{4 \cdot 3}$$
$$= \sqrt{4} \cdot \sqrt{3}$$
$$= 2\sqrt{3}$$

This is simplest radical from.

You try Example 14.

Example 14. Simplify $\sqrt{98}$.

Solution: We wish to write ____ as a product of factors, where at least one factor is a ________ ________

$$\sqrt{98} = \sqrt{49 \cdot 2}$$
$$= \sqrt{\_\_\_\_} \cdot \sqrt{\_\_\_\_}$$
$$= \_\_\_\_ \sqrt{\_\_\_\_}$$

Check your work on page 692. ▶

Example 15. Simplify $\sqrt{50x^3}$.

Solution

$$\sqrt{50x^3}$$
$$= \sqrt{25 \cdot 2 \cdot x^2 \cdot x}$$
$$= \sqrt{25} \cdot \sqrt{2} \cdot \sqrt{x^2} \cdot \sqrt{x}$$
$$= 5\sqrt{2} \cdot x\sqrt{x}$$
$$= 5 \cdot x \cdot \sqrt{2} \cdot \sqrt{x}$$
$$= 5x\sqrt{2 \cdot x}$$
$$= 5x\sqrt{2x}$$

You try completing Example 16.

Example 16. Simplify $\sqrt{108ab^4}$.

Solution

$$\sqrt{108ab^4}$$
$$= \sqrt{36 \cdot 3 \cdot a \cdot b^2 \cdot b^2}$$
$$= \sqrt{\_\_\_\_} \cdot \sqrt{3} \cdot \sqrt{\_\_\_\_} \cdot \sqrt{\_\_\_\_} \cdot \sqrt{\_\_\_\_}$$
$$= 6 \cdot \sqrt{3} \cdot \sqrt{a} \cdot b \cdot b$$
$$= 6 \cdot b \cdot b \cdot \sqrt{3} \cdot \sqrt{a}$$
$$= 6 \_\_\_\_ \sqrt{\_\_\_\_ \cdot \_\_\_\_}$$
$$= \_\_\_\_ \sqrt{\_\_\_\_}$$

Check your work on page 692. ▶

A square root radical is said to be in **simplest form** when the radicand contains no perfect square factors. If the exponent on a factor in the radicand is greater than 1, then some simplification of the square root is necessary.

Let's look at a radical in which the radicand is a fraction and try to simplify it.

$$\sqrt{\frac{4}{9}} = \frac{2}{3} \quad \text{and} \quad \frac{\sqrt{4}}{\sqrt{9}} = \frac{2}{3}$$

Therefore, it seems that

$$\sqrt{\frac{4}{9}} = \frac{\sqrt{4}}{\sqrt{9}}$$

Indeed, it is always possible to state that:

> The square root of a quotient is equal to the quotient of the square roots.
>
> $$\sqrt{\frac{a}{b}} = \frac{\sqrt{a}}{\sqrt{b}}$$ provided that a is not negative and b is positive.

This rule gives us another tool for simplifying radicals.

Try to complete Example 17.

Example 17. Simplify $\sqrt{\dfrac{25x^2}{49y^2}}$.

Solution

$$\sqrt{\frac{25x^2}{49y^2}} = \frac{\sqrt{25x^2}}{\sqrt{49y^2}}$$

$$= \frac{\sqrt{25} \cdot \sqrt{x^2}}{\sqrt{49} \cdot \sqrt{y^2}}$$

$$= \underline{\qquad}$$

Check your work on page 692. ▶

Example 18. Simplify $\sqrt{\dfrac{32x^3}{y^4}}$.

Solution

$$\sqrt{\frac{32x^3}{y^4}} = \frac{\sqrt{32x^3}}{\sqrt{y^4}}$$

$$= \frac{\sqrt{16 \cdot 2 \cdot x^2 \cdot x}}{\sqrt{y^2 \cdot y^2}}$$

$$= \frac{4x\sqrt{2x}}{y^2}$$

Our rule for quotients can also be used in reverse. That is, we may sometimes wish to make use of the fact that

$$\frac{\sqrt{a}}{\sqrt{b}} = \sqrt{\frac{a}{b}}$$

Suppose, for example, that we wish to simplify $\dfrac{\sqrt{50x^3}}{\sqrt{2x}}$.

$$\frac{\sqrt{50x^3}}{\sqrt{2x}} = \sqrt{\frac{50x^3}{2x}} \qquad \text{Rule for quotients of radicals.}$$

$$= \sqrt{25x^2} \qquad \text{Reduce the fraction as usual.}$$

$$= 5x \qquad \text{Simplify the radical.}$$

In general, we note that a quotient of two radicals should be written as a single radical whenever the numerator and denominator contain common factors. Once the numerator and denominator have no more common factors, the fraction should again be considered as the quotient of two separate radicals.

Example 19. Simplify $\dfrac{\sqrt{20x^3y^4}}{\sqrt{125xy}}$.

Solution: Since the numerator and denominator contain common factors, we write

$$\frac{\sqrt{20x^3y^4}}{\sqrt{125xy}}$$

$$= \sqrt{\frac{20x^3y^4}{125xy}} \qquad \text{Rewrite as one radical.}$$

$$= \sqrt{\frac{4x^2y^3}{25}} \qquad \text{Reduce the fraction.}$$

$$= \frac{\sqrt{4x^2y^3}}{\sqrt{25}} \qquad \text{Rewrite as separate radicals.}$$

$$= \frac{\sqrt{4 \cdot x^2 \cdot y^2 \cdot y}}{\sqrt{25}} \qquad \text{Look for perfect square factors.}$$

$$= \frac{2xy\sqrt{y}}{5} \qquad \text{Remove square roots of perfect square factors.}$$

Now you try Example 20.

Example 20. Simplify $\dfrac{\sqrt{18a^7c}}{\sqrt{6a^6c}}$.

Solution: Since the numerator and denominator contain common factors, we write

$$\frac{\sqrt{18a^7c}}{\sqrt{6a^6c}} = \sqrt{\frac{18a^7c}{6a^6c}}$$

$$= \sqrt{\underline{\qquad}}$$

Check your work on page 692. ▶

Trial Run

Write each radical in simplest form.

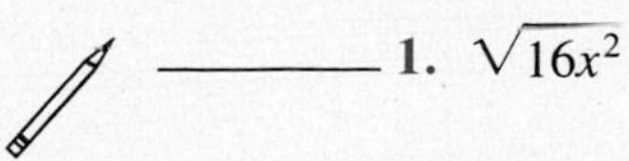

_______ 1. $\sqrt{16x^2}$

_______ 2. $\sqrt{18}$

_______ 3. $\sqrt{27a^3}$

_______ 4. $\sqrt{20x^3y^4}$

_______ 5. $\sqrt{\dfrac{9x^2}{64y^2}}$

_______ 6. $\sqrt{\dfrac{50x^3}{y^4}}$

_______ 7. $\dfrac{\sqrt{2x^5y}}{\sqrt{98xy}}$

_______ 8. $\dfrac{\sqrt{169a^4b^2}}{\sqrt{36c^2}}$

Answers are on page 692.

Examples You Completed

Example 1. Find $\sqrt{36}$.

Solution

$$\sqrt{36} = 6$$

because $(6)^2 = 36$.

Example 3. Find $\sqrt{1}$.

Solution

$$\sqrt{1} = 1$$

because $(1)^2 = 1$.

Example 6. Find $(\sqrt{36})^2$.

Solution

$$(\sqrt{36})^2 = (\sqrt{36})(\sqrt{36})$$
$$= (6)(6)$$
$$(\sqrt{36})^2 = 36$$

Example 8. Find $(\sqrt{53})^2$.

Solution

$$(\sqrt{53})^2 = 53$$

Example 10. Find $(\sqrt{x^2})^2$, $(\sqrt{y})^2$, and $(\sqrt{5x})^2$.

Solution

$$(\sqrt{x^2})^2 = x^2$$
$$(\sqrt{y})^2 = y$$
$$(\sqrt{5x})^2 = 5x$$

Example 12. Simplify $\sqrt{9a^2b^4}$.

Solution

$$\sqrt{9a^2b^4} = \sqrt{9 \cdot a^2 \cdot b^2 \cdot b^2}$$
$$= \sqrt{9} \cdot \sqrt{a^2} \cdot \sqrt{b^2} \cdot \sqrt{b^2}$$
$$= 3 \cdot a \cdot b \cdot b$$
$$= 3ab^2$$

Example 14. Simplify $\sqrt{98}$.

Solution: We wish to write 98 as a product of factors, where at least one factor is a perfect square.

$$\begin{aligned}\sqrt{98} &= \sqrt{49 \cdot 2}\\ &= \sqrt{49} \cdot \sqrt{2}\\ &= 7\sqrt{2}\end{aligned}$$

Example 16. Simplify $\sqrt{108ab^4}$.

Solution

$$\begin{aligned}\sqrt{108ab^4} &= \sqrt{36 \cdot 3 \cdot a \cdot b^2 \cdot b^2}\\ &= \sqrt{36} \cdot \sqrt{3} \cdot \sqrt{a} \cdot \sqrt{b^2} \cdot \sqrt{b^2}\\ &= 6 \cdot \sqrt{3} \cdot \sqrt{a} \cdot b \cdot b\\ &= 6b^2 \sqrt{3 \cdot a}\\ &= 6b^2\sqrt{3a}\end{aligned}$$

Example 17. Simplify $\sqrt{\dfrac{25x^2}{49y^2}}$.

Solution

$$\begin{aligned}\sqrt{\frac{25x^2}{49y^2}} &= \frac{\sqrt{25x^2}}{\sqrt{49y^2}}\\ &= \frac{\sqrt{25} \cdot \sqrt{x^2}}{\sqrt{49} \cdot \sqrt{y^2}}\\ &= \frac{5x}{7y}\end{aligned}$$

Example 20. Simplify $\dfrac{\sqrt{18a^7c}}{\sqrt{6a^6c}}$.

Solution: Since the numerator and denominator contain common factors, we write

$$\begin{aligned}\frac{\sqrt{18a^7c}}{\sqrt{6a^6c}} &= \sqrt{\frac{18a^7c}{6a^6c}}\\ &= \sqrt{3a}\end{aligned}$$

Answers to Trial Runs

page 686 **1.** 8 **2.** 10 **3.** 5 **4.** y **5.** $\frac{3}{x}$ **6.** $2y$

page 687 **1.** Rational **2.** Irrational **3.** Rational **4.** Rational

page 691 **1.** $4x$ **2.** $3\sqrt{2}$ **3.** $3a\sqrt{3a}$ **4.** $2xy^2\sqrt{5x}$ **5.** $\frac{3x}{8y}$ **6.** $\frac{5x\sqrt{2x}}{y^2}$ **7.** $\frac{x^2}{7}$ **8.** $\frac{13a^2b}{6c}$

Name ______________________________ Date ______________

EXERCISE SET 12.1

Decide if the following expressions represent rational or irrational numbers.

_______ 1. $\sqrt{25}$ _______ 2. $\sqrt{49}$

_______ 3. $\sqrt{5}$ _______ 4. $\sqrt{3}$

_______ 5. $\sqrt{1}$ _______ 6. $\sqrt{0}$

_______ 7. $\sqrt{8}$ _______ 8. $\sqrt{12}$

_______ 9. $\sqrt{\frac{25}{36}}$ _______ 10. $\sqrt{\frac{1}{4}}$

_______ 11. $\sqrt{65}$ _______ 12. $\sqrt{35}$

_______ 13. $\sqrt{\frac{2}{3}}$ _______ 14. $\sqrt{\frac{5}{7}}$

_______ 15. $\sqrt{225}$ _______ 16. $\sqrt{196}$

Simplify the following radical expressions.

_______ 17. $\sqrt{144}$ _______ 18. $\sqrt{169}$

_______ 19. $\sqrt{a^4}$ _______ 20. $\sqrt{a^6}$

_______ 21. $(\sqrt{8})^2$ _______ 22. $(\sqrt{10})^2$

_______ 23. $(\sqrt{x^5})^2$ _______ 24. $(\sqrt{a^3})^2$

_______ 25. $\sqrt{25b^2}$ _______ 26. $\sqrt{64c^2}$

_______ 27. $\sqrt{\frac{x^2}{36}}$ _______ 28. $\sqrt{\frac{y^2}{49}}$

_______ 29. $\sqrt{45}$ _______ 30. $\sqrt{40}$

_______ 31. $\sqrt{600}$ _______ 32. $\sqrt{128}$

_______ 33. $\sqrt{75}$ _______ 34. $\sqrt{108}$

_______ 35. $\sqrt{44}$ _______ 36. $\sqrt{80}$

_______ 37. $\sqrt{x^2}$ _______ 38. $\sqrt{x^4}$

_______ 39. $\sqrt{y^5}$ _______ 40. $\sqrt{y^3}$

_______ 41. $\sqrt{x^2y^3}$ _______ 42. $\sqrt{x^3y^2}$

_______ 43. $\sqrt{32a^3}$ _______ 44. $\sqrt{18b^3}$

_______ **45.** $\sqrt{4x^2y}$

_______ **46.** $\sqrt{9xy^2}$

_______ **47.** $\sqrt{81a^5b^4}$

_______ **48.** $\sqrt{100a^3b^6}$

_______ **49.** $\sqrt{192x^2y^4z^3}$

_______ **50.** $\sqrt{243x^4y^5z^2}$

_______ **51.** $\sqrt{180a^7b^3c}$

_______ **52.** $\sqrt{98a^5b^4c}$

_______ **53.** $\sqrt{\dfrac{4x^2}{25}}$

_______ **54.** $\sqrt{\dfrac{9x^2}{121}}$

_______ **55.** $\sqrt{\dfrac{32a^3}{b^4}}$

_______ **56.** $\sqrt{\dfrac{75a}{b^2}}$

_______ **57.** $\sqrt{\dfrac{40x^3}{9y^2}}$

_______ **58.** $\sqrt{\dfrac{27x^5}{4y^2}}$

_______ **59.** $\sqrt{\dfrac{20a^3b^3}{500ab}}$

_______ **60.** $\sqrt{\dfrac{35a^5b^5}{28a^3b^3}}$

_______ **61.** $\dfrac{\sqrt{75x^3}}{\sqrt{3x}}$

_______ **62.** $\dfrac{\sqrt{80x^5}}{\sqrt{5x}}$

_______ **63.** $\dfrac{\sqrt{15x^7}}{\sqrt{5x^3}}$

_______ **64.** $\dfrac{\sqrt{56x^5}}{\sqrt{7x^4}}$

_______ **65.** $\dfrac{\sqrt{3xy^3}}{\sqrt{363xy^2}}$

_______ **66.** $\dfrac{\sqrt{5xy^2}}{\sqrt{720y}}$

☆ Stretching the Topics

Simplify.

_______ **1.** $\sqrt{\dfrac{24x^{10}y^2}{y^6z^{28}}}$

_______ **2.** $-7xy^4\sqrt{63x^2y^7}$

_______ **3.** $\sqrt{0.16x^4y^2(a - b)^2}$

Check your answers in the back of your book.

If you can simplify the radical expressions in **Checkup 12.1**, you are ready to go on to Section 12.2.

Name ______________________ Date __________

✓ CHECKUP 12.1

Simplify the radical expressions.

______ 1. $\sqrt{49}$

______ 2. $\sqrt{9a^2}$

______ 3. $(\sqrt{12})^2$

______ 4. $\sqrt{80}$

______ 5. $\sqrt{75y^4}$

______ 6. $\sqrt{45a^3b^2}$

______ 7. $\sqrt{\dfrac{25x^2}{36y^2}}$

______ 8. $\sqrt{\dfrac{32x^5}{y^6}}$

______ 9. $\dfrac{\sqrt{50a^5b^3}}{\sqrt{2ab}}$

______ 10. $\dfrac{\sqrt{15a^5}}{\sqrt{5a}}$

Check your answers in the back of your book.

If You Missed Problems:	You Should Review Examples:
1	1–3
2	9
3	5–8
4–6	13–16
7, 8	17, 18
9, 10	19, 20

12.2 Operating with Radical Expressions

So far, we have spent time recognizing radical expressions and learning how to simplify them. As with any kind of algebraic expressions, we must now discuss the basic operations of addition, subtraction, and multiplication. We discuss division in Section 12.3.

Adding and Subtractng Radical Expressions

First we turn our attention to learning how to compute sums and differences of radical expressions such as $2\sqrt{3} + 5\sqrt{3}$. From our knowledge of combining like terms, it seems logical to conclude that the sum will be

$$2\sqrt{3} + 5\sqrt{3} = 7\sqrt{3}$$

The distributive property can be used to verify that this result is reasonable.

$$\begin{aligned} 2\sqrt{3} + 5\sqrt{3} &= (2 + 5)\sqrt{3} \\ &= 7\sqrt{3} \end{aligned}$$

The key fact to realize here is that the *radicals matched exactly*. As in our work with combining like terms, the rule for combining radical expressions demands that the radicals be *exactly alike*. They must both be square roots of the same radicand.

Combining Like Radicals. To combine **like radicals**, we combine the numerical coefficients and keep the same radical part.

Example 1. Simplify $17\sqrt{x} - 12\sqrt{x} + \sqrt{x}$.

Solution

$$\begin{aligned} &17\sqrt{x} - 12\sqrt{x} + \sqrt{x} \\ &= 17\sqrt{x} - 12\sqrt{x} + 1\sqrt{x} \\ &= 6\sqrt{x} \end{aligned}$$

You try Example 2.

Example 2. Simplify $3\sqrt{7} + 6\sqrt{7} - 10\sqrt{7}$.

Solution

$$\begin{aligned} &3\sqrt{7} + 6\sqrt{7} - 10\sqrt{7} \\ &= \_\_\_\_ \sqrt{7} \end{aligned}$$

Check your work on page 702. ▶

Let's try to simplify $\sqrt{18a} - \sqrt{8a} + \sqrt{2a}$. Notice that the radicals are all square roots, but the radicands are all *different*. Before we decide that these radicals cannot be combined, we must try to *simplify* each radical completely.

$$\begin{aligned} &\sqrt{18a} - \sqrt{8a} + \sqrt{2a} \\ &= \sqrt{9 \cdot 2a} - \sqrt{4 \cdot 2a} + \sqrt{2a} \\ &= 3\sqrt{2a} - 2\sqrt{2a} + 1\sqrt{2a} \\ &= 2\sqrt{2a} \end{aligned}$$

Try combining the radicals in Example 3.

Example 3. Simplify $\sqrt{50x} - 6\sqrt{32x}$.

Solution

$$\sqrt{50x} - 6\sqrt{32x}$$
$$= \sqrt{25 \cdot 2x} - 6\sqrt{16 \cdot 2x}$$
$$= 5\sqrt{2x} - 6 \cdot 4\sqrt{2x}$$

Check your work on page 702. ▶

Example 4. Simplify $6 + \sqrt{27} + \sqrt{48}$.

Solution

$$6 + \sqrt{27} + \sqrt{48}$$
$$= 6 + \sqrt{9 \cdot 3} + \sqrt{16 \cdot 3}$$
$$= 6 + 3\sqrt{3} + 4\sqrt{3}$$
$$= 6 + 7\sqrt{3}$$

Notice that these last two terms cannot be combined.

Trial Run

Simplify.

_______ 1. $2\sqrt{3} + 5\sqrt{3} - 9\sqrt{3}$

_______ 2. $8\sqrt{x} - 7\sqrt{x} + \sqrt{x}$

_______ 3. $\sqrt{32a} - 5\sqrt{18a}$

_______ 4. $11 + \sqrt{75} - \sqrt{12}$

_______ 5. $\dfrac{\sqrt{8}}{\sqrt{2}} + \sqrt{50} - \sqrt{72}$

Answers are on page 703.

Multiplying Radical Expressions

The methods needed for multiplying radical expressions are similar to those used for multiplying polynomials. As you remember, multiplying polynomials required use of the associative and distributive properties and the rules for combining like terms.

We discussed multiplying radicals in our work with simplifying radicals. At that time we agreed that

$$\sqrt{a} \cdot \sqrt{b} = \sqrt{a \cdot b} \quad \text{provided that } a \text{ and } b \text{ are not negative}$$

See if you can complete Example 5.

Example 5. Multiply $\sqrt{2} \cdot \sqrt{5} \cdot \sqrt{10}$.

Solution

$$\begin{aligned} &\sqrt{2} \cdot \sqrt{5} \cdot \sqrt{10} \\ &= \sqrt{2 \cdot 5 \cdot 10} \\ &= \sqrt{\_\_\_\_} \\ &= \_\_\_\_ \end{aligned}$$

Check your work on page 703. ▶

Example 6. Multiply $\sqrt{3a} \cdot \sqrt{3a}$.

Solution

$$\begin{aligned} &\sqrt{3a} \cdot \sqrt{3a} \\ &= (\sqrt{3a})^2 \\ &= 3a \end{aligned}$$

Example 6 illustrates a fact that we have used before and that will be even more useful to us shortly.

$\sqrt{a} \cdot \sqrt{a} = (\sqrt{a})^2 = a$ provided that a is not negative

Example 7. Multiply

$$\sqrt{2x - 1} \cdot \sqrt{2x - 1}$$

Solution

$$\begin{aligned} &\sqrt{2x - 1} \cdot \sqrt{2x - 1} \\ &= (\sqrt{2x - 1})^2 \\ &= 2x - 1 \end{aligned}$$

Now you complete Example 8.

Example 8. Multiply $\sqrt{y + 6} \cdot \sqrt{y + 6}$.

Solution

$$\sqrt{y + 6} \cdot \sqrt{y + 6}$$

Check your work on page 703. ▶

Example 9. Multiply $5\sqrt{2a} \cdot \sqrt{8a}$.

Solution

$$\begin{aligned} &5\sqrt{2a} \cdot \sqrt{8a} \\ &= 5\sqrt{2a \cdot 8a} \\ &= 5\sqrt{16a^2} \\ &= 5 \cdot 4a \\ &= 20a \end{aligned}$$

You complete Example 10.

Example 10. Multiply $(-3\sqrt{x})(-6\sqrt{x})$.

Solution

$$\begin{aligned} &(-3\sqrt{x})(-6\sqrt{x}) \\ &= -3(-6)\sqrt{x} \cdot \sqrt{x} \end{aligned}$$

Check your work on page 703. ▶

To multiply a single radical times a sum or difference, we must use the *distributive property* and then simplify.

Example 11. Multiply $\sqrt{3}(5 + \sqrt{3})$.

Solution

$$\begin{aligned} &\sqrt{3}(5 + \sqrt{3}) \\ &= \sqrt{3} \cdot 5 + \sqrt{3} \cdot \sqrt{3} \\ &= 5\sqrt{3} + 3 \end{aligned}$$

Example 12. Multiply $2\sqrt{2}(\sqrt{3} + \sqrt{8})$.

Solution

$$\begin{aligned} &2\sqrt{2}(\sqrt{3} + \sqrt{8}) \\ &= 2\sqrt{2} \cdot \sqrt{3} + 2\sqrt{2} \cdot \sqrt{8} \\ &= 2\sqrt{2 \cdot 3} + 2\sqrt{2 \cdot 8} \\ &= 2\sqrt{6} + 2\sqrt{16} \\ &= 2\sqrt{6} + 2 \cdot 4 \\ &= 2\sqrt{6} + 8 \end{aligned}$$

Trial Run

Simplify.

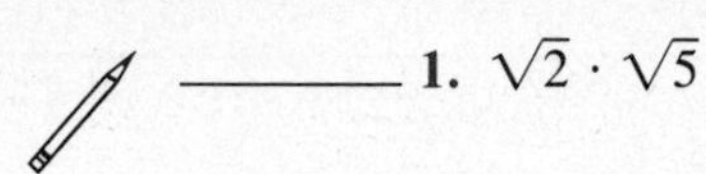

_______ 1. $\sqrt{2} \cdot \sqrt{5}$

_______ 2. $\sqrt{5 + a} \cdot \sqrt{5 + a}$

_______ 3. $4\sqrt{3a} \cdot \sqrt{12a}$

_______ 4. $(-2\sqrt{x})(5\sqrt{3x})$

_______ 5. $\sqrt{2}(3 - \sqrt{2})$

_______ 6. $3\sqrt{2}\,(\sqrt{5} + \sqrt{18})$

Answers are on page 703.

Let's spend a moment reviewing the multiplication of binomials to prepare ourselves for multiplying sums and differences of radicals. Recall how we found products of binomials using the FOIL method or formulas for special products.

FOIL Method	$(A + B)^2 = A^2 + 2AB + B^2$ $(A - B)^2 = A^2 - 2AB + B^2$	$(A + B)(A - B) = A^2 - B^2$
F, L, I, O $(x + 3)(x + 7)$	$(x + 5)^2$	$(x + 6)(x - 6)$
$= x^2 + 7x + 3x + 21$	$= x^2 + 2(5x) + 5^2$	$= x^2 - 6^2$
$= x^2 + 10x + 21$	$= x^2 + 10x + 25$	$= x^2 - 36$

We can use the same methods for multiplying sums and/or differences of radicals.

$$\overset{F}{}(2+\sqrt{3})(4+\sqrt{2}) = \overset{F}{2\cdot 4} + \overset{O}{2\cdot\sqrt{2}} + \overset{I}{\sqrt{3}\cdot 4} + \overset{L}{\sqrt{3}\cdot\sqrt{2}}$$

$$= 8 + 2\sqrt{2} + 4\sqrt{3} + \sqrt{6}$$

Here there were no like terms to combine.

Example 13. Find $(\sqrt{6}+1)^2$.

Solution

$$(\sqrt{6}+1)^2$$
$$= (\sqrt{6})^2 + 2\sqrt{6} + 1^2$$
$$= 6 + 2\sqrt{6} + 1$$
$$= 7 + 2\sqrt{6}$$

Now you try Example 14.

Example 14. Find $(8-\sqrt{y})^2$.

Solution

$$(8-\sqrt{y})^2$$
$$= 8^2 - 2\cdot 8\cdot\sqrt{y} + (\sqrt{y})^2$$
$$= \_\_\_\_ - \_\_\_\_\sqrt{y} + \_\_\_\_$$

Check your work on page 703. ▶

Example 15. Multiply

$$(5+\sqrt{x})\cdot(3+4\sqrt{x})$$

Solution

$$(5+\sqrt{x})(3+4\sqrt{x})$$
$$= 5\cdot 3 + 5\cdot 4\sqrt{x} + 3\sqrt{x} + 4\sqrt{x}\cdot\sqrt{x}$$
$$= 15 + 20\sqrt{x} + 3\sqrt{x} + 4x$$
$$= 15 + 23\sqrt{x} + 4x$$

Example 16. Multiply

$$(6+\sqrt{17})\cdot(6-\sqrt{17})$$

Solution

$$(6+\sqrt{17})(6-\sqrt{17})$$
$$= 6^2 - (\sqrt{17})^2$$
$$= 36 - 17$$
$$= 19$$

Example 16 is especially interesting because we multiplied two irrational quantities and our product was a *rational* number. There were no radicals left after we combined like terms. Notice that the original quantities were the *sum and difference of the same two terms*. Irrational numbers such as $6+\sqrt{17}$ and $6-\sqrt{17}$ are sometimes called **real conjugates** of each other. The product of an irrational number and its real conjugate is always a *rational* number. This fact will be useful to us in Section 12.3.

Example 17. Find the missing irrational numbers.

Irrational Number	Real Conjugate
$1+\sqrt{2}$	______
$3-\sqrt{5}$	______
______	$\sqrt{7}-1$
______	$2\sqrt{x}+\sqrt{13}$

Check your work on page 703. ▶

Example 18. Multiply

$$(\sqrt{5} + 2\sqrt{7}) \cdot (\sqrt{5} - 2\sqrt{7})$$

Solution

$$(\sqrt{5} + 2\sqrt{7})(\sqrt{5} - 2\sqrt{7})$$
$$= (\sqrt{5})^2 - (2\sqrt{7})^2$$
$$= 5 - 4(7)$$
$$= 5 - 28$$
$$= -23$$

You try Example 19.

Example 19. Multiply

$$(\sqrt{y} + 3) \cdot (\sqrt{y} - 3)$$

Solution

Check your work on page 703. ▶

Trial Run

Multiply

______ **1.** $(1 + \sqrt{2})(2 - \sqrt{3})$

______ **2.** $(5 - 2\sqrt{5})(2 + \sqrt{5})$

______ **3.** $(2 - 3\sqrt{7})(2 + 3\sqrt{7})$

______ **4.** $(\sqrt{x} + 3)(\sqrt{x} - 3)$

______ **5.** $(\sqrt{x} + 2)(\sqrt{x} - 4)$

______ **6.** $(\sqrt{x} - 5)^2$

Answers are on page 703.

▶ Examples You Completed

Example 2. Simplify $3\sqrt{7} + 6\sqrt{7} - 10\sqrt{7}$.

Solution

$$3\sqrt{7} + 6\sqrt{7} - 10\sqrt{7}$$
$$= (3 + 6 - 10)\sqrt{7}$$
$$= -1\sqrt{7}$$
$$= -\sqrt{7}$$

Example 3. Simplify $\sqrt{50x} - 6\sqrt{32x}$.

Solution

$$\sqrt{50x} - 6\sqrt{32x}$$
$$= \sqrt{25 \cdot 2x} - 6\sqrt{16 \cdot 2x}$$
$$= 5\sqrt{2x} - 6 \cdot 4\sqrt{2x}$$
$$= 5\sqrt{2x} - 24\sqrt{2x}$$
$$= -19\sqrt{2x}$$

Example 5. Multiply $\sqrt{2} \cdot \sqrt{5} \cdot \sqrt{10}$.

Solution

$$\begin{aligned} &\sqrt{2} \cdot \sqrt{5} \cdot \sqrt{10} \\ &= \sqrt{2 \cdot 5 \cdot 10} \\ &= \sqrt{100} \\ &= 10 \end{aligned}$$

Example 8. Multiply $\sqrt{y+6} \cdot \sqrt{y+6}$.

Solution

$$\begin{aligned} &\sqrt{y+6} \cdot \sqrt{y+6} \\ &= (\sqrt{y+6})^2 \\ &= y + 6 \end{aligned}$$

Example 10. Multiply $(-3\sqrt{x})(-6\sqrt{x})$.

Solution

$$\begin{aligned} &(-3\sqrt{x})(-6\sqrt{x}) \\ &= -3(-6)\sqrt{x} \cdot \sqrt{x} \\ &= 18(\sqrt{x})^2 \\ &= 18x \end{aligned}$$

Example 14. Find $(8 - \sqrt{y})^2$.

Solution

$$\begin{aligned} &(8 - \sqrt{y})^2 \\ &= 8^2 - 2 \cdot 8\sqrt{y} + (\sqrt{y})^2 \\ &= 64 - 16\sqrt{y} + y \end{aligned}$$

Example 17. Find the missing irrational numbers.

Irrational Number	Real Conjugate
$1 + \sqrt{2}$	$1 - \sqrt{2}$
$3 - \sqrt{5}$	$3 + \sqrt{5}$
$\sqrt{7} + 1$	$\sqrt{7} - 1$
$2\sqrt{x} - \sqrt{13}$	$2\sqrt{x} + \sqrt{13}$

Example 19. Multiply $(\sqrt{y} + 3)(\sqrt{y} - 3)$.

Solution

$$\begin{aligned} &(\sqrt{y} + 3)(\sqrt{y} - 3) \\ &= (\sqrt{y})^2 - 3^2 \\ &= y - 9 \end{aligned}$$

Answers to Trial Runs

page 698 **1.** $-2\sqrt{3}$ **2.** $2\sqrt{x}$ **3.** $-11\sqrt{2a}$ **4.** $11 + 3\sqrt{3}$ **5.** $2 - \sqrt{2}$

page 700 **1.** $\sqrt{10}$ **2.** $5 + a$ **3.** $24a$ **4.** $-10x\sqrt{3}$ **5.** $3\sqrt{2} - 2$ **6.** $3\sqrt{10} + 18$

page 702 **1.** $2 - \sqrt{3} + 2\sqrt{2} - \sqrt{6}$ **2.** $\sqrt{5}$ **3.** -59 **4.** $x - 9$ **5.** $x - 2\sqrt{x} - 8$
6. $x - 10\sqrt{x} + 25$

Name ______________________________ Date ____________

EXERCISE SET 12.2

Simplify each expression.

_______ 1. $2\sqrt{3} + \sqrt{3}$

_______ 2. $5\sqrt{6} - \sqrt{6}$

_______ 3. $5\sqrt{x} - 2\sqrt{x} + 8\sqrt{x}$

_______ 4. $9\sqrt{x} - 5\sqrt{x} - 6\sqrt{x}$

_______ 5. $\sqrt{27} - \sqrt{300}$

_______ 6. $\sqrt{32} - \sqrt{98}$

_______ 7. $\sqrt{54a} + \sqrt{24a}$

_______ 8. $\sqrt{125a} + \sqrt{80a}$

_______ 9. $\sqrt{81} - \sqrt{12} - \sqrt{48}$

_______ 10. $\sqrt{49} - \sqrt{45} - \sqrt{180}$

_______ 11. $4\sqrt{12x} + 3\sqrt{75x}$

_______ 12. $5\sqrt{72x} + 3\sqrt{18x}$

_______ 13. $5\sqrt{28} - 4\sqrt{50}$

_______ 14. $4\sqrt{27} - 3\sqrt{20}$

_______ 15 $\dfrac{\sqrt{288}}{\sqrt{4}} + 3\sqrt{50} - \sqrt{200}$

_______ 16. $\dfrac{\sqrt{360}}{\sqrt{8}} - 2\sqrt{405} + \sqrt{80}$

_______ 17. $\sqrt{75x^2} - \sqrt{300x^2}$

_______ 18. $\sqrt{72x^2} - \sqrt{32x^2}$

_______ 19. $\sqrt{3} \cdot \sqrt{7}$

_______ 20. $\sqrt{5} \cdot \sqrt{11}$

_______ 21. $\sqrt{2} \cdot \sqrt{12} \cdot \sqrt{6}$

_______ 22. $\sqrt{4} \cdot \sqrt{8} \cdot \sqrt{2}$

_______ 23. $\sqrt{11a} \cdot \sqrt{11a}$

_______ 24. $\sqrt{17a} \cdot \sqrt{17a}$

_______ 25. $\sqrt{x - 1} \cdot \sqrt{x - 1}$

_______ 26. $\sqrt{x + 5} \cdot \sqrt{x + 5}$

_______ 27. $\sqrt{15a} \cdot \sqrt{3a}$

_______ 28. $\sqrt{14a} \cdot \sqrt{2a}$

_______ 29. $3\sqrt{50} \cdot \sqrt{2}$

_______ 30. $\sqrt{27} \cdot \sqrt{3}$

_______ 31. $(5\sqrt{18x})(-2\sqrt{8x})$

_______ 32. $(-2\sqrt{27x})4\sqrt{3x})$

_______ 33. $(\sqrt{5x})(3\sqrt{10x})$

_______ 34. $(5\sqrt{3x})(\sqrt{21x})$

_______ 35. $\sqrt{5}(2 - \sqrt{5})$

_______ 36. $\sqrt{11}(2 - \sqrt{11})$

_______ 37. $2\sqrt{3}(\sqrt{2} + \sqrt{12})$

_______ 38. $3\sqrt{5}(\sqrt{2} + \sqrt{20})$

_______ 39. $-3(\sqrt{x} - 2\sqrt{5} + \sqrt{6})$

_______ 40. $-5(\sqrt{x} - 4\sqrt{2} + \sqrt{7})$

_______ 41. $\sqrt{x}(\sqrt{x} - 16)$

_______ 42. $\sqrt{y}(\sqrt{y} - 9)$

_______ 43. $\sqrt{6}(\sqrt{6} - 5) - 5(\sqrt{6} - 5)$

_______ 44. $\sqrt{2}(\sqrt{2} - 11) - 11(\sqrt{2} - 11)$

_______ 45. $\sqrt{x}(\sqrt{x} - 5) + 5(\sqrt{x} - 5)$

_______ 46. $\sqrt{a}(\sqrt{a} - 7) + 7(\sqrt{a} - 7)$

_______ 47. $4(3 - \sqrt{11}) - \sqrt{11}(3 - \sqrt{11})$

_______ 48. $2(7 - \sqrt{5}) + \sqrt{5}(7 - \sqrt{5})$

_______ 49. $2\sqrt{x}(\sqrt{x} - 1) - 3(\sqrt{x} + 1)$

_______ 50. $3\sqrt{x}(\sqrt{x} - 2) - 5(\sqrt{x} + 2)$

_______ **51.** $(5 - \sqrt{2})(1 - \sqrt{3})$

_______ **52.** $(4 - \sqrt{3})(1 + \sqrt{5})$

_______ **53.** $(7 - 2\sqrt{5})(3 + \sqrt{5})$

_______ **54.** $(4 - \sqrt{3})(9 + 2\sqrt{3})$

_______ **55.** $(2 - \sqrt{3x})(4 + \sqrt{3x})$

_______ **56.** $(5 - \sqrt{2x})(3 + \sqrt{2x})$

_______ **57.** $(\sqrt{3} - \sqrt{7})(\sqrt{3} + \sqrt{7})$

_______ **58.** $(\sqrt{10} - \sqrt{5})(\sqrt{10} + \sqrt{5})$

_______ **59.** $(\sqrt{x} - 10)(\sqrt{x} + 10)$

_______ **60.** $(\sqrt{x} - 9)(\sqrt{x} + 9)$

_______ **61.** $(\sqrt{y} - \sqrt{7})^2$

_______ **62.** $(\sqrt{y} - \sqrt{11})^2$

_______ **63.** $(a\sqrt{2} + 5)^2$

_______ **64.** $(a\sqrt{3} + 4)^2$

☆ Stretching the Topics

Simplify.

_______ **1.** $(5\sqrt{6} + 2\sqrt{3})^2$

_______ **2.** $(\sqrt{x + y} - \sqrt{x - y})(\sqrt{x + y} + \sqrt{x - y})$

_______ **3.** $(\sqrt{3} - \sqrt{5})^2 (\sqrt{3} + \sqrt{5})^2$

Check your answers in the back of your book.

If you can simplify the expressions in **Checkup 12.2**, you are ready to go on to Section 12.3.

Name ______________________________ Date ______________

CHECKUP 12.2

Simplify each expression.

______ 1. $\sqrt{24x} + \sqrt{54x}$

______ 2. $\sqrt{12} + \sqrt{75} - \sqrt{48}$

______ 3. $\dfrac{\sqrt{54}}{\sqrt{2}} - 5\sqrt{12} + \sqrt{300}$

______ 4. $\sqrt{15a} \cdot \sqrt{3a}$

______ 5. $\sqrt{y+6} \cdot \sqrt{y+6}$

______ 6. $\sqrt{5}(7 - \sqrt{10})$

______ 7. $2\sqrt{2}(\sqrt{3} + \sqrt{6})$

______ 8. $(\sqrt{x} - \sqrt{5})^2$

______ 9. $(3 + 2\sqrt{5})(4 - 3\sqrt{5})$

______ 10. $(\sqrt{x} - \sqrt{3})(\sqrt{x} + \sqrt{3})$

Check your answers in the back of your book.

If You Missed Problems:	You Should Review Examples:
1–3	1–4
4, 5	5–8
6, 7	11, 12
8	13, 14
9, 10	15–18

12.3 Rationalizing Denominators

Mathematicians generally agree that a fraction containing a radical in the denominator is not in simplest form. Fractions such as

$$\frac{1}{\sqrt{3}} \qquad \frac{2\sqrt{2}}{\sqrt{5}} \qquad \frac{5}{1+\sqrt{3}} \qquad \frac{3+\sqrt{7}}{5-\sqrt{10}}$$

are *not* in simplest form because each contains a radical in the denominator.

We must find a way to change each *irrational* denominator to a *rational* number. Recall that we are allowed to multiply the numerator and denominator of a fraction by the same nonzero number. Let's see if we can choose such a number of the fraction $\frac{1}{\sqrt{3}}$ to make its denominator *rational*. We want to multiply the irrational denominator ($\sqrt{3}$) by some number that will leave no radical in the denominator. Remembering that $\sqrt{3} \cdot \sqrt{3} = 3$, let's multiply numerator and denominator by $\sqrt{3}$ and simplify.

$$\begin{aligned} \frac{1}{\sqrt{3}} &= \frac{1}{\sqrt{3}} \cdot \frac{\sqrt{3}}{\sqrt{3}} \\ &= \frac{1 \cdot \sqrt{3}}{\sqrt{3} \cdot \sqrt{3}} \\ &= \frac{\sqrt{3}}{3} \end{aligned}$$

Notice that our denominator is now a rational number. Indeed, the numerator now contains a radical, but that is acceptable.

This process is called **rationalizing the denominator** because it changes the denominator into a rational number.

Rationalizing Denominators of One Term. To rationalize a denominator that is a single term containing a simplified radical, $\sqrt{a}$, we multiply numerator and denominator by that radical, $\sqrt{a}$.

See if you can rationalize the denominator in Example 1.

Example 1. Rationalize the denominator of $\frac{4}{\sqrt{2}}$.

Solution

$$\begin{aligned} \frac{4}{\sqrt{2}} &= \frac{4}{\sqrt{2}} \cdot \frac{\sqrt{2}}{\sqrt{2}} \\ &= \frac{4 \cdot \sqrt{2}}{\sqrt{2} \cdot \sqrt{2}} \end{aligned}$$

Did you remember to reduce your fraction? Check your work on page 713. ▶

Example 2. Rationalize the denominator of $\frac{2\sqrt{2}}{\sqrt{5}}$.

Solution

$$\begin{aligned} \frac{2\sqrt{2}}{\sqrt{5}} &= \frac{2\sqrt{2}}{\sqrt{5}} \cdot \frac{\sqrt{5}}{\sqrt{5}} \\ &= \frac{2\sqrt{2} \cdot \sqrt{5}}{\sqrt{5} \cdot \sqrt{5}} \\ &= \frac{2\sqrt{10}}{5} \end{aligned}$$

You complete Example 3.

Example 3. Rationalize the denominator of $\sqrt{\frac{5}{3}}$.

Solution

$$\sqrt{\frac{5}{3}} = \frac{\sqrt{5}}{\sqrt{3}}$$

$$= \frac{\sqrt{5}}{\sqrt{3}} \cdot \frac{\sqrt{\underline{\qquad}}}{\sqrt{\underline{\qquad}}}$$

Check your work on page 713. ▶

Example 4. Rationalize the denominator of $\frac{\sqrt{6} - 2}{\sqrt{6}}$.

Solution

$$\frac{\sqrt{6} - 2}{\sqrt{6}} = \frac{\sqrt{6} - 2}{\sqrt{6}} \cdot \frac{\sqrt{6}}{\sqrt{6}}$$

$$= \frac{(\sqrt{6} - 2)\sqrt{6}}{\sqrt{6} \cdot \sqrt{6}}$$

$$= \frac{\sqrt{6} \cdot \sqrt{6} - 2\sqrt{6}}{6}$$

$$= \frac{6 - 2\sqrt{6}}{6}$$

$$= \frac{\overset{1}{\cancel{2}}(3 - \sqrt{6})}{\underset{3}{\cancel{6}}}$$

$$= \frac{3 - \sqrt{6}}{3}$$

Notice we factored the numerator before reducing the fraction.

Trial Run

Rationalize each denominator and simplify.

_______ **1.** $\frac{1}{\sqrt{5}}$

_______ **2.** $\frac{5}{\sqrt{3}}$

_______ **3.** $\frac{2\sqrt{11}}{\sqrt{6}}$

_______ **4.** $\sqrt{\frac{7}{2}}$

_______ **5.** $\dfrac{1 + \sqrt{2}}{\sqrt{3}}$

_______ **6.** $\dfrac{\sqrt{10} - 4}{\sqrt{2}}$

Answers are on page 714.

Rationalizing denominators containing *two* terms requires a different kind of rationalizing factor. Suppose that we try to rationalize the denominator of $\dfrac{5}{7 + \sqrt{3}}$.

Recall from our multiplication examples that when we multiplied an irrational *sum* of two numbers (such as $7 + \sqrt{3}$) times an irrational *difference* of the same two numbers (such as $7 - \sqrt{3}$), the result was a *rational* number. Let's try rationalizing our denominator by multiplying numerator and denominator by $7 - \sqrt{3}$ (the real *conjugate* of $7 + \sqrt{3}$).

$$\frac{5}{7 + \sqrt{3}} = \frac{5}{7 + \sqrt{3}} \cdot \frac{7 - \sqrt{3}}{7 - \sqrt{3}}$$ Multiply numerator and denominator by conjugate of $7 + \sqrt{3}$.

$$= \frac{5(7 - \sqrt{3})}{(7 + \sqrt{3})(7 - \sqrt{3})}$$ Rewrite product as one fraction.

$$= \frac{5(7 - \sqrt{3})}{7^2 - (\sqrt{3})^2}$$ Multiply factors in denominator. Remember that $(A + B)(A - B) = A^2 - B^2$.

$$= \frac{5(7 - \sqrt{3})}{49 - 3}$$ Simplify terms in denominators.

$$= \frac{5(7 - \sqrt{3})}{46}$$ Simplify denominator.

$$= \frac{35 - 5\sqrt{3}}{46}$$ Remove parentheses in numerator.

We have successfully rationalized the denominator.

Rationalizing Denominators of Two Terms. To rationalize a denominator containing a sum or difference of two terms, multiply the numerator and denominator by the real conjugate of the original denominator.

In rationalizing denominators, it is wise to wait until the denominator is a rational number before multiplying the factors in the numerator. Otherwise, you might overlook a common factor that would allow you to reduce the fraction.

Now complete Example 5.

Example 5. Rationalize the denominator of $\frac{\sqrt{3}}{5 - \sqrt{2}}$.

Solution: The real conjugate of $5 - \sqrt{2}$ is $5 + \sqrt{2}$.

$$\frac{\sqrt{3}}{5 - \sqrt{2}} = \frac{\sqrt{3}}{5 - \sqrt{2}} \cdot \frac{5 + \sqrt{2}}{5 + \sqrt{2}}$$

$$= \frac{\sqrt{3}(5 + \sqrt{2})}{(5 - \sqrt{2})(5 + \sqrt{2})}$$

$$= \frac{\sqrt{3}(5 + \sqrt{2})}{5^2 - (\sqrt{2})^2}$$

Check your work on page 713. ▶

Example 6. Rationalize the denominator of $\frac{\sqrt{3}}{9 + \sqrt{3}}$.

Solution: The real conjugate of $9 + \sqrt{3}$ is $9 - \sqrt{3}$.

$$\frac{\sqrt{3}}{9 + \sqrt{3}} = \frac{\sqrt{3}}{9 + \sqrt{3}} \cdot \frac{9 - \sqrt{3}}{9 - \sqrt{3}}$$

$$= \frac{\sqrt{3}(9 - \sqrt{3})}{(9 + \sqrt{3})(9 - \sqrt{3})}$$

$$= \frac{\sqrt{3}(9 - \sqrt{3})}{9^2 - (\sqrt{3})^2}$$

$$= \frac{\sqrt{3}(9 - \sqrt{3})}{81 - 3}$$

$$= \frac{\sqrt{3}(9 - \sqrt{3})}{78}$$

$$= \frac{9\sqrt{3} - 3}{78}$$

$$= \frac{\overset{1}{\cancel{3}}(3\sqrt{3} - 1)}{\underset{26}{\cancel{78}}}$$

$$= \frac{3\sqrt{3} - 1}{26}$$

Example 7. Rationalize the denominator of $\frac{8}{\sqrt{5} - \sqrt{7}}$.

Solution: The real conjugate of $\sqrt{5} - \sqrt{7}$ is $\sqrt{5} + \sqrt{7}$.

$$\frac{8}{\sqrt{5} - \sqrt{7}} = \frac{8}{\sqrt{5} - \sqrt{7}} \cdot \frac{\sqrt{5} + \sqrt{7}}{\sqrt{5} + \sqrt{7}}$$

$$= \frac{8(\sqrt{5} + \sqrt{7})}{(\sqrt{5} - \sqrt{7})(\sqrt{5} + \sqrt{7})}$$

$$= \frac{8(\sqrt{5} + \sqrt{7})}{(\sqrt{5})^2 - (\sqrt{7})^2}$$

$$= \frac{8(\sqrt{5} + \sqrt{7})}{5 - 7}$$

$$= \frac{\overset{-4}{\cancel{8}}(\sqrt{5} + \sqrt{7})}{\underset{1}{\cancel{-2}}}$$

$$= -4(\sqrt{5} + \sqrt{7})$$

$$= -4\sqrt{5} - 4\sqrt{7}$$

You complete Example 8.

Example 8. Rationalize the denominator of $\frac{\sqrt{7} - 2}{\sqrt{7} + 2}$.

Solution: The real conjugate of $\sqrt{7} + 2$ is $\sqrt{7} - 2$.

$$\frac{\sqrt{7} - 2}{\sqrt{7} + 2} = \frac{\sqrt{7} - 2}{\sqrt{7} + 2} \cdot \frac{\sqrt{7} - 2}{\sqrt{7} - 2}$$

$$= \frac{(\sqrt{7} - 2)^2}{(\sqrt{7} + 2)(\sqrt{7} - 2)}$$

$$= \frac{(\sqrt{7})^2 - 2 \cdot 2\sqrt{7} + 2^2}{(\sqrt{7})^2 - 2^2}$$

Check your work on page 713. ▶

Trial Run

Rationalize each denominator and simplify.

_______ 1. $\frac{2}{7 + \sqrt{5}}$

_______ 2. $\frac{-3}{\sqrt{2} + 1}$

_______ 3. $\frac{\sqrt{5}}{4 - \sqrt{5}}$

_______ 4. $\frac{3 - \sqrt{5}}{3 + \sqrt{5}}$

Answers are on page 714.

Examples You Completed

Example 1. Rationalize the denominator of $\frac{4}{\sqrt{2}}$.

Solution

$$\frac{4}{\sqrt{2}} = \frac{4}{\sqrt{2}} \cdot \frac{\sqrt{2}}{\sqrt{2}}$$

$$= \frac{4 \cdot \sqrt{2}}{\sqrt{2} \cdot \sqrt{2}}$$

$$= \frac{\overset{2}{\cancel{4}}\sqrt{2}}{\underset{1}{\cancel{2}}}$$

$$= 2\sqrt{2}$$

Example 3. Rationalize the denominator of $\sqrt{\frac{5}{3}}$.

Solution

$$\sqrt{\frac{5}{3}} = \frac{\sqrt{5}}{\sqrt{3}}$$

$$= \frac{\sqrt{5}}{\sqrt{3}} \cdot \frac{\sqrt{3}}{\sqrt{3}}$$

$$= \frac{\sqrt{5} \cdot \sqrt{3}}{\sqrt{3} \cdot \sqrt{3}}$$

$$= \frac{\sqrt{15}}{3}$$

Example 5. Rationalize the denominator of $\frac{\sqrt{3}}{5 - \sqrt{2}}$.

Solution: The real conjugate of $5 - \sqrt{2}$ is $5 + \sqrt{2}$.

$$\frac{\sqrt{3}}{5 - \sqrt{2}} = \frac{\sqrt{3}}{5 - \sqrt{2}} \cdot \frac{5 + \sqrt{2}}{5 + \sqrt{2}}$$

$$= \frac{\sqrt{3}(5 + \sqrt{2})}{(5 - \sqrt{2})(5 + \sqrt{2})}$$

$$= \frac{\sqrt{3}(5 + \sqrt{2})}{5^2 - (\sqrt{2})^2}$$

$$= \frac{\sqrt{3}(5 + \sqrt{2})}{25 - 2}$$

$$= \frac{\sqrt{3}(5 + \sqrt{2})}{23}$$

$$= \frac{5\sqrt{3} + \sqrt{6}}{23}$$

Example 8. Rationalize the denominator of $\frac{\sqrt{7} - 2}{\sqrt{7} + 2}$.

Solution: The real conjugate of $\sqrt{7} + 2$ is $\sqrt{7} - 2$.

$$\frac{\sqrt{7} - 2}{\sqrt{7} + 2} = \frac{\sqrt{7} - 2}{\sqrt{7} + 2} \cdot \frac{\sqrt{7} - 2}{\sqrt{7} - 2}$$

$$= \frac{(\sqrt{7} - 2)^2}{(\sqrt{7} + 2)(\sqrt{7} - 2)}$$

$$= \frac{(\sqrt{7})^2 - 2 \cdot 2\sqrt{7} + 2^2}{(\sqrt{7})^2 - 2^2}$$

$$= \frac{7 - 4\sqrt{7} + 4}{7 - 4}$$

$$= \frac{11 - 4\sqrt{7}}{3}$$

Answers to Trial Runs

page 710 **1.** $\frac{\sqrt{5}}{5}$ **2.** $\frac{5\sqrt{3}}{3}$ **3.** $\frac{\sqrt{66}}{3}$ **4.** $\frac{\sqrt{14}}{2}$ **5.** $\frac{\sqrt{3} + \sqrt{6}}{3}$ **6.** $\sqrt{5} - 2\sqrt{2}$

page 713 **1.** $\frac{7 - \sqrt{5}}{22}$ **2.** $-3\sqrt{2} + 3$ **3.** $\frac{4\sqrt{5} + 5}{11}$ **4.** $\frac{7 - 3\sqrt{5}}{2}$

Name ______________________________ Date ____________

EXERCISE SET 12.3

Rationalize the denominators. Write your answers in simplest form.

______ 1. $\frac{3}{\sqrt{3}}$

______ 2. $\frac{5}{\sqrt{5}}$

______ 3. $\frac{2}{\sqrt{7}}$

______ 4. $\frac{5}{\sqrt{6}}$

______ 5. $\frac{4}{\sqrt{2}}$

______ 6. $\frac{9}{\sqrt{3}}$

______ 7. $\frac{12}{\sqrt{10}}$

______ 8. $\frac{21}{\sqrt{6}}$

______ 9. $\frac{9}{\sqrt{32}}$

______ 10. $\frac{8}{\sqrt{27}}$

______ 11. $\frac{\sqrt{5}}{\sqrt{8}}$

______ 12. $\frac{\sqrt{7}}{\sqrt{12}}$

______ 13. $\frac{\sqrt{10}}{\sqrt{5}}$

______ 14. $\frac{\sqrt{6}}{\sqrt{2}}$

______ 15. $\sqrt{\frac{1}{2}}$

______ 16. $\sqrt{\frac{1}{3}}$

______ 17. $\sqrt{\frac{4}{5}}$

______ 18. $\sqrt{\frac{9}{7}}$

______ 19. $\sqrt{\frac{3}{x}}$

______ 20. $\sqrt{\frac{2}{x}}$

______ 21. $\frac{1 + \sqrt{2}}{\sqrt{5}}$

______ 22. $\frac{2 + \sqrt{3}}{\sqrt{2}}$

______ 23. $\frac{\sqrt{8} - 6}{\sqrt{2}}$

______ 24. $\frac{\sqrt{10} - 5}{\sqrt{5}}$

______ 25. $\frac{\sqrt{2} + \sqrt{5}}{\sqrt{10}}$

______ 26. $\frac{\sqrt{3} + \sqrt{2}}{\sqrt{6}}$

______ 27. $\frac{1}{2 + \sqrt{3}}$

______ 28. $\frac{1}{5 - \sqrt{2}}$

______ 29. $\frac{-12}{3 - \sqrt{6}}$

______ 30. $\frac{-6}{2 + \sqrt{7}}$

______ 31. $\frac{\sqrt{2}}{1 - \sqrt{2}}$

______ 32. $\frac{\sqrt{3}}{2 - \sqrt{3}}$

_______ 33. $\frac{\sqrt{8}}{\sqrt{3} + 1}$

_______ 34. $\frac{\sqrt{12}}{\sqrt{5} - 1}$

_______ 35. $\frac{2\sqrt{3}}{\sqrt{5} + \sqrt{2}}$

_______ 36. $\frac{2\sqrt{5}}{\sqrt{11} - \sqrt{3}}$

_______ 37. $\frac{2 + \sqrt{2}}{\sqrt{2} + 1}$

_______ 38. $\frac{3 + \sqrt{5}}{\sqrt{5} - 2}$

_______ 39. $\frac{\sqrt{6} + 2}{\sqrt{6} - 2}$

_______ 40. $\frac{\sqrt{7} + 1}{\sqrt{7} - 1}$

☆ Stretching the Topics

Rationalize the denominators.

_______ 1. $\sqrt{\frac{a^2}{4(a + b)}}$

_______ 2. $\frac{3\sqrt{x}}{5\sqrt{x} - 2\sqrt{y}}$

_______ 3. $\frac{2\sqrt{x - y}}{\sqrt{x - y} - 2}$

Check your answers in the back of your book.

If you can rationalize the denominators in **Checkup 12.3**, you are ready to go on to Section 12.4.

Name ________________________________ Date ______________

✓ CHECKUP 12.3

Rationalize the denominators. Write your answers in simplest form.

______ **1.** $\dfrac{6}{\sqrt{6}}$

______ **2.** $\dfrac{20}{\sqrt{10}}$

______ **3.** $\dfrac{7}{\sqrt{12}}$

______ **4.** $\dfrac{\sqrt{6}}{\sqrt{15}}$

______ **5.** $\sqrt{\dfrac{8}{3}}$

______ **6.** $\dfrac{\sqrt{12} - 10}{\sqrt{2}}$

______ **7.** $\dfrac{\sqrt{5} + \sqrt{15}}{\sqrt{20}}$

______ **8.** $\dfrac{\sqrt{3}}{6 - \sqrt{3}}$

______ **9.** $\dfrac{5 + \sqrt{11}}{\sqrt{11} - 3}$

______ **10.** $\dfrac{\sqrt{7} - 2}{\sqrt{7} + 2}$

Check your answers in the back of your book.

If You Missed Problems:	You Should Review Examples:
1–4	1, 2
5	3
6, 7	4
8	5–7
9, 10	8

12.4 Solving More Quadratic Equations

Recall from Chapter 7 that a **quadratic equation** is an equation of the form

$$ax^2 + bx + c = 0 \qquad \text{where } a \neq 0$$

In that chapter we learned to solve such equations by factoring and using the zero product rule.

$x^2 + 3x - 4 = 0$			$x^2 - 9 = 0$		
$(x + 4)(x - 1) = 0$			$(x + 3)(x - 3) = 0$		
$x + 4 = 0$	or	$x - 1 = 0$	$x + 3 = 0$	or	$x - 3 = 0$
$x = -4$	or	$x = 1$	$x = -3$	or	$x = 3$

Every quadratic equation in Chapter 7 could be solved by means of factoring to yield *rational* solutions. Unfortunately, it is not true that all quadratic equations have solutions that are rational numbers. We turn our attention now to methods that can be used for solving *any* quadratic equation.

Solving Quadratic Equations of the Form $ax^2 = c$

Some quadratic equations do not contain a first-degree term. For example,

$$x^2 - 2 = 0$$

is a quadratic (second-degree) equation containing *no first-degree term*. We would like to factor the left-hand side of this equation, but no rational factors will "work." Let's rewrite our equation by adding 2 to both sides.

$$x^2 - 2 = 0$$
$$x^2 = 2$$

Now we are looking for a number x whose square is the number 2.

Certainly, $x = \sqrt{2}$ is a solution because it satisfies the original equation.

$$x^2 = 2$$
$$(\sqrt{2})^2 \stackrel{?}{=} 2$$
$$(\sqrt{2})(\sqrt{2}) \stackrel{?}{=} 2$$
$$2 = 2$$

But another solution is $x = -\sqrt{2}$ because it also satisfies the original equation.

$$x^2 = 2$$
$$(-\sqrt{2})^2 \stackrel{?}{=} 2$$
$$(-\sqrt{2})(-\sqrt{2}) \stackrel{?}{=} 2$$
$$2 = 2$$

Our solutions for $x^2 = 2$ are

$$x = \sqrt{2} \qquad \text{or} \qquad x = -\sqrt{2}$$

As before, we discover that a second-degree equation has *two* solutions.
You try Example 1.

Example 1. Solve $x^2 = 7$.

Solution

$$x^2 = 7$$
$$x = \sqrt{} \qquad \text{or}$$
$$x = -\sqrt{}$$

Check your work on page 724. ▶

Complete Example 2.

Example 2. Solve $2x^2 - 22 = 0$.

Solution

$$2x^2 - 22 = 0$$
$$2x^2 = \underline{}$$
$$x^2 = \underline{}$$
$$x = \underline{} \qquad \text{or} \qquad x = -\underline{}$$

Check your work on page 724. ▶

Solving Equations of the Form $x^2 = C$. If $x^2 = C$ and C is *not* negative, the solutions are

$$x = \sqrt{C} \quad \text{or} \quad x = -\sqrt{C}$$

If $x^2 = C$ and C is negative, there is no real number solution.

To save time and space when solving quadratic equations of the form $x^2 = C$ (where C is not negative), we may start by writing both solutions at the same time, using the notation

$$x = \pm\sqrt{C}$$

This shortcut allows us to simplify both solutions at once, but we shall always conclude by stating the solutions separately.

Example 3. Solve $3x^2 - 5 = 0$.

Solution

$$3x^2 - 5 = 0$$

$$3x^2 = 5$$

$$x^2 = \frac{5}{3}$$

$$x = \pm\sqrt{\frac{5}{3}} \qquad \text{Solve for } x.$$

$$= \frac{\pm\sqrt{5}}{\sqrt{3}} \qquad \text{Separate the radicals.}$$

$$= \frac{\pm\sqrt{5}}{\sqrt{3}} \cdot \frac{\sqrt{3}}{\sqrt{3}} \qquad \text{Rationalize the denominator.}$$

$$x = \frac{\pm\sqrt{15}}{3} \qquad \text{Multiply.}$$

$$x = \frac{\sqrt{15}}{3} \quad \text{or} \quad x = \frac{-\sqrt{15}}{3}$$

Write both solutions.

You try completing Example 4.

Example 4. Solve $2x^2 - 11 = 2$.

Solution

$$2x^2 - 11 = 2$$

$$2x^2 = 13$$

$$x^2 = \frac{13}{2}$$

$x =$ ____ or $x =$ ____

Check your work on page 724. ▶

Example 5. Solve $x^2 + 1 = 0$.

Solution

$$x^2 + 1 = 0$$

$$x^2 = -1$$

There is no real number solution.

Complete Example 6.

Example 6. Solve $x^2 + 7 = 5$.

Solution

Check your work on page 724. ▶

Trial Run

Solve.

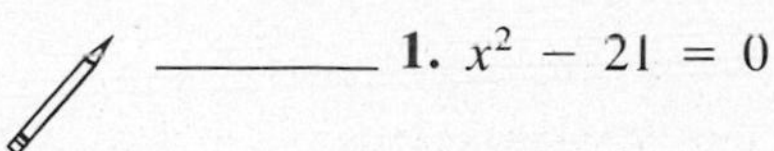

______ 1. $x^2 - 21 = 0$

______ 2. $3x^2 = 15$

______ 3. $2x^2 - 16 = 0$

______ 4. $5 = 2x^2 - 6$

Answers are on page 724.

Using the Quadratic Formula

We know how to solve a quadratic equation of the form

$$ax^2 + bx + c = 0$$

as long as the left-hand side can be factored, but suppose that it cannot be factored by our usual methods. Fortunately, mathematicians have derived a formula that allows us to solve any quadratic equation.

> **Quadratic Formula.** If $ax^2 + bx + c = 0$, the solutions can be found by
>
> $$x = \frac{-b \pm \sqrt{b^2 - 4ac}}{2a}$$

Let's practice using this formula with an equation whose solutions can also be found by factoring; then we can be more convinced that the formula "works." Solve the equation

$$x^2 + 2x - 3 = 0$$

Approach 1. Use factoring.

$$x^2 + 2x - 3 = 0$$

$$(x + 3)(x - 1) = 0$$

$$x + 3 = 0 \quad \text{or} \quad x - 1 = 0$$

$$x = -3 \quad \text{or} \quad x = 1$$

Approach 2. Use the quadratic formula.

$$x^2 + 2x - 3 = 0$$

We carefully note that here

$$a = 1 \quad b = 2 \quad \text{and} \quad c = -3$$

and we substitute these values into the formula.

$$x = \frac{-b \pm \sqrt{b^2 - 4ac}}{2a}$$

$$x = \frac{-2 \pm \sqrt{2^2 - 4(1)(-3)}}{2(1)}$$

$$= \frac{-2 \pm \sqrt{4 + 12}}{2}$$

$$= \frac{2 \pm \sqrt{16}}{2}$$

$$= \frac{-2 \pm 4}{2}$$

$$x = \frac{-2 + 4}{2} \quad \text{or} \quad x = \frac{-2 - 4}{2}$$

$$= \frac{2}{2} \quad \text{or} \quad = \frac{-6}{2}$$

$$x = 1 \quad \text{or} \quad x = -3$$

Notice that both approaches yielded the same two solutions. Of course, factoring was a much quicker method to use in this example. We shall *always solve by factoring when possible*; otherwise, we shall use the quadratic formula.

Example 7. Solve $x^2 + 5x + 1 = 0$.

Solution: We cannot factor $x^2 + 5x + 1$, so we must use the quadratic formula. Here

$$a = 1 \qquad b = 5 \qquad \text{and} \qquad c = 1$$

$$x = \frac{-b \pm \sqrt{b^2 - 4ac}}{2a}$$

$$x = \frac{-5 \pm \sqrt{5^2 - 4(1)(1)}}{2(1)}$$

$$= \frac{-5 \pm \sqrt{25 - 4}}{2}$$

$$x = \frac{-5 \pm \sqrt{21}}{2}$$

$$x = \frac{-5 + \sqrt{21}}{2} \quad \text{or} \quad x = \frac{-5 - \sqrt{21}}{2}$$

Try using the quadratic formula in Example 8.

Example 8. Solve $2x^2 - 7x + 2 = 0$.

Solution

$$2x^2 - 7x + 2 = 0$$

We note that

$$a = 2$$

$$b = -7$$

$$c = 2$$

Example 9. Solve $x^2 + 2x = 1$.

Solution

$$x^2 + 2x = 1$$

$$x^2 + 2x - 1 = 0$$

We note that

$$a = 1$$

$$b = 2$$

$$c = -1$$

and substitute into the quadratic formula.

$$x = \frac{-b \pm \sqrt{b^2 - 4ac}}{2a}$$

$$x = \frac{-(-7) \pm \sqrt{(-7)^2 - 4(2)(2)}}{2(2)}$$

$x =$ ________ or $x =$ ________

Check your work on page 724. ▶

and we substitute into the quadratic formula.

$$x = \frac{-b \pm \sqrt{b^2 - 4ac}}{2a}$$

$$x = \frac{-2 \pm \sqrt{2^2 - 4(1)(-1)}}{2(1)}$$

$$= \frac{-2 \pm \sqrt{4 + 4}}{2}$$

$$= \frac{-2 \pm \sqrt{8}}{2}$$

$$= \frac{-2 \pm \sqrt{4 \cdot 2}}{2}$$

$$= \frac{-2 \pm 2\sqrt{2}}{2}$$

$$= \frac{\overset{1}{\cancel{2}}(-1 \pm \sqrt{2})}{\underset{1}{\cancel{2}}}$$

$$x = -1 \pm \sqrt{2}$$

$$x = -1 + \sqrt{2}$$

or $\quad x = -1 - \sqrt{2}$

In Example 9, notice that we simplified the radical and then *factored* the numerator in order to reduce the fraction.

Trial Run

Write each equation in the form $ax^2 + bx + c = 0$. Then give the values of a, b, and c.

________ **1.** $2x^2 = 3x - 2$

________ **2.** $x^2 - 5x + 2 = 7x + 5$

Use the quadratic formula to solve each equation.

________ **3.** $x^2 + x - 3 = 0$

________ **4.** $2x^2 - 3x - 1 = 0$

________ **5.** $2y^2 + 4y = -1$

Answers are on page 724.

▶ Examples You Completed

Example 1. Solve $x^2 = 7$.

Solution

$$x^2 = 7$$

$$x = \sqrt{7} \quad \text{or} \quad x = -\sqrt{7}$$

Example 2. Solve $2x^2 - 22 = 0$.

Solution

$$2x^2 - 22 = 0$$
$$2x^2 = 22$$
$$x^2 = 11$$
$$x = \sqrt{11} \quad \text{or} \quad x = -\sqrt{11}$$

Example 4. Solve $2x^2 - 11 = 2$.

Solution

$$2x^2 - 11 = 2$$
$$2x^2 = 13$$
$$x^2 = \frac{13}{2}$$
$$x = \pm\sqrt{\frac{13}{2}}$$
$$= \pm\frac{\sqrt{13}}{\sqrt{2}}$$
$$= \pm\frac{\sqrt{13}}{\sqrt{2}} \cdot \frac{\sqrt{2}}{\sqrt{2}}$$
$$x = \pm\frac{\sqrt{26}}{2}$$
$$x = \frac{\sqrt{26}}{2} \quad \text{or} \quad x = \frac{-\sqrt{26}}{2}$$

Example 8. Solve $2x^2 - 7x + 2 = 0$.

Solution

$$2x^2 - 7x + 2 = 0$$

We note that

$$a = 2$$
$$b = -7$$
$$c = 2$$

and substitute into the quadratic formula

$$x = \frac{-b \pm \sqrt{b^2 - 4ac}}{2a}$$
$$x = \frac{-(-7) \pm \sqrt{(-7)^2 - 4(2)(2)}}{2(2)}$$
$$= \frac{7 \pm \sqrt{49 - 16}}{4}$$
$$x = \frac{7 \pm \sqrt{33}}{4}$$
$$x = \frac{7 + \sqrt{33}}{4}$$
$$\text{or} \quad x = \frac{7 - \sqrt{33}}{4}$$

Example 6. Solve $x^2 + 7 = 5$.

Solution

$$x^2 + 7 = 5$$
$$x^2 = -2$$

There is no real number solution.

Answers to Trial Runs

page 721 **1.** $x = \sqrt{21}$ or $x = -\sqrt{21}$ **2.** $x = \sqrt{5}$ or $x = -\sqrt{5}$ **3.** $x = 2\sqrt{2}$ or $x = -2\sqrt{2}$ **4.** $x = \frac{\sqrt{22}}{2}$ or $x = \frac{-\sqrt{22}}{2}$

page 723 **1.** $2x^2 - 3x + 2 = 0$; $a = 2$, $b = -3$, $c = 2$ **2.** $x^2 - 12x - 3 = 0$; $a = 1$, $b = -12$, $c = -3$ **3.** $x = \frac{-1 \pm \sqrt{13}}{2}$ **4.** $x = \frac{3 \pm \sqrt{17}}{4}$ **5.** $y = \frac{-2 \pm \sqrt{2}}{2}$

Name ____________________ Date __________

EXERCISE SET 12.4

Solve each of the following quadratic equations. Write all answers in simplest radical form and reduce all answers to lowest terms.

______ **1.** $x^2 = 12$

______ **2.** $x^2 = 32$

______ **3.** $9a^2 = 5$

______ **4.** $4a^2 = 7$

______ **5.** $5x^2 - 1 = 0$

______ **6.** $6x^2 - 5 = 0$

______ **7.** $6y^2 - 15 = 0$

______ **8.** $8y^2 - 28 = 0$

______ **9.** $3y^2 - 8 = 7$

______ **10.** $5y^2 - 17 = 8$

______ **11.** $9 = 11 - a^2$

______ **12.** $8 = 13 - a^2$

______ **13.** $4x^2 + 3 = 3$

______ **14.** $6x^2 + 8 = 8$

______ **15.** $a^2 - 80 = 0$

______ **16.** $a^2 - 75 = 0$

______ **17.** $x^2 + x - 3 = 0$

______ **18.** $x^2 + 6x + 1 = 0$

______ **19.** $a^2 = 5a + 5$

______ **20.** $a^2 = 2a + 2$

______ **21.** $x^2 - 3 = 4x$

______ **22.** $x^2 - 2 = 3x$

______ **23.** $y^2 = 2y + 1$

______ **24.** $y^2 = 4y + 2$

______ **25.** $a^2 + 7a = 1$

______ **26.** $a^2 + 5a = 4$

______ **27.** $3x^2 - 2x - 2 = 0$

______ **28.** $5x^2 - 3x - 1 = 0$

______ **29.** $4 - 6y = -y^2$

______ **30.** $4z - 1 = -z^2$

______ **31.** $2z = 15 - 2z^2$

______ **32.** $7z = 6z^2 - 1$

☆ Stretching the Topics

Solve.

______ **1.** $(x - 3)(3x + 5) = (2x + 7)(x + 1)$

______ **2.** $x^2 - 3\sqrt{2}\,x - 4 = 0$

______ **3.** $\sqrt{3}\,x^2 - 6x + 2\sqrt{3} = 0$

Check your answers in the back of your book.

If you can solve the quadratic equations in **Checkup 12.4**, you are ready to go on to Section 12.5.

Name ______________________ Date ____________

CHECKUP 12.4

Solve.

______ 1. $x^2 = 2$

______ 2. $4a^2 = 20$

______ 3. $y^2 - 32 = 0$

______ 4. $9x^2 - 80 = 100$

______ 5. $9 - 3x^2 = 9$

______ 6. $x^2 + 4x - 4 = 0$

______ 7. $x^2 = 3x - 1$

______ 8. $3x^2 - 6x + 2 = 0$

______ 9. $2x^2 - 5x = 1$

______ 10. $9x^2 - 7 = 6x$

Check your answers in the back of your book.

If You Missed Problems:	You Should Review Examples:
1–3	1, 2
4, 5	3–6
6	7
7–10	8, 9

12.5 Switching from Word Statements to Quadratic Equations

Now that we have the tools for solving any quadratic equation, let's practice switching from words to quadratic equations. Do not be surprised if our equations have irrational solutions. To find approximations for such irrational numbers, we may use the table of square roots inside the back cover of this book.

Example 1. A carpenter wishes to construct a square tabletop with an area of 8 square feet. How long should each side of the tabletop be?

Solution: We can let

$$x = \text{length of each side} \quad \text{(feet)}$$
$$x^2 = \text{area of square top} \quad \text{(square feet)}$$

Then

$$\begin{aligned} x^2 &= 8 \\ x &= \pm\sqrt{8} \\ &= \pm\sqrt{4 \cdot 2} \\ &= \pm 2\sqrt{2} \end{aligned}$$

$$x = 2\sqrt{2} \quad \text{or} \quad x = -2\sqrt{2}$$

We reject the negative value for length and the solution is $x = 2\sqrt{2}$ ft. If we wish an approximation for the length, we may use $\sqrt{2} \doteq 1.414$. Then

$$\begin{aligned} x &\doteq 2(1.414) \\ &\doteq 2.818 \end{aligned}$$

Each side of the tabletop should be approximately 2.818 feet long.

You try completing Example 2.

Example 2. This spring Martina enlarged her square garden by doubling the width and tripling the length. If the area of the enlarged garden is 72 square yards, what were the approximate dimensions of her square garden?

Solution: We do not know the length of the sides of the square garden, so we can let

$$x = \text{length of side of square garden} \quad \text{(yards)}$$

Let's illustrate both gardens. Remember that she doubled the width and tripled the length.

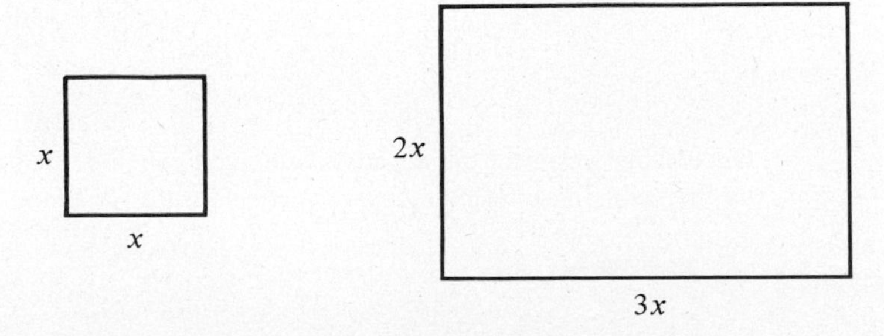

The area of her new rectangular garden can be found by $A = l \cdot w$.

$$\begin{aligned} \text{Area} &= (2x)(3x) \\ &= 6x^2 \end{aligned}$$

So we must solve the equation

$6x^2 = 72$	New area is 72 square yards.
$x^2 = \frac{72}{6}$	Isolate x^2.
$x^2 = 12$	Simplify.
$x = \pm\sqrt{12}$	Solve for x.
$x = \pm\sqrt{\_\_\_\_ \cdot \_\_\_\_}$	Factor the radicand.
$x = \pm \_\_\_\_ \sqrt{3}$	Remove square root.

$x = \_\_\_\_ \sqrt{3}$ or $x = - \_\_\_\_ \sqrt{3}$

We reject the negative value for the length of a side and approximate the positive value using $\sqrt{3} \doteq 1.732$.

$$x = \_\_\_\_ \sqrt{3}$$
$$\doteq \_\_\_\_ (1.732)$$
$$\doteq \_\_\_\_$$

The original garden measured approximately ____ yards on each side.

Check your work on page 732. ▶

Let's solve the problem stated at the beginning of this chapter using the Pythagorean Theorem.

Example 3. A ladder that is 20 feet long is leaning against the wall of a building. If the foot of the ladder is 10 feet from the bottom of the wall, how far up the wall is the top of the ladder?

Solution: We can let

$$x = \text{distance up wall} \quad \text{(feet)}$$

From the Pythagorean formula we know that

$$x^2 + 10^2 = 20^2$$
$$x^2 + 100 = 400$$
$$x^2 = 300$$
$$x = \pm\sqrt{300}$$
$$= \pm\sqrt{100 \cdot 3}$$
$$x = \pm 10\sqrt{3}$$

20 feet

x

10 feet

$$x = 10\sqrt{3} \quad \text{or} \quad x = -10\sqrt{3}$$

Once again we reject the negative value and conclude that the top of the ladder is $10\sqrt{3}$ feet up the wall. If we wish an approximation for this distance, we may use $\sqrt{3} \doteq 1.732$. Then

$$x = 10\sqrt{3}$$
$$\doteq 10(1.732)$$
$$x \doteq 17.32$$

The top of the ladder is approximately 17.32 feet up the wall.

Example 4. One irrational number is 6 less than another irrational number. The product of the two numbers is -7. Find the irrational numbers.

Solution: We can let

$$x = \text{first irrational number}$$
$$x - 6 = \text{second irrational number}$$
$$x(x - 6) = \text{product of numbers}$$

Our equation must be

$$x(x - 6) = -7$$
$$x^2 - 6x = -7$$
$$x^2 - 6x + 7 = 0$$

We solve by the quadratic formula.

$$x = \frac{-b \pm \sqrt{b^2 - 4ac}}{2a}$$

Here $a = 1$, $b = -6$, and $c = 7$.

$$x = \frac{-(-6) \pm \sqrt{(-6)^2 - 4(1)(7)}}{2(1)}$$
$$= \frac{6 \pm \sqrt{36 - 28}}{2}$$
$$= \frac{6 \pm \sqrt{8}}{2}$$
$$= \frac{6 \pm \sqrt{4 \cdot 2}}{2}$$
$$= \frac{6 \pm 2\sqrt{2}}{2}$$
$$= \frac{\overset{1}{\cancel{2}}(3 \pm \sqrt{2})}{\underset{1}{\cancel{2}}}$$
$$x = 3 \pm \sqrt{2}$$

There are two pairs of irrational numbers.

First pair	*Second pair*
$x = 3 + \sqrt{2}$	$x = 3 - \sqrt{2}$
$x - 6 = 3 + \sqrt{2} - 6$	$x - 6 = 3 - \sqrt{3} - 6$
$= -3 + \sqrt{2}$	$= -3 - \sqrt{2}$

So the pairs of irrational numbers that satisfy the conditions of our problem are

$$3 + \sqrt{2} \quad \text{and} \quad -3 + \sqrt{2}$$

or

$$3 - \sqrt{2} \quad \text{and} \quad -3 - \sqrt{2}$$

▶ The Example You Completed

Example 2

Solution

$$6x^2 = 72$$

$$x^2 = \frac{72}{6}$$

$$x^2 = 12$$

$$x = \pm\sqrt{12}$$

$$x = \pm\sqrt{4 \cdot 3}$$

$$x = \pm 2\sqrt{3}$$

$$x = 2\sqrt{3} \quad \text{or} \quad x = -2\sqrt{3}$$

$$x \doteq 2(1.732)$$

$$\doteq 3.464$$

The original garden measured approximately 3.464 yards on each side.

Name ______________________________ Date ____________

EXERCISE SET 12.5

Solve and give the answers in simplest radical form, unless directed otherwise.

________ **1.** The area of a square is 50 square centimeters. Find the length of a side of the square.

________ **2.** The area of a square is 98 square inches. Find the length of a side of the square.

________ **3.** The length of the Latham's patio is the same as the width. If it takes 432 square feet of outdoor carpet to cover the patio, find the length of one side of the patio. (Approximate answer to nearest tenths.)

________ **4.** Rosa's family room is a square. If it takes 675 square feet of carpeting to cover it, find the length of one side of the family room. (Approximate answer to nearest tenths.)

________ **5.** In a right triangle, the hypotenuse is 8 feet and one side is 4 feet. Find the length of the other side.

________ **6.** If a flower bed in the shape of a right triangle has one side that measures 6 feet and the hypotenuse is 12 feet, find the length of the other side.

________ **7.** One leg of a right triangle is twice as long as the other leg and the hypotenuse is 15 meters long. Find the length of each leg of the triangle. (Approximate answer to nearest hundredths.)

________ **8.** One leg of a right triangle is 3 times as long as the other leg and the hypotenuse is 50 meters long. Find the length of each leg of the triangle. (Approximate answer to nearest hundredths.)

________ **9.** The width of a rectangle is 6 meters less than the length. If the area of the rectangle is 10 square meters, find the length of the rectangle.

________ **10.** The length of a rectangle is 2 meters more than the width. If the area of the rectangle is 15 square meters, find the width of the rectangle.

________ **11.** One irrational number is 2 less than another irrational number. The product of the two numbers is 9. Find the irrational numbers.

________ **12.** One irrational number is 2 more than another irrational number. The sum of the squares of the two numbers is 5. Find the irrational numbers.

________ **13.** A rectangle is 3 feet by 5 feet. If both dimensions are increased by the same amount, the area will be 30 square feet. Find the number of feet by which each dimension is increased.

________ **14.** A rectangle is 4 feet by 6 feet. If both dimensions are increased by the same amount, the area will be 50 square feet. Find the number of feet by which each dimension is increased.

_______ **15.** The dimensions of a picture including the frame are 8 inches and 10 inches. Find the width of the frame if the area of the picture inside the frame is 50 square inches. (Approximate answer to nearest tenths.)

_______ **16.** The dimensions of a rectangular linen tablecloth including the border are 5 feet by 7 feet. Find the width of the lace if the area of the linen before adding the border was 27 square feet. (Approximate answer to nearest tenths.)

☆ Stretching the Topics

_______ **1.** One leg of a right triangle is twice as long as the other and the hypotenuse is 5 feet more than the longer leg. Find the length of each side of the triangle.

_______ **2.** Each side of a square is 2 inches shorter than a diagonal of the square. Find the length of the side and the area of the square.

Check your answers in the back of your book.

If you can do **Checkup 12.5**, you are ready to do the **Review Exercises for Chapter 12.**

Name ______________________ Date __________

CHECKUP 12.5

_______ **1.** The area of a square is 32 square inches. Find the length of a side of the square.

_______ **2.** If one side of a square is decreased by 2 meters and the other side is increased by 3 meters, the resulting rectangle has an area of 32 square meters. Find the length of a side of the original square. (Approximate to nearest hundredths.)

_______ **3.** One irrational number is 4 more than another irrational number. The product of the two numbers is 1. Find the irrational numbers.

_______ **4.** In a right triangle one leg is 12 feet and the hypotenuse is twice the other leg. Find the length of the other leg. (Approximate to nearest tenths.)

Check your answers in the back of your book.

If You Missed Problem:	You Should Review Examples:
1	1
2	2
3	4
4	3

Summary

In this chapter we discussed square roots of numbers and agreed to call the positive square root of a positive number a the **principal square root of a**, denoted by $\sqrt{a}$. We noted that the square root of zero is zero and the square root of a negative number is *not* a real number. The square root of a positive number is always real, but it may be rational ($\sqrt{81} = 9$) or irrational ($\sqrt{13}$).

To multiply, divide and/or simplify radical expressions we used two important properties of radicals.

Property	Symbols	Examples
Multiplication property for radicals	$\sqrt{a} \cdot \sqrt{b} = \sqrt{ab}$	$\sqrt{12a} \cdot \sqrt{3a} = \sqrt{36a^2} = 6a$ $\sqrt{98a^3b^2} = \sqrt{49 \cdot 2 \cdot a^2 \cdot a \cdot b^2}$ $= 7ab\sqrt{2a}$
Division property for radicals	$\dfrac{\sqrt{a}}{\sqrt{b}} = \sqrt{\dfrac{a}{b}} \quad (b \neq 0)$	$\dfrac{\sqrt{300a^3}}{\sqrt{3a}} = \sqrt{\dfrac{300a^3}{3a}}$ $= \sqrt{100a^2} = 10a$ $\sqrt{\dfrac{9x^2}{4}} = \dfrac{\sqrt{9x^2}}{\sqrt{4}} = \dfrac{3x}{2}$

We found that square roots can be combined in a sum or difference only if their radicands match exactly. Then we combine them in the same way that we combine like terms.

We learned that fractions containing square roots in the denominator are not considered to be in simplest form. Ridding the denominator of an irrational square root is called **rationalizing the denominator**.

If the Denominator Is:	We Can Rationalize the Denominator by:	Examples
A single simplified radical term, $\sqrt{a}$	Multiplying numerator and denominator by the same radical, $\sqrt{a}$	$\dfrac{3}{\sqrt{5}} = \dfrac{3}{\sqrt{5}} \cdot \dfrac{\sqrt{5}}{\sqrt{5}}$ $= \dfrac{3\sqrt{5}}{5}$
An irrational sum or difference of two terms	Multiplying numerator and denominator by the real conjugate of the original denominator	$\dfrac{\sqrt{7}}{\sqrt{7}+2} = \dfrac{\sqrt{7}}{\sqrt{7}+2} \cdot \dfrac{\sqrt{7}-2}{\sqrt{7}-2}$ $= \dfrac{\sqrt{7}(\sqrt{7}-2)}{7-4}$ $= \dfrac{7-2\sqrt{7}}{3}$ $\dfrac{3}{6-\sqrt{5}} = \dfrac{3}{6-\sqrt{5}} \cdot \dfrac{6+\sqrt{5}}{6+\sqrt{5}}$ $= \dfrac{3(6+\sqrt{5})}{36-5}$ $= \dfrac{18+3\sqrt{5}}{31}$

Finally, we developed methods for solving second-degree (or quadratic) equations that cannot be solved by factoring.

Equation	Solution	Examples
$x^2 = c$ (c is not negative)	$x = \sqrt{c}$ or $x = -\sqrt{c}$	$3x^2 = 15$ $x^2 = 5$ $x = \sqrt{5}$ or $x = -\sqrt{5}$
$ax^2 + bx + c = 0$	$x = \dfrac{-b \pm \sqrt{b^2 - 4ac}}{2a}$	$x^2 + 3x + 1 = 0$ $x = \dfrac{-3 \pm \sqrt{3^2 - 4(1)(1)}}{2(1)}$ $x = \dfrac{-3 \pm \sqrt{5}}{2}$ $x = \dfrac{-3 + \sqrt{5}}{2}$ or $x = \dfrac{-3 - \sqrt{5}}{2}$

We found we could use these methods to solve word problems that might not have rational solutions.

❑ Speaking the Language of Algebra

1. The positive square root of a positive number is called the __________ __________ __________ of that number.

2. The square root of a negative number is not a __________ number.

3. In the expression $\sqrt{a}$, the symbol $\sqrt{}$ is called the __________ , and a is called the __________ .

4. The number $\sqrt{7}$ is an example of a number that is real but __________ .

5. The process by which we rid a denominator of radicals is called __________ __________ __________ .

6. For the irrational number $\sqrt{3} + 2$, the irrational number $\sqrt{3} - 2$ is called the __________ __________ .

7. If $x^2 = c$ (and c is not negative), the two solutions are $x =$ __________ or $x =$ __________ .

8. If $ax^2 + bx + c = 0$, the solutions can be found by $x =$ __________ . This is called the __________ __________ .

Check your answers in the back of your book.

Name ______________________ Date ____________

REVIEW EXERCISES for Chapter 12

Simplify the following radical expressions.

______ **1.** $\sqrt{144}$

______ **2.** $\sqrt{64a^2}$

______ **3.** $(\sqrt{13})^2$

______ **4.** $\sqrt{45y^4}$

______ **5.** $\sqrt{48a^3b^2}$

______ **6.** $\sqrt{\dfrac{27x^5}{y^4}}$

______ **7.** $\dfrac{\sqrt{98a^7b^3}}{\sqrt{2ab}}$

______ **8.** $\dfrac{\sqrt{63a^7}}{\sqrt{7a}}$

______ **9.** $\sqrt{20x} + \sqrt{80x} - \sqrt{45x}$

______ **10.** $\sqrt{21a} \cdot \sqrt{7a}$

______ **11.** $(3\sqrt{10})(-4\sqrt{15})$

______ **12.** $\sqrt{2}(3 - \sqrt{14})$

______ **13.** $5\sqrt{3}(\sqrt{2} + \sqrt{8})$

______ **14.** $\sqrt{5}(\sqrt{3} - 2) - 4(\sqrt{5} + 3)$

______ **15.** $(4 - 3\sqrt{6})(1 + 2\sqrt{6})$

______ **16.** $(\sqrt{a} - \sqrt{5})(\sqrt{a} + \sqrt{5})$

Rationalize the denominators. Write each answer in simplest form.

______ **17.** $\dfrac{15}{\sqrt{3}}$

______ **18.** $\dfrac{\sqrt{15}}{\sqrt{12}}$

______ **19.** $\sqrt{\dfrac{7}{5}}$

______ **20.** $\dfrac{\sqrt{12} - 21}{\sqrt{3}}$

______ **21.** $\dfrac{\sqrt{7} + \sqrt{21}}{\sqrt{14}}$

______ **22.** $\dfrac{\sqrt{11}}{5 - \sqrt{11}}$

______ **23.** $\dfrac{4 + \sqrt{3}}{\sqrt{3} - 2}$

______ **24.** $\dfrac{\sqrt{13} - 2}{\sqrt{13} + 2}$

Solve the equations.

______ **25.** $y^2 - 48 = 0$

______ **26.** $9a^2 = 18$

______ **27.** $7 - 4x^2 = 9$

______ **28.** $x^2 + 6x - 3 = 0$

______ **29.** $x^2 = 4x - 1$

______ **30.** $5x^2 - 10x + 2 = 0$

______ **31.** $2x^2 - 3x = 1$

______ **32.** $3x^2 = 4x + 2$

______ **33.** In a right triangle, the lengths of the two legs are the same and the hypotenuse is 6 centimeters. Find the length of each leg.

______ **34.** The length of a rectangle is 4 feet more than the width. If the area of the rectangle is 2 square feet, find its length.

Check your answers in the back of your book.

If You Missed Exercises:	You Should Review Examples:	
1, 2	SECTION 12.1	1, 2, 9
3		5, 6
4, 5		15, 6
6		18
7, 8		19, 20
9	SECTION 12.2	3
10, 11		9, 10
12–14		11, 12
15, 16		13–16
17, 18	SECTION 12.3	1, 2
19		3
20, 21		4
22		5, 6
23, 24		8
25–27	SECTION 12.4	1–4
28, 29		7, 9
30–32		8, 9
33	SECTION 12.5	3
34		4

If you have completed the **Review Exercises** and corrected your errors, you are ready to take the **Practice Test for Chapter 12**.

Name ______________________ Date __________

PRACTICE TEST for Chapter 12

Simplify the radical expressions.	Section	Examples
______ 1. $\sqrt{\dfrac{16}{25}}$	12.1	2
______ 2. $(\sqrt{3a})^2$	12.1	10
______ 3. $\sqrt{81a^2}$	12.1	11
______ 4. $\sqrt{32}$	12.1	13, 14
______ 5. $\sqrt{48a^3b^4}$	12.1	16
______ 6. $\sqrt{\dfrac{32x^3}{y^4}}$	12.1	18
______ 7. $\dfrac{\sqrt{500x^7y^3}}{\sqrt{5x^5y}}$	12.1	19, 20
______ 8. $2\sqrt{160z} + \sqrt{40z} - 6\sqrt{10z}$	12.2	3
______ 9. $(-5\sqrt{10a})(\sqrt{30a})$	12.2	9, 10
______ 10. $4\sqrt{6}(\sqrt{12} - \sqrt{3})$	12.2	11, 12
______ 11. $(\sqrt{y} - \sqrt{10})^2$	12.2	14
______ 12. $(4 + \sqrt{x})(5 - 3\sqrt{x})$	12.2	15
______ 13. $(7 - 2\sqrt{6})(7 + 2\sqrt{6})$	12.2	16
______ 14. $(\sqrt{a} - 5)(\sqrt{a} + 5)$	12.2	19
Rationalize the denominators and simplify.		
______ 15. $\dfrac{20}{\sqrt{10}}$	12.3	1
______ 16. $\sqrt{\dfrac{5}{3}}$	12.3	3
______ 17. $\dfrac{3 + \sqrt{5}}{\sqrt{5}}$	12.3	4
______ 18. $\dfrac{\sqrt{7}}{4 + \sqrt{7}}$	12.3	5, 6
______ 19. $\dfrac{6}{\sqrt{2} - \sqrt{3}}$	12.3	7

		Section	Example
______	20. $\dfrac{1 - \sqrt{10}}{1 + \sqrt{10}}$	12.3	8

Solve the equations.

		Section	Example
______	21. $x^2 = 72$	12.4	1
______	22. $7x^2 - 21 = 0$	12.4	2
______	23. $x^2 + 5x - 1 = 0$	12.4	7
______	24. $x^2 + 8x = -10$	12.4	9
______	25. The width of a rectangular tablecloth is 2 feet less than the length. The area of the tablecloth is 11 square feet. Find the length of the tablecloth.	12.5	2, 4

Check your answers in the back of your book.

Name ________________________________ Date ____________

SHARPENING YOUR SKILLS after Chapters 1–12

	Problem	SECTION
______	**1.** Find 115% of \$630.	1.3
______	**2.** Compare -3.98 and -4.21 using $<$ or $>$.	2.1
______	**3.** Evaluate $\dfrac{5b(a^2 - 3a + 1)}{(b - c)^2}$ when $a = 1$, $b = -1$, and $c = 2$.	4.1
______	**4.** Is -2 a solution for $3(2x - 1) + 5 = 7 - 5x$?	4.1
______	**5.** Find the restrictions on the variable for the expression $\dfrac{x^2 - 3x - 28}{2x^2 - 15x + 7}$.	8.1
______	**6.** Does the ordered pair $\left(\dfrac{1}{2}, \dfrac{-5}{4}\right)$ satisfy $3x - 2y = 1$?	10.1

Simplify.

	Problem	SECTION
______	**7.** $\dfrac{-8(7 - 2) - 4(-2)}{-4(-12 + 10)}$	2.3
______	**8.** $-[5x - 5(x - 4)] - 5[2(x - 2) - 2x]$	3.3
______	**9.** $7a^3b(-ab^2)^4$	5.1
______	**10.** $\left(\dfrac{-x^{-2}y^3}{2xy^{-4}}\right)^{-3}$	5.3
______	**11.** $-x^2y^3(2x + 4xy^4) + 5x^3y(y^2 - 2y^6)$	6.3
______	**12.** $(3x - 2y)^2$	6.3
______	**13.** $(x - 5)(x^2 + 5x + 25)$	6.3
______	**14.** $(x^4 - 16) \div (x - 2)$	6.4
______	**15.** $\dfrac{18x^2}{34y^3} \div \dfrac{9y^2}{17x}$	8.2
______	**16.** $\dfrac{15x^2 - 47x + 36}{5x^2 - 24x + 27} \cdot \dfrac{x^2 + 5x - 24}{3x^2 + 2x - 8}$	8.2
______	**17.** $\dfrac{5x - 15}{5x^2 - 16x + 3} - \dfrac{2}{5x^2 - x}$	8.4

Factor completely.

	Problem	SECTION
______	**18.** $a^3b^2 - 14a^2b^3 + 4ab^4$	7.3
______	**19.** $63m^2 + 38mn - 16n^2$	7.3

Solve.

______ **20.** $3[x - (2x - 1)] = -2(x - 3) - 6$ 4.2

______ **21.** $4x(x - 1) - 40 = 3(x^2 + 5) + 2x$ 7.5

______ **22.** $\frac{x}{x-9} + \frac{1}{9} = \frac{9}{x-9}$ 8.6

______ **23.** $\frac{-x}{2} + 1 \geq \frac{2x}{5} - 2$ 9.2

______ **24.** Solve by substitution. 11.2

$$3x - 9y = -13$$
$$2x + y = 3$$

______ **25.** Solve by elimination. 11.3

$$9x - 5y = 8$$
$$4x + 3y = 14$$

26. Graph $y = 3x + 4$ using the slope-intercept method. 10.4

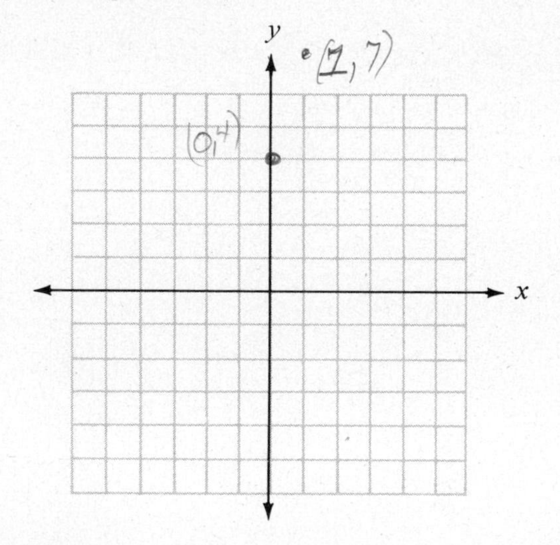

27. Graph $x + 2 = 0$ by inspection. 10.3

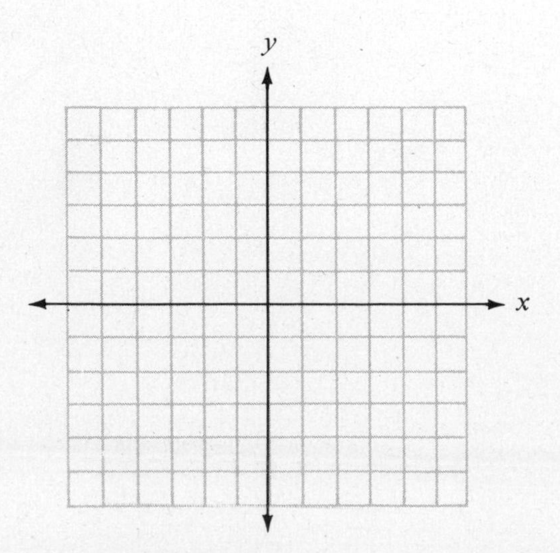

_______ **28.** The support beams of a barn form a right triangle. The longest beam is 17 feet and the shortest beam is 7 feet less than the other. Find the length of the piece of lumber that would be needed to cut all three beams. 7.6

_______ **29.** A car rental agency charges customers $35.50 per day plus $0.25 per mile driven. If Lena's car rental bill was $134.25 for 1 day, find how many miles she drove the car. 4.3

_______ **30.** Ray and Glenn are planning to drive to Canada, sharing the time at the wheel. Ray drives at one constant rate and Glenn drives at another constant rate. If Ray drives for 7 hours and Glenn for 4 hours, they can travel 680 miles. If Ray drives 5 hours and Glenn for 6 hours, they can travel 690 miles. How fast does each one drive? 11.4

Check your answers in the back of your book.

Answers

This section includes the answers to Odd-Numbered Exercises, Stretching the Topics, Checkups, Speaking the Language of Algebra, Review Exercises, Practice Tests, and Sharpening Your Skills.

CHAPTER 1

Exercise Set 1.1 (page 17)

1. 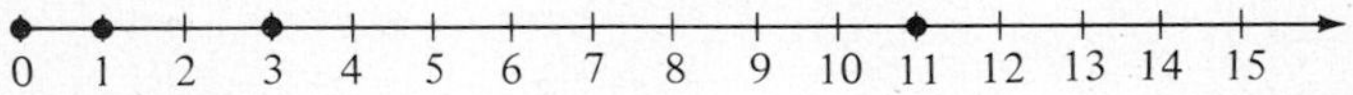

3. $A(2)$; $B(6)$: $C(9)$; $D(12)$ **5.** $12 > 7$ **7.** $0 < 9$ **9.** $22 < 55$ **11.** $32 = 15 + 17$
13. $35 = 7 \cdot 5$ **15.** 6 **17.** 12 **19.** 48 **21.** 160 **23.** 2 **25.** 47 **27.** 5
29. 64 **31.** 5 **33.** 30 **35.** 64 **37.** 37 **39.** 2 **41.** 28 **43.** Undefined
45. 28 **47.** 37 **49.** 96 **51.** 0 **53.** 31 **55.** 40 **57.** 28 **59.** 0

Stretching the Topics

1. 56 **2.** 1 **3.** 0

Checkup 1.1 (page 21)

1. 80 **2.** 100 **3.** 40 **4.** $\frac{11}{16}$ **5.** 6 **6.** 23 **7.** 92 **8.** 85 **9.** 0
10. Undefined

Exercise Set 1.2 (page 35)

1. $\frac{1}{5}$ **3.** $\frac{7}{5}$ **5.** $\frac{9}{16}$ **7.** $\frac{5}{7}$ **9.** $\frac{3}{2}$ **11.** $\frac{1}{7}$ **13.** $\frac{2}{21}$ **15.** 3 **17.** $\frac{2}{5}$ **19.** $\frac{1}{15}$
21. 9 **23.** $\frac{14}{15}$ **25.** $\frac{7}{3}$ **27.** $\frac{1}{15}$ **29.** $\frac{55}{4}$ **31.** $\frac{9}{10}$ **33.** $\frac{42}{49}$ **35.** $\frac{96}{36}$ **37.** $\frac{36}{99}$
39. $\frac{51}{45}$ **41.** $\frac{63}{72}$ **43.** $\frac{45}{5}$ **45.** $\frac{93}{252}$ **47.** $\frac{4}{5}$ **49.** $\frac{1}{3}$ **51.** $\frac{5}{13}$ **53.** $\frac{1}{11}$ **55.** 1
57. 0 **59.** $\frac{8}{15}$ **61.** $\frac{2}{9}$ **63.** $\frac{11}{5}$ **65.** 1 **67.** $\frac{47}{90}$ **69.** $\frac{19}{15}$

Stretching the Topics

1. $\frac{43}{60}$ **2.** $\frac{43}{150}$ **3.** $\frac{1}{10}$

Checkup 1.2 (page 39)

1. $\frac{7}{9}$ **2.** $\frac{40}{56}$ **3.** $\frac{2}{21}$ **4.** $\frac{3}{5}$ **5.** 15 **6.** $\frac{117}{80}$ **7.** $\frac{10}{21}$ **8.** $\frac{13}{3}$ **9.** $\frac{29}{56}$ **10.** $\frac{41}{35}$

Exercise Set 1.3 (page 51)

1. $\frac{4}{5}$ **3.** $\frac{11}{20}$ **5.** $\frac{7}{40}$ **7.** $\frac{3}{250}$ **9.** 2.300 **11.** $15.00 **13.** 13.200 **15.** 0.75
17. 0.63 **19.** 0.83 **21.** 0.53 **23.** 0.36 **25.** 1.53 **27.** $25.80 **29.** 5.285
31. 2.614 **33.** 0.02106 **35.** 0.02616 **37.** 3.9 **39.** 0.203 **41.** 0.19 **43.** 0.371
45. 1.50 **47.** 0.005 **49.** 9% **51.** 710% **53.** 3.2% **55.** 175% **57.** 61.1
59. 6.48 **61.** 388.8 **63.** 498 **65.** 222 **67.** 937.8 **69.** 312.5

Stretching the Topics

1. 3873.75 **2.** $625 **3.** $1.50

Checkup 1.3 (page 53)

1. $\frac{3}{125}$ **2.** 43.500 **3.** 0.875 **4.** 0.0325 **5.** 12.5% **6.** 5.862 **7.** $7.02
8. 0.224 **9.** 0.008 **10.** 27.625

Exercise Set 1.4 (page 61)

1. $49.17 **3.** 285 miles **5.** $1495 **7.** $7000 **9.** 430 **11.** 21 **13.** $\frac{3}{10}$
15. $7\frac{8}{15}$ **17.** 7 m **19.** 251.25 **21.** $21.25 **23.** 5343.61 sq m **25.** $5.39
27. 35.75 **29.** 1800

Stretching the Topics

1. $105.66 **2.** $13.39

Checkup 1.4 (page 67)

1. $\frac{13}{20}$ **2.** $673.69 **3.** 29.8 **4.** $148.47 **5.** $101.75

Speaking the Language of Algebra (page 68)

1. Whole numbers **2.** Sum; difference **3.** Factors; product **4.** Dividend; divisor; quotient
5. Proper; improper **6.** Common factors **7.** Invert; multiply **8.** Same denominator
9. Whole number; fractional **10.** Per hundred; two; left

Review Exercises (page 69)

1. $3 < 19$ **2.** $43 = 32 + 11$ **3.** $91 = 13 \cdot 7$ **4.** 57 **5.** 72 **6.** 76 **7.** 72
8. 35 **9.** 8 **10.** Undefined **11.** 0 **12.** $\frac{3}{5}$ **13.** $\frac{1}{8}$ **14.** $\frac{1}{27}$ **15.** $\frac{3}{8}$ **16.** $\frac{32}{75}$
17. $\frac{4}{9}$ **18.** 0.3125 **19.** 13.1 **20.** 17.1665 **21.** $177.49 **22.** 0.088 **23.** 5.2
24. 9% **25.** 0.127 **26.** $660 **27.** $50 **28.** $7275 **29.** $\frac{13}{20}$ **30.** 6.5 lb

Practice Test for Chapter 1 (page 71)

1. $24 = 13 + 11$ **2.** $55 = 11 \cdot 5$ **3.** 192 **4.** 102 **5.** 37 **6.** Undefined **7.** 0
8. $\frac{10}{21}$ **9.** $\frac{1}{18}$ **10.** 15 **11.** $\frac{31}{36}$ **12.** $\frac{11}{12}$ **13.** 0.375 **14.** 7.14 **15.** 19.485
16. 0.8225 **17.** 5.02 **18.** $22.95 **19.** $5.47 **20.** 4.05

CHAPTER 2

Exercise Set 2.1 (page 85)

1.

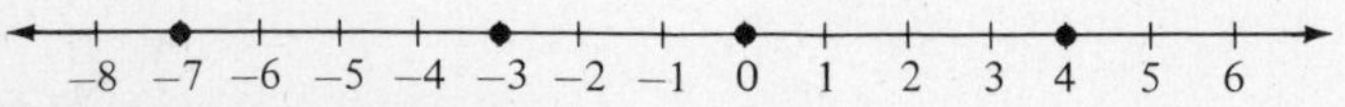

3. $A(-5)$; $B(-2)$; $C(0)$; $D(3)$ **5.** **(a)** $5 < 12$ **(b)** $-4 > -10$ **(c)** $8 > -11$
(d) $-3 < 7$ **7.** **(a)** $2 + (-7)$ **(b)** $-9 + 6$ **(c)** $-6 + (-20)$ **(d)** $10 + 8$
9. **(a)** $6 + [3 + (-2)]$ **(b)** $(-5 + 4) + (-3)$ **(c)** $-4 + [9 + (-2)]$

(d) $[7 + (-2)] + 11$ **11.** 5 **13.** -8 **15.** 3 **17.** 12 **19.** 11 **21.** 22
23. -12 **25.** 12 **27.** -6 **29.** -7 **31.** 13 **33.** 20 **35.** 8 **37.** -17
39. -14 **41.** 18 **43.** -21 **45.** 0 **47.** -11 **49.** 10 **51.** -12.7 **53.** -12
55. 24 **57.** 3 **59.** 3 **61.** Undefined **63.** -7 **65.** 0 **67.** -8 **69.** 0

Stretching the Topics
1. 65 **2.** 36 **3.** 120

Checkup 2.1 (page 89)
1. $3 > -15$ **2.** $-5 + 17$ **3.** $[-9 + (-3)] + 7$ **4.** -12 **5.** -5 **6.** 8
7. -1 **8.** -15 **9.** 0 **10.** 1

Exercise Set 2.2 (page 95)
1. $12 + (-9) = 3$ **3.** $11 + (-27) = -16$ **5.** $-15 + (-7) = -22$ **7.** $12 + 3 = 15$
9. $-9 + 7 = -2$ **11.** 10 **13.** 25 **15.** -11 **17.** -9 **19.** 3 **21.** -12
23. -10 **25.** -6 **27.** 0 **29.** 1 **31.** 15 **33.** -2 **35.** -6 **37.** 40 **39.** 3
41. -16 **43.** -7 **45.** 13 **47.** 2 **49.** 2 **51.** Undefined **53.** -21 **55.** 0
57. -26

Stretching the Topics
1. -28 **2.** 78 **3.** 18.9

Checkup 2.2 (page 97)
1. 7 **2.** 21 **3.** -22 **4.** -10 **5.** -18 **6.** 6 **7.** 1 **8.** -8 **9.** -1
10. 23

Exercise Set 2.3 (page 111)
1. (a) $-8(-9)$ **(b)** $-12(5)$ **(c)** $6 \cdot 0$ **(d)** $7(-3)$ **3. (a)** $3(4 \cdot 2)$ **(b)** $(6 \cdot 5)(-7)$
(c) $[-7(-3)](10)$ **(d)** $9[-2(-8)]$ **5. (a)** $-3(7) + (-3)(2)$ **(b)** $5(-8) - 5(3)$
(c) $4 \cdot 7 - 4(-2)$ **(d)** $-10(-9) + (-10)(8)$ **7.** -30 **9.** 42 **11.** -24 **13.** 150
15. 0 **17.** 45 **19.** -80 **21.** 37 **23.** 64 **25.** -27 **27.** -288 **29.** -4
31. -2 **33.** 5 **35.** -2 **37.** -7 **39.** -9 **41.** -12 **43.** Undefined
45. -1 **47.** 1 **49.** -4 **51.** -7 **53.** $\frac{-1}{6}$ **55.** -13 **57.** -3 **59.** 51
61. 200 **63.** 1 **65.** -71

Stretching the Topics
1. 63 **2.** -256 **3.** -34

Checkup 2.3 (page 115)
1. $5(-3)$ **2.** $[(-3)(2)](-7)$ **3.** $-5(-2) + (-5)(8)$ **4.** -30 **5.** -5 **6.** -2
7. -60 **8.** -4 **9.** -2 **10.** $\frac{-2}{3}$

Exercise Set 2.4 (page 125)
1. 8.3 **3.** 5.7 **5.** $\frac{-2}{3}$ **7.** $\frac{-3}{4}$ **9.** $\frac{8}{21}$ **11.** $\frac{-14}{15}$ **13.** -6 **15.** 4.31
17. $\frac{-3}{7}$ **19.** -8.8786 **21.** 30.104.3 **23.** $\frac{-27}{64}$ **25.** 1.44 **27.** $\frac{1}{625}$ **29.** -0.7
31. 0.22 **33.** $\frac{1}{7}$ **35.** $\frac{5}{2}$ **37.** 1.47 **39.** $\frac{2}{5}$

Stretching the Topics
1. $\frac{16}{25}$ **2.** -0.0002854 **3.** $\frac{-7}{48}$

Checkup 2.4 (page 127)
1. $\frac{4}{5}$ **2.** $\frac{-1}{3}$ **3.** 1.71 **4.** $\frac{-17}{30}$ **5.** $\frac{1}{40}$ **6.** 4.3 **7.** $\frac{-27}{125}$ **8.** 0.1 **9.** -2.8 **10.** -1.58

Exercise Set 2.5 (page 133)

1. $87 - 14 = 73$ **3.** $1605 - 312 = 1293$ **5.** $\frac{\$5.86}{2} = \$2.93$ **7.** $53(4.5) = 238.5$ miles

9. $\frac{2}{3}(25) = 16\frac{2}{3}$ **11.** $3(75¢) - 7(25¢) = 50¢$ won **13.** $\$1.75 + \$1.10(7) = \$9.45$

15. $2(\$-5) + \$7 - \$3 = \$-6$; she owes him \$6.

17. $2\left(150\frac{3}{4} + 250\frac{2}{3}\right) - 3\frac{1}{3} - 6\frac{1}{2} = 793$ **19.** $2(\$3.85) + 4(\$4.90) = \$27.30$

21. $0.055(\$1.69 + \$1.25 + \$0.45) = \$0.19$; $(\$1.69 + \$1.25 + \$0.45) + \$0.19 = \$3.58$; $\$10.00 - \$3.58 = \$6.42$ **23.** $27(2) - 2(13) - 1.5(10) = 13$

Stretching the Topics

1. \$61.59 loss **2.** \$12,668

Checkup 2.5 (page 137)

1. $\$785.28 - (\$119.61 + \$28.75) = \$636.92$ **2.** $3(20) - 2(10) = 40$ ft

3. $\frac{\$625{,}823 - \$125{,}729}{3} = \$166{,}698$ **4.** $48 + \frac{2}{9} + \frac{1}{8} + 0 - \frac{1}{6} - \frac{2}{3} = 47\frac{37}{72}$ ft

5. $\frac{\$-5(3) + \$7(2) - \$2.50(4)}{9} = \$-1.22$; loss of \$1.22

Speaking the Language of Algebra (page 141)

1. Integers **2.** Integer; nonzero integer **3.** Irrational **4.** Absolute value of *a*; distance; 0
5. Negative **6.** Positive; negative **7.** Zero
8. Addition, multiplication; subtraction, division **9.** Opposite **10.** Positive; negative
11. Positive; positive; negative **12.** *A*; *n*

Review Exercises (page 143)

1.

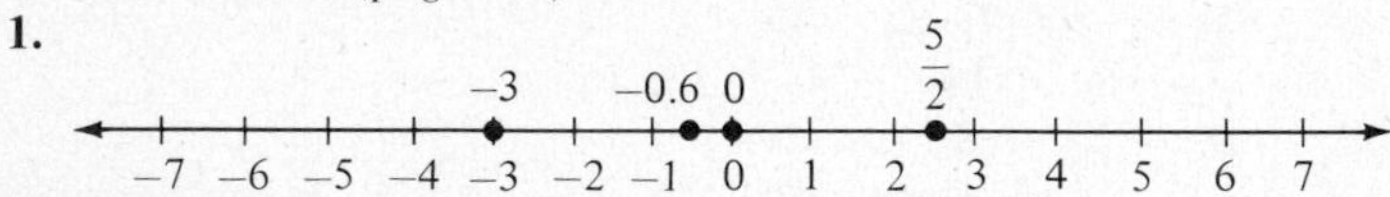

2. $-6 + (-5)$ **3.** $-3 + (-2 + 5)$ **4.** $5(-4)$ **5.** $[-7(-8)](9)$
6. $-2(-6) + (-2)(9)$ **7. (a)** $4 < 14$ **(b)** $-5 > -11$ **(c)** $9 > -12$ **(d)** $-9 < 3$
8. 13 **9.** −18 **10.** 3 **11.** −5 **12.** 0 **13.** 38 **14.** −16 **15.** 9 **16.** 120
17. 1 **18.** 105 **19.** −9 **20.** −7 **21.** 14 **22.** 3 **23.** 0 **24.** 8
25. Undefined **26.** 1.38 **27.** $\frac{-13}{12}$ **28.** −0.21 **29.** −42.2 **30.** $\frac{-5}{6}$
31. Won \$3 **32.** \$232.42 **33.** $\$102\frac{1}{4}$ **34.** $\frac{25}{72}$ **35.** Above 0.33 in.

Practice Test for Chapter 2 (page 147)

1.

−1.6 $-\frac{2}{3}$ 0 0.7 $\frac{9}{4}$

−7 −6 −5 −4 −3 −2 −1 0 1 2 3 4 5 6 7

2. $5(-3)$ **3.** $(-4 + 9) + (-2)$ **4.** $-4(11) + (-4)(-2)$ **5. (a)** $-15 < -1$
(b) $0 > -3$ **(c)** $6 > 5.9$ **(d)** $\frac{1}{3} < \frac{1}{2}$ **6.** −8 **7.** −13 **8.** 7 **9.** 0 **10.** 42
11. −30 **12.** −16 **13.** −2 **14.** $\frac{-23}{3}$ **15.** 0 **16.** Undefined **17.** −5.45
18. $\frac{-35}{24}$ **19.** −0.96 **20.** −140 **21.** $\frac{189}{640}$ **22.** 1.75 lb **23.** $\frac{1}{3}$ c.

Chapter 3

Exercise Set 3.1 (page 157)

1. $y + 10$ **3.** $-3m$ **5.** $a - 6.5$ **7.** $\frac{t}{-15}$ **9.** $x + (-9)$ **11.** $3.2q$

13. $n - 6.3$ **15.** $\frac{2x}{7}$ **17.** $\frac{1}{6}k + 9$ **19.** $\frac{p}{7} - 5$ **21.** $9 - 4x$ **23.** $-4(x + 1.5)$
25. 7; -6 **27.** -3; 0 **29.** 42; -49 **31.** -32; 40 **33.** -4; 0 **35.** 1; -3
37. \$2.50y; \$25; \$62.50 **39.** $75 - x$; 66.3; 61.8 **41.** $200 - 3n$; 180.5; 110
43. \$50 + \$12n; \$122; \$194 **45.** \$587.39 − $6x$; \$437.39; \$287.39 **47.** $\frac{x}{32}$; \$3; \$5.50
49. $2N - 3$; 27; 47

Stretching the Topics
1. $\frac{4y - 3x}{15}$ **2.** -7 **3.** $2108 - \frac{1}{3}x$; 2084 sq ft

Checkup 3.1 (page 161)
1. $x + (-3)$ **2.** $11 - a$ **3.** $-6m$ **4.** $2b + 5$ **5.** -3; 0 **6.** -4; undefined
7. -21; 0 **8.** 1; 3 **9.** $\frac{x}{3}$; \$50,000; \$120,000 **10.** 0.06(\$2500) + \$500x; \$1650; \$7650

Exercise Set 3.2 (page 167)
1. $3x, -5$ **3.** $\frac{1}{2}a, -7b, 9$ **5.** $1, -5m, 6m$ **7.** $13x, -5y, 2z, -7$
9. -19 (constant); $4x$ (variable) **11.** $2x, 3y$ (variable)
13. 5.2 (constant); $-4a, -7b$ (variable) **15.** $1, -4$ (constant); $7m, -5n, 14p$ (variable)
17. 3 is coefficient of x **19.** $\frac{1}{4}$ is coefficient of x; -12 is coefficient of y
21. 3 is coefficient of a; 4 is coefficient of b
23. 0.3 is coefficient of m; -1 is coefficient of n; -2 is coefficient of p **25.** $8a$ and $-3a$
27. $3h$ and $5h$; $-2k$ and $5k$ **29.** $\frac{2}{5}a$ and $3a$; -6 and -9
31. $3x$ and $-1.2x$; $-0.24y$ and $4y$; 1 and -7 **33.** $-2a$ **35.** $3x$ **37.** $9 - 0.5y$
39. $15m + 3$ **41.** $7h + k - 4$ **43.** $11x + 5y + 7z$ **45.** $-8a + 10b + 7$
47. $m - 11n + 4p$ **49.** $-v - 6w$

Stretching the Topics
1. $-0.8a + 1.79b + 0.22c$ **2.** $\frac{-7}{10}x - \frac{17}{3}y + \frac{7}{4}z$ **3.** $20m + 20n - 5p$

Checkup 3.2 (page 169)
1. $3, -7x, 2y$ **2.** $11, -13$ (constant); $-4a, 7a$ (variable)
3. 3 is coefficient of x, -1 is coefficient of y, -4 is coefficient of z.
4. $5x$ and $-7x$; $-2y$ and $2y$; 4 and -1 **5.** $-4a$ **6.** $4x$ **7.** $5 + 8y$ **8.** $m + 2$
9. $-x + 7y - 6$ **10.** $-5b + 2c$

Exercise Set 3.3 (page 179)
1. $8x$ **3.** -1.5 **5.** -9 **7.** $6x$ **9.** $-3y$ **11.** $30k$ **13.** $-5x$ **15.** $5x$ **17.** a
19. m **21.** $-5y$ **23.** $-n$ **25.** $11x$ **27.** $7x$ **29.** $\frac{1}{4}a - \frac{1}{4}b$ **31.** $-3x - 6y$
33. $9x - 3y$ **35.** $-20a + 28b$ **37.** $-18m - 27n$ **39.** $3x - 12y$ **41.** $\frac{-1}{2}x + 5y$
43. $3a + 7b - 5.4$ **45.** $x + 1$ **47.** $7a + 5b$ **49.** $-6m$ **51.** $9x - 4y$
53. $11m - 20$ **55.** $-7x + 10y$ **57.** $4a + 14$ **59.** $10x - 15y$ **61.** $-38a + 6b$
63. $-20x$

Stretching the Topics
1. $3.2x - 0.7y + 0.3$ **2.** $-9a + 11b$ **3.** $90x + 135$

Checkup 3.3 (page 181)
1. $-32x$ **2.** $40y$ **3.** $-4x$ **4.** $-a$ **5.** $-5x$ **6.** $-10x + 15y$ **7.** $2x - 8$
8. $12a - 15b$ **9.** $-6a + 6$ **10.** $49n$

Speaking the Language of Algebra (page 183)

1. Numerical coefficient; numerical coefficient; constant **2.** Numerical coefficients
3. Like terms **4.** Distributive **5.** Innermost

Review Exercises (page 185)

1. $6x$ **2.** $y + 9$ **3.** $3m + 10$ **4.** $23.7 - 4k$ **5.** $9(x + 7)$ **6.** -7; -21
7. -8; undefined **8.** -6; 0 **9.** 17 **10.** 0; -4 **11.** $4h$; 14 miles **2.** $\frac{x}{3}$; \$70,000
13. \$200 + \$19.50d; \$258.50 **14.** $3x$ (variable); -12 (constant) **15.** $-2x$, $7y$ (variable)
16. $4a$, $-3b$ (variable); 2 (constant) **17.** $3m$, $4n$, $-p$ (variable); -2, -9 (constant)
18. 3 is coefficient of x **19.** 7 is coefficient of x; -2 is coefficient of y
20. -3 is coefficient of a; 1 is coefficient of b
21. 1 is coefficient of m; -1 is coefficient of n; 2 is coefficient of p **22.** $-2x$ **23.** $5m + 2$
24. $-7a - 13b + 3$ **25.** $4x + 5z$ **26.** $-20x$ **27.** $24y$ **28.** $-6x$ **29.** a
30. $-3x$ **31.** $28a - 35b$ **32.** $-7x + 2y$ **33.** $-4x + 7$ **34.** $-16m$
35. $-9a - 6$ **36.** $20x - 50y$

Practice Test for Chapter 3 (page 187)

1. $m - 5$ **2.** $3x + 9$ **3.** $10(b + 4.6)$ **4.** 2; -16 **5.** -14; -16 **6.** $\frac{19}{3}$
7. \$33.79n; \$168.95 **8.** \$200 + \$4.10x; \$323 **9.** $\frac{\$180}{m}$; \$4 **10.** $-4n$ **11.** $-3x - 7$
12. $-4y + 2z$ **13.** $-16y$ **14.** $6x$ **15.** $-12n$ **16.** $6x + 2$ **17.** $-18x + 12y$
18. $4y - 7x$ **19.** $-12a + 12$ **20.** $-42x - 14y$

Sharpening Your Skills after Chapters 1–3 (page 189)

1. $\frac{7}{8}$ **2.** 0.83 **3.** $\frac{8}{27}$ **4.** 3.15 **5.** $2 \cdot 3 - 7 > -8$ **6.** Undefined **7.** $\frac{61}{40}$
8. 0.5664 **9.** 9 **10.** 27 **11.** 12 **12.** 4 **13.** 16 **14.** 6
15. $\frac{-13 + 12}{7} = -\frac{1}{7}$

Chapter 4

Exericse Set 4.1 (page 203)

1. Yes **3.** Yes **5.** Yes **7.** No **9.** Yes **11.** No **13.** $x = 7$ **15.** $x = 12$
17. $y = 2.6$ **19.** $y = -8$ **21.** $m = 5$ **23.** $a = 10$ **25.** $k = 6$ **27.** $x = 0$
29. $m = \frac{-5}{2}$ **31.** $y = -8$ **33.** $a = -7$ **35.** $x = 0$ **37.** $x = 7$ **39.** $x = -36$
41. $x = \frac{-35}{2}$ **43.** $x = 0.6$ **45.** $x = 0$ **47.** $x = -6.5$ **49.** $x = 12$ **51.** $x = -4$
53. $x = -56$ **55.** $x = 0$ **57.** $x = -40$ **59.** $x = 63$ **61.** $x = -6$ **63.** $x = \frac{-27}{2}$

Stretching the Topics

1. $x = \frac{7}{12}$ **2.** $x = -1.5$ **3.** $y = d - c$ **4.** $y = \frac{d}{a}$

Checkup 4.1 (page 205)

1. Yes **2.** $x = -1$ **3.** $x = 7$ **4.** $x = 1.2$ **5.** $x = 0$ **6.** $x = 9$ **7.** $x = -12$
8. $x = 0$ **9.** $x = 24$ **10.** $x = -60$

Exercise Set 4.2 (page 217)

1. $x = 7$ **3.** $x = 10$ **5.** $y = -5$ **7.** $x = 1$ **9.** $a = -9$ **11.** $m = \frac{-13}{2}$
13 $x = -13$ **15.** $x = -3$ **17.** $x = 13$ **19.** $y = 0$ **21.** $a = 8$
23. $x = 10$

25. $x = 6$ **27.** $x = 0$ **29.** $x = -10$ **31.** $x = 12$ **33.** $a = -4$ **35.** $y = 16$
37. $x = -4$ **39.** $x = 5$ **41.** $x = 16$ **43.** $x = -3$ **45.** $x = \frac{35}{2}$ **47.** $x = 1$
49. $x = 5$ **51.** $x = 2$ **53.** $x = 0.7$ **55.** $x = -2$ **57.** $x = \frac{15}{7}$ **59.** $x = -4$
61. $x = -2$ **63.** $x = -4$ **65.** $x = 0$ **67.** $x = -11$ **69.** $x = 3$ **71.** $x = -2$
73. $x = 1$ **75.** $x = -3$ **77.** $d = \frac{C}{\pi}$ **79.** $y = \frac{3 - x}{2}$ **81.** $t = \frac{I}{pr}$
83. $a = P - b - c$ **85.** $p = \frac{k - c}{2c}$ **87.** $x = b$ **89.** $n = \frac{l - a + d}{d}$

Stretching the Topics
1. $x = -0.5$ **2.** $x = -2$ **3.** $M = \frac{6V - hB - hb}{4h}$

Checkup 4.2 (page 221)
1. $x = 5$ **2.** $x = -1$ **3.** $y = -6$ **4.** $x = -3$ **5.** $x = 4$ **6.** $x = -8$
7. $x = 6$ **8.** $y = \frac{13}{6}$ **9.** $y = 8$ **10.** $r = \frac{A - P}{Pt}$

Exercise Set 4.3 (page 229)
1. Lydia, 5 lb; Lyle 7 lb **3.** 9 hr **5.** Julia, 15 tickets; Sara, 60 tickets **7.** 13 m
9. $0.55 **11.** $36 **13.** 12 lb **15.** 31 by 36 in. **17.** $1200 **19.** 66 **21.** $1980
23. $420,000 **25.** $415

Stretching the Topics
1. 14, 13, 12, 11 in. **2.** $43,000

Checkup 4.3 (page 231)
1. 14 **2.** 9 by 27 in. **3.** 24 **4.** 80 **5.** 180 bu

Speaking the Language of Algebra (page 233)
1. Mathematical sentence **2.** Solution; satisfies **3.** Isolate **4.** Substitute; original equation

Review Exercises (page 235)
1. Yes **2.** No **3.** No **4.** Yes **5.** $x = 10$ **6.** $y = -14$ **7.** $m = 4$
8. $a = 2$ **9.** $x = 8$ **10.** $x = -18$ **11.** $x = -7$ **12.** $x = 0$ **13.** $x = -3$
14. $y = -14$ **15.** $x = \frac{19}{2}$ **16.** $x = 1$ **17.** $x = -3$ **18.** $x = 4$ **19.** $x = -6$
20. $x = 4$ **21.** $x = 4$ **22.** $x = -10$ **23.** $y = 5$ **24.** $x = \frac{17}{4}$ **25.** $a = 2$
26. $x = -19$ **27.** $x = -3$ **28.** $x = 0$ **29.** $x = 11$ **30.** $a = \frac{P - c}{2}$ **31.** 45
32. 4 by 12 ft **33.** 4

Practice Test for Chapter 4 (page 237)
1. Yes **2.** No **3.** $y = 13$ **4.** $x = 13$ **5.** $x = -10$ **6.** $x = 4$ **7.** $x = 8$
8. $x = -7$ **9.** $x = \frac{9}{2}$ **10.** $x = 1$ **11.** $x = -15$ **12.** $y = 15 - 10x$
13. Harvey, 37 tickets; Martha, 148 tickets **14.** 2 by 5 in. **15.** 83 mi

Sharpening Your Skills after Chapters 1–4 (page 239)
1. -86 **2.** $\frac{8}{15}$ **3.** 104 **4.** $\frac{15}{8}$ **5.** -11.37 **6.** 16 **7.** -7 **8.** $-2x - 3y$
9. $-6x$ **10.** $12x + 28y$ **11.** 49 **12.** -3 **13.** $\frac{2}{3} < \frac{3}{4}$ **14.** -10
15. 500n$, $6000

Chapter 5

Exercise Set 5.1 (page 251)

1. $3x^2y^2$ **3.** $-a^4$ **5.** $(-2x)^3$ **7.** $3x^3 - 4y^3$ **9.** $8 \cdot a \cdot a \cdot a$ **11.** $(3a)(3a)(3a)$ **13.** $-1x \cdot x \cdot x \cdot x$ **15.** $\frac{2}{3}x \cdot x \cdot x \cdot y \cdot y \cdot y$ **17.** -27 **19.** -16 **21.** 4 **23.** 64 **25.** 1 **27.** x^5 **29.** a^7 **31.** y^9 **33.** $-2a^{12}$ **35.** x^7y^5 **37.** $2m^8$ **39.** $\frac{1}{2}x^9y^4$ **41.** $0.5x^4y^4z^5$ **43.** $3x^9yz^2$ **45.** a^6 **47.** x^5y^5 **49.** $49x^2$ **51.** $\frac{1}{9}y^4$ **53.** $-64x^3$ **55.** x^{18} **57.** $81x^4y^6$ **59.** $16x^4y^{12}$ **61.** $\frac{1}{4}x^5y^6$

Stretching the Topics

1. $\frac{-1}{8}x^{14}y^{16}z^{18}$ **2.** $42x^4y^3$ **3.** $x^{12}y^5z^9$ **4.** $72a^{19}b^{14}c^4$

Checkup 5.1 (page 253)

1. 16 **2.** $\frac{-1}{3}x^4$ **3.** $0.5x \cdot x \cdot x \cdot y \cdot y$ **4.** -216 **5.** x^8 **6.** $-y^6$ **7.** $-5x^3y^7$ **8.** a^{18} **9.** $\frac{4}{9}x^2$ **10.** $-8x^3y^3$

Exercise Set 5.2 (page 263)

1. x^7 **3.** a^4 **5.** 1 **7.** $\frac{1}{y^2}$ **9.** $4x^3$ **11.** $-3y^5$ **13.** $\frac{1}{3x^2}$ **15.** $\frac{-3}{a^2}$ **17.** $\frac{-x^2}{5y^3}$ **19.** $\frac{7}{ab^2}$ **21.** $\frac{-x}{y^6}$ **23.** $\frac{-2z^2}{y^2}$ **25.** $\frac{7a^3b^2}{6c^2}$ **27.** $\frac{y^4}{x}$ **29.** $\frac{-y^7}{3x}$ **31.** $\frac{x^3}{8}$ **33.** $\frac{-27}{y^3}$ **35.** $\frac{x^8}{y^4}$ **37.** $\frac{-x^{10}}{y^{15}}$ **39.** $\frac{16x^8}{y^4}$ **41.** $\frac{25a^6}{4b^4}$ **43.** $\frac{-32x^{10}y^5}{243z^5}$ **45.** $\frac{9b^6}{a^4}$ **47.** $\frac{-z^9}{27x^6}$ **49.** $4x^2$

Stretching the Topics

1. $\frac{xy^2}{2z^{13}}$ **2.** $-ab$ **3.** $\frac{11}{12}x^2$

Checkup 5.2 (page 265)

1. x^5 **2.** $\frac{-1}{a^3}$ **3.** $\frac{3}{x^2}$ **4.** $\frac{-a^2}{2b^4}$ **5.** $\frac{x}{z^3}$ **6.** xy^4 **7.** $\frac{x^5}{32}$ **8.** $\frac{16}{a^4}$ **9.** $\frac{-x^{12}}{8y^3}$ **10.** $\frac{64a^9}{b^6}$

Exercise Set 5.3 (page 279)

1. $8x^6$ **3.** a^3 **5.** $\frac{5}{y^3}$ **7.** $\frac{-27a^6}{b^3}$ **9.** $\frac{1}{9x^2}$ **11.** $\frac{-1}{125}$ **13.** $\frac{25}{16}$ **15.** $\frac{1}{36x^2}$ **17.** $\frac{6}{x^2}$ **19.** 4 **21.** $\frac{1}{a^3}$ **23.** $2x^4$ **25.** $\frac{-5}{x^3}$ **27.** $\frac{1}{a^{12}}$ **29.** y **31.** $\frac{9}{x^3}$ **33.** $\frac{1}{64x^3}$ **35.** $\frac{27}{a^{12}}$ **37.** $\frac{y^{10}}{64}$ **39.** $\frac{25a^6}{b^{10}}$ **41.** $\frac{x^{10}y^6}{81}$ **43.** $\frac{1}{7}$ **45.** x^{15} **47.** $\frac{3}{5a^{10}}$ **49.** $\frac{1}{3a^7b^{12}}$ **51.** $\frac{-2x^2}{3}$ **53.** $\frac{27}{8x^3}$ **55.** a^2x^4 **57.** $\frac{x^{12}}{y^6}$ **59.** $\frac{9}{x^8y^2}$

Stretching the Topics

1. $\frac{3y^8}{2x^8}$ **2.** $\frac{4b^{20}}{27a^{16}}$ **3.** y^{10}

Checkup 5.3 (page 281)

1. $\frac{1}{32x^5}$ **2.** a^4 **3.** $\frac{3}{y^2}$ **4.** 2 **5.** $-3x^7$ **6.** $\frac{a^2}{2}$ **7.** $\frac{-x^4}{3}$ **8.** $\frac{3}{x^8}$ **9.** $9x^4$
10. $\frac{1}{27a^9b^6}$

Speaking the Language of Algebra (page 284)

1. Base; exponent; a; n **2.** Parenthesis **3.** Add **4.** Multiply **5.** Factor **6.** Subtract
7. Numerator; denominator **8.** Reciprocal **9.** Numerator **10.** 1

Review Exercises (page 285)

1. $-2x^2y^3$ **2.** $\left(\frac{1}{3}a\right)^3 = \frac{a^3}{27}$ **3.** $-3m \cdot m \cdot m \cdot n \cdot n$ **4.** $(0.5a)(0.5a)(0.5a)$ **5.** -27
6. $\frac{1}{16}$ **7.** -2 **8.** 1 **9.** x^{10} **10.** $-2a^9$ **11.** $9x^5y^8$ **12.** m^6 **13.** $-27y^3$
14. $x^6y^4z^2$ **15.** $3a^{13}b^6$ **16.** $\frac{1}{27}x^3$ **17.** $\frac{1}{x^3}$ **18.** $-5x^9$ **19.** $\frac{-3x^3}{y^6}$ **20.** $\frac{-y^2}{x^3}$
21. $\frac{8x^7}{y}$ **22.** $\frac{-27a^6}{8b^3}$ **23.** $\frac{16a^{12}}{b^{20}}$ **24.** $\frac{1}{25x^2}$ **25.** $\frac{4}{y^3}$ **26.** $\frac{1}{x^4}$ **27.** $4a^2$ **28.** $\frac{8}{x^{11}}$
29. $\frac{1}{9x^2}$ **30.** $-7x^8$ **31.** $\frac{a^3}{4}$ **32.** $\frac{16}{81x^4}$ **33.** $4y^2$ **34.** $\frac{x^4}{4}$

Practice Test for Chapter 5 (page 287)

1. -64 **2.** -3 **3.** 4 **4.** x^9 **5.** $-2y^8$ **6.** $4x^7y^3z^4$ **7.** a^{12} **8.** $x^{12}y^6$
9. $-27x^3$ **10.** $16a^{16}y^{12}$ **11.** $5a^7b^3$ **12.** $\frac{1}{x^8}$ **13.** $-2x^5$ **14.** $\frac{-3x^5}{y^4}$ **15.** $\frac{-b^2}{a^3}$
16. $-32x^2y^{11}$ **17.** $\frac{-27a^{15}}{64b^6}$ **18.** $\frac{1296x^{12}}{y^{16}}$ **19.** $\frac{7}{x^4}$ **20.** $\frac{-5}{w^5}$ **21.** $\frac{1}{81a^4}$ **22.** $\frac{x^8}{25y^{12}}$
23. $\frac{-3}{x^4}$ **24.** $\frac{4x^5}{y^5}$ **25.** $\frac{-3a^5b^5}{4}$

Sharpening Your Skills after Chapters 1–5 (page 289)

1. 2 **2.** $-20b$ **3.** $2x - y + z$ **4.** 0.85 **5.** -0.96 **6.** $\frac{2}{21}$ **7.** No
8. $8(x + 13)$ **9.** $x = 22$ **10.** $x = 40$ **11.** $x = 8$ **12.** $x = 3$ **13.** $w = \frac{P - 2l}{2}$
14. Ethel, 80 cans; Sam, 210 cans **15.** 4 by 16 ft

Chapter 6

Exercise Set 6.1 (page 297)

1. Monomial **3.** Binomial **5.** Trinomial **7.** Binomial **9.** Binomial **11.** Trinomial
13. Polynomial of four terms **15.** $4x^2 - 2x + 4$ **17.** $5a - 3c$ **19.** $-4x^2 - 2x - 9$
21. $-12x^2$ **23.** $-8xy$ **25.** $6a$ **27.** $-0.8ab^3$ **29.** $-6x^5y^4$ **31.** $5x^2 - 15x + 10$
33. $-6a + 4b - c$ **35.** $-8a^2 + 8a - 4$ **37.** $3x^2 - 2x + 1$ **39.** $-x + 2y$
41. $4x - 17y$ **43.** $0.4x^2 - 2.7x + 1$ **45.** $-5y^2 - 12y + 21$ **47.** 1 **49.** $-y^2 - 3$

Stretching the Topics

1. $-7h + 30k$ **2.** $-3xy - 3xz$ **3.** $4a + 2b$

Checkup 6.1 (page 299)

1. Trinomial **2.** Monomial **3.** $3x^2 - 9$ **4.** $12xy$ **5.** $12a^3b^2$ **6.** $2x^2 - 4x + 1$
7. $-3a + 4b + 7c$ **8.** $-5x^2 + 11x + 20$ **9.** $-2x^2 - 5x + 6$ **10.** $2y^2 - 4y + 10$

Exercise Set 6.2 (page 305)

1. $10x^5$ **3.** $2x^8$ **5.** $10a^{12}$ **7.** $-0.9y^4$ **9.** $-6m^8$ **11.** $-12x^7y^5$ **13.** $2x^7y^4$
15. $42x^4y^3$ **17.** $35x^3y^4z^5$ **19.** $-24x^5yz^2$ **21.** $20x^7$ **23.** $-200y^{23}$ **25.** $-24a^4b^7$

27. $10x + 15$ **29.** $-14x^2 + 28$ **31.** $6x^2 - 12x$ **33.** $2y^3 - 5y^2$ **35.** $-a^4 + 2a^2$
37. $36x^4 - 16x^2$ **39.** $-21a^5 + 9a^4$ **41.** $10x^2y + 4xy^2$ **43.** $-11x^3y^2 + 4x^2y^3$
45. $14a^5b^2 - 2a^3b^4$ **47.** $2m^3 - 10m^2 + 12m$ **49.** $15x^4y^4 - 5x^3y^3 - 20x^2y^2$
51. $-3a^2b + 2a^3b^2 - 4a^4b^3$ **53.** $32m^4n - 8m^3n^2 + 4m^2n^3 - 12mn^4$ **55.** $4x^3y$
57. $x^4 + 2x^2 + 24$ **59.** $9m^2 - n^2$ **61.** $4ax - 5axy$ **63.** $7x^2y - 10xy^2$

Stretching the Topics
1. $-3m^5n^2 - 7m^4n^3 + 2m^3n^4$
2. $3x^3y^2 - 7x^2y^3 + 7x^2y^4$ **3.** $4a^4 - 4a^2b^2 + 22b^3$

Checkup 6.2 (page 307)
1. $-21x^{11}$ **2.** $8a^{12}$ **3.** $-2x^4y^4z^3$ **4.** $-2a^4 + 14a^3 - 10a^2$ **5.** $15x^2y - 6xy^2 + 12xy$
6. $12x^4y^4 - 18x^3y^3 + 24x^2y^2$ **7.** $-8m^4n + 2m^3n^2 - 4m^2n^3 + 2mn^4$ **8.** $-x^3y$
9. $15x^2 + xy + y^2$ **10.** $7ax$

Exercise Set 6.3 (page 315)
1. $x^2 + x - 6$ **3.** $x^2 + 13x + 36$ **5.** $x^2 - 0.8x + 0.15$ **7.** $y^2 - 18y + 81$
9. $a^2 + 0.4a + 0.04$ **11.** $x^2 - 16$ **13.** $y^2 - \frac{1}{9}$ **15.** $12y^2 - 23y + 5$
17. $-6a^2 - 11a + 35$ **19.** $36x^2 - 25$ **21.** $18 - 13y + 2y^2$ **23.** $2a^2 - 11ab + 5b^2$
25. $2x^2 - 5xy - 3y^2$ **27.** $-2x^2 + 5xy - 3y^2$ **29.** $a^2 + 6ab + 9b^2$ **31.** $0.04 - 0.25x^2$
33. $\frac{1}{4}x^2 - xy + y^2$ **35.** $ac + ad - 2bc - 2bd$ **37.** $a^3 - 3a^2 - 4a + 12$
39. $6a^4 - 31a^2 + 5$ **41.** $25a^4 - 20a^2 + 4$ **43.** $a^6 - 2a^3 - 3$ **45.** $x^2y^2 - 5xy - 50$
47. $a^2b^2 - 4c^2$ **49.** $9y^2 - 12yz + 4z^2$ **51.** $36a^4 - b^2$ **53.** $100 + 10x^2 + \frac{1}{4}x^4$
55. $x^3 - 5x^2 + 7x - 3$ **57.** $b^3 - 125$ **59.** $x^4 + 3x^3 - 2x^2 - 12x - 8$
61. $2 + x + x^2 + 6x^3$ **63.** $0.3x^3 - 2.1x^2 + 3x$ **65.** $-49y^2 + y^4$

Stretching the Topics
1. $6x^3 - 19x^2y + 16xy^2 - 4y^3$ **2.** $x^3 - 3x^2 + 3x - 1$ **3.** $2x^2 - xy - 3y^2$

Checkup 6.3 (page 317)
1. $x^2 + 3x - 10$ **2.** $2y^2 - 15y + 18$ **3.** $25 - 20y + 4y^2$ **4.** $4a^2 + 20ab + 25b^2$
5. $25 - 4a^2$ **6.** $y^2 - \frac{9}{25}$ **7.** $a^3 - 6a^2 + 7a - 42$ **8.** $3x^4 - 7x^2 - 20$ **9.** $x^3 + 8$
10. $-3x + 0.4x^2 + 0.2x^3$

Exercise Set 6.4 (page 325)
1. $x + 5$ **3.** $8x - 7$ **5.** $4 - 3x$ **7.** $5x^2y - 2$ **9.** $2d - 3 - \frac{1}{4d}$
11. $-y^2 + ay + b$ **13.** $-m^2 - 3mn + 5n^2$ **15.** $3b + 4c - \frac{d}{2}$ **17.** $x + 7$
19. $a - 4$ **21.** $z^2 - 3z + 2$ **23.** $x^2 - 3x + 9 + \frac{-54}{x + 3}$ **25.** $4x - 7$
27. $3x^3 + x^2 + 4x + 1$ **29.** $3x^2 + 3x - 1 + \frac{1}{x - 1}$

Stretching the Topics
1. $x^4 + 2x^3y + 4x^2y^2 + 8xy^3 + 16y^4$ **2.** $3a^3 + 4a^2b + 5ab^2 + 2b^3$ **3.** $4x$, $8x + 6xy^3$

Checkup 6.4 (page 327)
1. $2x^2 - \frac{1}{3}$ **2.** $-a + 4b + \frac{3b^2}{a}$ **3.** $2y + 3 + \frac{-3}{y + 4}$ **4.** $x^2 + 2x - 1$
5. $3x^3 + x^2 + 3x + 1$ **6.** $a - 3$

Exercise Set 6.5 (page 335)

1. $4l - 14,\ l^2 - 7l$ **3.** $12x,\ 9x^2$ **5.** $3x + 15,\ \dfrac{x^2 + 7x}{2}$ **7.** $5x^2$ **9.** $x^2 + x$
11. $x^2 + 12x$ **13.** $x^2 + 5x - 45$ (or $x^2 + 23x + 81$) **15.** $x^2 + 4x$
17. $(w^2 + 15w)$ sq rods **19.** $4w + 30$ **21.** $(x^2 + 8x - 20)$ sq ft **23.** $2w^2 + 6w$
25. $\dfrac{x^2 - 9x}{2}$ **27.** $5x^2 + 20x + 25$ **29.** $5x^2 + 12x$

Stretching the Topics

1. $\dfrac{1}{4}l^2 - 3l + 9$ **2.** $2t - \dfrac{1}{t}$ **3.** $n^2 + 4n + 3$

Checkup 6.5 (page 337)

1. $12x,\ 9x^2$ **2.** $16w - 6,\ 7w^2 - 3w$ **3.** $4x + 10,\ \dfrac{x^2 + 7x}{2}$ **4.** $2x^2 + 14x + 40$
5. $4x^2 + 2x + 1$

Speaking the Language of Algebra (page 340)

1. Monomial, one; binomial, two; trinomial, three **2.** Add like terms **3.** Distributive
4. Binomials **5.** $A^2 + 2AB + B^2$ **6.** $A^2 - B^2$ **7.** Monomial **8.** Long division

Review Exercises (page 341)

1. Binomial **2.** Binomial **3.** Monomial **4.** Trinomial **5.** $3x^2 + 3x - 3$
6. $-2a^2 - 3a$ **7.** $-14xy$ **8.** $6x^2$ **9.** $-3.2a^2b^3$ **10.** $-5x^2 - 15x + 10$
11. $-x^2 + 2x - 1$ **12.** $1.6x^2 - 0.6x + 0.4$ **13.** $2x - 6y$ **14.** $-7a - 2b$
15. $-3x^2 - 4x - 14$ **16.** x^{10} **17.** $-6a^8$ **18.** $-18x^5y^8$ **19.** m^6 **20.** $81y^4$
21. $x^6y^4z^2$ **22.** $3a^{13}b^6$ **23.** $\dfrac{1}{8}x^3$ **24.** $4x - 2.4$ **25.** $21y^2 + 14y$
26. $3m^4 - 12m^3n - 15m^2n^2$ **27.** $35x^2y - 55xy^2$ **28.** $-6a^2b + 3a^3b^2 - 4a^4b^3$
29. $2x^2y + xy^2$ **30.** $x^2 + x - 72$ **31.** $12y^2 - 5y - 2$ **32.** $4a^2 - 9b^2$
33. $x^2 - 10xy + 25y^2$ **34.** $4x^2y^2 - 4xy + 1$ **35.** $m^4 + 2m^2 - 15$
36. $6x^4 + 2x^2y^2 - 20y^4$ **37.** $27x^3 - 36x^2 + 12x$ **38.** $2x^3 - 18x$ **39.** $-36a^2 + 16a^4$
40. $y^3 - 4y^2 - y + 4$ **41.** $x^2 - 2x - \dfrac{2}{3}$ **42.** $-x^2 - 15xy + 5y$ **43.** $x + 7$
44. $x - 6 + \dfrac{29}{x + 4}$ **45.** $x^2 - 4x + 5 + \dfrac{-1}{x + 1}$ **46.** $4x^3 + 2x^2 + \dfrac{1}{2x - 1}$
47. $y - 1 + \dfrac{2}{y^2 + y + 1}$ **48.** $l^2 - 10l$ **49.** $5x^2 + 8x + 4$ **50.** $-2x^2 + x$

Practice Test for Chapter 6 (page 343)

1. Monomial **2.** Trinomial **3.** $6x + 6y - z$ **4.** $-2.1a^2$ **5.** $9x^2 - 2x + 16$
6. $-15y^{10}$ **7.** $-20x^5y^3$ **8.** $-5a^2c + 15ac^2 + 5c^3$ **9.** $12x^2y - 24xy^2 + 24x^2y^2$
10. $x^2 + x - 110$ **11.** $12x^2 - 23x + 10$ **12.** $a^2 - 14a + 49$ **13.** $x^2 - \dfrac{4}{9}$
14. $20a^2 + 7ac - 3c^2$ **15.** $y^4 + 0.2y^2 - 0.35$ **16.** $16x^2 - 8xz + z^2$
17. $x^3 + 8$ **18.** $-2y^3 + 162y$ **19.** $-48x^2 + 24xy - 3y^2$ **20.** $3x^2 - 5x - 4$
21. $\dfrac{-1}{y} + \dfrac{2}{x} + \dfrac{5}{6x^2}$ **22.** $2x - 7$ **23.** $x^2 + 3x + 4 + \dfrac{3}{x - 3}$ **24.** $10x^2 + 12x + 4$
25. $x^2 + 6x + 9$

Sharpening Your Skills after Chapters 1–6 (page 345)

1. 0.625 **2.** $-3 + [4 + (-10)]$ **3.** 3 **4.** Yes **5.** -6 **6.** $\dfrac{31}{36}$ **7.** $\dfrac{8}{7}$
8. $6a + 4b + c$ **9.** $-36x + 42y$ **10.** $9x^6y^5$ **11.** $\dfrac{-8x^2}{9y^3z^3}$ **12.** -7 **13.** $x = 5$
14. $x = 8$ **15.** 6 by 9 ft

Chapter 7

Exercise Set 7.1 (page 355)

1. $7(x + 2)$ **3.** $7(3x - 4)$ **5.** $6(7 - y)$ **7.** $6(a^2 - 6)$ **9.** $-6(2x - 3y)$
11. $15(2x^2 + y^2)$ **13.** $7(x^2 - 3x - 1)$ **15.** $5(2y^2 + 3y - 7)$ **17.** $7(4 - a + 5a^2)$
19. $10(5x^2 - xy + 2y^2)$ **21.** $2(x^3 - 3x^2 + 2x + 4)$ **23.** $x(x + 1)$
25. $-y(y^2 - y + 2)$ **27.** $2b(b^2 - 3b - 2)$ **29.** $b^3(7 + 2b^2 - 3b^4)$
31. $3x(x^2 - 2xy + 4y^2)$ **33.** $-a^2b(a^2 - 6ab + 5b^2)$ **35.** $2xy(2x^2y^2 - 10xy + 1)$
37. $x^3y^2z(1 - 2xy - x^2y^2)$ **39.** $9x(7x + 8)$ **41.** $20(5a^3 - 3b^3)$ **43.** $16b^3(1 + 4b)$
45. $6a(a^2 - 7a - 13)$ **47.** $5xy(x^2 + 16y^2)$ **49.** $-7b^2(1 + 6b - 5b^2)$
51. $xy(9x^2 - 12xy + 20y^2)$ **53.** $8a^2(9a^2 - 6a - 1)$ **55.** $4xy(3x^2 - 2xy + y^2)$
57. $9xyz(x^3 + 2x^2y - 3xy^2 - 5y^3)$ **59.** $(5 + c)(a + b)$ **61.** $(x - 5)(x^2 + 6)$
63. $(7a - 4b)(10a + 3)$ **65.** $(y - 2)(3x - 1)$ **67.** $(y^2 + 4)(2x + 3)$
69. $(x - 2)(3x^2 + 1)$

Stretching the Topics

1. $(m - n)(2m - 2n - 1)$ **2.** $3x^{2m}(x^5 + 3)$ **3.** $\frac{1}{6}h(B + b + 4M)$

Checkup 7.1 (page 357)

1. $3(x + 5)$ **2.** $7(5 - 2y)$ **3.** $-x(x - 1)$ **4.** $5(3x^2 - 2xy + 5y^2)$
5. $8a(a^2 - 2a - 5)$ **6.** $3ab(3a^2b^2 - 9ab + 1)$ **7.** $-8m(m^2 + 3)$
8. $x^2yz(x^2 - xy + y^2)$ **9.** $(m + n)(5 + p)$ **10.** $(x + 2)(3x - 1)$

Exercise Set 7.2 (page 363)

1. $(x + 5)(x - 5)$ **3.** $(3y + 1)(3y - 1)$ **5.** $(7 + a)(7 - a)$ **7.** $(2a + 11)(2a - 11)$
9. $\left(5b + \frac{2}{3}\right)\left(5b - \frac{2}{3}\right)$ **11.** $(x + y)(x - y)$ **13.** $(a + 0.6b)(a - 0.6b)$
15. $(2x + 15y)(2x - 15y)$ **17.** $(mn + 12)(mn - 12)$ **19.** $(3ab + c)(3ab - c)$
21. $(0.8xy + 0.5z)(0.8xy - 0.5z)$ **23.** $(x^2 + 9)(x + 3)(x - 3)$
25. $(5x^2 + 11)(5x^2 - 11)$ **27.** $(15 + a^2b^2)(15 - a^2b^2)$ **29.** $3(x + 5)(x - 5)$
31. $11(x + 2y)(x - 2y)$ **33.** $y(4 + 3x)(4 - 3x)$ **35.** $6m(m + 5)(m - 5)$
37. $b\left(7a + \frac{1}{2}b\right)\left(7a - \frac{1}{2}b\right)$ **39.** $2y^2(x + 6y)(x - 6y)$ **41.** $3(m^2 + 9)(m + 3)(m - 3)$
43. $2xy(x + 2y)(x - 2y)$ **45.** $3xy(3xy + 2)(3xy - 2)$ **47.** $a^2(a^2 + 5)(a^2 - 5)$
49. $z^2(17xy + 1)(17xy - 1)$

Stretching the Topics

1. $(a + b + 3c)(a + b - 3c)$ **2.** $(2x + 3z)(2x - 2y - 3z)$ **3.** $(x^a + 5y^b)(x^a - 5y^b)$

Checkup 7.2 (page 365)

1. $(x + 8)(x - 8)$ **2.** $(5 + y)(5 - y)$ **3.** $(3a + 10)(3a - 10)$ **4.** $\left(x + \frac{5}{7}y\right)\left(x - \frac{5}{7}y\right)$
5. $(x^2 + 4)(x + 2)(x - 2)$ **6.** $3(x + 0.5y)(x - 0.5y)$ **7.** $7y^2(x + 2y)(x - 2y)$
8. $5(m^2 + 9m^2)(m + 3n)(m - 3n)$ **9.** $4xy(x + 10y)(x - 10y)$ **10.** $3a(1 + 9a)(1 - 9a)$

Exercise Set 7.3 (page 379)

1. $(x - 4)(x - 3)$ **3.** $(x + 0.5)(x + 0.4)$ **5.** $(x - 12)^2$ **7.** $(a - 7)(a + 2)$
9. $(x + 4)^2$ **11.** $(x - 0.7)(x + 0.5)$ **13.** $(x + 9)(x - 8)$ **15.** $(x - 2y)^2$
17. $\left(x - \frac{1}{5}y\right)^2$ **19.** $(m + 3n)(m - 2n)$ **21.** $(13 - m)(3 - m)$
23. $(xy + 0.5)(xy - 0.3)$ **25.** $(x^2 - 3)(x^2 - 2)$ **27.** $3(x - 3)^2$
29. $-2y(y - 5)(y + 2)$ **31.** $4(m^2 + 3n^2)(m^2 - 2n^2)$ **33.** $xy(x - 10y)(x - 8y)$
35. $2a^2b(a - 16b)(a + 2b)$ **37.** $(2x + 1)(x + 3)$ **39.** $(3a - 5)(3a - 5)$
41. $(5y + 4)(2y - 3)$ **43.** $(4m - 3)(2m + 5)$ **45.** $(3x + 8y)(2x - 7y)$
47. $(9x - 2y)^2$ **49.** $(11 + 6a)^2$ **51.** $5(3c - d)(2c - 9d)$
53. $-2(2y - 7)(2y - 7)$ **55.** Prime **57.** $a^2b^2(4b - 1)(b - 6)$
59. $2xy(3x + 7)(4x - 5)$ **61.** $(0.3k - 5h)^2$ **63.** $(x^2 - 5)(x + 2)(x - 2)$

Stretching the Topics

1. $(a + b - 7)(a + b + 2)$ **2.** $xy(x + 2y)(x - 2y)(x + 7y)(x - 7y)$
3. $(5a^x + 7b^y)(a^x - 3b^y)$

Checkup 7.3 (page 381)

1. $(x - 8y)(x + 5y)$ **2.** $(7 - x)(3 - x)$ **3.** $(x + 1.1)(x - 0.2)$ **4.** $5(x + y)(x - 3y)$
5. Prime **6.** $x\left(x - \frac{1}{3}\right)^2$ **7.** $2xy(x + 8y)(x - y)$ **8.** $(5x - 2)(x - 3)$
9. $-y(4 - 3y)(2 + y)$ **10.** $2ab(2a + 5b)(3a - b)$

Exercise Set 7.4 (page 385)

1. $(x - 6)(x - 5)$ **3.** $6(c - d)(7c - 4d)$ **5.** $(5x + y)(4a - 3)$ **7.** $(7x - 3y)^2$
9. $(4b + 11)(4b - 11)$ **11.** $4a(3a^2b - 4a + 5b^3)$ **13.** $2a(a + 10)(a - 10)$
15. $(ab - 12c)(ab + c)$ **17.** $x^2(3x^2 + 5y^2)^2$ **19.** $(13 + 2a)^2$ **21.** $-8(3x^2 - x + 7)$
23. $(a^2 + 9)(a - 3)$ **25.** $(x^2 + 7)(x + 3)(x - 3)$ **27.** $14a(a - 1)^2$
29. $xyz(x - 5y)(x - 3y)$ **31.** $-x^2(x - 17)(x - 2)$ **33.** $3(7x^2 - 3x + 1)$ **35.** $(x - 9y)^2$
37. $(5x - 9y)(2x + 3y)$ **39.** $-5(6y^2 + 45y - 8)$ **41.** $a^2b^2(2b - 7)^2$
43. $(x - 4)(x + 3)(x - 3)$ **45.** $4x(x - 4y)(x - y)$
47. $a^2(a^2 + 4)(a + 2)(a - 2)$ **49.** $-y^2(11 + 4x)(11 - 4x)$

Stretching the Topics

1. $-ab^2(4a^2 + 3)(3a^2 - 7)$ **2.** $(a + b - 5c)^2$ **3.** $(5x - y)^2$

Checkup 7.4 (page 387)

1. $(3x - 7)^2$ **2.** $9(2x^2 - 7x - 1)$ **3.** $4(x + 3)(x - 3)$ **4.** $(9x^2 + y^2)(3x + y)(3x - y)$
5. $(x - 1)(x + 3)(x - 3)$ **6.** $(y - 9)(y + 6)$ **7.** $(3m + 2)(3m - 8)$
8. $b(8a + 5b)(8a - 5b)$ **9.** $3xy(10x^2 - 6x - 33)$ **10.** $10y^2(x^2 + 12y^2)(x + 2y)(x - 2y)$

Exercise Set 7.5 (page 399)

1. $x = 3, x = 2$ **3.** $x = -0.2, x = 0.6$ **5.** $a = -13, a = -5$ **7.** $x = 0, x = \frac{7}{8}$
9. $x = 7, x = -9$ **11.** $y = 3, y = -1$ **13.** $x = 4, x = 2$ **15.** $x = \frac{5}{4}, x = \frac{-5}{4}$
17. $y = -6, y = -3$ **19.** $m = 0, m = -1.5$ **21.** $z = -9, z = 6$ **23.** $x = 0, x = 5$
25. $z = 15$ **27.** $y = 10, y = -10$ **29.** $n = 0, n = 2$ **31.** $y = -11, y = 3$
33. $x = \frac{2}{5}, x = -3$ **35.** $x = 0.3, x = -0.6$ **37.** $y = 9, y = -9$ **39.** $x = 0, x = \frac{5}{2}$
41. $x = 9, x = -8$ **43.** $a = -5, a = 2$ **45.** $x = 8$ **47.** $x = \frac{1}{5}, x = 2$
49. $a = 12, a = -12$ **51.** $y = 0, y = -17$ **53.** $y = 20, y = -5$
55. $x = 0.9, x = 0.5$ **57.** $x = -2, x = -3$ **59.** $y = \frac{2}{3}, y = \frac{-2}{3}$ **61.** $x = 3$
63. $y = 0, y = 1$ **65.** $x = 9, x = -3$ **67.** $y = 6, y = -6$
69. $x = 5, x = -5$ **71.** $a = \frac{2}{7}, a = \frac{1}{3}$ **73.** $x = 8, x = -6$ **75.** $x = 5$
77. $x = -1$ **79.** $y = 0, y = \frac{4}{5}$

Stretching the Topics

1. $m = \frac{-3}{5}, m = -2$ **2.** $x = \frac{-1}{3}, x = -2$ **3.** $x = 3, x = -3$

Checkup 7.5 (page 401)

1. $x = 0, x = 13$ **2.** $a = 9, a = -7$ **3.** $y = \frac{4}{5}, y = \frac{-4}{5}$ **4.** $x = -7, x = -3$
5. $x = 0, x = 0.4$ **6.** $x = 8, x = -5$ **7.** $x = 3, x = 2$ **8.** $a = 3$
9. $x = 8, x = -8$ **10.** $y = -5, y = -2$

Exercise Set 7.6 (page 409)

1. 4, 9 or -7, -2 **3.** 8 by 9 ft **5.** 5 or -7 **7.** 9 cm **9.** 8 cm, 15 cm **11.** 8 by 16 ft **13.** 150 by 200 ft **15.** 5 ft, 12 ft **17.** 7, 8 or -8, -7 **19.** -6, -5 or 5, 6 **21.** 4, 6

Stretching the Topics

1. 16 in. **2.** $S = 2\pi r(r + h)$, 828.96 sq in. **3.** 2, 4

Checkup 7.6 (page 411)

1. 5 cm by 7 cm **2.** 15 ft **3.** 60 ft **4.** -6, -2 or 4, 8 **5.** -3, -2 or 2, 3

Speaking the Language of Algebra (page 413)

1. Factoring **2.** Common factor **3.** Difference of two squares **4.** perfect square trinomial **5.** Prime **6.** Quadratic, second-degree **7.** Zero product

Review Exercises (page 415)

1. $5(2x - 3)$ **2.** $y^2\left(2y - \frac{3}{5}\right)$ **3.** $2xy(5x + 2y)$ **4.** $-a^2b(3 - 2ab + 4a^2b^2)$
5. $(x + y)(4 + z)$ **6.** $(5 + y)(5 - y)$ **7.** $3(m + 4n)(m - 4n)$
8. $2x^2(3x + 2y)(3x - 2y)$ **9.** $(a - 8b)^2$ **10.** $(x - 5)(x - 3)$ **11.** $(xy - 5)(xy + 2)$
12. $2(2a^2 - 10ab + b^2)$ **13.** $(3x - 7)(2x + 1)$ **14.** $-2(x - 3y)(x + y)$
15. $5(2m + 11)(m - 4)$ **16.** $(13 + 2a)(13 - 2a)$ **17.** $(8x - 3)(x - 1)$
18. $3(y^2 + 25)(y + 2)(y - 2)$ **19.** $(x - 4)(x + 1)(x - 1)$ **20.** $x = 7, x = -8$
21. $x = 13, x = -13$ **22.** $x = -7$ **23.** $m = 0, m = \frac{-2}{3}$ **24.** $a = 9, a = -9$
25. $z = -13, z = -4$ **26.** $y = -3, y = -8$ **27.** $y = 0, y = -2.3$
28. $x = 21, x = 3$ **29.** $x = 9, x = -9$ **30.** $x = 10, x = -3$ **31.** 6 by 10 ft
32. 15 cm **33.** 9, 13 or -13, -9 **34.** 12 ft **35.** 5, 7 or -3, -1

Practice Test for Chapter 7 (page 417)

1. $-5(2x - 5)$ **2.** $11x(3x - 7)$ **3.** $3y^2(2y^3 + y^2 - 5)$ **4.** $-2ab(4a^2 - 7ab + 4b^2)$
5. $(a - 3)(y + 5)$ **6.** $(4x + 5a)(4x - 5a)$ **7.** $(0.6 + ab)(0.6 - ab)$
8. $7(x + 1)(x - 1)$ **9.** $(a + 4b)(a + 5b)$ **10.** $(x - 2)(x - 11)$ **11.** $(10 - x)(2 + x)$
12. $(2x - 3)(10x + 1)$ **13.** $-3b(a - 4b)^2$ **14.** $(x + 3)(x - 3)(x + 1)(x - 1)$
15. $x = 0, x = -11$ **16.** $x = 0, x = -1.4$ **17.** $n = 5, n = -5$
18. $a = -7, a = -10$ **19.** $y = \frac{5}{11}, y = \frac{-5}{11}$ **20.** $x = 4$ **21.** $y = 4, y = -2$
22. $x = 10, x = -2$ **23.** 4 by 10 m **24.** 13 yd

Sharpening Your Skills after Chapters 1–7 (page 419)

1. $\frac{-64}{125}$ **2.** 15.336 **3.** $\frac{8}{9}$ **4.** -6 **5.** $-7.2k$ **6.** $9x + 8$ **7.** $-8x^8y^4$
8. $3y^2 - 8yz + 4z^2$ **9.** $-45x^7y^7$ **10.** $-5a^2b + 2a^3b^2 - 7a^4b^3$ **11.** $8a^2 - 2ab - 15b^2$
12. $-y^4 - 2y^3 + 15y^2$ **13.** $-5x^2 + 4xy - \frac{4}{3}y^2$ **14.** $4x + 3 + \frac{1}{x - 5}$ **15.** $x = 36$
16. $x = -2$ **17.** $x = 40$ **18.** 432 **19.** $3l^2 - 10l$, 525 sq m **20.** $t = \frac{A - p}{pr}$

CHAPTER 8

Exercise Set 8.1 (page 431)

1. $x \neq 0$ **3.** $x \neq -1$ **5.** $x \neq 0$ **7.** $y \neq 8$ **9.** $x \neq \frac{1}{2}$ **11.** $2x^3$ **13.** $\frac{-a}{4}$
15. $\frac{3x^2}{y^2}$ **17.** $\frac{-3b^2}{4a^2}$ **19.** $\frac{5}{4x^2z}$ **21.** $\frac{x + y}{3}$ **23.** $a + 5b$ **25.** $9y - 7$ **27.** $\frac{2}{16 + x}$
29. $\frac{6(a^2 + b^2)}{5}$ **31.** $5(x^2 + y^2)$ **33.** $\frac{7x^2}{x + 1}$ **35.** $\frac{-4a - 5c}{6}$ **37.** $\frac{2x^2 + 3x + 1}{10}$

39. $\frac{a+b}{x+y}$ 41. $\frac{x-y}{2(x+y)}$ 43. $\frac{1}{2}$ 45. $\frac{-1}{3}$ 47. $\frac{x-y}{3}$ 49. $\frac{x-7}{4}$ 51. $\frac{3}{x+y}$
53. $\frac{-7}{3(a-b)}$ 55. $\frac{x-2y}{3}$ 57. $\frac{x+5}{x-4}$ 59. $\frac{x+4}{x+5}$ 61. $\frac{5(x+3)}{x-7}$ 63. $\frac{x+5y}{x+y}$
65. $\frac{2(a+b)}{a-3b}$ 67. $\frac{(x+5)(x-5)}{x^2+25}$ 69. $\frac{1}{2b-1}$ 71. $\frac{(x^2+4y^2)(x+2y)}{x+4y}$ 73. $\frac{-1}{9}$
75. $-(x+5)$

Stretching the Topics

1. $m \neq 0, m \neq \frac{-3}{2}, m \neq -4$ 2. $\frac{-2b(a-c)}{3(a+c)}$ 3. $2x(b+c)^{n-1}$

Checkup 8.1 (page 435)

1. $x \neq 0$ 2. $x \neq \frac{1}{2}$ 3. $\frac{x^3}{y^2}$ 4. $\frac{-1}{4x}$ 5. $4y-5$ 6. $\frac{1}{5x-4y}$ 7. $\frac{a+b}{4(a-b)}$
8. $\frac{-1}{3(x+y)}$ 9. $\frac{m+2n}{2}$ 10. $\frac{x+4}{x+5}$

Exercise Set 8.2 (page 443)

1. $\frac{x}{y^2}$ 3. $\frac{-27z^2}{28x^2y}$ 5. $\frac{105y}{x^2}$ 7. b 9. $\frac{-x}{9}$ 11. $4xz$ 13. $\frac{5b^5}{a^2}$ 15. $\frac{18y^2}{x^2}$
17. $\frac{4b^4}{9a}$ 19. $\frac{5x^6}{6y^{11}}$ 21. $\frac{2}{x+3}$ 23. $\frac{3}{8x}$ 25. $\frac{2b}{a}$ 27. $\frac{7}{3}$ 29. $x-3y$
31. $\frac{-2(x-2)}{3(x+2)}$ 33. $\frac{5y(x+3)}{8(x-3)}$ 35. $\frac{x+4}{x+6}$ 37. $\frac{x+5y}{5}$ 39. $y(x-7)$ 41. $\frac{x+5}{49}$
43. $\frac{1}{2(x-9)}$ 45. $\frac{-3}{x-2}$ 47. $\frac{1}{x+3}$ 49. $\frac{5(a-b)}{a^2b^2}$ 51. 1

Stretching the Topics

1. $\frac{243x^7}{40}$ 2. $\frac{(a+b)^2(a-b)^2}{a^3}$ 3. $\frac{20x(x+y)}{3y(x+2y)(2x+3y)}$

Checkup 8.2 (page 445)

1. $\frac{-14x^4}{z}$ 2. $\frac{10a^3c^2}{7b}$ 3. $\frac{-27z^2}{28x^2y}$ 4. $\frac{2}{y}$ 5. $\frac{2(x-3)}{3(x+3)}$ 6. -1 7. $\frac{x-8}{y}$
8. $\frac{1}{3(x+6)}$ 9. $\frac{(x-11)(x-8)}{(x+3)^2}$ 10. $\frac{2n-1}{n-6}$

Exercise Set 8.3 (page 451)

1. $\frac{2x}{3x}$ 3. $\frac{-3x}{15x^2}$ 5. $\frac{8xy}{2x^2y^2}$ 7. $\frac{9xy^2}{21y^3}$ 9. $\frac{-30a^2}{15abc}$ 11. $\frac{3xy}{3y}$ 13. $\frac{-3x^2y}{3y^3}$
15. $\frac{-8a^2b^4}{4a^2b^2}$ 17. $\frac{8x-8y}{12(x-y)}$ 19. $\frac{25x^2}{10x^3}$ 21. $\frac{-4a-4b}{a(a+b)}$ 23. $\frac{3x^2}{x(x+y)}$
25. $\frac{2ab-2a}{(b+1)(b-1)}$ 27. $\frac{6x^3+12x^2}{(x-5)(x+2)}$ 29. $\frac{x^3+5x^2}{x^2(x+3)}$ 31. $\frac{a^2-5a+4}{(a-3)(a-4)}$
33. $\frac{-3xy+3y}{x(x-1)^2}$ 35. $\frac{x^2-4xy+3y^2}{2(x-3y)(x+y)}$

Stretching the Topics

1. $\frac{3x^3-x^2-10x}{x^2(2x-3)(3x+5)}$ 2. $\frac{5x-5y}{(a+b)(x-y)}$ 3. $\frac{-3x^3-6x^2-3x-6}{(x^2+1)(x+2)(x-2)}$

Checkup 8.3 (page 453)

1. $\frac{54a^2b}{9a^3b^2}$ 2. $\frac{-25xyz^2}{10y^2z^3}$ 3. $\frac{-3xy^2}{18y^4}$ 4. $\frac{16a^4b^2}{16a^2b^2}$ 5. $\frac{x^2-3x-10}{(x+2)(x+2)}$ 6. $\frac{2a^4-a^3b}{a^3(a-3b)}$
7. $\frac{2a^2b}{ab(a+b)}$ 8. $\frac{5x^2-30x}{(x+6)(x-6)}$ 9. $\frac{x^2+2x-8}{(x+4)(x-1)}$ 10. $\frac{a^2-b^2}{5(a+b)^2}$

Exercise Set 8.4 (page 463)

1. $3x$ 3. $\frac{5a-6}{7}$ 5. $\frac{1}{x}$ 7. $\frac{2x-1}{y}$ 9. $\frac{-2}{b}$ 11. $\frac{9y}{x+3y}$ 13. 3 15. $\frac{2x}{2x+5}$
17. $\frac{x}{x-2}$ 19. $3(x-5)$ 21. $\frac{2a-3}{a}$ 23. $\frac{1}{x-2y}$ 25. $\frac{5x-3y}{30}$ 27. $\frac{15x+23}{5x^2}$
29. $\frac{7y-1}{y}$ 31. $\frac{2(2-x^2)}{x}$ 33. $\frac{5y+2x}{xy}$ 35. $\frac{2(18y-25x)}{15xy}$ 37. $\frac{4y-5x}{x^2y^2}$
39. $\frac{-3x^2-8y^2}{12x^2y^2}$ 41. $\frac{7x+9}{(x+2)^2}$ 43. $\frac{-2x-32}{(x+4)(x-4)}$ 45. $\frac{x^2+x-6}{3(x+1)}$ 47. $\frac{-11}{3(x+1)}$
49. $\frac{2x+24}{(x+2)(x-2)}$ 51. $\frac{x^2-x-5}{x(x+5)(x-5)}$ 53. $\frac{1}{x^2}$ 55. $\frac{3x+35}{(x+7)^2(x-7)}$
57. $\frac{4x}{(x+4)(x-2)}$ 59. $\frac{2x-5}{x(2x-1)}$ 61. $\frac{2x-36}{(x-9)^2(x+9)}$ 63. $\frac{x^2-2}{(2x-3)(x+2)(x+1)}$

Stretching the Topics

1. $\frac{-5n+24}{(n-5)(n-3)}$ 2. $\frac{3x^3+21x^2-36}{x(x+7)(x+3)(x-3)}$ 3. $\frac{-2}{x-3y}$

Checkup 8.4 (page 467)

1. $\frac{-x}{2}$ 2. $\frac{2a-1}{b}$ 3. $x-8$ 4. $\frac{5}{2x}$ 5. $\frac{3y+2x}{xy^2}$ 6. $\frac{11x-21}{x(x-3)}$ 7. $\frac{15}{2(x+1)}$
8. $\frac{3x+25}{(x+5)^2(x-5)}$ 9. $\frac{1}{x^2}$ 10. $\frac{17}{(x-9)(x+7)}$

Exercise Set 8.5 (page 475)

1. $\frac{4}{7}$ 3. $\frac{x^5}{4y}$ 5. $\frac{3x}{x+3y}$ 7. $\frac{5}{4(x-4)}$ 9. $\frac{x^2}{x-7}$ 11. $\frac{3}{x^2}$ 13. $\frac{2(5x-2)}{x^2}$
15. 6 17. $\frac{x-7}{x}$ 19. $\frac{1}{x-1}$ 21. $\frac{x}{5x+1}$ 23. $\frac{3}{4}$ 25. $\frac{3+x}{3x}$ 27. $\frac{5(2x-5)}{x}$
29. $\frac{9(x+5)}{x(x+9)}$

Stretching the Topics

1. $\frac{5x(x-y)}{(x-4y)(x+2y)}$ 2. $\frac{8(y-2)^2}{(y+3)(y-3)}$ 3. $\frac{-(x+1)}{2(x-12)}$

Checkup 8.5 (page 477)

1. $\frac{3}{x^3y}$ 2. $\frac{9x^3y^4}{2}$ 3. $\frac{4x^2}{3x+5}$ 4. $\frac{(x+3)(x+1)}{3}$ 5. $\frac{3x}{x-8}$ 6. $\frac{x+4}{x}$
7. $\frac{5a+4}{a}$ 8. $\frac{x+9}{9x}$ 9. $\frac{4(x-1)}{x}$ 10. $\frac{4(x-6)}{x(x-4)}$

Exercise Set 8.6 (page 485)

1. $x=\frac{7}{2}$ 3. $x=4$ 5. $x=\frac{-14}{5}$ 7. $y=-15$ 9. $x=\frac{90}{11}$ 11. $x=\frac{33}{2}$
13. $y=-1$ 15. $x=3$ 17. $x=-5$ 19. $z=-3$ 21. $x=3$ 23. $x=\frac{13}{3}$
25. $x=-8$ 27. $x=\frac{44}{9}$ 29. No solution 31. $x=\frac{-23}{4}$ 33. $x=\frac{-1}{10}$
35. $a=\frac{38}{7}$ 37. $b=\frac{-1}{5}$ 39. $x=\frac{-2}{5}$ 41. $x=-41$ 43. $x=\frac{-17}{3}$
45. $a=\frac{3}{5}$ 47. No solution 49. $x=0$ 51. $x=5, x=8$ 53. $x=5, x=\frac{-3}{2}$
55. $y=4, y=-3$ 57. $x=2, x=\frac{1}{3}$ 59. $x=3$

Stretching the Topics

1. $x = -11$ **2.** $x = \dfrac{abck}{bc - 2ac + 3ab}$ **3.** $n = \dfrac{Ir}{E - IR}$

Checkup 8.6 (page 489)

1. $y = 18$ **2.** $x = -2$ **3.** $x = -11$ **4.** $y = -23$ **5.** $a = \dfrac{23}{2}$ **6.** $x = \dfrac{55}{17}$

7. No solution **8.** $x = 3$ **9.** $x = -1$ **10.** $x = 1, x = -2$

Exercise Set 8.7 (page 499)

1. \$2800 **3.** Louise, 56 mph; husband, 50 mph **5.** José 20, Rick 16 **7.** 165 mi

9. 2500 **11.** 12 by 18 ft **13.** 3 **15.** $3\frac{1}{3}$ hr **17.** $3\frac{3}{7}$ hr **19.** \$900

Stretching the Topics

1. 1 **2.** 30 days **3.** 2 mph

Checkup 8.7 (page 501)

1. \$240,000 **2.** 100 **3.** Faye, 6 mph; daughter, 4 mph **4.** 2 **5.** $2\frac{2}{5}$ hr

Speaking the Language of Algebra (page 505)

1. Rational algebraic expression **2.** 0; undefined **3.** Numerators; denominators
4. Denominator **5.** Single fraction **6.** Least common denominator **7.** Extraneous
8. Proportion

Review Exercises (page 507)

1. $\frac{1}{2}$ **2.** $\frac{-1}{5}$ **3.** $x \neq -4$ **4.** $x \neq \frac{1}{2}$ **5.** $2x^5$ **6.** $\dfrac{5x^2y^2}{4}$ **7.** $a + 3b$ **8.** $\dfrac{8x^4}{x^2 + 1}$

9. $\frac{-1}{5}$ **10.** $\dfrac{x - 4}{5x}$ **11.** $\dfrac{1}{5a - 6b}$ **12.** $\dfrac{-3}{x + y}$ **13.** $\dfrac{x - 5}{2x - 3}$ **14.** $x + y$ **15.** $\dfrac{7y^2}{6}$

16. $9abc$ **17.** $10b^4$ **18.** $\dfrac{y^2(x + 4)}{8z(x - 4)}$ **19.** $\dfrac{5x + 2}{5x - 2}$ **20.** $\frac{1}{2}$ **21.** $\dfrac{15xy}{35y^2}$

22. $\dfrac{3y^2 + 6y}{(y + 2)(y - 2)}$ **23.** $\dfrac{2x + 3}{y}$ **24.** $\dfrac{x - 5}{3x - 1}$ **25.** $\dfrac{32x + 1}{4x^2}$ **26.** $\dfrac{-7x + 24}{6x}$

27. $\dfrac{-4x^2 - 10xy + 3y^2}{12x^2y^2}$ **28.** $\dfrac{-2x}{3(x - 3)}$ **29.** $\dfrac{2a + 3}{(a + 2)(a + 1)}$ **30.** $\dfrac{-3x + 17}{(x + 6)(x - 3)}$

31. $\dfrac{5}{2ab^5c^2}$ **32.** $\dfrac{x + 7}{7x}$ **33.** $x = 16$ **34.** $x = -15$ **35.** No solution **36.** $a = 1$

37. $a = -6$ **38.** $x = 1, x = 3$ **39.** 12 gal **40.** 25 **41.** $1\frac{1}{5}$ hr **42.** 65 mph

Practice Test for Chapter 8 (page 511)

1. $\frac{7}{8}$ **2.** $y \neq \frac{-1}{3}$ **3.** $\dfrac{-y}{4x^3}$ **4.** $\dfrac{6x^2}{x + 2}$ **5.** $\dfrac{2x}{x + 5}$ **6.** $\dfrac{-(x - y)}{3}$ **7.** $\dfrac{2b^3}{5a^2}$

8. $\dfrac{18}{25x^2y^4}$ **9.** $\dfrac{2x}{5}$ **10.** 1 **11.** $\dfrac{(x + 1)(2x + 1)}{3x^4}$ **12.** $\dfrac{x - 1}{a}$ **13.** $\dfrac{x - 1}{x + 2}$

14. $\dfrac{5a - 9ab + 60b}{15a^2b^2}$ **15.** $\dfrac{-60}{(y + 5)(y - 5)}$ **16.** $\dfrac{a - 8}{a(a - 3)(a + 4)}$ **17.** $x = \dfrac{21}{2}$

18. $x = 5$ **19.** $x = -1$ **20.** 12 **21.** 90 **22.** $2\frac{2}{9}$ hr

Sharpening Your Skills after Chapters 1–8 (page 513)

1. Undefined **2.** 2 **3.** $5x + 85y$ **4.** $-8x^8y^9$ **5.** $\dfrac{729b^{10}}{784a^{10}}$ **6.** $-36a^5b^5$

7. $4a^2 - 12ab + 9b^2$ **8.** $x^2 - 0.25y^2$ **9.** $-2x^3 + 6x^2y + 20xy^2$

10. $5xy(3x - y)(x - 4y)$ **11.** $2(x + 7)(x - 7)$ **12.** $(x - 5y)^2$ **13.** $-5(3x - 2)(x + 6)$
14. $x = 0$ **15.** $y = \frac{13}{7}$ **16.** $y = 0, y = 3$ **17.** 5 in. and 12 in. **18.** 38

CHAPTER 9

Exercise Set 9.1 (page 521)

1. $<$ **3.** $>$ **5.** $=$ **7.** $<$ **9.** $<$ **11.** $>$ **13.** $>$ **15.** $-11 < -5 < 0$
17. $-2 < 3 < 7$ **19.** $\frac{-4}{5} < \frac{-2}{5} < 0$

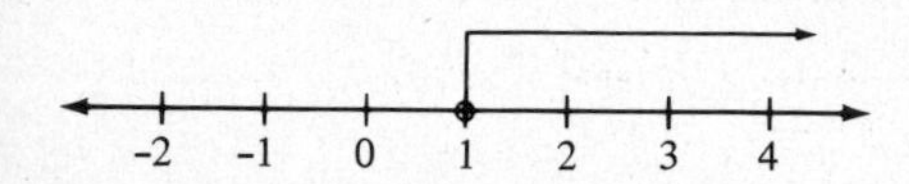

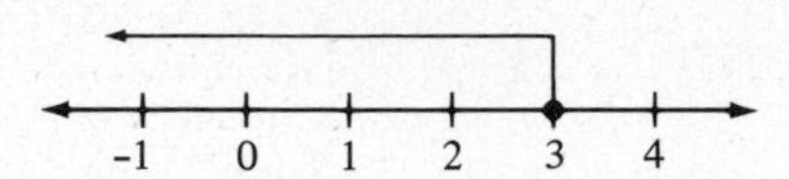

25.

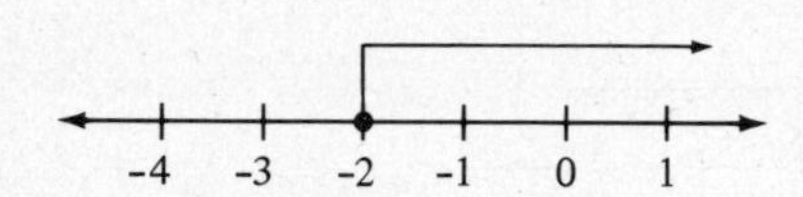

27.

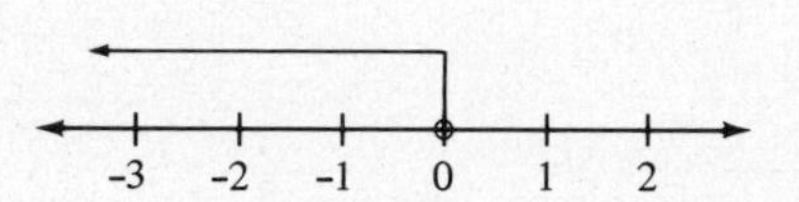

29.

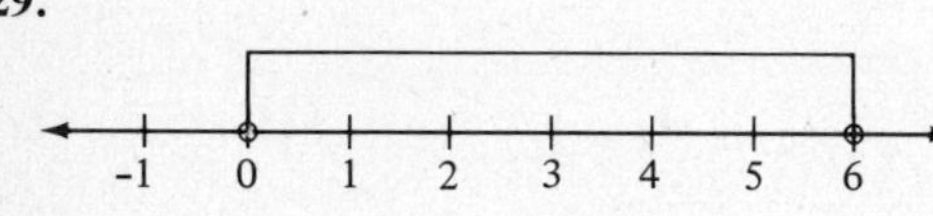

31.

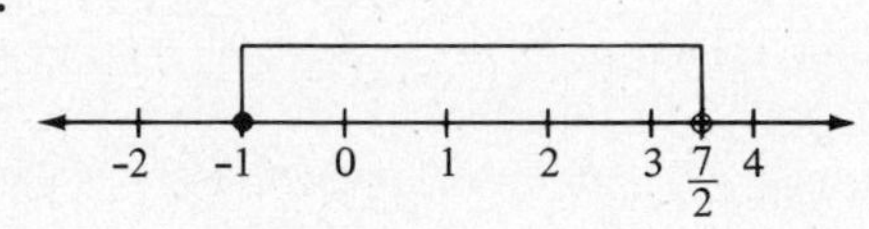

33.

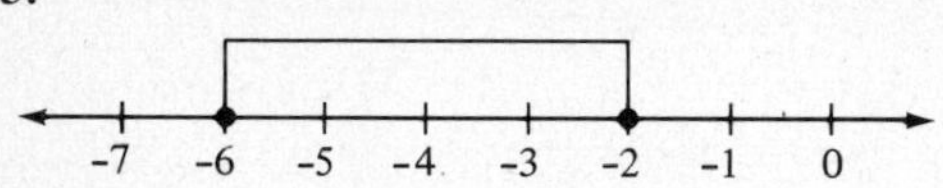

35.

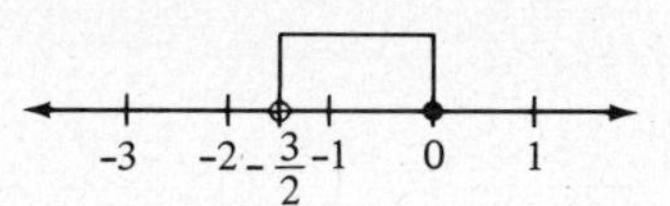

37.

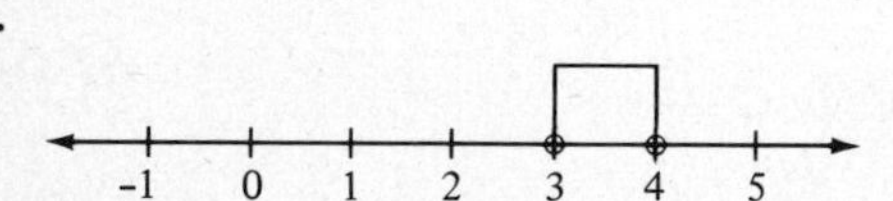

39.

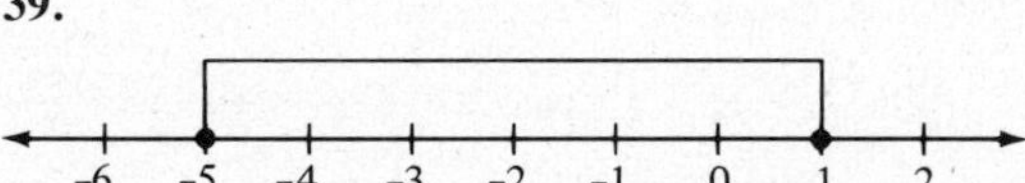

Stretching the Topics

1.

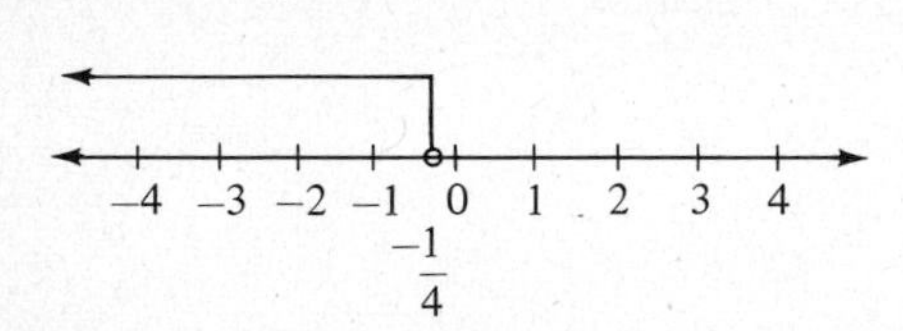

2.

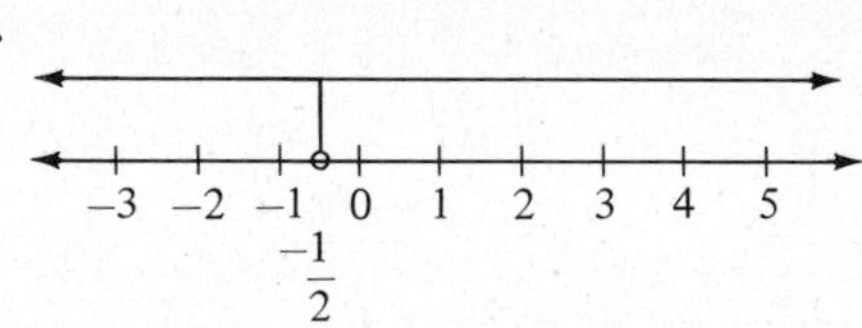

3.

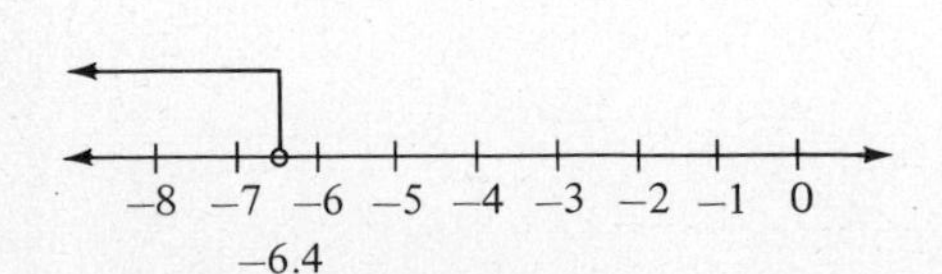

Checkup 9.1 (page 523)

1. $<$ **2.** $=$ **3.** $>$ **4.** $>$ **5.** $-5 < -2 < 0$ **6.** $\frac{3}{2} < 3 < \frac{7}{2}$

7.

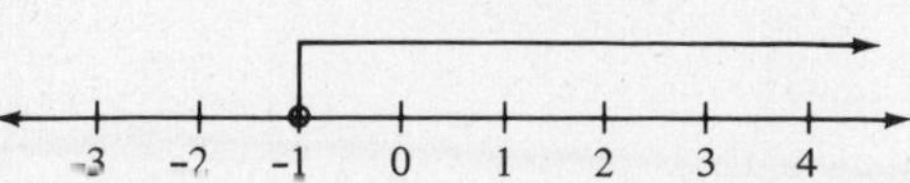

8.

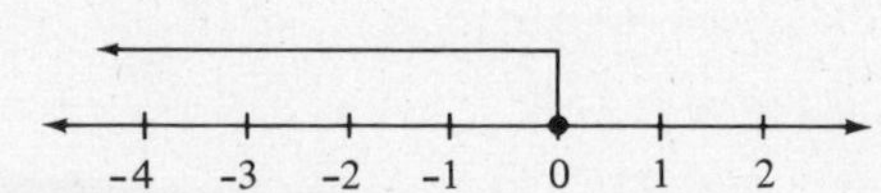

9.

10.

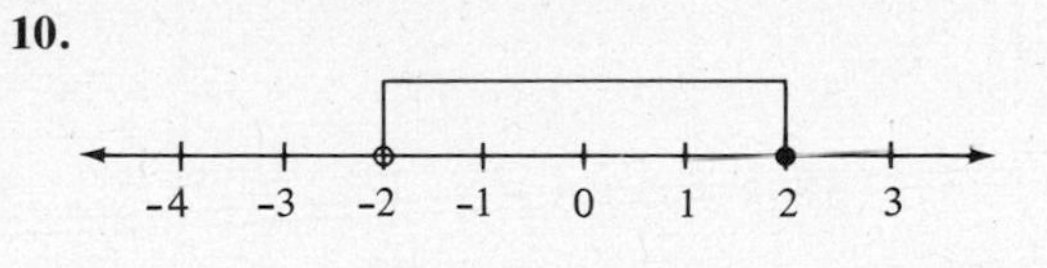

Exercise Set 9.2 (page 533)

1. $1 < 9$ **3.** $6 > -5$ **5.** $8 > -12$ **7.** $-6 < -2$ **9.** $-7 > -14$ **11.** $-2 < 1$
13. $13 > 1$ **15.** $4 > -8$

17. $x \leq 4$

19. $x > -5$

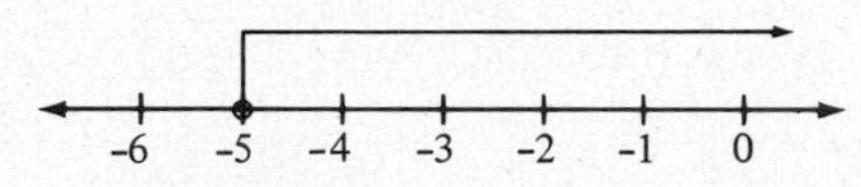

21. $a < \frac{-7}{2}$

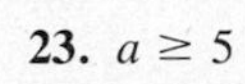

23. $a \geq 5$

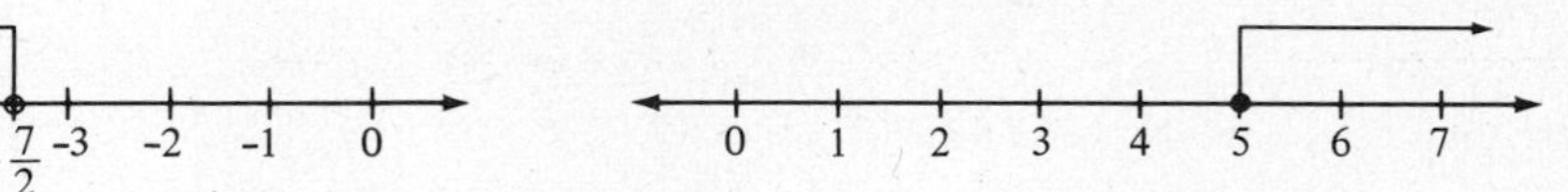

25. $x < -3$

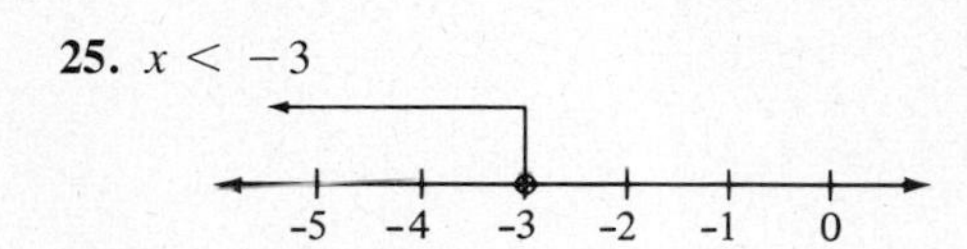

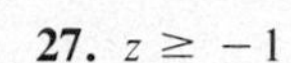

27. $z \geq -1$

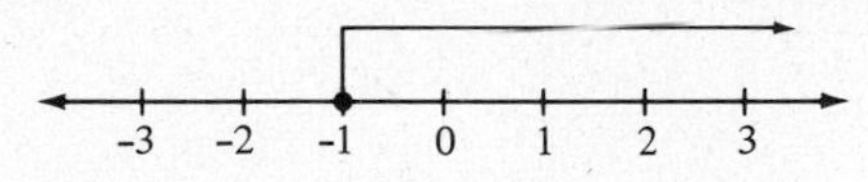

29. $y < 2$

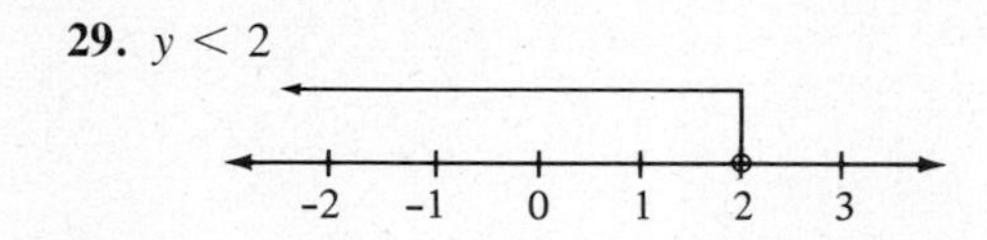

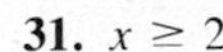

31. $x \geq 2$

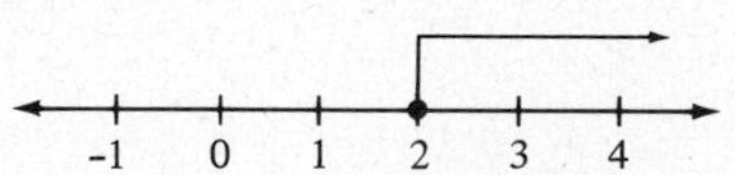

33. $x \geq \frac{9}{2}$ **35.** $y < 0$ **37.** $x > 12$ **39.** $a < -1$ **41.** $x \leq \frac{1}{2}$ **43.** $x < \frac{-5}{4}$

45. $z > -3$ **47.** $x \leq 12$ **49.** $x \leq \frac{-1}{5}$ **51.** $-2 < x < 3$ **53.** $-4 \leq x \leq \frac{1}{2}$

55. $-3 \leq x \leq 4$ **57.** $-4 \leq x < 16$ **59.** $-1 \leq x \leq \frac{2}{3}$

Stretching the Topics

1. $x \geq \frac{8}{5}$ **2.** $y < \frac{51}{7}$ **3.** $\frac{-31}{2} \leq x < \frac{23}{2}$

Checkup 9.2 (page 535)

1. $3 > -4$ **2.** $-2 < 3$

3. $x \geq -2$

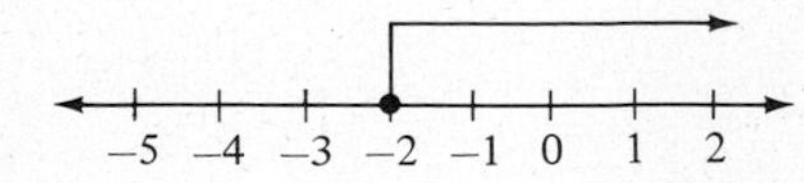

4. $a \geq -3$

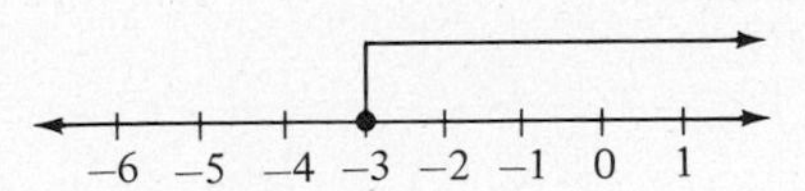

5. $x > 3$

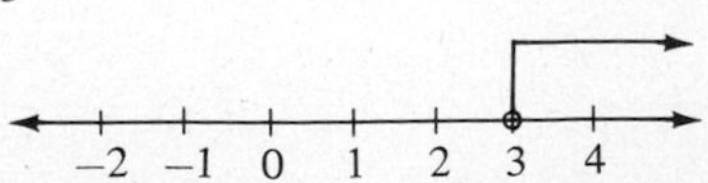

6. $-1 \leq y \leq 2$

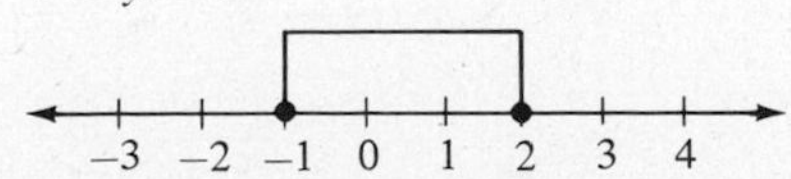

7. $a \leq \frac{5}{3}$ **8.** $x > \frac{-5}{4}$ **9.** $y > \frac{-1}{2}$ **10.** $2 \leq x \leq 6$

Exercise Set 9.3 (page 543)

1. $-2 \le x \le 3$ **3.** $x > 0$ **5.** $x < 5$ **7.** $x > 1$ or $x < -2$ **9.** No solution
11. $\frac{1}{2} < x < 4$ **13.** $x > 0$ **15.** $-2 < x < 2$ **17.** $2 \le x < 3$ **19.** $x \ge -5$
21. $x < -1$ or $x > 0$ **23.** $x > 4$ **25.** $x < 3$ **27.** $x \le 3$ **29.** $x > 3$
31. $5 \le x < 11$ **33.** $x < -4$ or $x > 13$ **35.** $x \ge 3$ **37.** $x > -8$

Stretching the Topics

1. No solution **2.** $x \ge \frac{9}{16}$ or $x \le \frac{7}{11}$ **3.** $x > 4$

Checkup 9.3 (page 545)

1. $x > 3$ **2.** $x > -2$ **3.** $a \le -2$ or $a \ge 4$ **4.** No solution **5.** All real numbers
6. $0 < x < 4$ **7.** $-4 < x < 4$ **8.** $-1 < x < \frac{3}{2}$ **9.** $x > 5$ **10.** $x < \frac{3}{2}$ or $x > 4$

Exercise Set 9.4 (page 551)

1. At least 8 ft **3.** No more than 176 **5.** No more than $14\frac{1}{2}$ ft **7.** $2 < n < 5$
9. At least 350 **11.** $w \le 24$ **13.** At least 2400 **15.** No more than 19 by 26 ft
17. $w \le 5\frac{2}{3}\,m$ **19.** $4 \le j \le 10$

Stretching the Topics

1. $\$10{,}000 < S \le \$25{,}000$ **2.** $w \le 54$ ft

Checkup 9.4 (page 553)

1. No more than 4 **2.** At least 4 **3.** No more than $54\frac{1}{2}$ by 109 ft **4.** At least 950
5. At least $50.75

Speaking the Language of Algebra (page 556)

1. Inequalities **2.** $2 < x < 5$ **3.** Left **4.** Multiply; divide; negative **5.** $3n \ge 18.7$

Review Exercises (page 557)

1. $<$ **2.** $>$ **3.** $-10 < -7 < 3$ **4.** $0 < \frac{5}{3} < 4$

5.

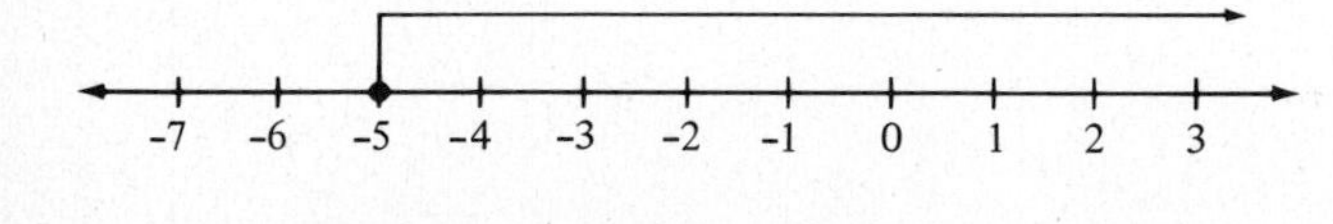

6.

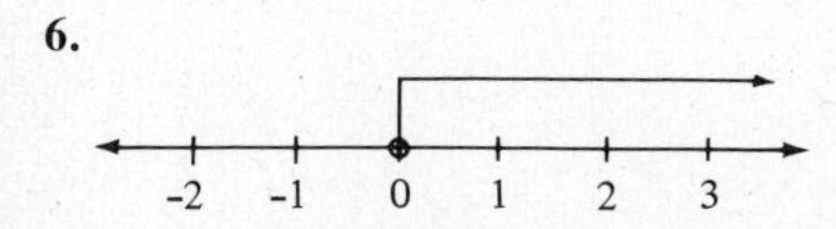

7.

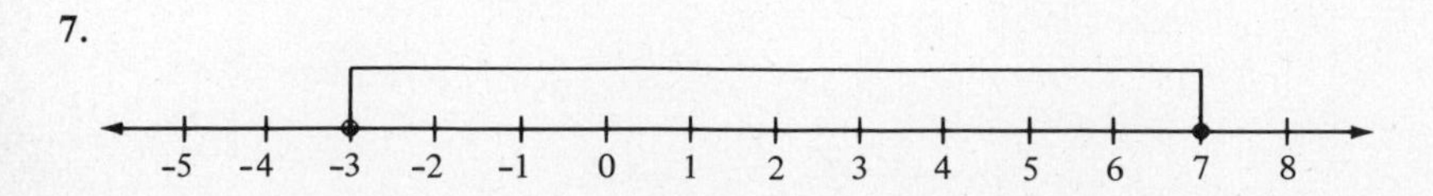

8.

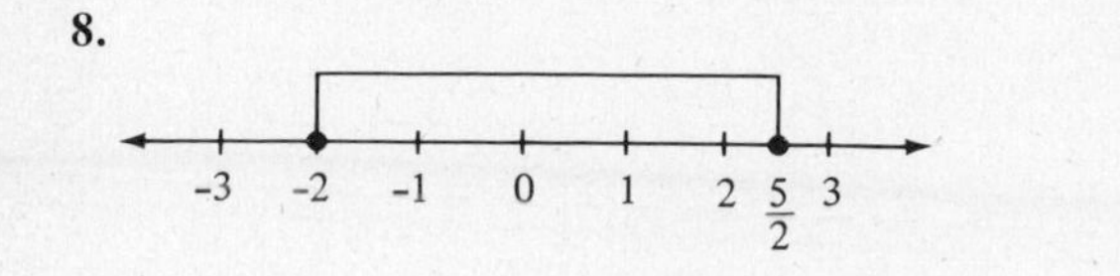

9. $4 < 13$ **10.** $-21 > -30$ **11.** $-5 < 2$ **12.** $-9 > -15$

13. $x \leq 4$

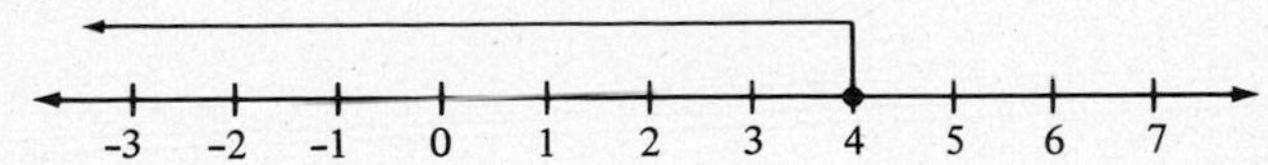

14. $y < 5$

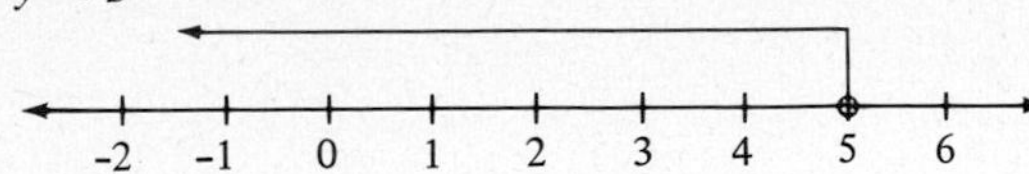

15. $a \geq -8$

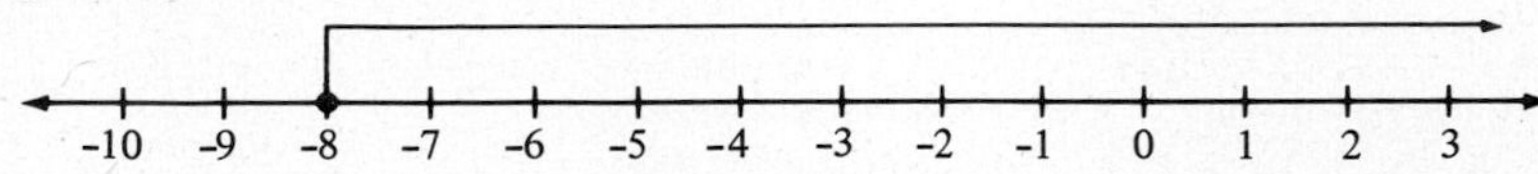

16. $x > 6$

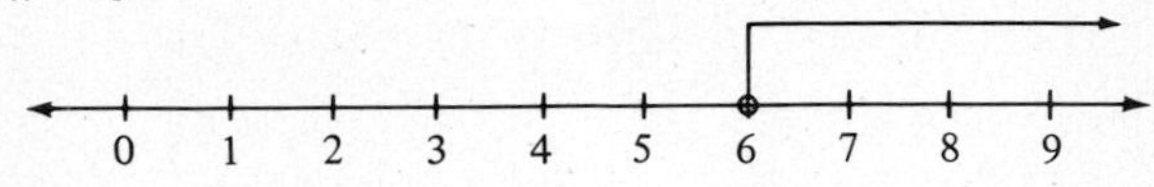

17. $x > -5$ **18.** $x \geq 2$ **19.** $x > \frac{-9}{5}$ **20.** $x \leq \frac{18}{5}$ **21.** $y \leq -6$ **22.** $\frac{-2}{3} \leq y \leq 2$

23. $2 < x \leq 6$ **24.** $-3 \leq x \leq 3$ **25.** $x > 3$ **26.** $x \geq 6$ **27.** $x < -1$ or $x > 3$

28. At least 15 **29.** No more than 12 by 24 ft **30.** At least 300

Practice Test for Chapter 9 (page 559)

1. $-1 < \frac{-1}{2} < \frac{1}{4}$

2.

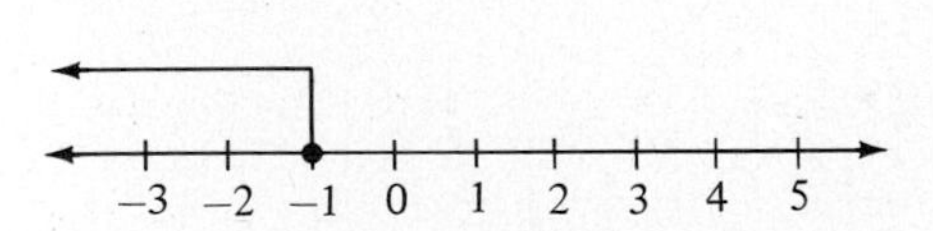

3.

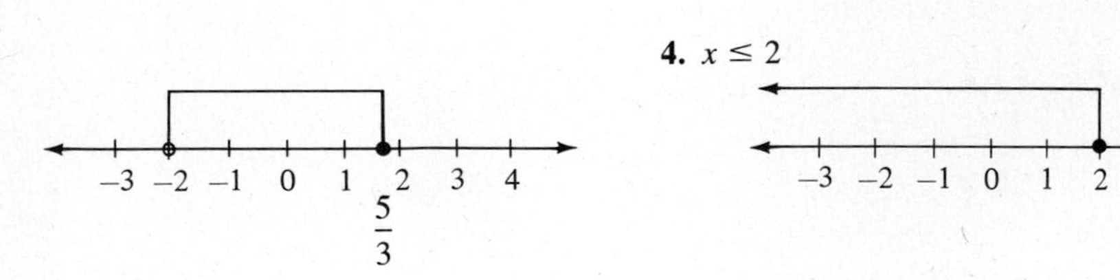

4. $x \leq 2$

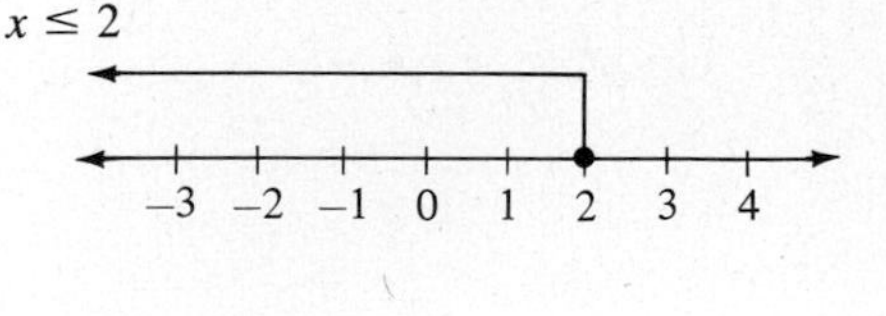

5. $x > 5$

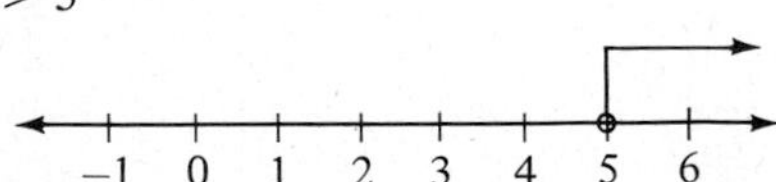

6. $x > -9$ **7.** $x > 0$ **8.** $x < \frac{1}{2}$

9. $a \leq -17$ **10.** $y \leq \frac{-72}{29}$ **11.** $-2 \leq x \leq -1$ **12.** $1 < x < 4$ **13.** $-8 \leq x \leq 0$

14. $x < -5$ **15.** $x < -11$ or $x > 5$ **16.** No more than 80

Sharpening Your Skills after Chapters 1–9 (page 561)

1. 62.5% **2.** $\frac{4}{5} > \frac{5}{7}$ **3.** \$442.50 **4.** 33 **5.** $x \neq 4, x \neq -4$ **6.** $2x - 2y$

7. $18x^4y^6$ **8.** $x^3 - 1$ **9.** $-24x + 4x^2 + 4x^3$ **10.** $\frac{m - 3n}{2}$ **11.** $\frac{3(x - 7)}{4(x + 7)}$

12. $\frac{13}{4(x + 1)}$ **13.** $\frac{1}{xy^3(x + 2)}$ **14.** $8ab(2ab - 1)(ab - 1)$

15. $6(m^2 + 4)(m + 2)(m - 2)$ **16.** $-xy(x + 3y)(x + 5y)$ **17.** $x = 9$ **18.** $y = 4$

19. $x = 3, x = -13$ **20.** $x = 0, x = 9$ **21.** $x = -2$ **22.** $a = \frac{-16}{3}$ **23.** 7 by 21 ft

24. 9, 10 or -10, -9 **25.** Louise, 56 mph; husband, 50 mph

CHAPTER 10

Exercise Set 10.1 (page 571)

1. Yes **3.** No **5.** Yes **7.** No **9.** No **11.** Yes **13.** Answers will vary.
15. Answers will vary. **17.** Answers will vary. **19.** Answers will vary.
21. Answers will vary. **23.** Answers will vary. **25.** Answers will vary.
27. Answers will vary. **29.** $y = 4x + 8$ **31.** $x + y = 28$ **33.** $l = 2w - 5$
35. $x + 5y = 100$ **37.** $2l + 2w = 48$ **39.** $x + y = 12{,}500$
41. $3.20x + 4.80y = 147.20$ **43.** $c = 5d - 2$ **45.** $3x + 2y = 20$

Stretching the Topics

1. Yes **2.** $x = 4, y = 13; x = 6, y = 19; y = 3x + 1$ **3.** $0.085x + 0.12y = 0.10(x + y)$

Checkup 10.1 (page 573)

1. No **2.** No **3.** Yes **4.** No **5.** Answers will vary. **6.** Answers will vary.
7. Answers will vary. **8.** $y = 2x - 500$ **9.** $2x + 3y = 17$ **10.** $2l + 2w = 196$

Exercise Set 10.2 (page 585)

1–8.

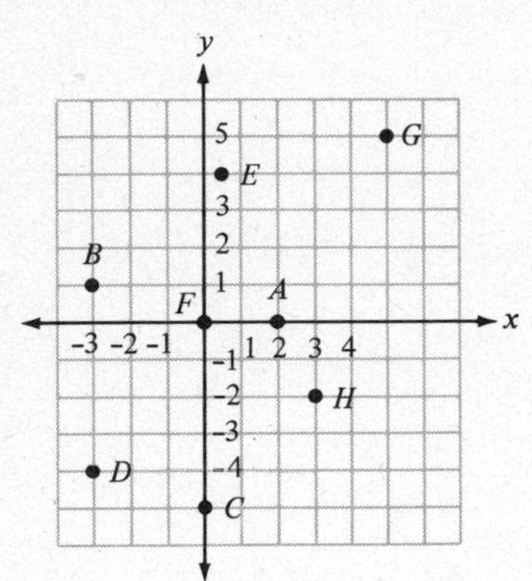

9. $A(2, 5)$ **11.** $C(2, -1)$ **13.** $E(0, 0)$
15. $G(-3, -3)$

17.

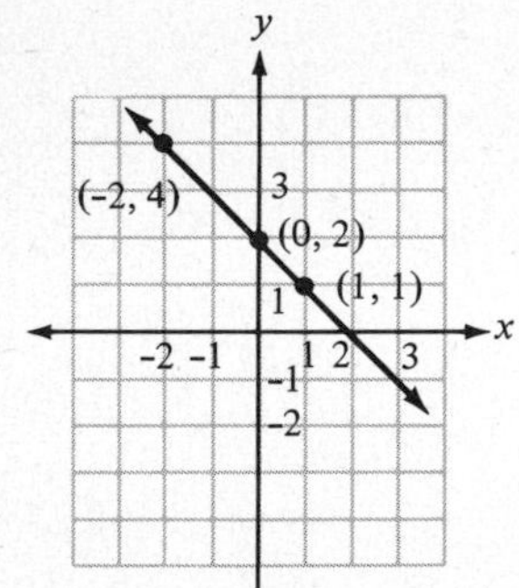

19.

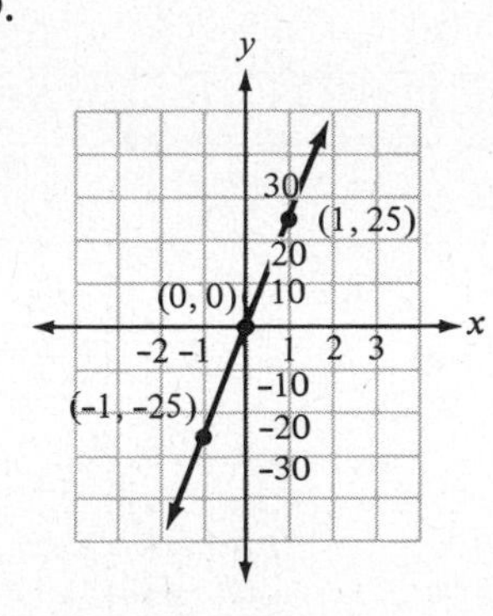

21.

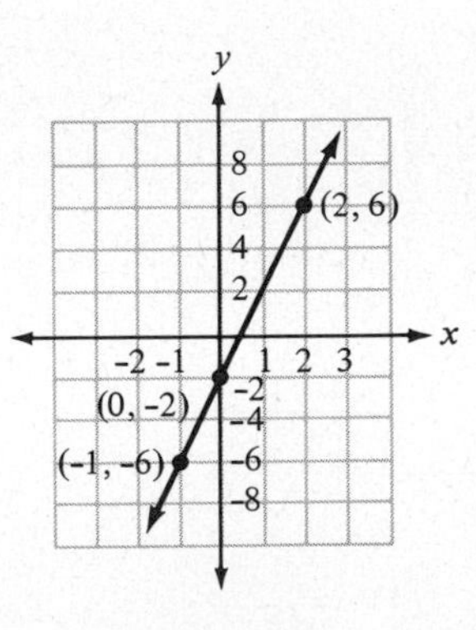

23.

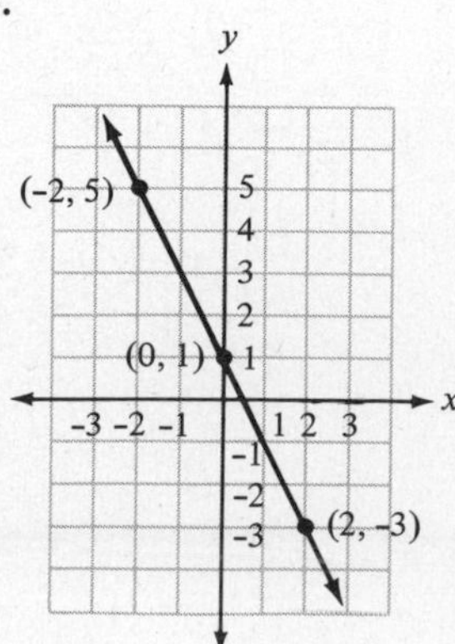

25.

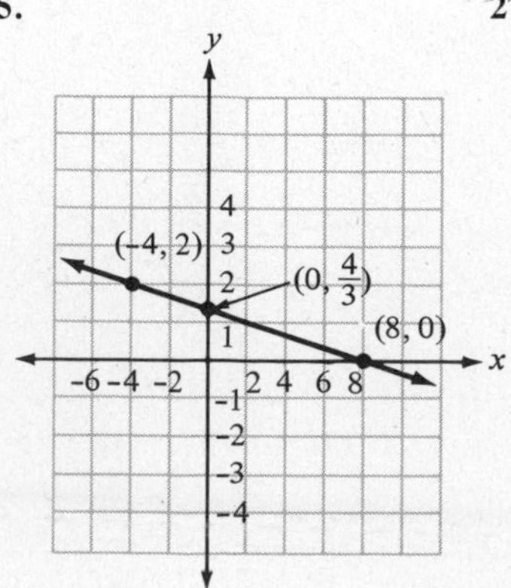

27.

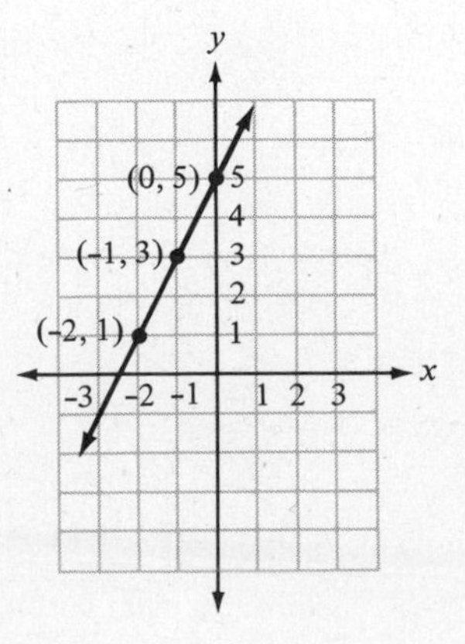

Stretching the Topics

1.

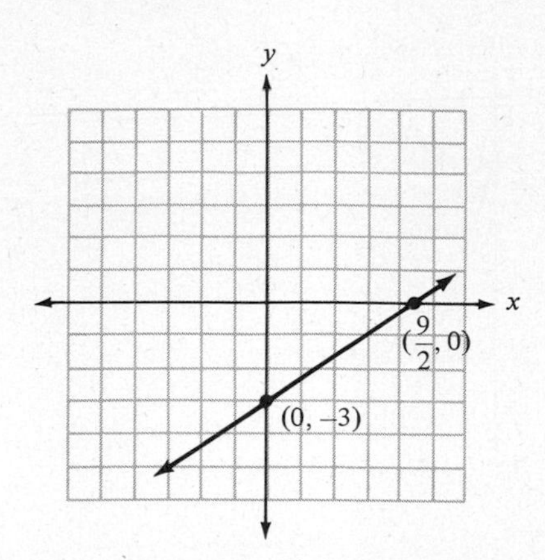

2.

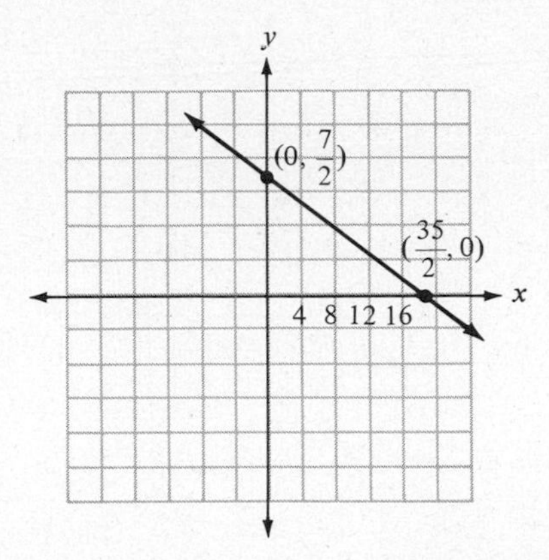

3.

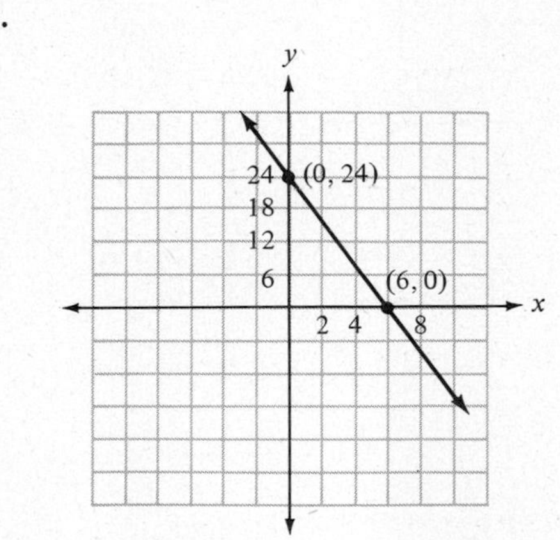

Checkup 10.2 (page 587)

1–3.

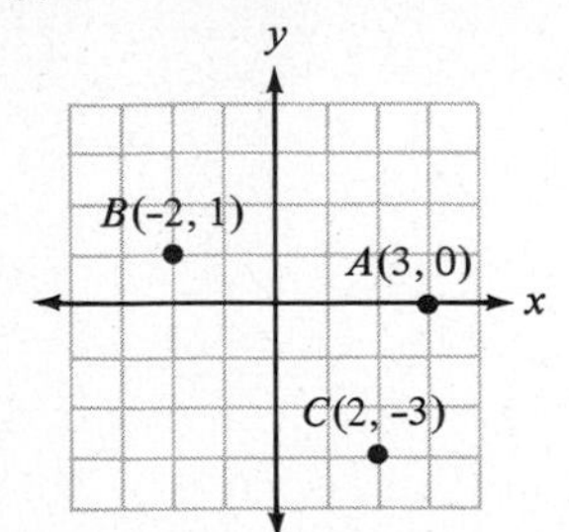

4. $A(-2, 0)$
5. $B(4, 2)$
6. $C(3, -2)$

7.

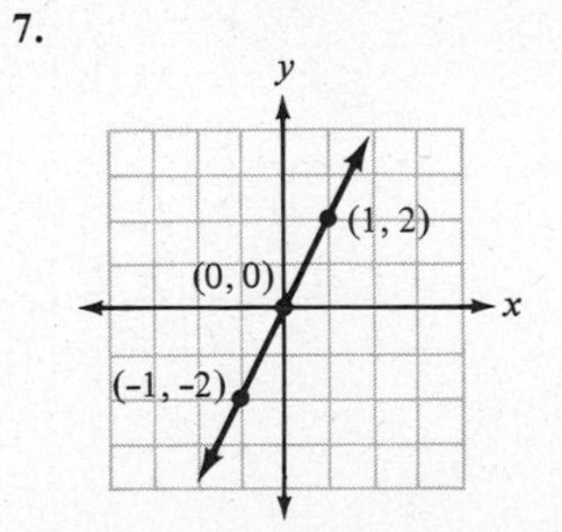

8.

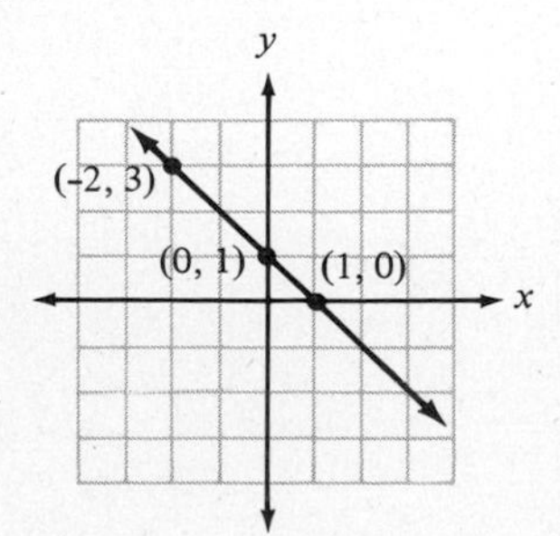

9.

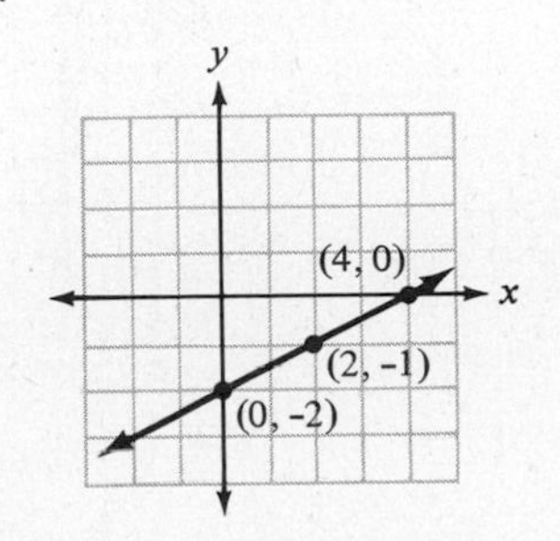

10.

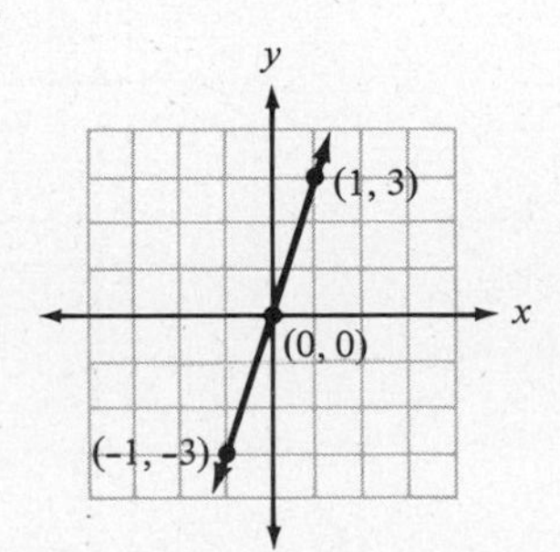

Exercise Set 10.3 (page 599)

1.

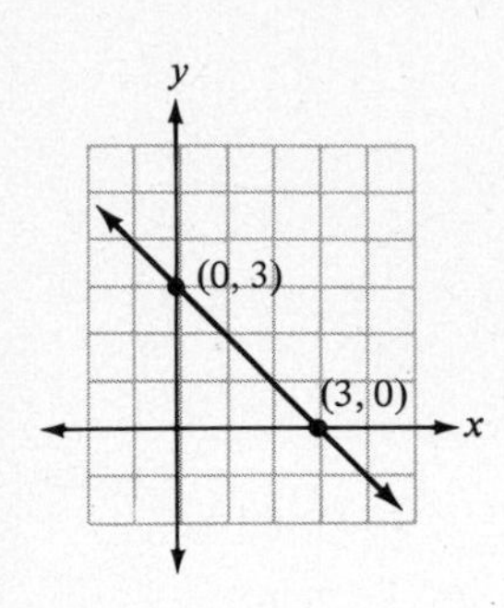

3.

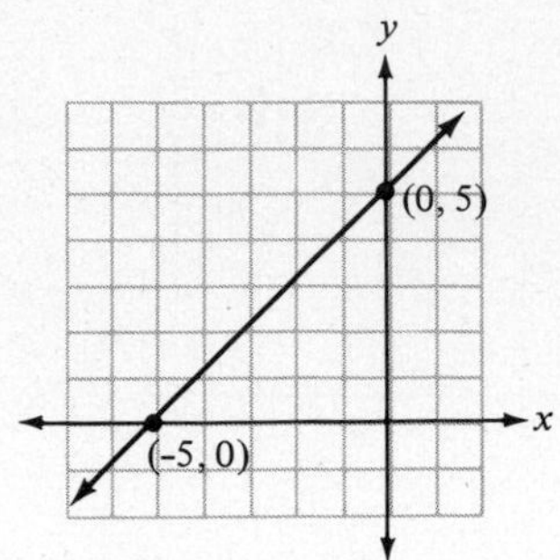

5.

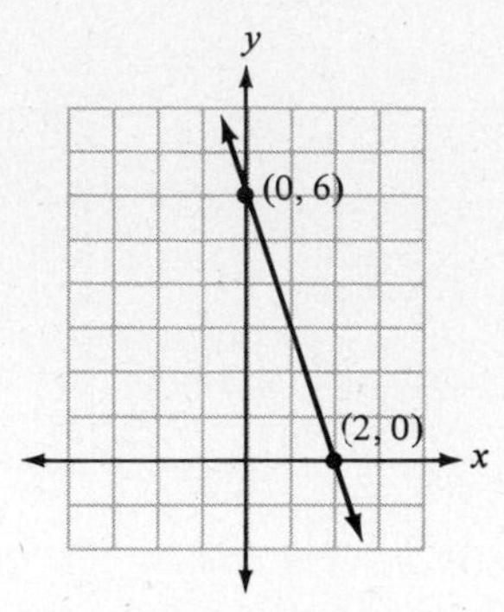

7.

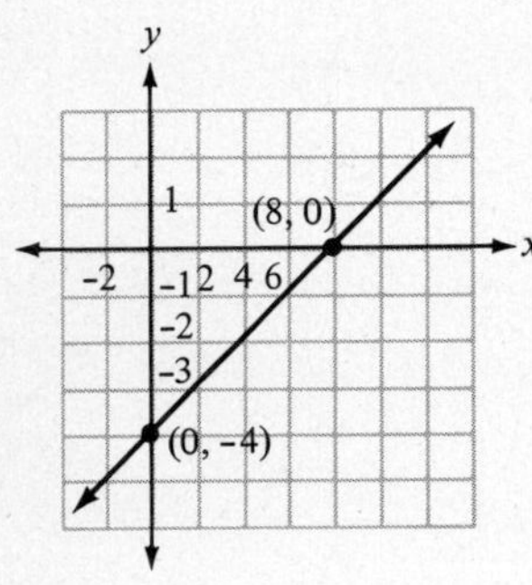

9.

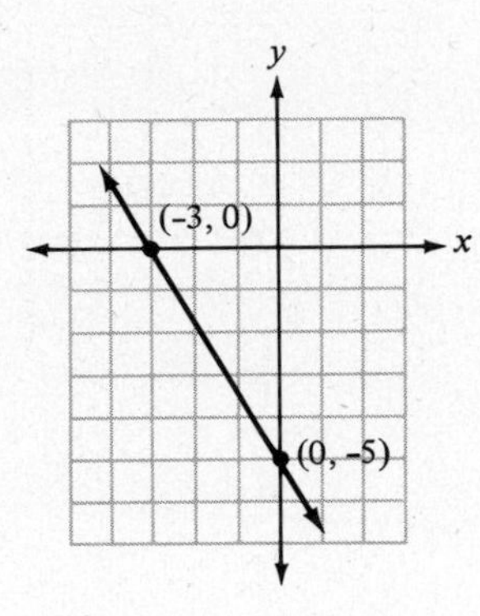

11.

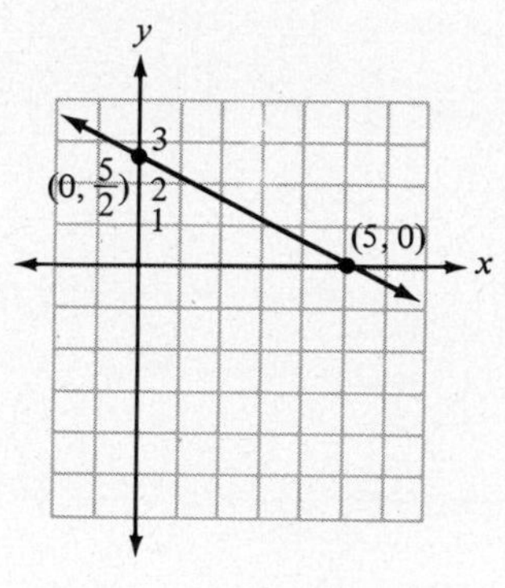

13.

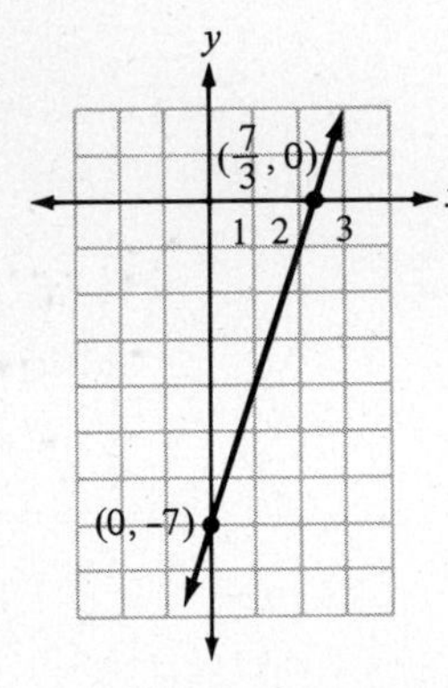

15.

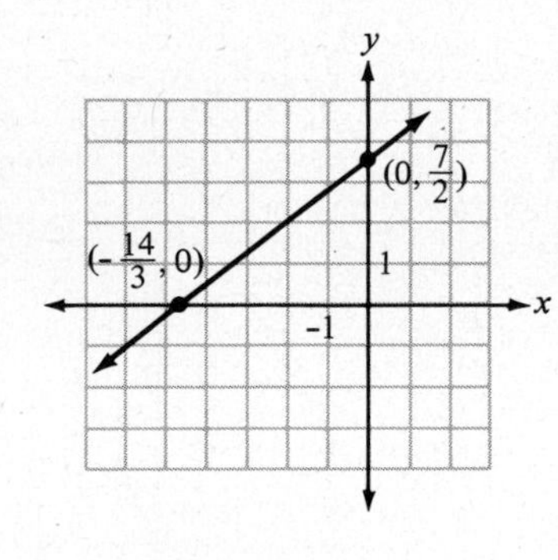

17.

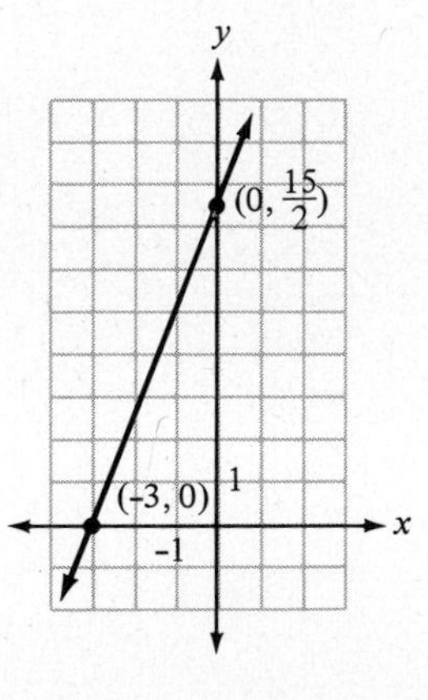

19.

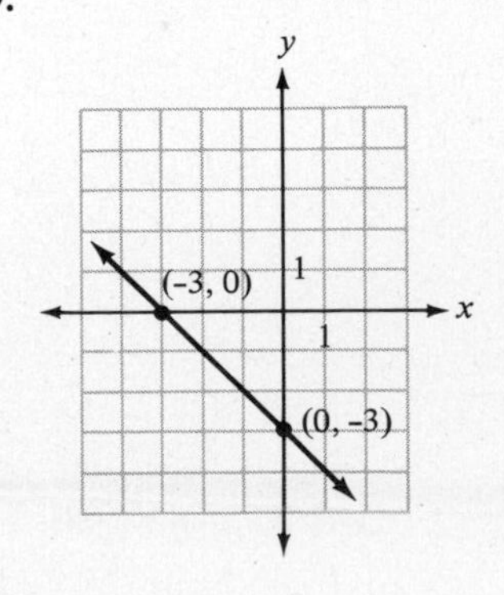

21.

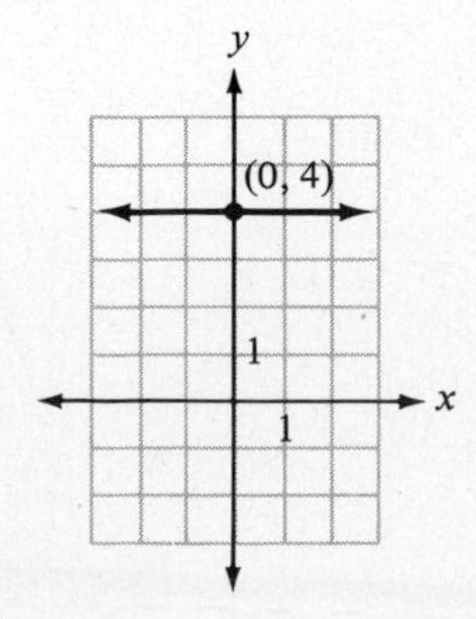

23.

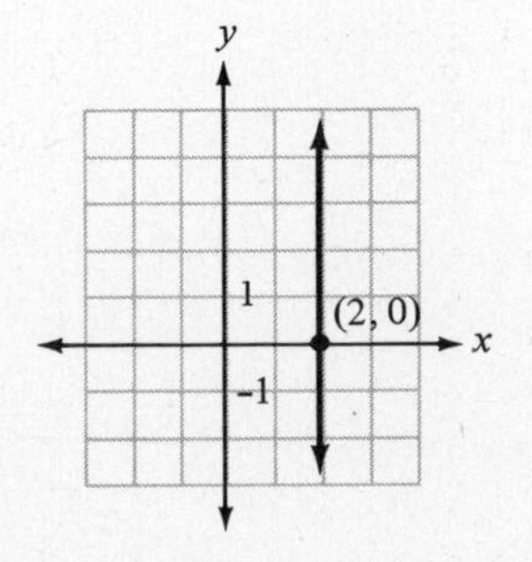

25.

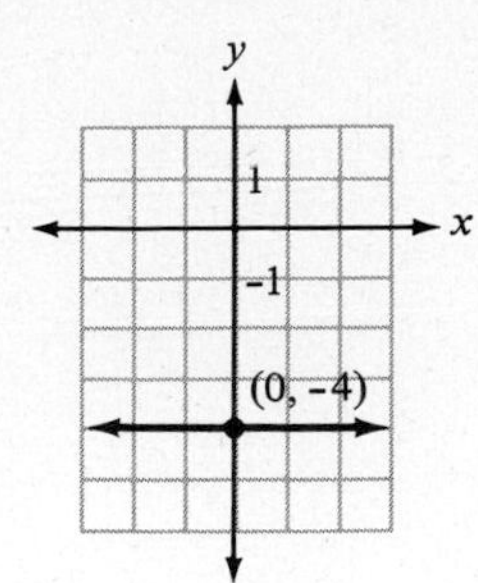

27.

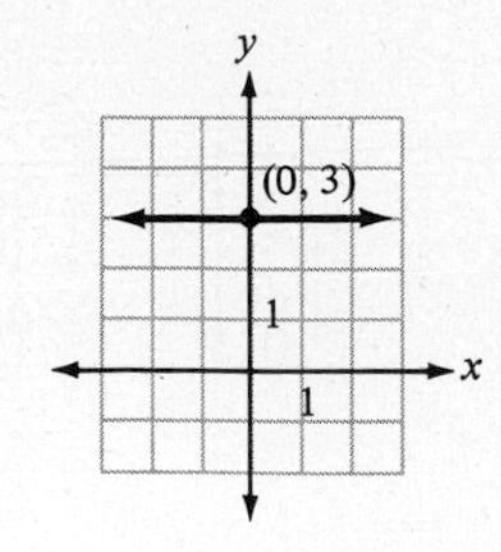

29.

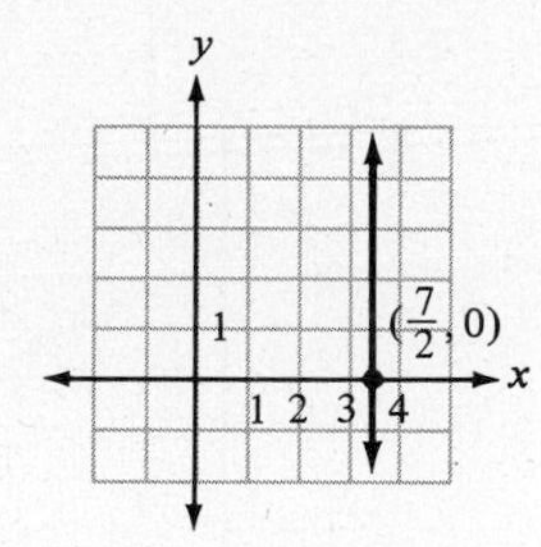

31.

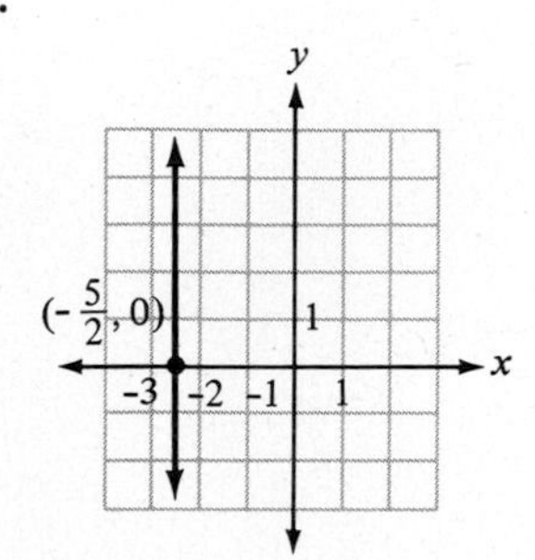

33.

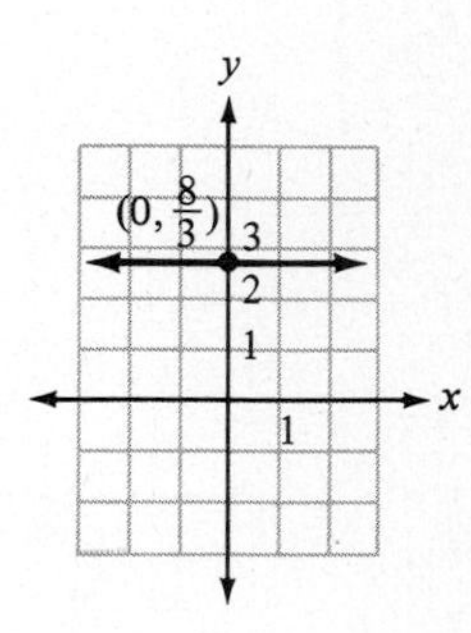

35.

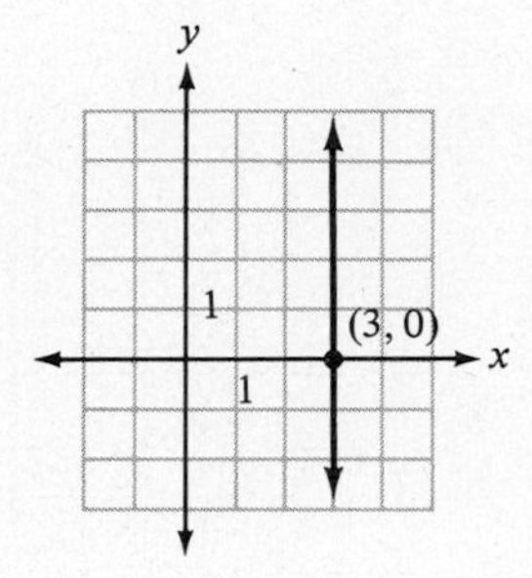

Stretching the Topics

1.

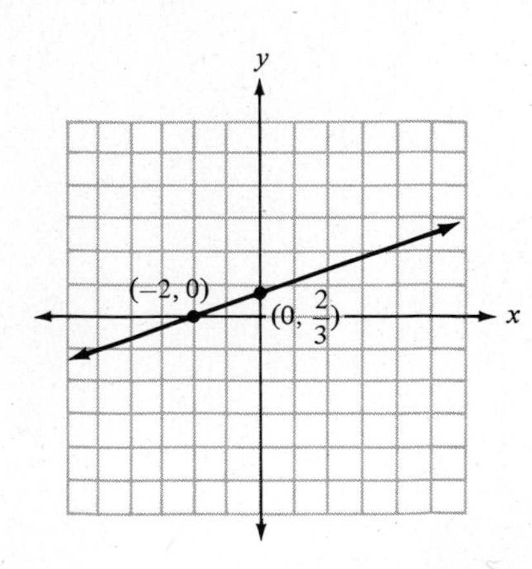

2.

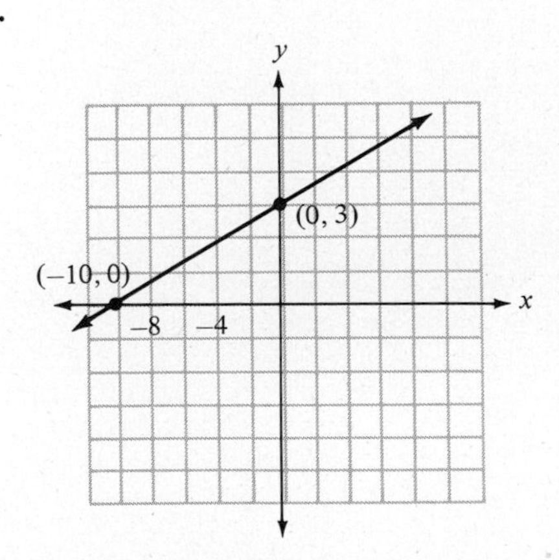

3.

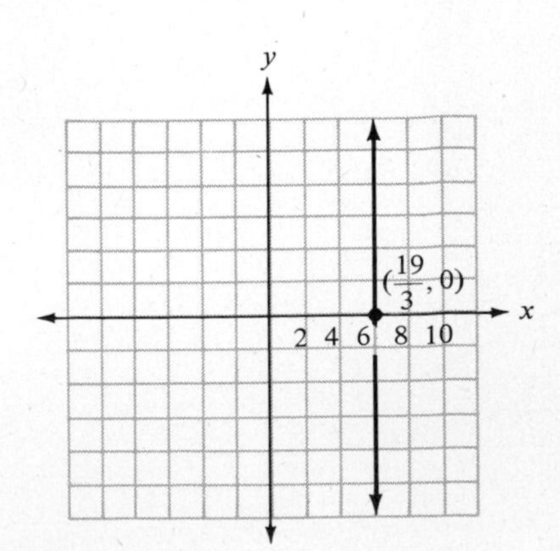

Checkup 10.3 (page 601)

1.

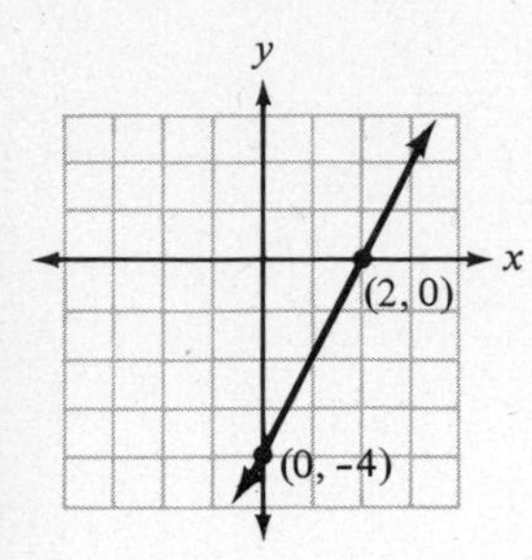

2.

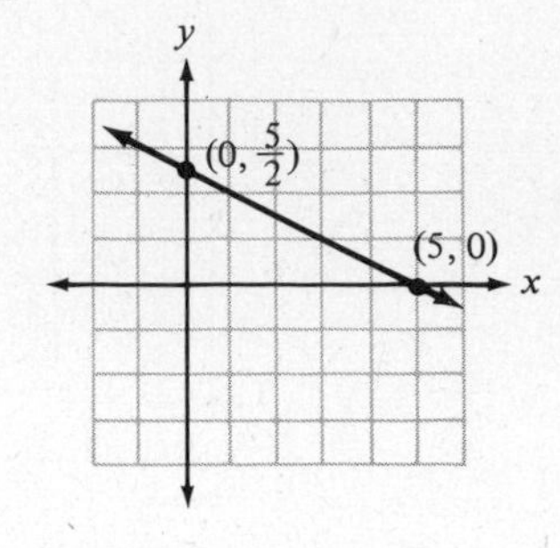

3.

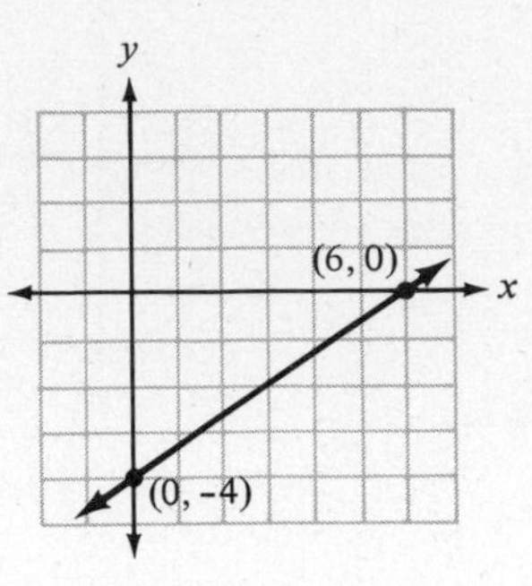

4.

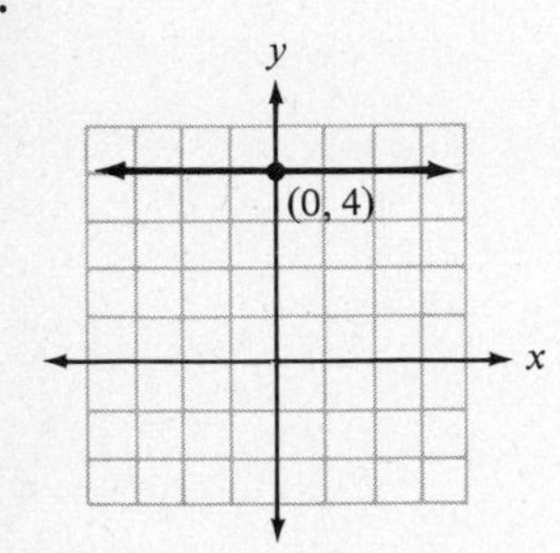

5.

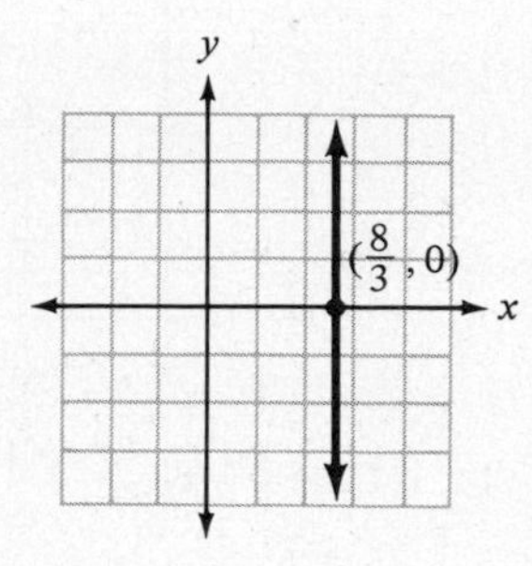

6.

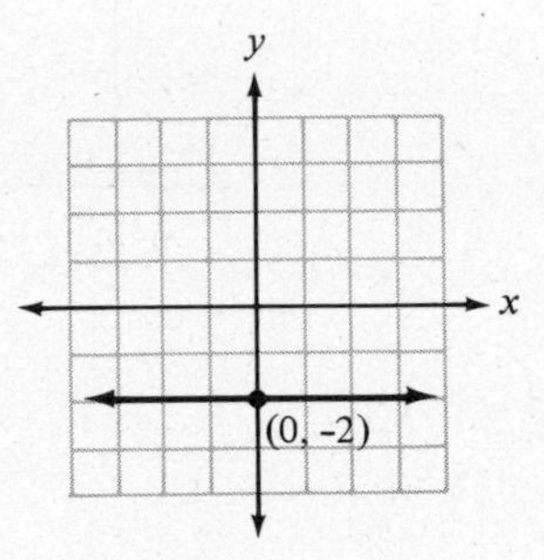

Exercise Set 10.4 (page 617)

1. $\frac{5}{3}$ **3.** Undefined **5.** -1 **7.** 0 **9.** 3

11.

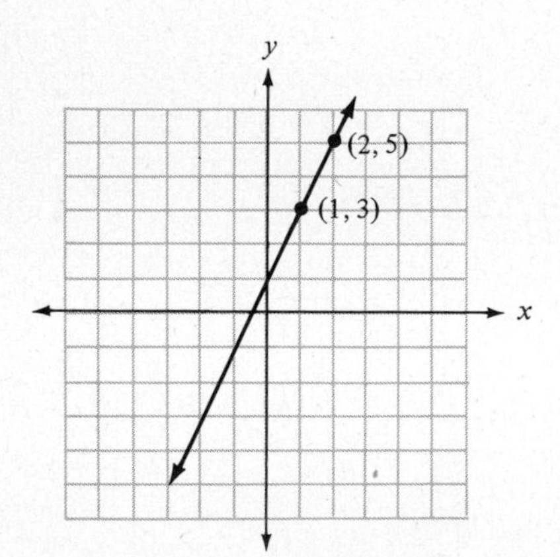

13.

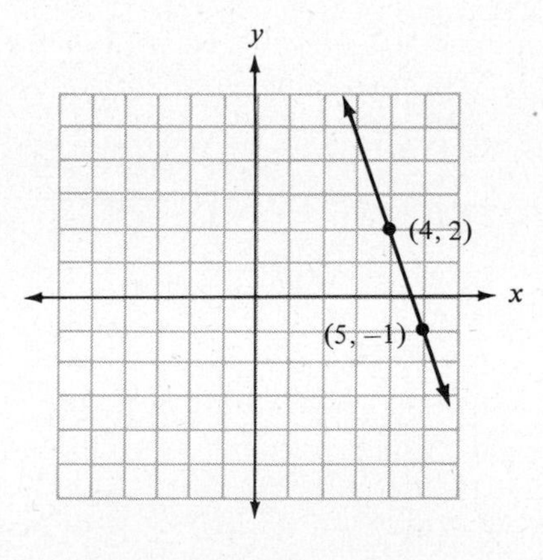

15.

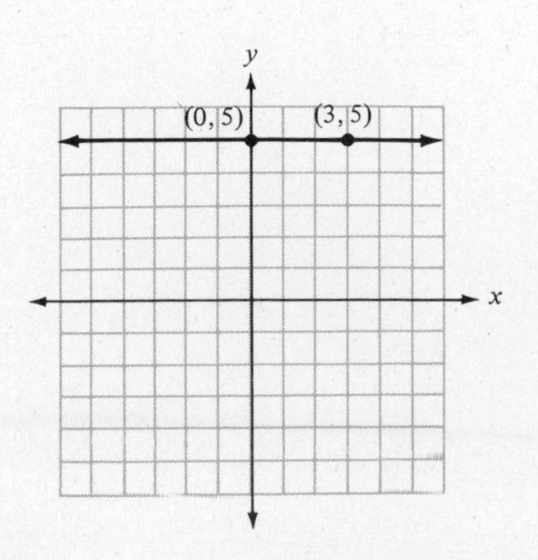

17.

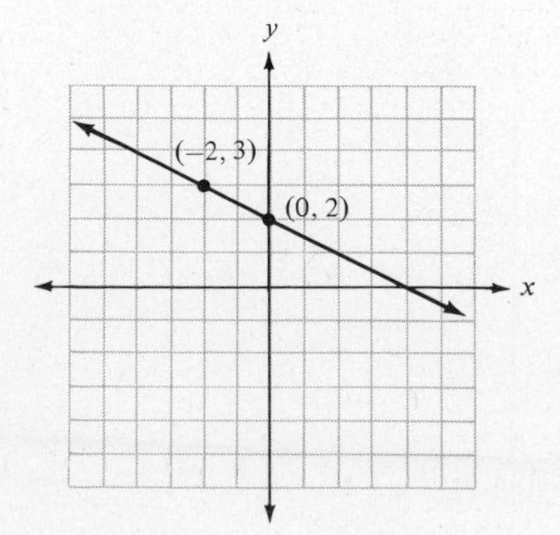

19.

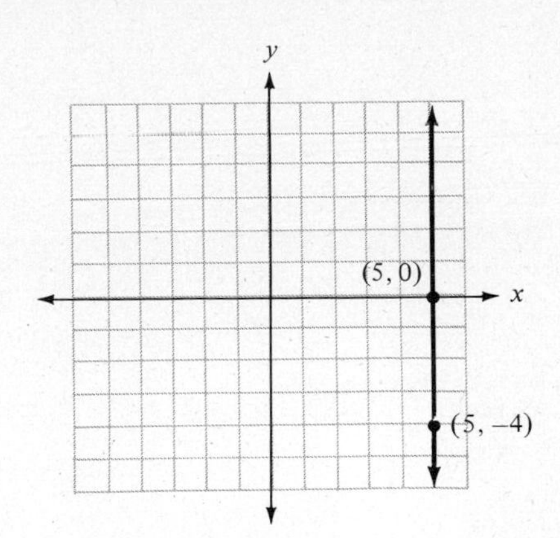

21.

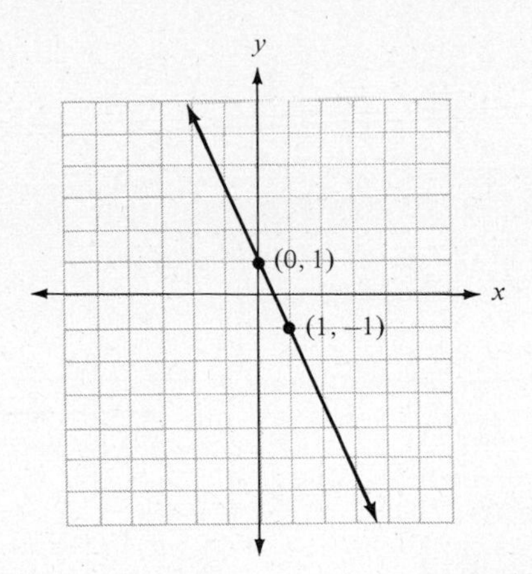

23.

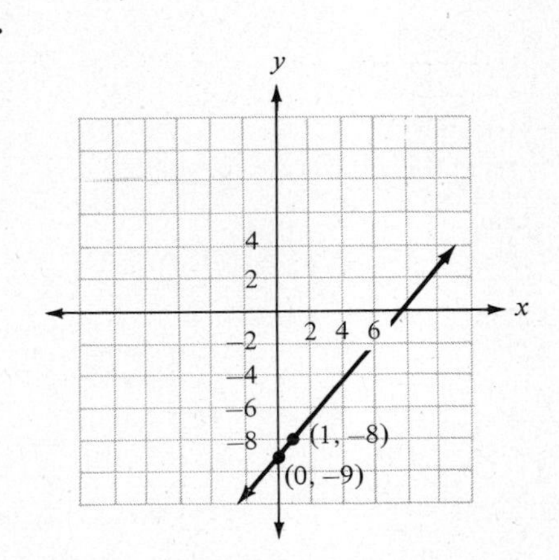

25.

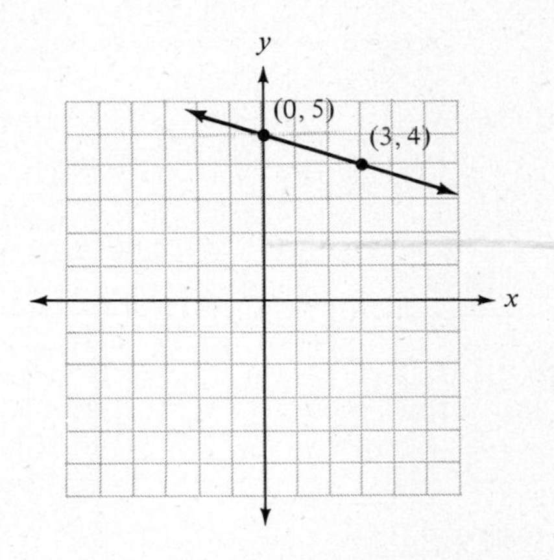

27.

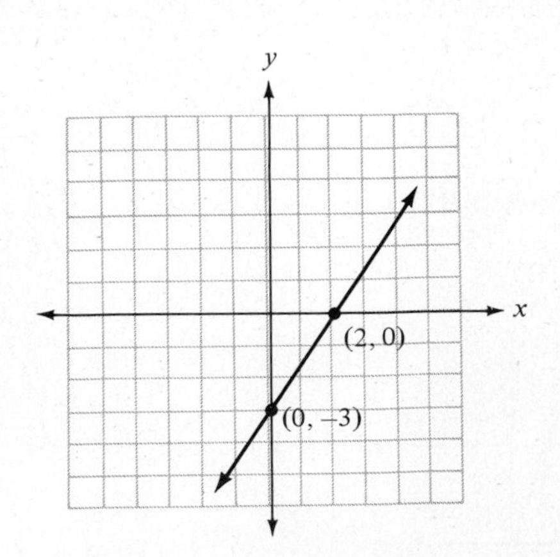

29.

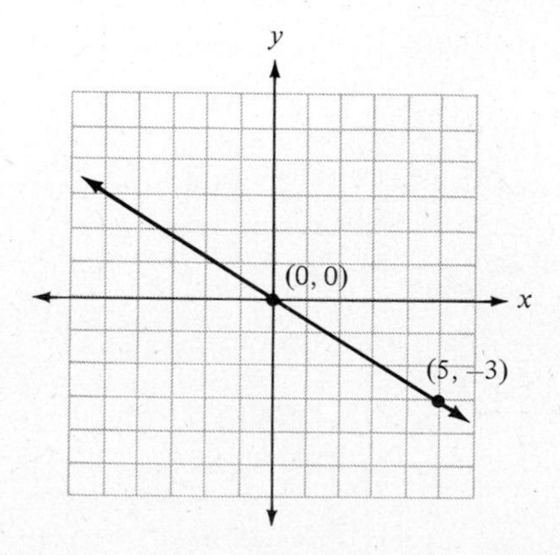

Stretching the Topics

1. $m = \frac{-2}{3}$ **2.** $m = \frac{7c}{4k}$ **3.** $x = 5$

Checkup 10.4 (page 619)

1. -1 **2.** 0

3.

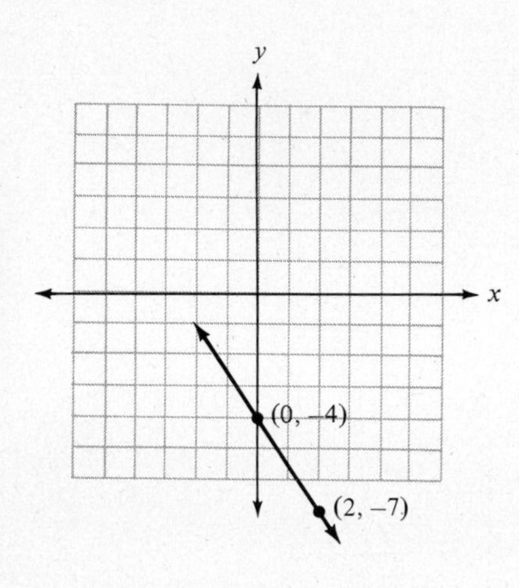

4.

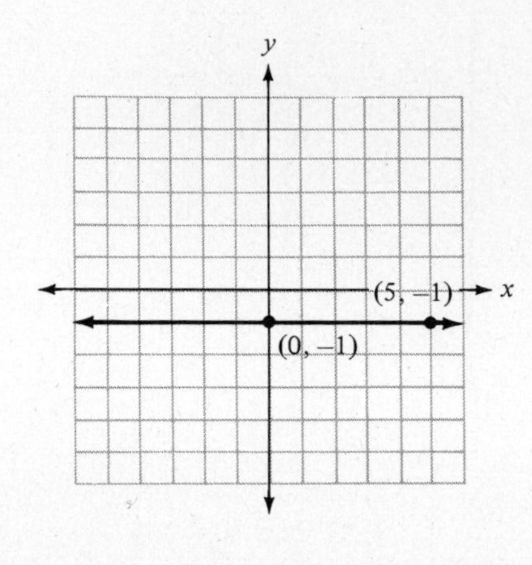

5.

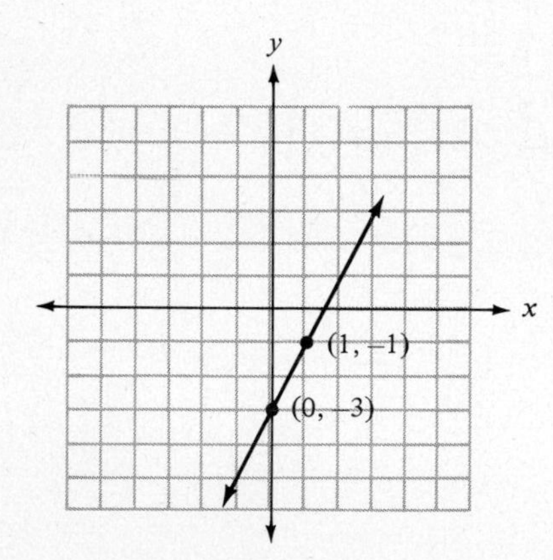

6.

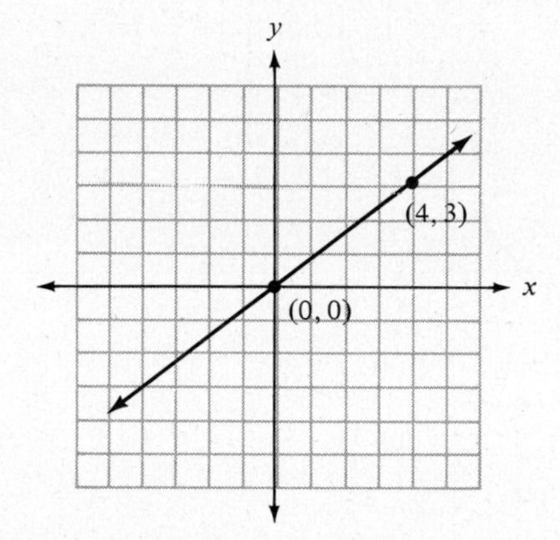

Speaking the Language of Algebra (page 621)

1. 3; -2 **2.** x-intercept; y-intercept **3.** Origin **4.** Straight line **5.** $\dfrac{y_2 - y_1}{x_2 - x_1}$
6. 0; $y = k$, a constant **7.** Undefined; $x = k$, a constant **8.** Slope; y-intercept

Review Exercises (page 623)

1. Yes **2.** Yes **3.** Answers will vary. **4.** Answers will vary. **5.** Answers will vary.
6. Answers will vary. **7.** $l = 3w - 5$; answers will vary.
8. $5x + 3y = 45$; answers will vary.  **10.** $A(-4, 1)$, $B(2, 0)$, $C(3, -4)$, $D(4, 3)$ 

9.

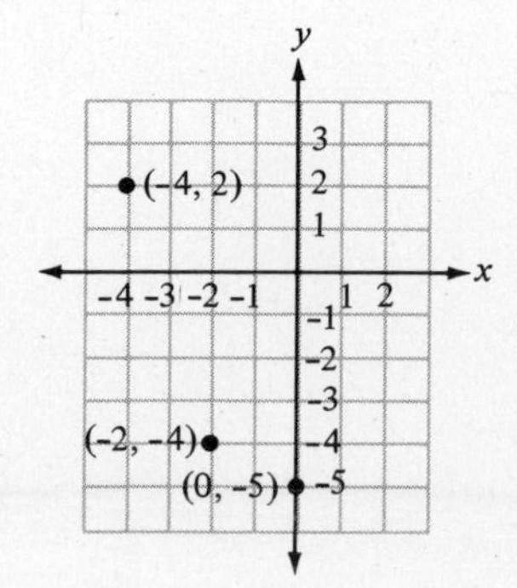

11.

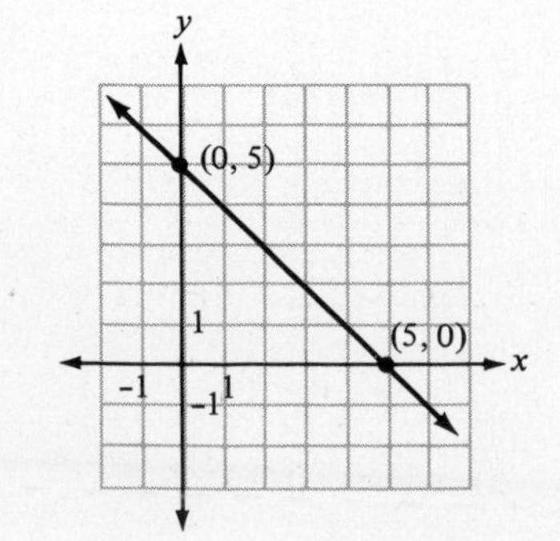

12.

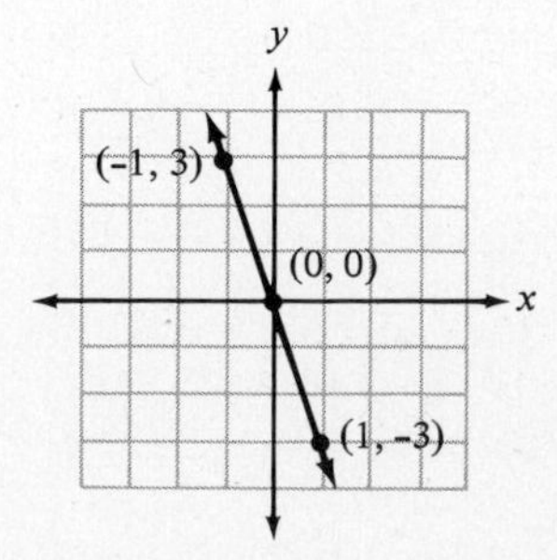

13.

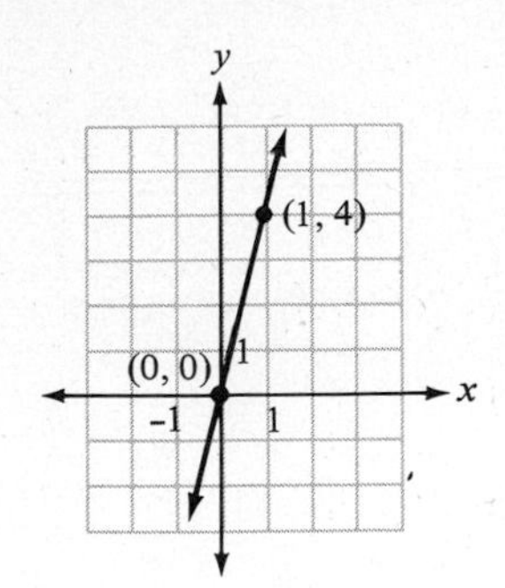

14.

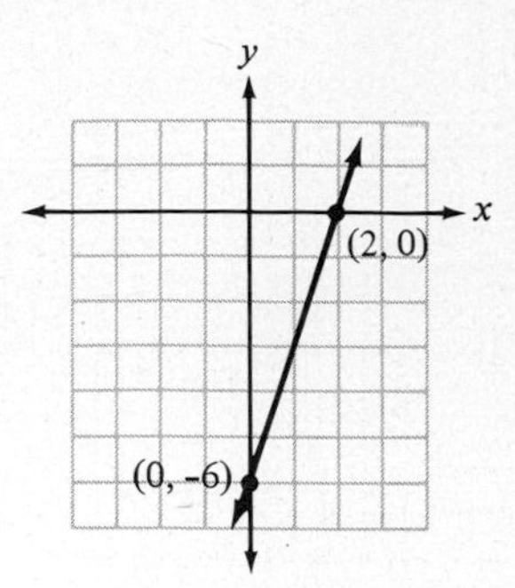

15.

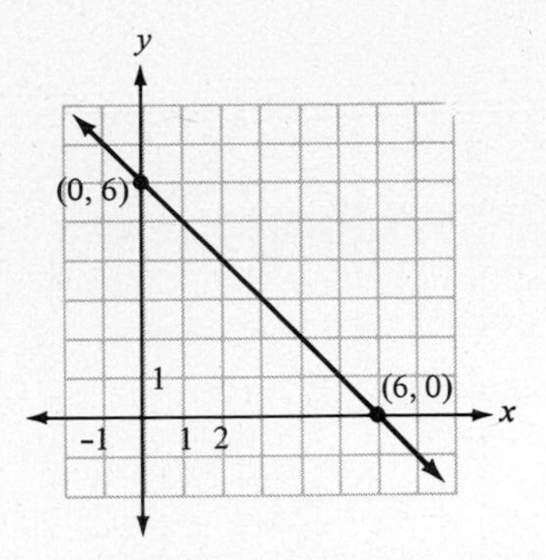

16.

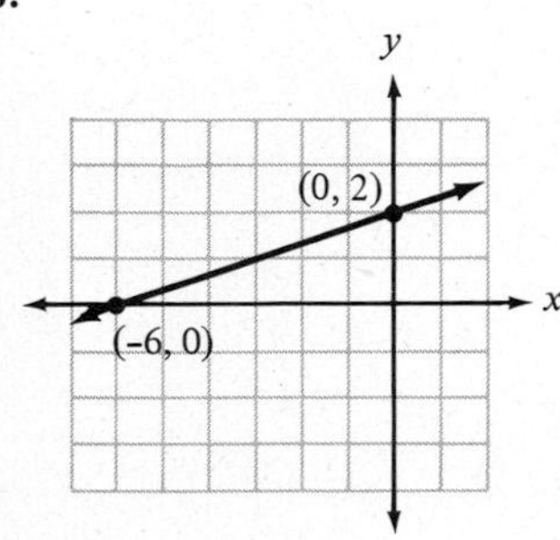

17.

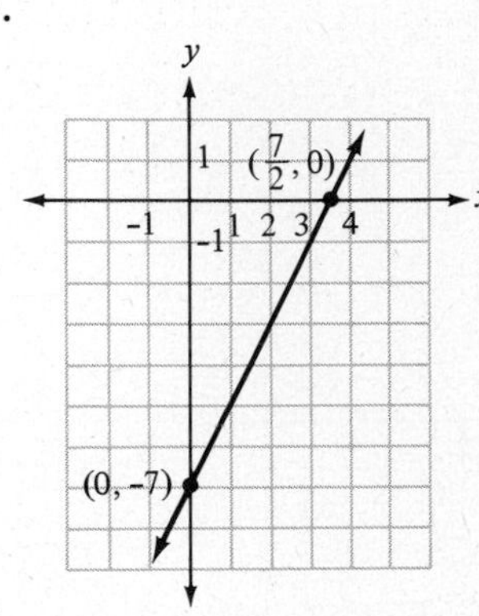

18.

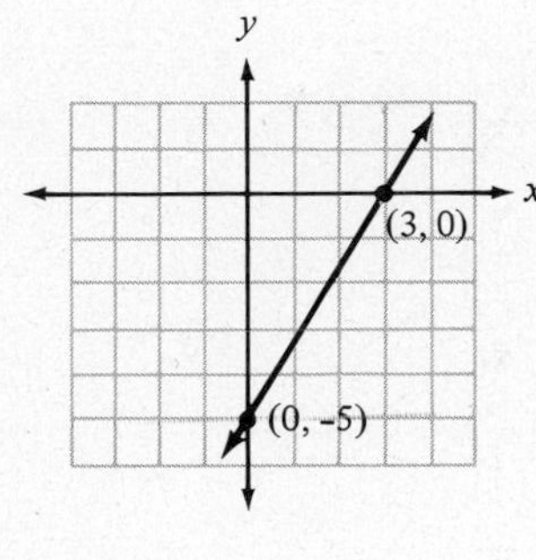

19.

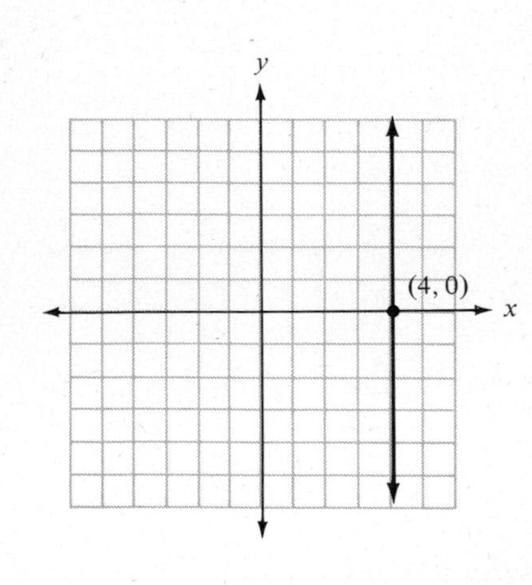

20.

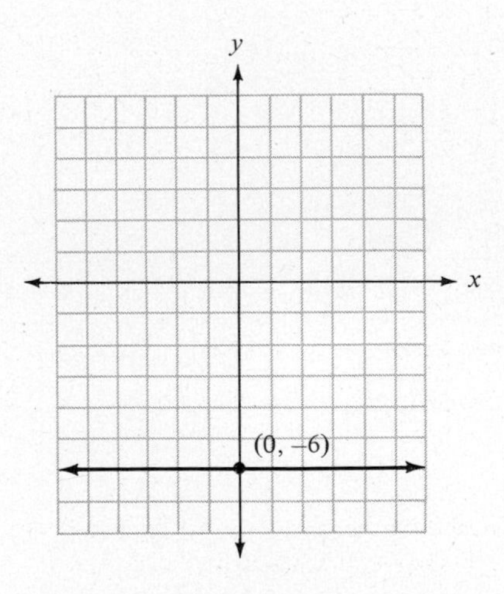

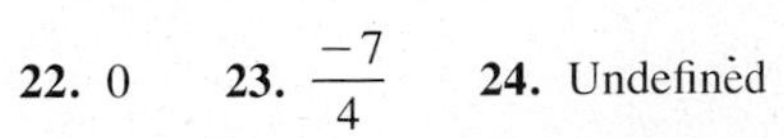

21. 1 **22.** 0 **23.** $\frac{-7}{4}$ **24.** Undefined

25.

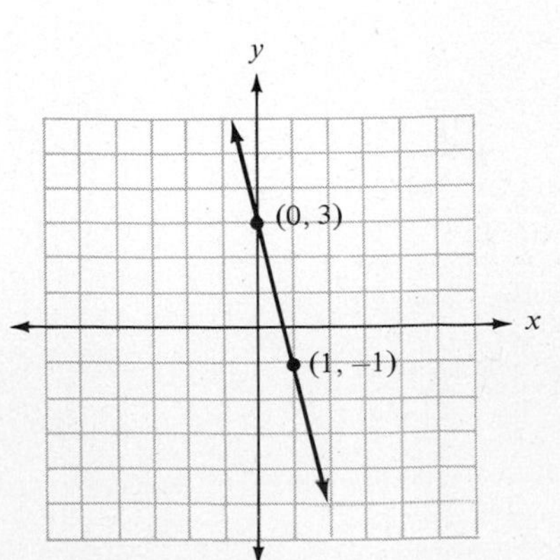

26.

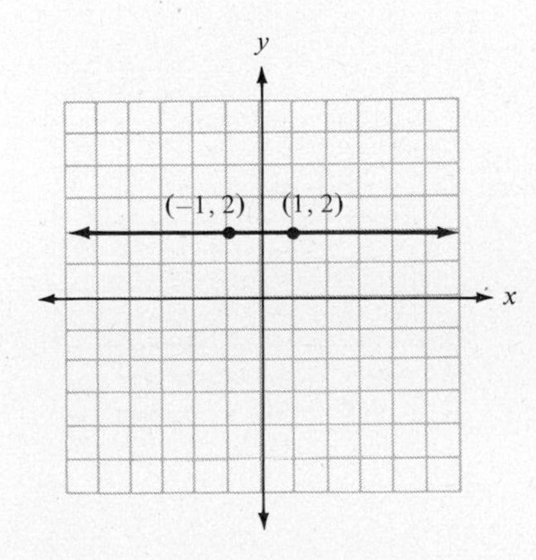

27.

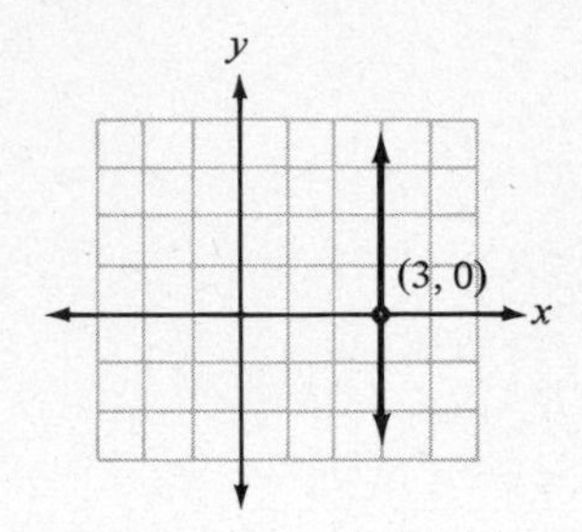

28.

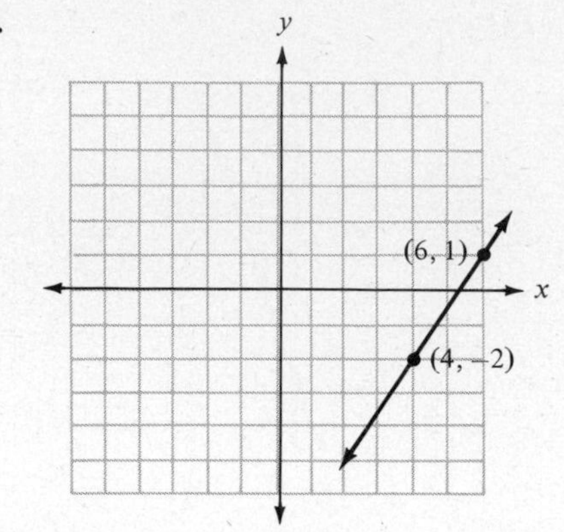

29.

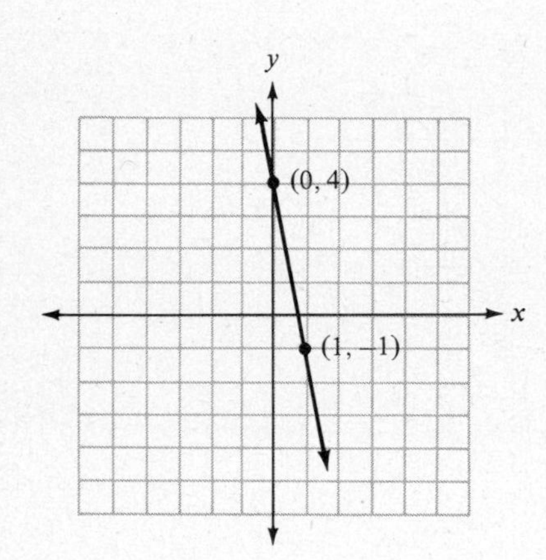

30.

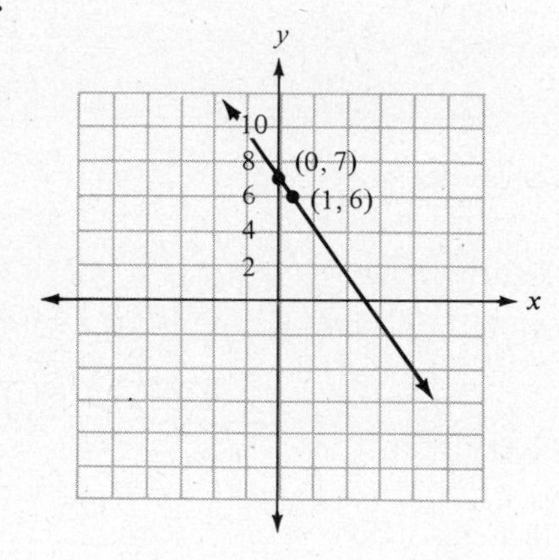

31.

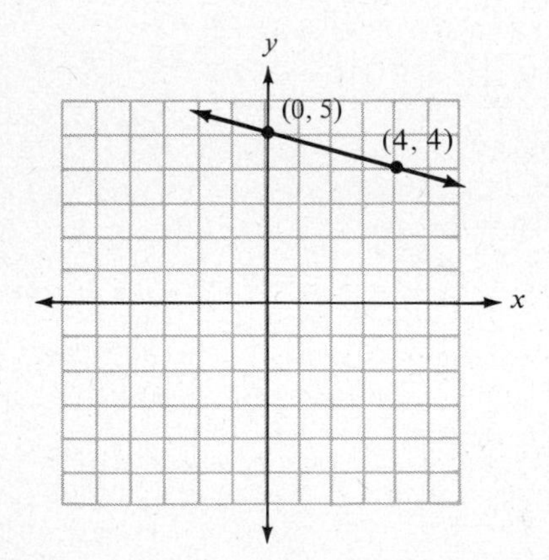

32.

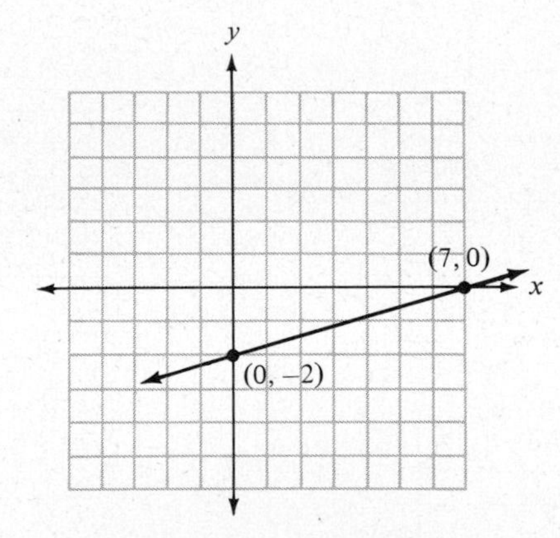

33.

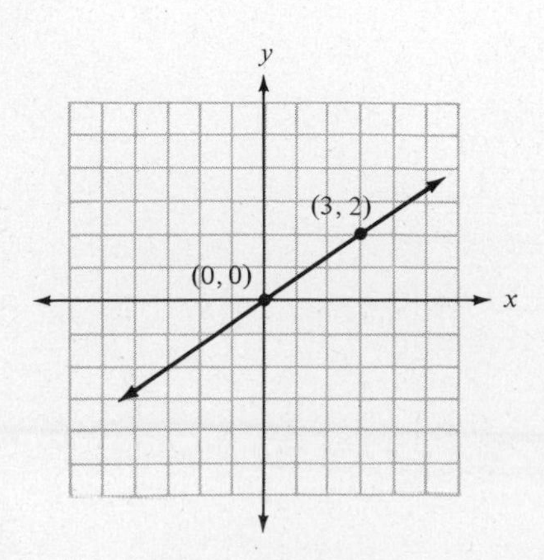

34.

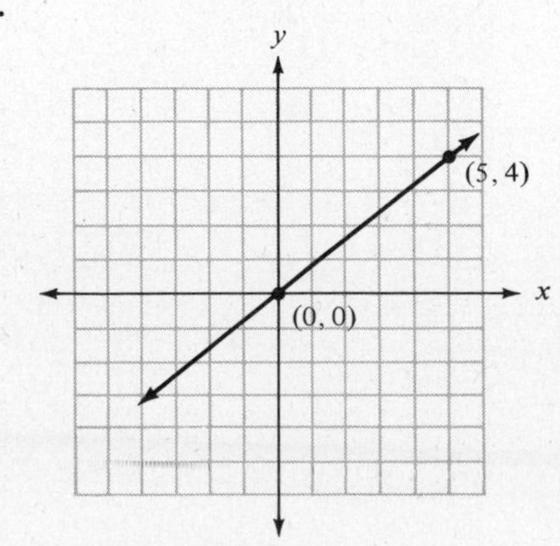

Practice Test for Chapter 10 (page 625)

1. No **2.** Yes **3.** $y = \frac{1}{2}x + 15$ **4.** $0.22x + 0.14y = 15.40$

5.

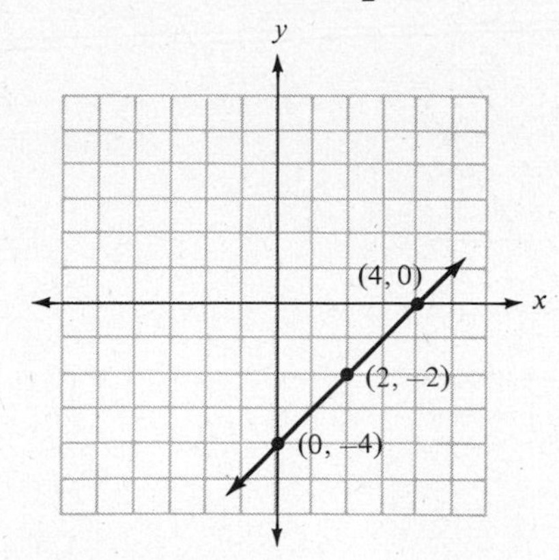

6.

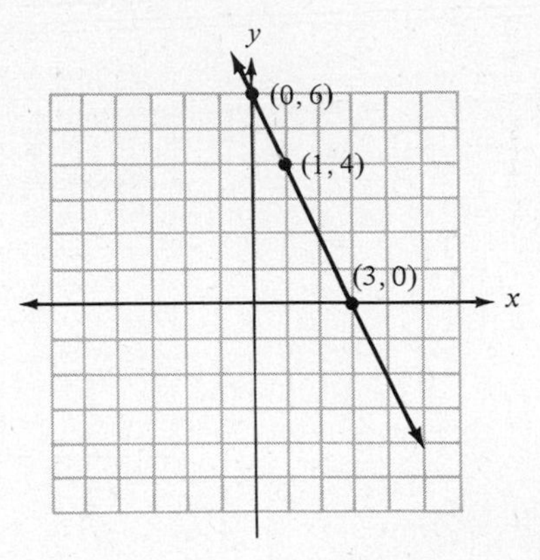

7.

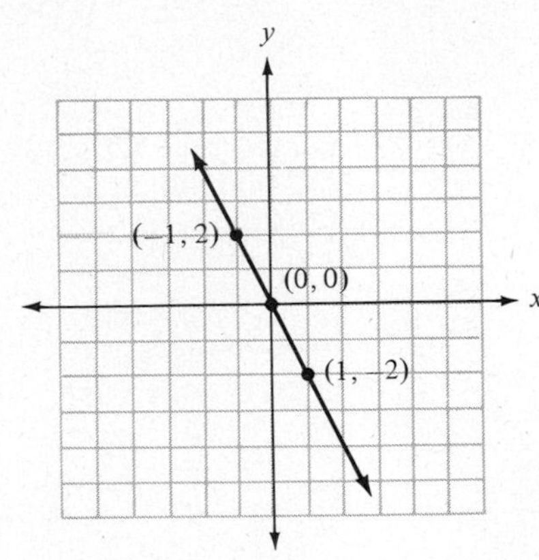

8.

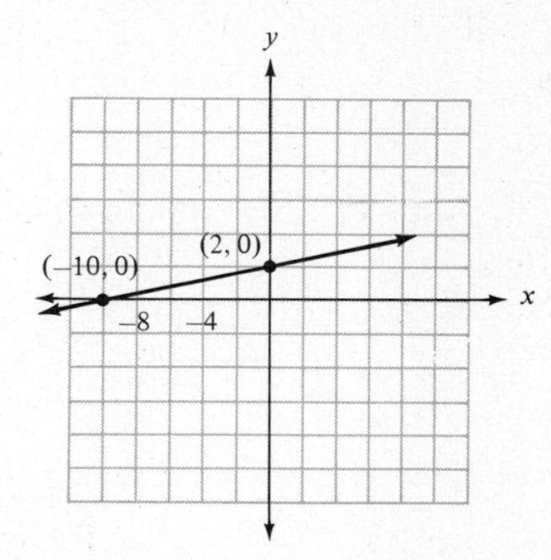

9.

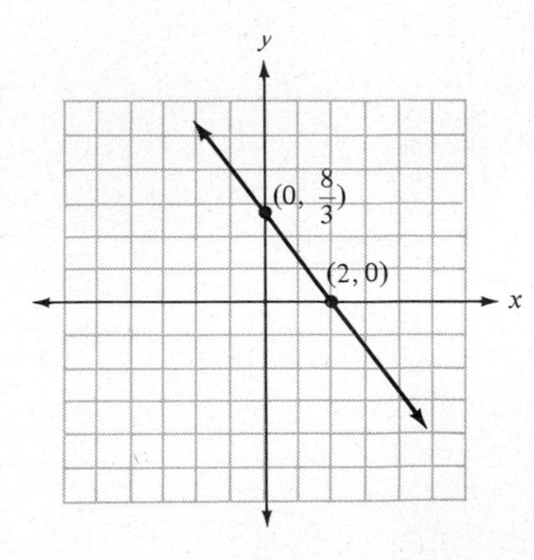

10

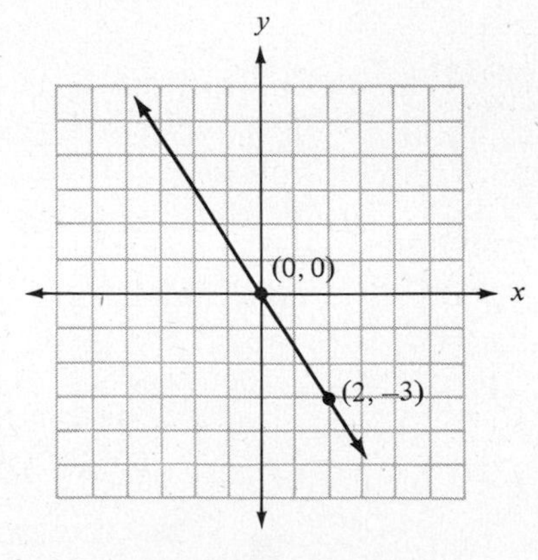

11.

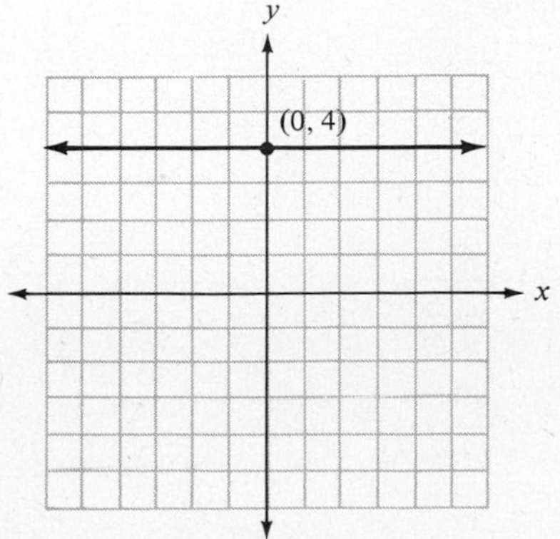

12.

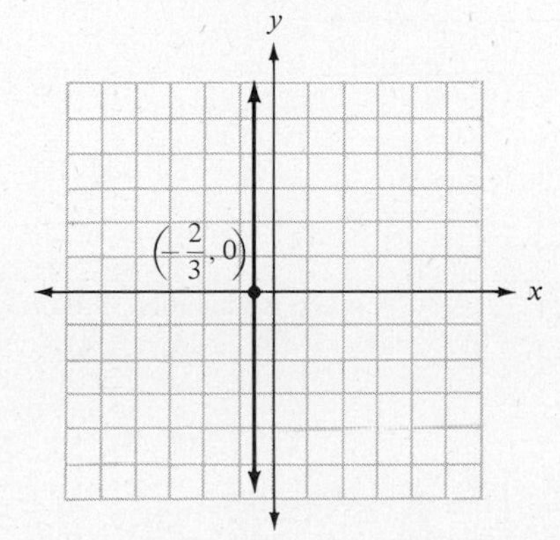

13. $\frac{1}{8}$

14.

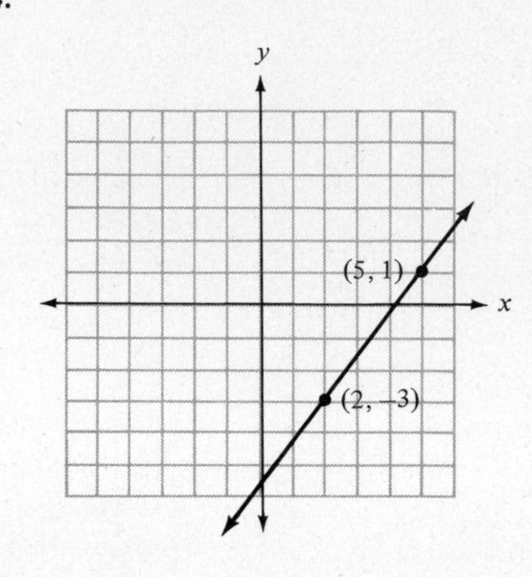

15.

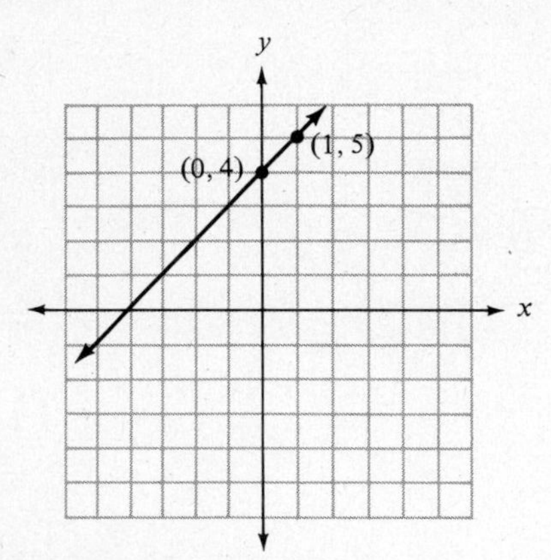

16.

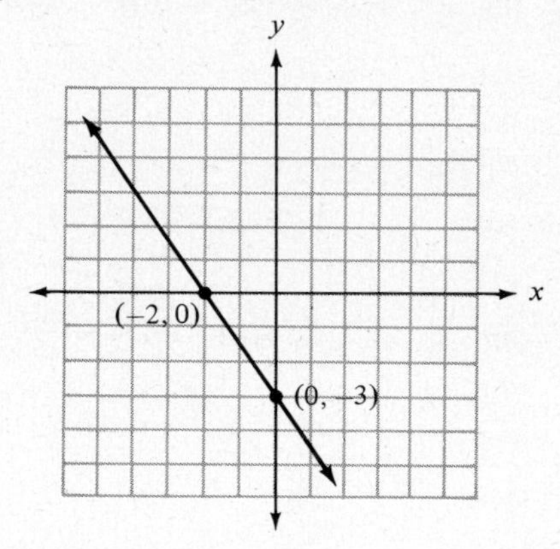

Sharpening Your Skills after Chapters 1–10 (page 629)

1. 0.193 **2.** 0.09 **3.** $\frac{-1}{18}$ **4.** 0.15 **5.** $16x - 12y$ **6.** $\frac{-3}{2}y$ **7.** $\frac{16b^{14}}{a^7}$
8. $\frac{-8y^3}{x^5}$ **9.** $40x^{11}y^3$ **10.** $-5axy$ **11.** $x^4 - 9$ **12.** $x - 5 - \frac{4}{x}$
13. $x^2 - 2x + 3 + \frac{-1}{x + 2}$ **14.** $\frac{1}{2x + 3y}$ **15.** $\frac{3(3y - 1)}{2xy(3y + 1)}$ **16.** $\frac{-12y^2z^3}{5x^2}$
17. $\frac{13x - 8}{5x(x + 4)}$ **18.** $y = 4$ **19.** $x = 7, x = -3$ **20.** $y = -4, y = -5$ **21.** $a = 1$
22. $b = \frac{-1}{11}$ **23.** $x > \frac{-4}{5}$ **24.** $1 \leq y \leq 4$ **25.** $x < 3$ **26.** $P = \frac{A}{1 + rt}$
27. $x \geq -2$

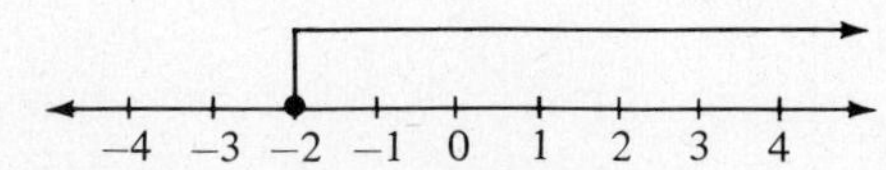

28. 7 **29.** 5 by 9 cm **30.** $x = 2$

CHAPTER 11

Exercise Set 11.1 (page 639)

1. (1, 3) **3.** (−2, 6) **5.** (−2, −12) **7.** No solution (parallel lines) **9.** (2, 0)
11. All points on the line $2x - y = 2$ **13.** (2, 3) **15.** (7, 4) **17.** (4, 1)
19. No solution (parallel lines)

Stretching the Topics

1. (5, −3) **2.** All points on the line $2x - 3y = 0$

Checkup 11.1 (page 641)
1. $(1, -1)$ **2.** No solution (parallel lines) **3.** $(-3, 4)$
4. All points on the line $-x + 2y = 4$

Exercise Set 11.2 (page 649)
1. $(2, 3)$ **3.** $(-1, -2)$ **5.** $(6, 19)$ **7.** $(17, 5)$ **9.** Ordered pairs satisfying $y = 2x - 4$
11. $(-5, -9)$ **13.** $\left(8, \frac{-9}{2}\right)$ **15.** $\left(\frac{24}{5}, \frac{6}{5}\right)$ **17.** $\left(3, \frac{7}{3}\right)$ **19.** No solution
21. $(1, 3)$ **23.** $(-23, -10)$ **25.** $\left(\frac{1}{10}, \frac{17}{10}\right)$

Stretching the Topics
1. $(0.05, -0.1)$ **2.** $\left(-6, \frac{13}{3}\right)$ **3.** $(2a, a)$

Checkup 11.2 (page 651)
1. $(3, 6)$ **2.** No solution **3.** $\left(6, \frac{5}{2}\right)$ **4.** $(2, 2)$ **5.** $\left(\frac{5}{3}, \frac{4}{3}\right)$

Exercise Set 11.3 (page 661)
1. $(1, -1)$ **3.** $(-1, -1)$ **5.** $\left(4, \frac{7}{4}\right)$ **7.** $(2, -2)$ **9.** $\left(\frac{-1}{2}, -1\right)$ **11.** $\left(\frac{1}{2}, \frac{3}{4}\right)$
13. $(2, 1)$ **15.** $\left(\frac{-11}{3}, 4\right)$ **17.** No solution **19.** $\left(0, \frac{-5}{2}\right)$ **21.** $\left(\frac{-2}{3}, \frac{11}{4}\right)$
23. $(1, 1)$ **25.** Ordered pairs satisfying $14x + 7y = 7$ **27.** $\left(1, \frac{-1}{3}\right)$ **29.** $(1, 1)$

Stretching the Topics
1. $(4, 9)$ **2.** $\left(\frac{-5}{7}, \frac{-36}{7}\right)$ **3.** $k = 6$

Checkup 11.3 (page 663)
1. $(2, 0)$ **2.** $(3, -2)$ **3.** No solution **4.** $\left(\frac{4}{3}, \frac{-1}{3}\right)$ **5.** $(2, 7)$

Exercise Set 11.4 (page 671)
1. -1 and 7 **3.** 8 by 21 ft **5.** 13 five-dollar bills, 12 one-dollar bills
7. 9 books at \$2, 11 books at \$3 **9.** \$3000 in Abell's, \$5000 in Betti's
11. \$16 for jeans, \$12 for tops **13.** 24°, 66° **15.** \$5 for records, \$7 for tapes
17. blouse \$21, skirt \$28 **19.** 100 of each

Stretching the Topics
1. \$10,000 at 8%, \$6000 at 10% **2.** 170

Checkup 11.4 (page 673)
1. \$5 as waitress, \$4 as window washer **2.** shirts \$13, ties \$5 **3.** 40 by 60 in.
4. 5 and -12 **5.** 1176 advance tickets, 1649 at door

Speaking the Language of Algebra (page 675)
1. Satisfy **2.** Substitution **3.** No **4.** One

Review Exercises (page 677)
1. $(2, 3)$ **2.** $(2, 1)$ **3.** $(-14, -7)$ **4.** $(-1, 2)$ **5.** $(-1, 3)$ **6.** $\left(\frac{1}{2}, \frac{-3}{2}\right)$
7. $(-2, -8)$ **8.** No solution **9.** $(13, 3)$ **10.** $(-4, 1)$ **11.** $(5, 2)$ **12.** $(0, 3)$
13. No solution **14.** $\left(\frac{55}{31}, \frac{-15}{31}\right)$ **15.** Ordered pairs satisfying $-6x + 5y = 9$ **16.** $(1, 2)$

17. (0, 8) **18.** (−10, −5) **19.** (3, −2) **20.** $\left(\frac{55}{23}, \frac{6}{23}\right)$ **21.** 10 and 12
22. 24 by 34 m **23.** \$1250 at Union, \$450 at National **24.** 8 bottles at \$5, 12 bottles at \$3

Practice Test for Chapter 11 (page 679)

1. (1, 2) **2.** (−5, 3) **3.** (2, 5) **4.** No solution **5.** $\left(\frac{1}{2}, -2\right)$
6. Ordered pairs satisfying $3x - 3y = 6$ **7.** (10, 4) **8.** (2, 3) **9.** (−1, −2)
10. $\left(\frac{19}{71}, \frac{-2}{71}\right)$ **11.** \$3 for mowing, \$4.50 for digging **12.** 60 by 100 yd

Sharpening Your Skills after Chapters 1–11 (page 681)

1. $-2.39 < -2.184$ **2.** $\frac{8}{5}$ **3.** $y \neq 3, y \neq \frac{-5}{2}$ **4.** $-1 < \frac{-2}{3} < \frac{-1}{5}$ **5.** -3
6. $\frac{56}{405}$ **7.** $18a - 4$ **8.** $-4x^5y^5z^5$ **9.** $2a^4 - 10a^3 + 4a^2 - 20a$ **10.** $\frac{6}{x - 6}$
11. $\frac{7x}{20(x - 3)}$ **12.** $(9x^2 + 4)(3x + 2)(3x - 2)$ **13.** $7xy(x + 5y)(x - 5y)$
14. $(a + b)(2x - 3y)$ **15.** $(xy - 9z)(xy + z)$ **16.** $x = -8$ **17.** $x = 0, x = 9$
18. $y = 0, y = 3$ **19.** $x = 42$ **20.** $y > 0$ **21.** $x > 6$
22. **23.** $m = \frac{3}{5}$ **24.** 3 **25.** 20

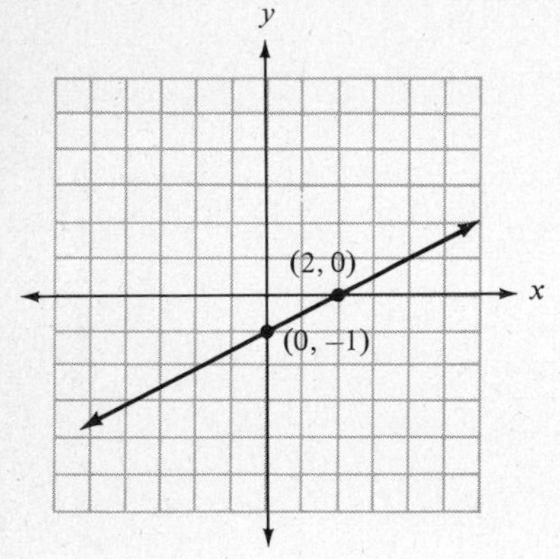

CHAPTER 12

Exercise Set 12.1 (page 693)
1. Rational **3.** Irrational **5.** Rational **7.** Irrational **9.** Rational **11.** Irrational
13. Irrational **15.** Rational **17.** 12 **19.** a^2 **21.** 8 **23.** x^5 **25.** $5b$ **27.** $\frac{x}{6}$
29. $3\sqrt{5}$ **31.** $10\sqrt{6}$ **33.** $5\sqrt{3}$ **35.** $2\sqrt{11}$ **37.** x **39.** $y^2\sqrt{y}$ **41.** $xy\sqrt{y}$
43. $4a\sqrt{2a}$ **45.** $2x\sqrt{y}$ **47.** $9a^2b^2\sqrt{a}$ **49.** $8xy^2z\sqrt{3z}$ **51.** $6a^3b\sqrt{5abc}$ **53.** $\frac{2x}{5}$
55. $\frac{4a\sqrt{2a}}{b^2}$ **57.** $\frac{2x\sqrt{10x}}{3y}$ **59.** $\frac{ab}{5}$ **61.** $5x$ **63.** $x^2\sqrt{3}$ **65.** $\frac{\sqrt{y}}{11}$

Stretching the Topics
1. $\frac{2x^5\sqrt{6}}{y^2z^{14}}$ **2.** $-21x^2y^7\sqrt{7y}$ **3.** $0.4x^2y(a - b)$

Checkup 12.1 (page 695)

1. 7 **2.** $3a$ **3.** 12 **4.** $4\sqrt{5}$ **5.** $5y^2\sqrt{3}$ **6.** $3ab\sqrt{5a}$ **7.** $\frac{5x}{6y}$ **8.** $\frac{4x^2\sqrt{2x}}{y^3}$
9. $5a^2b$ **10.** $a^2\sqrt{3}$

Exercise Set 12.2 (page 705)

1. $3\sqrt{3}$ **3.** $11\sqrt{x}$ **5.** $-7\sqrt{3}$ **7.** $5\sqrt{6a}$ **9.** $9 - 6\sqrt{3}$ **11.** $23\sqrt{3x}$
13. $10\sqrt{7} - 20\sqrt{2}$ **15.** $11\sqrt{2}$ **17.** $-5x\sqrt{3}$ **19.** $\sqrt{21}$ **21.** 12 **23.** $11a$
25. $x - 1$ **27.** $3a\sqrt{5}$ **29.** 30 **31.** $-120x$ **33.** $15x\sqrt{2}$ **35.** $2\sqrt{5} - 5$
37. $2\sqrt{6} + 12$ **39.** $-3\sqrt{x} + 6\sqrt{5} - 3\sqrt{6}$ **41.** $x - 16\sqrt{x}$ **43.** $31 - 10\sqrt{6}$
45. $x - 25$ **47.** $23 - 7\sqrt{11}$ **49.** $2x - 5\sqrt{x} - 3$ **51.** $5 - 5\sqrt{3} - \sqrt{2} + \sqrt{6}$
53. $11 + \sqrt{5}$ **55.** $8 - 2\sqrt{3x} - 3x$ **57.** -4 **59.** $x - 100$ **61.** $y - 2\sqrt{7y} + 7$
63. $2a^2 - 10a\sqrt{2} + 25$

Stretching the Topics

1. $162 + 60\sqrt{2}$ **2.** $2y$ **3.** 4

Checkup 12.2 (page 707)

1. $5\sqrt{6x}$ **2.** $3\sqrt{3}$ **3.** $3\sqrt{3}$ **4.** $3a\sqrt{5}$ **5.** $y + 6$ **6.** $7\sqrt{5} - 5\sqrt{2}$
7. $2\sqrt{6} + 4\sqrt{3}$ **8.** $x - 2\sqrt{5x} + 5$ **9.** $-18 - \sqrt{5}$ **10.** $x - 3$

Exercise Set 12.3 (page 715)

1. $\sqrt{3}$ **3.** $\frac{2\sqrt{7}}{7}$ **5.** $2\sqrt{2}$ **7.** $\frac{6\sqrt{10}}{5}$ **9.** $\frac{9\sqrt{2}}{8}$ **11.** $\frac{\sqrt{10}}{4}$ **13.** $\sqrt{2}$ **15.** $\frac{\sqrt{2}}{2}$
17. $\frac{2\sqrt{5}}{5}$ **19.** $\frac{\sqrt{3x}}{x}$ **21.** $\frac{\sqrt{5} + \sqrt{10}}{5}$ **23.** $2 - 3\sqrt{2}$ **25.** $\frac{2\sqrt{5} + 5\sqrt{2}}{10}$
27. $2 - \sqrt{3}$ **29.** $-4(3 + \sqrt{6})$ **31.** $-\sqrt{2} - 2$ **33.** $\sqrt{6} - \sqrt{2}$ **35.** $\frac{2\sqrt{15} - 2\sqrt{6}}{3}$
37. $\sqrt{2}$ **39.** $5 + 2\sqrt{6}$

Stretching the Topics

1. $\frac{a\sqrt{a + b}}{2(a + b)}$ **2.** $\frac{15x + 6\sqrt{xy}}{25x - 4y}$ **3.** $\frac{2x - 2y + 4\sqrt{x - y}}{(x - y) - 4}$

Checkup 12.3 (page 717)

1. $\sqrt{6}$ **2.** $2\sqrt{10}$ **3.** $\frac{7\sqrt{3}}{6}$ **4.** $\frac{\sqrt{10}}{5}$ **5.** $\frac{2\sqrt{6}}{3}$ **6.** $\sqrt{6} - 5\sqrt{2}$ **7.** $\frac{1 + \sqrt{3}}{2}$
8. $\frac{2\sqrt{3} + 1}{11}$ **9.** $13 + 4\sqrt{11}$ **10.** $\frac{11 - 4\sqrt{7}}{3}$

Exercise Set 12.4 (page 725)

1. $x = \pm 2\sqrt{3}$ **3.** $a = \pm \frac{\sqrt{5}}{3}$ **5.** $x = \pm \frac{\sqrt{5}}{5}$ **7.** $y = \pm \frac{\sqrt{10}}{2}$ **9.** $y = \pm \sqrt{5}$
11. $a = \pm \sqrt{2}$ **13.** $x = 0$ **15.** $a = \pm 4\sqrt{5}$ **17.** $x = \frac{-1 \pm \sqrt{13}}{2}$
19. $x = \frac{5 \pm 3\sqrt{5}}{2}$ **21.** $x = 2 \pm \sqrt{7}$ **23.** $y = 1 \pm \sqrt{2}$ **25.** $a = \frac{-7 \pm \sqrt{53}}{2}$
27. $x = \frac{1 \pm \sqrt{7}}{3}$ **29.** $y = 3 \pm \sqrt{5}$ **31.** $z = \frac{-1 \pm \sqrt{31}}{2}$

Stretching the Topics

1. $x = \frac{13 \pm \sqrt{257}}{2}$ **2.** $x = \frac{3\sqrt{2} \pm \sqrt{34}}{2}$ **3.** $x = \sqrt{3} \pm 1$

Checkup 12.4 (page 727)

1. $x = \pm\sqrt{2}$ **2.** $a = \pm\sqrt{5}$ **3.** $y = \pm 4\sqrt{2}$ **4.** $x = \pm 2\sqrt{5}$ **5.** $x = 0$
6. $x = -2 \pm 2\sqrt{2}$ **7.** $x = \frac{3 \pm \sqrt{5}}{2}$ **8.** $x = \frac{3 \pm \sqrt{3}}{3}$ **9.** $x = \frac{5 \pm \sqrt{33}}{4}$
10. $x = \frac{1 \pm 2\sqrt{2}}{3}$

Exercise Set 12.5 (page 733)

1. $5\sqrt{2}$ cm **3.** 20.8 ft **5.** $4\sqrt{3}$ ft **7.** 6.71 m, 13.42 m **9.** $(3 + \sqrt{19})$ m
11. $1 + \sqrt{10}$ and $-1 + \sqrt{10}$ or $1 - \sqrt{10}$ and $-1 - \sqrt{10}$ **13.** $(-4 + \sqrt{31})$ ft
15. 0.9 in.

Stretching the Topics

1. $(10 + 5\sqrt{5})$ ft, $(20 + 10\sqrt{5})$ ft, $(25 + 10\sqrt{5})$ ft **2.** $(2 + 2\sqrt{2})$ in., $(12 + 8\sqrt{2})$ sq in.

Checkup 12.5 (page 735)

1. $4\sqrt{2}$ in. **2.** 5.68 m **3.** $-2 + \sqrt{5}$ and $2 + \sqrt{5}$ or $-2 - \sqrt{5}$ and $2 - \sqrt{5}$ **4.** 13.8 ft

Speaking the Language of Algebra (page 738)

1. Principal square root **2.** Real **3.** Radical; radicand **4.** Irrational
5. Rationalizing the denominator **6.** Real conjugate **7.** $x = \sqrt{c}$ or $x = -\sqrt{c}$
8. $\dfrac{-b \pm \sqrt{b^2 - 4ac}}{2a}$; quadratic formula

Review Exercises (page 739)

1. 12 **2.** $8a$ **3.** 13 **4.** $3y^2\sqrt{5}$ **5.** $4ab\sqrt{3a}$ **6.** $\dfrac{3x^2\sqrt{3x}}{y^2}$ **7.** $7a^3b$ **8.** $3a^3$
9. $3\sqrt{5x}$ **10.** $7a\sqrt{3}$ **11.** $-60\sqrt{6}$ **12.** $3\sqrt{2} - 2\sqrt{7}$ **13.** $15\sqrt{6}$
14. $\sqrt{15} - 6\sqrt{5} - 12$ **15.** $-32 + 5\sqrt{6}$ **16.** $a - 5$ **17.** $5\sqrt{3}$ **18.** $\dfrac{\sqrt{5}}{2}$
19. $\dfrac{\sqrt{35}}{5}$ **20.** $2 - 7\sqrt{3}$ **21.** $\dfrac{\sqrt{2} + \sqrt{6}}{2}$ **22.** $\dfrac{5\sqrt{11} + 11}{14}$ **23.** $-(11 + 6\sqrt{3})$
24. $\dfrac{17 - 4\sqrt{13}}{9}$ **25.** $y = \pm 4\sqrt{3}$ **26.** $a = \sqrt{2}, a = -\sqrt{2}$ **27.** No solution
28. $x = -3 \pm 2\sqrt{3}$ **29.** $x = 2 \pm \sqrt{3}$ **30.** $x = \dfrac{5 \pm \sqrt{15}}{5}$ **31.** $x = \dfrac{3 \pm \sqrt{17}}{4}$
32. $x = \dfrac{2 \pm \sqrt{10}}{3}$ **33.** $3\sqrt{2}$ cm **34.** $(2 + \sqrt{6})$ ft

Practice Test for Chapter 12 (page 741)

1. $\dfrac{4}{5}$ **2.** $3a$ **3.** $9a$ **4.** $4\sqrt{2}$ **5.** $4ab^2\sqrt{3a}$ **6.** $\dfrac{4x\sqrt{2x}}{y^2}$ **7.** $10xy$ **8.** $4\sqrt{10z}$
9. $-50a\sqrt{3}$ **10.** $12\sqrt{2}$ **11.** $y - 2\sqrt{10y} + 10$ **12.** $20 - 7\sqrt{x} - 3x$ **13.** 25
14. $a - 25$ **15.** $2\sqrt{10}$ **16.** $\dfrac{\sqrt{15}}{3}$ **17.** $\dfrac{3\sqrt{5} + 5}{5}$ **18.** $\dfrac{4\sqrt{7} - 7}{9}$
19. $-6(\sqrt{2} + \sqrt{3})$ **20.** $\dfrac{2\sqrt{10} - 11}{9}$ **21.** $x = \pm 6\sqrt{2}$ **22.** $x = \pm \sqrt{3}$
23. $x = \dfrac{-5 \pm \sqrt{29}}{2}$ **24.** $x = -4 \pm \sqrt{6}$ **25.** $(1 + 2\sqrt{3})$ ft

Sharpening Your Skills after Chapters 1–12 (page 743)

1. \$724.50 **2.** $-3.98 > -4.21$ **3.** $\dfrac{5}{9}$ **4.** No **5.** $x \neq 7, x \neq \dfrac{1}{2}$ **6.** No **7.** -4
8. 0 **9.** $7a^7b^9$ **10.** $\dfrac{-8x^9}{y^{21}}$ **11.** $3x^3y^3 - 14x^3y^7$ **12.** $9x^2 - 12xy + 4y^2$
13. $x^3 - 125$ **14.** $x^3 + 2x^2 + 4x + 8$ **15.** $\dfrac{x^3}{y^5}$ **16.** $\dfrac{x + 8}{x + 2}$ **17.** $\dfrac{5x - 2}{x(5x - 1)}$
18. $ab^2(a^2 - 14ab + 4b^2)$ **19.** $(9m + 8n)(7m - 2n)$ **20.** $x = 3$ **21.** $x = 11, x = -5$

22. No solution **23.** $x \leq \frac{10}{3}$ **24.** $\left(\frac{2}{3}, \frac{5}{3}\right)$ **25.** (2, 2)

26.

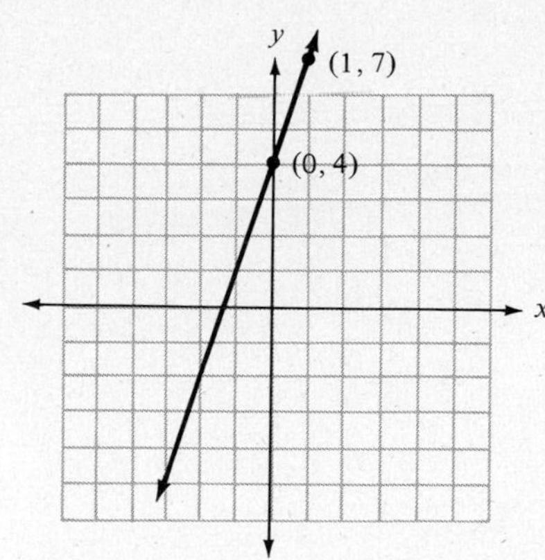

27.

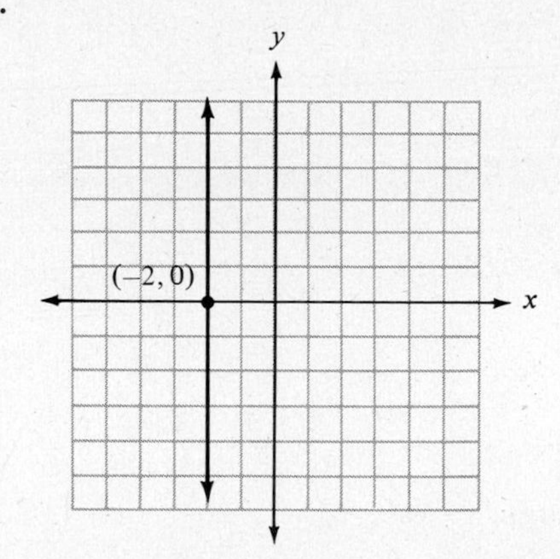

28. 40 ft **29.** 395 **30.** Ray, 60 mph; Glenn, 65 mph

Index

Absolute value, 75
Addition
decimals, 44
fractions, 28
integers, 76
polynomials, 293
radicals, 697
rational expressions, 455
property of equality, 194
property of zero, 6, 81
whole numbers, 3
Additive inverse, 81
Algebraic expression, 150
evaluating, 151
switching to, 153
Associative property
for addition, 80
for multiplication, 102
Average, 58

Base, 11, 242
Binomial
definition, 292
factor, common, 350
multiplication, 309
product by FOIL, 311
squaring, 310

Cartesian coordinate plane, 575
Closure properties, 122
Coefficient, numerical, 163
Commutative property
for addition, 79
for multiplication, 101
Conjugate, real, 701
Constant, 149
equation, 592
Coordinate of point, 2, 576

Decimals
adding, subtracting, 44
dividing, 45
multiplying, 44
and percents, 46
place value, 41
repeating, 42
rounding, 43
Denominator
LCD, 30, 459
rationalizing, 709
Difference of two squares, 310
factoring, 359
Distance, 491
Distributive property, 173, 294
Division
definition, 123
integers, 105
by monomial, 319
monomial by constant, 172
by polynomial, 321
rational expressions, 437
whole numbers, 5
with zero, 8

Equations
conditional, 192
constant, 592
first-degree, 192
fractional, 479
graph of, 578
linear, 580
second-degree (quadratic), 391, 719
switching from words to, 223, 403, 491, 564, 729
systems of, 631
in two variables, 566
Exponents 11, 104, 242
laws of, 245–260
negative, 267–275
one and zero, 244
whole number, 242

Factor
common, 23

Factor *(cont.)*
  common binomial, 350
  common monomial, 348
Factoring
  ac-method, 375
  common monomial, 348
  difference of two squares, 359
  fourth-degree trinomials, 377
  grouping, 350
  perfect square trinomials, 371
  trinomials, 367–372
Fractions
  adding and subtracting, 28, 455
  building, 27, 447
  complex, 469
  dividing, 26, 437
  equality, 493
  equations with, 479
  improper, 23
  multiplying, 24, 437
  proper, 23
  reducing, 23, 424

Graphing
  arbitrary point method, 578
  constant equations, 592
  equations, 578
  inequalities, 517
  intercept method, 589
  linear equations, 579
  number line, 2
  slope-intercept method, 609

Identity
  additive, 6, 81
  equation, 192
  multiplicative, 9, 103
Inequalities, 2, 516
  combining, 537
  graphing, 517
  solving first-degree, 425
  switching from words to, 547–549
  variable, 517
Integers, 74
  adding, 76
  consecutive, 331
  dividing, 105
  multiplying, 99
  subtracting, 91
Intercepts, 589
Irrational numbers, 118, 687

Lines
  graphing, 578–611
  horizontal, 593
  parallel, 633
  vertical, 595

Mixed number, 23
Monomial, 292
Multiplication
  associative property for, 102
  binomial, 309
  commutative property, 101
  distributive property, 103, 173
  FOIL method, 311
  integers, 99
  monomials, 171, 301
  property of equality, 196
  property of one (identity), 9, 103
  property of zero, 7
  rational expressions, 437
  whole numbers, 4
Multiplicative inverse, 123

Natural numbers, 2
Number line, 2

Opposites, 74
Order of operations, 12
Ordered pairs, 564
  graphing, 575
Origin, 575

Percents, 46
Plotting points, 2, 575
Polynomials
  addition, 293
  definition, 292
  division by monomial, 317
  long division, 321
  multiplication, 302
  prime (irreducible), 360
  switching from words to, 329–332
Powers
  definition, 11, 104, 242
  dividing, 255
  multiplying, 245
  of products, 247
  of quotients, 258
  raising to power, 247
Proportion, 493
Pythagorean theorem, 330

Quadratic equations
  incomplete, 719
  solving by factoring, 391
  solving by formula, 721